the human organism

RUSSELL MYLES DeCOURSEY

Professor Emeritus of Zoology
Biological Sciences Group
University of Connecticut

the human organism

FOURTH EDITION

McGRAW-HILL BOOK COMPANY

New York   *St. Louis*   *San Francisco*   *Düsseldorf*   *Johannesburg*   *Kuala Lumpur*   *London*   *Mexico*

Montreal   *New Delhi*   *Panama*   *Paris*   *São Paulo*   *Singapore*   *Sydney*   *Tokyo*   *Toronto*

This book was set in Laurel by York Graphic Services, Inc.
The editors were Thomas A. P. Adams and Shelly Levine Langman;
the designer was Barbara Ellwood;
the production supervisor was Thomas J. LoPinto.
New drawings were done by Eric G. Hieber Associates Inc.
Von Hoffmann Press, Inc., was printer and binder.

Library of Congress Cataloging in Publication Data

DeCoursey, Russell Myles, date
    The human organism.

    Bibliography: p. 612
    1. Human physiology. 2. Anatomy, Human. I. Title.
[DNLM: 1. Anatomy. 2. Physiology. QT4 D297h 1974]
QP34.5.D38 1974   612   73-23016
ISBN 0-07-016234-4

**the human
organism**

4 5 6 7 8 9 0 VHVH 7 9 8 7

contents

PREFACE   xi

UNIT 1   THE CELL

Chapter 1   *Basic principles*   2
Matter and energy   3
Solutions   14
Questions   14
Suggested reading   14

Chapter 2   *Functional aspects of cellular metabolism*   15
Movement through membranes   15
Homeostasis   19
Energy sources   19
Utilization of energy   21
Cellular enzymes and coenzymes   22
Summary   26
Questions   26
Suggested reading   26

Chapter 3   *Cellular structure and function*   28
Some physiological aspects of living matter   28
Structure of a cell   29
Mitotic division of cells   43
Genetics   47
Summary   52
Questions   54
Suggested reading   54

UNIT 2   STRUCTURAL ELEMENTS

Chapter 4   *Epithelial and connective tissues*   56
Epithelial tissues   56
Connective tissues   64
Albinism   71
Dermis   71
Cartilage   72
Bone   72
Terms of reference   81
Body cavities   83
Planes of reference   84
Summary   85
Questions   86
Suggested reading   86

Chapter 5   *The skeleton*   88
The axial skeleton   89
The appendicular skeleton   102
Sprains, fractures, and bone deficiencies   116

v

Synovial bursae 117
Articulations 118
Summary 119
Questions 120

Chapter 6 *Physiology of muscles* 121
Physiological characteristics of muscle tissue 121
The physiology of muscle 124
Summary 139
Questions 140
Suggested reading 141

Chapter 7 *Skeletal muscles* 142
Muscles of the head, face, and neck 144
Anterior muscles of the shoulder girdle and muscles of the thorax 148
Muscles of the back 151
Muscles of the shoulder and arm 154
Muscles of the abdominal wall 164
Muscles of the pelvic floor 168
Posterior muscles of the hip; the gluteal muscles 169
Muscles of the thigh 171
Muscles of the leg and foot 175
Muscles of the foot 180
Summary 182
Questions 185
Suggested reading 185

UNIT 3 THE NERVOUS SYSTEM AND THE SENSE ORGANS

Chapter 8 *Neural integration* 188
The neuron 188
Nerves 194
The nerve impulse 195
White matter and gray matter 201
Structure of the spinal nerve 202
Reflex activity 204
Summary 206
Questions 207
Suggested reading 207

Chapter 9 *The brain and spinal cord* 208
Basic divisions 208
Meninges 208
Ventricles 212
Neuroglia 213
Cerebrum 215
Learning and memory processes 223
Chemical compounds affecting the brain 224
Brain waves 226
Headache 227
The brainstem 227
Cerebellum 228
The medulla 230
Cranial nerves 230

	Spinal cord	233
	Spinal nerves	237
	Summary	238
	Questions	239
	Suggested reading	239

Chapter 10	*The autonomic nervous system*	*241*
	Sympathetic, or thoracolumbar, division	243
	Parasympathetic, or craniosacral, division	248
	Physiology of the autonomic nervous system	252
	Chemical transmitters	255
	Summary	256
	Questions	257
	Suggested reading	257

Chapter 11	*Special senses*	*259*
	The sensory unit	259
	Classification of receptors	259
	Cutaneous sensations	260
	The gustatory sense	262
	The olfactory sense	263
	Hearing and equilibrium	267
	Summary	280
	Questions	281
	Suggested reading	281

Chapter 12	*Special senses: vision*	*283*
	Accessory structures	288
	Lacrimal structures	284
	Extrinsic eye muscles	284
	Structure of the eyeball	286
	The formation and movement of fluids within the eyeball	300
	Mechanism of vision	301
	Binocular or stereoscopic vision	305
	Color vision	307
	The afterimage	309
	Color blindness	310
	Summary	312
	Questions	313
	Suggested reading	314

| UNIT 4 | THE CIRCULATORY SYSTEM | |

Chapter 13	*The blood*	*318*
	Functions	318
	Quantity, or volume, of blood	320
	Plasma	320
	Red cells: erythrocytes	322
	White cells: leukocytes	327
	Blood platelets	333
	The clotting mechanism	334
	Hemophilia	336
	Serum	336
	Storage of whole blood and plasma	336

Contents

vii

Blood grouping 337
The Rh factor 339
Other factors 341
Radiation hazards affecting the blood 342
The spleen and blood supply 342
Summary 344
Questions 345
Suggested reading 345

Chapter 14 *The heart and general circulation* 346
The heart 346
Arteries, veins, and capillaries 355
Principal systemic arteries 357
Arteries of the pelvis and lower extremities 362
The venous system 363
Fetal circulation 367
Summary 369
Questions 371
Suggested reading 371

Chapter 15 *Physiology of circulation* 372
Blood pressure 375
Hypertension 382
The lymphatic system 383
Summary 388
Questions 390
Suggested reading 390

UNIT 5 THE RESPIRATORY SYSTEM AND INTERNAL RESPIRATION

Chapter 16 *External and internal respiration* 392
The trachea 393
The lungs 394
Respiratory movements 397
Respiratory centers 399
Factors directly affecting the respiratory center 401
Internal respiration and cellular metabolism 406
Effects of rapid decompression 410
Artificial respiration 410
Summary 412
Questions 414
Suggested reading 414

UNIT 6 DIGESTION

Chapter 17 *The digestive system* 416
Mouth 416
Tonsils 416
Tongue 417
Teeth 417
Salivary glands 421
Pharynx 423
Act of swallowing 423
Esophagus 424

	Abdomen	425
	Stomach	426
	Duodenum	431
	Pancreas	432
	Regulation of carbohydrate metabolism	434
	Liver	438
	Jejunum and ileum	445
	Large intestine	449
	Summary	451
	Questions	452
	Suggested reading	453

Chapter 18	*Foods, nutrition, and metabolism*	*454*
	Carbohydrates	455
	Fats and related compounds	456
	Proteins	457
	Inorganic requirements in the diet	460
	Metabolism	475
	Summary	488
	Questions	490
	Suggested reading	491

UNIT 7 EXCRETION

Chapter 19	*The kidneys*	*494*
	Embryonic origin of the urinary system	494
	Kidneys	497
	Ureters and urinary bladder	509
	Other excretory organs	511
	Summary	511
	Questions	511
	Suggested reading	511

UNIT 8 GLANDS OF INTERNAL SECRETION

Chapter 20	*The endocrine system*	*516*
	Functions of hormones	516
	Hypophysis	519
	Thyroid gland	528
	Parathyroid glands	533
	Adrenal glands	535
	Hormones of the gonads	541
	Thymus	546
	The pineal gland	547
	Prostaglandins	547
	Hormones associated with the digestive tract	547
	Summary	548
	Questions	548
	Suggested reading	548

UNIT 9 REPRODUCTION

Chapter 21	*The reproductive system*	*552*
	Female reproductive system	552
	Male reproductive system	560

Sterilization   564
Venereal diseases   565
Summary   567
Questions   568
Suggested reading   568

UNIT 10 DEVELOPMENTAL ANATOMY AND PHYSIOLOGY

Chapter 22 *Human development*   570
Gametogenesis   570
Meiosis   570
The behavior of the sex chromosomes in gametogenesis   573
Spermatogenesis   577
Oogenesis   579
Fertilization   583
Cleavage   584
Implantation   585
Germ layers and their derivatives   586
Extraembryonic, or fetal, membranes   587
Twinning   590
Development of the human embryo   592
Growth   600
Childbirth   607
Summary   609
Questions   610
Suggested reading   611

REFERENCE BOOKS   612

GLOSSARY   615

INDEX   630

preface

The fourth edition of "The Human Organism" presents much new material, incorporating the results of recent research and newer attitudes toward teaching anatomy and physiology. In this regard, there seems to be a trend away from detailed factual anatomy and more toward functional anatomy, with greater interest in physiology. Additional anatomical detail, therefore, has been limited except for the musculature of the forearm and hand, which had been requested by several users of the text. Medical terminology, medical detail, and pathology have not been stressed, in the belief that beginning students are often confused by such terminology. By the same token, a minimal amount of chemistry is presented, since many first-year students have had little chemistry and most of them have had no classwork in biochemistry.

A brief discussion of introductory genetics has been added, with examples in appropriate places throughout the book. The chapter on muscle physiology has been largely rewritten, and new material has been introduced throughout the text. Greater attention has been given to such subjects as negative-feedback mechanisms, hormone receptors and receptor sites on target organs, hormone-releasing factors, and the role of cyclic AMP as a "second messenger" for many hormone functions. The counter-current concept of kidney functions is discussed also. More than thirty new tables have been added as a learning and review aid for the student.

It is well recognized that much of the research in human physiology depends on animal experimentation. The text includes a little histology, some comparative anatomy, and an appropriate amount of developmental anatomy, in order that the reader may gain some knowledge of essential related fields in biology.

An attempt has been made to keep related subject matter localized, not scattered throughout the book. The best method of presentation, we believe, involves integration of structure and function of an organ system in its relation to the whole organism.

Though designed primarily for a one-semester course, the text should be readily adaptable to a term system or a two-semester course. The book has been used as a text in liberal arts colleges and universities, schools of nursing, physical therapy, physical education, and pharmacy, and in the rapidly developing community colleges. A new edition of "A Laboratory Manual of Human Anatomy and Physiology," by DeCoursey and Dolyak, McGraw-Hill Book Company, New York, has also been prepared. And an instructor's manual is available.

The suggested readings at the ends of chapters include recent publications affording more detailed or more extensive information. They may be assigned as outside reading with written reports or as oral seminar reports, or they may be used for discussion classes. Generally, they have been selected from readily available publications.

I should like to thank those colleagues who have read parts of the manuscript. I am indebted to the illustrator and to the publishing staff of the McGraw-Hill Book Company for their work. In particular, I should like to thank the senior editor, Mr. Thomas A. P. Adams, for his part in supervising this new edition.

Russell Myles DeCoursey

unit one

the cell

chapter 1
basic principles

It is difficult to comprehend the complexity of the human organism with all its variety of structure and intricacy of function. Chemically and physically complex, it operates with amazing efficiency. Probably there are thousands of chemical reactions taking place within our bodies even as we sit quietly studying.

The living world about us is composed of plants and animals of almost infinite variety. There are relatively simple forms, such as the algae among the plants or the protozoa of the animal kingdom; there are also very complex organisms, such as insects, trees, and man.

Living organisms have certain structural and functional characteristics that serve to identify all living matter. The structural unit of living matter is the cell. It is difficult to give a comprehensive definition of a cell, because individual cells vary widely in structure and function. The usual concept is that a cell is a microscopic entity composed of protoplasm and containing a nucleus and many minute structures called organelles. The components of a cell are contained within a plasma membrane. The cell is capable of performing the fundamental functions of living organisms.

Our knowledge of the fine structure of a cell is closely connected with the development of the microscope. Early in the nineteenth century various workers had observed cells in plant and animal life, but they had failed to develop their ideas into a clear concept of the structural nature of living matter. Finally a new principle evolved: all tissues are composed of cells.

However, the exact nature and function of structures within the nucleus and cytoplasm of the cell remained largely a mystery until recent years. When the light microscope became perfected as a modern research instrument, more and more features of the cell were observed in greater detail. Better methods of fixation and staining helped to delineate structures within the nucleus and cytoplasm. The phase microscope enabled observers to see more clearly the structures in living cells under contrasting light. Nevertheless, there is a strict limitation to the degree of magnification obtainable with a microscope using a light source.

The resolving power of the unaided human eye is about 0.1 millimeter. Dots closer together than this are seen as single dots or are blurred. Compound microscopes, of course, have a much greater resolving power, and magnifications of a thousand times are commonly achieved, but the best obtainable magnification of the light microscope is not great enough to reveal the minute details of cell structure.

The electron microscope uses a beam of electrons instead of light waves. The resolving power is very greatly increased, and electron micrographs with magnifications of 50,000 to 60,000 times are common. Much higher magnifications can be obtained by enlarging the original picture, although the resolution cannot be increased by this means.

The electron beam passes through electromagnetic fields which control or refract the electrons in much the same way that glass lenses and condensers direct light waves in the light microscope. Since electrons cannot be seen, their effect must be observed on a photographic plate or on a fluorescent screen. There are some disadvantages to this method of studying biological material, even though the contributions to science have been enormous. The material to be studied must be extremely dry, and ultramicrotomes are required for cutting the sections sufficiently thin. Materials are not colored with biological dyes or stains but are "shadowed" by treatment with solutions of metals such as osmium, tungsten, or chromium in order to obtain contrast.

The physiology of the cell largely involves the activities of the living protoplasm of the cell. The physiology of a complex organism such as the human body includes not only the physiology of cells and tissues but also the function of organs and organ systems.

All physiology is based on chemical and physical principles. It has become difficult to distinguish clearly between biochemical, biophysical, and biological studies of function, but these studies are basic to physiology. Before we are ready to study the broad field of physiology, therefore, we shall survey briefly certain aspects of elementary chemistry and physics.

MATTER AND ENERGY

From a physical standpoint, matter occupies space and has mass and weight. It is described as gaseous, liquid, or solid. Oxygen and carbon dioxide are gases concerned with respiration; the blood and other body fluids are liquids, while much of the skeleton exists as solid matter. Matter is composed of chemical *elements*, commonly referred to by their chemical symbols, as carbon, C; oxygen, O; hydrogen, H; and nitrogen, N. Usually these symbols are taken from the first letter or the first two letters of the English or Latin name of the element, but some are more difficult to learn than others. The following are often misunderstood: sodium, Na; potassium, K; copper, Cu; iron, Fe; and mercury, Hg. There are many chemical elements; some 90 naturally occurring elements are included in the periodic table of chemical elements. In this table, the elements are arranged in order of increasing atomic numbers. Hydrogen has an atomic number of 1, and uranium has an atomic number of 92. There are, in addition, 10 or more transuranium elements which extend the list from 93 to over 100, but these elements are produced artificially, by prolonged atomic bombardment of uranium, and are not stable; that is, they exist for only a short period of time. A few examples from the periodic table are shown in Figure 1.1, but they are not arranged, as in the periodic table, to show chemical relationships.

ATOMS All chemical elements have a precise atomic structure. The atom, in turn, is composed of elementary particles called *electrons, protons,* and *neutrons*. Atomic structure consists of electrons revolving in orbitals around a nucleus. Although a diagram of atomic orbitals commonly shows electrons in their shells or subshells revolving around the nucleus in planetary fashion, this is not entirely in keeping with modern theory. Electrons move in certain volumes of space and exhibit wavelike properties—for example, the beam of electrons used in the electron microscope.

The nucleus of the atom consists of protons, each carrying a positive charge, and neutrons, which are not charged. Electrons are negatively charged. When the positive protons and the negative electrons are equal in number, the positive and negative charges are equally balanced, and the atom is neutral (Figure 1.2).

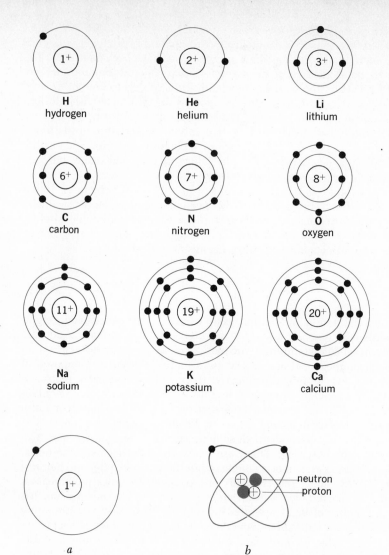

FIGURE 1.1
Diagrams of a few elements selected from the periodic table to show the structure of various atoms. The number in the center of each diagram represents the positive nuclear charge which is balanced by the negatively charged electrons.

FIGURE 1.2
Diagrammatic representation of *a* the hydrogen atom and *b* the helium atom. These are not pictures of atoms but merely symbols representing their structure.

Protons or neutrons have mass or weight, but electrons have so little mass that it is considered negligible. Since the electrons moving in their orbitals are of a minute size, there is a relatively large amount of space within the atom. It has been said that if the nucleus of an atom could be enlarged to the size of a baseball, the electron would be just a speck 2,000 feet away. Different kinds of atoms have different numbers of electrons revolving around the nuclei, and the electrons may be in one to seven different orbitals.

The periodic table indicates that the elements can be arranged according to the number of electrons (or protons) in the atoms. The hydrogen atom has one electron and one proton, and the atomic number of hydrogen is 1. The helium atom has two electrons, carbon six, nitrogen seven, and oxygen eight (Figure 1.1). The examples pictured also show that the electrons are located in certain prescribed orbitals, or shells, indicated diagrammatically by circles. Each shell has a fixed maximum number of electrons that it can

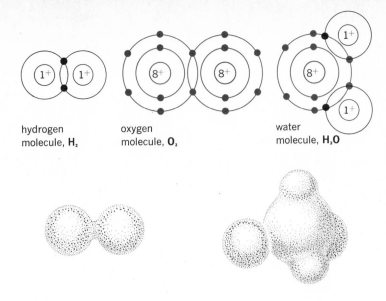

hydrogen
molecule, **H₂**

oxygen
molecule, **O₂**

water
molecule, **H₂O**

FIGURE 1.3

The formation of a molecule of water. The covalent bond between the oxygen atom and each hydrogen atom represents a pair of electrons shared between the two. The two bulges in the oxygen atom below are exaggerated representations of the unshared pairs of electrons.

hold. The first shell is limited to two electrons in the orbital nearest the atomic nucleus. The second shell has a maximum of eight electrons. We shall not include the subshells, which contain electrons with very slightly different energies, but the maximum number of electrons is known for every shell and subshell.

When all the shells of an atom contain their maximum number of electrons, the atom is nonreactive. Again let us consider the atom of helium, an inert gas (Figure 1.1). The first shell contains the maximum number of two electrons; the atom is stable electronically, and it is almost completely inert from a chemical standpoint. Hydrogen, however, with one orbital electron may act as an electron donor, as in the reaction $H + Cl \rightarrow H^+Cl^-$, or pairs of hydrogen atoms may share electrons to form a hydrogen molecule, H_2. Oxygen, with an incomplete outer shell of six electrons, may complete the outer shell by sharing electrons with another oxygen atom to form a molecule of oxygen, O_2. In the formation of a molecule of water, H_2O, an oxygen atom shares its six outer electrons with two hydrogen atoms and in this way completes its outer shell of eight electrons. The hydrogen atoms at the same time acquire an inner stable shell of two electrons each (Figure 1.3).

MOLECULES We are all familiar with many kinds of matter, or substances, such as water, sugar, and salt. Water, as we have just seen, is composed of hydrogen and oxygen, two atoms of hydrogen and one of oxygen comprising a molecule of water, H_2O. A *molecule* is a unit of matter with a definite structure. Some molecules, such as oxygen, O_2, have two identical atoms, whereas carbon dioxide, CO_2, has two different kinds, and the sugar glucose, $C_6H_{12}O_6$, has three different kinds of atoms. Some molecules, such as those of proteins, have a very complex atomic structure. In molecules composed of atoms of the shared-electron type, the atoms are connected by covalent chemical bonds. There is no actual transfer of electrons. Certain types of atoms, usually in the nonmetal group, commonly share electrons and form

Matter and energy

5

covalent compounds. These atoms include hydrogen, oxygen, nitrogen, and carbon. Hydrogen, however, may share its electron, or it may transfer it. Referring again to Figure 1.1, we see that oxygen can accept two electrons in its outer shell and therefore has a valence of 2; nitrogen can accept three electrons and has a valence of 3; a carbon atom can accept four electrons and ordinarily has a valence of 4 (see Table 1.1).

We may very well ask this question: How did living organisms arise on the earth? What were the first compounds of biological significance? Atoms of hydrogen, oxygen, carbon, and nitrogen probably became available to form simple compounds as the earth cooled. Hydrogen and oxygen formed water, H_2O; at the same time hydrogen may very well have combined with nitrogen to form ammonia, NH_3. The flammable gas methane, CH_4, could have been formed by the combination of hydrogen and carbon atoms. Further cooling of the earth permitted great rain clouds to form, and centuries of rain produced the oceans. The water on the earth's surface then contained atoms from atmospheric gases, O, H, N, C, as well as dissolved minerals and salts, and the stage was set for the formation of more complex carbon compounds. Radiation from the sun and lightning from rain clouds probably played a part in providing energy for the synthesis of more complex molecules.

methane, CH_4 ammonia, NH_3

TABLE 1.1
Some common chemical elements

Element	Symbol	Atomic number	Atomic weight*	Common valences
Hydrogen	H	1	1	1
Carbon	C	6	12	4
Nitrogen	N	7	14	3
Oxygen	O	8	16	2
Sodium	Na	11	23	1
Magnesium	Mg	12	24	2
Phosphorus	P	15	31	3, 5
Sulfur	S	16	32	2
Chlorine	Cl	17	35	1
Potassium	K	19	39	1
Calcium	Ca	20	40	2
Iron	Fe	26	56	2, 3
Copper	Cu	29	63	1, 2
Iodine	I	53	127	1

*The atomic weights shown here are rounded off to the nearest whole number.

The atomic number indicates the number of protons in the nucleus of the atom. The atomic weight is equal to the sum of the protons and neutrons.

At earth level, mass and weight are the same, but in outer space, for example, with the loss of gravity from the earth a condition of weightlessness develops while the mass remains the same.

Molecular weight is equal to the combined atomic weights of all the atoms composing the molecule. Water, H_2O, with two hydrogen atoms and one oxygen atom, affords a simple example. Its molecular weight is 18.

CARBON COMPOUNDS   The carbon atom is of special significance since it is the key to the formation of the molecules characteristic of living matter. In the methane formula, we see that carbon has a valence, or bonding capacity, of 4. This means that it can be linked with other atoms, such as those of hydrogen, or that carbon atoms can be linked with each other. In this way the structural formulas of carbon compounds are written as rings or as chains.

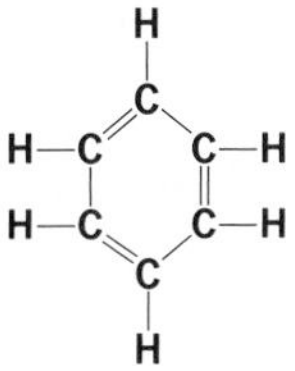

benzene, C_6H_6

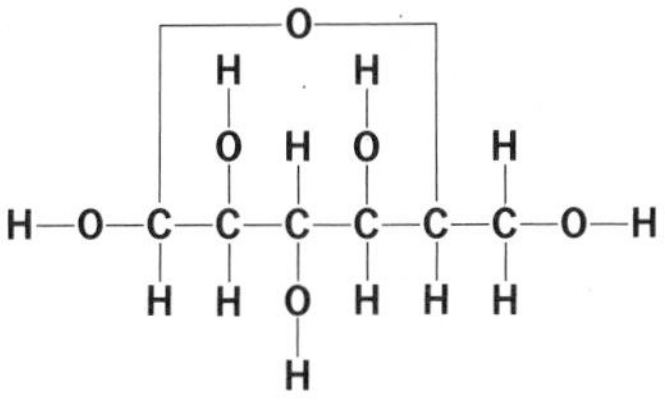

glucose, $C_6H_{12}O_6$

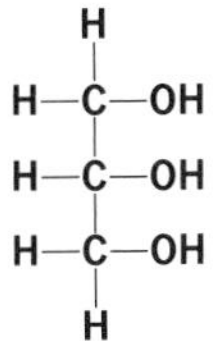

glycerol, $C_3H_8O_3$

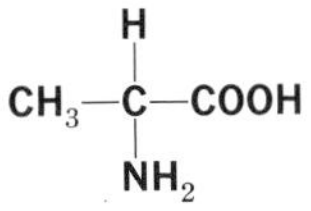

an amino acid
(alanine, $C_3H_7NO_2$)

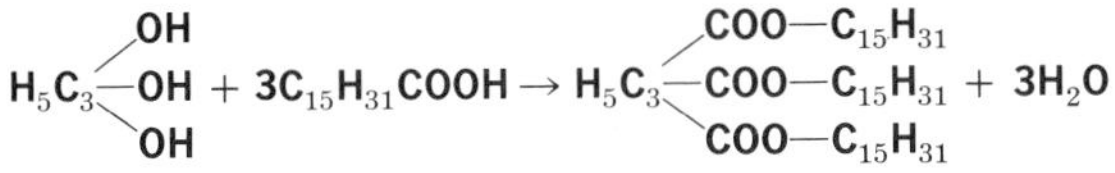

$$H_5C_3 \overset{OH}{\underset{OH}{-OH}} + 3C_{15}H_{31}COOH \rightarrow H_5C_3 \overset{COO-C_{15}H_{31}}{\underset{COO-C_{15}H_{31}}{-COO-C_{15}H_{31}}} + 3H_2O$$

glycerol   a fatty acid   a fat
                (palmitic)   (tripalmitin)

Since such linkage of carbon atoms in rings and chains is principally associated with the products of living cells or material that was at one time alive, these complex structures are called *organic compounds*. The organic substances regarded as essential in the formation of living matter include sugars (carbohydrates), lipids, amino acids, and proteins.

ATOMIC WEIGHT   Since atoms are unbelievably light, many millions together would have no effect even on sensitive scales. It is not possible actually to weigh atoms, but chemical elements do have *atomic mass*, which has been assigned to them as relative rather than absolute weight. The hydrogen atom is the lightest of all atoms. It was early suggested that hydrogen be given the atomic weight of 1 and that other elements be given atomic weights on a relative basis. There were difficulties, however, in using hydrogen as a standard. Consequently, oxygen, with a value of 16, was chosen as the standard. The weights of equal volumes of hydrogen and oxygen, taken under equal conditions of temperature and pressure, show the atomic mass of hydrogen to be about one-sixteenth that of oxygen, or, more specifically, in the proportion of 1.008 for hydrogen to 16 for oxygen. More recently the carbon atom has been selected as the standard, with a relative atomic weight of 12. This, however, represents the mass of a particular carbon atom; the actual value, based on carbon atoms of different weights, is 12.01115 (see isotopes below). Other examples are calcium, Ca, 40.08; iron, Fe, 55.847; nitrogen, N, 14.0067; oxygen, O, 15.9994; potassium, K, 39.102; and sodium, Na, 22.9898 (see Table 1.1). The atomic weights of the elements are given

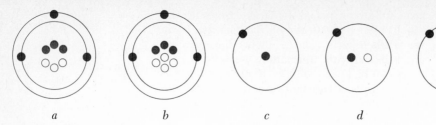

a   b   c   d   e

FIGURE 1.4

Lithium and hydrogen isotopes.
a Lithium isotope, atomic weight
6, the nucleus containing three
protons and three neutrons.
b Lithium isotope, atomic weight
7, the nucleus containing three
protons and four neutrons.
c Hydrogen, ^{1}H. *d* Deuterium, ^{2}H.
e Tritium, ^{3}H. Tritium-labeled
compounds are used extensively
in research.

in the accepted table of international atomic weights, which is available in chart form and in chemistry textbooks.

Isotopes It is assumed that a given compound is made up of countless similar molecules. Thus water is composed of many molecules that have the chemical formula H_2O. Since each molecule of water has the same proportion of hydrogen and oxygen, the mass proportion of hydrogen and oxygen in a certain quantity of water is constant. Water in bulk is made up of 8 parts of oxygen to 1.008 parts of hydrogen. Chemists have been able to determine the number of atoms that constitute individual molecules of chemical compounds. Some elementary forms of matter that appear to have the same chemical properties have been found to differ in atomic weight. Such forms of matter are called *isotopes* (Figure 1.4). Thus sulfur has four stable isotopes, with atomic weights of 32, 33, 34, and 36. Chlorine has two stable isotopes, with atomic weights of 35 and 37. Although chlorine has the atomic weight of 35.457, a given quantity of chlorine is made up of a mixture of isotopes, some with the atomic weight of 35 and some with that of 37. The difference in the atomic weights of isotopes of a given element results from the varying numbers of neutrons in the differing forms. Thus chlorine atoms, with a constant number of 17 protons in the atom, can vary as to the number of neutrons. Some chlorine atoms have 18 neutrons, and others have 20.

Numbers of isotopes vary with different elements. Oxygen has three stable isotopes: ^{16}O, ^{17}O, and ^{18}O; carbon has three: ^{12}C, ^{13}C, and ^{14}C. Of these, ^{12}C and ^{13}C are stable; ^{14}C occurs naturally and is radioactive. Hydrogen has isotopes ^1H, ^2H, and ^3H; the last occurs naturally and is unstable. Calcium has six stable isotopes: ^{40}Ca, ^{42}Ca, ^{43}Ca, ^{44}Ca, ^{45}Ca, and ^{48}Ca; iron has four: ^{54}Fe, ^{56}Fe, ^{57}Fe, and ^{58}Fe. Although there are physical differences in the isotopes of a certain element, the chemical properties are the same. This fact has been of great value in the study of physiological processes, for the action of carbon, for example, is the same whether the investigator is studying the effect of the stable carbon element or its radioactive isotope. When radioactive isotopes are introduced into the living body, they emit energized particles, or rays, and their progress can be traced by means of a Geiger counter.

IONIZATION Molecules of all substances are constituted of definite combinations of atoms. An atom may be shown diagrammatically by drawing two or more concentric circles (Figure 1.1), the inner circle to represent the nucleus of the atom and the outer circle or circles to represent the orbital or orbitals containing the electrons. An atom contains as many electrons in its orbital as there are protons in its nucleus. If atoms gain or lose electrons, they become charged particles called *ions*, and the process is called *ionization*. Many inorganic compounds that dissolve in aqueous solution tend to

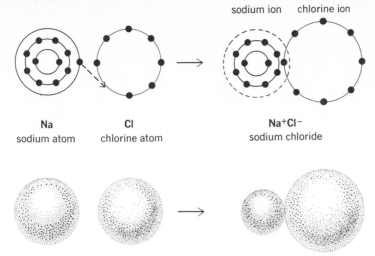

sodium ion   chlorine ion

Na   Cl
sodium atom   chlorine atom

Na$^+$Cl$^-$
sodium chloride

FIGURE 1.5
The formation of sodium chloride by electron transfer. The sodium atom is shown with a single electron in its highest normal energy level. The chlorine atom is shown with seven electrons in its highest normal energy level (only the highest, or third, energy level is shown). If the chlorine atom gains one electron, it becomes a negatively charged chlorine ion. If the sodium atom loses an electron, it becomes a positively charged ion.

ionize. If the atom gains electrons, it has a negative charge and is called an *anion;* if it loses an electron or electrons, it has a positive charge and is termed a *cation* (cat-eye-on). Since an atom is normally neutral, the positive charge on the nucleus is balanced by the negative charges on the electrons. If the atom gains a negative charge by acquiring an electron or electrons, the particle becomes a negative ion. If an electron is lost, the balance between nucleus and electron changes in favor of the positive charge on the nucleus. The particle then becomes a positive ion. In an electric field, anions move toward the anode, or positive pole, whereas cations, being positively charged, move toward the cathode, or negative pole. Particles having the same charge will exert a force to repel each other. The force or work required is a positive force, and a positive potential is indicated. A negative potential is developed when the particles have opposite charges and attract each other. Sodium chloride, NaCl, for example, is ionized in water into sodium ions, Na$^+$, and chloride ions, Cl$^-$. The plus and minus symbols indicate the sign of the electric charge. Such an ion-containing solution is capable of conducting an electric current and is therefore an *electrolytic solution* or electrolyte. Acids, bases, and salts that dissociate into ions when dissolved in water are electrolytes.

This type of chemical reaction involves the transfer of electrons, which usually occurs between metals and nonmetals. If the outer energy shell has few electrons, the atom will probably act as an electron donor. For example, if the outer shell lacks only one, two, or three electrons in order to become stable, the atom may act as an electron acceptor. In this kind of reaction, metals ordinarily are donors, and nonmetals are acceptors. In the transfer reaction of the sodium atom and the chlorine atom to a molecule of sodium chloride, sodium, the donor, is a metal, and chlorine, the acceptor, is a nonmetal (Figure 1.5).

Valence is determined by the number of electrons an atom contributes in chemical union. The valences of some of the elements are given in Table 1.1. It will be noted that some elements have two valences. They usually form more stable compounds at one valence than at the other. In such compounds with metals, for example, ferrous chloride and ferric chloride,

Matter and energy

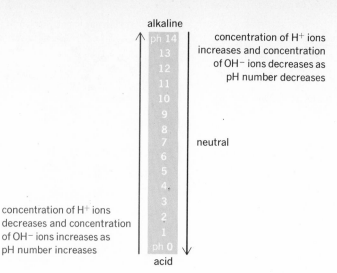

alkaline

concentration of H+ ions
increases and concentration
of OH− ions decreases as
pH number decreases

neutral

concentration of H+ ions
decreases and concentration
of OH− ions increases as
pH number increases

acid

FIGURE 1.6
Diagram illustrating the range of hydrogen-ion concentration of solutions. A solution with a pH value of 7 is neutral.

the term for the lower valence state ordinarily has the ending *-ous*, whereas that for the upper valence state bears the ending *-ic*. These valences may also be expressed as Fe^{2+} and Fe^{3+}.

Hydrogen-ion concentration (pH) The symbol pH refers to the hydrogen-ion concentration of a solution. In pure water, the hydrogen ions, H^+, and the hydroxyl ions, OH^-, are in equal concentration. Pure water, therefore, is neutral in reaction and has a pH of 7. Alkaline solutions range from pH 7 to pH 14, and acid solutions from pH 7 to pH 0 (Figure 1.6). As the pH number decreases, the hydrogen-ion concentration rises, and the solution becomes more acid; that is, a solution of pH 3 is more acid than a solution of pH 5. Tissue fluids are generally close to pH 7 in their reaction. Saliva is usually slightly acid (pH 6.8); blood is normally about pH 7.4 and therefore slightly alkaline. The gastric fluid in the stomach is the most acid substance in the body. Samples of gastric juice as usually taken are diluted with water and contain mucin, enzymes, some organic matter, and electrolytes, but the mixture has a pH range of about 1.6 to 2.6.

The hydrogen-ion concentration of water under ordinary conditions is close to 1/10,000,000, or 0.0000001, or 10^{-7} mole per liter. (A mole, or gram molecule, is the weight of a substance in grams which is equivalent to its molecular weight.) The pH of a solution is derived according to the formula pH = log 1/(H^+ concentration). [The logarithm of a number is the power to which 10 must be raised in order to equal the number. The logarithm of 10 is 1; $10^2 = 100$, and so the logarithm of 100 is 2; the logarithm of 1,000 (log 10^3) is 3.] The pH system was devised to simplify the terminology involved in expressing hydrogen-ion concentration in powers of 10 (see Table 1.2).

ACIDS, BASES, AND SALTS An acid may be defined as a substance that dissociates when dissolved in water, releasing hydrogen ions, H^+. The hydrogen atom, as we have seen, consists of a single proton and a single electron. When the hydrogen atom loses an electron, the proton, unable to exist alone in a water solution, tends to combine with some other ion or molecule. In a water solution it is really a hydrated hydrogen ion (that is, attached to a

TABLE 1.2
Scale of pH, or hydrogen-ion concentration,
in moles per liter

Decimal	Exponent	pH
1.	10^0	0
0.1	10^{-1}	1
0.01	10^{-2}	2
0.001	10^{-3}	3
0.0001	10^{-4}	4
0.00001	10^{-5}	5
0.000001	10^{-6}	6
0.0000001	10^{-7}	7
0.00000001	10^{-8}	8
0.000000001	10^{-9}	9
0.0000000001	10^{-10}	10
0.00000000001	10^{-11}	11
0.000000000001	10^{-12}	12
0.0000000000001	10^{-13}	13
0.00000000000001	10^{-14}	14

water molecule), but for most purposes it is convenient to consider it simply as a hydrogen ion with the symbol H^+.

There are strong acids and weak acids. Acids such as hydrochloric acid, HCl, or sulfuric acid, H_2SO_4, which ionize almost completely and produce hydrogen ions in considerable quantity, are strong acids.

$$HCl \rightarrow H^+ + Cl^-$$

Sulfuric acid ionizes in two stages, since the bisulfate ion (HSO_4) itself ionizes as follows:

$$H_2SO_4 \rightarrow H^+ + HSO_4^-$$
$$HSO_4^- \rightarrow H^+ + SO_4^-$$

Weak acids do not ionize as completely as strong acids, but all acids ionize to some extent. Such acids as acetic acid, $C_2H_4O_2$; citric acid, $C_6H_8O_7$; lactic acid, $C_3H_6O_3$; or carbonic acid, H_2CO_3, are considered weak acids. The formula for acetic acid is usually written CH_3COOH, and it dissociates in the following manner:

$$CH_3COOH \rightarrow CH_3COO^- + H^+$$

Acetic acid is only moderately ionized, meaning that relatively little of it is dissociated as hydrogen ions, H^+, and acetate ions, CH_3COO^-, at any one time and also that the ions readily recombine to form CH_3COOH.

Blue litmus paper turns red when dipped into an acid solution, indicating that hydrogen ions are present.

A compound that dissociates in aqueous solution to yield hydroxyl ions, OH^-, is called a *base*. A strong base readily dissociates into ions, as, for example, sodium hydroxide:

$$NaOH \rightarrow Na^+ + OH^-$$

Strong bases combine readily with hydrogen ions, H^+, whereas weak bases combine weakly. Other substances, such as carbonates, proteinates, or hemoglobinates, can act as bases. Since the acid forms of these compounds are relatively weak acids, the basic ions are relatively strong. These substances dissociate as follows:

$$NaHCO_3 \longrightarrow Na^+ + HCO_3^-$$
sodium bicarbonate   base

$$KHb \longrightarrow K^+ + Hb^-$$
potassium hemoglobinate   base

$$NaPr \longrightarrow Na^+ + proteinate^-$$
sodium proteinate   base

Bases have a soapy feeling, and they turn red litmus paper to blue.

When an acid and a base interact chemically in solution, a salt is produced, as in the following reaction:

$$HCl + NaOH \rightarrow NaCl + H_2O$$

The metallic ion, Na^+, replaces the H^+ ion of the hydrochloric acid, and the salt sodium chloride is formed. The H^+ and OH^- ions combine to form water, H_2O.

BUFFERS   A buffer is a substance that, when added to a solution, tends to maintain the hydrogen-ion concentration against the action of either excess acid or alkali. Within the body the buffering action of the blood affords the clearest illustration of the way a buffering system operates. In the first place, all acids ionize to a certain extent, and the blood contains relatively weak acids, such as carbonic and lactic acids. Also present in the blood are bases such as sodium bicarbonate, $NaHCO_3$.

If a small quantity of HCl is mixed with pure water, the hydrogen-ion concentration increases considerably in the absence of a buffering system. If the same amount of HCl is introduced into a system buffered by sodium bicarbonate, $NaHCO_3$, as in the blood, more carbonic acid, H_2CO_3, is produced, the amount of $NaHCO_3$ is reduced somewhat, and the neutral salt NaCl is formed, but the hydrogen-ion concentration increases only moderately because H_2CO_3 is a much weaker acid than HCl.

$$NaHCO_3 + H^+ \rightarrow H_2CO_3 + Na^+$$

Most of the carbon dioxide of the blood is present in the form of bicarbonate ions, HCO_3^-:

$$CO_2 + H_2O \rightleftharpoons H_2CO_3 \rightleftharpoons H^+ + HCO_3^-$$

If a small quantity of a base such as sodium hydroxide, NaOH, is added to a blood solution containing carbonic acid, H_2CO_3, sodium bicarbonate and water are formed, and the hydrogen-ion concentration is lowered only moderately.

$$H_2CO_3 + Na^+ + OH^- \rightarrow NaHCO_3 + H_2O$$

In addition to the bicarbonate buffer system, two other buffer systems are important in body fluids. They are *phosphate* and *protein buffer systems*.

Phosphate buffers   Phosphates are present in body fluids as the more acid monobasic sodium phosphate, NaH_2PO_4, and as the more alkaline dibasic sodium phosphate, Na_2HPO_4. Since the monobasic form has two hydrogen atoms and one of them is readily ionized in body fluids, the monobasic form, NaH_2PO_4, acts as a weak acid, whereas the dibasic form, Na_2HPO_4, acts as a weak base. However, there is about four times as much alkaline sodium phosphate in body fluids as there is of the monobasic form. If a small amount of hydrochloric acid is introduced into the dibasic buffer system, the neutral salt NaCl is formed, and the hydrogen-ion concentration changes only slightly.

$$Na_2HPO_4 + HCl \rightarrow NaH_2PO_4 + NaCl$$

The addition of a base such as sodium hydroxide to a monobasic sodium phosphate buffer (the monobasic form acting as a weak acid) gives the following reaction:

$$NaOH + NaH_2PO_4 \rightarrow Na_2HPO_4 + H_2O$$

Mixtures of monobasic and dibasic phosphates are nearly neutral in their reaction.

Protein buffers A protein molecule is often very complex, since it may contain as many as 20 different amino acids in its structural organization. Furthermore, a very large molecule such as hemoglobin may contain around 140 amino acids in each of its four polypeptide chains. The proteins of blood plasma are weak acids, capable of dissociating to form hydrogen ions and conjugate basic proteinate ions. Plasma proteins are important buffers, and their buffering action extends over a wider range than that of inorganic buffers.

Hemoglobin is best known as the red coloring matter of the blood, which has a chemical affinity for oxygen. Actually hemoglobin is composed of (1) a protein, globin, and (2) heme, the red coloring matter. When it is carrying oxygen in loose chemical combination, it is called *oxyhemoglobin*, $HHbO_2$. After the oxygen has been given up to the tissues, the compound is referred to as *reduced hemoglobin*, HHb. Oxyhemoglobin is the stronger acid of the two, but both are capable of dissociating.

$$H^+ + Hb^- \rightarrow HHb$$
$$\text{base}$$

Hemoglobin is confined to the red cells of the blood, where its concentration is very high. It is regarded as one of the most important buffers of the blood. Hydrogen ions pass readily through the membrane of the red cell. Inside the red cell, hemoglobin reacts with potassium ions, K^+, to form alkaline potassium hemoglobinate, KHb, and buffers the carbonic acid, which passes into the red cells from the surrounding plasma. A more complete account of the function of hemoglobin will be found in Chapter 13. The discussion deals with the movement of ions through selectively permeable membranes.

$$H_2CO_3 + KHb \rightarrow HHb + K^+ + HCO_3^-$$

The pH of the blood and body fluids is regulated not only by buffer systems but also by the action of the excretory organs of the body. As the hydrogen-ion concentration of the blood increases, the respiratory system is stimulated to remove more CO_2. Also, the kidneys can excrete excess acids or alkalies and so aid in maintaining the normal pH of the blood (see Chapters 16 and 19).

SOLUTIONS

Physiological solutions commonly consist of a solid dissolved in a liquid. The liquid (solvent) may dissolve the solid substance without any chemical change, as sugar dissolves in water, or the solvent may react chemically with a solute to form a solution. The dissolved substance in a solution is the *solute*.

The concentration of a solution may be indicated in several different ways. It is commonly expressed as percentage or parts per hundred, such as the number of grams of the substance present in 100 milliliters of solution.

A mole or gram molecular weight is defined as the weight of a substance in grams equivalent to its molecular weight. If a substance has a molecular weight of 35, one mole would be 35 grams of the substance. A molar solution contains one mole or one molecular weight of substance per liter. Note that in preparing a molar solution, one dissolves 1 gram molecular weight of the solute in a volume of solvent sufficient to make 1 liter of solution. In a *molal* solution one starts with 1,000 grams of solvent and adds the gram molecular weight of the substance to be dissolved. This may result in a volume greater than 1 liter.

The millimole (0.001) is often used in biological preparations where much smaller amounts are required.

QUESTIONS

1 How can isotopes be used in the study of physiological processes?

2 Which is more alkaline, a solution of pH 8 or one of pH 10? Which is more acid, a solution of pH 2 or one of pH 5?

3 Explain the value of buffers and how they function.

4 What characterizes a strong acid?

5 Define a base.

6 What happens during ionization? Distinguish between an anion and a cation. In an electric field, the anions move toward which pole, the anode or the cathode? Why do they do this?

7 How is a salt produced?

8 Give examples of matter as it exists in liquid, gaseous, or solid states.

9 Discuss the structure of atoms.

10 Discuss the sharing of electrons and the transfer of electrons. Which types of atoms commonly share electrons, and which types usually transfer electrons?

11 What do you understand the terms covalence and electrovalence to mean?

12 In all probability, what were the first chemical compounds of biological significance formed on the earth?

13 Why is the carbon atom of special significance?

14 Define mole or gram molecular weight.

15 What differences of procedure should one consider in preparing a molar solution and a molal solution?

SUGGESTED READING

There are a number of good introductory textbooks of chemistry and physics that should prove helpful. In addition, the following paperbacks and articles are cited:

COULT, D. A.: "Molecules and Cells," chap. 2, Basic Chemistry for the Study of Living Matter, Houghton Mifflin Company, Boston, 1967.

GOLDSBY, RICHARD A.: "Cells and Energy," chap. 1, An Introduction to the Language of Chemistry, The Macmillan Company, New York, 1967.

KLEINSMITH, L. J.: Molecular Mechanisms for the Regulation of Cell Function, *BioScience*, **22**:343–347 (1972).

SPEAKMAN, J. C.: "Molecules," chap. 1, The Molecular Concept; chap. 2, The Atom; chap. 3, Valence and Its Electronic Interpretation, McGraw-Hill Book Company, New York, 1966.

chapter 2
functional aspects of cellular metabolism

All living cells must maintain the individual organization of their protoplasm, but at the same time they must provide for the intake of food materials and oxygen and also for excretion. An impermeable membrane would contain the protoplasm of the cell and all its constituents, but the cell could not survive because it could not receive nourishment or rid itself of waste products. A selectively permeable membrane restrains the larger molecules of substances such as colloids, of which protoplasm is composed, but permits the passage of smaller molecules of certain nutrient materials necessary for maintaining cellular metabolism. A selectively permeable membrane also permits the outward movement of excretory products. Cells, of course, are dependent upon a constant interchange of substances from their environment. Living membranes often act differently from what would be expected of purely physical forces providing passage of materials through nonliving membranes. The cell membrane, then, appears to be selective, permitting the passage of some substances but excluding others. We shall discuss some methods by which materials pass through membranes.

MOVEMENT THROUGH MEMBRANES

DIFFUSION   When we detect the odor of an orange being peeled, we have an example of the *diffusion*, or spreading, of gases through the air. Substances can also diffuse through a liquid medium, as when a crystal of copper sulfate or some potassium permanganate is dropped into water and the color can be seen to spread away from its source. The diffusing substance goes into solution in water, which in this case is its *solvent*. The molecules of the solute spread slowly from the area of greatest concentration and eventually spread evenly and reach an equilibrium throughout the solvent (Figure 2.1).

OSMOSIS   Molecules of a great variety of substances move in and out through cell membranes, but not all molecules are able to pass through. The membrane is called *semipermeable* or *selectively permeable*. In general, cell membranes must enclose the protoplasm of the cell but still permit oxygen to pass in and carbon dioxide to pass out. Nutritive materials also must enter the cell, and excretory materials must be disposed of.

Strictly speaking, *osmosis* refers to the passing of the solvent, not the solute, through a selectively permeable membrane. In living organisms the solutions are aqueous; therefore, the solvent is water passing through living cell membranes. In a common laboratory demonstration, a membrane is tied across the larger open end of a thistle tube. The thistle tube, filled with sugar solution or protein solute, is inverted and suspended in a beaker of water. If the membrane is a so-called "perfect membrane," the larger molecules

FIGURE 2.1

Diagrams illustrating diffusion and osmosis. Diffusion through a permeable membrane permits molecules of the solute to spread from the area of greatest concentration within the inverted thistle tube until they reach an equilibrium throughout the solvent. In osmosis, the movement of larger molecules out of the thistle tube is restricted by the selectively permeable membrane. In the case of sugar solutions, the greater movement of water is inward toward the solution. As a consequence, the diluted sugar solution rises in the thistle tube.

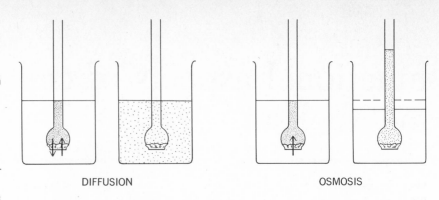

DIFFUSION   OSMOSIS

cannot pass out, but smaller water molecules can pass in either direction. The greater movement of water molecules will be inward toward the solution; there is a smaller concentration of H_2O molecules in the solution than in pure water. Consequently, the water level will rise in the neck of the thistle tube. The greater tendency of the solvent molecules to flow toward the higher concentration of solute substance (for example, the greater flow of H_2O molecules into the thistle tube than out of it) is measured as *osmotic pressure*. It is this pressure that causes the water level to rise in the thistle tube. In general, osmotic pressure can be determined for any soluble substance, but it varies with a number of factors, such as the concentration of the substance in solution and the temperature. Eventually, the solutions in the osmosis demonstration reach an equilibrium. In living systems, however, osmosis does not ordinarily reach an equilibrium, because materials passing through cell membranes are continually assimilated by the metabolic activities of protoplasm and excretory products are continually removed by the blood (Figure 2.2).

Living cells of the body are generally surrounded by a tissue fluid that has the same osmotic pressure as the protoplasm within the cell membrane. In such cases, a water balance exists between the contents of the cell and the surrounding medium, which is said to be *isotonic*. Blood is ordinarily isotonic to the red cells that it carries. A physiological saline solution (0.9 percent sodium chloride) is isotonic to mammalian red cells and is often used to dilute blood for study under the microscope.

If blood cells are placed in distilled water or any other *hypotonic solution*, water molecules move toward the higher concentration of salts in the protoplasm, and the cell swells and bursts. *Hypertonic solutions*, on the other hand, contain a greater concentration of salts or other solute, and water moves out of the protoplasm toward the greater concentration of solute. The cell then shrinks and presents a shriveled appearance (Figure 2.3).

FIGURE 2.2

Schematic diagram of the chemical makeup of the cell membrane.

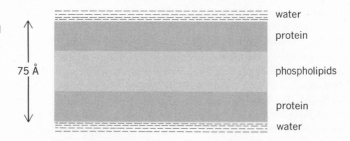

75 Å

water

protein

phospholipids

protein

water

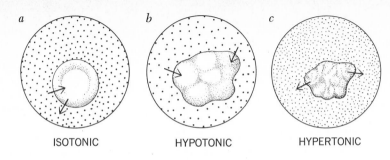

ISOTONIC   HYPOTONIC   HYPERTONIC

FIGURE 2.3
Passive transport. Effect of placing cells in solutions of various concentrations. *a* A cell placed in a fluid that has the same osmotic pressure as the contents of the cell, neither shrinks nor expands. The fluid is isotonic with the fluid of the cell. *b* If the cell is placed in a fluid of lower concentration than that existing within the cell, water enters the cell and the cell expands. The fluid is hypotonic with respect to the content of the cell. *c* A fluid of higher concentration than that found within the cell is a hypertonic solution. A cell placed in a hypertonic solution shrinks. The cell membrane is considered to be selectively permeable.

DIALYSIS Evidently cell membranes are not perfect membranes, because many substances in solution pass through. Artificial membranes having pores of different sizes can be prepared. These membranes permit crystalloids to pass but withhold the larger molecules of colloids. A simple experiment can be made by filling a piece of dialysis tubing with some colloid, such as gelatin in solution, and including a salt solution. The end of the tube is tied and the bag immersed in a beaker of water. The crystalloids (salts) pass through the membrane (*dialyze*), while the larger colloid molecules remain inside the bag. Water molecules move inward to replace salt molecules. It would be possible to remove the salt completely from the bag by continually replacing the water in the beaker, for, of course, the salt concentration will reach an equilibrium unless fresh water is added. The process of separating crystalloids from colloids in a solution containing both by permitting the crystalloids to pass through a membrane is called *dialysis*.

Filtration is a mechanical process whereby fluid and dissolved substances are passed through a physical barrier such as a membrane. Suspended substances may not be able to pass through the barrier because the particles are too large. The fluid that passes through is the *filtrate*. Filtration requires force; a good example is the filtration of solutes through the capillary wall as a result of higher pressure within the capillary than outside in the tissue spaces. Filtration through the glomeruli of the kidneys is another example.

TRANSPORT ACROSS MEMBRANES The protoplasm of the cell is colloidal in nature, and it is contained by the cell membrane. The cell is not a perfect osmotic system because many substances in solution are able to pass through the plasma membrane. Molecules of water, oxygen, and carbon dioxide pass through cellular membranes readily, but larger molecules, such as those of glucose, do not and therefore require some special mechanism for transportation through the membrane. It has become evident that simple diffusion and osmosis alone cannot adequately explain the movement of all substances in and out of the cell.

The concept of a porous cellular membrane persists, although pores have not been identified, even under electron microscopy. If present, they must be very minute, since a diameter of 7 angstroms or less (an angstrom is 0.0000001 millimeter) has been suggested for them. There is a hypothesis that the pores may open and close, the open pores compressing the membrane into pillars between the pores. Also that the pores would be likely to be closed by the preparation for electron microscopy and therefore would not be observed. At any rate, if the membrane were porous, the passage of small molecules and the exclusion of larger ones would be more readily explained. Since living membranes are permeable to water, minute channels containing water molecules may possibly exist. Porosity, even if it is a

Movement through membranes

property of the cell membrane, would not explain all the problems involved in transport. Membranes are selectively permeable and ordinarily do not permit the passage of ionized particles except under certain conditions.

It has been determined that cell membranes are composed of phospholipids and proteins. The outer limiting membrane is about 75 angstroms wide and consists of a double layer of proteins which partially constitute two dense lines, with a pale zone of lipids between. J. D. Robertson has called this the *unit membrane*, since it seems applicable to all cellular membranes (see Figure 2.3). The lipoprotein composition may very well account for the membrane's selective permeability. Some fat-soluble substances, such as alcohol and ether, pass through cellular membranes readily, but most of the larger, more complex molecules require some aid if they are to pass at all.

ACTIVE TRANSPORT It is generally accepted that the movement of ions such as sodium and potassium ions across plasma membranes against a concentration gradient requires some sort of active transport and involves so-called "biological pumps." Sodium ions outside the cell in the extracellular fluid have a concentration some 10 times greater than the concentration for sodium ions inside the cell, whereas the concentration of potassium ions inside the cell is around 30 times that existing outside. Sodium and potassium ions have the same positive charge. These ions evidently are forced or pumped through the cell membrane, since they are continually moving in and out and do not reach an equilibrium.

Changes in electric potential or chemical changes in the membrane facilitate the passage of some substances while inhibiting others. Changes in electric potential are clearly demonstrated in the initiation of muscular contraction and in conduction of the nerve impulse along nerve fibers. These examples are taken up later under the appropriate organ systems. Theories of chemical change within the membrane are usually based on the premise that enzyme systems within the membrane provide the energy for binding ions to a diffusible substance or that molecules may be chemically combined with a lipid-soluble substance which would then diffuse through the lipids of the cell membrane.

The concept of transport by a mobile carrier also may help to explain the movement of glucose and other substances against a concentration gradient. This concept postulates a chemical carrier that would combine with a monosaccharide, for example, on one side of the membrane, transport it through the membrane, and release it chemically unchanged on the other side.

The transportation of amino acids presents a similar problem. The molecules are too large to pass readily through the cell membrane. Though the mechanism of their transport is not fully understood, it is most likely that they are subject to active transport by chemical carriers.

OTHER METHODS OF TRANSPORT It is evident that our knowledge of transport across membranes is far from complete. It may be found that permeability of membranes is related to cell metabolism as a whole, rather than being just a property of the membrane. Apparently a considerable variety of enzymes and coenzymes are involved, and adenosine triphosphate (ATP) is available as an energy source. Furthermore, there is some evidence that the endoplasmic reticulum, the mitochondria, and the electron-transport system may influence the permeability of the plasma membrane (these are discussed in the following chapter). In bacterial cells, and quite possibly in other cells as well, there are enzymatic active-transport systems, called *permeases,* which seem to be highly specific in the transport of organic nutrient materials necessary for cellular metabolism. There is also the possibility that *ATPase* may play some part in membrane transport. ATPase is an enzyme that causes the breakdown of ATP. It has been considered that, in active-transport systems using ATP as an energy source, ATPase may also act as an enzyme involved in the active transport of materials through cell membranes.

HOMEOSTASIS

Free-living cells constantly struggle to maintain themselves in a changing environment. They must adapt to changes in temperature, hydrogen-ion concentration, lack of oxygen, lack of food materials, and many other varying conditions. In the higher organisms the entire body is subjected to changes in the external environment, but within the body the cells are protected by many different mechanisms that tend to stabilize the internal environment. Warm-blooded animals have an elaborate mechanism for maintaining a constant temperature. We have seen that buffers in the blood and tissues help to regulate the acid-base balance. There is also a chemical mechanism that functions to supply an adequate amount of oxygen to cells deep within the tissues of the body and to remove the waste products of metabolism. The tissue fluid is the liquid environment of cells in the body. This environment is kept in a state of relative constancy by the operation of regulatory mechanisms. The term *homeostasis* refers to this constant state of the internal environment.

Chemical and physical processes such as osmosis, filtration, or active transport across cell membranes would reach a state of equilibrium if tissues were not living and in an active state of metabolism. Fluids constantly move in and out of cells as food materials are supplied and used up, with a consequent release of energy. In a living organism, a state of *dynamic equilibrium* exists, in which there is a constant but balanced movement of materials in and out of the tissues.

ENERGY SOURCES

We read in the study of physics that thermodynamics is concerned with energy in its different forms. The first law of thermodynamics applies to the functions of living matter as well as to the nonliving. In general terms this important law states that the total amount of energy in the universe is constant, that is, energy cannot be created or destroyed. It can be converted from one form to another, but it is never lost or gained. This is also known as the law of the conservation of energy. The ball bearing which we encounter in physics laboratory classes exhibits potential energy even at rest, since it is not performing work. But when it rolls down an inclined plane, kinetic energy of motion is released and work is accomplished. The metal ball rolling against resistance also produces heat, another form of energy. Chemical energy is released when substances such as ATP or sugar are broken down into simpler components. Kinetic energy may appear in many forms such as electrical, chemical, physical, but it is always concerned with matter in motion. When a fire burns, combustion takes place in the presence of oxygen, dehydrogenation occurs, and energy in the form of heat is liberated. Both the combustion of fuels and the metabolic process known as cellular respiration represent the breaking down of high-energy compounds. In the case of fire and in the process of respiration, oxygen can act as a hydrogen acceptor, and water is formed as a by-product. When fuel is burned by fire, heat is liberated suddenly; in cellular respiration, food substances are broken down step by step, and their energy is liberated slowly. This energy is not set free, as in combustion, but is incorporated in other chemical substances to be released in stepwise fashion as the substance is degraded. In the redistribution of energy to more usable forms, 50 percent or more is trapped. When molecules in substances that can be used as fuel are built up (wood, coal, oil), chemical bonds are formed, and the energy that goes into the formation of these bonds comes directly or indirectly from the sun. The same can be said of food substances; the energy incorporated in the molecules of food substances represents stored solar energy.

Photosynthesis is the process in which light energy is used to form carbohydrate from carbon dioxide and water in the presence of chlorophyll, the green coloring matter in plants. Transforming light energy into chemical energy is a very complex process. Light energy excites electrons in chlorophyll molecules so that, in a series of steps or phases, energy in the form of adenosine triphosphate (ATP) is built up. In another phase, light energy is used to build up chemical energy involving an electron acceptor, nicotinamide adenine dinucleotide (NAD). A related coenzyme is nicotinamide adenine dinucleotide phosphate (NADP). The reduced form (NADPH) constitutes the source of hydrogen for the formation of carbohydrate. The chemical energy of ATP and NADPH is utilized in the formation of glucose from carbon dioxide and water. Another series of steps is involved, but the overall summary of events is expressed as follows:

$$6CO_2 + 12H_2O \rightarrow C_6H_{12}O_6 + 6O_2 + 6H_2O$$

Most botany and biology textbooks provide adequate descriptions of the various phases of photosynthesis; see also the references to articles in *Scientific American* at the end of this chapter.

OXIDATION-REDUCTION   The terms *oxidation* and *reduction* often are not clearly understood. Essentially, oxidation involves the removal of an electron from a molecule, whereas reduction involves the addition of an electron. When we recall that the hydrogen atom contains one proton and one electron, we can see why oxidation often means removing hydrogen and reduction means adding hydrogen. However, oxidation is accomplished not only by the release of hydrogen from a compound but by the addition of oxygen, or both. Either process involves the loss of electrons and the release of energy. Oxidations, therefore, are described as *exergonic;* that is, energy is given off as the molecule is "degraded."

Cellular respiration affords an example of oxidation in the use of glucose as a source of energy:

$$C_6H_{12}O_6 + 6O_2 \xrightarrow{\text{respiration}} 6CO_2 + 6H_2O + ATP + energy$$

In this reaction, glucose is oxidized and loses electrons; energy is released as the glucose molecular structure is broken down and is trapped in the terminal-bond phosphate of ATP. Respiration, then, is essentially the reverse of photosynthesis.

Reduction is *endergonic.* It is accomplished by the loss of oxygen or by the addition of hydrogen, or both. It includes the addition of electrons and requires the utilization of energy. Whenever an oxidation reaction occurs, there is an accompanying reduction reaction in some other substance. Since electrons are never set free but must always be accepted by another atom, it follows that, whenever an atom or group of atoms is oxidized, another atom or group of atoms is reduced.

Oxidation occurs as hydrogen atoms are released even though no oxygen is added. *Dehydrogenases*, the enzymes that catalyze the release of hydrogen from a substance, often require coenzymes which act as hydrogen acceptors. The coenzymes nicotinamide adenine dinucleotide (NAD), flavin adenine dinucleotide (FAD), and nicotinamide adenine dinucleotide phosphate (NADP) commonly act as hydrogen acceptors in biological oxidations. The names of NAD and NADP are new designations for diphosphopyridine nucleotide (DPN) and triphosphopyridine nucleotide (TPN), respectively. However, DPN and TPN are still in use. $NADH_2$ is reduced nicotinamide adenine dinucleotide.

The following reaction will illustrate some of the points under consideration:

$$NAD + CH_3CHOHCOOH \underset{+2H}{\overset{-2H}{\underset{\text{dehydrogenase}}{\overset{\text{lactic}}{\rightleftharpoons}}}} CH_3COCOOH + NADH + H^+$$

lactic acid   pyruvic acid

Here, lactic acid loses hydrogen and electrons and is oxidized to form pyruvic acid. The electrons and one hydrogen atom are accepted by NAD to form NADH, and H^+ is released. However, pyruvate—not lactate—is the primary product of cellular metabolism. A simplified sequence is as follows:

$$glucose \xrightarrow{\text{glycolysis}} pyruvate \xrightarrow{\text{tricarboxylic acid cycle}} CO_2 + H_2O$$

$$\Updownarrow$$

lactate

UTILIZATION OF ENERGY

Respiration in man involves not only the exchange of respiratory gases in breathing (O_2 and CO_2) and their transport to and from the tissues by way of the bloodstream but also a most important series of cellular reactions in which energy is obtained from nutrient substances such as glucose. The step-by-step dehydrogenation is enzymatically controlled so that the energy released is never really free, and only about half of it is given off as heat. The energy bound up in molecules of glucose or other foodstuffs is not directly available for the metabolic functions of cells. Cellular energy requirements are met by withdrawing stored or bound energy from a high-energy compound called ATP, adenosine triphosphate. Adenosine is composed of a purine called adenine (a nitrogenous base) and a five-carbon sugar, ribose. If one phosphoric acid group is added, one phosphate bond is added to yield adenosine monophosphate (AMP), a relatively low-energy compound. But oxidative metabolism permits adenosine monophosphate to form one or two additional high-energy bonds with phosphate, thereby increasing its bound energy tremendously. With one additional phosphate bond, usually indicated ~P, AMP becomes adenosine diphosphate (ADP) (Figure 2.4). An additional phosphate bond gives adenosine triphosphate (ATP). The great value of this substance lies in the ability of enzymes to transfer phosphate groups, including their energy capacity, to other substances. We shall see later that energy from ATP plays the key role in muscle contraction, provides for transmission

FIGURE 2.4

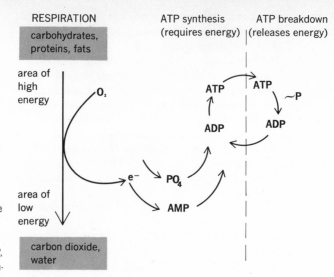

Diagram illustrating some aspects of cellular metabolism. Foodstuffs represent a high-energy level. Eventually they break down to CO_2 and H_2O, representing a low-energy level. The energy cycle indicates that some of the energy derived from the breakdown of foodstuffs contributes to the synthesis of ATP. The breakdown of ATP to ADP yields energy for cellular metabolism. The synthesis of ATP is illustrated at the left of the broken vertical line, whereas its breakdown to ADP with release of energy is shown on the right. ADP, adenosine diphosphate; AMP, adenosine monophosphate; ATP, adenosine triphosphate; PO_4, phosphate radical. The high-energy phosphate bond $\sim$P contributes energy for cellular metabolism.

of the nerve impulse, and takes part in probably hundreds of cellular processes. Energy also goes into the resynthesis of many compounds, including ATP reserves.

The energy supplied to the ATP energy-carrying mechanism must ultimately come from foods. The oxidation of certain breakdown products of carbohydrates, fats, and proteins provides the needed energy.

adenine +   ribose   → adenosine
 a purine   a pentose sugar, 5C

adenosine diphosphate (ADP)

adenosine triphosphate (ATP)

Phosphate groups are derivatives of phosphoric acid, H_3PO_4, but in equations representing energy transfer the phosphate group, PO_4, is commonly symbolized P. The high-energy phosphate bond is written $\sim$P.

CELLULAR ENZYMES AND COENZYMES

The cell is able to carry on the processes of metabolism with the aid of enzymes, which are organic catalysts. A catalyst initiates or regulates the rate of a chemical reaction without itself being changed by the reaction.

Cellular enzymes are biological catalysts aiding chemical reactions within the cell. A few enzymes, such as the digestive enzymes, function outside the cell or at some distance from their place of origin, but most enzymes are intracellular. A convention of naming provides that the ending -*ase* be applied to names of enzymes. An enzyme that catalyzes an oxidation-reduction reaction is therefore termed an *oxidoreductase*, and an enzyme concerned with the digestion of proteins is a *protease*. While perhaps a hundred different enzymes are necessary to carry on the activities of the cell, the majority bring about chemical changes by oxidation or by hydrolysis. Oxidative reactions, as we have seen, generally involve the addition of oxygen in a chemical reaction or removal of hydrogen atoms, or both. Hydrolysis, on the other hand, refers to the chemical changes brought about by the addition of one or more molecules of water to a compound and catalyzed by hydrolase. These changes result in the breaking of chemical bonds in the process of forming simpler substances. The digestive enzymes are hydrolases capable of breaking down the larger food molecules into smaller molecules. Hydrolases include the carbohydrases, proteases, and lipases as well as the phosphatases and nucleases. Enzymes are for the most part quite specific in their action; that is, each enzyme works best in catalyzing a single chemical. It is this greater "efficiency" in catalyzing one particular reaction, rather than a number of different reactions with equal efficiency, that is referred to by the term *enzyme specificity*. Thus a certain enzyme may be essential for one step in a chemical reaction, but a different enzyme may be equally essential for any further reaction.

Probably all enzymes are proteins. Many have been prepared in a highly purified state, and some have been crystallized, retaining their full activity. Since enzymes are protein in nature, they are inactivated by heat. Some enzymes have a metal or a chemical compound associated with the protein. These metals or compounds constitute the cofactor or coenzyme which must be associated with the protein if the enzyme is to retain its active state. The substance acted upon by the enzyme is called the *substrate*. Each enzyme catalyzes a chemical reaction most rapidly with a particular substrate but also catalyzes reactions with other substrates of the same general type, although perhaps more slowly.

The chemical reactions of metabolism are often intricate and involved. The process of deriving energy from food substances requires numerous enzymes and proceeds through an elaborate series of steps. A knowledge of the function of some of the enzymes, such as the decarboxylases, dehydrogenases, and oxidases, is helpful in understanding these steps.

In the initial phases of carbohydrate metabolism, glycogen and glucose are broken down to less complex substances. Low-energy phosphorylation accompanies the breakdown, and net energy is expended in the process. These steps, leading to the formation of pyruvic acid, represent an anaerobic phase. A later phase involving the tricarboxylic acid cycle is an oxidative phase, wherein a large amount of energy is produced and retained in the form of ATP. This is a most important phase of cellular energy metabolism.

Let us consider briefly some of the steps involved in the anaerobic breakdown of glycogen and glucose to pyruvic acid in the process called *glycolysis*. The first step involves a phosphorylase that catalyzes the transfer of a phosphate bond when glycogen is broken down to glucose 1-phosphate. The enzyme phosphoglucomutase catalyzes the conversion of glucose 1-phosphate to glucose 6-phosphate. The conversion of glucose to glucose 6-phosphate is catalyzed by the enzyme hexokinase. The phosphate group comes from ATP as this energy source is broken down to ADP. The glycolytic phase of carbohydrate breakdown ends with the production of pyruvic acid, as indicated in the following sequence:

$$\text{glycogen} \xrightarrow[\text{PO}_4]{\text{phosphorylase}} \text{glucose 1-phosphate}$$

$$\downarrow \text{phosphoglucomutase}$$

$$\text{glucose} + \text{ATP} \xrightarrow[\text{ADP}]{\text{hexokinase}} \text{glucose 6-phosphate}$$

$$\downarrow \text{glycolysis}$$

$$\text{pyruvic acid}$$

An understanding of the oxidative metabolism of pyruvic acid requires a study of the tricarboxylic acid cycle (Figure 2.5). In this cycle a great amount of energy is liberated from food substances and used in the formation of ATP. The details of the cycle will be considered in Chapter 16, but the cycle itself provides a good example of the essential action of enzymes and coenzymes. As pyruvic acid enters the cycle, CO_2 is given off and also hydrogen, in packets of 2H. The removal of CO_2 from substances undergoing oxidation is accomplished by a group of enzymes called *decarboxylases*. Certain vitamins may act as coenzymes to some of the decarboxylases. They are *cocarboxylases* and are associated with a derivative of the vitamin substance thiamine. Breaking up carbon chains by decarboxylation is a common method of oxidation. The carbon dioxide that is exhaled comes, in part, from the breakdown of pyruvic acid by this process.

At the beginning of the TCA cycle, pyruvic acid, $C_3H_4O_3$, is decarboxylated and oxidized, releasing H_2. The remaining two-carbon portion unites with coenzyme A and as acetyl CoA enters the cycle combining with four-carbon oxaloacetic acid to form six-carbon citric acid. There are two

FIGURE 2.5

The citric acid cycle is the energy cycle of cellular respiration. Here the breakdown products of food metabolism (carbohydrates, fats, and proteins) become further degraded stepwise, with the release of hydrogen or electron pairs as NAD is reduced to $NADH_2$ and FAD to $FADH_2$. These reductions occur at four places. There are also two decarboxylations where CO_2 is released within the cycle. One molecule of glucose produces two molecules of pyruvic acid. Therefore the cycle would be traversed twice in the breakdown of one molecule of glucose. The yield, then, is 16H, 2ATP, and $4CO_2$ molecules.

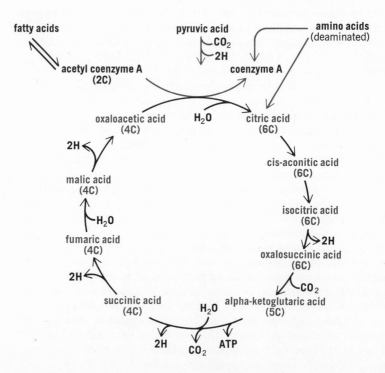

24

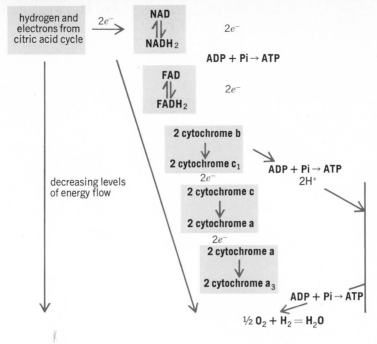

FIGURE 2.6

Hydrogen or electrons removed from food end products by the citric acid cycle are passed along a series of carrier substances called the electron-transport system. In this series of coenzymes and cytochromes there are three places where ATP is formed. This means that an electron pair passing along the system in oxidative phosphorylation produces three molecules of ATP. The terminal oxidation with cytochrome oxidase a_3 produces electrons which unite with molecular oxygen to form H_2O.

In the transfer of hydrogen from the citric acid cycle, NAD is commonly the hydrogen acceptor, but NADP also may act as a hydrogen acceptor in some cases. NAD carries either electrons or hydrogen atoms. FAD and coenzyme Q transport only hydrogen atoms, whereas the cytochromes transfer only one electron per cycle. In this case there is a dissociation of the hydrogen atom into H^+ and e^-.

more decarboxylations as the compounds are degraded to five- and four-carbon forms. The hydrogen or electrons released are readily accepted by the electron transport system where much of the energy produced is used in the formation of ATP.

Dehydrogenases catalyze the removal of hydrogen. In the cytochrome series the released hydrogen is accepted immediately by the coenzyme NAD. This coenzyme acts as an electron or hydrogen acceptor and passes electrons along to another acceptor called FAD. The vitamin riboflavin forms part of this group of flavoproteins. FAD passes electrons to heme-containing proteins called *cytochromes* and finally to cytochrome oxidase. This is a true *oxidase*, since it transfers electrons to molecular oxygen, resulting in the formation of water. Most of the energy of the electron-transport series goes into the formation of ATP. We shall consider the cytochrome system in greater detail later (see discussion of cytochrome system, Chapter 16). Figure 2.6 illustrates the electron-transport series.

In summary, physiological oxidations are catalyzed not only by oxidases but also by dehydrogenases. Dehydrogenation represents a complex series of steps by which hydrogen ions (or electrons) are transferred from one coenzyme to another until finally accepted by molecular oxygen.

Coenzymes, as we have seen, play an important part as catalysts in biological oxidations. The better-known coenzymes are those combined with vitamins of the B group, although it may be determined finally that all vitamins function as coenzymes. *Coenzyme A*, for example, is a phosphorylated nucleotide in combination with the vitamin pantothenic acid (see the section on vitamins in Chapter 18). It is the coenzyme of certain transfer reactions associated with the tricarboxylic acid cycle, and it functions not only in carbohydrate and protein metabolism but also in the degradation and synthesis of fatty acids (see Figure 2.5).

SUMMARY Materials are transported across membranes in several ways, including diffusion, osmosis, and active transport. The pressure resulting from the movement of water molecules toward the greater concentration of the soluble substance is called osmotic pressure. Dialysis is the process of separating crystalloids from colloids in a solution containing both by permitting the crystalloids to diffuse through a membrane that retains the colloids. Isotonic, hypotonic, and hypertonic solutions are defined and illustrated. Filtration is the removal of substances from suspension by passage of materials through a physical barrier, such as a membrane.

Active transport of materials across membranes may be accomplished in various ways. It may involve chemical combination of some substances with lipoproteins within the membrane or maintaining ionic equilibrium by pumping sodium ions out of the cell. It may mean employing enzyme systems to supply the energy for active transport. Transport by mobile carrier means that a chemical carrier combines with the substance to be transported, moves it through the cell membrane, and releases it, chemically unchanged, on the other side. Enzymatic active transport may include permeases, which have been demonstrated in bacterial cells. In general, chemical or physical aid in passing substances through a membrane may be considered as a kind of active transport.

Oxidation occurs when an atom or group of atoms loses one or more electrons. Reduction takes place when electrons are gained.

The utilization of energy is illustrated by the process of degrading carbohydrates or other foodstuffs and releasing energy. Phosphate bonds contribute to the building up of stored energy in ADP and ATP.

Cellular enzymes and coenzymes are discussed. An enzyme is a biological catalyst. The names of enzymes generally have the ending -ase. The substance acted upon by an enzyme is a substrate. The functions of oxidases, dehydrogenases, hydrolases, and decarboxylases are considered. Cytochrome oxidase functions as a part of the cytochrome series. Coenzymes are cofactors associated with enzymes. They activate enzymes.

QUESTIONS

1 Discuss the different methods of transport across membranes.
2 Describe a selectively permeable membrane.
3 What is meant by osmotic pressure?
4 How does filtration differ from osmosis?
5 Explain dialysis.
6 Discuss the concept of transport by mobile carrier.
7 If blood cells are placed in a hypotonic solution, what will happen to them?
8 In what ways can a substance be oxidized?
9 Discuss the utilization of energy by cells.
10 Explain the action of enzymes and coenzymes.
11 When muscles contract and a part is moved, what kind of energy is used? If the arm is simply flexed at the elbow and no load or weight is held in the hand, is any work accomplished? Is any heat produced?

SUGGESTED READING   BAILER, J. C., Jr.: Some Coordination Compounds in Biochemistry, *Am. Scientist,* **59**(5):586–592 (1971).
Discusses the role of metal ions in life processes and in metal-containing enzymes.
HENDRICKS, STERLING B.: Salt Transport across Cell Membranes, *Am. Scientist,* **52**:306–333 (1964).
RABINOWITCH, E. I., and GOVINDJEE: The Role of Chlorophyll in Photosynthesis, *Sci. Am.,* **213**:74–83 (1965).
An article on chlorophyll as catalyst for the transfer of hydrogen atoms.

ROBERTSON, J. DAVID: The Membrane of the Living Cell, *Sci. Am.*, **206:**64-72 (1962).

An excellent popular article.

SOLOMON, ARTHUR K.: Pumps in the Living Cell, *Sci. Am.*, **207:**100-108 (1962).

A discussion of active transport in the excretion of sodium.

WHITTAM, R., and K. P. WHEELER: Transport across Cell Membranes, *Ann. Rev. Physiol.*, **32:**21-60 (1970).

A technical discussion of various aspects.

Suggested reading

27

chapter 3
cellular structure and function

The cell affords one of the most engaging subjects for study in the entire field of biology. Numerous books and articles have been written about it. Modern scientific technology enables us to look into its minute structures as never before, while physiological and biochemical studies have unraveled life processes hardly imagined only a few years ago.

In the first two units we have considered some physical and chemical aspects of cell physiology and of cellular metabolism. We turn now to the cell itself and study the structure and function of its component parts.

SOME PHYSIOLOGICAL ASPECTS OF LIVING MATTER

The living matter of the cell is commonly called *protoplasm*, although this has become a very general term. In a broad sense it refers to the living matter within the cell or plasma membrane. The fluid substance of the nucleus may be called *nucleoplasm*, and the viscid material outside the nucleus, *cytoplasm*. Various structures within the cytoplasm are sometimes called, in a general way, *cytoplasmic inclusions*. However, it is more usual to refer to the specific membranes or organelles by name, since their individual structures and functions are fairly well known.

A chemical analysis of cellular contents indicates that cellular material contains oxygen, hydrogen, carbon, nitrogen, and many other elements found in nonliving substances, but within the cell this complex material has the unique distinction of being alive and performing all the activities of living matter.

In the living cell these physiological activities include irritability, motility, the various aspects of metabolism, growth, and reproduction.

Irritability This refers to the ability of a cell or organism to respond to external stimuli, which may include the effects of a change in the cell's environment. In an individual cell, the stimulus may cause a change to occur at the cell membrane or within the cell itself. The stimulus may induce an excitatory state in the cell or membrane, as in nerve or muscle cells. All cells are capable of responding to a proper stimulus, but nerve cells exhibit a high degree of irritability and respond by conducting a nerve impulse (*conductivity*). Muscle cells respond by contracting and exhibit a high degree of *contractility*.

Motility The function of movement, of which there are various forms, is called motility. The streaming of protoplasm in the ameba and in the cells of certain plants can be demonstrated. Ciliated cells afford another example of motility. The male sex cells, or spermatozoa, are capable of locomotion by swimming movements. Certain white blood cells are capable of ameboid movement. The contractility of muscle demonstrates a very high degree of motility in the contraction of muscle fibers.

Metabolism This term is applied to life processes in general. It involves all the chemical activities of the cell or the organism. The metabolic process concerned with building new materials and storing energy is called *anabolism;* chemical activity concerned with the breakdown of materials into simpler substances and releasing energy is referred to as *catabolism.* Examples of metabolic processes are respiration, digestion, assimilation, secretion, and excretion.

Metabolic processes utilize energy. We have seen that green plants use light energy in the presence of chlorophyll to synthesize carbohydrates, proteins, and fats. Animal cells break down these food substances and release their energy in the process of respiration. However, the energy obtained from foods is not used directly. It goes into the formation of ATP. The release of energy from the breakdown of ATP supplies the energy for the metabolic processes that represent the work of the cell.

Growth and multiplication of cells Growth of a cell is characterized by an increase in its volume or mass and is closely associated with metabolism. If anabolism is greater than catabolism, growth takes place. Growth occurs when there is a synthesis of more complex substances from simpler ones, as in the synthesis of proteins from amino acids. When a cell attains its growth, it may undergo mitotic division; the resulting smaller cells eventually grow to full size. Tissues grow by the multiplication of cells, normally through mitotic division and the subsequent expansion of these cells to normal size.

Reproduction The ability of cells to multiply asexually by mitotic division, as in the growth of tissues, should not be confused with the ability of an organism to reproduce new individuals sexually and so perpetuate the species. Sexual reproduction in the higher animals and man refers to the production of germ cells, spermatozoa and ova, that are capable of forming a new individual by fertilization. The development of the germ cells is discussed in Chapter 22.

STRUCTURE OF A CELL

There are many kinds of cells, and they assume many shapes. In general, a cell is a microscopic mass of protoplasm containing a nucleus and more or less spherical in shape. But there are also the spindle-shaped cells of visceral muscle tissue, the thin flat cells of squamous epithelium, the cube-shaped cells of cuboidal epithelium, and many others. Though most cells are microscopic in size, some egg cells with their nutritive material may be quite large. The human red blood cell is 0.008 millimeter in diameter, but voluntary-muscle cells may be as long as 4 centimeters. The axon process of a motor-nerve cell may extend from the lower part of the spinal cord to a muscle in the foot and be well over 3 feet long. The diameter of the nerve-cell body, however, is only about 125 to 130 microns (0.125 to 0.130 millimeter) even for the larger motor-cell bodies of the spinal cord (Figure 3.1).

THE CELL MEMBRANE The contents of a cell are contained by a thin, flexible cell membrane or plasma membrane. The membrane as seen in an electron micrograph is approximately 75 angstroms thick and consists of two dense portions composed of proteins with a less dense lipid layer in between (Figure 3.2). It is a selectively permeable membrane and osmotically active. The contour of the membrane varies considerably in different cells. It may project outward in a series of minute folds, called *microvilli* (Figure 4.4), or it may

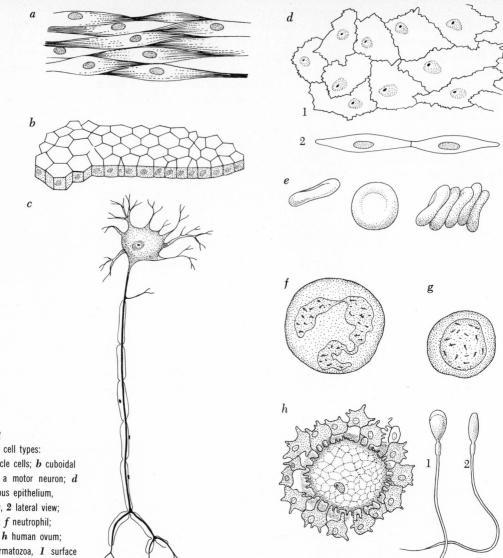

FIGURE 3.1
Representative cell types:
a smooth-muscle cells; *b* cuboidal
epithelium; *c* a motor neuron; *d*
simple squamous epithelium,
1 surface view, *2* lateral view;
e erythrocytes; *f* neutrophil;
g lymphocyte; *h* human ovum;
i human spermatozoa, *1* surface
view, *2* lateral view.

invaginate to form *pinocytic vesicles. Pinocytosis* means "drinking" by cells and appears to be still another way in which fluids and some more solid substances may enter cells. The invaginations become pinched off to form vesicles within the cytoplasm, where vesicles and vacuoles are commonly observed. The vesicles may contain food substances, but it is not known exactly how these substances pass out of a vesicle into the cytoplasm; apparently the membrane and the materials enclosed are absorbed by the cytoplasm.

Deeper invaginations of the surface membrane form a network throughout the cytoplasm. This series of irregular channels is called the *endoplasmic reticulum,* (Figure 3.3). The membranes, with a few exceptions, are studded on the sides facing the cytoplasm with minute dark-staining granules called *ribosomes.* Granular endoplasmic reticulum is also called

Cellular structure and function

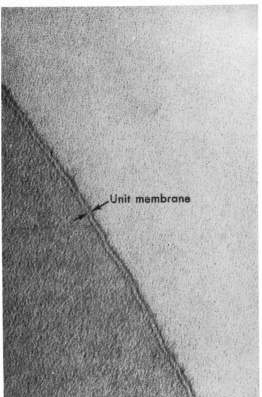

FIGURE 3.2
The unit membrane. An electron micrograph showing the three-layered structure of an erythrocyte plasma membrane. ×280,000. [*Courtesy of J. D. Robertson, from Roy O. Greep, "Histology," McGraw-Hill Book Company, New York (1973).*]

Unit membrane

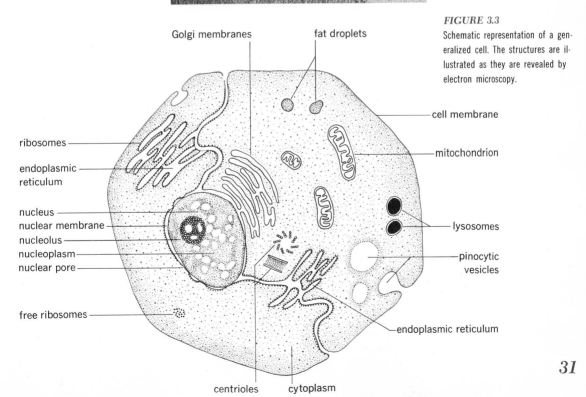

FIGURE 3.3
Schematic representation of a generalized cell. The structures are illustrated as they are revealed by electron microscopy.

Golgi membranes

fat droplets

ribosomes

endoplasmic reticulum

nucleus

nuclear membrane

nucleolus

nucleoplasm

nuclear pore

free ribosomes

cell membrane

mitochondrion

lysosomes

pinocytic vesicles

endoplasmic reticulum

centrioles

cytoplasm

ergastoplasm and is sometimes referred to as *rough reticulum*. Ribosomes contain ribonucleic acid (RNA) and play an important part in the synthesis of proteins.

The *Golgi apparatus* is an agranular, or smooth, laminated membranous structure. This specialized complex is usually located near the nucleus. Some other parts of the endoplasmic reticulum are also smooth but are not considered to be a part of the Golgi complex. The Golgi apparatus consists of stacks of saccules, usually eight to ten, piled on top of one another, not unlike a stack of miniature pancakes in appearance. Secretory products enter the bottom layer, and the chemically prepared secretions emerge from the top layer. It has been determined that in a variety of secretory cells the Golgi apparatus is the place where carbohydrate is added to the protein to produce large molecules of glycoprotein. In outline, after amino acids are built into proteins on ribosomes, the protein molecules become associated with the endoplasmic reticulum and are conveyed toward the Golgi apparatus, where they are later found in the basal saccules. Here the protein molecules are combined with carbohydrate to form glycoprotein. In goblet cells, which are mucus secreting, the upper or distal saccules become transformed into globules of mucus, which are released at the top of the cell. It appears, then, that the Golgi apparatus occupies a key position in the membrane system of the cell, regulating the packaging and flow of materials into the cytoplasm or, in the case of secretory cells, outward through the cell membrane.

The endoplasmic reticulum also forms the outer portion of the nuclear membrane. The inner membrane appears to be separate from the outer membrane, but at intervals the two appear to fuse, forming thin single layers. These thin areas are called *pores*. In many photomicrographs the pores appear to be wide open, permitting the nucleoplasm to be in direct contact with the cytoplasm; this may indeed be true.

THE CYTOPLASM The clear, viscid, more or less fluid portion of the cell exclusive of the nucleus is called the *cytoplasm*. It contains food substances, such as glycogen or glucose, proteins, and lipids. There are also enzymes and various electrolytes, such as potassium, magnesium, chloride, phosphate, and calcium, and, in lesser amounts, sodium and bicarbonate. Often the cytoplasm close to the cell membrane is in more of a gel state than liquid and is sometimes referred to as the *ectoplasm*. The more liquid cytoplasm in the interior of the cell is the *endoplasm*. We have already discussed the endoplasmic reticulum and the Golgi complex. There are also numerous cytoplasmic inclusions, or organelles, in what may be called the *cytoplasmic matrix*. The more clearly defined of these minute structures or inclusions are the mitochondria, lysosomes, and centrioles.

Mitochondria The cytoplasm of the cell contains several hundred to several thousand minute bodies called *mitochondria*. They vary somewhat in shape but are usually elongate, sausage-shaped structures approximately 15,000 angstroms long and 5,000 angstroms wide. They are visible, when properly stained, under the high power of the compound microscope, but their detailed structure is revealed only by the electron microscope. The outer wall, or covering, is composed of an outer and inner membrane with a fluid-filled space between (see Figure 3.4). The inner membrane extends into the interior, forming a series of shelflike projections, called *cristae*, which are sacs filled with fluid. The liquid is believed to be gel-like, rather than a thin fluid. The inner chamber into which the cristae project is filled with

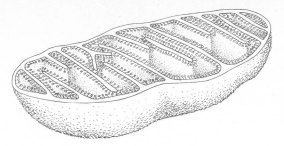

FIGURE 3.4

The mitochondrion, magnification about ×400,000. Outer surface particles have been observed on rat-liver mitochondria but not on other types.

a somewhat dense material called the *matrix*. The outer membrane in most cases appears to be smooth. Cristae of the inner membrane are studded with spherical particles mounted on stalks (Figure 3.5). The stalks, in turn, appear to be attached to globular base pieces within the membrane. These tripartite structures and the inner membrane are thought to be concerned with oxidative phosphorylation and electron transport of the cell, but the evidence is not conclusive. The enzymes of the citric or tricarboxylic acid cycle (Figure 2.5) generally have been considered to be located in or on the inner membrane, rather than in the outer membrane.

The mitochondria have been called the powerhouses of the cells, since nearly all the ATP is synthesized in these organelles. This means that enzymes associated with the ultimate breakdown of food substances by way of the tricarboxylic acid cycle are located here. These include the dehydrogenases and decarboxylases. Glycolysis, however, takes place in the cytoplasm and is not directly associated with the particles within the mitochondria.

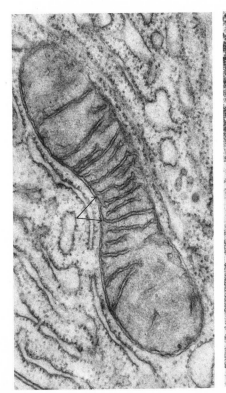

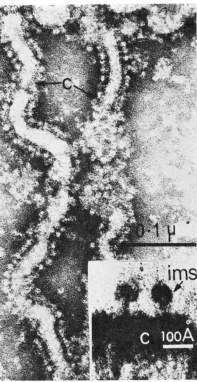

FIGURE 3.5

a Electron micrograph of a mitochondrion; arrows indicate cristae. ×64,000. [*Courtesy of James A. Freeman (see bibliography).*] *b* Stalked subunits projecting from cristae. ×192,000. Inset: subunits at higher magnification. Spacing between units is 100 angstroms. C, christae; ims, inner membrane subunits. [*Courtesy of Donald F. Parsons, Science, 140:985–987 (1963). Copyright, 1963, by the American Association for the Advancement of Science.*]

Structure of a cell

Electrons released from the tricarboxylic acid cycle are accepted by the coenzyme NAD (nicotinamide adenine dinucleotide). NAD is thereby reduced to NADH + H+ (Figure 2.6). Electrons are passed along a series of catalysts including NAD, FAD (flavin adenine dinucleotide), and the cytochrome series, which ends with cytochrome oxidase and the transfer of electrons and hydrogen to molecular oxygen in the formation of water (see discussion of the cytochrome system, Chapter 16).

Cellular respiration, therefore, is largely a function of the mitochondria. As hydrogen and electrons are passed along the series of catalysts, the high-energy phosphorylated nucleotide, adenosine triphosphate (ATP), is formed. This process requires oxygen and is referred to as *oxidative phosphorylation*. The phosphorylation occurs at certain places along the electron-transfer chain; at these places there is a coupling of inorganic phosphate to adenosine diphosphate to produce ATP. A large amount of energy released by oxidation is trapped and stored in the high-energy bonds of ATP. The mitochondria, then, represent the energy source of the cell.

There is a hypothesis that mitochondria do not arise *de novo* but that they come from preexisting mitochondria; that is, they are self-replicating bodies. There seems to be evidence that mitochondria (and centrioles) contain deoxyribonucleic acid (DNA) of their own, not nuclear DNA. If this evidence proves to be correct, these structures may be units of cytoplasmic inheritance.

Some electron micrographs suggest that mitochondria may arise from membranes of the endoplasmic reticulum, but there seems to be considerable doubt about this interpretation at the present time.

The microsome fraction This is a term commonly applied to the particles obtained from centrifugation of disintegrated cells. The microsome fraction may consist of fragments of the endoplasmic reticulum, some fragments with ribosomes still attached, ribosomal fragments smaller than mitochondria, and ribosomal particles.

Lysosomes Lytic bodies in the cell enclosed in a limiting membrane are called *lysosomes*. These dark-staining bodies were originally known as the "dense bodies" of the cytoplasm. The enzymes they contain are largely hydrolases, which function in breaking down large molecules by the addition of water. The enzymes include phosphatases, glycosidases, sulfatases, cathepsins (protein-digesting enzymes), ribonuclease, and deoxyribonuclease. Lysosomes are organelles of intracellular digestion. They exist in various forms; some are filled with enzyme-storage granules; others are active digestive vacuoles, while a third type consists of inactive residual bodies filled with indigestible material. Phagocytic white blood cells contain large digestive granules, which are similar to, if not identical with, lysosomes. In general, it appears that lysosomes in various kinds of cells digest materials taken into the cell by pinocytosis or phagocytosis. If lysosomes break down, their enzymes are capable of digesting the cell itself.

The contents of the primary lysosome, or storage granule, are thought to arise from ribosomes and to pass through the endoplasmic reticulum to the Golgi apparatus, where the mature lysosomes appear, each containing a specific enzyme. It is the secondary lysosome, or digestive vacuole, that contains a variety of enzymes.

Centrioles The high power of the microscope reveals near the nucleus a somewhat dense area in the cytoplasm commonly referred to as the *centrosome* (central body). It is usually associated with the Golgi complex and contains one or two dark bodies called *centrioles*. The electron microscope has revealed the rather surprising detail of the centriole. Each centriole is paired, and the two parts usually lie at right angles to each other [that is, in electron micrographs one centriole usually appears in cross section and the other in longitudinal section or position (Figure 3.3)]. Each consists of nine groups of cylindrical bodies arranged in a circle, each group commonly containing two or three cylindrical fibers. Centrioles are approximately 0.5 micron long and 0.15 micron in diameter. They are self-replicating bodies, and there is no limiting membrane. They appear to play some part in the formation of the spindle during mitotic division. However, they have not been demonstrated in the cells of flowering plants.

It is of interest that the basal bodies of cilia, flagella, and sperm tails have a structure similar to centrioles. In this case there are the typical nine groups of cylindrical fibers plus a pair of cylindrical fibers in the middle of the circle. In motile sperm, the groups of fibers extend from the basal body into the sperm tail.

THE NUCLEUS   The nucleus is a vital part of the cell organization. Located near the center of the cell, it is usually a spherical body, although it may assume various shapes. The *nucleoplasm* is the fluid protoplasm, or nuclear sap, surrounded by the nuclear membrane. As we have seen, the nuclear membrane is composed of two layers with more or less open spaces called *pores*. The outer membrane is continuous with the endoplasmic reticulum. The nucleoplasm contains a small spherical body, the *nucleolus* (little nucleus). The nucleus may contain two or more nucleoli. In fixed and stained preparations the nucleus appears as a dark body in the pale cytoplasm and the nucleolus as a still darker body within the nucleoplasm. The nucleolus has no membrane surrounding it. Internally it is composed of dense granules and fibrils of ribonuclear protein, which stains darkly. There are large amounts of ribonucleic acid in the nucleolus, and there is good evidence that ribosomal RNA is synthesized there. There is evidence also that the nucleolus plays a part in the biogenesis of cytoplasmic ribosomes. The nucleus contains the chromosomes, which are in a very extended, threadlike form during interphase, when the cells are not dividing. During cell division they are evident as thickened, compact bodies.

Chromosomes   The human cell has 46 chromosomes. These chromosomes contain genes, which are the essential hereditary factors of the cell and of the organism. Genes are responsible also for the development and regulation of the metabolic processes performed by the cell. For example, genes determine protein synthesis. Enzymes are protein in nature, and therefore genes control the formation of hundreds of enzymes. These enzymes are thought to be specific for catalyzing hundreds and perhaps thousands of chemical reactions which characterize all phases of cellular metabolism.

The principal chemical substances in chromosomes are basic proteins, called *histones*, and deoxyribonucleic acid (DNA). Chemical analysis of nuclei indicates that RNA is present in chromosomes in lesser amounts. Alkaline phosphatase is also revealed.

The DNA molecule   The DNA molecule is a most interesting structure in that it is the key to life processes and is an example of a complex molecule able to replicate itself. This nucleic acid is constructed as a double, rather than single, helix chain of nucleotides, each strand being the molecular complement of the other, which would help to explain its replication. The DNA

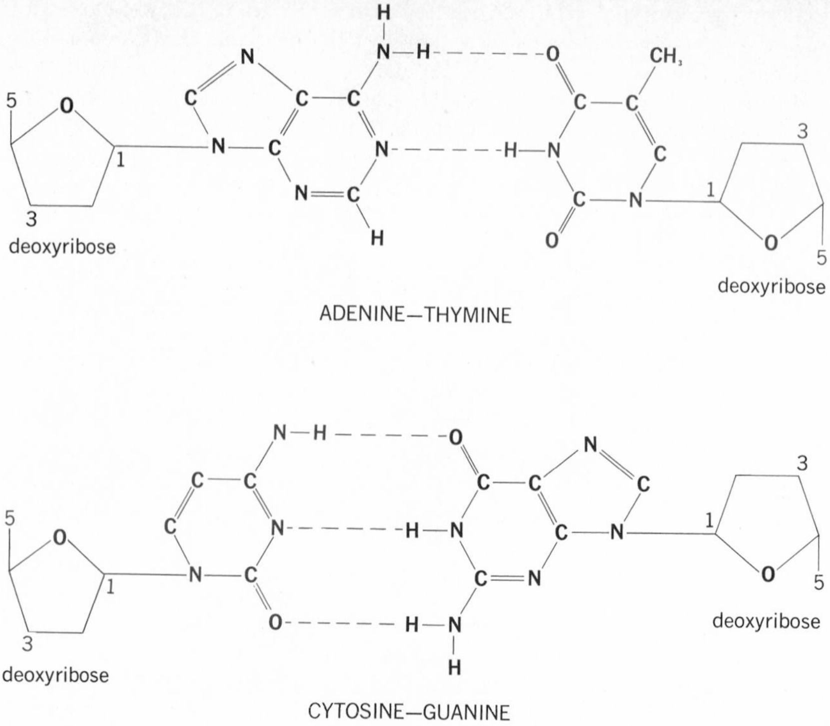

ADENINE—THYMINE

CYTOSINE—GUANINE

FIGURE 3.6

Structural formulas for adenine and thymine, cytosine and guanine.

TABLE 3.1

Comparison of bases and sugars in DNA and RNA

Bases	DNA	RNA
	CG	CG
	AT	AU

Sugars	Deoxyribose	Ribose
	CG	CG
	GC	GC
	AT	AU
	TA	UA

NOTE: A, adenine; C, cytosine; G, guanine; T, thymine; U, uracil.

Cellular structure and function

36

molecule is so constructed that it resembles a ladder twisted into a helix (a spiral, somewhat like the thread of a screw). The sides of the ladder are composed of deoxyribose sugar groups alternating with phosphate groups. The two sides of the ladder show a reverse polarity with reference to the sugar and phosphate units. The rungs of the ladder, which extend between two sugar groups, contain nitrogenous bases held together by hydrogen bonds. The nitrogenous bases are adenine, thymine, guanine, and cytosine (Figure 3.6 and Table 3.1). The bases are arranged so that adenine is paired with thymine and guanine is paired with cytosine. In 1953 Watson and Crick developed a proposed structure of the DNA molecule while they were working together at the Cavendish Laboratory in Cambridge, England. Models show only a small part of the DNA chain of nucleotide units, which may be well over 10,000 units in length (Figure 3.7).

The ratio of adenine to thymine or of guanine to cytosine is constant in any given organism; therefore the sequence of these bases must provide the genetic code for each organism. For example, the code specifies which amino acids are to be linked into polypeptide chains in the formation of proteins.

Adenine and guanine are purines, whereas cytosine, thymine, and uracil are pyrimidines. These bases will be considered further in the discussion of DNA and RNA. Probably adenine is more familiar in the form of adenosine. When the purine adenine is combined with a pentose sugar (ribose), it forms adenosine, a nucleo*side*. If a phosphate group is added to the sugar, the RNA nucleo*tide*, adenosine monophosphate (AMP, adenylic acid), is formed. Polymerization of nucleotides produces nucleic acids, such as DNA and RNA.

Replication of DNA occurs at interphase, just preceding mitotic divi-

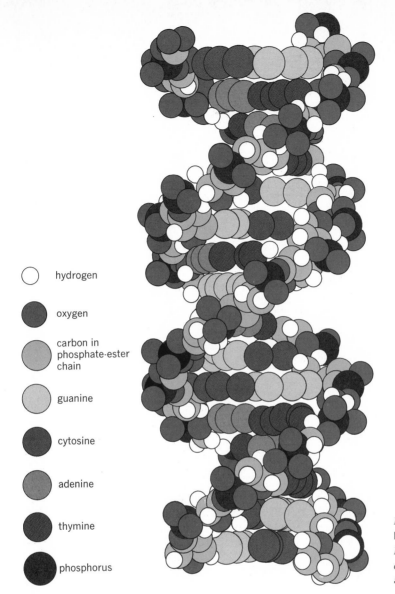

○	hydrogen
●	oxygen
●	carbon in phosphate-ester chain
●	guanine
●	cytosine
●	adenine
●	thymine
●	phosphorus

FIGURE 3.7
Molecular model of DNA. [*From L. D. Hamilton, A Bulletin of Cancer Progress, CA,* **5:***159 (1955).*]

a nitrogen base + a pentose sugar → a nucleoside

pyrimidines or purines + pentose + phosphate → nucleotides

hundreds or thousands of nucleotides $\xrightarrow{\text{polymerization}}$ nucleic acids

sion, in all cells that are capable of dividing. The genetic code, then, is duplicated at each cell division, the original DNA molecule acting as a template for the new. The 46 chromosomes in the human cell comprise 23 pairs. For a gene located at one point on one chromosome, there exists another gene resembling it on the other chromosome of the pair, for the male parent contributes 23 chromosomes in the sperm, and the female parent contributes 23 chromosomes in the ovum. The chromosome pairs are

Structure of a cell

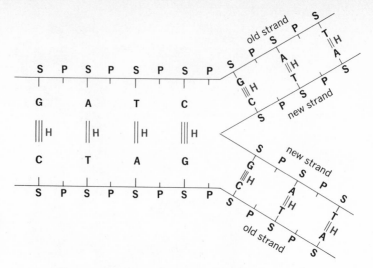

FIGURE 3.8
Diagram illustrating the structure
of the DNA molecule and its
method of replication. A, adenine;
C, cytosine; G, guanine; T, thymine;
H, hydrogen; P, phosphate; S,
sugar.

restored at the time of fertilization of the ovum. In every mitotic cell division thereafter each chromosome divides (see Figure 3.12).

At this point we shall digress briefly to consider the structure and classification of some of these bases. Pyrimidines and purines are nitrogenous bases containing in their ring structure both carbon and nitrogen atoms. Pyrimidines are represented by a six-sided ring with four carbon atoms and two nitrogen atoms. Purines retain the pyrimidine ring but include a side ring consisting of one carbon atom and two nitrogen atoms (Figure 3.6). These bases combine with phosphorylated pentose sugars (five carbon atoms) to form more complex molecules called *nucleotides*. Nucleotides are the subunits of nucleic acids.

Functions of DNA and RNA If the DNA molecule is to serve as the bearer of hereditary characteristics in the genes, it must be able to perform at least two functions: It must be able to replicate itself, maintaining the integrity of its genetic code, and it must provide a code for the production of proteins. It is thought that the replication of the DNA molecule probably begins at one end, where it unwinds and forms new strands complementary to the old strands, as shown in Figure 3.8. The gradual uncoiling and rotation would produce two molecules exactly alike. Note that adenine and thymine form two hydrogen bonds between them, whereas guanine and cytosine are held together by three hydrogen bonds.

The DNA molecule, in addition to replicating itself, must be able to act as a template in the formation of ribonucleic acid in order to provide for the synthesis of proteins. In the same general way that the enzyme DNA polymerase is essential to catalyze the polymerization of DNA from its four deoxyribonucleotides, so the enzyme RNA polymerase is necessary to catalyze the synthesis of RNA from its DNA template. The four essential triphosphates are ATP, UTP, CTP, and GTP. RNA is a complementary copy of only one of the two DNA strands. Probably only a part of the DNA strand is coded at any one time. Whereas the DNA molecule is remarkably stable, RNA breaks down in a short time and new RNA is resynthesized. It has been suggested that the RNA molecule resembles a hairpin loop, with the single strand twisted on itself but with some regions unpaired. The base pairing is believed to be incomplete, probably because there is a lack of

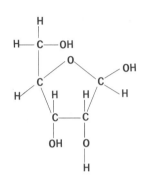

URACIL   PHOSPHORIC ACID

DEOXYRIBOSE   RIBOSE

FIGURE 3.9
Structural formulas of uracil, phosphoric acid, deoxyribose, and ribose.

complementary bases on an opposing strand. The single strand consists of repeating units of the pentose sugar ribose (instead of deoxyribose, found in DNA) and phosphate. One of the four bases is attached to each sugar unit. Three of the bases are identical with those found in DNA. The fourth base is uracil, which replaces the thymine of the DNA molecule. It is closely related chemically to thymine (Figure 3.9).

Approximately 75 percent of the cytoplasmic RNA in animal cells is found in the ribosomes (ribosomal RNA). Some RNA is found also in the nucleus. Nuclear RNA in animal cells constitutes about 15 percent of the total cellular RNA. It is found principally in the chromosomes and in the nucleolus. There is also a small amount in the nuclear sap, but this portion is not clearly defined. It is of interest that nuclear RNA varies in amount in different kinds of cells and that therefore the amount of RNA is not directly related to the amount of DNA in the cell.

Three kinds of RNA are commonly recognized; these are called, provisionally, *ribosomal* RNA (*r*RNA), *messenger* RNA (*m*RNA), and *transfer* or *soluble* RNA (*t*RNA or *s*RNA).

Ribosomes and ribosomal RNA Earlier we noted that the endoplasmic reticulum was studded with ribosomes on its outer surfaces. Ribosomes are spheroid bodies containing about equal amounts of RNA and protein. As we have indicated, the synthesis of ribosomal RNA occurs in the nucleolus. The structural RNA as part of the ribosome apparently takes no part in the genetic coding for the synthesis of proteins. Ribosomes are commonly considered to be composed of two subunits, which are usually described in Svedberg units. A Svedberg unit is based on the sedimentation rate of

Structure of a cell

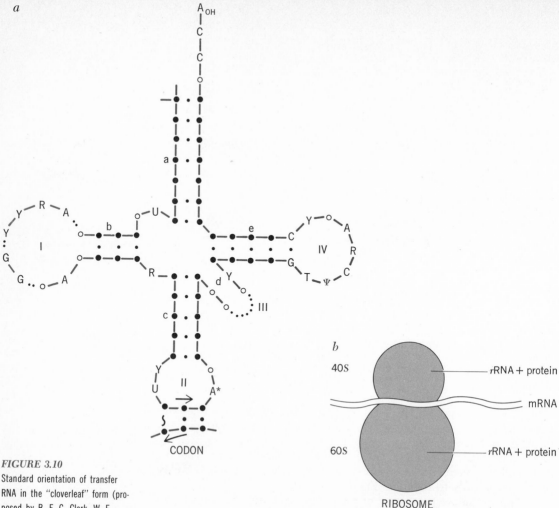

a

FIGURE 3.10

Standard orientation of transfer RNA in the "cloverleaf" form (proposed by B. F. C. Clark, W. E. Cohn, and T. H. Jukes). I, II, III, and IV are "loops" or unpaired base regions; a, b, c, d, and e represent helical (base-paired) regions or "arms." The upper arrow indicates position of the anticodon. Loop III is not always present. [*Courtesy of T. H. Jukes and Lila Gatlin, in Recent Studies Concerning the Coding Mechanism, Prog Nucleic Acid Res Mol Bio, 11:303–350 (1971)*] *b* A ribosome. 40S and 60S are sedimentation constants for eukaryotic cells with 80S ribosomes.

40

particles thrown down in the ultracentrifuge. The sedimentation constant for bacterial ribosomes is 70S, and for the two subunits including protein it is 30S and 50S (Figure 3.10*b*).

Protein synthesis takes place on groups of ribosomes called *polyribosomes* or *polysomes*, rather than on individual ribosomes. Electron microscopy reveals a fine filament connecting ribosomes and believed to be messenger RNA. There seems to be some evidence of an opening between the subunits of ribosomes. This has been interpreted as a possible opening for *m*RNA to pass through. It might also permit *t*RNA to reach *m*RNA at this point, but this is hypothetical. Another interesting suggestion, backed by some experimental evidence, proposes that ribosomal subunits come together for the synthesis of proteins but that at other times they may exist as individual subunits.

Messenger RNA A second type, messenger RNA, is considered to be nuclear RNA that moves out of the nucleus and into the cytoplasm, carrying information necessary for the synthesis of proteins. It then becomes attached to ribosomes. The *m*RNA formed on a strand of DNA represents a transcription of the genetic information contained in a certain portion of the DNA molecule. This complementary copy transcribes the genetic code from DNA to *m*RNA. The sequence of bases in *m*RNA specifies the arrangement of amino acids in the protein to be synthesized.

It has been demonstrated many times that RNA is capable of moving from the nucleus into the cytoplasm. Presumably it moves out through nuclear pores. It is surprising that *m*RNA in some bacteria has a very rapid turnover, in a matter of minutes. This may not be true for the cells of higher animals, but the attachment of *m*RNA to ribosomes and the formation of polyribosomes is quite rapid.

Transfer RNA   Transfer RNA is the third form of RNA to be considered. It is a molecule of low molecular weight and contains around 80 nucleotides, as compared with *m*RNA, which is estimated to contain hundreds of subunits. The small molecule consists of a chain of nucleotides arranged in a cloverleaf pattern with some base pairings between the usual RNA four bases A-U, C-G (Figure 3.10*a*). In addition, there are several unusual bases such as inosinic acid (which can substitute for guanine in base pairing), pseudouracil, methylguanine, and others.

Structurally we find certain characteristics that are now accepted as applying to all *t*RNA molecules. One end of the chain of nucleotides always ends with the bases CCA. This is the 3′ end where the amino acid is attached. Guanine is found at the other free end or 5′ end. The loop opposite the free ends contains the anticodon, which consists of three unpaired bases. The anticodon is complementary to the three bases which designate an amino acid on *m*RNA. Referring to Figure 3.10, the solid circles represent bases in the helical regions. They are usually paired and held together by hydrogen bonds, which are indicated by the small dots centered between them. The open circles are also bases but usually unpaired. R and Y, purine-pyrimidine nucleosides; T, ribothymidine; psi, pseudouridine;* a modified base. Intensive research has revealed the base sequence of several amino acids. The first to be completed was alanine, by Holley and his associates. The anticodon for alanine was found to be IGC. This would base-pair with CCG on *m*RNA.

The function of *t*RNA is to carry activated amino acids to a complementary coded area on *m*RNA. The amino acids are first activated by ATP to form a higher-energy compound. The activating enzyme is amino acyl synthetase, and the energy released by the breakdown of ATP enables the amino acid and *t*RNA to form a chemical bond. The synthetase enzymes are considered to be specific for each amino acid. It is proposed that in order to accomplish specificity, specific binding sites would be necessary, (1) to enable the enzyme to recognize the conformation of a certain amino acid, and (2) to permit recognition of the correct *t*RNA (Figure 3.11).

After the *t*RNA becomes located on complementary bases in a strand of *m*RNA, the amino acids carried there form peptide bonds with other amino acids, and a polypeptide chain of amino acids begins to grow. Having deposited their amino acids in correct sequence, it is assumed that the *t*RNA molecules move away but that the polypeptide chain is always attached to a *t*RNA until the complete sequence on *m*RNA has been read. A current concept suggests that the ribosome moves along the *m*RNA while *t*RNA molecules assemble amino acids according to the genetic code. The formation of polypeptide chains is remarkably rapid, requiring only a few seconds or perhaps a minute for longer chains. It would be well to consider that the

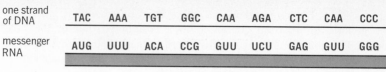

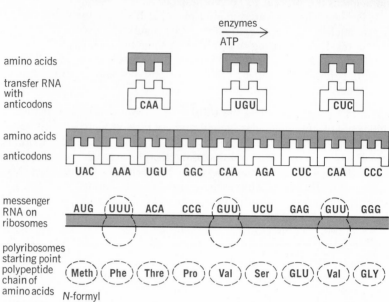

FIGURE 3.11
Diagram illustrating the function of messenger and transfer RNA. The genetic code of DNA is transcribed to *m*RNA, which then combines with protein and moves into the cytoplasm, becoming associated with polyribosomes. Amino acids activated by specific enzymes and ATP are carried to coded areas on *m*RNA by transfer RNA. Anticodons carried by transfer RNA recognize their complementary codons on *m*RNA, resulting in the formation of a chain of polypeptides. It is assumed that *t*RNA units become released after amino acids are located on *m*RNA and polypeptide chains are formed.

newly formed proteins are essential to the formation of enzymes and many cellular products and tissues.

The "letters" of the genetic code It is well established now that the base pairs of the DNA molecule determine the arrangement of amino acids in the formation of proteins. We have seen how DNA, acting as a template, transcribes triplet complementary codons to *m*RNA. Since only four bases, as compared with 20 amino acids, are involved in protein synthesis, more than one base must be used to designate an amino acid. The use of two bases would provide only 16 different combinations; therefore the minimum number of bases must be three. All possible combinations using three bases would number 64, far more than the number needed, but it was found that in some cases as many as four combinations of bases will code for the same amino acid. Such a code is said to be "degenerate."

It has been many years since Nirenberg and Matthei (1961) prepared a synthetic "messenger" RNA containing only the base uracil. When this synthetic polymer of uridylic acid (poly U) was added to an active cell-free mixture derived from colon bacilli (*Escherichia coli*), it was found that only the amino acid phenylalanine was produced. It seemed evident that the coding for phenylalanine should be UUU. Actually UUC also will code for phenylalanine. Various methods have been used to work out the complete codons for all 20 amino acids (Table 3.2).

A few triplet combinations do not produce amino acids; these were originally termed "non-sense" codons. It now appears that UAA (ochre), UGA and UAG (amber) are terminator codons that indicate the end of synthesis for certain polypeptide chains. A release factor is required also.

TABLE 3.2
The amino acid code

UUU Phenylalanine	UCU Serine	UAU Tyrosine	UGU Cysteine
UUC Phenylalanine	UCC Serine	UAC Tyrosine	UGC Cysteine
UUA Leucine	UCA Serine	UAA Chain Termn.	UGA Chain Termn.
UUG Leucine	UCG Serine	UAG Chain Termn.	UGG Tryptophan
CUU Leucine	CCU Proline	CAU Histidine	CGU Arginine
CUC Leucine	CCC Proline	CAC Histidine	CGC Arginine
CUA Leucine	CCA Proline	CAA Glutamine	CGA Arginine
CUG Leucine	CCG Proline	CAG Glutamine	CGG Arginine
AUU Isoleucine	ACU Threonine	AAU Asparagine	AGU Serine
AUC Isoleucine	ACC Threonine	AAC Asparagine	AGC Serine
AUA Isoleucine	ACA Threonine	AAA Lysine	AGA Arginine
AUG Methionine	ACG Threonine	AAG Lysine	AGG Arginine
GUU Valine	GCU Alanine	GAU Aspartic acid	GGU Glycine
GUC Valine	GCC Alanine	GAC Aspartic acid	GGC Glycine
GUA Valine	GCA Alanine	GAA Glutamic acid	GGA Glycine
GUG Valine	GCG Alanine	GAG Glutamic acid	GGG Glycine

Codons which initiate synthesis have been more difficult to determine, but AUG has been identified as the starting codon in bacteria. It is likely that these codons will apply to eukaryote cells as well.

MITOTIC DIVISION OF CELLS

The process of cell division by mitosis involves a series of nuclear changes that result in the accurate division of the cell's chromosome complement. Actually the process involves a continuous series of changes, but for convenience it is usually divided into four phases: prophase, metaphase, anaphase, and telophase. The stages of a typical mitotic cell division are shown in Figures 3.12 and 3.13.

A cell that is not dividing is in the stage called *interphase*. Cells that show no outward signs of division are frequently called *resting cells*, although this term is not very appropriate since it implies that such a cell is not active so far as its metabolism is concerned. Actually, a most important event occurs at interphase: The DNA of the nucleus replicates itself. It can be demonstrated that the DNA content of the cell doubles before any mitotic changes are visible under the microscope. It is of interest that, whereas diploid somatic cells vary considerably in their chemical composition, for any given species most of them possess the same amount of DNA in their nuclei. Furthermore, the germ cells, which are haploid (that is, contain half the normal number of chromosomes), also have half the amount of DNA compared with the diploid body cells. However, certain cells, such as some liver cells, may be tetraploid, containing twice the normal number of chromosomes. Other specialized cells may have even higher ploidy.

Rapidly growing tissues contain many cells in various stages of mitotic division, but in older tissues most of the cells may be in a more or less permanent interphase. It has been estimated that in the adult human being most of the somatic cells have passed through 50 mitotic divisions during the growth of the individual. Some cells (for example, brain cells) do not continue to divide once the system is mature.

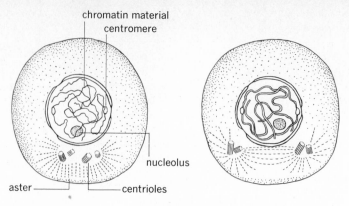

chromatin material
centromere

nucleolus

aster

centrioles

INTERPHASE

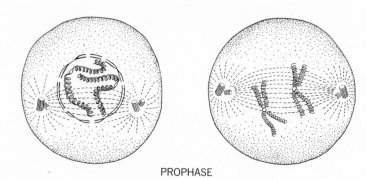

PROPHASE

METAPHASE

ANAPHASE

Cellular structure and function

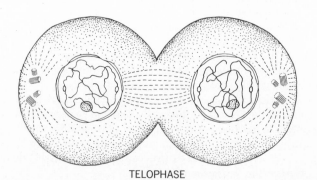

TELOPHASE

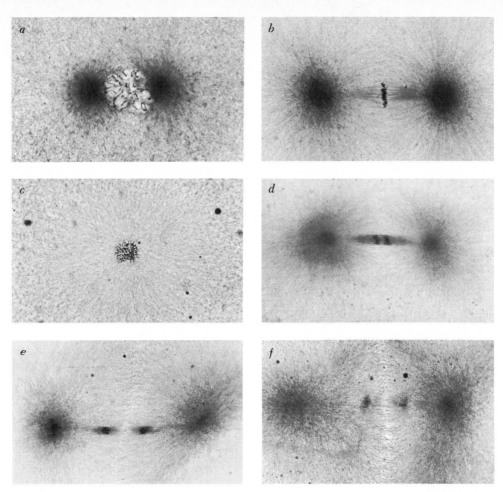

FIGURE 3.13

Photomicrographs of animal mitosis: *a* early prophase; *b* metaphase; *c* metaphase, polar view;
d early anaphase; *e* late anaphase;
f telophase. Note plane of cleavage
and formation of daughter nuclei.
(*Courtesy of General Biological Supply House, Inc., Chicago.*)

The nucleoprotein material is greatly elongated at interphase, but
individual chromosomes are not visible. During this phase the DNA, in
replicating itself, forms two strands. Apparently one strand acts as a template
in the formation of the new strand, which would mean that the genes are
duplicated at this time as the DNA replicates.

STAGES OF MITOSIS *Prophase* The appearance of the cell changes as it enters
the early stages of mitotic division. There is considerable activity in the
cytoplasm around the centrioles, where lines of flow radiate outward. These
lines form the *aster*. The cylindrical nature of the paired centrioles has been
described. They are self-replicating structures. The replication occurs at the
end of the mitotic division, in the telophase, when each centriole forms a
new centriole, making two pairs in the cell. In the late interphase or early
prophase the pairs of centrioles start to move away from each other, and
filaments of the spindle form between them. The chromosomes, still in the
extended state, are now in two strands. By late prophase, when the centrioles
have migrated to opposite poles, a fully developed spindle lies between them.
The spindle appears to be a fibrous structure, protein in nature, and the
filaments are often described as microtubules. The chromosome strands now

Mitotic division of cells

become tightly coiled, and the chromatin material becomes condensed around them so that the coils are no longer visible. The chromosomes are divided longitudinally into two *chromatids* lying side by side at the earliest stage of visibility. A minute pale area in the chromatid is called the *centromere* or *kinetochore*, indicating the point of attachment to the spindle fiber. This area remains unstained by the dyes ordinarily used for staining chromosomes. The two areas, one in each chromatid, lie side by side and face opposite poles of the spindle. The nuclear membrane and nucleolus ordinarily disappear at about this time, and mitosis progresses to the metaphase.

Metaphase This is usually a brief but important stage leading to the separation of the chromatids. The chromosomes become arranged with their centromeres in the equatorial plane of the spindle. The centromere is not necessarily located in the center of the chromosome; this part is arranged around the periphery of the equator, and the ends of the chromosomes can be in various positions. The centromeres divide in the late metaphase, but the actual separation of the chromatids is considered to occur in the anaphase.

Anaphase The chromatids start to move away from each other toward the opposite centrioles. They separate at the centromeres first, the ends often presenting a V-shaped appearance as though the chromosomes were being pulled by some force at the middle. As the two chromatids move away from each other, they properly may be called chromosomes, since each will be an individual chromosome in the newly formed daughter cells. The late anaphase is indicated when the centromeres have migrated far apart toward their respective spindle poles. Mazia estimates that the chromosomes travel toward opposite poles during anaphase at about 1 micron per minute.

Telophase The two groups of chromosomes never quite reach the poles of the spindle because at the completion of their migration the poles and spindle tend to disappear as the two new nuclei are organized. The nuclear membranes and the nucleoli again become visible, but the matrices of the chromosomes become invisible as the chromosomal filaments uncoil in the newly formed nuclei. As stated earlier, a new centriole is produced by each centriole, and the two new cells assume the characteristics of cells in interphase.

A cytoplasmic constriction forms between the two developing nuclei and progresses until the cell is completely divided. The constriction forms exactly where the equator of the spindle was located at metaphase. The two daughter cells grow rapidly until they have attained their typical size, after which they may, in turn, undergo mitotic division.

The time necessary for the accomplishment of a mitotic division varies greatly in different tissues. An average time may be about 3 hours, although some tissues have been observed to complete mitosis in 30 minutes and others may require several hours. An optimum temperature facilitates the process. The prophase usually requires the longest period of time, whereas the metaphase is one of the shortest phases.

The significance of the elaborate process of mitosis appears to lie in the provision for equal quantitative and qualitative division of the DNA chromatin material. The result is that each daughter cell has identical hereditary characteristics.

Meiosis refers to the behavior of chromosomes in reproductive cells whereby homologous chromosome pairs undergo a reductional division which provides that the reduced or haploid number of chromosomes (23) will be present in the oocyte and the same number will be present in the spermatocyte. Upon fertilization of the ovum, the chromosome number is restored to the diploid condition (46). Meiosis is considered in detail in Chapter 22 on human development (see Figures 22.2 and 22.3).

In earlier times, before the nineteenth century, animal breeders produced crosses which they considered to be intermediate in appearance between the two parents. Many people also had crossed plants having different characteristics (for example, flowers of different colors) and had produced "hybrids" which were different from the parents, but it remained for the Austrian monk Gregor Mendel (1865) to discover some of the fundamental mechanisms which govern the laws of inheritance. It would be well to realize that at that time nothing was known about genes and chromosomes.

Mendel worked with peas in his garden at the monastery at Brünn. He observed many differences in the plants and believed them to be inherited. He tested reddish flowers, white flowers, tall plants, short plants, yellow seeds and green seeds, as well as smooth seeds and wrinkled seeds. He selected pure-line plants for his experiments and observed that when he crossed parent plants that were characterized by a single trait, the offspring in the first generation all resembled the parent having the dominant trait. In reality, Mendel dealt with two traits in his experiments with the flowers of peas. He crossed violet-red flowers having long axes with white flowers having short axes, a dihybrid cross.

A simpler experiment with flower color consists of a monohybrid cross between plants having red flowers and plants bearing white flowers. If we are dealing with dominant pure-line strains, the first-generation plants (F_1) should have flowers that are all red. The second-generation cross (F_2) between

FIGURE 3.14
A simple monohybrid cross with complete dominance.

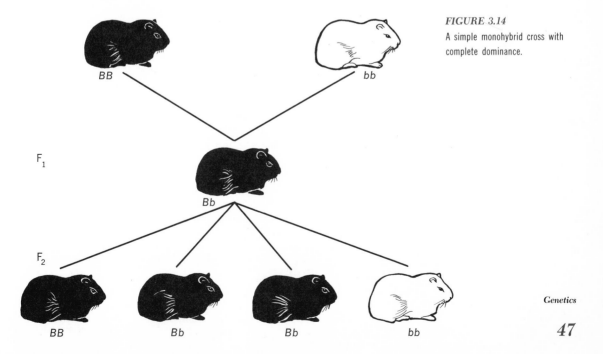

these "hybrid" red plants should produce about three-fourths red-flowered plants and one-fourth white-flowered. This is, of course, the characteristic 3:1 ratio. Strictly speaking, true hybrids now are considered to be the result of crosses between two different species.

Mendel's pure-line parents would now be called *homozygous* and, as in Figure 3.14, would be written *BB*, indicating black dominant. The recessive character would be written in small letters *bb* or *ww* for homozygous white. The F_1 (first filial) generation, which is a mixture of parental genes, would be heterozygous and would be written *Bb*. It is evident that in guinea pigs, black is dominant over white. The ratio for the F_2 would be written *BB Bb Bb bb*. The phenotypic ratio, then, would be three black animals to one white.

It was not known in Mendel's day that in meiosis the chromosome pairs are separated and that this leads to the segregation of allelic genes. (Alleles refer to genes having the same locus on homologous chromosomes.) Nevertheless, Mendel's first law of inheritance was the *law of segregation*. He assumed that there were two units of inheritance for each trait in the parent, and he called them *determiners*. If these determiners segregate or separate in the parent, they must come together again at the time of fertilization of the ovum. We know now that in meiosis the paired chromosomes do separate and that in the development of the sex cells (gametes) each ovum or spermatozoon receives only one chromosome of a pair.

Mendel's second law is called the law of independent assortment. It refers to the segregation of each pair of genes independently from other pairs. If we select a dihybrid cross illustrating complete dominance as an example, the F_1 generation exhibits the dominant characteristics of one parent. The F_2 generation demonstrates independent assortment or free distribution and recombination of genes (Figure 3.15). The ratio in this case

FIGURE 3.15
A dihybrid cross illustrating the inheritance of two pairs of contrasting characters, rough or smooth coat and color.

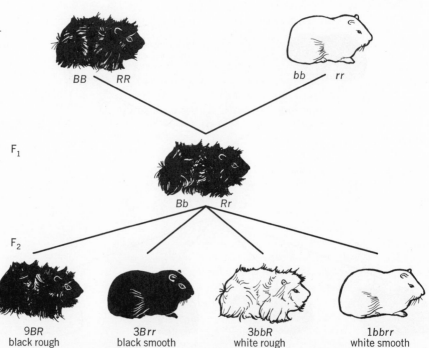

is $9:3:3:1$. This is a phenotypic ratio consisting of nine black rough animals, three black smooth, three white rough, and one white smooth. These results indicate complete dominance, but dominance is not always complete.

The inheritance of flower color and leaf form in the snapdragon plant is a common example of incomplete dominance. In crossing red-flowered plants *RR* with white-flowered plants *rr*, the F_1 generation *Rr* consists of pink flowers rather than red, illustrating incomplete dominance. Also, crossing broad-leafed snapdragon plants with narrow-leafed snapdragon plants produces a leaf of intermediate form in the F_1 generation.

LINKAGE AND CROSSING-OVER Genes located in the same chromosome may be inherited as a group. They are linked together and therefore cannot segregate independently. It is well known now that each pair of homologous chromosomes will include numerous gene pairs. Linkage, then, would prove an exception to Mendel's second law.

SEX-LINKED CHARACTERS Human sex chromosomes are labeled XY for males and XX for females. The peculiarities of sex-linked inheritance stem from the fact that the controlling character is located in the X chromosome. The gene is passed to the female, where it does not gain expression. The female as a carrier, however, passes the gene in the X chromosome to some of her sons. The condition is expressed in the sons because the Y chromosome has no normal counterpart (gene) to pair with the gene in the X chromosome. These sex-linked characters are not inherited directly from male parent to son. They skip a generation. Examples in the human being are the inheritance of color blindness (Figure 12.17) and hemophilia (Chapter 13).

Crossing-over occurs at an early stage of meiosis when homologous chromosomes pair in synapsis. Bivalent chromosomes are formed at this time, which divide lengthwise to form chromatids (Figure 3.16). The four chromatids are called *tetrads*. In this thin, threadlike stage, crossing-over may occur. The results obtained from crossing-over indicate that genes are, in fact, arranged in linear order on chromosomes. The process of crossing-over merely leads to an interchange of sections of homologous chromosomes and results in a recombination of linked genes. Since two chromatids exchange equal segments, there is no loss or gain of genes (Figure 3.17).

Phenotype and *genotype* are useful terms of quite different meaning. Phenotype refers simply to a group of individuals of similar appearance and does not consider genetic composition. Genotype refers to individuals having the same hereditary or genetic constitution.

MUTATIONS Inherited changes in the germ plasm are called mutations. They may be naturally occurring mutations or may be induced, as by x-ray or chemical means. They may result from changes in chromosome numbers or from changes in the structure of chromosomes. These are known as chromosomal mutations. There are also gene mutations, caused by changes in single genes or at a single locus.

Occasionally diploid gametes occur during gametogenesis. If the spermatozoon is diploid instead of haploid and the ovum is also diploid, then the resulting fertilized ovum or zygote will be tetraploid. Tetraploidy ($4n$) is common in cultivated plants, but it is rare in animals.

Down's syndrome (mongolism) is a good example of a serious disability associated with the addition of one chromosome, autosome number 21, a chromosomal mutation (Figure 22.8). In Klinefelter's syndrome the individual

Genetics

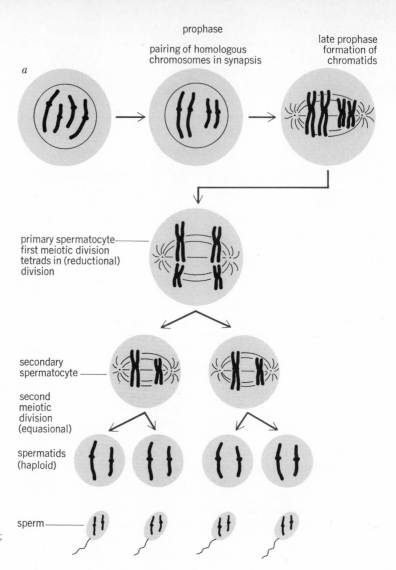

prophase

pairing of homologous
chromosomes in synapsis

late prophase
formation of
chromatids

a

primary spermatocyte
first meiotic division
tetrads in (reductional)
division

secondary
spermatocyte

second
meiotic
division
(equasional)

spermatids
(haploid)

sperm

FIGURE 3.16
Gametogenesis: *a* spermatogenesis;
b oogenesis.

is male but the sex chromosomes are XXY; $2n = 47$ (Figure 22.6). Turner's syndrome is characterized by sexually immature females who have only one X chromosome (Chapter 22).

An authenticated case of gene translocation occurs in sickle-cell anemia, where the amino acid valine by mutation has replaced glutamic acid in one of the two polypeptide chains of the hemoglobin molecule (Chapter 13). If we consider that the genetic code for glutamic acid is either GAA or GAG and that two of the code words for valine are GUA or GUG, it appears that the mutation which results in sickle-cell hemoglobin may very well have been caused by the change in one base of a triplicate nucleotide.

Newer terminology questions the older definition of the gene as the unit of inheritance. If the gene represents a portion of a DNA molecule, then it may include many nucleotide pairs. The term *cistron*, coined by Benzer, would better represent a portion of a DNA molecule having iden-

Cellular structure and function

50

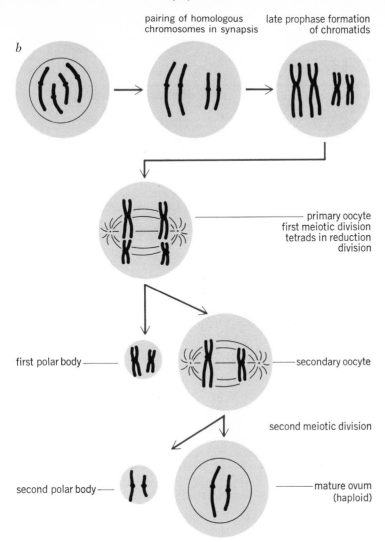

prophase

pairing of homologous chromosomes in synapsis

late prophase formation of chromatids

b

primary oocyte
first meiotic division
tetrads in reduction division

first polar body

secondary oocyte

second meiotic division

second polar body

mature ovum (haploid)

tifiable functional significance and carrying genetic information. Genes, in general, do not function independently. Geneticists speak of modifier genes and suppressor genes in bacteria, meaning that the action of a certain gene may be modified or suppressed by another gene. The operator gene is considered to be necessary in order that another gene, perhaps a structural gene, may be able to act. The operator gene may control the action of several structural genes. The operon is thought to represent a group of closely linked genes, including a structural gene and an operator gene.

It would seem, then, that hereditary traits are controlled by a number of genes acting together, not by the action of individual genes. Scientists now have the ability to transfer small portions of DNA from one cell to another. Being able to insert genes into mammalian cells should open up inviting new areas for experimentation and for understanding of gene function in higher animals and man. The cell displays an extremely complex

Genetics

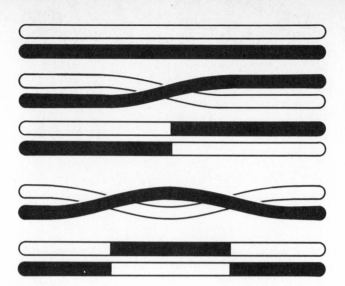

organization, and it is likely that we are only beginning to discover its secrets.

Cancer results from the uncontrolled growth of cells. Normally cells grow and divide in a controlled manner, although very little is known about how the control is exerted. Some cells grow and replace themselves throughout life, while other cells, such as muscle and nerve cells, do not divide at all in the adult. Some sort of feedback system may control the normal growth and multiplication of cells.

It is known that x-rays and other ionizing radiation as well as certain chemicals can cause cancerous growths to develop. Viruses causing cancer in animals have been isolated, but no virus has as yet been isolated in the case of human cancer, although it remains a distinct possibility that viruses may be involved in at least some human cancers. There is also the possibility, as some investigators strongly believe, that a change in the genetic composition of the cell may lead to uncontrolled cell growth and cancer.

Cancers are able to invade surrounding tissues by a process called *metastasis*. Cancer cells may spread throughout the body by way of blood or lymph vessels. At present there is no cure for cancer, but early surgery is often effective.

SUMMARY   A living cell is a microscopic mass composed of a nucleus and cytoplasm surrounded by a cell membrane. A cell displays physiological activities such as irritability, motility, metabolism, growth, and reproduction.

A study of the structure of the cell includes the characteristics of cell membranes and their associated structures, among which are microvilli, pinocytic vesicles, the endoplasmic reticulum, ribosomes, and the Golgi complex.

The cytoplasm contains a number of minute structures. Among them are the mitochondria, lysosomes, and the centrioles. The mitochondria are the sites of the energy metabolism of the cell, including cellular respiration. Lysosomes contain digestive enzymes and are organelles of digestion within the cell. The electron microscope has revealed the centrioles as paired cylindrical structures commonly placed at right angles to each other. They are active during mitotic division.

The nucleus is a small body, usually spherical in shape, located near the center of the cell. The nucleoplasm is the nuclear protoplasm and is surrounded by the nuclear membrane. The nucleus contains at least one nucleolus, or little nucleus. Chromosomes are located in the nucleus.

Chromosomes contain genes, which control hereditary characteristics. There are 46 chromosomes, or 23 pairs, in the human cell. Chemically, the body of chromosomes is composed of nucleoprotein, and the principal nucleic acid is DNA, although RNA is present in lesser amounts. By enzymatic action genes determine protein synthesis.

The DNA molecule, constructed in the form of a helix, is a double-stranded chain of nucleotides. It may be compared to a ladder, in which the sides are composed of deoxyribose sugar groups alternating with phosphate groups; the rungs extend between sugar groups and contain nitrogenous bases held together by hydrogen bonds. There are four such bases: adenine, thymine, guanine, and cytosine. Adenine is always paired with thymine and guanine with cytosine in the DNA molecule. The sequence of these bases provides the genetic code for the synthesis of proteins. Adenine and guanine are purines; cytosine, thymine, and uracil are pyrimidines.

Ribonucleic acid (RNA) is closely related chemically to DNA. It is single-stranded but twisted on itself to form a double helix. The single strand contains repeating units of ribose sugar and phosphate. There are four bases in RNA, as in DNA; three of the bases are the same as those found in DNA, but the fourth base is uracil, which replaces the thymine of the DNA molecule.

It is generally accepted that there are three kinds of RNA: ribosomal, messenger, and transfer RNA. About 75 percent of the RNA found in the cytoplasm is ribosomal. Another 15 percent occurs in the nucleus associated with the chromosomes and the nucleolus.

It is believed that a strand of DNA acts as a template for the synthesis of *m*RNA. According to this hypothesis, *m*RNA leaves the nucleus and enters the cytoplasm, where it becomes attached to ribosomes. The chemical bases of *m*RNA are complementary to those in one strand of DNA, uracil replacing thymine. The sequence of bases in *m*RNA determines the arrangement of amino acids in the polypeptide chains of the proteins to be synthesized.

A third form of RNA, called *transfer* or *soluble RNA*, transfers activated amino acids to a complementary coded area on *m*RNA. Transfer RNA is a single-stranded molecule with the nucleotides arranged in a cloverleaf pattern. One free end of the chain of nucleotides where amino acids are attached ends with the bases CCA. The other free end contains guanine. The loop opposite the free ends carries the three unpaired bases of the anticodon. These bases are complementary to three bases on *m*RNA designating an amino acid.

The "letters" of the genetic code are formed by using three bases in differing combinations. In this way amino acids may be arranged according to a coding of bases on *m*RNA such as UUU for phenylalanine, GUA for valine, or CAU for histidine.

Protein synthesis probably takes place on groups of ribosomes rather than on individual ones. Groups of ribosomes are called polyribosomes or polysomes.

Cells divide by the process of mitosis, which includes the following stages: prophase, metaphase, anaphase, and telophase. A cell that is not dividing is in interphase.

A brief study of mendelian genetics concludes the chapter.

1 Describe some of the physiological activities performed by living cells.

2 Discuss some of the various aspects of cellular metabolism. Describe some of the factors involved in cellular respiration.

3 In what way are the mitochondria involved in energy metabolism? Where is ATP formed?

4 What kinds of enzymes are found in lysosomes? Would it be possible for them to digest the cell itself?

5 Discuss the different forms and functions of cellular membranes. Briefly, what is the function of ribosomes?

6 Describe the structure of the centriole.

7 How is the DNA molecule constructed? How does RNA differ from DNA?

8 What is the significance of the replication of DNA? When does the replication occur? What is the nature of the chromatin material at this time?

9 Discuss the function of the three kinds of RNA.

10 How would it be possible for the "letters" of the genetic code to determine the arrangement of amino acids in the synthesis of proteins?

11 Trace the chromosomes through the stages of mitosis.

12 What is the significance of homeostasis?

13 Explain Mendel's two laws of inheritance.

14 Discuss sex-linked inheritance and give examples.

15 Explain the differences between mitosis and meiosis.

SUGGESTED READING

BOURNE, G. H.: "Division of Labor in Cells," 2d ed., Academic Press, Inc., New York, 1970.

CRICK, F. H. C.: The Genetic Code III, *Sci. Am.*, **215**:55-62 (1966).

De BUSK, A. GIB: "Molecular Genetics," The Macmillan Company, New York, 1968.

FREEMAN, JAMES A.: "Cellular Fine Structure," McGraw-Hill Book Company, New York, 1964.

GRATZER, W. B.: "Readings in Molecular Biology," selected from *Nature*, The MIT Press, Cambridge, Mass., 1971.

GREEN, DAVID E.: The Mitochondrion, *Sci. Am.*, **210**:63-74 (1964).

HARDMAN, J. G., G. A. ROBISON, and E. W. SUTHERLAND: Cyclic Nucleotides, *Ann. Rev. Physiol.*, **33**:311-336 (1971).

HOLLEY, R. W.: The Nucleotide Sequence of a Nucleic Acid, *Sci. Am.*, **214**:30-39 (1966).

KORNBERG, ARTHUR: The Synthesis of DNA, *Sci. Am.*, **219**(4):64-78 (1968).

KURLAND, C. G.: Ribosome Structure and Function Emergent, *Science*, **169**:1171-1177 (1970).

Mc ELROY, WILLIAM D.: "Cell Physiology and Biochemistry," 3d ed., Prentice-Hall, Inc., Englewood Cliffs, N.J., 1971.

MAZIA, DANIEL: How Cells Divide, *Sci. Am.*, **205**:101-120 (1961).

MENDELL, G.: Experiments in Plant Hybridization, translation in Sinnott, E. W., L. C. Dunn, and T. Dobzhansky, "Principles of Genetics," McGraw-Hill Book Company, New York, 1958.

MILLER, O. L., Jr.: The Visualization of Genes in Action, *Sci. Am.*, **228**:34-42 (1973).

NEUTRA, M., and C. P. LEBLOND: The Golgi Apparatus, *Sci. Am.*, **220**(2):100-107 (1969).

NOMURA, M.: Ribosomes, *Sci. Am.*, **221**(4):28-35 (1969).

NORTHCOTE, D. H.: The Golgi Apparatus, *Endeavor*, **30**(109):26-33 (1971).

PARSONS, DONALD F.: Mitochondrial Structure: Two Types of Subunits on Negatively Stained Mitochondrial Membranes, *Science*, **140**:985-987 (1963).

WAGNER, R. P.: Genetics and Phenogenetics of Mitochondria, *Science*, **163**:1026-1031 (1969).

WOESE, CARL R.: The Problem of Evolving a Genetic Code, *Bio Science*, **20**(8):471-480 (1970).

unit two

structural elements

chapter 4
epithelial and connective tissues

Tissues are composed of organized groups of cells that are similar in origin, structure, and function. Associated with the cells is an intercellular substance of various types. A group of muscle cells, therefore, may be referred to as muscle tissue, or a group of nerve cells may form nervous tissue. There are not so many kinds of distinct tissues, however, as one might suppose. They are usually limited to muscular, nervous, epithelial, and connective tissues. Muscle and nervous tissues will be discussed in Chapters 6 and 8. Blood and lymph may be considered as a type of connective tissue in which the intercellular substance is fluid.

EPITHELIAL TISSUES

Epithelial tissues cover the surface of the body (the skin) or form the delicate linings of body cavities that open directly or indirectly to the surface. Since the tissues that line the thoracic and abdominal cavities have a different embryological origin, they are referred to as *mesothelial tissues* or *mesothelium*. Likewise, delicate tissues that line the heart and blood vessels are called *endothelial tissues* or *endothelium*. The word *epithelium* may be used in an elementary sense to cover all these kinds of tissues.

The cells of epithelial tissues are placed close to one another, there being very little intercellular substance between them. The tissues formed by the cells are usually delicate, especially when they line a cavity, and are often only one cell layer thick.

All epithelial tissues have a free surface and a basal surface. The basal surface lies on tissue that is formed by the condensation of the ground substance of the underlying connective tissue. It is usually considered that this membrane supports the epithelium and serves to attach it to the underlying tissue.

Epithelial tissues are classified according to the shape of the cells as squamous, cuboidal, and columnar. If the cells are arranged in a single layer, it is called *simple epithelium*. If there is more than one layer of cells, it is called *stratified epithelium*.

SIMPLE SQUAMOUS EPITHELIUM This type of epithelium consists of flat cells in a single layer, somewhat like tiling in a mosaic pattern (Figure 4.1). When the cell is viewed from above, the nuclei usually appear oval in shape, but when viewed from the side, the nuclei present a raised area on the free surface of a very thin cell. This type of epithelium forms the thin serous linings of the alveoli of the lungs. It is found also in the lens of the eye and in the lining of the membranous labyrinth of the inner ear. The mesothelium, which lines the body cavities, and the delicate endothelial lining of the heart and blood vessels may be classified as simple squamous epithelium.

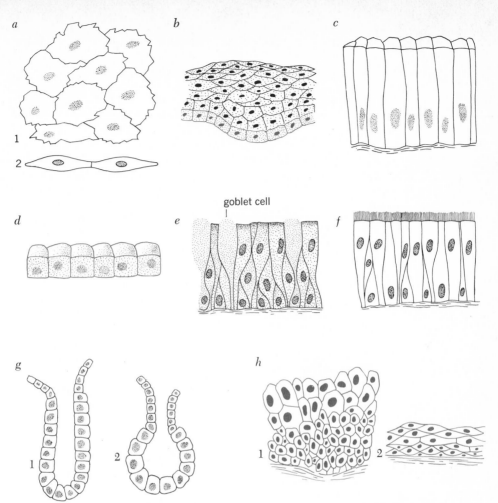

goblet cell

FIGURE 4.1

Diagrammatic representation of various types of epithelial tissues: *a* simple squamous epithelium, *1* surface view, *2* lateral section; *b* stratified squamous epithelium in longitudinal section; *c* simple columnar; *d* cuboidal; *e* columnar epithelium with goblet cells; *f* ciliated pseudostratified columnar epithelium; *g* cuboidal epithelium in *1* simple tubular and *2* alveolar glands; *h* transitional epithelium, as in the urinary bladder, *1* relaxed, *2* extended.

STRATIFIED SQUAMOUS EPITHELIUM The cells of stratified squamous epithelium tend to be cuboidal in the deeper strata but flatten as the various layers approach the free surface. If one were to scrape the lining of the cheek with a toothpick and place the scrapings under a microscope, one would see the thin, flat, squamous cells of the surface layer. The cells of the deeper layers multiply by mitotic divisions so that the surface layer is constantly replenished as it wears away. The basement membrane in this type of tissue is not a straight line, and the basal layer of cells is not arranged in a regular pattern. Underlying tissues project up into the epithelial layer in numerous rounded elevations called *papillae*. The basement membrane therefore appears as an irregular line in histological sections. Stratified squamous epithelium forms the external protective layer of the skin; it also lines the mouth, pharynx, esophagus, vagina, and anal canal.

A variation of stratified epithelium is found in the lining of the urinary bladder, the ureters, and the basal parts of the urethrae. This is the so-called "transitional epithelium," which is capable of being relaxed or extended. In the relaxed condition the cells tend to be spherical in shape, but when

Epithelial tissues

the bladder or tube is distended, the cells of the upper layers become flattened and elongated, resembling stratified squamous epithelium. No distinct basement membrane is present in this kind of tissue (Figure 4.1h).

CUBOIDAL EPITHELIUM This type of epithelium consists of cube-shaped cells and can be found lining the ducts of many glands. When one looks down on a row of these cells, the surfaces appear square, representing one side of a cube. The nuclei are spherical and located near the center of each cell. Cuboidal epithelium is found in the lining of the smaller ducts and tubules of the salivary glands, pancreas, liver, and kidneys. Although cuboidal epithelium is not considered a distinct type by many authorities, it is a useful designation for the characteristic cells lining the ducts of glands.

SIMPLE COLUMNAR EPITHELIUM When the cells in a single row are laterally compressed so that they are taller than their width, they form simple columnar epithelium. The oval-shaped nucleus is located toward the base of the cell. This type of tissue is commonly found throughout the digestive tract, notably in the lining of the stomach and intestine, in the gallbladder, and in the larger ducts of digestive glands. These cells are concerned with the secretion of digestive fluids and the absorption of food materials.

Goblet cells Present in columnar epithelium as enlarged flask-shaped cells, goblet cells are interspersed among typical columnar cells. They may be considered as unicellular glands that produce a mucoid secretion. They are well represented in the epithelial lining of the intestine, becoming abundant in the large intestine (Figures 4.1 and 4.2a and b).

VARIATIONS IN EPITHELIAL CELLS *Pseudostratified epithelium* Pseudostratified epithelium is a modification of simple columnar epithelium. As a result of compression, the cells are of different heights, and their nuclei therefore are in irregular layers. The majority of the cells are long and slender and extend from the basement membrane to the free surface, but a considerable number of cells are confined to an area close to the basement membrane. The distribution of the nuclei at different levels is characteristic. This form of epithelium lines the nose, the trachea, and the larger bronchial tubes of the lungs. It also lines portions of the male and female reproductive structures.

Ciliated epithelium Columnar epithelium is often ciliated. Hairlike processes called *cilia* project from the free surface of cells and wave continuously, producing currents in the surface fluids. Ciliated epithelium lines the trachea and bronchial tubes. Movements of the cilia produce a current moving upward and assist in clearing the lungs of foreign matter. Ciliated cells are also found in the oviducts, where they help to move the egg downward toward the uterus. The epididymis and the vas deferens are lined with ciliated epithelium.

Electron micrographs reveal the minute structure of cilia (Figures 4.1f and 4.3). In cross section they are seen to be composed of a central pair of filaments surrounded by nine groups of filaments. Each group usually contains two filaments, but sometimes there are three. Centrioles have a similar structure, as we have seen, but lack the central pair of filaments. In longitudinal sections of tissue showing the detailed structure of cilia, a basal body or basal corpuscle can be seen at the base of each cilium. The basal body resembles a centriole. Numerous mitochondria are located below the basal body.

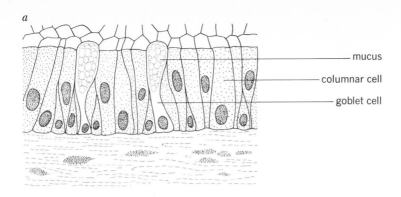

mucus

columnar cell

goblet cell

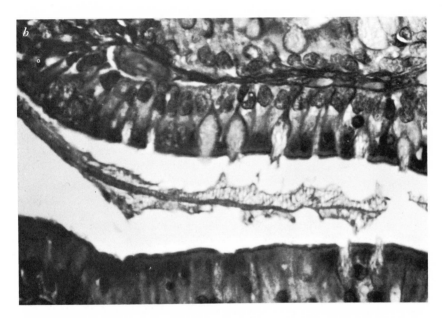

FIGURE 4.2

a Columnar epithelium of the intestine, including goblet cells. *b* Photomicrograph of columnar epithelium. (*Courtesy of General Biological Supply House, Inc., Chicago.*)

Microvilli The minute projections on the free-surface membrane in epithelial cells, shown by electron microscopy, are called *microvilli* (Figure 4.4). They are present especially in cells concerned with absorption, such as the columnar epithelium lining the intestine and in the "brush border" of cells in the proximal convoluted tubules of the kidney. Wherever present, microvilli greatly increase the absorption surface.

Desmosomes The junctions of certain epithelial cells, especially those in the squamous epithelium of the skin, represent firm attachments of one cell to another. Electron micrographs indicate a disk of dense material on either side of opposing cell membranes at these places (Figure 4.5). The dense areas are called *desmosomes*. Fine filaments enter the desmosome from the cytoplasm but do not cross the intercellular space of the desmosome.

MULTICELLULAR GLANDS Multicellular glands often become exceedingly complex in their structure and function. A large number of epithelial cells are involved and locally become secretory cells. Structurally, two types are

Epithelial tissues

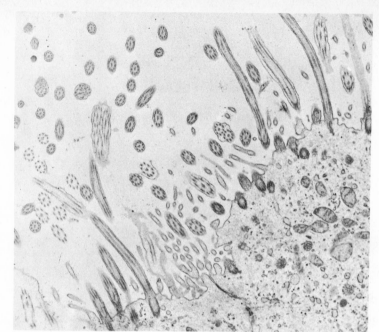

FIGURE 4.3
Electron micrograph showing cilia on the surface of dog tracheal epithelium. ×11,500. Note the structure of cilia in cross sections. (*From James A. Freeman, "Cellular Fine Structure," McGraw-Hill Book Company, New York, 1964.*)

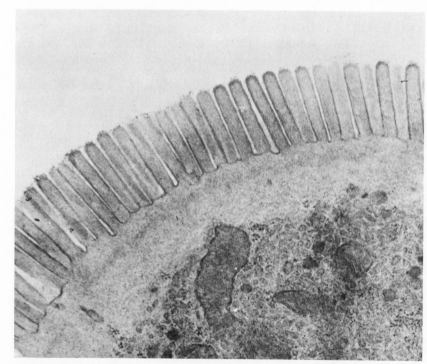

FIGURE 4.4
Electron micrograph of microvilli in the absorptive epithelial cells of rat ileum. ×37,000. (*From James A. Freeman, "Cellular Fine Structure," McGraw-Hill Book Company, New York, 1964.*)

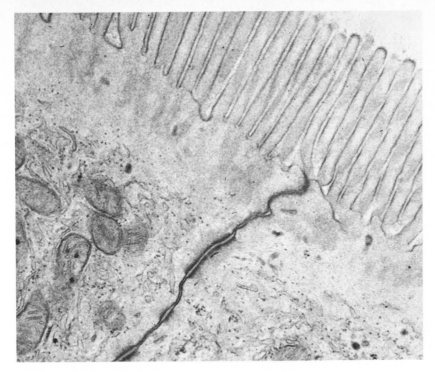

FIGURE 4.5
Electron micrograph showing desmosomes between squamous epithelial cells. The desmosome consists
of fine fibrils which converge into
a dense thickening of each plasma
membrane. ×39,000. (*From
James A. Freeman, "Cellular
Fine Structure," McGraw-Hill
Book Company, New York,
1964.*)

usually recognized, resulting from simple infolding of the epithelium; they
are simple tubular and simple alveolar glands. The first is an in-pocketing,
shaped like a test tube; the second is slightly more complex, being enlarged
at the base and shaped more like a flask. Glandular structure becomes more
complex by the branching of these simple multicellular types, and often the
tubular and alveolar types are combined in a single gland (Figure 4.1).

Sweat glands and *sebaceous glands* are examples of simple multicellular
glands. There are two kinds of sweat glands, *merocrine* (eccrine) and *apocrine*
glands (Figure 4.6). The unbranched tubular gland with a long duct opening
at a pore in the skin is an eccrine gland. The coiled tubular portion is the
secretory part; the long narrow duct conveys the secretion to the surface.
Sweat glands are found over almost all the skin surfaces of the body but
are most abundant on the forehead, in the axillae of the arms, the palms
of the hands, and the soles of the feet. The eccrine glands pour water (sweat)
onto the skin surface, which by evaporation has a cooling effect. They
function also to moisten the palms and soles since these are friction surfaces
and a little moisture provides a better grip. In addition to water, salt and
some other electrolytes are lost in perspiration. Sweat glands are larger and
generally more active in men than in women.

Apocrine glands of the skin are large, specialized sweat glands. Probably they were derived originally from scent glands in their evolution, and
they are still partly responsible for the odorous substance in sweat. The glands
are located predominantly in the axillae and anogenital areas. Unlike the
eccrine glands, they are not directly concerned with temperature regulation
but respond to emotional states, stress, and sexual excitement. The glands

Epithelial tissues

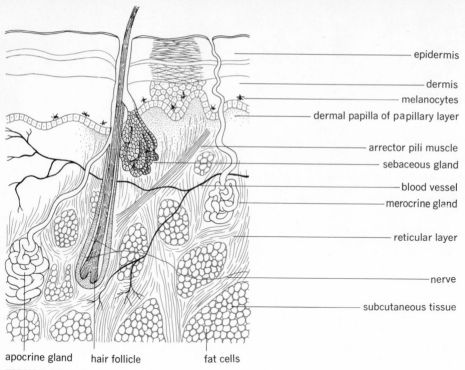

epidermis

dermis

melanocytes

dermal papilla of papillary layer

arrector pili muscle

sebaceous gland

blood vessel

merocrine gland

reticular layer

nerve

subcutaneous tissue

apocrine gland   hair follicle   fat cells

FIGURE 4.6

Section through the skin, illustrating a merocrine (eccrine) gland, an apocrine gland, and a hair follicle. Detail of the subepithelium is shown only in part. The melanocytes are pigment cells.

become enlarged at puberty, and in the female they are active during menstruation. The apocrine gland is usually larger than the eccrine gland but produces only a small amount of fluid. Profuse sweating in the armpits tends to spread the apocrine fluid. The action of bacteria produces most of the objectionable odor of perspiration.

Sebaceous glands are ordinarily associated with hair follicles. Their ducts open into the upper part of the hair follicle. They supply an oily or waxy secretion called *sebum* to the hair and skin. The sebum serves to protect the skin against becoming too dry and also reduces absorption through the skin. They are holocrine glands (Table 4.1).

Glands located in the skin of the external canal of the ear secrete earwax and are called *ceruminous glands.* They are modified sweat glands.

Apocrine glands, such as the mammary glands, will be considered in Chapter 21 on the reproductive system.

Endocrine glands, or glands of internal secretion, do not have ducts and ordinarily deliver their secretions directly into the bloodstream for distribution over the body. They will be considered under appropriate chapter headings.

EPITHELIAL MEMBRANES Epithelial membranes are of two general types: *mucous membranes* and *serous membranes.* They are composed of various types of epithelial cells upon a layer of connective tissue. Mucous membranes line the structures of organ systems that open to the exterior. These are the respiratory, digestive, and urinogenital organ systems. A mucous membrane

TABLE 4.1
Types of multicellular glands

Type	Functional description	Examples
Endocrine	Internally secreting and ductless	Endocrine pancreatic glands, adrenal glands
Exocrine		
Merocrine	Secretion does not include secretory cells.	Salivary glands, exocrine pancreatic glands, eccrine sweat glands
Apocrine	Secretions accumulate near free ends of glands and contain parts of cells but basal portions remain intact and functional.	Mammary glands and large, specialized sweat glands
Holocrine	Secretions accumulate within the cell itself. The cell and the secretion are discharged together. The secretory cells are replaced by new cells.	Sebaceous glands

does not necessarily secrete mucus; for example, the lining of the urinary bladder is not a mucus-secreting type. However, the surface epithelium of most organs of these systems is kept moist by the secretion of glands of various types.

A serous membrane is composed of two layers, the *parietal layer*, which lines a cavity, and the *visceral layer*, which covers an organ. This membrane lines the body cavities that do not open to the exterior, and it covers the organs that lie within these cavities. An exception is the peritoneal cavity in the female. It is not a closed cavity, since the uterine tubes open into it, but it is lined with a serous membrane. Serous membranes produce a thin fluid which serves to lubricate and moisten the areas and the organs covered.

The serous membrane lining the abdominal cavity is the *peritoneum*. Its visceral layer covers most of the organs of the abdomen. It is composed of simple squamous epithelium on a base of connective tissue. The epithelial layer is called a *mesothelium* because it is derived from mesoderm. The peritoneum forms the *mesentery*, a thin double-layered membrane which supports the intestine, and an apronlike fold, extending downward from the stomach, called the *omentum*, which lies over the intestine.

The *pleura* is a serous membrane lining the thoracic cavity and covering the lungs. The pleural cavity is a potential space between the parietal and visceral layers. It contains a minute amount of fluid, which enables the two moist and lubricated surfaces to move against each other freely and smoothly.

The *pericardium* is a serous membrane that covers the heart with a double layer.

A serous membrane also lines the interior of the heart, the blood vessels, and lymphatic vessels. This simple squamous epithelium is called an *endothelium*. It is a very thin layer forming the smooth lining of the chambers of the heart and lining the larger blood vessels. In the capillaries it functions as a selectively permeable membrane between the blood and the tissues.

Connective tissues are found throughout the body. They help to form the framework of organs, and they pervade tissue spaces, filling in and connecting organs and various other structures.

It is characteristic of connective tissues that the number of cells is minimal, but the intercellular substance is ordinarily abundant. The connective tissues contrast sharply with epithelial tissues in this respect. The intercellular substance, which varies considerably in different kinds of connective tissues, makes it possible to classify these tissues as areolar, adipose, fibrous, cartilage, or bone. In this section we shall discuss the first three types.

Typical embryonic connective-tissue cells, with the exception of cartilage and bone cells, are large, star-shaped cells with numerous processes. They are called *fibroblasts,* and they give rise to the fibers found in adult connective tissues. Fibroblasts are thought to arise from an early embryonic tissue known as *mesenchyme,* in which irregular cells with their protoplasmic processes form a network enclosing a fluid material or matrix. There is a greater number of cells in the young embryonic material than in differentiated connective tissues, and there is relatively less intercellular substance.

As connective tissues develop, the cells lose their stellate appearance and are often separated by more abundant intercellular material. This material varies from a fluid, or semifluid, mucoid substance to a firm ground substance, as in cartilage or the rigid matrix of bone.

Fibers that are characteristic of connective tissues lie within the intercellular material or matrix. They are of two general types: *collagenous* and *elastic fibers.* Collagenous, or white, fibers are found in wavy bundles. They contain collagen, an albuminoid protein. The individual fibrils do not branch and are typically found in relatively large bundles. They show very little elasticity and are therefore used abundantly in strong structures such as ligaments and tendons.

Elastic fibers occur singly and branch to unite with similar fibers. They are larger, individually, than collagenous fibrils, are slightly yellow when grouped in fresh material, and are straight, not undulating, in their course through the tissue. They contain the protein, elastin, which is responsible for their elasticity. Since they have a highly developed property of elasticity and since they differ in appearance from collagenous fibers, they are commonly referred to as *yellow elastic fibers* (Figures 4.7, 4.8, and 4.9).

Collagen, the principal constituent of connective tissue, is the major structural protein of the body. It is synthesized primarily by fibroblasts, and collagenous fibers are found abundantly in the connective tissues of the skin and in ligaments, tendons, cartilage, and bone. The electron microscope shows that collagen fibrils are cross-striated (Figure 4.10). Each fibril is composed of long-chain polypeptide molecules of tropocollagen. It is thought that tropocollagen molecules are synthesized by the fibroblasts and extruded from the cell, where they form collagen fibrils by polymerization. The tropocollagen molecules in linear array overlap slightly; this gives rise to their striated appearance. When newly formed molecules are dissolved in a cold salt solution, they will re-form into typical striated collagen upon being warmed.

Epithelial and connective tissues

AREOLAR CONNECTIVE TISSUE Areolar tissue is the most abundant of the connective tissues and the most widely distributed. It is composed of rather

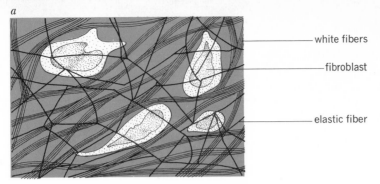

white fibers

fibroblast

elastic fiber

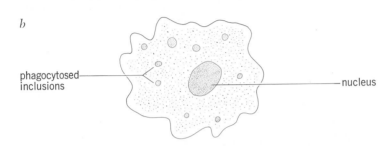

b

phagocytosed inclusions

nucleus

FIGURE 4.7
a Areolar connective tissue, including elastic fibers, white fibers, and fibroblasts. *b* Macrophage with cellular inclusions by phagocytosis.

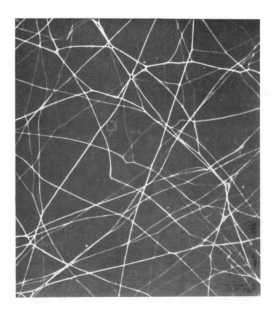

FIGURE 4.8
Network of elastic fibers in rat mesentery. [*From Roy O. Greep (ed.), "Histology," McGraw-Hill Book Company, New York, 1973.*]

primitive cells in a semifluid base or matrix. Also within the matrix are wavy bundles of white fibers and straight yellow elastic fibers forming an interlacing network (Figure 4.7). The spaces (*areolae*) between the fiber bundles vary with the type of tissue. Loose areolar tissue consists of relatively few fibers with wide spaces between the fibers. When removing the skin from

Connective tissues

65

FIGURE 4.9
Collagenous fibers in rat mesentery.
[*From Roy O. Greep (ed.),*
"Histology," McGraw-Hill
Book Company, New York,
1973.]

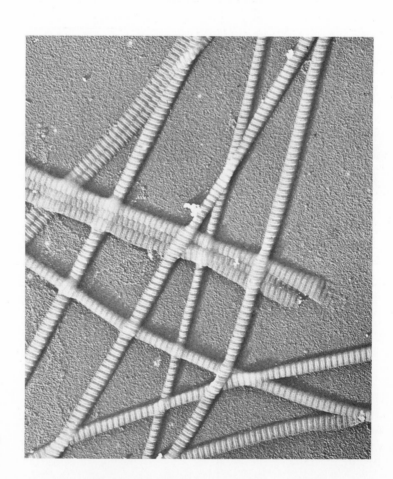

FIGURE 4.10
Electron micrograph of collagen.
Note cross-striations. (*Courtesy*
of Dr. Jerome Gross.)

an animal, one observes that the skin is attached to the body by strands of loose areolar tissue. Dense areolar tissue is found in the deeper layer of the skin itself, forming the dermis. Areolar tissue is found throughout the body, surrounding organs, supporting them, and attaching them to various structures. It invests blood vessels and nerves and is found between organs and other structures. Areolar tissue is continuous throughout the body and may therefore serve as a pathway for the spreading of infections. It is also concerned with the water balance of tissues, since it can absorb and hold a very considerable amount of water. Under certain abnormal conditions this may result in swelling (edema) of the ankles or other parts of the body, because of the accumulation of water in the tissues.

ADIPOSE TISSUE Fat cells are associated with loose connective tissue. Subepidermal areolar tissue is commonly filled with fat cells. Within a framework of fibers are single cells or groups of cells, each of which is filled with fat. The cells become so distended with fat that the cytoplasm consists of only a thin ring encircling each cell. The nucleus lies in the ring of cytoplasm, which resembles the setting in a signet ring (Figure 4.11a and b).

Although some adipose tissue is differentiated in the embryo and appears to function as a fat-storage organ, fat cells of the areolar tissues of the skin and mesenteries are considered to be fibroblasts that have been modified by the accumulation of oil droplets within their cytoplasm. As more fat is deposited, the cell increases in size until groups of large fat cells make up the greater part of adipose tissue. The matrix, containing fibers, becomes a slender framework between masses of fat-laden cells. If the accumulated fat is used up in body metabolism, these cells resume the fibroblast appearance.

Fat is not deposited indiscriminately over the body. It is found primarily as a subcutaneous layer or as a deposition on membranes such as the mesenteric, or greater omentum (see Chapter 17). It is usually found around the kidneys, helping to hold them in place, around the intestine, and in the furrows of the heart. The eyeball lies embedded in a padding of fat. There is a layer of fat around the joints of the skeletal system, and fat is present in the marrow of long bones. It fills in between muscles and helps to support various structures such as blood vessels and nerves.

The subepidermal layer of fat acts as an insulating layer which prevents excessive heat loss from internal organs. It may be assumed that those individuals with a thick subepidermal fat layer, other factors being favorable, are better insulated for a cold environment. Women commonly have a somewhat thicker layer of adipose tissue under the skin than men. One can observe that fat tends to be deposited in certain areas, for example, in the ventral abdominal area.

Adipose tissue is an important source of reserve energy, since stored fat can be oxidized and used as food. When we discuss the utilization of foods, we shall find that a great deal more heat can be obtained from fats than from any other kind of food.

Fat becomes something of a liability, especially in older persons, since an aging circulatory system is forced to carry blood to nourish all the extra pounds of fat. The action of the heart can be hampered somewhat by a deposition of fat on its surface. One may be sure that in seriously overweight persons fat has been deposited internally around organs as well as in the subepidermal layers.

white fibers

fat cell

nucleus

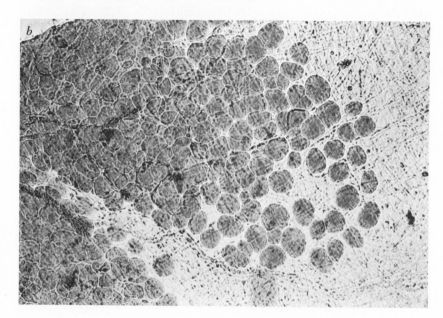

The control of one's weight is largely a matter of diet. If a person
eats more than he can expend in energy, the excess is likely to be stored
as fat.

FIBROUS CONNECTIVE TISSUE Although connective tissues generally contain
fibers, the fibrous portion of such structures as ligaments and tendons is so
great that they are commonly classified as examples of dense fibrous tissue.
In tendons, which are the best examples, large wavy bundles of white
collagenous fibers are seen in longitudinal section, and between the bundles
of fibers are rows of tendon cells (Figure 4.12*a* and *b*). Cytoplasmic layers
extend outward from cells and tend to encircle bundles of fibers. In cross
section, slender processes of tendon cells may be observed branching out
and anastomosing around bundles of fibers.

Fibrous connective tissue is a very strong, tough tissue formed into
cords of glistening white. It is pliant, yet has very little elasticity. The blood

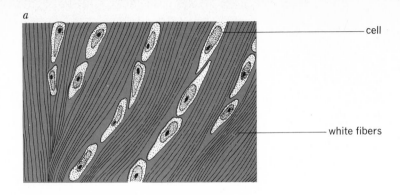

cell

white fibers

FIGURE 4.12
a Diagram of tendinous tissue.
b Photomicrograph of fibrous
connective tissue. (*Courtesy of
Carolina Biological Supply
Company.*)

supply is not abundant. If this tissue is injured, as in sprains, it may take a long time for the injury to heal.

Tendons attach muscles to bones. They can be observed in the wrist or ankle and also along the back of the hand or in the foot, where they are often long cords extending out to the digits.

Ligaments help to hold bones together at the joints. They may be in the form of flat bands or capsules. Though most ligaments are composed largely of white fibers, some ligaments contain elastic fibers in abundance.

Aponeuroses are thin tendinous sheets of fibrous tissue that ordinarily attach to flat muscles.

Fasciae cover muscles and help to hold them in place. They are thin sheets of white fibrous connective tissue.

RETICULAR TISSUE Reticular tissue contains delicate white fibers that form a network with primitive connective-tissue cells. The reticular cells have numerous protoplasmic processes, and the nuclei are flat, oval disks. In lymphoid or adenoid tissue, lymphocytes are found in the fibrous network. Reticular tissue is found in lymph nodes, in the spleen, and in bone marrow.

Elastic connective tissue Sometimes considered as a distinct type, this is a fibrous tissue in which elastic fibers are more abundant than white collagenous fibers. Cells are located between the fibers in the ground substance. The predominance of yellow fibers gives the tissue a slightly yellow color. Elastic tissue has the properties of elasticity and extensibility. It is found in the walls of blood vessels, in respiratory structures such as the lungs and bronchioles, in the vocal folds, and between the cartilages of the larynx. It is present in a few elastic tendons.

Mast cells are fairly large cells present in fibrous connective tissues. They are characterized by dense cytoplasmic granules and resemble granular white blood cells to a certain extent. But the granules are somewhat smaller in mast cells, and since the granules are soluble in water, they are not rendered visible with ordinary aqueous staining techniques.

Mast cells are fairly well represented in subcutaneous areolar connective tissues. They contain an anticoagulant of the blood called *heparin*. They also release *histamine*, as a result of injury, which stimulates vasodilatation in small arterioles and capillaries and also increases the permeability of capillaries to protein. Histamine is an amine derived from histidine. It is broken down by histaminase, an enzyme common to most tissues.

Macrophages, or *histiocytes*, are active phagocytic cells of loose fibrous connective tissues. They resemble monocytes of the blood and are closely related to them. However, macrophages are not present in the blood; they are wandering cells in the tissues or fixed cells lining passageways in such structures as lymph nodes and spleen. In healthy connective tissues, the cells are characterized by cytoplasmic processes extending outward, giving the cell a somewhat stellate appearance, but when they are active phagocytic cells, as in an area of inflammation, they lose their cytoplasmic processes and appear as oval-shaped cells. The scanning electron microscope shows these cells as having a rough, membranous surface with microvilli.

Macrophages in certain organs are often given special names, such as the *Kupffer cells* of the liver and "dust cells" in the alveoli of the lungs. In lymph nodes, spleen, and bone marrow they have commonly been called *reticuloendothelial cells*. All macrophages seem to be closely related and capable of ingesting large amounts of foreign material (Figure 4.7b).

THE INTEGUMENT   The skin forms a remarkable protective layer over the outer surface of the body. The skin folds inward and becomes continuous with the mucous linings of the mouth, nose, urethra, vagina, and anus at mucocutaneous junctions. There are two principal layers in the skin, the *epidermis*, formed by stratified squamous epithelium, and the *dermis*, or *corium*, which is a fibrous connective-tissue layer. The *hypodermis*, a loose connective-tissue layer beneath the dermis, often is filled with subcutaneous fatty tissue but is not a part of the skin. It has fibrous connections with such covering membranes beneath it as the fasciae that cover muscles.

The skin in its many functions helps to maintain body fluids and aids in maintaining blood pressure and regulating body temperature. Sensory receptors in the skin include those reacting to tactile stimuli, pressure, and temperature. The skin has the ability to repair its own minor wounds. Scratches and minor abrasions are commonplace occurrences, but without the healing ability of the skin, man could not long survive.

The outer epidermal cells are dead cells. but in dying they produce a surface layer that is protective against air, water, most foreign substances, and microorganisms. They tend also to ward off excessive ultraviolet radiation. As the outer, dead cells wear away, the living, deeper layers move more cells toward the surface to take their place. These cells produce the fibrous protein *keratin*, which is the main constituent of the surface layer of the skin and also of such skin derivatives as hair and nails.

The epidermis is ordinarily a thin flexible layer, but on the palms of the hands and soles of the feet it is much thicker and rougher. A ridged or rough surface affords a better grip. The ridges, whorls, and arches of the fingertips are distinctive for each individual and establish the validity of fingerprinting as a means of identification.

The skin becomes more active at puberty, the hair follicles enlarge, and body hair becomes more prominent. The sebaceous glands and sweat glands also become more active. The number of hair follicles does not differ greatly between men and women, but body hair on women is usually soft, short, and essentially colorless and so escapes notice. In old age the skin becomes thin, dry, and wrinkled.

The color of the skin depends largely upon the amount of melanin present. Melanin is the dark-colored pigment normally present in stellate cells called *melanocytes* (Figure 4.6). These cells lie in the deeper layers of the epidermis. People with fair complexions produce a small amount of melanin in the skin; those with dark complexions produce a much greater amount. Skin pigment is not evenly distributed; it occurs in much greater concentration in such areas as the areoles of the breasts, where melanocytes are also present, and in decreased amounts in the palms of the hands and soles of the feet. It is protective against the ultraviolet rays of the sun and undoubtedly has had survival value for those peoples exposed to the intense sunlight of the tropics.

The pigment carotene is present also and gives "white" skin a yellowish color. The pink blush of healthy white skin is caused by the blood circulating through the deeper layers.

ALBINISM

Because they are unable to produce melanin, human albinos have pale straw-colored hair, pink eyes, and fair, reddish skin. Albinism is inherited as a homozygous recessive character. Individuals with normal pigmentation will be either homozygous *AA* or heterozygous *Aa* for this character. The recessive homozygous condition *aa* produces the albino. Albinos are unable to form melanin because a tyrosinase enzyme system is lacking. They have a normal number of melanocytes, but the recessive gene apparently does not provide for the formation of the enzyme tyrosinase which is essential for the production of melanin.

DERMIS

The *dermis*, or *corium*, is composed of fibrous connective tissue. It extends from a layer of dermal papillae just underneath the epidermis to a reticular layer which is continuous with the subcutaneous layer. The papillary layer contains more elastic than collagenous fibers. The reticular layer is composed

Dermis

of a network of collagenous fibers. The dermis contains blood vessels, nerves, sweat glands, hair follicles, and cutaneous sensory receptors (see Figure 4.6).

CARTILAGE

Consisting of modified connective-tissue cells surrounded by massive amounts of intercellular material, cartilage is a dense white connective tissue. This material is composed of white fibers, yellow fibers, and a ground substance (or matrix). The ground substance, chondrin, is found in a limited area around the cells. The great bulk of the intercellular substance is composed of a dense mass of white fibers. The individual fibers are not visible under ordinary light.

Cartilage is ordinarily found in relatively thin sheets covered by a connective tissue called the *perichondrium*. Since cartilage itself is essentially nonvascular, the perichondrium, being well supplied with blood vessels, acts as a nourishing membrane. Lymphatics and nerves are also confined to the perichondrium. Nutrient fluids diffuse slowly through the dense intercellular substance to nourish the cells.

There are three forms of cartilage in the body. They are largely differentiated on the basis of variations in the structure of their intercellular substances into hyaline, yellow elastic, and white fibrocartilage.

HYALINE CARTILAGE This is the most abundant and most widely distributed type of cartilage. Its structure corresponds to the general description of cartilage given above. The relatively few cells are located in cavities called *lacunae*. The intercellular substance is composed of white collagenous fibers. It has a glistening white or pearl-like appearance, but in thin sections there is a slightly bluish tinge. This type of cartilage, in the embryo, is the forerunner of bone. Hyaline cartilage is found in the nose, in the larynx, and in the rings of the trachea and bronchial tubes. It is the type of cartilage attached to the ribs and covers the ends of bones at joints. In the latter case, cartilage promotes the smooth action of joints and helps to absorb some of the shock of walking, running, and jumping (Figure 4.13a and b).

YELLOW ELASTIC CARTILAGE Developing from hyaline cartilage, yellow elastic cartilage is somewhat similar in appearance, but yellow elastic fibers are present in great numbers within a network of white collagenous fibers. The cells lie in lacunae, and a perichondrium is always present. The elastic fibers impart a certain degree of elasticity to this tissue. Elastic cartilage is found in the external ear, in the epiglottis, in the eustachian tube, and in some of the small laryngeal cartilages (Figure 4.14a and b).

WHITE FIBROCARTILAGE In general appearance white fibrocartilage resembles fibrous connective tissue. It is composed of bundles of wavy white collagenous fibers, the bundles more or less parallel to one another. Cartilage cells lie in lacunae in the matrix, surrounded by bundles of fibers (Figure 4.15a and b).

Fibrocartilage is found notably between the vertebrae in the intervertebral disks and in the cartilage between the two pubic bones. It is placed around the outer borders of movable joints, where it tends to make the joint cavities deeper. The tissue is often not well defined, grading off into areolar or fibrous connective tissue. When fibrocartilage is associated with hyaline cartilage in a joint or in the intervertebral disks, hyaline

Epithelial and connective tissues

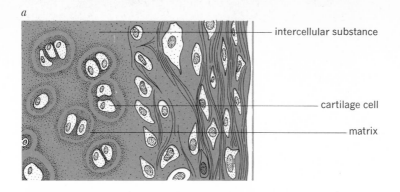

intercellular substance

cartilage cell

matrix

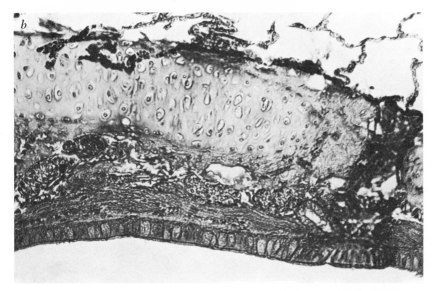

FIGURE 4.13
a Hyaline cartilage. *b* Photomicrograph of hyaline cartilage. (*Courtesy of Carolina Biological Supply Company.*)

cartilage always makes contact with the bone and gradually grades into fibrocartilage. The plates of fibrocartilage between the vertebrae form a padding that absorbs much of the shock transmitted through the vertebral column as the result of walking or jumping.

SYNOVIAL MEMBRANES   These membranes are composed of connective tissues; unlike mucous membranes, they have no epithelial layer. They line the articular capsules of joints and secrete synovial fluid into the articular cavity. Synovial fluid is a thin, clear fluid which lubricates the joints. *Bursae* and *tendon sheaths*, the sacs inserted to cushion parts that move on each other, such as tendons moving over bones, also are formed by connective-tissue membranes lined by synovial membranes, and they contain synovial fluid.

BONE

Bone, or osseous tissue, is the hardest of the connective tissues. Its hardness is due largely to the deposition of inorganic salts in an organic matrix called *ossein.* The two most abundant salts are calcium phosphate and calcium carbonate. The organic materials include fibers embedded in a matrix containing protein substances (collagen), bone cells, blood vessels, and cartilagi-

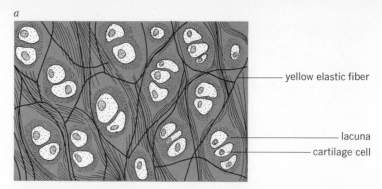

— yellow elastic fiber

— lacuna
— cartilage cell

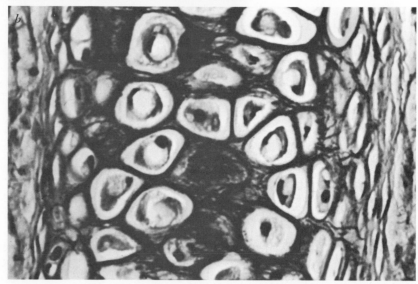

FIGURE 4.14
a Elastic cartilage. *b* Photomicrograph of elastic cartilage. (*Courtesy of Carolina Biological Supply Company.*)

nous substances. The crystalline mineral found in bone is hydroxyapatite. The mineral apatite is calcium phosphate fluoride, $CaFCa_4(PO_4)_3$, commonly found in rocks. In hydroxyapatite, the fluorine atoms are replaced by hydroxyl groups, OH. The crystals of mineral matter are embedded in an organic matrix composed largely of crystalline collagen. Hyaluronic acid also is present. Bone can be decalcified by placing it in vinegar or some other weak acid solution. The inorganic salts are then dissolved out, while the organic structure remains. The bone retains its shape but becomes pliable and can be bent or twisted. It is still strong but flexible. In another experiment the organic material may be driven off by exposing a bone to intense heat. Again, the bone retains its shape, but now it appears as bone ash—chalky white and very brittle.

STRUCTURE Bone is not the solid homogeneous material that it may appear to be upon superficial examination. Only the walls of the shaft are composed of hard bone; the ends are filled with a porous, spongy network called *cancellous bone*. Within the shaft is a long medullary canal containing bone marrow, of which there are two types. The type in the medullary canal is yellow marrow. It contains connective tissue, blood vessels, immature blood cells, and a great many fat cells, which give it a yellow color (Figure 4.16).

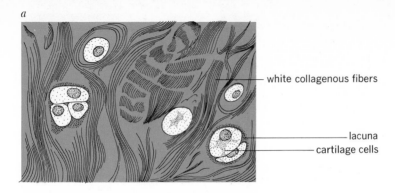

a

white collagenous fibers

lacuna
cartilage cells

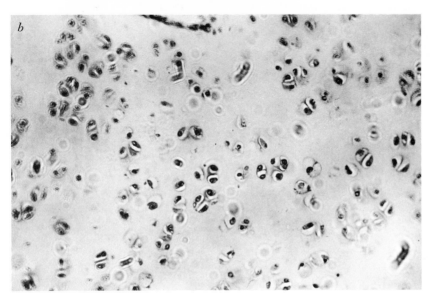

b

FIGURE 4.15
a Fibrocartilage. *b* Photomicrograph
of fibrocartilage. (*Courtesy of
Carolina Biological Supply
Company.*)

Red marrow is found in the ends of long bones in the cancellous tissue. It is more abundant in the larger flat bones and in the ribs. There are fat cells in the red marrow, but they are limited to individual cells distributed among a great number of developing blood cells. After birth, the red cells and granular white cells of the blood originate in the red marrow. The breast bone, or sternum, is a convenient bone for obtaining samples of red marrow.

A study of a very thin transverse section taken from the shaft of a bone shows that this osseous tissue is composed of concentric rings (concentric lamellae). The centers of the rings are canals, called *haversian*, or *central*, *canals*, that contain blood vessels and small amounts of connective tissue. Between the concentric layers of bone are cavities called *lacunae*, in which bone cells lie. Radiating out in all directions from the lacunae are tiny canals (*canaliculi*), which branch and anastomose until they form a network throughout the bone. Minute protoplasmic filaments grow outward from osteocytes into the canaliculi. Thus bone itself becomes a living tissue, containing cells and connecting strands of cytoplasm even in the hardest and most compact portions of its structure. Haversian canals are not isolated; they are joined by communicating canals. The concentric arrangement of layers of bone and bone cells around a canal is called a *haversian system*

Bone

75

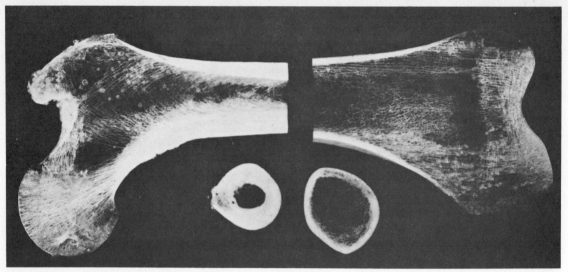

FIGURE 4.16
Section of proximal portion of the human femur showing internal structure. (*Courtesy of Otto Kampmeier, Ph.D., M.D., and the Upjohn Company.*)

or an *osteon*. The areas between haversian systems are filled in with lamellae that do not have a concentric arrangement. These are the interstitial lamellae (Figure 4.17).

The *periosteum* is a tough fibrous membrane that covers bone except at the articulating surfaces. There are two layers within the periosteum, although they are not clearly defined. The outer layer is composed of connective tissue filled with blood vessels, lymphatic vessels, and nerves. The inner layer contains elastic fibers, cells, and blood vessels, although it is not as vascular as the outer layer. It is generally considered that the cells of the inner layer are capable of becoming bone cells. In the young there are many osteoblasts in the inner layer of the periosteum next to the surface of growing bone, but in an adult the cells remain inactive unless an injury occurs. If a bone is fractured, this osteogenic layer supplies new bone to replace the old and aids in repairing the break (Figure 4.18) (Table 4.2).

Blood vessels and nerves from the periosteum penetrate into the hard substance of bone through canals (*Volkmann's canals*). These minute passageways pass through the lamellae as secondary vascular canals. Some of them connect with blood vessels in haversian canals, and others follow an irregular path through solid bone. Nutrient arteries supply the bone marrow. They commonly penetrate the shaft of a long bone near the middle and divide, sending branches into the marrow toward either end. There are veins also in the marrow cavity, many of them emerging through the cancellous tissue at the ends of long bones.

GROWTH   Most of the bones of the cranium and face develop embryonically between membranes and are called *membrane bones*. These bones do not preform in cartilage. The embryonic appendicular skeleton is formed of cartilage that is later replaced by bone. Bones formed in this way are commonly called *cartilage bones*. There are, therefore, two methods of bone formation in the embryo: intramembranous, as in the case of skull bones, and intracartilaginous, referring to the replacement of cartilage by bone as in the appendicular skeleton.

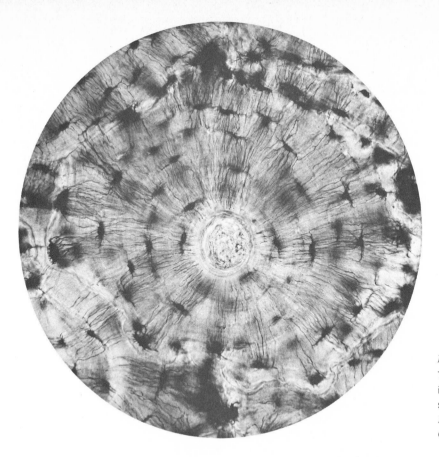

FIGURE 4.17

Thin cross section of compact bone, illustrating an osteon, or haversian system. (*Courtesy of General Biological Supply House, Inc., Chicago.*)

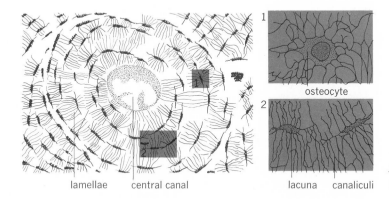

1

osteocyte

2

lamellae   central canal

lacuna   canaliculi

FIGURE 4.18

Microscopic structure of bone.

Development of membrane bone   The bones of the cranium develop from centers of ossification in connective tissue. Fibroblasts, located in fibrous connective tissue, change to *osteoblasts,* or *bone-forming cells.* The intercellular substance becomes gelatinous in its consistency. Spicules of bone appear, and calcium salts are deposited in the matrix. Finally, long, thin plates surrounded by osteoblasts appear in the connective tissue. Some of the osteoblasts become located individually in lacunae and are then called *osteocytes* or

Bone

77

TABLE 4.2

Epithelial and connective tissues

Type	Where found	Description
Epithelial		
Simple squamous	Thin serous linings of alveoli of lungs	Flat oval cells in a single layer
	The mesothelium which lines body cavities	
	The endothelial lining of the heart and blood vessels	
Stratified squamous	Lining of cheek	Several layers of cells; the surface layer consists of flat cells, the deeper layers tend toward cuboidal
	The external protective layer of the skin	
Cuboidal	Lining of glandular ducts and tubules	Somewhat cube-shaped
Simple columnar	Lining of organs of digestive tract (inner surface layer)	A column of cells, each cell taller than wide; goblet cells present, especially in intestine
Ciliated columnar	Lining the trachea and bronchial tubes; the oviducts, epididymis, and vas deferens	Hairlike processes called *cilia* project from free surfaces
Connective		
Areolar or loose	Around organs; between organs and other structures	Fibers and fibroblasts in a semifluid base or matrix
Adipose or fat	Subepidermal, on membranes and around organs	Signet-ring cells, the nucleus lies in a thin ring of cytoplasm
Dense fibrous	Ligaments, tendons, aponeuroses, and fasciae	Mostly collagenous white fibers with rows of cells between bundles of fibers
Elastic	Walls of blood vessels; respiratory structures such as lungs and bronchioles, vocal folds	Collagenous fibers but with a greater amount of elastic fibers
Cartilage		
Hyaline	Covers the ends of bones at joints. Larynx, nose, rings of trachea, and bronchi (the forerunner of bone)	Intercellular substance composed of white collagenous fibers. The cells lie in lacunae. (All types of cartilage covered by perichondrium.)
Yellow elastic	External ear Auditory tube	Yellow elastic fibers in a network of white fibers. Cells in lacunae
White fibrocartilage	Intervertebral disks. Symphysis between pubic bones	Bundles of wavy white fibers. Cells in lacunae
Bone	Skeletal structures	A hard connective tissue covered with periosteum. Bone cells arranged in a haversian system

bone cells. They maintain connections with each other and with the nourishing membrane through the canaliculi.

The connective tissue around the developing bone also forms the periosteum, a nourishing membrane capable of producing bone. The inner layer of this fibrous covering is lined with osteoblasts and is known as the *osteogenic layer.* More compact bone is deposited immediately under the periosteum, whereas the intermediate portion is composed of spicules of bone in loose spongy arrangement. The skull bones have not completed their ossification at birth. There are membranous areas called *fontanels* between them. The thickest portion of the bone is at the center of ossification; the bones are paper thin at the edges (Figure 4.19).

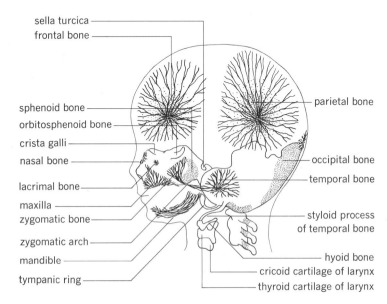

sella turcica
frontal bone
sphenoid bone
orbitosphenoid bone
crista galli
nasal bone
lacrimal bone
maxilla
zygomatic bone
zygomatic arch
mandible
tympanic ring

parietal bone
occipital bone
temporal bone
styloid process of temporal bone
hyoid bone
cricoid cartilage of larynx
thyroid cartilage of larynx

FIGURE 4.19

Skull of human embryo of about 12 weeks. *(After Hertwig and Patten, in Bradley M. Patten, "Human Embryology," 3d ed., McGraw-Hill Book Company, New York, 1968.)*

Development of cartilage bone The skeleton of the developing embryo, with the exception of most of the skull bones, is formed of hyaline cartilage. After the second month of development, cartilage is gradually replaced by bone. The perichondrium becomes the periosteum and forms a cylinder of bone at the middle of the shaft in the long bones. The cartilage within this ring of bone begins to degenerate. The cartilage cells appear swollen, and calcification occurs in this area. The cartilage cells die as they are released by the degeneration of cartilage. Endochondral bone then begins to replace the degenerating cartilage at the center of ossification. The shaft is gradually converted into bone as ossification proceeds away from the center toward the ends of the bone. There are evidently two kinds of bone formation in cartilage bones: perichondral, in which bone forms around cartilage, and endochondral, in which bone replaces degenerating cartilage. However, bone always arises from connective-tissue membranes, even in the case of cartilage bone.

While ossification progresses from the center of the shaft, the cartilage ends continue to grow and the bone increases in length. There is necessarily considerable reorganization inside the bone. Primary marrow cavities are formed by the inward growth of perichondral tissue into degenerating

Bone

79

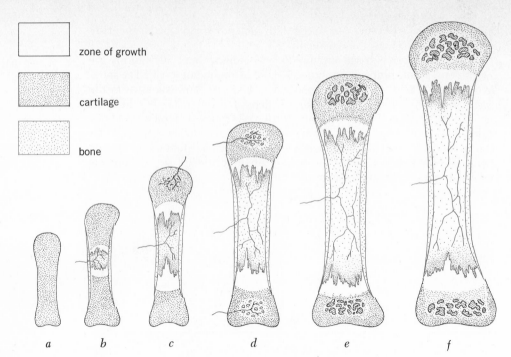

zone of growth

cartilage

bone

FIGURE 4.20

a to *f* stages in the ossification of a long bone.

a   *b*   *c*   *d*   *e*   *f*

cartilage. Blood vessels accompany the periosteal buds. Cells of perichondral tissue form the primary bone marrow, which fills the marrow cavities. Finally the primary marrow spaces are united to form the single medullary cavity of long bones (Figure 4.20). A thin connective-tissue membrane called the *endosteum* lines the marrow cavity. It is thinner than the periosteum and is more evident in the larger yellow-marrow cavities.

The marrow cavity in the bone of an adult is considerably larger than its original cartilage antecedent in the embryo. If a metal band is placed around a bone of a young animal, the metal band will be located in the bone marrow of the adult. This indicates that the marrow cavity is enlarged as the bone grows and that all the original cartilage and the bone that was formed around it has been resorbed during the growth of the bone. So far as growth in diameter is concerned, bones increase in size as successive layers of bone are produced by the osteogenic layer of the periosteum.

Osteoclasts are giant cells, multinucleated and associated with areas in which bone is being resorbed. It is generally assumed that they function in breaking down bony structure, as, for example, in the enlargement of the marrow cavity of long bones during growth. However, resorption of bone does not appear to depend entirely on the activity of osteoclasts. The secretion of the parathyroid glands increases the withdrawal of calcium from bones. It also stimulates the multiplication and activity of osteoclasts. This may indicate that the parathyroid hormone accomplishes the removal of calcium from bones by increasing the activity of the osteoclasts.

The shaft of a long bone is called the *diaphysis*. The ends are the *epiphyses*. There may be several centers of ossification. The ossification of the shaft progresses toward the epiphyses. The epiphyses are mostly cartilaginous at birth; they begin to ossify at various times after birth. Ossification can proceed from more than one center in the epiphyses of the larger bones. Bone-forming tissue invades the cartilage, and spicules of bone appear, forming a network of spongy bone (Figure 4.21).

Both the diaphysis and the epiphyses change to bone before growth is completed, but a cartilage plate, the epiphyseal plate, separates them during the entire period of growth. Proliferation of cartilage cells and the formation of new matrix proceed on the side of the plate toward the epiphysis. The cartilage toward the diaphysis is gradually replaced by bone. These zones of growth are maintained in the long bones of the female until between the seventeenth and nineteenth years, when they are entirely replaced by bone, and no further growth in the length of bones or in the height of the individual occurs. However, young women commonly attain essentially all their growth in height by the time they are fifteen to sixteen years old. In the male the zones of growth are replaced by bone between the nineteenth and twenty-first years. For practical purposes the period of growth in height is largely completed at about sixteen to eighteen years of age, although complete fusion of all epiphyses may not occur until twenty-five years of age (Figure 4.22).

Living bones may react and change their shape somewhat in response to undue demands on them for support. On the other hand, if they are not used they may become greatly weakened. Bones require a certain amount of stress in order to maintain their strength.

Elderly people may experience a deficiency in the maintenance of a hard bony structure in the skeleton, the bones exhibiting a greater degree of porosity than normal. The condition known as *osteoporosis* occurs when the body fails to maintain a firm bony matrix. This physiological breakdown in bony tissue commonly affects the vertebrae and pelvis. In women it is more common after menopause.

TERMS OF REFERENCE

It is necessary to consider a few terms of reference in order to understand directions and positions in regard to the human body. In animals that walk on four legs, the head may be said to be anterior and the tail posterior or caudal. The back is dorsal and the underneath side is ventral. The human being, however, walks in an upright position and the terms applied to the quadruped may not apply so well to the biped. The human figure, in anatomy, is studied in the anatomical position, which means erect and facing the viewer (a frontal view) or with the back to the viewer (a posterior view). The hands are held at the side with the thumbs pointing away from the body. In this position, the head is *superior* or *cranial* and the feet are at the inferior end of the body. *Cephalic* may be used also in referring to the head. *Anterior* (or *ventral*) refers to the frontal aspect of the body while *posterior* (or *dorsal*) refers to the back.

Among other directional terms in common use are *proximal* and *distal*. Proximal refers to the area closest to the body or to the midline within the body; *distal* designates the portion farthest away (Table 4.3). *Medial* indicates structures in or close to the midline; *lateral* means to the side. *Superficial* structures are those that are close to the body surface; *deep* structures are those farther away from the surface. *Parietal* usually refers to a body wall or to a membrane lining a wall, for example, the outer membrane lining the thoracic or abdominal cavities. *Visceral* indicates a membrane covering

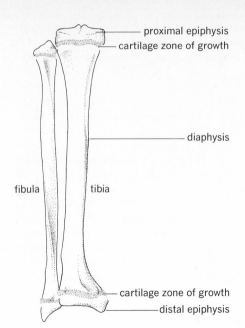

proximal epiphysis
cartilage zone of growth

diaphysis

fibula   tibia

cartilage zone of growth
distal epiphysis

FIGURE 4.21

The tibia and fibula of a child about five years old.

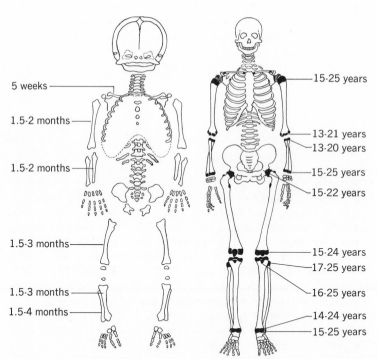

5 weeks

1.5-2 months

1.5-2 months

1.5-3 months

1.5-3 months
1.5-4 months

15-25 years

13-21 years
13-20 years

15-25 years
15-22 years

15-24 years
17-25 years

16-25 years

14-24 years
15-25 years

FIGURE 4.22

Differences between the fetal and the adult skeleton. The figures for the fetal skeleton indicate the stage of development in the fetal months; epiphyses begin to appear about this time. The figures for the adult skeleton indicate a wide range of years during which the zones of growth disappear and the epiphyses are completely fused with the diaphysis. [*After Graca and Noback in Barry J. Anson (ed.), "Morris' Human Anatomy," 12th ed., McGraw-Hill Book Company, New York, 1966.*]

an organ, as the visceral peritoneum covers the viscera of the abdominal cavity.

Directional terms relating to the hand and foot may cause some difficulty in interpreting their position. The palm of the hand is the *volar* or *palmar* surface; the back of the hand is posterior or dorsal. The sole of the foot is the *plantar surface*, whereas the opposite side is dorsal.

Epithelial and connective tissues

82

TABLE 4.3
Terms of reference

Terms	Definition	Example
Superior	Toward the head	The head is superior to the thorax.
Inferior	Toward the feet	The feet are inferior to the knees.
Anterior, or ventral	Front	The nose is anterior to the ear.
Posterior, or dorsal	Refers to the back	The vertebral column is posterior to the sternum (breast bone).
Medial or mesal	The midline	The linea alba lies along the mesal, or midsagittal, plane.
Lateral	Toward the side	The arms are lateral to the trunk.
Proximal	Nearer the midline or nearer the trunk	The shoulder is proximal to the elbow.
Distal	Farther from the midline or trunk	The fingers are distal to the wrist.
Parietal	Refers to a body wall or to the membrane lining the wall	The thoracic wall is lined by a membrane, the parietal pleura.
Visceral membrane	A membrane covering an organ	The visceral pleura covers the lungs.
Peripheral	Toward the external surface	The skin contains peripheral sensory receptors.
Superficial	Structures closer to the surface, such as superficial muscles	The external oblique muscle is superficial to the internal abdominal oblique muscle.
Deep	Structures more internal	The interossei are deep muscles of the hand.

The cavities of the body may be considered as *dorsal* and *ventral* cavities. Each of these cavities may, in turn, be divided into two parts. It should be understood that these cavities are filled completely with organs and body fluids. The dorsal cavity is subdivided into a *cranial cavity* containing the brain and into a *spinal cavity* containing the spinal cord. The ventral cavity is subdivided into the *thoracic* and *abdominal-pelvic* cavities. The thoracic cavity contains the lungs, the heart with its blood vessels, and several other structures, such as the thymus, esophagus, trachea, nerves, and lymphatic vessels.

The abdominal cavity is separated from the thoracic cavity by the thoracic diaphragm. This large cavity contains the stomach, intestine, liver, pancreas, kidneys, and spleen. There is no distinct structural separation between the abdominal and pelvic cavities, and they are commonly considered together as the abdominopelvic cavities. Within the pelvic cavity we find the urinary bladder, the internal reproductive organs, and the terminal portion of the large intestine.

BODY CAVITIES

Body cavities

83

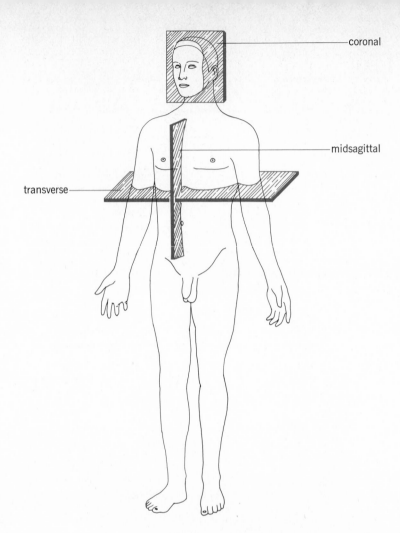

coronal

midsagittal

transverse

FIGURE 4.23
Anatomical planes of reference.

PLANES OF REFERENCE

It is often useful to consider regions of the body in the light of planes of reference. Organs and various body structures may be conveniently located when viewed in certain planes. We speak of *sagittal sections*, meaning that the body or organ has been cut into right and left portions. A *midsagittal* plane divides the body into equal right and left halves. *Parasagittal* refers to a plane parallel to the median plane (Figure 4.23).

Coronal or *frontal* sections divide the body along a coronal plane into anterior (or ventral) portions and posterior (or dorsal) parts. A coronal section of the head divides the head into an anterior portion which includes the face and a posterior portion which includes the back or occipital region.

A *horizontal* or *transverse plane* divides the body into superior and inferior portions. It lies at right angles to the longitudinal planes. Similar sections of organs and tissues are commonly called cross sections. If cut along the longitudinal axis, they are longitudinal sections. Oblique sections also may be made.

Tissues are groups of cells that are similar in origin, structure, and function. **SUMMARY** Epithelial tissues are thin and usually delicate and contain very little intercellular substance. Types of epithelial tissues include simple and stratified squamous, cuboidal, and columnar. Columnar epithelium may be ciliated, as in the lining of the trachea and bronchial tubes; it may contain goblet cells, as in the lining of the intestine. The electron microscope reveals the minute structure of cilia to consist of nine groups of filaments surrounding a central pair. Mitochondria are numerous at the base of each cilium.

Epithelial cells may have minute projections from their free surface. These are microvilli, and they increase the absorptive surface.

Junctions between epithelial cells have been demonstrated. These disks of dense material are desmosomes.

There are two kinds of sweat glands, merocrine (eccrine) and apocrine. Merocrine glands are concerned with temperature regulation, and their ducts emerge in the pores of the skin. Apocrine glands are stimulated to secrete under certain conditions such as emotional states, stress, and sexual excitement. The sebaceous glands of hair follicles are holocrine glands.

Mucous membranes are epithelial membranes that line the respiratory, digestive, and urogenital tracts. Various types of glands in these membranes keep the surface moist. Mucus is commonly secreted but not in all cases.

Serous membranes line body cavities that do not open to the exterior, such as the abdominal and the thoracic cavities. The pleural cavity is a potential cavity between the parietal and visceral layers of the pleura. The pericardium is a double layer covering the heart. Serous membranes also line the interior of the heart, the blood vessels, and the lymphatic vessels. Serous fluid is thin and watery.

Synovial membranes are composed of connective tissues and, unlike mucous membranes, have no epithelial layer. They line the articular capsules of joints and secrete synovial fluid. Bursae and tendon sheaths also are lined by synovial membranes and contain synovial fluid.

Connective tissues are tough and fibrous and usually contain an abundance of intercellular substance. Areolar, adipose, dense fibrous, and reticular tissues and cartilage and bone are types of connective tissues. Ligaments and tendons are examples of dense fibrous connective tissue. There are three kinds of cartilage: hyaline cartilage, yellow elastic cartilage, and white fibrocartilage.

Collagen has been called the major structural protein of the body. Collagenous fibers are abundant in the skin and in ligaments, tendons, cartilage, and bone. When highly magnified, collagen has a distinctly striated appearance.

Mast cells are fairly large cells found in fibrous connective tissues. The cytoplasm is granular, and they resemble somewhat the eosinophils of the blood. They produce heparin, an anticoagulant of blood and also release histamine, a vasodilator.

Macrophages are phagocytic cells found in loose connective tissues. They may be wandering cells or fixed cells lining passageways in such structures as lymph nodes and spleen. In the liver they are called Kupffer cells. They constitute the reticuloendothelial cells of lymph nodes, spleen, and bone marrow. They resemble the monocytes of the blood and are closely related to them, but macrophages are not found in the blood.

The skin, or integument, is composed of two principal layers, the epidermis and the dermis. The epidermis is composed of stratified squamous epithelium, whereas the dermis is a fibrous connective-tissue layer. The hypodermis lies beneath the dermis but is not a part of the skin.

Epidermal cells produce the fibrous protein keratin, which is the principal constituent of the surface layer and also of nails and hair.

The color of the skin depends largely on the amount of pigment present. The pigment is melanin and is produced by cells called melanocytes.

The deeper layers of the skin composing the dermis, or corium, contain blood vessels, nerves, sweat glands, hair follicles, and the cutaneous sensory receptors.

The bones of the cranium that do not preform as cartilage are membrane bones. The bones of the appendicular skeleton preform as cartilage and are called cartilage bones. The haversian system of bone structure consists of concentric layers of bone and bone cells around the haversian canal. Bone cells lie in cavities called lacunae, from which minute canals (canaliculi) radiate outward. The periosteum, a tough fibrous membrane, is the protective membrane that covers bone except at the articulating surfaces. The inner layer of the membrane is osteogenic.

The marrow cavity is lined by a thin connective-tissue membrane called the endosteum. Large, multinucleated cells, called osteoclasts, are found in areas in which bone is being resorbed. They are concerned with the breaking down of bony structure and therefore aid in enlarging the marrow cavity. The osteogenic layer of the periosteum provides for the increase in diameter in bones by producing successive layers on the outside.

Membrane bones develop directly from centers of ossification in connective tissue. Cartilage bones develop from hyaline cartilage, which is gradually replaced by bone. Cartilage zones of growth remain between the diaphysis and the epiphyses in growing bones. Zones of growth in the long bones ordinarily change to bone between the seventeenth and twenty-first years, after which there is no further growth in height.

QUESTIONS	
1	Describe the outstanding structural characteristics of epithelial tissues. How do they differ from connective tissues?
2	How are bones nourished?
3	Where would you find squamous, cuboidal, or columnar epithelium?
4	Where is ciliated epithelium found?
5	Compare the minute structure of cilia with that of centrioles.
6	Discuss the differences in structure and function between eccrine and apocrine glands.
7	Where are sebaceous glands found, and what is their function?
8	Describe the differences between mucous, serous, and synovial membranes.
9	What use is made of collagen?
10	How do macrophages function?
11	Discuss the structure and function of the skin.
12	Describe an osteon.
13	Discuss some of the functions of adipose tissue.
14	What are the individual uses of tendons, ligaments, aponeuroses, and fasciae?
15	Discuss three types of cartilage, indicating differences in structure and function. Locate various types of cartilage in the body.
16	Discuss the structure of bone, including the embryonic development of different kinds of bone. How do bones grow?
17	Where do the red cells of the blood originate in the adult?

SUGGESTED READING   Several textbooks on histology are listed at the back of the book.

CONGDON, C. C.: Bone Marrow Transplantation, *Science*, **171**:1116–1124 (1971).

FRASER, R. D. B.: Keratins, *Sci. Am.*, **221**:86–96 (1969).

GROSS, JEROME: Collagen, *Sci. Am.*, **204:**121-130 (1961).

MONTAGNA, WILLIAM: The Skin, *Sci. Am.*, **212:**56-66 (1965).

ROSS, R.: Wound Healing, *Sci. Am.*, **220:**40-50 (1969).

——— and PAUL BORNSTEIN: Elastic Fibers in the Body, *Sci. Am.*, **224:**44-52 (1971).

RUSHMER, ROBERT F., K. J. K. BUETTNER, J. M. SHORT, and G. F. ODLAND: The Skin, *Science*, **154:**343-348 (1966).

WARFEL, A. H., and S. S. ELBERG: Macrophage Membranes Viewed through a Scanning Electron Microscope, *Science*, **170:**446-447 (1970).

Suggested reading

chapter 5
the skeleton

The skeleton forms the bony framework of the body; it is the principal support for the various parts and affords protection for vital organs. The skeletal muscles are attached to bones by tendons. Muscular contractions provide the power of movement to skeletal parts.

Bones are structures that have undergone many adaptive changes to perfect them for their present use. The bones in the arm of man, for example, are homologous with the bones in the wing of a bird or the wing of a bat. Bones are not to be regarded, therefore, as mere pieces of framework, but as living, active organs which have had a long phylogenetic history of development and which in the history of the individual are subject to considerable change during growth and even in later life.

It is most important that the skeletal system be regarded as a living organ system, not as a collection of dried bones. It is here that calcium and phosphorus are stored for use in maintaining metabolic functions. Calcium, as we shall see later, is essential for the proper functioning of the nervous system, the contraction of muscles, and the clotting of the blood. We have noted before that the red cells of the blood are formed in the red marrow of bones.

Hormones from the thyroid and parathyroid glands, together with vitamin D, regulate the calcium level of the blood and largely control the deposition of calcium in bones. Under certain conditions, calcium may be withdrawn from bones in order to maintain the calcium level of the blood, or additional calcium may be absorbed from the intestine under the influence of vitamin D.

Phosphate is essential in carbohydrate metabolism and also in energy metabolism, resulting in the formation of adenosine triphosphate (ATP) by a process called *oxidative phosphorylation*. Consider also the role of phosphorus in the formation of various organic compounds such as phosphocreatine, phospholipids, and phosphoproteins.

There are 206 named bones in the body. They may be classified according to shape as long, short, flat, and irregular. The larger bones of the arms and legs are examples of long bones. The bones of the wrist and ankle are short bones. The skull bones and shoulder blades are good examples of flat bones. Irregular bones include those not readily classified, such as the vertebrae.

The few small round bones that develop in a joint or tendon are called *sesamoid bones*. The *patella* (or kneecap) is the largest of these bones, which are usually embedded in tendons of the hand or foot. An example would be the two sesamoid bones which are placed under the metatarsophalangeal joint of each great toe. Occasionally, isolated bones are formed along sutures of the skull. These are *Wormian bones*.

Projections from the surface of bones are called *processes*. There are various types of projections; one is the *condyle*, which is defined as a rounded articulating surface forming a joint at the end of a bone. A *tubercle* is a slightly rounded elevation from the surface of a bone that affords attachment

The skeleton

to a muscle or ligament. Several other types of projections are given specific names.

Cavities may also be of various types. A *foramen* is an opening through which blood vessels and nerves pass. *Sinus* is a term used for a cavity in a bone. A *fossa* is a depression in the surface of a bone.

The skeleton is commonly considered to consist of two divisions. The axial skeleton includes the head, the hyoid bone, and the vertebral column, sternum, and ribs. The bones of the arms and legs plus the shoulder and pelvic girdles comprise the appendicular skeleton (Color Plate A).

THE AXIAL SKELETON

THE SKULL The skull bones form the skeletal structure of the head and face. For purposes of study the bones that are included in the cranium may be considered separately from those that form the skeletal structure of the face.

The cranial bones The parietal and temporal bones are paired, and the others are single, making a total of eight bones in the cranium.

The cranial bones
frontal
parietal
occipital
temporal
sphenoid
ethmoid

The cranial bones are joined by an irregular line of fibrous tissue that unites the edges of the bones and forms a *suture*. Such joints are known as *immovable joints* (Figure 5.1). The principal sutures are the *coronal*, between the frontal and parietal bones; the *lambdoidal*, between the occipital and parietal bones; and the *sagittal*, which extends along a midsagittal line from the base of the nose to the posterior fontanel.

In the skull of a newborn infant the sagittal suture divides the frontal bone into two parts. The space between the frontal bones and the parietal bones is the *anterior fontanel*. This is the largest of several spaces that occur at the angles of union of the bones of the skull. There is membranous tissue between the cranial bones at the fontanels and sutures. The anterior fontanel normally closes at about the age of eighteen months. The posterior fontanel is located at the posterior end of the sagittal suture between the occipital and parietal bones. It is much smaller than the anterior fontanel and closes a few months after birth.

The *frontal bone* forms the anterior part of the cranium, that is, the forehead. It also forms the upper part of the orbits (or eye sockets) and a

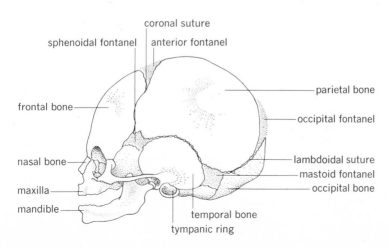

FIGURE 5.1

Skull of a newborn infant, showing development of bones and fontanels.

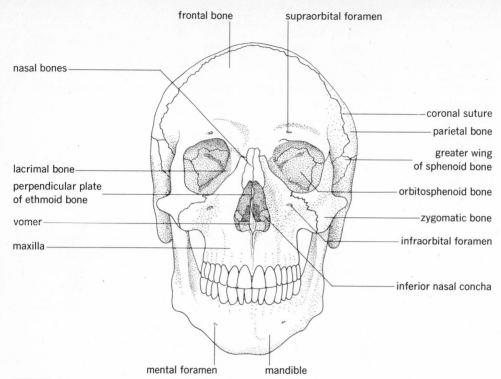

nasal bones

frontal bone

supraorbital foramen

coronal suture

parietal bone

greater wing
of sphenoid bone

lacrimal bone

perpendicular plate
of ethmoid bone

orbitosphenoid bone

vomer

zygomatic bone

maxilla

infraorbital foramen

inferior nasal concha

mental foramen

mandible

FIGURE 5.2

Frontal views of a mature skull.

small anterior part of the nasal cavity. Above the orbits are the supraorbital ridges, and beneath these ridges lie two cavities, the *frontal sinuses* (Figure 5.2).

The *occipital bone* forms the back of the cranium and a considerable part of the base of the skull. The large opening at the base of the skull through which the spinal cord passes is the *foramen magnum* of the occipital bone. The two *occipital condyles* located on either side of the foramen present a smooth articular surface where the skull rests on the first cervical vertebra. The external occipital protuberance can be felt along the median line posteriorly.

Between the frontal and occipital bones are the right and left *parietal bones*. They form a considerable part of the sides and top of the cranium.

The *temporal bones*, located below the parietal bones on either side, form a part of the sides and base of the skull. The thick petrous portion contains the hearing apparatus. The styloid process projects downward from the undersurface. Above and anterior to the petrous portion, the bone becomes relatively thin and platelike. This is the squamous part of the temporal bone. From its lower lateral surface the zygomatic process extends out to meet the temporal process of the zygomatic bone and form the zygomatic arch. Below the base of the zygomatic process is a depression, the mandibular fossa, into which the condyle of the mandible, or lower jaw, articulates.

The mastoid portion of the temporal bone lies posterior to the external auditory canal. The mastoid process projects downward from this portion. This part of the bone contains air spaces, the mastoid air cells, which are

The skeleton

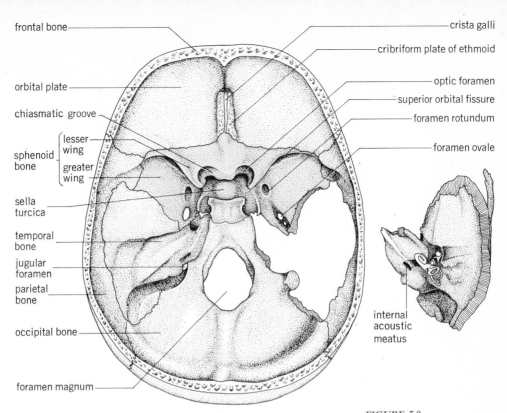

frontal bone

orbital plate

chiasmatic groove

sphenoid bone
- lesser wing
- greater wing

sella turcica

temporal bone

jugular foramen

parietal bone

occipital bone

foramen magnum

crista galli

cribriform plate of ethmoid

optic foramen

superior orbital fissure

foramen rotundum

foramen ovale

internal acoustic meatus

FIGURE 5.3
Base of the skull viewed from above and showing the floor of the cranial cavity. At right is the temporal bone, including the ear structures.

connected with each other and with the cavity of the middle ear. Infections of the middle ear can enter these air spaces, causing the condition known as *mastoiditis.* Since there is only a thin plate of bone between the mastoid air cells and the membrane covering the brain, an uncontrolled infection can spread to this membrane and cause a much more serious inflammation. The mastoid operation consists of cutting into the mastoid process to permit external drainage of the infection (Figure 5.3).

Within the middle ear there are three small bones that are a part of the hearing mechanism. These bones are the malleus, incus, and stapes. They will be considered later under a discussion of the ear.

The *sphenoid bone* is somewhat more difficult to locate. It is wedged in between other cranial bones to form a part of the anterior cranial structure. The name *sphenoid* is taken from a Greek word that means a wedge. The bone consists of a median body portion and three sets of lateral paired processes: the greater wings, the lesser wings, and the pterygoid processes (Figure 5.4).

The body of the sphenoid lies between the ethmoid and occipital bones. It contains two large air spaces, the *sphenoidal sinuses,* which are separated by a thin layer of bone. The sinuses drain into the nasal cavity. On the upper part of the body there is a depression called the *sella turcica,* meaning Turk's saddle. The hypophysis, or pituitary body, rests in this fossa (Table 5.1).

The greater wings of the sphenoid spread out from the body and form a part of the anterolateral floor of the braincase. They then turn upward, where they can be seen from the outside and form a part of the side wall of the skull just anterior to the temporal bone on either side (Figure 5.5).

The axial skeleton

91

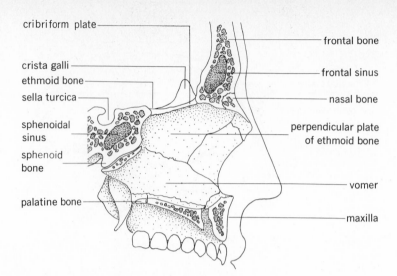

cribriform plate

crista galli

ethmoid bone

sella turcica

sphenoidal sinus

sphenoid bone

palatine bone

frontal bone

frontal sinus

nasal bone

perpendicular plate of ethmoid bone

vomer

maxilla

FIGURE 5.4
A sagittal section of the anterior part of the skull.

The lesser wings are somewhat anterior and on a higher plane than the greater wings. They are best seen as forming a part of the floor of the braincase, but they also form the posterior part of the roof of the orbit.

The pterygoid processes project downward from the body of the sphenoid, where they are located in the posterior part of the side walls of the nasal cavity.

The *ethmoid bone* is a cranial bone of lightweight construction, located anterior to the body of the sphenoid bone. Its lateral masses contain labyrinths of air spaces, the *ethmoidal sinuses*. Its perpendicular plate forms the upper part of the nasal septum. The *cribriform plates* are horizontal portions of the ethmoid bone that lie in the floor of the braincase anteriorly. They also form the roof of the nasal cavity. They are perforated by tiny openings that permit the passage of the fibers of the olfactory nerve from the nasal mucosa to the olfactory bulbs of the brain. Between the cribriform plates a crest of bone rises to furnish attachment for the falx cerebri, a portion of the membrane covering the brain. The crest of bone is called the *crista galli* (cock's comb). The ethmoid bone forms a part of the orbit. From the sides of the labyrinth portion, two pairs of thin scroll-shaped bones project into the nasal cavity. These are the superior and middle nasal conchae.

The facial bones

mandible
maxillae
palatine bones
vomer
zygomatic bones
lacrimal bones
nasal bones
inferior nasal conchae

The facial bones   The bones of the face are all paired except the mandible and the vomer.

The *mandible* is the lower jaw bone. It has a U-shaped horizontal part and two vertical parts, known as the *rami*. Each ramus is divided into an anterior *coronoid process* and a posterior *condyloid process*. The temporal muscle is attached to the coronoid process. The condyle of the condyloid process articulates on the temporal bone. The teeth of the lower jaw are embedded in cavities in the upper part of the alveolar border. There is an opening anteriorly on either side of the mandible, called the *mental foramen*, which permits the passage of the inferior mental nerve (Figure 5.5 *a* and *b*).

At birth the right and left sides of the mandible are united anteriorly by fibrous tissue. The two sides start to grow together during the first year, and the union is usually complete by the end of the second year. At birth

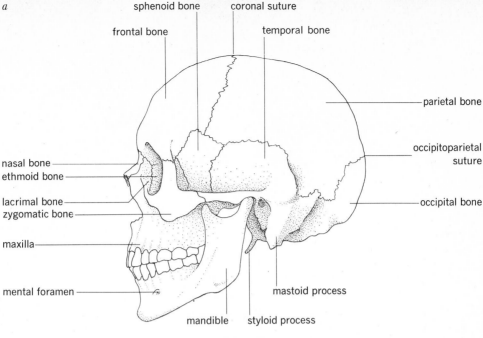

a

sphenoid bone

coronal suture

frontal bone

temporal bone

nasal bone

ethmoid bone

lacrimal bone

zygomatic bone

maxilla

mental foramen

parietal bone

occipitoparietal suture

occipital bone

mastoid process

mandible   styloid process

LATERAL VIEW

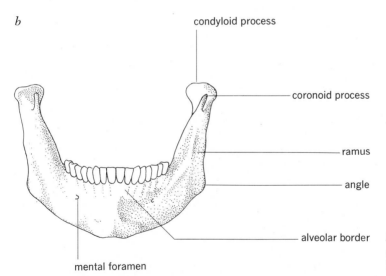

b

condyloid process

coronoid process

ramus

angle

alveolar border

mental foramen

FIGURE 5.5
a Lateral aspect of the skull; *b* the adult mandible.

the rami are short and slope away from the body instead of standing at a sharp angle as they do in an adult. The body of the bone lengthens as a child grows, in order to accommodate the molar teeth. It also deepens as the upper alveolar border, which supports the teeth, is built up. In old age the alveolar border is absorbed after the teeth are lost, and the chin then becomes more prominent. Aged persons are therefore able to raise the chin until it comes close to the nose.

The axial skeleton

TABLE 5.1

Cranial bones

Name	Description
Single bones	
Frontal	Forms the anterior part of the cranium or forehead; also the upper part of the orbits or eye sockets. Contains the frontal sinuses.
Occipital	Forms the posterior part of the cranium and part of the base of the skull. Contains a large opening, the foramen magnum.
Sphenoid	Constitutes a part of the base of the skull; also a part of the cranial wall. The body of the bone is wedged in just anterior to the occipital, and the greater wings form a part of the side wall anterior to the temporal bone. Contains the sella turcica.
Ethmoid	Located anterior to the body of the sphenoid. Contains the ethmoidal sinuses. The crista galli and cribriform plates are a part of this bone.
Paired bones	
Parietal	Lie between the frontal and occipital bones on either side and form part of the sides and top of the cranium.
Temporal	Located below the parietal bones on either side. Contains the inner ear.
Auditory ossicles *6 paired bones* Malleus (2) Incus (2) Stapes (2)	Tiny inner ear bones located in the temporal bones. They function in the transmission of sound.
Hyoid 1 U-shaped bone	Suspended from the styloid process of the temporal bone, affords attachment for the tongue and several tongue muscles. (Not a skull bone.)

The *maxillae* are paired bones that unite to form the upper jaw. There are four processes arising from each bone: the palatine, alveolar, zygomatic, and frontal processes. The maxillae form the greater part of the floor of the orbits, the side walls and floor of the nasal passage, and the anterior part of the roof of the mouth. The palatine processes form a part of the floor of the nasal passage and the anterior part of the hard palate. The condition known as *cleft palate* occurs when the two palatine processes fail to unite before birth. This condition varies from a simple form, in which there is a divided uvula, to a much more complicated form, in which the cleft may extend through the soft palate and the hard palate and involve the lips. A cleft in the lip, sometimes referred to as *harelip*, may be on either side or both but is not medial.

The alveolar process supports the teeth of the upper jaw.

The zygomatic process articulates with the zygomatic bones.

The frontal process arises almost vertically from the body of the maxilla and forms the greater part of the skeleton of the outer nose.

Two large cavities located on either side of the nasal passage are called the *maxillary sinuses*. They are located in the body of each bone, and they open into the nasal cavity.

The *palatine bones* form the posterior part of the hard palate; vertical

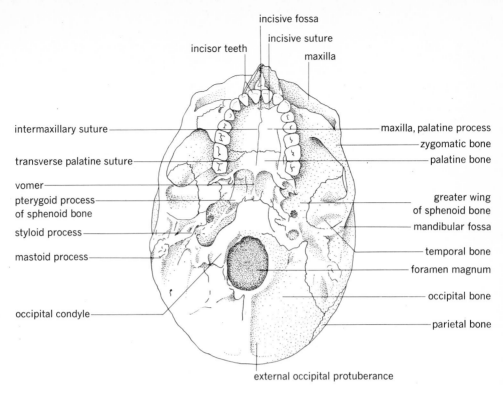

incisive fossa
incisor teeth
incisive suture
maxilla

intermaxillary suture
maxilla, palatine process
zygomatic bone
palatine bone

transverse palatine suture

vomer
pterygoid process
of sphenoid bone
greater wing
of sphenoid bone
mandibular fossa

styloid process
temporal bone
foramen magnum

mastoid process

occipital bone

parietal bone

occipital condyle

external occipital protuberance

VENTRAL VIEW

portions form a part of the lateral walls of the nasal cavity, posterior to the maxilla. The vertical portions extend upward as far as the orbits, where the orbital processes form a small part of the floor of the orbits (Figure 5.6).

The *vomer* (plowshare bone) is a thin, single bone that forms the lower part of the nasal septum. Its lower anterior border is grooved to receive the cartilage septum of the nose; its upper anterior border fuses with the perpendicular plate of the ethmoid bone. The vomer often turns to one side or the other, making the nasal passages of unequal size (Figure 5.4).

The *zygomatic,* or *malar, bones* are known commonly as the *cheek-bones.* They form a prominence below the orbits and also form a part of the outer border and floor of the orbits. The temporal process of each bone projects posteriorly to meet the zygomatic process of the temporal bone. These two processes form the zygomatic arch on either side of the skull (Figure 5.2).

The *lacrimal bones* are located at the inner angle of the orbits directly posterior to the frontal processes of the maxillae. These are the smallest bones of the face. Each bone is grooved opposite a similar groove in the frontal process of the ethmoid bone, thus forming the opening through which the tear duct passes into the nasal cavity (Figure 5.7).

The *nasal bones* are two small, elongate bones that lie side by side to form the upper part of the bridge of the nose. The nasal cartilages form the lower part of the structural framework and are largely responsible for its shape (Table 5.2).

The *inferior nasal conchae* (turbinated bones) are thin, shelflike plates

The axial skeleton

95

TABLE 5.2

The facial bones

Name	Description
Single bones	
Mandible	The lower jaw. Holds the lower set of teeth. There are two processes, an anterior coronoid and a posterior condyloid.
Vomer	Forms the lower and posterior part of the nasal septum.
Paired bones	
Maxilla	The upper jaw. The alveolar process supports the upper set of teeth. Contains the maxillary sinus.
Palatine	Forms the posterior part of the hard palate, a part of the floor and lateral wall of the nasal cavity, and a portion of the orbital floor.
Zygomatic	The cheekbone; forms the anterior part of the zygomatic arch and a part of the lateral wall of the orbit.
Lacrimal	Located at the inner angle of the orbit. A thin, slender bone. A groove provides for the passage of the tear duct.
Nasal	Forms the bony part of the nasal bridge.
Inferior nasal concha	A thin, scroll-like bone projecting into the nasal cavity below the superior and middle conchae of the ethmoid.

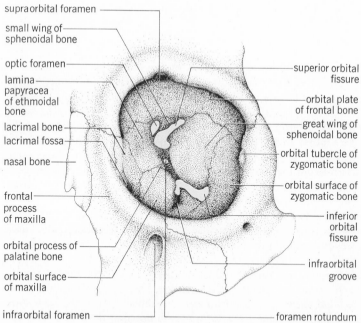

FIGURE 5.7
Left orbital cavity.

projecting into the nasal cavity on either side below the superior and middle conchae of the ethmoid bone. They are often described as resembling a scroll of parchment, since they turn on themselves. They are covered with mucous membrane in the living state. All the nasal conchae perform a useful service in warming and moistening the air that passes over their warm, moist

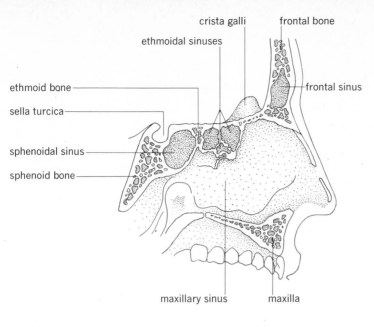

FIGURE 5.8
Lateral section of the skull, including paranasal sinuses.

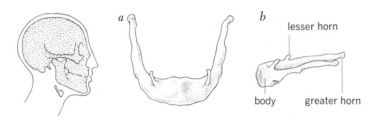

FIGURE 5.9
The hyoid bone: *a* frontal or anterior view; *b* lateral view.

membranes. The air is also partially cleansed, since some of the dust particles adhere to the mucous membrane.

The paranasal sinuses of the frontal, ethmoid, sphenoid, and maxillary bones also open into the nasal cavity or into the nasopharynx by means of small openings. The nasal mucous membrane that lines the nose extends through these openings and lines the cavities of these bones. The sinuses are ordinarily air cavities; but when infections of the upper part of the respiratory tract, such as colds, cause the mucous membrane to secrete abnormally, the sinuses may become filled with secretions. Inflammation of the nasal mucosa involving the sinuses is *sinusitis*. Inflammation of the lining of the small openings from the sinus into the nasal cavity can completely block the drainage of the sinus, and the secretions accumulating under pressure may cause severe headaches. In such cases drainage is usually established by shrinking the inflamed membrane by means of drugs or by forcing an opening into the sinus to permit drainage (Figure 5.8).

The *hyoid bone* is shaped like the letter U. It is located in the neck just above the larynx (Adam's apple) and affords attachment for the tongue and for several of the tongue muscles. The greater horns extend posteriorly from the body of the bone and are suspended by ligaments from the styloid processes of the temporal bones (Figure 5.9). The hyoid bone arises in the embryo from the cartilages of the second and third visceral arches. It is,

The axial skeleton

97

therefore, not strictly a skull bone but a bone derived from the embryonic visceral skeleton, or splanchnocranium.

THE BONES OF THE TRUNK *The vertebral column* The vertebral column is composed of individual bones called *vertebrae*. The column is divided into 7 cervical, 12 thoracic, 5 lumbar, 5 sacral, and 4 coccygeal vertebrae as they develop in the embryo. Before the sacral vertebrae fuse to form the sacrum, and before the coccygeal vertebrae fuse, the total number of vertebrae is 33. In an adult the sacrum is a single bone, and the coccyx is usually more or less fused, thus making a total of 26 bones.

The vertebral column, or backbone, supports the trunk and affords protection for the spinal cord. The individual vertebrae vary, but in general they consist of a body (or centrum), which supports the weight of the trunk, and an arch, which with the body forms the vertebral foramen, the opening through which the spinal cord passes.

Each vertebra has a spinous process projecting dorsad. The process affords attachment for ligaments and for muscles of the back (Figure 5.10). There are two transverse processes projecting laterad and dorsad from a typical vertebra that also afford attachment for muscles.

Articular processes, which form joints, join the vertebrae into a series or column. The articulating surfaces are covered with cartilage, which promotes smoothness in action.

There are typically seven *cervical vertebrae* in the neck region of mammals. The first cervical vertebra is the *atlas*. It is specialized, with large smooth surfaces that articulate with the occipital condyles for the support of the skull. The body of the atlas is not well developed, and it appears more nearly as a circular bone with two transverse processes. There is only a posterior tubercle in place of a dorsal spinous process (Figure 5.11).

The *axis* is the second of the cervical vertebrae. From its upper surface there arises a strong bony process, the *odontoid process*, upon which the atlas pivots when the head is turned. This process represents a striking specialization to permit freedom of head movement. The dorsal spine of the axis is thick and strong compared with the spines of other cervical vertebrae (Figure 5.12).

The seventh cervical vertebra has a long, narrow dorsal spine, which is not forked. In the cervical vertebrae the transverse processes on either side form a foramen, which transmits the vertebral artery.

The 12 *thoracic vertebrae* are larger and stronger than the cervical vertebrae. The dorsal spinous processes are long and project downward. The transverse processes are large and help to support the ribs. These transverse processes, except the eleventh and twelfth vertebrae, have facets that articulate with the tubercles of the ribs. There are also demifacets on the body of each vertebra for the articulation of a rib head on either side.

The *lumbar vertebrae* are large, and their dorsal processes are shaped somewhat like an ax. The transverse processes are thin. The fifth lumbar vertebra is not typical, since it is modified for articulation with the sacrum.

The *sacrum* is formed by the fusion of the five sacral vertebrae. It is somewhat triangular in shape, with the apex downward, and is held between the hipbones by very strong ligaments. The inner surface is smooth and concave. The outer surface is very irregular and convex. The sacral canal is the passageway for the posterior end of the spinal cord, where the roots of the sacral and coccygeal nerves lie.

The *coccyx* consists of four rudimentary vertebrae. The first coccygeal

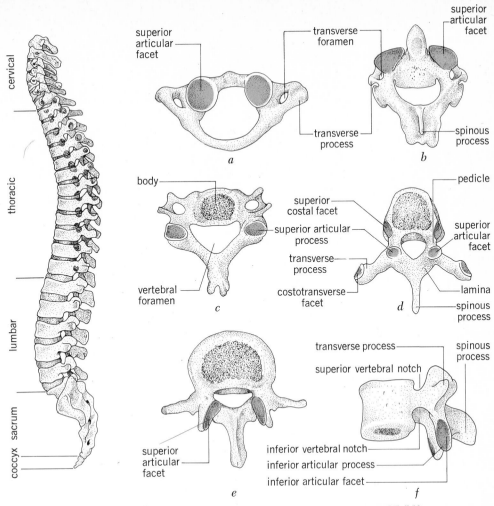

a

superior articular facet

transverse foramen

transverse process

superior articular facet

b

spinous process

body

superior costal facet

superior articular process

transverse process

costotransverse facet

vertebral foramen

c

pedicle

superior articular facet

lamina

spinous process

d

superior articular facet

e

transverse process

superior vertebral notch

spinous process

inferior vertebral notch

inferior articular process

inferior articular facet

f

cervical

thoracic

lumbar

sacrum

coccyx

FIGURE 5.10

The vertebral column, illustrating individual vertebrae: *a* atlas; *b* axis; *c* fourth cervical; *d* sixth thoracic; *e* third lumbar, superior view; *f* lateral view.

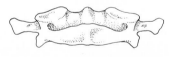

ATLAS

FIGURE 5.11

The atlas, anterior view.

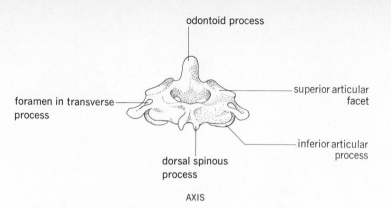

odontoid process

foramen in transverse process

superior articular facet

inferior articular process

dorsal spinous process

AXIS

FIGURE 5.12
The axis.

vertebra usually remains free until past middle age, when all four vertebrae may become fused. The coccyx in man may be regarded as a vestigial tail.

The *vertebral column* is the dorsal support for the trunk, but it is not a rigid support. The individual vertebrae are separated by cartilage disks that absorb the shocks of walking and jumping. The vertebrae are bound together by ligaments. Muscles are attached to the dorsal spines and to transverse processes that act as levers and permit limited movement. The column of vertebrae can be bent forward or backward or to either side, and considerable rotation is possible.

We are often admonished to keep the spine straight, but such advice is given with lateral curvatures in mind. There are natural dorsoventral curvatures, and the column normally is not straight. There are two primary curvatures, thoracic and sacral; both are concave anteriorly and are present at birth. The cervical curvature is a secondary one and develops as the child begins to hold its head up and sit upright. It is convex anteriorly. The lumbar curvature is also a secondary, or compensating, curvature and is convex anteriorly. It forms the hollow of the back just above the sacrum in the lumbar region (Figure 5.10).

Good posture requires that the body be held erect so that none of the curves of the vertebral column is exaggerated. When a person stands with head erect, chest up, and lower part of the abdomen held in, there should be a vertical alignment of the ear, shoulder, hip, knee, and ankle (Figure 5.13).

Poor posture permits the head to tip forward, the shoulders and chest to sag downward, and the abdomen to protrude. Under these conditions the thoracic curvature may be accentuated and cause a round-shouldered or even a hunchbacked condition, known as *kyphosis*.

The lumbar curvature may be greater than normal, a condition called *lordosis*. This curvature is normally more pronounced in the female than in the male.

Scoliosis means a lateral curvature. The vertebral column often shows a slight curvature toward the right, but this curvature can become exaggerated by poor postural conditions, such as writing at a desk that is too low. A condition in which one shoulder is lower than the other or in which the hipbones are not level may be the result of lateral curvature of the vertebral column.

The thorax The thorax is formed by the thoracic vertebrae, the ribs, the costal cartilages, and the sternum. It affords protection for the heart, thoracic blood

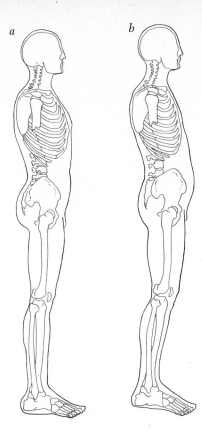

vessels, and lungs and aids in supporting the bones of the shoulder girdle. The diaphragm forms the floor of the thoracic cavity.

There are 12 pairs of ribs in both male and female. The true ribs articulate with thoracic vertebrae and attach by means of the costal cartilages to the sternum. The first 7 pairs are *true ribs.* The next 3 pairs—the eighth, ninth, and tenth—are *vertebrochondral ribs.* They attach indirectly to the sternum anteriorly by cartilages that join each other. The eleventh and twelfth pairs are tipped with cartilage, but they do not attach anteriorly. They are *vertebral,* or *floating, ribs* (Figure 5.14).

A typical rib is a long, slender, curved bone. It articulates posteriorly with two vertebrae. The first, tenth, eleventh, and twelfth ribs each articulate with a single vertebra. There is a tubercle just beyond the neck region of the rib, where it articulates with the transverse process of the vertebra. The eleventh and twelfth ribs have no tubercle for articulation with a transverse process. At the anterior end the rib is attached by costal cartilage. The eleventh and twelfth ribs again are exceptions (Figure 5.15).

The *sternum* is the breastbone. It is a thin, flat bone located in the anterior part of the thorax along a median line. It is composed of three parts: the manubrium, body, and xiphoid, or ensiform, process. The manubrium lies at the top of the thorax; the clavicle (or collarbone) and the cartilages of the first pair of ribs attach to it. The cartilages of the second pair of ribs attach partly to the manubrium and partly to the body. The body of the sternum is the longest part. The costal cartilages of the second to the seventh pairs of ribs attach to it. The xiphoid process projects downward from the

The axial skeleton

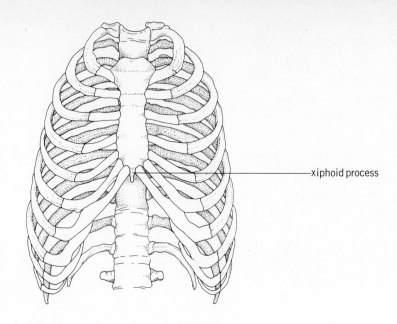

FIGURE 5.14
The thoracic cage.

xiphoid process

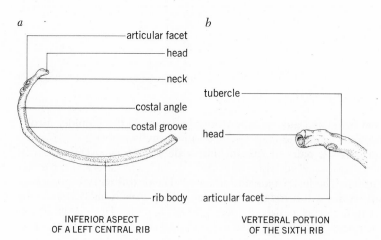

a   *b*

articular facet

head

neck

tubercle

costal angle

costal groove

head

rib body   articular facet

FIGURE 5.15
a A typical rib; *b* articulation
of the sixth rib.

INFERIOR ASPECT
OF A LEFT CENTRAL RIB

VERTEBRAL PORTION
OF THE SIXTH RIB

body of the sternum. It varies considerably in shape and in youth and until middle age is composed of cartilage. During middle age it gradually ossifies until in old age it is entirely bone and tends to fuse with the body (Figure 5.14) (Table 5.3).

THE APPENDICULAR SKELETON

THE SHOULDER GIRDLE AND ITS APPENDAGES The vertebrate skeleton of an animal that walks on all four legs may be described as a bridge of vertebrae supported by four pillars, the legs. The fact that man walks erect and that the arms have become modified has not altered the basic skeletal plan. The arms articulate with the shoulder girdle; the legs support the pelvic girdle (Figure 5.16).

TABLE 5.3

Bones of the vertebral column and thorax

	No. of bones	Description
Vertebral column		
Cervical vertebrae	7 (the atlas, axis, and 5 others)	The atlas is the first cervical (neck) vertebra. It turns on the second cervical vertebra, or axis. The first two are highly specialized to permit turning the head.
Thoracic vertebrae	12	Have facets for the attachment and support of the 12 pairs of ribs
Lumbar vertebrae	5	The largest vertebrae, located between the thorax and pelvis, in the small of the back
Sacrum	5	Fused vertebrae
Coccyx	4	Fused, small, poorly developed
Thorax		
Ribs	12 pairs	True ribs, 7 pairs attached to sternum by costal cartilages; the 8th, 9th, and 10th pairs attached upward by costal cartilages; the 11th and 12th ribs are not attached anteriorly and are floating ribs
Sternum	1	The breast bone, composed of three parts: the manubrium, body, and xiphoid or ensiform process

The shoulder girdle is composed of two clavicles and two scapulae. It is not a complete girdle, since the sternum separates the clavicles anteriorly and muscles attach the scapulae to the trunk posteriorly (Figure 5.17).

The *clavicles* are the collarbones. They can be felt anteriorly just above the thorax on either side. They are slender, curved brace bones forming the anterior part of the shoulder girdle. Anteriorly they articulate with the manubrium, and posteriorly with the acromion processes of the scapulae.

The *scapulae* present broad, flat surfaces for the attachment of shoulder muscles (Figure 5.18). The inner surface of each bone is somewhat concave. The outer, or dorsal, surface is divided unequally by a ridge called the *spine*. The spine terminates as the acromion process. The clavicle articulates with this process. Another process, the coracoid, projects upward and forward underneath the clavicle. Below these two processes there is a smooth, circular, depressed area: the glenoid fossa, or cavity. The head of the humerus (the bone of the arm above the elbow) articulates in this cavity.

The *humerus* is the long bone of the upper arm. Its rounded head fits into the glenoid cavity of the scapula and forms a *ball-and-socket joint* at the shoulder. Such a joint permits rotating movements. Just beyond the constricted neck region there are two bony prominences, the greater tuberosity and the lesser tuberosity. The greater tuberosity is a lateral projection and can be felt as the point of the shoulder. The lesser tuberosity projects anteriorly when the arm is resting at the side (Figure 5.19*a*, *b*, *c*).

The area for a short distance below the tuberosities has been called the *surgical neck* of the bone because many fractures occur there.

The tendon for the long head of the biceps muscle lies in the groove between the tubercles (intertubercular groove).

The appendicular skeleton

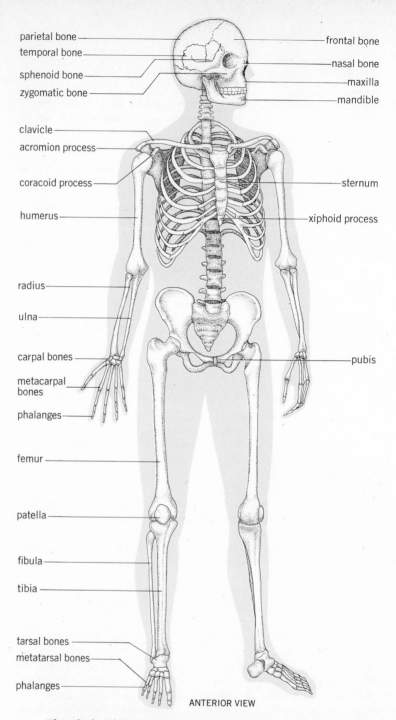

parietal bone

temporal bone

sphenoid bone

zygomatic bone

frontal bone

nasal bone

maxilla

mandible

clavicle

acromion process

coracoid process

humerus

sternum

xiphoid process

radius

ulna

carpal bones

metacarpal bones

phalanges

femur

patella

fibula

tibia

tarsal bones

metatarsal bones

phalanges

pubis

ANTERIOR VIEW

The shaft of the humerus is long and rather slender. The deltoid tuberosity is located about midway on the lateral margin. The deltoid muscle is inserted there.

The distal extremity of the humerus widens and flattens. There are two smooth articular surfaces. The outer capitulum articulates with the radius,

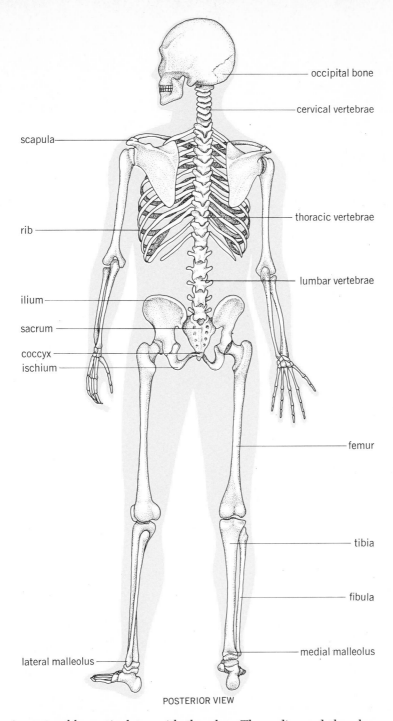

occipital bone

cervical vertebrae

scapula

thoracic vertebrae

rib

lumbar vertebrae

ilium

sacrum

coccyx

ischium

femur

tibia

fibula

medial malleolus

lateral malleolus

POSTERIOR VIEW

and the inner trochlea articulates with the ulna. The radius and the ulna are the bones of the forearm. Above the capitulum on the anterior side is a shallow fossa, which receives the margin of the head of the radius when the arm is fully flexed.

On the posterior side the olecranon fossa receives the olecranon process

The appendicular skeleton

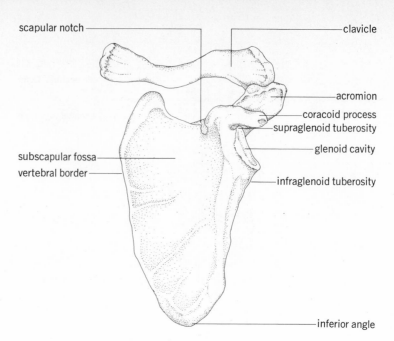

scapular notch

clavicle

acromion

coracoid process

supraglenoid tuberosity

subscapular fossa

glenoid cavity

vertebral border

infraglenoid tuberosity

inferior angle

THE LEFT SCAPULA, ANTERIOR VIEW

FIGURE 5.17
Left shoulder girdle (anterior view).

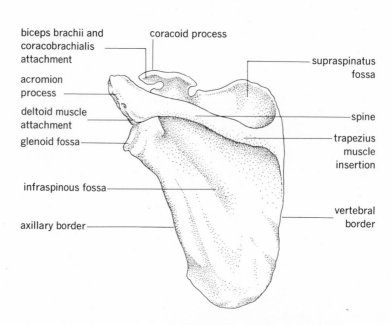

biceps brachii and coracobrachialis attachment

coracoid process

supraspinatus fossa

acromion process

deltoid muscle attachment

spine

glenoid fossa

trapezius muscle insertion

infraspinous fossa

vertebral border

axillary border

FIGURE 5.18
Left scapula (posterior view).

of the ulna when the arm is fully extended. The olecranon fossa is separated from the coronoid fossa by a thin bony partition.

The *ulna* is the longer bone of the forearm. It can be felt along the back of the forearm from the elbow to the wrist. The proximal end of the bone presents two processes, the olecranon and the coronoid; between them is the trochlear notch. The olecranon is the point of the elbow, and although it is a strong process, it is often broken. When the arm is extended, the

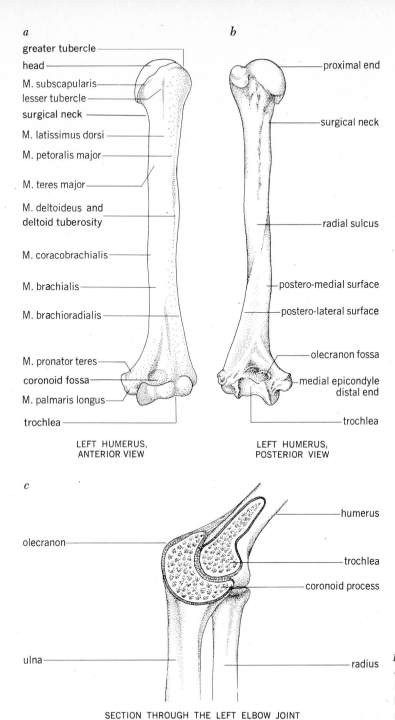

a

greater tubercle
head
M. subscapularis
lesser tubercle
surgical neck
M. latissimus dorsi
M. petoralis major
M. teres major
M. deltoideus and deltoid tuberosity
M. coracobrachialis
M. brachialis
M. brachioradialis
M. pronator teres
coronoid fossa
M. palmaris longus
trochlea

LEFT HUMERUS,
ANTERIOR VIEW

b

proximal end
surgical neck
radial sulcus
postero-medial surface
postero-lateral surface
olecranon fossa
medial epicondyle distal end
trochlea

LEFT HUMERUS,
POSTERIOR VIEW

c

olecranon
humerus
trochlea
coronoid process
ulna
radius

SECTION THROUGH THE LEFT ELBOW JOINT

FIGURE 5.19
a Left humerus (anterior view) indicating muscle attachments; *b* left humerus (posterior view); *c* section through left elbow joint.

anterior tip of the process is received by the olecranon fossa of the humerus. The trochlear notch articulates against the trochlea of the humerus and forms a hinge joint. A *hinge joint* permits movement in only one plane. On the inner side of the coronoid process there is a smooth articular surface that

The appendicular skeleton

107

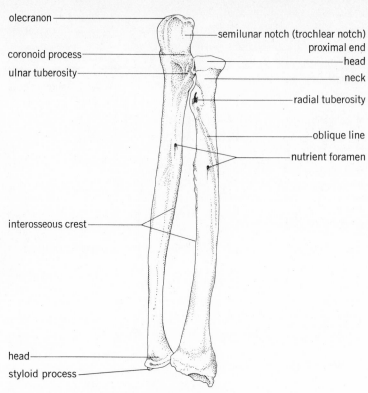

ulna   radius

olecranon

semilunar notch (trochlear notch)
proximal end
coronoid process   head

ulnar tuberosity   neck

radial tuberosity

oblique line

nutrient foramen

interosseous crest

head
styloid process

ANTERIOR VIEW OF THE BONES OF THE LEFT FOREARM

FIGURE 5.20
Left radius and ulna (anterior
view).

articulates with the head of the radius. This is the radial notch. The distal
extremity consists of the head of the ulna and a small lateral process called
the *styloid process.* It can be felt just above the wrist on the little-finger
side.

The *radius* is a cylindrical bone resembling the spoke of a wheel. It
lies laterad to the ulna on the same side of the forearm as the thumb. The
head presents a smooth, hollowed-out disk, which articulates with the capitulum of the humerus. As the hand is rotated, the radius pivots on the
capitulum. This is a good example of a *pivot joint.* Below the constricted
neck of the bone on the medial side, the radial tuberosity affords attachment
for the tendon of the biceps brachii muscle. The lower extremity is flattened
and presents two articular surfaces, one for articulation with wrist bones
and the other, the ulnar notch, to permit rotating movement on the ulna
(Figure 5.20).

There are eight small bones in the wrist, known as *carpal bones.* (For
a study of the individual wrist bones the reader is referred to a reference
text on human anatomy.) The bones of the carpus (or wrist) are bound
together by ligaments, yet their arrangement in two rows permits some
movement.

The *metacarpal bones* make up the skeletal framework of the hand.
There are five bones in each hand. At their distal ends they articulate with
the proximal row of finger bones, the *phalanges.* There are 14 phalanges,
three for each finger and two for the thumb (pollex) (Figure 5.21) (Table
5.4).

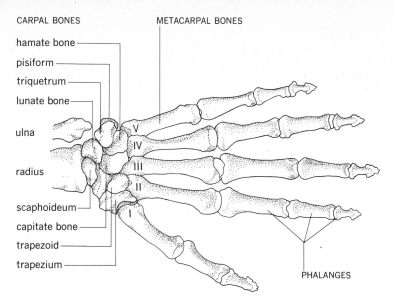

CARPAL BONES   METACARPAL BONES

hamate bone ——
pisiform ——
triquetrum ——
lunate bone ——
ulna ——
radius ——
scaphoideum ——
capitate bone ——
trapezoid ——
trapezium ——

V
IV
III
II
I

PHALANGES

FIGURE 5.21
Bones of the wrist and hand
(palmar surface of hand).

TABLE 5.4
Bones of the shoulder girdle and the upper extremity

Location	Name	Total no. of bones	Description
Shoulder	Clavicle(2)	4	The "collarbone"
	Scapula(2)		The "shoulder blade," a large, thin, flat bone
Arm	Humerus(2)	2	The long bone of the arm
Forearm	Radius(2)	4	The bone that rotates; at the base it is on the thumb side
	Ulna(2)		Forms hinge joint at elbow; at the base it is on the little-finger side
Wrist	Scaphoid(2) Lunate(2) Triquetrum(2) Pisiform(2) Trapezium(2) Trapezold(2) Capitate(2) Hamate(2)	16	Two transverse rows of small bones in the wrist
Hand	Metacarpal(10)	38	Long bones of the palm
	Phalanges(28)		Bones of the fingers; two in the thumb, three in each finger

THE PELVIC GIRDLE AND ITS APPENDAGES   The *hipbones* (ossa coxae) are two large, flat bones that form the pelvic girdle. They are the broadest bones in the skeleton. They form a symphysis anteriorly, and the sacrum completes the girdle posteriorly. The posterior appendages are attached to the pelvic girdle (Figure 5.22).

The appendicular skeleton

109

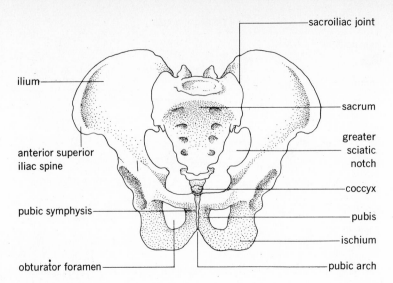

FIGURE 5.22
Pelvic girdle: male pelvis (anterior
view).

Each hipbone develops in the fetus as three separate bones, the *ilium*,
ischium, and *pubis*. The ilium is the uppermost and largest of the three parts.
Its crest can be felt along the upper border of the hipbone. The anterior
superior spine, from which anatomical and surgical measurements are made,
can be located by following the crest forward to the most prominent anterior
projection. Before the three parts of the hipbone fuse, the ilium is separated
from the ischium and pubis by a suture that passes through the *acetabulum*,
the cavity that receives the head of the femur (Figure 5.23).

The *ischium* is the lower posterior portion of the hipbone. The ischial
tuberosities are large, bony prominences that support the body in a sitting
position. The obturator foramen, formed by the ischium and the pubis on
either side, is the largest foramen in the skeleton.

The lower anterior portion of each hipbone is the *pubis*. The pubic
bones meet anteriorly along a median line and form a slightly movable joint,
the symphysis pubis. There is a disk of fibrocartilage between the two bones,

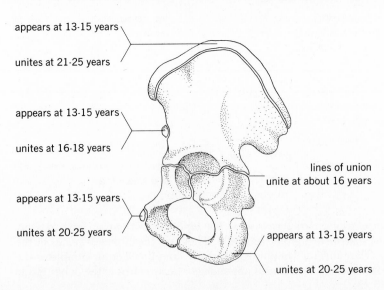

FIGURE 5.23
The hipbone during growth (lat-
eral view). [*Redrawn from
J. Parsons Schaeffer (ed.),
"Morris' Human Anatomy,"
12th ed., McGraw-Hill Book
Company, New York, 1966.*]

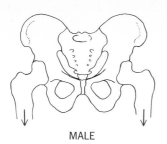
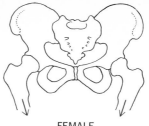

MALE   FEMALE

FIGURE 5.24
Differences between the pelves of the human female and the male. Note that the femur is nearly vertical in the male but slopes inward from the wider female pelvis. The pubic arch is wider in the female pelvis than in the male, and the opening of the true pelvis of the female is larger and more oval shaped than that of the male.

and they are held together by ligaments. The symphysis is a useful landmark for pelvic measurements. Below the symphysis the pubic bones form the pubic arch.

The *pelvis* is considered to consist of two parts: an upper, or greater, pelvis, bounded laterally by the broad upper portions of the ilia, and the true pelvis, which lies below the brim. The true pelvis is bounded by the lower parts of the hipbones and posteriorly by the sacrum and coccyx.

The female pelvis differs from that of the male in many respects. The differences are due largely to the adaptation of the true pelvis of the female to childbearing. It is wider, and the cavity is larger and shorter. The sacrum is shorter and broader, and the coccyx is more freely movable. The ischial tuberosities as well as the acetabula are farther apart. The pubic arch is much wider in the female pelvis (Figure 5.24).

The sacroiliac joint is a symphysis at the back of the pelvis. Tall, slender persons are especially subject to strain at this joint between the sacrum and the ilium, which may cause severe backache. The surface area above the sacrum is marked off by the posterior superior iliac dimple in the skin on either side.

The *femur*, or thigh bone, is the bone of the leg above the knee. It is a strong, heavy bone and the longest in the body. At the upper end of the bone there is a rounded head, which forms a ball-and-socket joint with the acetabulum in the hipbone. Below the long, constricted neck region there are two processes, the greater and lesser trochanters, that afford attachment for some of the muscles of the thigh and gluteal regions. The intertrochanteric crest lies between the two trochanters.

Along the posterior side of the shaft there is a ridge, the linea aspera, to which muscles are attached. The lower end of the femur widens, and at the extremity there are two bony prominences, the lateral and medial condyles. The smooth articular surfaces form a *hinge joint* with the tibia. The patella also articulates with the lower extremity of the femur on the anterior patellar surface. The femur is inclined inward because of the width of the pelvis. The degree of inclination is greater in women than in men (Figure 5.25*a* and *b*).

The *patella*, or kneecap, is a flat sesamoid bone embedded in the tendon of the quadriceps femoris muscle. It is the largest sesamoid bone in the body, measuring about 2 inches in both length and width, and is surrounded by bursae. The patella and the bursae serve to protect the knee joint. When one bends the knee, the patella becomes fixed against the intercondyloid notch of the femur. The ligamentum patellae is a strong ligament that extends from the patella to the tibia. When one kneels, the weight of the body rests partly on this stretched ligament and partly on the tubercle of the tibia rather than on the patella. Persons who work on their knees are subject to inflam-

The appendicular skeleton

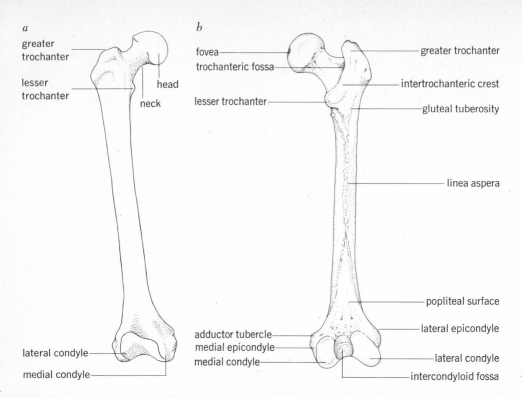

a

greater trochanter

lesser trochanter

head

neck

lateral condyle

medial condyle

RIGHT FEMUR, ANTERIOR VIEW

b

fovea

trochanteric fossa

lesser trochanter

greater trochanter

intertrochanteric crest

gluteal tuberosity

linea aspera

popliteal surface

lateral epicondyle

adductor tubercle

medial epicondyle

medial condyle

lateral condyle

intercondyloid fossa

THE RIGHT FEMUR, POSTERIOR VIEW

FIGURE 5.25

The femur (thigh bone): *a* right femur (anterior view); *b* posterior view.

mations of the bursae (bursitis). The condition commonly referred to as water on the knee can result from injury to the bursae (Figure 5.26*a* and *b*).

The *tibia*, or shin bone, is the larger bone of the leg below the knee. The upper end of the bone widens and forms the lateral and medial condyles. Between the condyles there is a sharp projection, the intercondyloid eminence. The articular surfaces that form the knee joint with the condyles of the femur are slightly concave. The shaft of the bone is roughly triangular, with the anterior apex sharp and well defined. The sharp anterior border of the shaft is known as the *shin*. It can be felt as a ridge along the anterior part of the leg. The lower end of the tibia articulates with the *talus*, an ankle bone. The lower end is not so wide as the upper extremity. There is a long process on the medial side known as the *medial malleolus*. It can be felt as the inner prominence of the ankle. On the opposite side there is a depressed articular surface where the lower extremity of the fibula articulates.

The *fibula* is a long, slender bone lateral to the tibia. Its head articulates with the tibia, but the articulation is below the knee joint. The lower end projects downward below the tibia and forms the *lateral malleolus*, the outer ankle bone. The talus is confined between the lateral malleolus of the fibula and the medial malleolus of the tibia (Figure 5.27).

The tarsus Seven bones form the tarsus. They are the *calcaneus* (or heel bone), the *talus* (astragalus), *navicular bone, cuboid bone,* and the first, second, and third *cuneiform bones.* The tarsal bones are larger and more specialized than the carpal bones. The calcaneus, the largest of the tarsal bones, helps to

The skeleton

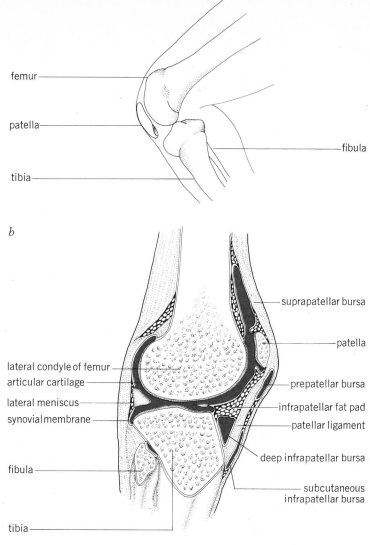

a

femur

patella

tibia

fibula

b

suprapatellar bursa

patella

lateral condyle of femur

articular cartilage

lateral meniscus

synovial membrane

prepatellar bursa

infrapatellar fat pad

patellar ligament

deep infrapatellar bursa

fibula

subcutaneous
infrapatellar bursa

tibia

FIGURE 5.26
The knee joint: *a* lateral view; *b*
section through knee joint.

support the weight of the body and affords attachment for the large muscles of the calf of the leg. The talus, which lies above the calcaneus and articulates with the tibia and fibula, is at the top of the longitudinal arch of the foot. The navicular bone is immediately anterior to the talus on the inner side of the foot. The cuboid bone is just anterior to the calcaneus on the lateral side. The three cuneiform bones form a row in front of the navicular with the first cuneiform bone on the inner side (Figure 5.28*a* and *b*).

The metatarsus   The five bones that comprise the metatarsus are homologous with the metacarpal bones of the hand. The metatarsal bones articulate at their proximal ends with the tarsal bones to form the longitudinal arch of the foot. The metatarsal bones also form the transverse arch. Distally the metatarsal bones articulate with the phalanges.

The phalanges   The phalanges are the bones of the toes. They are homologous

The appendicular skeleton

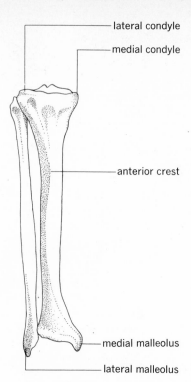

lateral condyle

medial condyle

anterior crest

medial malleolus

lateral malleolus

RIGHT FIBULA — RIGHT TIBIA
(ANTERIOR VIEW)

FIGURE 5.27
Right tibia and fibula (anterior
view).

TABLE 5.5
Bones of the pelvic girdle and lower extremity

Location	Name	Total no. of bones	Description
Hipbones	Os coxae (innominate)(2)	2	The ilium, ischium, and pubis fuse to form the hipbone. The skeletal pelvis includes the sacrum and coccyx.
Thigh	Femur(2)	2	The longest bone in the body
Knee	Patella(2)	2	The "kneecap"
Leg	Tibia(2)	4	Anteromedial bone of leg, the "shin bone"
	Fibula(2)		Lateral leg bone
Ankle	Calcaneus(2) Talus(2) Cuboid(2) Navicular(2) Cuneiform(6)	14	Ankle and heel bones
Foot	Metatarsal(10)	38	Bones that help to form the arches; form the framework of the foot
	Phalanges(28)		Toe bones; two in great toe, three in each of the others

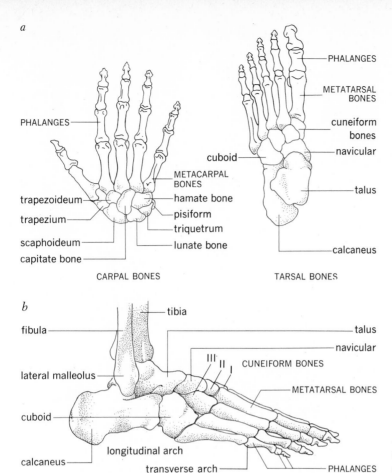

FIGURE 5.28
a Homology of hand and foot; *b* arches of the foot (lateral view).

with the bones of the fingers. There are two phalanges in the great toe and three in each of the other toes (Table 5.5).

The arches of the foot The foot has been modified to support the body in an erect position. The bones are so arranged that they form two arches, *longitudinal* and *transverse*. The arches are not rigid but yield as weight rests upon them, springing back into place as the weight is lifted. A well-functioning arch, therefore, gives a certain spring to the step (Figure 5.28).

The bones of the foot that form the arches are bound together by ligaments. The arches are also supported by the muscles and tendons of the foot. If the muscles and tendons become weakened from malnutrition or occupational conditions that require long hours of standing without much exercise, the ligaments may allow the arches to sag and produce the condition known as *flatfoot*. The tarsal bones are often forced toward the inner side of the foot in a pronated position (Figure 5.29).

Muscular exercises, such as standing on the toes, can do much to strengthen the musculature of the foot and can aid in restoring strength to the arch. Artificial arch supports should be used only if there seems to be no chance of restoring the normal condition of the arch by strengthening the muscles.

Weakening of the transverse arch often causes pain at the base of the

The appendicular skeleton

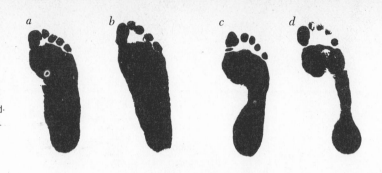

toes. Wearing high-heeled shoes forces most of the weight of the body onto the ends of the metatarsal bones. They may enlarge and push sideways out of line. The inner line of the shoe should be straight, not curved outward at the toe. The painful enlargement at the joint between the first metatarsal bone and the great toe is called a *bunion*.

SPRAINS, FRACTURES, AND BONE DEFICIENCIES

Sprain indicates that the ligaments and tendons of a joint have been injured. They may be stretched or torn away from their attachment to a bone. Sprains heal slowly and may require the attention of a physician for several weeks.

Fracture is a term applied to a break in a bone. The bones of elderly persons can be very brittle and can break easily. The bones of children contain a great deal of organic matter, and when a fracture occurs, the bone may not break apart but some of the fibers may be broken. Such a fracture is called a *greenstick fracture* because it resembles the way a green stick breaks. If the broken ends of a bone protrude through the skin, the fracture is classified as a *compound fracture*. If the skin is not broken, it is called a *simple fracture* (Figure 5.30).

Fractures require the services of a physician. It is often advisable not to move a patient until the arrival of a physician for fear of causing greater injury. If the patient must be moved, first aid consists of applying splints and simple bandaging to afford support to the injured parts. In the case of a compound fracture the greatest care should be exercised to prevent infections from entering the wound.

A physician should set the bones, preferably with the aid of x-rays so that the ends of the bones can be placed together properly. A callus of new bone forms around the broken ends and holds them together. The periosteum plays an important part in the regeneration of new bony tissue.

Dislocations occur when bones are forced from their proper position at joints. A dislocation usually involves strained or torn ligaments and considerable inflammation. Reduction of the dislocated bones should be performed by a physician except perhaps in the case of dislocated fingers, which can be put back into place by a firm pull.

Rickets is a vitamin-deficiency condition that results in a disturbance in the normal calcification of bones. Vitamin D is the vitamin concerned. In severe cases the long bones of the body do not grow straight. Common

THE

HUMAN

BODY

THE HUMAN BODY

An accurate knowledge of the architecture of the human body is c
great importance to students in hygiene, nursing, and physiology. Th
anatomical plates on the following pages have been specially prepare
and included in this text to provide a ready reference source not onl
of the various structures of the body, but also of their relationships t
one another.

The various body systems are shown in the natural color of health
tissues, and special attention has been paid to labeling the structur
and organs that are usually discussed in college-level courses. It is sug
gested that frequent reference to this section will prove invaluable i
helping the student visualize more clearly the body as a function
organism. To facilitate their use, certain illustrations have been place
on facing pages for easier comparison, and the same general body ou
lines have been used throughout. Furthermore, all structures are show
in proportionate size to afford a clearer understanding of their size an
the space they occupy in the body.

These illustrations were prepared by Robert J. Demarest, Medic
Illustrator, in consultation with a number of medical authorities.

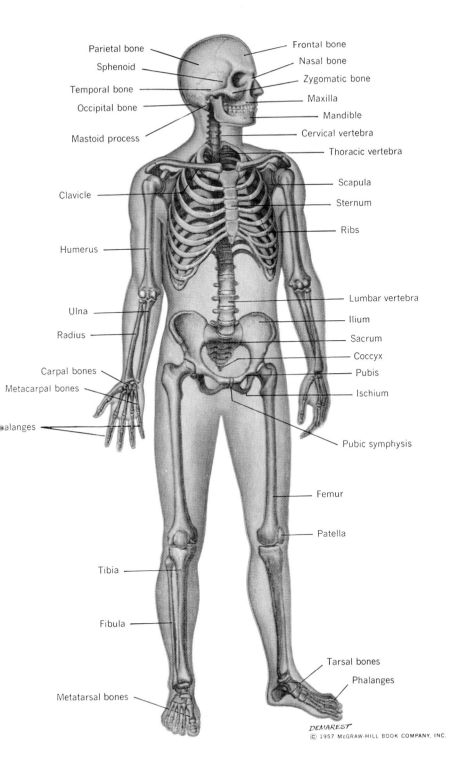

Parietal bone

Sphenoid

Temporal bone

Occipital bone

Mastoid process

Clavicle

Humerus

Ulna

Radius

Carpal bones

Metacarpal bones

alanges

Frontal bone

Nasal bone

Zygomatic bone

Maxilla

Mandible

Cervical vertebra

Thoracic vertebra

Scapula

Sternum

Ribs

Lumbar vertebra

Ilium

Sacrum

Coccyx

Pubis

Ischium

Pubic symphysis

Femur

Patella

Tibia

Fibula

Metatarsal bones

Tarsal bones

Phalanges

DEMAREST

© 1957 McGRAW-HILL BOOK COMPANY, INC.

PLATE A   SKELETAL SYSTEM (anterior view)

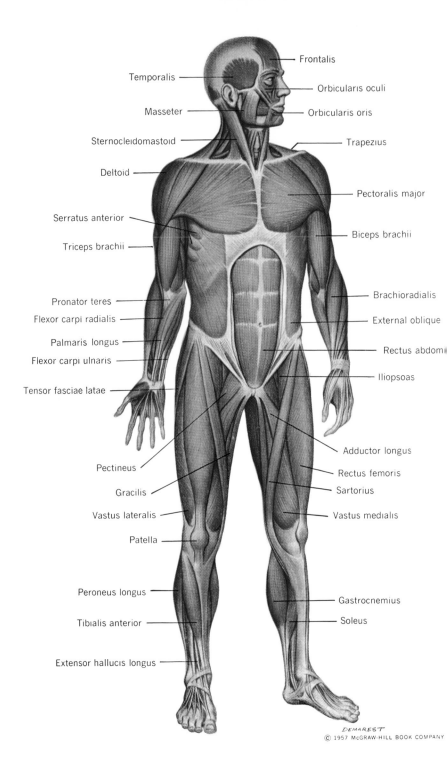

Frontalis

Temporalis

Orbicularis oculi

Masseter

Orbicularis oris

Sternocleidomastoid

Trapezius

Deltoid

Pectoralis major

Serratus anterior

Biceps brachii

Triceps brachii

Pronator teres

Brachioradialis

Flexor carpi radialis

External oblique

Palmaris longus

Rectus abdomi

Flexor carpi ulnaris

Iliopsoas

Tensor fasciae latae

Adductor longus

Pectineus

Rectus femoris

Gracilis

Sartorius

Vastus lateralis

Vastus medialis

Patella

Peroneus longus

Gastrocnemius

Tibialis anterior

Soleus

Extensor hallucis longus

DEMAREST
© 1957 McGRAW-HILL BOOK COMPANY

PLATE B   MUSCULAR SYSTEM (anterior view)

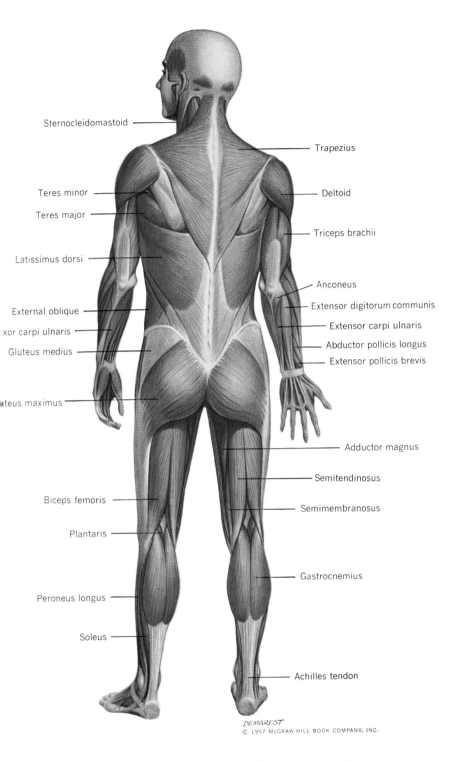

Sternocleidomastoid

Teres minor

Teres major

Latissimus dorsi

External oblique

xor carpi ulnaris

Gluteus medius

teus maximus

Biceps femoris

Plantaris

Peroneus longus

Soleus

Trapezius

Deltoid

Triceps brachii

Anconeus

Extensor digitorum communis

Extensor carpi ulnaris

Abductor pollicis longus

Extensor pollicis brevis

Adductor magnus

Semitendinosus

Semimembranosus

Gastrocnemius

Achilles tendon

DEMAREST
© 1957 McGRAW-HILL BOOK COMPANY, INC.

PLATE C   MUSCULAR SYSTEM (posterior view)

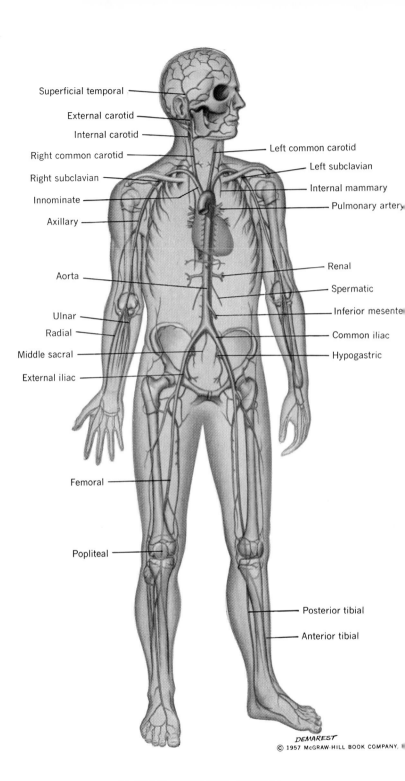

Superficial temporal

External carotid

Internal carotid

Right common carotid

Right subclavian

Innominate

Axillary

Aorta

Ulnar

Radial

Middle sacral

External iliac

Left common carotid

Left subclavian

Internal mammary

Pulmonary artery

Renal

Spermatic

Inferior mesenteric

Common iliac

Hypogastric

Femoral

Popliteal

Posterior tibial

Anterior tibial

DEMAREST

PLATE D   ARTERIAL SYSTEM

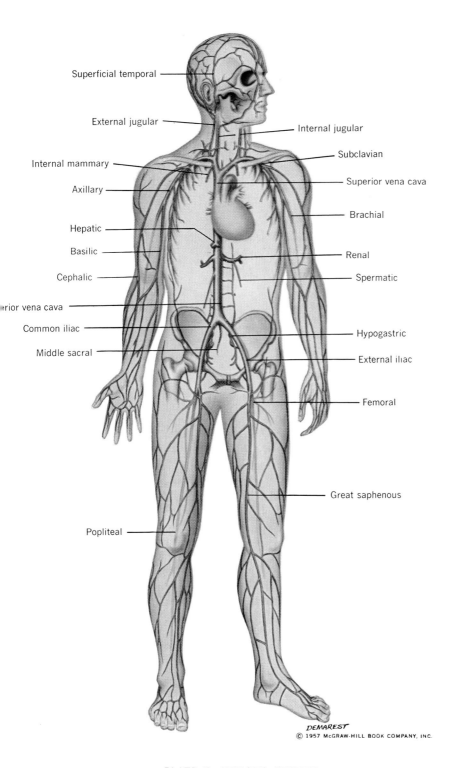

Superficial temporal

External jugular

Internal jugular

Internal mammary

Subclavian

Axillary

Superior vena cava

Hepatic

Brachial

Basilic

Cephalic

Renal

rior vena cava

Spermatic

Common iliac

Middle sacral

Hypogastric

External iliac

Femoral

Great saphenous

Popliteal

DEMAREST
© 1957 McGRAW-HILL BOOK COMPANY, INC.

PLATE E VENOUS SYSTEM

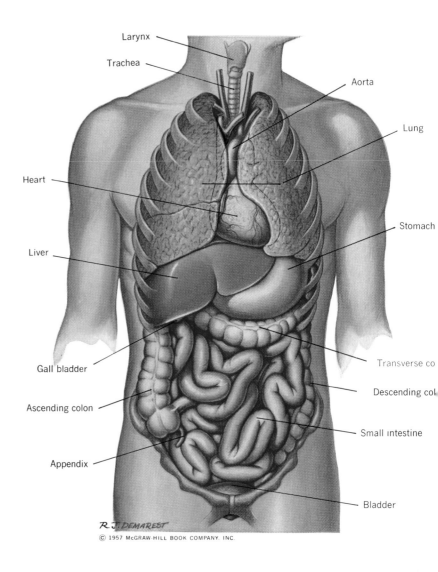

Larynx

Trachea

Aorta

Lung

Heart

Stomach

Liver

Gall bladder

Transverse co

Descending col

Ascending colon

Small intestine

Appendix

Bladder

R J DEMAREST

© 1957 McGRAW-HILL BOOK COMPANY. INC.

PLATE F VISCERA

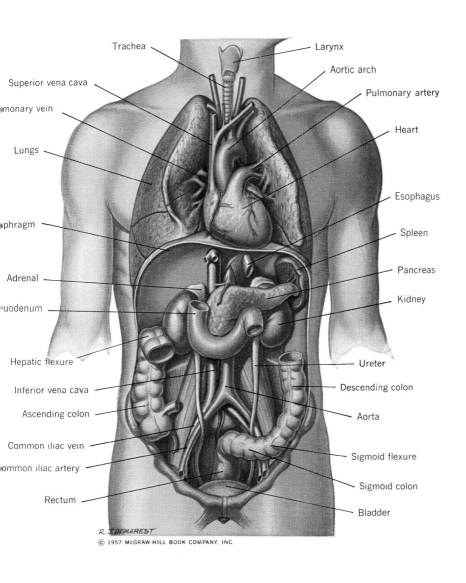

Trachea

Larynx

Aortic arch

Pulmonary artery

Superior vena cava

monary vein

Heart

Lungs

Esophagus

Spleen

phragm

Pancreas

Adrenal

Kidney

uodenum

Hepatic flexure

Ureter

Inferior vena cava

Descending colon

Ascending colon

Aorta

Common iliac vein

Sigmoid flexure

ommon iliac artery

Sigmoid colon

Rectum

Bladder

R. J. DEMAREST

© 1957 McGRAW-HILL BOOK COMPANY, INC.

PLATE G   VISCERA (deep structures)

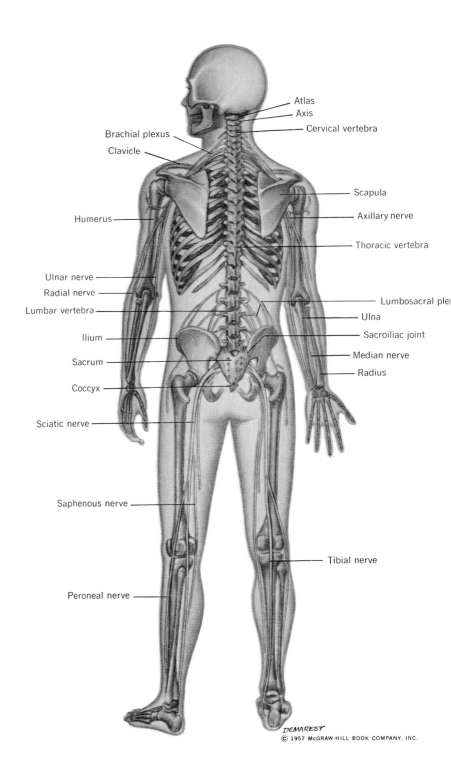

Atlas
Axis
Cervical vertebra

Brachial plexus
Clavicle

Scapula
Axillary nerve

Humerus

Thoracic vertebra

Ulnar nerve
Radial nerve
Lumbar vertebra

Lumbosacral ple
Ulna

Ilium

Sacroiliac joint

Sacrum

Median nerve
Radius

Coccyx

Sciatic nerve

Saphenous nerve

Tibial nerve

Peroneal nerve

DEMAREST
© 1957 McGRAW-HILL BOOK COMPANY, INC.

PLATE H   SKELETAL SYSTEM (posterior view showing spinal nerves)

INDEX

ctor pollicis longus, C
les' tendon, C
ctor longus, B
ctor magnus, C
ıal, G
neus, C
ʼior tibial vein, E
ı, D, F, G
ıc arch, G
ndix, F
ıding colon, F, G
ıry artery, D
ıry nerve, H
ıry vein, E

ıc vein, E
ıs brachii, B
ler, F, G
ıial plexus, H
ıial vein, E
ıioradialis, B

ıal bones, A
alic vein, E
ıcal vertebrae, A, H
cle, A, H
yx, A, H
ı, F, G
ımon carotid artery, D, G
ımon iliac artery, D
ımon iliac vein, E, G

ıid, B, C
ending colon, F, G
ıhragm, G
lenum, G

ısor carpi ulnaris, C
ısor digitorum communis,

ısor hallucis longus, B
ısor pollicis brevis, C
ʼnal carotid artery, D
ʼnal iliac artery, D
ʼnal iliac vein, E
ʼnal jugular vein, E
ʼnal oblique, B, C

ʼral artery, D
ʼral vein, E
ır, A, H
ıa, A, H
ʼr carpi radialis, B
ʼr carpi ulnaris, B, C
ital bone, A
talis, B

ılladder, F
ʼrocnemius, B, C
eus maximus, C
eus medius, C
ilius, B
ıt saphenous vein, E

Heart, F, G
Hepatic flexure, G
Hepatic vein, E
Humerus, A, H
Hypogastric artery, D
Hypogastric vein, E

Iliopsoas muscle, B
Ilium, A, H
Inferior mesenteric artery, D
Inferior vena cava, E, G
Innominate artery, D
Internal carotid artery, D
Internal jugular vein, E
Internal mammary artery, D
Internal mammary vein, E
Ischium, A

Jugular veins, E

Kidney, G

Larynx, F, G
Latissimus dorsi muscle, C
Liver, F
Lumbar vertebrae, A, H
Lumbosacral plexus, H
Lung, F, G

Mammary artery, D
Mammary vein, D
Mandible, A
Masseter muscle, B
Mastoid process, A
Maxilla, A
Median nerve, H
Metacarpal bones, A
Metatarsal bones, A
Middle sacral artery, D
Middle sacral vein, E

Nasal bone, A

Occipital bone, A
Orbicularis oculi muscle, B
Orbicularis oris muscle, B

Palmaris longus muscle, B
Pancreas, G
Parietal bone, A
Patella, A, B
Pectineus muscle, B
Pectoralis major muscle, B
Peroneal nerve, H
Peroneus longus muscle, B, C
Phalanges, A
Phrenic arteries, D
Plantaris muscle, C
Popliteal artery, D
Popliteal vein, E
Posterior tibial vein, E
Pronator teres muscle, B
Pubic symphysis, A
Pubis, A

Pulmonary artery, D, G
Pulmonary vein, G

Radial artery, D
Radial nerve, H
Radius, A, H
Rectum, G
Rectus abdominis muscle, B
Rectus femoris muscle, B
Renal artery, E
Renal vein, E
Ribs, A

Sacroiliac joint, H
Sacrum, A, H
Saphenous nerve, H
Saphenous veins, E
Sartorius muscle, B
Scapula, A, H
Sciatic nerve, H
Semimembranosus muscle, C
Semitendinosus muscle, C
Serratus anterior muscle, B
Sigmoid colon, G
Sigmoid flexure, G
Small intestine, F
Soleus muscle, B, C
Spermatic artery, D
Spermatic vein, E
Sphenoid, A
Sternocleidomastoid muscle, B, C
Sternum, A
Stomach, F
Subclavian artery, D
Subclavian vein, E
Superficial temporal artery, D
Superficial temporal vein, E
Superior vena cava, E, G

Tarsal bones, A
Temporal bone, A
Temporalis muscle, B
Tensor fasciae latae, B
Teres major muscle, C
Teres minor muscle, C
Thoracic vertebrae, A, H
Tibia, A
Tibial nerve, H
Tibialis anterior muscle, B
Trachea, F, G
Transverse colon, F
Trapezius muscle, B, C
Triceps brachii muscle, B, C

Ulna, A, H
Ulnar artery, B
Ulnar nerve, H
Ureter, G

Vastus lateralis muscle, B
Vastus medialis muscle, B

Zygomatic bone, A

ABOUT THE ARTIST

ROBERT J. DEMAREST's medical illustrations have appeared in many texts and medical journals as well as in several popular weekly magazines. Such a variety of uses clearly indicates the appeal of the life-like medical illustrations this young artist does. Trained at the Art Students' League and the School for Medical Illustrators, Demarest on graduation was honored as the recipient of the Saunders' Fellowship in Medical Art at the University of Pennsylvania. Mr. Demarest is currently associated with the College of Physicians and Surgeons, Columbia University.

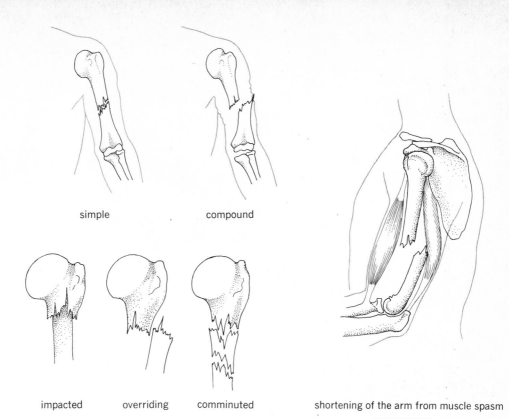

simple   compound

impacted   overriding   comminuted   shortening of the arm from muscle spasm

FIGURE 5.30

Various types of fractures. (*After William W. Stiles, "Individual and Community Health," McGraw-Hill Book Company, New York, 1953.*)

manifestations of rickets are bowlegs, knock-knees, or a very narrow chest with deformities of the sternum and costal cartilages.

The diet of children should include foods rich in vitamin D, such as egg yolk, butter, and whole milk. It may be advisable to supplement the diet with fish-liver oils or vitamin D concentrates. Fresh vegetables should be included in the diet to provide calcium and phosphorus, which are necessary for the proper hardening of bone.

Finally, the diet can be supplemented by exposing the skin to the direct rays of the sun. When the surface of the body is irradiated with ultraviolet light, it is capable of manufacturing its own vitamin D. For further discussion, see Chapter 18, section on vitamins.

SYNOVIAL BURSAE

Bursae are sacs of fibrous connective tissue lined with synovial membrane. They are commonly located at joints to prevent the friction of one surface moving upon another. They may be subcutaneous, such as the bursa over the patella or the olecranal bursa. They are found also between tendons or between muscles and bones, wherever friction is likely to occur. The synovial membrane secretes a viscid lubricating fluid. Mechanical injuries to bursae or infections can cause the fluid to accumulate. Inflammation of the bursa is called *bursitis*. Common locations of bursitis are subdeltoid (above the shoulder joint), at the elbow, around the calcaneus tendon, and around the knee (Figure 5.31).

Synovial bursae

117

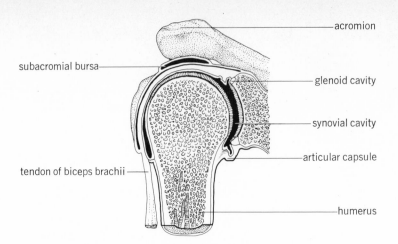

acronion

subacromial bursa

glenoid cavity

synovial cavity

articular capsule

tendon of biceps brachii

humerus

FIGURE 5.31
A synovial joint. Section through the shoulder joint, an example of a freely movable ball-and-socket type.

ARTICULATIONS

Various kinds of joints or articulations have been mentioned in considering the skeletal system. Joints are commonly classified according to the amount or kind of movement associated with the articulation (see Table 5.6). The study of articulations is called *arthrology*. The better-known term *arthritis* refers to inflammation of the joints.

Joints may be described as immovable, slightly movable, and freely movable. They may be classified also, in the same order, as fibrous, cartilaginous, and synovial joints (Table 5.6).

IMMOVABLE JOINTS   There are several kinds of immovable joints. The sutures between cranial bones are an example. These bones are fitted together by an interlocking arrangement and are held together by fibrous connective tissue. The sutures often become more ossified in the later years.

SLIGHTLY MOVABLE JOINTS   These are articulations that have a pad of fibrocartilage between two bones. The intervertebral disks have a core of more pliable tissue, which permits some flexibility in the vertebral column. The symphysis between the pubic bones is also of this type, although there is practically no movement. The sacroiliac articulation is a similar example. At the termination of pregnancy there is some slight flexibility in these pelvic articulations occasioned by structural changes in the fibrocartilage.

TABLE 5.6
Classification of joints or articulations

Scientific name	Type	Kind of articulation	Example
Synarthrosis	Immovable	Fibrous	Sutures between cranial bones
Amphiarthrosis	Slightly movable	Fibrocartilaginous	Joints between the vertebrae
Diarthrosis	Freely movable	Synovial	Hinge joints, ball-and-socket joints, pivot joints

TABLE 5.7

Types of synovial joints

Type	Description	Example
Gliding joint	One smooth, flat surface moving over another plane surface	Wrist and ankle bones
Hinge joint (ginglymus)	A rounded convex surface fitting into a concave surface; movement in one plane only	Elbow joint between the humerus and the ulna
Pivot joint	A shallow depression at the end of one bone rotating on a rounded surface	The radius pivots on the capitulum of the humerus
Ball-and-socket joint	The ball-shaped head of one bone fitting into a socket in another bone	The head of the humerus fits into the scapula to form the shoulder joint
Condyloid joint	A modified ball-and-socket type with limited movement	The joints between the metacarpals and the first phalanges of the fingers
Saddle joint	A concave-convex joint; biaxial, permitting movement from side to side and back and forth	The carpal-metacarpal joint of the thumb

FREELY MOVABLE JOINTS   Articulations of the ball-and-socket type are the most freely movable, but hinge joints permit free movement in a single plane. Pivot joints permit rotary movements, such as the atlas turning on the axis or the proximal end of the radius pivoting around the ulna. The articulating ends of bones are covered with hyaline cartilage. The joint is enclosed in a fibrous articular capsule, which is supported by the capsular ligament. The capsule is lined by synovial membrane, which secretes synovial fluid to lubricate the articulation. The small space between the articular cartilages and enclosed by the synovial membrane is called the *articular cavity*. It contains synovial fluid. The articular cavity often is continuous with the cavities of bursae (Table 5.7).

SUMMARY

The cranial bones are the frontal, parietal, occipital, temporal, sphenoid, and ethmoid bones. The occipital bone contains the foramen magnum and the occipital condyles. The anterior fontanel is a space located medially between the frontal and parietal bones. The posterior fontanel is located at the posterior end of the sagittal suture between the occipital and parietal bones. The temporal bones contain the middle and internal ear structures, and each bone has a mastoid process. Outstanding features of the sphenoid bones are the sella turcica and the greater and lesser wings. The crista galli, cribriform plate, perpendicular plate, and superior and middle nasal conchae are ethmoidal structures.

The facial bones are the mandible, maxilla, palatine bone, vomer, zygomatic bone, lacrimal bone, nasal bone, and inferior nasal concha. These are all paired bones except the mandible and the vomer. The mandible is the lower jaw; the maxillae form the upper jaw. The palatine bones form the posterior part of the bony roof of the mouth. The vomer is the lower portion of the median nasal septum. The zygomatic bones are the cheekbones. The lacrimal bones are fragile, grooved bones located medially, just inside the bony orbit. The nasal bones form the bridge of the nose, and the inferior nasal conchae are scroll-like bones located below the ethmoidal nasal conchae in the nasal passageway. The bones containing the paranasal sinuses are the frontal, ethmoid, and sphenoid bones and the maxillae.

The vertebral column is composed of 33 vertebrae: 7 cervical, 12 thoracic, 5 lumbar, 5 sacral, and 4 coccygeal. The first cervical vertebra is the atlas, and the second is the axis. The sacral vertebrae are fused to form the sacrum. The coccyx consists of four rudimentary vertebrae more or less fused.

The thoracic cage is formed by 12 thoracic vertebrae, 12 pairs of ribs, the costal cartilages, and the sternum.

The shoulder girdle consists of two clavicles and two scapulae. The clavicles are the collarbones. The head of the humerus articulates with the scapula.

The bones of the arm are the humerus, radius, and ulna. There are eight carpal bones in the wrist. Metacarpal bones form the framework of the hand. Finger bones are phalanges.

The hipbones form the pelvic girdle. Each hipbone develops as three separate bones: the ilium, ischium, and pubis.

The thigh bone is the femur. The patella, or kneecap, protects the knee joint anteriorly. The tibia and fibula are bones of the leg. There are seven tarsal bones. Of these, the heel bone is the calcaneus, and the bone that articulates the ankle is the talus. The five metatarsal bones of the foot are homologous with the metacarpal bones of the hand. The phalanges are the toe bones. The tarsal and metatarsal bones form the longitudinal arch of the foot. The metatarsal bones form the transverse arch.

Bursae are sacs of fibrous connective tissue lined with synovial membrane. They contain synovial fluid and function between moving parts to prevent friction.

Articulations are classified as immovable, slightly movable, and freely movable. The sutures between cranial bones are immovable joints. The intervertebral disks permit some movement between the vertebrae and are slightly movable joints. Hinge, ball-and-socket, and pivot joints are classified as freely movable.

QUESTIONS

1 Describe the skull of a newborn infant.
2 Discuss calcium and phosphorus metabolism as applied to bone.
3 Locate the fontanels.
4 Which bones contain paranasal sinuses?
5 Name and locate the natural curvatures of the vertebral column.
6 What bone forms the bony prominence above the shoulder joint? At the elbow?
7 When the hand is turned from palm side up to palm side down, which bone rotates at the elbow?
8 What are some of the differences between male and female pelves?
9 Of what use are the bursae?
10 What bone forms the large bony prominence on the inner side of the ankle?
11 Compare the structure of the hand and that of the foot. How do they differ?
12 Discuss the structure and function of the two arches of the foot.

The skeleton

Several textbooks on histology and anatomy are listed at the back of the book.

chapter 6
physiology of muscles

The muscles of the body are responsible for movement. *Skeletal muscles* cause the inert skeletal framework to move, as in walking, running, and jumping. Supporting this activity, there is a special kind of musculature in the heart—*cardiac muscle* tissue—which by its constant contracting and relaxing enables the heart to beat and to supply blood to all parts of the body. Still a third kind of muscular tissue makes up the viscera: the organs of the body such as the stomach, intestine, and uterus. *Visceral* or *smooth muscle* is responsible for such internal activity as churning the food in the stomach and passing it through the intestine. Muscular tissue accounts for 40 to 50 percent of the body weight, the percentage being slightly greater in men than in women (Table 6.1).

PHYSIOLOGICAL CHARACTERISTICS OF MUSCLE TISSUE

Muscular tissue is composed of specialized cells. The characteristic functions of the tissue are essentially the same as the functions of the cells that make up the tissue. One characteristic of living tissue is *excitability,* the quality of reacting to a stimulus. Muscular tissue responds to stimuli. Another characteristic of muscle tissue is *contractility.* A muscle, when stimulated, has the ability to become shorter and thicker. Work is accomplished by the contraction of muscles. Muscle tissue also exhibits *extensibility.* Skeletal muscles are usually arranged in antagonistic pairs so that when one muscle contracts, the opposing muscle is extended. The action of muscles in flexing the arm or leg illustrates this principle (Figure 6.1). Cardiac muscle tissue exhibits extensibility when the chambers of the heart are distended as they fill with blood. Visceral muscle shows great extensibility, as when the stomach is distended by food or when the urinary bladder is distended by accumulated urine. *Elasticity* is the ability of muscle tissue to return to its normal length after the force applied to it has been relieved.

TABLE 6.1
Muscle types and characteristics

Type	Contraction	Muscle cell	Striations	Nuclei
Skeletal	Voluntary	Long cylindrical	Strongly striated	Multiple nuclei around periphery of fiber
Cardiac	Involuntary	Fibers short and branching	Finely striated	Elongate-oval between intercalated disks
Smooth	Involuntary	Spindle-shaped	Nonstriated	Single nuclei, usually in middle of cell

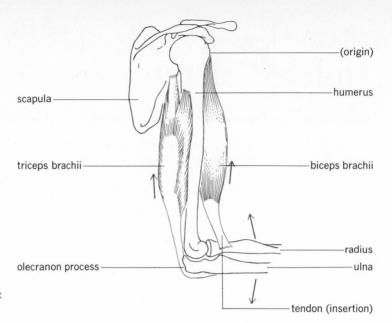

FIGURE 6.1
The biceps brachii and the triceps brachii as examples of antagonistic muscles.

SKELETAL MUSCLE The muscles that are designed primarily to give movement to the skeletal framework are called *skeletal muscles,* and they represent a considerable portion of the body weight. A piece of beefsteak is largely skeletal muscle. These muscles are ordinarily attached to bones by tendons. Since the gross action of skeletal muscles may be directed by higher centers of the brain, their action is voluntary, and they are called *voluntary muscles.*

Structure of a fiber Muscle tissue itself is very complex. Under a microscope, a bit of skeletal muscle tissue appears to be composed of many fibers extending along the length of muscle. Each fiber is finely striated with alternating light and dark bands (Figures 6.2 and 6.3). The fibers represent the structural unit of a muscle. They are cylinders of small diameter, 0.01 to 0.1 millimeter, about the same size as human hair. They are 1 to 40 or 50 millimeters in length, and some may become inserted in the muscle tendon. Others, however, are located only in the body of the muscle itself. The muscle fiber is enclosed in a delicate but strong sheath called the *sarcolemma,* which is essentially the cell membrane. This sheath encloses the semifluid protoplasm of the fiber, the *sarcoplasm.* There are many nuclei in the sarcoplasm and they lie just underneath the sarcolemma. Numerous mitochondria are present, functioning in the production of ATP. Muscle mitochondria are often called *sarcosomes,* especially in the flight muscles of insects, where they are very large and extend in long strands between the muscle fibrils. *Myofibrils,* or muscle fibrils, are threadlike structures, cross-striated like the fiber in which they lie. They have no surface membrane, since they are composed of protein molecules. The electron microscope reveals that a myofibril is composed of still smaller filaments. These filaments are of two kinds, thick and thin. Thick filaments are about 0.01 micron in diameter and 1.5 microns in length. Thin filaments are about 0.005 micron in diameter and about 2 microns long. (A micron is 0.001 millimeter, or $\frac{1}{25,000}$ of an inch.) In cross section it may be seen that each thick filament is surrounded by six thin filaments in a hexagonal arrangement.

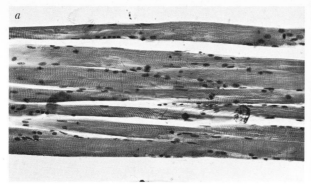
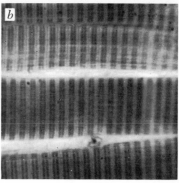

FIGURE 6.2
Skeletal muscle tissue. Note striations and nuclei. *a* Striations
shown moderately magnified.
(*Courtesy of General Biological Supply House, Inc., Chicago.*) *b* The resolution is much
greater, and various zones and
bands can be identified. Compare
with Figure 6.6. (*Courtesy of
Dr. W. M. Copenhaver.*)

Thick and thin filaments overlap for part of their length, a factor that explains the dark, dense, *anisotropic*, or A, band (Figure 6.4). A lighter colored, less dense area, called the *isotropic* or I band, is composed of thin filaments only, in relaxed muscle, whereas the narrow H band contains thick filaments only. A narrow zone of dense material, known as the Z line, lies in the middle of each I zone and divides the fiber into segments called *sarcomeres*. In other words, the segment limited by a Z membrane at either end is a sarcomere.

According to a theory developed by A. F. Huxley, H. E. Huxley, and J. Hanson, contraction of the myofibril is caused by the two sets of filaments sliding over each other. It is observed that the A band, composed of thick filaments, retains a constant length during either contraction or extension of the myofibril. This would indicate that the thick filaments do not contract or lengthen. Neither do the thin filaments change in length. Since the thin filaments are attached to the Z line, the sarcomere shortens in contraction of the myofibril as the thin filaments move further into the H band of the A zone (Figure 6.5). It appears that the thin filaments divide at the Z line,

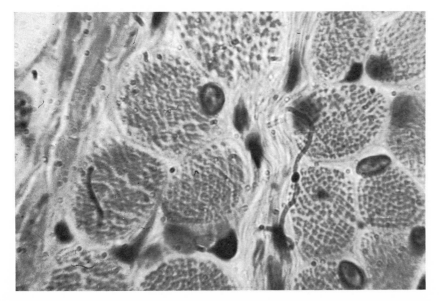

FIGURE 6.3
Cross section of skeletal muscle
tissue. Note muscle fibers, fibrillae,
and nuclei. (*Courtesy of Dr.
W. M. Copenhaver.*)

*Physiological
characteristics
of muscle tissue*

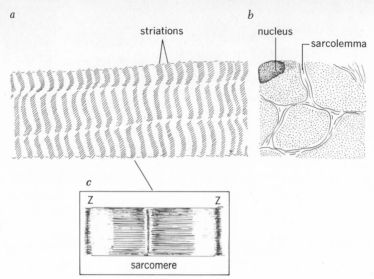

striations   nucleus   sarcolemma

a   b   c

Z   Z

sarcomere

FIGURE 6.4

Diagrammatic representation of
skeletal muscle greatly enlarged:
a longitudinal section; *b* cross
section; *c* diagram illustrating a
sarcomere.

presenting a Y-shaped, staggered appearance at their ends. Each branch of
the Y makes a connection with another thin filament in the next sarcomere.

As we have seen, during contraction the Z lines come closer together.
The various bands and zones do not present the same appearance during
contraction as in the resting state. The I band and H zone are somewhat
lighter in color, since there is no overlap of thick and thin filaments in these
areas when the myofibril is stretched (Figure 6.6). But during contraction
the I-band area becomes greatly constricted and the H zone also narrows,
almost to the point of disappearing, as thin filaments invade this region. In
a maximum contraction the ends of thin filaments come together and may
overlap each other.

In the middle of the A band there appears a thin line which apparently
is due to a slight widening of the thick filaments at their middle. This is
called the M line. A study of the fine structure of the M line reveals that
it contains short M filaments. It is suggested that the structures of the M
line may be concerned with guiding the sliding filaments and assuring their
alignment. Apparently there is a newly recognized protein associated with
the M line, called, provisionally, M substance.

THE PHYSIOLOGY OF MUSCLE

MUSCLE PROTEINS   Scientists have worked for many years to unravel the
mystery of muscle contraction, but now the sliding filament theory seems
to be generally accepted. Several proteins are involved, but actin of the thin
filaments and myosin of the thick filaments certainly play a principal part.
These two may be extracted individually from muscle and purified. When
they are placed together in a proper solution, they combine to form a more
complex protein called *actomyosin.* Fibers of actomyosin made artificially
and placed in a solution of proper ionic composition will shorten when
adenosine triphosphate (ATP) is introduced. However, this in vitro experiment is not thought to represent the method by which muscles contract in
vivo.

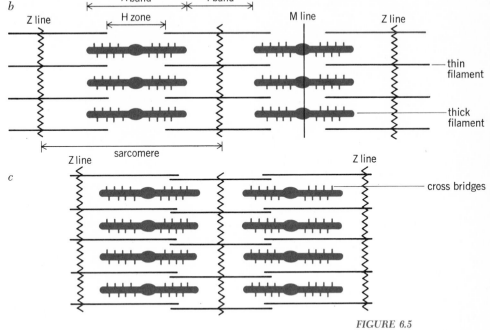

FIGURE 6.5
Diagram of muscle zones and filaments: *a* a section of striated
muscle fiber showing zones and
bands; *b* resting muscle; *c* contracted muscle; *d* schematic representation of myosin molecules
indicating that the heads point in
one direction in one half of a
thick filament and in the opposite
direction in the other half. The
heads of these molecules represent the cross bridges projecting
out from thick filaments to thin
filaments. In contraction the two
sets of filaments slide over each
other and the Z lines move closer
together. A sarcomere is the area
between two Z lines. The Z lines
pass through I bands. H zones are
located in A bands. (*After
Bernard Katz, "Nerve, Muscle, and Synapse,"
McGraw-Hill Book Company,
New York, 1966.*)

Three proteins compose the structure of thin filaments, *actin* being the more abundant one. The actin filament consists essentially of two strands of globular molecules wound around each other in a helical arrangement. Although actin is the principal protein of the thin filaments, two additional proteins are present in much smaller amounts. These are *tropomyosin* and *troponin*. Tropomyosin appears to be an integral part of the actin filament, while troponin molecules seem at least to be associated with it. These two proteins are concerned with the calcium sensitivity of the myofibrils, a subject to be discussed later.

Myosin is the protein found in the thick filaments of the A band. The myosin molecule, with a globular head and a linear tail, somewhat resembles a minute golf club. The molecule is composed of two subunits or meromysins, based on its sedimentation properties in the ultracentrifuge. The light mero-mysin (LMM) portion of the molecule forms a part of the filament itself, while the heavy meromysin forms the cross bridge and the globular head.

Myosin molecules are placed so that they lie in opposite directions within the thick filament, that is, the heads point in one direction in one half and in the opposite direction in the other half. There are no cross bridges

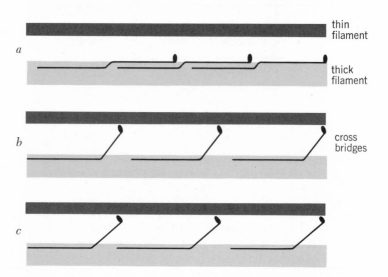

FIGURE 6.6

Electron micrograph of striated muscle tissue from rabbit (original magnification 24,000 diameters). The broad, dense areas are A bands. The less dense area passing through the middle of the A band is the H zone. The wide, light-colored areas are the I bands. The black band passing through the I band is the Z line. (*Courtesy of H. E. Huxley.*)

FIGURE 6.7

Proposed action of cross bridges: *a* in resting muscle; *b* during contraction (the cross bridges are thought to reach out from myosin filaments and make contact with actin filaments); *c* the head of the myosin molecule slants with the movement of the filaments. (*After Hugh Huxley.*)

thin filament

a

thick filament

b

cross bridges

c

in the middle of the myosin filament. The structure shown diagrammatically as the head of the molecule is misleading (Figure 6.5*e*). Actually two globular units form the head.

Myosin has a most interesting characteristic in that it may act as an enzyme in the dephosphorylation of adenosine triphosphate (ATP). It catalyzes the removal of a phosphate group from ATP (ATP–ADP) in the process of transforming chemical into mechanical energy. It is, therefore, an ATPase. The ATPase activity takes place in the globular head of the molecule, which also contains the actin-binding site.

Considerable interest now centers on the cross bridges of the myosin filaments and their function with reference to contraction. Originally, electron micrographs portrayed them as short, stiff projections spirally arranged

Physiology of muscles

126

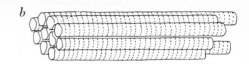

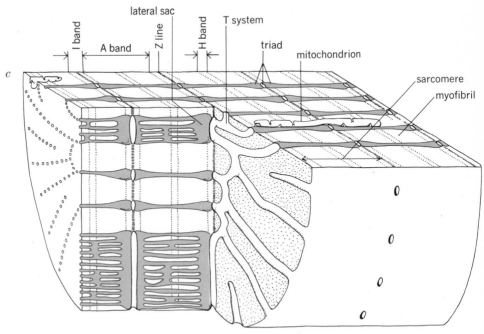

FIGURE 6.8

A diagram of a frog muscle fiber illustrating the relationship between the sarcoplasmic reticulum and the transverse system: *a* a muscle; *b* a group of muscle fibers enlarged; *c* detail of a muscle fiber as revealed by electron microscopy. The sarcoplasmic reticulum is shown in color. [*After Keith Porter, The Sarcoplasmic Reticulum, Sci. Am.,* **212:**73–80 (1965). *Copyright 1965, by Scientific American, Inc. All rights reserved.*]

along the thick filament. It now appears that the cross bridges may be far more flexible than originally proposed, reaching out to make contact with a reactive site on the thin actin filament (Figure 6.7).

THE NATURE OF THE CONTRACTION PROCESS Probably we should say at once that the exact method of contraction is still undetermined. As a first step, the heavy meromysin of the cross bridge binds to actin of the thin filament, and this results in activating myosin ATPase. In the next step ATPase breaks down ATP to ADP, thus releasing energy. The manner in which this release of energy causes movement at the cross bridges is not known. There is some indication that the cross bridges pull the sliding filaments over each other in a movement that resembles the motion of oars in rowing a boat.

THE INTERNAL MEMBRANE SYSTEMS *The sarcoplasmic reticulum* is a part of the internal membrane system of muscle. It consists of a network of channels and vesicles within the muscle fiber, in many ways similar to the endoplasmic reticulum of other cells. The network extends along the myofibrils in the sarcoplasm of the muscle cell. It can be demonstrated in cardiac tissue and reaches its highest development in skeletal muscle but does not seem to be present in smooth visceral muscle. In addition to the tubules that run parallel to the myofibrils there exists a system of *transverse tubules* which make up the second part of the internal membrane system (Figure 6.8). The transverse

The physiology of muscle

tubular system is a direct infolding of the surface membrane, and in frog muscle the transverse channels enter the fiber at the Z line. They are not necessarily at the same position in all classes of animals. *Lateral sacs* of the sarcoplasmic reticulum associated with the transverse system form the triads, which consist of sacs or vesicles on either side of the transverse tubule near the surface membrane. Lateral sacs are also called lateral cisternae or outer vesicles by various authors. Apparently the sarcoplasmic reticulum and the channels of the transverse system are not continuous; there is no indication that they open into each other. Nevertheless, they evidently cooperate in aiding the contraction of the myofibrils.

It is necessary now to consider the effect of the nerve impulse on muscle. The impulse is conveyed to the muscle by a motor neuron (Figure 3.1c). When a motor nerve is stimulated, the nerve impulse is transmitted to the muscle membrane, where chemical and physical changes occur, beginning at the neuromuscular junction, or end plate (Figures 6.9 and 6.10). This is the end-plate potential, characterized by local depolarization of the surface membrane. Each fiber is covered by an electrically polarized membrane; the outside of the resting membrane is positive; the inside is about 90 millivolts negative.

Depolarization of the plasma membrane produces an action potential which results in contraction. As the action potential spreads rapidly, the membrane exhibits increased permeability to sodium and potassium ions, the movement of these ions being responsible for the depolarization.

It is commonly accepted that the nerve impulse causes the release of a transmitter substance at the neuromuscular junction (Figure 6.9). In the peripheral nervous system this substance is acetylcholine and is responsible for a change in the permeability at the motor end plate, initiating the motor end-plate potential. If the end-plate potential reaches threshold level, depolarization of the muscle-fiber membrane occurs and an action potential spreads over the fiber, resulting in contraction. An enzyme, acetylcholinesterase, is responsible for the rapid breakdown of acetylcholine.

One of the problems in the development of muscle physiology involved explaining how the electrical signal could reach the interior of the muscle fiber so that the myofibrils located deep inside the fiber could be affected by the action potential. The discovery of the transverse tubular system has helped to explain this problem, but it is much more complicated than a simple depolarization of the transverse tubular membrane. The depolarization wave provides for the release of calcium ions, presumably stored in the lateral sacs of the sarcoplasmic reticulum, and a rapid building up of free calcium ions.

We recall that the thin filaments contain troponin and tropomyosin as well as actin. This complex of proteins acts to prevent the interaction of actin and myosin and so maintains a state of relaxation. Troponin is capable of binding calcium ions specifically, and when enough free calcium is present, it interacts with tropomyosin and releases the inhibition to the combining of actin and myosin. The first step is the activation of myosin ATPase by actin, which then permits actin to combine with myosin ATP. Magnesium ions are essential for this reaction. It is important to understand that calcium ions do not act directly to cause contraction but rather act indirectly, as a releasing factor. Upon activation of myosin ATPase, ATP is broken down to ADP, releasing energy for contraction. The exact way in which a chemical reaction causes movement of the cross bridges is not known.

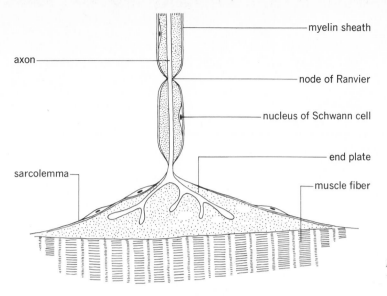

axon

myelin sheath

node of Ranvier

nucleus of Schwann cell

end plate

sarcolemma

muscle fiber

FIGURE 6.9
The myoneural junction.

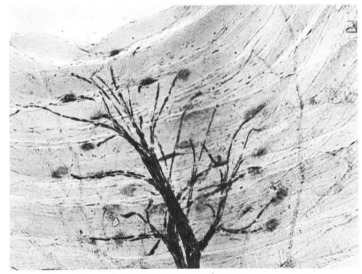

FIGURE 6.10
Motor end plates in striated muscle. (*Courtesy of Dr. Mac V. Edds.*)

SUMMARY OF EXCITATION-CONTRACTION The nerve impulse conveyed by a motor neuron arrives at the motor end plate or neuromuscular junction (Figure 6.10), releasing a transmitter substance. The motor end-plate potential, thus formed, initiates depolarization of the muscle-fiber membrane, and an action potential spreads over the fiber. The transverse tubule membrane is depolarized, releasing calcium ions from lateral sacs in the sarcoplasmic reticulum. Calcium ions are bound by the troponin-tropomyosin complex, releasing the inhibition to the interaction of actin and myosin ATPase. Activation of myosin ATPase permits actin to combine with myosin ATP, releasing energy as ATP breaks down to ADP, resulting in electromechanical coupling and contraction.

The physiology of muscle

RELAXATION Forces providing relaxation of the myofibrils are not well understood. A relaxing factor is postulated but has not been determined. As the calcium level falls, relaxation occurs, but the reaction is a complicated one. Presumably calcium is withdrawn from the troponin-tropomyosin complex, and a state of inhibition to contraction is established. It has been shown that magnesium and ATP play some part in relaxation. Intact ATP is thought to break the bond between actin and myosin, a necessary step in relaxation. In rigor mortis, for example, it is considered that no more ATP can be produced and that therefore the myofibrils are unable to break the bond between actin and myosin. This results in the continued contraction and eventual hardening of the muscles which exists for some time after death. When muscles have been contracting strongly or are greatly fatigued just before death, rigor mortis sets in more promptly.

SOURCES OF ENERGY The immediate source of energy for the contraction of muscle comes from the breakdown of ATP.

$$ATP + H_2O \rightarrow ADP + Pi + energy$$

The amount of ATP in muscle is not great, and a source of supply from anaerobic and aerobic metabolism would probably be too slow to support rapid movement. There is, however, another high-energy phosphate compound capable of releasing energy; this is creatine phosphate (CP). Much of this energy is used in the resynthesis of ATP, thus making up for any loss occurring during exercise.

$$ADP + CP \rightleftharpoons ATP + creatine$$

The enzyme involved in this reaction is creatine kinase or creatine phosphoryltransferase. This is a reversible reaction and provides for the rebuilding of creatine phosphate by ATP during periods of rest or at least during periods of reduced activity.

Glycogen is usually abundant in muscle tissue. In the process of glycolysis, glycogen is broken down to pyruvic acid. This is an anaerobic phase, and a number of steps are involved. The energy released during this phase is relatively low. Mitochondria are present in muscle tissue in considerable numbers. When pyruvic acid is formed, it enters mitochondria and is oxidized in the tricarboxylic acid cycle. ATP is formed in the passage of hydrogen and electrons along a series of catalysts and cytochromes in the electron-transport series, and finally hydrogen combines with molecular oxygen to form water. Also, CO_2 is given off during the tricarboxylic acid cycle.

During muscle contraction, when ATP breaks down to ADP, the ADP enters the mitochondria, where it acts as a phosphate acceptor and is resynthesized to ATP.

Glycogen is stored in the liver, and muscle glycogen is replenished from this source. Lactic acid, the reduced form of pyruvic acid, does not accumulate in muscle except during severe sustained exercise or during extreme muscular effort, when a deficiency of oxygen may occur. (Under ordinary conditions muscular activity is supported by an adequate supply of oxygen.) Pyruvic acid enters the tricarboxylic acid cycle and is oxidized.

$$ \underset{\text{lactic acid}}{\begin{array}{c} CH_3 \\ | \\ H-C-OH \\ | \\ COOH \end{array}} \quad \underset{+2H}{\overset{-2H}{\underset{\longleftarrow}{\xrightarrow{\text{lactic dehydrogenase}}}}} \quad \underset{\text{pyruvic acid}}{\begin{array}{c} CH_3 \\ | \\ C=O \\ | \\ COOH \end{array}} $$

About 10 times as much energy is liberated from glucose when pyruvate is oxidized by way of the mitochondria than is obtained during the anaerobic glycolysis of glucose, but muscles can contract for some time without an immediate supply of oxygen. The recovery phase in muscle is the time when oxidation occurs and when carbon dioxide is formed. If lactic acid is produced, it is carried by the blood to the liver and resynthesized to liver glycogen.

In the breakdown of these chemical compounds the energy produced can be measured in calories, since heat is a form of energy. There are two phases of heat production in contracting muscles; one is the *initial heat* of contraction, and the other phase is called *recovery heat*. Initial heat develops quickly, and since the process is nonoxidative, it comes largely from the breakdown of organic phosphates rather than from the breakdown of glycogen. Recovery heat develops more slowly after relaxation and results primarily from glycolysis and the resynthesis of ATP and CP.

The mechanical efficiency of muscle compares favorably with the mechanical efficiency of gas engines, which is only about 25 percent. Muscles have been estimated to convert 25 to 30 percent of the energy into movement (work); the rest is given off as heat. In an engine the energy given off as heat is largely a loss, but in the body the heat produced helps to maintain a body temperature of about 98.6°F. A whole series of events, to be discussed later, is concerned with the maintenance and control of body temperature, but one factor should be mentioned here. As one becomes cold, the muscles become tense and finally break into uncontrolled contractions, called *shivering*. These rapid contractions produce heat and so help to warm the body. On a very hot day muscles are apt to be relaxed and flaccid, and one moves about as little as possible.

OXYGEN DEBT Human beings are entirely dependent upon a steady intake of oxygen. About 200 to 300 cubic centimeters of oxygen per minute is required when the body is at rest, and many times this amount is needed for vigorous exertion. The trained athlete can utilize 4 or 5 liters of oxygen per minute. The oxygen requirement for vigorous exercise may be 16 to 20 liters per minute or higher, depending on the degree of exertion. It is apparent that in extreme exertion oxygen intake will be considerably less than the oxygen requirement of the muscles. One must, therefore, go into debt for oxygen. The runner breathes heavily as he rests after the race, until the oxygen debt is repaid.

It is fortunate that the energy necessary for the immediate contraction of muscle is supplied by nonoxidative reactions. However, one could not perform strenuous exercise for more than 1 or 2 minutes if one were deprived of oxygen. We have seen that oxygen is necessary in the recovery phase of muscle. If there is not enough oxygen available, lactic acid accumulates after strenuous muscular activity and is then transported to the liver, where enzyme systems convert the greater portion of it to glycogen. Oxidative energy is necessary also to rebuild organic phosphates. One can go into debt for oxygen, but the debt must be paid before normal exercise is resumed.

Moderate exercise, such as walking or working in the garden, may incur no oxygen debt. Physiological adjustments are such that a balance is attained between oxygen needs and energy requirements. The heart and breathing rates become adjusted to exercise, and one reaches a condition described

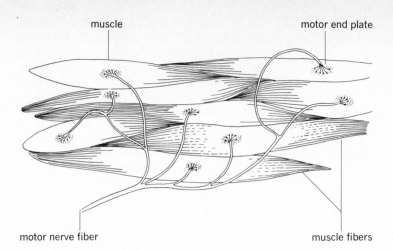

muscle   motor end plate

motor nerve fiber   muscle fibers

FIGURE 6.11
A motor unit.

as a *steady state*. An athlete in training can perform moderate exercise without additional lactic acid appearing in the blood.

Myoglobin is an important muscle protein capable of storing oxygen. The molecule is built around a heme group, which has an iron atom at the center. The heme group combines with oxygen and holds it in loose chemical combination until it can be utilized by contracting muscle. The molecular structure of this protein has been worked out by Kendrew and his associates at Cambridge. The molecule contains 2,600 atoms, and the single polypeptide chain consists of around 150 amino acids. It is a relatively small molecule when compared with the complexity of its chemical relative, hemoglobin. The function of myoglobin has not been fully understood. Apparently it releases its store of oxygen only during conditions of exhausting exercise when intracellular oxygen pressure may become extremely low. It has been suggested also that it may function to promote diffusion of oxygen rather than to store it.

MOTOR UNIT   It is fundamental to an understanding of muscle-nerve physiology that we establish the concept of the motor unit (Figure 6.11). A motor unit consists of a motor neuron and its terminal filaments which innervate individual muscle fibers. Skeletal muscle fibers are innervated in groups of perhaps a hundred by the branches of a single motor neuron. With such a common innervation these fibers then respond as a unit. The muscle fibers of a motor unit are not all localized in one part of the muscle but are spread out through the body of the muscle. This arrangement tends to equalize the contraction and prevents a strong contraction in just one part of the muscle.

STUDY OF SKELETAL MUSCLE CONTRACTION   Experimentally, much can be learned about the nature of contraction in an isolated skeletal muscle by arranging for a muscle to record its contractions on the revolving drum of an instrument called a *kymograph*. The drum is covered with smoked or waxed paper that will record movements of a writing lever. A muscle commonly used is taken from the hind leg of a frog. It is fitted in a clamp at one end, and the other end is attached to the writing lever.

Since the muscle has been removed from the body, artificial stimuli must be employed to cause contraction. Artificial stimuli may be of several different types, namely, mechanical, as striking the muscle or pinching it;

Physiology of muscles

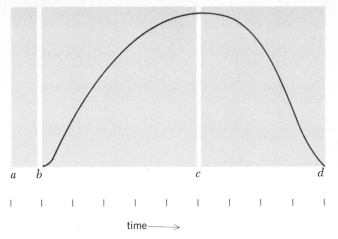

a b   c   d

time ⟶

FIGURE 6.12
The phases of a simple muscular contraction, a muscle "twitch": *a* the initiation of the stimulus; *b* beginning of contraction; *a–b* latent period, about $\frac{1}{100}$ second in frog muscle; *b–c* contraction phase, about $\frac{4}{100}$ second; *c–d* relaxation phase, about $\frac{5}{100}$ second. Each division of the time scale below represents $\frac{1}{100}$ second.

chemical, applying chemicals or as the result of chemical changes within the muscle; thermal, applying heat or cold; and electrical. Electricity is commonly used because it is readily available, its strength is easily controlled, and it does not damage the tissues unduly.

The current used can be obtained from a battery and led through an induction coil. Wires from the coil go to electrodes inserted in the muscle. A signal magnet is used to record the instant that the stimulus is given. A bioelectric stimulator is now commonly used in place of the battery and coil, since it affords better control of voltage and frequency.

Response to stimuli A great deal can be learned about the contraction of muscle by observing its response under various conditions. Having the muscle prepared for experimental procedure, we are now ready to stimulate it and watch the action of the muscle. As the muscle is stimulated by the electric current, it contracts, but the mark made by the signal magnet seems to indicate that it did not start to contract the very instant that it was stimulated. There is a brief period of perhaps 0.01 second in frog muscle from the time the stimulus is received until the muscle starts to contract. This is the *latent period*. It is a much shorter period in mammalian muscle, averaging about 1 millisecond, that is, 0.001 second. The contraction phase is pictured on the drum by the upward motion of the writing lever. If the drum is revolving slowly, the tracing will slant sharply upward; but if the drum is revolving rapidly, it will be an upward curve. This represents the period of contraction of the muscle and lasts about 0.04 second.

The downward tracing represents the relaxation phase and lasts approximately 0.05 second. The muscle accomplishes work by its contraction, but the relaxation phase is important too. Relaxation enables the muscle to protect itself against fatigue. It is not now regarded as an entirely passive stage but rather as an active process (Figure 6.12).

Suppose we now introduce a series of stimuli of constant strength into the muscle. We must allow time for the muscle to contract and relax, but this will take only about 0.1 second in frog muscle. If we repeat the stimuli slowly at 0.5-second intervals, we find that the muscle makes a series of tracings on the revolving drum as it contracts and relaxes. The first few tracings may show an increasing height with each contraction. This is the so-called *staircase phenomenon*, or *treppe*, which is associated with increasing efficiency of muscular contraction.

The physiology of muscle

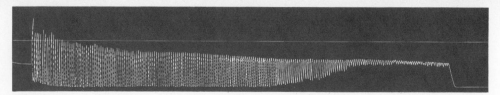

FIGURE 6.13

Fatigue curve of an isolated skeletal muscle.

The muscle, having reached the peak of its performance, contracts and relaxes regularly in response to a series of regularly spaced stimuli. After numerous contractions the muscle does not relax completely, and therefore the tracings become shorter. The lever is not lifted so high, and with the shortening of the muscle the lever does not go so far down. This type of shortening of muscle is called *contracture* rather than *contraction*.

Fatigue We see in the gradual shortening of the isolated muscle a loss of efficiency as fatigue sets in. The response is slower, the latent period is increased a slight extent, the time necessary for contraction increases, and the relaxation period becomes slower and slower. A study of the relaxation phase deserves special attention. In fatigue it is important to observe not only that the muscle gradually fails to contract but also that it fails to relax (Figure 6.13).

Summation When the rate of stimulation is increased to such an extent that the second stimulus is received by the muscle while it is still contracting in response to the first stimulus, a *summation* effect is noted. The muscle then contracts to a greater extent than it would in response to a single stimulus (Figure 6.14*b*).

Summation of subminimal stimuli A summation effect may also be obtained with subminimal (or subliminal) stimuli. A subminimal stimulus is one below the threshold level, that is, one that is too weak to cause a response. If a series of subminimal stimuli is introduced rapidly into a muscle, however, the muscle may be stimulated to contract. The original stimulus brings about a physiological change in the muscle. The succeeding stimuli, if frequent enough, produce an additive effect that results in contraction.

Tetanus If frog muscle is stimulated at a rate of 20 or 30 stimuli per second, there is time for only a partial relaxation between stimuli and the muscle maintains a state of contraction called *incomplete tetanus* (Figure 6.14*c, d*). Stimulation at an increased rate of 35 to 50 stimuli per second should result in complete tetanus.

A single muscular contraction (a muscle twitch) would be of little use in coordinated movement. Those muscles that enable us to stand up, walk, run, or bend over maintain a state of physiological tetanus for long periods of time with comparatively little fatigue. The condition of tetanus can apply to only a portion of a muscle rather than to an entire muscle. The tension in posture muscles is maintained by a series of nervous stimuli to groups of nerve endings in the parts of the muscles involved. Flexing the biceps slowly is a coordinated contraction involving many fibers and is much slower than a muscle twitch. This action may therefore be considered as a tetanus of those fibers that are in a state of contraction.

Refractory period Experimentally a muscle can be stimulated at a rate so rapid that it fails to respond to each stimulus. There is a brief period in muscle when it is nonirritable and, therefore, will not react to stimuli. This period

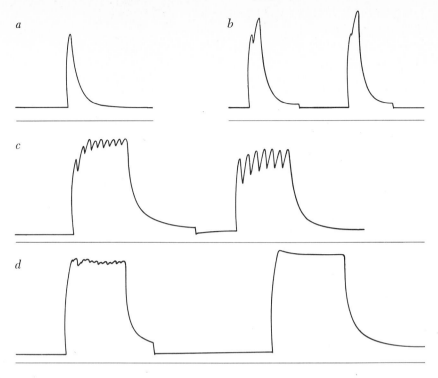

FIGURE 6.14
A tracing or myogram of muscle contractions: *a* a simple muscle twitch following a single stimulus; *b* summation, when two stimuli are introduced in rapid succession; *c* incomplete tetanus; *d* complete tetanus.

of about 0.001 second in warm-blooded skeletal muscle is known as the *refractory period*. Since it is of such short duration, it probably does not limit activity under normal conditions. The stimuli that result in tetanus of skeletal muscle are evidently spaced to the extent that they do not fall within the brief refractory period. The refractory period in heart muscle extends practically throughout the systolic contraction phase. Furthermore, a relative refractory period lasts well into diastole, the relaxing period of the heartbeat. Therefore, tetanus does not occur in cardiac tissue. Visceral muscle has the longest refractory period of any muscle tissue. Recent investigations indicate that the refractory period in muscles should be interpreted as being directly dependent upon the refractory period in nerves.

Types of contraction   In the foregoing experiments with isolated muscle, we observed contractions of muscle under load. Muscles contract more efficiently if they have an appropriate load. When a load is first placed on isolated muscle and the muscle stimulated, a slight extension occurs; this slight stretching produces tension, and the muscle responds by shortening. The muscle functions more efficiently after it has adjusted to the load. The type of contraction exhibited when a muscle becomes shorter and thicker is called an *isotonic contraction*. The contraction of the biceps brachii is an example. The tension remains essentially constant, but the muscle shortens, performing work.

In another type of contraction, the tension increases with the load, but the muscle does not shorten. This type is called *isometric*, from the Greek words that mean "the same measurement." If one carries a bucket of water with the arm extended, the muscles of the arm and trunk support the additional weight, but there is little or no change in the length of the muscles

The physiology of muscle

involved, only an increase in tension. In isometric contraction, energy is utilized and heat is produced, as it is in isotonic contraction. The "fixed" postural muscles of the trunk and neck are isometric.

SMOOTH MUSCLE *Structure* Smooth muscle is found in various organs of the body, in the internal eye muscles, and in the walls of blood vessels. It is nonstriated, since it is not composed of myofibrils as in skeletal muscle. It receives its innervation from the autonomic nervous system rather than from the conscious part of the brain and is, therefore, involuntary.

There are two kinds of smooth-muscle tissue, a mass-unit type composed of thin sheets or circular bands, as in the wall of the intestine, and a multiunit type composed of individual fibers, as in the intrinsic muscles of the eye and in the walls of blood vessels. Most of the smooth muscle is of the mass-unit type, which includes a number of involuntary sphincter muscles such as the pyloric sphincter at the distal end of the stomach.

The thin flat sheets of muscle or circular bands of muscle are composed of spindle-shaped cells of microscopic size, commonly ranging from around 6 microns in diameter to 50 to 100 microns in length, although they may stretch to greater lengths. The cells usually lie in a longitudinal position with the slender ends fitted in against the thick middle portions of adjacent cells (Figure 6.15). Some thin strands of connective tissue lie between the cells, but the membranes are in close association and this may have some functional significance. Specialized areas in the membranes between adjacent cells have been observed which may permit an action potential to progress from one cell to another. In this way many cells may be stimulated to contract in unison. The cells receiving the initial stimulus are considered to be the pacemaker cells.

Smooth-muscle cells of this type, therefore, form a functional syncytium, though probably not a true syncytium, which would require a confluence of protoplasm between the cells. This is the kind of muscle found in the walls of the stomach, intestine, uterus, ureters, and urinary bladder—in other words, in the hollow viscera—and is therefore called *visceral muscle*. It is capable of rhythmic waves of contraction and relaxation, referred to as *peristalsis*. There are no specialized nerve endings equivalent to the motor end plates of skeletal muscle. Nerve action appears to be more regulatory in nature. Visceral muscle contracts and relaxes much more slowly than skeletal muscle; a single contraction may require a few seconds or even minutes.

Multiunit smooth muscle This type is found in the iris and ciliary muscles of the eye, the nictitating membrane of vertebrates, the walls of blood vessels, and the pilomotor muscles attached to hairs. The individual muscle fibers may be innervated by more than one nerve fiber. The muscle fibers are capable of more rapid contraction than is found in visceral muscle, and they do not exhibit peristalsis. A stimulus of threshold level is followed by membrane depolarization, a rather long latent period, and then a fairly rapid contraction. In pilomotor muscles, the fibers are arranged longitudinally to form a very small muscle in the skin attached to the base of the hair. These are the arrector muscles that cause the hair to stand on end as a reaction to fear or cold.

Chemical components The contractile elements of smooth muscle are not well understood. Actin, myosin, troponin, and tropomyosin can be extracted, indicating that there is evidently some relationship to the chemistry of

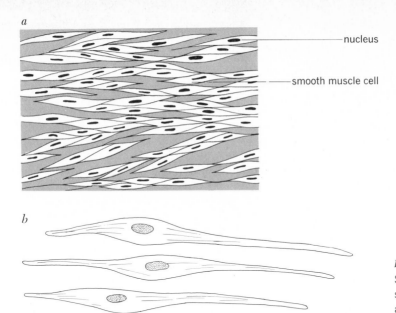

a

nucleus

smooth muscle cell

b

FIGURE 6.15
Smooth-muscle tissue showing long spindle-shaped cells: *a* cells moderately contracted; *b* cells relaxed.

contraction exhibited by skeletal muscle. However, there is no evident sarcoplasmic reticulum or system of transverse tubules. Some thin actin filaments appear to be present, and recently some filaments resembling myosin filaments have been revealed in certain vertebrates. It would be difficult, however, to explain contraction in smooth muscle on the same basis as the well-known methods of contraction in skeletal muscle. Calcium ions probably are essential to initiating contraction, as in skeletal muscle, possibly being released from binding sites in or on the cell membrane.

The resting potential of smooth-muscle cells is considered to be around 60 millivolts, the inside of the cell membrane being negative and the outside positive. Slow depolarizations occur in the pacemaker cells, followed by rhythmic contraction of the whole muscle.

Tonus in smooth muscle Smooth muscle, unlike skeletal muscle, is able to maintain a state of mild contraction or tone for considerable periods of time even though experimentally isolated from its nerve supply. An earlier hypothesis that all skeletal muscle exhibits tone even when it is relaxed is apparently in error. Tracings of action potentials show no electrical activity in relaxed skeletal muscle. Good tone in visceral muscle is, of course, very important in maintaining a good state of health.

Plasticity The visceral muscles of hollow organs such as the stomach, urinary bladder, and uterus maintain tension even though subjected to considerable extension. In this way, the urinary bladder may be greatly extended but still be capable of voiding urine because the muscles retain tension even though stretched. The muscles still exert tension even when greatly shortened, and one is capable of voiding urine when the bladder is nearly empty. The ability of smooth muscle to shorten or stretch and still maintain effective tension is an example of its plasticity.

Autonomic innervation, neurohumors, and hormones The activity of smooth muscle is exceedingly complex, since it is under the regulatory control of the

The physiology of muscle

autonomic nervous system, certain neurohumors, and hormones. The autonomic system is composed of two divisions: a sympathetic, or thoracolumbar, division and a parasympathetic, or craniosacral, division. The sympathetic division consists of a chain of ganglia and nerves on either side of the spinal cord through the thoracic and lumbar regions, whereas the parasympathetic division is associated with certain cranial and sacral nerves (see Chapter 10 on the autonomic nervous system). Adrenergic postganglionic fibers of the sympathetic nervous system release *norepinephrine* as a transmitter substance. It is assumed that there are reactive sites on the muscle-cell membrane. Norepinephrine and *epinephrine* (Adrenalin) are classified chemically as catecholamines. Epinephrine, a hormone released by the adrenal glands, is blood-borne and has a pronounced effect on the smooth muscles of blood vessels (see Chapter 20 on the endocrine system).

The role of adenosine 3′,5′-monophosphate (cyclic AMP) is not well known as yet. It is formed from ATP in a reaction catalyzed by the enzyme adenyl cyclase. It is suggested that it acts as a beta-adrenergic receptor, and it has been characterized as a "second messenger" in adrenergic reactions. In smooth muscle, it may function to release calcium ions, but this has not been proved.

The neurotransmitter released by the parasympathetic postganglionic fibers is *acetylcholine*, which ordinarily has an opposing effect to that produced by epinephrine. These effects are taken up in detail under the discussion of the autonomic nervous system.

CARDIAC MUSCLE   The muscular wall of the heart is composed of specialized cardiac muscle tissue, which is considered to be intermediate in structure between smooth and skeletal muscle. The tissue consists of a branching network of fibers, giving strength to the musculature. The cells are elongate with large nuclei. Heart muscle is finely striated and involuntary.

Cardiac muscle resembles skeletal muscle in possessing thin actin filaments and thick myosin filaments. Therefore it is considered to have a sliding filament type of contraction. Sarcomeres are plainly marked by prominent Z lines. The sarcoplasmic reticulum is evident, and transverse tubules invaginate from the surface membrane in the region of the Z lines.

Like visceral muscle, heart muscle contracts in a rhythmic manner, the action potential spreading from one cell to another through specialized areas of the membrane called "tight" junctions. At the ends of cardiac muscle cells are thickened areas commonly known as *intercalated disks*. Under electron magnification the intercalated disks appear as irregular structures affording firm support for the attachment of myofibrils. Dense, heavily pigmented regions of the disk resemble desmosomes (Figure 6.16).

Heart muscle has an inherent ability to contract rhythmically for all the years of life. Nerve impulses to the heart have a regulatory function, but heart muscle is not entirely dependent on its nerve supply. The heart of a frog or turtle may continue to beat even though removed from the body, and a small piece of cardiac tissue, with proper care, will continue to pulsate for some time. This automaticity results from a slow depolarization toward a threshold level which, when reached, is followed by contraction. During contraction, the muscle will not respond to additional stimuli for the duration of its long, absolute refractory period, which may last from 1 to 5 seconds. Therefore there is no summation and no complete tetanus is possible in cardiac muscle.

Following the initiation of contraction, a slow repolarization of the

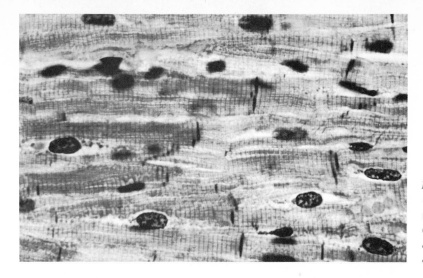

FIGURE 6.16
Cardiac tissue, showing striations, nuclei, and intercalated disks. (*Courtesy of General Biological Supply House, Inc., Chicago.*)

membrane begins, but the contraction phase is completed and the muscle is in its relaxing phase before the membrane is completely repolarized. The long refractory period provides time for the chambers of the heart to fill with blood before the next contraction phase starts.

Similarities in structure and function within the sarcomere would seem to indicate that basically the mechanism of contraction in skeletal muscle and that in cardiac muscle bear a close relationship. In cardiac muscle, action potentials are thought to cause the release of calcium ions, presumably from the sarcoplasmic reticulum, and it is likely that actin combines with myosin to release energy as in skeletal muscle. But cardiac muscle has not been studied as extensively as skeletal muscle, and there may be considerable difference in its mode of action. There are numerous large mitochondria or sarcosomes in cardiac tissue, and this would be consistent with its need for ATP as a source of energy. Myoglobin also is present. Heart muscle requires an abundant supply of oxygen. The action of the heart will be considered further in Chapter 14 on the heart.

SUMMARY

There are three kinds of muscle tissue: smooth, cardiac, and skeletal. Smooth muscle is described as unitary, that is, contracting as a unit, or it may be multiunit, consisting of individual fibers, each with its own motor nerve ending. Smooth muscle is nonstriated and involuntary. The term *visceral muscle* is used for the smooth muscle found in the walls of hollow organs of the viscera, where it is capable of contracting in rhythmic waves, an action described as *peristalsis*. Such organs include the stomach, intestine, and urinary bladder. Multiunit smooth muscle is found in the intrinsic muscles of the eye, in the walls of blood vessels, and in the pilomotor muscles. All smooth muscle exhibits a mild state of contraction called *tonus*. The action of smooth muscle is regulated by the autonomic nervous system, neurohumors, and hormones.

Cardiac muscle is found in the muscular walls of the heart. It is striated and involuntary and contains *intercalated disks*. Heart muscle has developed a high degree of automaticity which enables it to pulsate rhythmically. Since it maintains a long refractory period, it does not respond to summation and there is no complete tetanus.

Skeletal muscle is striated and voluntary. It is composed of cylindrical muscle fibers, each of which is enclosed in a sheath called the *sarcolemma*. The protoplasm of a fiber is the sarcoplasm. Within the sarcoplasm are threadlike *myofibrils*. The myofibril is composed of still smaller filaments described as thin, actin filaments and thick, myosin filaments. Contraction occurs when the two sets of filaments slide over each other and produce a shortening of the fiber.

Two additional proteins are associated with actin in the thin filament. These are *tropomyosin* and *troponin*. They appear to be concerned with calcium ion sensitivity of the myofibril.

The myosin molecule has a globular head and a linear tail. It is composed of two subunits called *meromysins*. Light meromysin forms a part of the thick filament itself, while heavy meromysin forms the cross bridge and the globular head. Myosin molecules are placed so that they lie in opposite directions within the thick filament, that is, the heads point in one direction in one half and in the opposite direction in the other half of the filament. Cross bridges extend out and make contact with reactive sites on the thin filaments.

The *sarcoplasmic reticulum* is a part of the internal membrane system of muscle. There are also transverse tubules which develop from a direct infolding of the surface membrane. The transverse tubules apparently carry an electric signal to the myofibrils located deep inside the fiber.

When a nerve impulse reaches the *myoneural junction* or motor end plate in skeletal muscle, a transmitter substance, acetylcholine, is released and an end-plate potential develops. Depolarization of the muscle-fiber membrane then occurs, and an action potential spreads over the fiber, resulting in contraction.

A *motor unit* innervates groups of muscle fibers, enabling these fibers to contract as a unit. The immediate source of energy in contraction of muscle is the breakdown of ATP. Creatine phosphate is another high-energy phosphate compound capable of releasing energy. Much of this energy is used in the resynthesis of ATP. Glycolysis, through several steps, results in the formation of pyruvic acid. This is an anaerobic phase and releases a small amount of energy.

Myoglobin is a muscle protein capable of holding oxygen in loose chemical combination. Vigorously contracting muscle may utilize this store of oxygen.

Muscle contraction is characterized as isotonic or isometric. In isotonic contraction, the muscle thickens and shortens. This is the type of contraction involved in body movement. In isometric contraction, the tension increases with the load, but the muscle does not shorten. The muscles concerned with maintaining posture often exhibit this type of contraction.

ᵉˢTIONS

1 Discuss the structure and function of smooth muscle, indicating differences between the two types.
2 Why is it that a state of complete tetanus does not occur in cardiac muscle?
3 Describe the sarcoplasmic reticulum. Suggest a possible function for transverse tubules.
4 How are the myosin cross bridges constructed?
5 What is a motor unit?
6 Make a drawing similar to the curve produced on a kymograph record, indicating the response of muscle to single stimuli of constant strength. Label and explain. About how long is the latent period in frog muscle?

7 What is meant by threshold stimulus?

8 If subminimal stimuli of just below threshold level are repeated rapidly, the muscle may respond by contracting. Explain.

9 Experimentally, how could you obtain complete tetanus in isolated frog muscle?

10 Make a drawing to illustrate the tracing made on a kymograph by a muscle as it reaches a state of fatigue. Is it fully contracted? Is it completely relaxed?

11 Explain the chemical nature of contraction in skeletal muscle.

12 What is meant by oxygen debt?

13 Does shivering have any value? If so, what is gained?

14 Distinguish between isotonic and isometric contraction.

15 What is the advantage of having numerous mitochondria associated with muscle fibers?

16 What is the chemical difference between rigor and normal contraction?

SUGGESTED READING

ASHLEY, C. C.: Calcium and the Activation of Skeletal Muscle, *Endeavour,* **30**(109):18-25 (1971).

AXELSSON, J.: Catecholamine Functions, *Ann. Rev. Physiol.,* **33**:1-30 (1971).

BRECKENRIDGE, BRUCE McL.: Cyclic AMP and Drug Actions, *Ann. Rev. Pharmacol.,* **10**:19-34 (1970).

DeLANGE, R. J., R. G. KEMP, W. D. RILEY, R. A. COOPER, and E. G. KREBS: Activation of Skeletal Muscle Phosphorylase Kinase by ATP and Adenosine 3'5' monophosphate, *J. Biol. Chem.,* **243**:2200-2208 (1968).

GIBBS, C. L., W. F. H. M. MOMMAERTS, and N. V. RICCHIUTTI: Energetics of Cardiac Contraction, *J. Physiol.,* **191**:25-46 (1967).

HURWITZ, LEON, and AMIN SURIA: The Link between Agonist Action and Response in Smooth Muscle, *Ann. Rev. Pharmacol.,* **11**:303-326 (1971).

HUXLEY, H. E.: The Mechanism of Muscular Contraction, *Sci. Am.,* **213**:18-27 (1965).

————: The Mechanism of Muscular Contraction, *Science,* **164**:1356-1366 (1969).

KATZ, BERNARD: "Nerve, Muscle and Synapse," McGraw-Hill Book Company, New York. 1966.

MERTON, P. A.: How We Control the Contraction of Our Muscles, *Sci. Am.,* **226**:30-37 (1972).

PEACHY, L. D.: Muscle, *Ann. Rev. Physiol.,* **30**:201-440 (1968).

PORTER, KEITH R., and CLARA FRANZINI-ARMSTRONG: The Sarcoplasmic Reticulum, *Sci. Am.,* **212**:73-80 (1965).

RASMUSSEN, H., and A. TENENHOUSE: Cyclic Adenosine Monophosphate, Ca^{++} and Membranes, *Proc. Nat. Acad. Sci., USA,* **59**:1364-1370 (1968).

REEDY, M. K.: Cross-bridges and Periods in Muscle, *Am. Zool.,* **7**(3):465-481 (1967).

SANDOW, A.: Skeletal Muscle, *Ann. Rev. Physiol.,* **32**:87-138 (1970).

SUTHERLAND, E. W., and G. A. ROBINSON: The Role of Cyclic 3'5' AMP in Responses to Catecholamines and Other Hormones, *Pharmacol. Revs.,* **18**:145-161 (1966).

VAN DER KLOOT, W. G.: Membrane Depolarization and the Metabolism of Muscle, *Am. Zool.,* **7**:661-669 (1967).

VON EULER, U. S.: Pieces in the Puzzle, *Ann. Rev. Pharmacol.,* **11**:1-12 (1971).

WILKIE, D. R.: "Muscle," St. Martin's Press, Dunmore, Pa., 1968.

chapter 7
skeletal muscles

Muscles, as we have seen, contract and relax, and in the process, convert chemical energy into mechanical energy. A skeletal muscle may be considered as an organ of the muscular system. It has a nerve supply and an adequate distribution of blood vessels and lymphatic vessels. There is also considerable connective tissue in muscles. We should not overlook the very important functions of cardiac and smooth muscles, but in this chapter we shall consider only skeletal muscles.

A muscle is composed of bundles of microscopic fibers, each bundle separated by connective tissue. The word *fiber* is often used to denote whole bundles of fibers, which represent small strands within a large muscle. In this sense, fiber refers to the visible strands. Striated muscles account for a large portion of body weight. They constitute around 35 percent of the body weight in women and a little over 40 percent in men. Muscular tissue is red or pink, as one can observe in the raw meat at a butcher shop.

Over four hundred muscles comprise the organs of the muscular system of the human body. They must have connections with the circulatory and nervous systems in order to perform their functions. One of these functions is to enable the skeletal structure to move. Groups of muscles attached to bone by tendons apply force or power by their contraction. The attachment to bone may be a direct attachment, in which the connective tissue surrounding muscle fibers appears to fuse directly with the periosteum covering the bone, or it may be a tendinous attachment. *Tendons* are ordinarily long cords of white fibrous connective tissue; when flattened into a broad sheet, they are called *aponeuroses*. Each muscle is covered with a thin connective-tissue sheath called the *fascia*. There is often considerable blending of these connective tissues to strengthen the attachment.

Tendons are often protected by a connective-tissue covering called a *tendon sheath.* These sheaths may be lined with a serous membrane and consist of two layers with a synovial cavity between them. The cavity contains synovial fluid, a viscid fluid something like the white of an uncooked egg. Serous membranes are secreting membranes. When they form protective linings, sheaths, or sacs for bones and muscles, they are called *synovial membranes.* Synovial sheaths, for example, are present around the tendons of the flexor and extensor muscles of the fingers and toes.

Synovial bursae are sacs between tendons, between tendon and bone, or over joints to protect the movements of tendons and prevent friction. They too contain synovial fluid. There is a large bursa over the patella at the knee joint, and others are found at the shoulder, elbow, hip, heel, and under various muscles.

ORIGIN AND INSERTION Ordinarily the more stationary attachment of a muscle is called the *origin,* and the more movable attachment is called the *insertion.* In most cases the origin is on the more fixed part, and the insertion is on the more movable part. It is often the case, especially in the appendages, that the origin is proximal and the insertion distal.

LEVERS OF THE BODY   Muscles contract and force is applied to accomplish work through a system of levers. The bones of the skeleton are the levers, the joints are the fulcrums, and the contraction of muscles supplies the force so that a weight can be lifted or a part can be moved and overcome resistance. This concept does not apply equally well to all muscles, however.

Levers are of three classes, depending on the relative position of the fulcrum, the force applied, and the resistance to be overcome or balanced. We shall consider only levers of the third class, since this is the common type of leverage used in body mechanics. In levers of the third class, force is applied across a joint or between the weight and the fulcrum. A common example is the application of force from the biceps and brachialis muscles, which are located anteriorly in the forearm. The elbow is the fulcrum, and the weight is at the hand. The force of contraction is applied across the elbow joint in flexing the arm. In this type of leverage, force is sacrificed in order to obtain quickness of movement and motion over a greater area.

ACTION OF MUSCLES   Ordinarily muscles do not act singly, but in groups. Even the simplest act requires muscular coordination. Taking a step forward is an example of very complicated coordination, not only of the muscles of the legs and feet but of the trunk as well. Muscles are usually so placed that they operate as antagonistic groups. This means that, as one group of muscles contracts, the opposing group relaxes. A common example is the flexing of the arm at the elbow by the contraction of the biceps and brachialis muscles. These muscles are so placed that their attachment is across the elbow joint anteriorly. They are inserted in the bones of the forearm just distal to the elbow. The opposing muscle is the triceps, located on the posterior side of the arm; its tendon is inserted in the olecranon process of the ulna. When the triceps contracts, the biceps and brachialis relax, and the arm is extended at the elbow. Muscles that cause a part to bend at a joint are referred to as *flexors;* those muscles that cause a part of the body to be extended are called *extensors.* The antagonistic action of muscles requires reciprocal innervation; that is, as nervous impulses are directed to one group of muscles stimulating them to contract, the opposing group must be permitted to relax. Muscular movements aid circulation mechanically by literally squeezing blood and lymph through the vessels during exercise. A person may "freshen up" or "warm up" by muscular exercise if the circulation has become sluggish from remaining quiet for too long a time.

The muscles described on the following pages were selected largely because they illustrate some principle of contraction or leverage or because of the importance of their functions. Though it is difficult to say that one muscle is more important than another, certain ones are *prime movers* and offer better illustrations of muscle performance than others. For this reason, certain muscles may be described more completely and others mentioned only incidentally. It is not the purpose of this book to give undue emphasis to the exact origin and insertion of muscles. The reader is referred to one of the larger works on human anatomy for more detailed descriptions.

It is misleading in many cases to mention only the most evident function of a muscle, since movement depends upon the coordinated activity of a number of muscles. One function of the biceps brachii is the flexing of the lower arm upon the upper; but if opposing muscles are fixed, the biceps can be used to turn the hand to a palm-upward position. Since the heads of the biceps arise from the scapula, this muscle can also aid in drawing

Skeletal muscles

143

the upper arm forward or in pulling the shoulder forward if the arm is in a fixed position.

The muscles concerned with maintaining posture are good examples of *fixators*. They must maintain the body position against the pull of gravity. Muscles located around joints often act as fixators or stabilizers, thus freeing other muscles to act as prime movers. Muscles which aid the prime mover by contracting in unison with it are *synergists*.

Following are a number of terms, perhaps of somewhat limited application, which may prove useful. Movement of a part away from the midline of the body is *abduction*, whereas movement toward the body or toward the midline is *adduction*. Movement of the arm away from the body is abduction. Movement of the leg toward the midline is adduction. Rotating movements holding the arm or leg straight are *circumduction* when the arm or leg circumscribes a cone by this movement. Rotation of the hand so that the palm is upward is *supination;* moving the palm downward illustrates *pronation*. The foot can be turned inward so that the sole is toward the midline, a movement called *inversion. Eversion* is a much more limited movement, but the sole can be turned slightly outward.

The study of the muscular system is termed *myology*. This term, however, ordinarily refers to a study of skeletal muscles. *Kinesiology* is concerned with voluntary muscles and their action in man.

MUSCLES OF THE HEAD, FACE, AND NECK

Muscles of the head, face, and neck
epicranius
orbicularis oculi
orbicularis oris
levator palpebrae
  superioris
corrugator
zygomaticus
buccinator
masseter
temporalis
pterygoideus medialis
platysma
sternocleidomastoideus
splenius capitus

Facial muscles are of interest partly because their contractions alter facial expression. While the facial muscles may not present an especially attractive appearance (Figure 7.1), beauty being only skin deep, the mechanism of movement is interesting. The principal muscle of the scalp is the *epicranius,* or *occipitofrontalis, muscle.* It is composed of paired muscles at the back of the head that arise from tendinous attachments to the occipital bone and paired frontal muscles attached to the skin at the eyebrows. Between these two sets of muscles is a large epicranial aponeurosis, a fibrous membrane extending across the top of the cranium. The occipitalis tends to fix and tighten the aponeurosis. With the aponeurosis fixed, the frontalis wrinkles the forehead horizontally, raises the eyebrows, and aids in raising the eyelids. It plays a part in facial expressions denoting perplexity, surprise, horror, or visual concentration. When the lower attachment is fixed, the frontalis pulls the scalp forward.

The principal movements of the face center about the eyes and mouth, where there are circular, or sphincter, muscles. Around the eye is the *orbicularis oculi,* and about the oral cavity, or mouth, is the *orbicularis oris* (Figure 7.1). These are highly reflex, but voluntary, sphincter muscles.

The *orbicularis oculi* has numerous functions, since different parts of this muscle can be brought into play. When the medial and lateral fibers shorten, a wink or prolonged closure of the lid results; strong contraction of the entire muscle results in squinting or lowering of the eyebrows. The muscle that pulls the eyelid up is the *levator palpebrae superioris.* It lies within the orbit just under the roof and is inserted into the eyelid. A small muscle arising on the frontal bone above the nasal bones and inserted into the skin of the eyebrow is called the *corrugator.* Its contraction causes the forehead to become wrinkled perpendicularly, as in frowning. Its action is closely associated with that of the orbicularis oculi.

Skeletal muscles

144

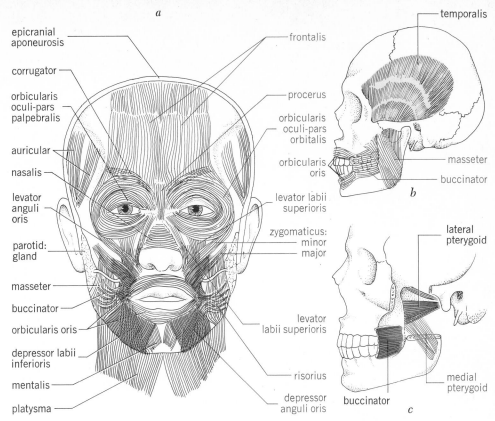

a
epicranial
aponeurosis
corrugator
orbicularis
oculi-pars
palpebralis
auricular
nasalis
levator
anguli
oris
parotid
gland
masseter
buccinator
orbicularis oris
depressor labii
inferioris
mentalis
platysma

frontalis
procerus
orbicularis
oculi-pars
orbitalis
orbicularis
oris
levator labii
superioris
zygomaticus:
minor
major
levator
labii superioris
risorius
depressor
anguli oris

temporalis
masseter
buccinator
b

lateral
pterygoid
medial
pterygoid
buccinator
c

FIGURE 7.1
a Muscles of the head and face
(anterior view), *b* and *c* muscles
of mastication (lateral view).

There are six muscles within the orbit, attached to the eyeball. They are concerned with moving the eyeball and will be discussed in connection with a study of the eye.

The lips, lower jaw, and tongue are other movable parts of the face. The sphincter muscle, the *orbicularis oris*, closes the mouth opening. Acting with other muscles the orbicularis oris may purse the lips, as in whistling, or the lower lip may be drawn upward, as in pouting. Several other muscles attach to the outer margins of this sphincter and serve to open the lips by their combined contraction. Among them is the *zygomaticus*, which extends from the zygomatic bone downward to the corner of the mouth, where the right and left muscles acting together pull the corners of the mouth upward into a smile. Contracting further, these muscles raise the cheek and lower eyelid, producing the little wrinkles at the outer corner of the eye so commonly associated with mirth and a kindly disposition.

The *buccinator muscle* has been called the trumpeter's muscle, because it draws the corners of the mouth back and flattens the lips against the teeth. The cheeks are kept taut by its contraction, an essential in the correct playing of a trumpet or bugle. More important is its function during chewing; it cooperates with muscles of the tongue in keeping food between the teeth. That this cooperation is effective is shown by the infrequency of biting either the tongue or the cheek.

The forcible movement of the lower jaw is upward for chewing food. Among the several muscles responsible for this action is the *masseter*, which

Muscles of the head, face, and neck

145

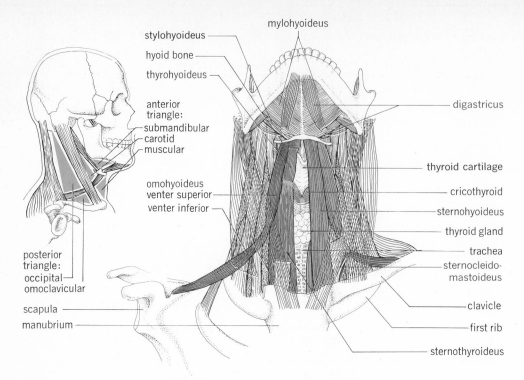

FIGURE 7.2

Superficial muscles of the neck.

extends downward from the zygomatic bone to the angle of the lower jaw, where it attaches to the outer surface. The muscle can be observed in action if you watch someone chewing or can be felt if you place a hand over the angle of the jaw and close the jaws tightly. Cooperating with the masseter in elevating the jaw are the *temporalis* and *pterygoideus medialis* (*internal pterygoid*) muscles (Figure 7.1*b* and *c*).

Several muscles that move the hyoid bone are listed in Table 7.2 (see also Figure 7.2).

The downward sag of the mouth in grief and exhaustion is caused by the contraction of the *platysma,* which inserts from below on the corners of the mouth. This is an exceedingly thin superficial muscle arising over a wide area at about the level of the second rib. It forms a thin sheet of muscle anterolaterally over the neck and is inserted partly into the connective tissue over the mandible and partly into the skin of the cheek and at the corner of the mouth. In extreme exertion, as in the case of a runner at a track meet, the platysma is tense, the lips are set, and the skin of the neck is wrinkled vertically. Since the platysma is attached to the mandible, it may aid in depressing the jaw, particularly in yawning (Table 7.1).

Several muscles are concerned with movements of the tongue. They are paired muscles and include four pairs of muscles within the tongue, the *intrinsic muscles,* and three pairs of muscles having their origin outside and below the tongue, the *extrinsic muscles.* These muscles are responsible for the exceedingly intricate movements of the tongue during mastication and swallowing and for the highly coordinated movements of the tongue and lips necessary for speech.

The *sternocleidomastoideus,* as indicated by its name, arises by two heads, one from the top of the sternum (the manubrium) and the other from the clavicle (*cleido-* is a combining form pertaining to the clavicle) (Figure 7.2). The muscle is inserted by a short tendon on the mastoid process of

TABLE 7.1

Principal muscles of the head and face, including muscles of mastication

Muscle	Origin	Insertion	Action	Nerve
Buccinator	Alveolar processes of mandible and maxilla	Orbicularis oris	Compresses cheek, retracts angle of mouth	Facial
Corrugator supercilii	Arch of frontal bone	Skin of forehead	Draws eyebrows toward each other	Facial
Depressor labii inferioris	Mandible near mental foramen	Lower lip	Draws lower lip down	Facial
Epicranius				
Frontalis	Epicranial aponeurosis	Skin of forehead	Elevates eyebrows, wrinkles skin of forehead horizontally	Facial
Occipitalis	Superior nuchal line of occiput	Epicranial aponeurosis	Fixes and tightens aponeurosis	Facial
Levator anguli oris	Maxilla	Angle of mouth	Raises angle of mouth	Facial
Levator labii superioris	Zygomatic arch	Upper lip and border of nostril	Raises lip and dilates nostril	Facial
Levator palpebri superioris	Roof of orbit	Skin of upper eyelid	Raises upper eyelid	Oculomotor
Mentalis	Incisor fossa of mandible	Skin of chin	Protrudes lower lip, wrinkles chin	Facial
Orbicularis oculi	Medial aspect of orbit	Skin around eyes	Closes eyelids	Facial
Orbicularis oris	Muscles about mouth	In muscles surrounding mouth	Closes and purses the lips	Facial
Platysma	Fasciae of upper thorax	Mandible and skin at angle of mouth	Depresses mandible and corners of mouth	Facial
Procerus	Nasal bone and nasal cartilage	Skin above nose	Wrinkles skin over bridge of nose	Facial
Risorius	Fasciae over masseter muscle	Skin at corner of mouth	Raises corner of mouth and draws it back	Facial
Zygomaticus major and minor	Zygoma	Skin at angle of mouth	Raises corner of mouth upward and backward	Facial
Muscles of mastication				
Masseter	Zygomatic process and zygomatic arch	Coronoid process, angle and ramus of mandible	Raises jaw and holds it tight	Trigeminal, mandibular branch
Pterygoideus lateralis	Greater wing of sphenoid and lateral pterygoid plate	Condyloid process of mandible	Protrudes and opens jaw, moves it from side to side	Trigeminal, mandibular branch
Pterygoideus medialis	Maxilla and pterygoid fossa	Ramus and angle of mandible	Aids in closing jaw	Trigeminal, mandibular branch
Temporalis	Temporal fossa and fascia	Coronoid process of mandible	Raises and retracts jaw	Trigeminal, mandibular branch

TABLE 7.2

Muscles of the neck, the anterior triangle

Muscle	Origin	Insertion	Action	Nerve
Digastricus				
Venter anterior	Mandible	Hyoid bone	Elevates hyoid, depresses mandible	Trigeminal
Venter posterior	Mastoid notch	Hyoid, greater horn	Draws hyoid back	Facial
Mylohyoideus	Inner surface of mandible	Hyoid bone	Elevates hyoid, base of tongue, and floor of mouth; depresses mandible	Trigeminal
Omohyoideus, venter superior, venter inferior	Body of hyoid	Suprascapular notch	Depresses hyoid	Ansa cervicalis
Sternocleidomastoideus	Upper portion of manubrium and medial third of clavicle	Mastoid process	Draws head toward shoulder on same side; acting together they flex the neck and extend the head	Cervical 2 and 3; spinal accessory XI
Sternohyoideus	Posterior aspect of manubrium	Body of hyoid	Depresses hyoid	Ansa cervicalis
Sternothyroideus	Posterior aspect of manubrium	Thyroid cartilage	Depresses larynx	Ansa cervicalis
Thyrohyoideus	Thyroid cartilage	Greater horn of hyoid	Depresses hyoid, elevates larynx	Ansa cervicalis

the temporal bone. The right and left muscles form the anterior triangle of the neck. Acting individually, they rotate the head toward the opposite side. They are capable of bending the head forward when acting together, but only against increased resistance, as when one is lying on his back and wishes to raise his head. This is the muscle commonly affected when one is said to have a wryneck or a stiff neck. The trapezius muscle may be affected also (Table 7.2).

The *splenius capitus* is one of a group of deeper, dorsal, cervical muscles. Arising from the dorsal spinal processes of the first to fourth thoracic vertebrae and from the seventh cervical vertebra, it is inserted on the occipital bone and on the mastoid process of the temporal bone underneath the insertion of the sternocleidomastoid muscle. When right and left muscles act in unison, the head is pulled back (extended). When they act singly, the head is rotated or inclined toward the muscle attached on the right or left side. Since this is one of the deeper muscles, it is not illustrated.

ANTERIOR MUSCLES OF THE SHOULDER GIRDLE AND MUSCLES OF THE THORAX

Muscles of the thorax
  pectoralis major
  pectoralis minor
  serratus anterior
  external intercostals
  internal intercostals
  diaphragm

The *pectoralis major* (Plate B and Figure 7.3) is the large triangular muscle covering the upper part of the chest anteriorly. It arises from attachments to the clavicle, the sternum, and the cartilages of the upper six ribs. The lower part of the muscle arises from the aponeurosis of the external oblique muscle. From this broad area of attachment the muscle fibers converge and are inserted by a short flat tendon into the lateral margin of the intertubercular groove of the humerus. When the arm is raised, the pectoralis

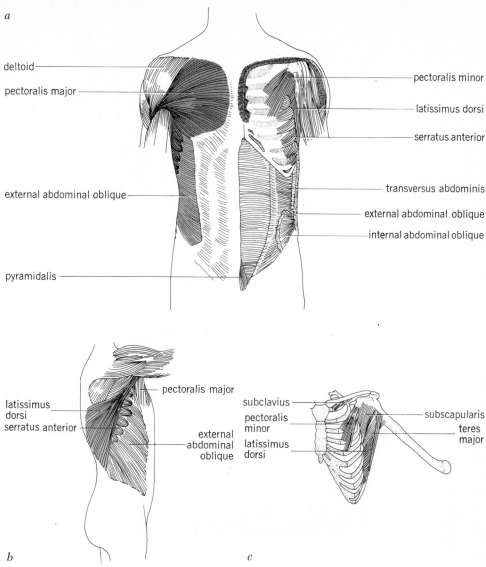

a

deltoid

pectoralis major

external abdominal oblique

pyramidalis

pectoralis minor

latissimus dorsi

serratus anterior

transversus abdominis

external abdominal oblique

internal abdominal oblique

latissimus dorsi
serratus anterior

pectoralis major

external abdominal oblique

subclavius

pectoralis minor

latissimus dorsi

subscapularis

teres major

b

c

FIGURE 7.3

Muscles of the pectoral region and anterior abdominal wall.

major, in cooperation with other muscles, pulls the arm down toward the chest. It also draws the arm across the chest and rotates it inward. With the arm fixed, as in climbing, it pulls the chest upward. It is therefore of value in forced inspiration or in artificial respiration.

The *pectoralis minor* lies directly underneath the pectoralis major. It is flat, thin, and more slender than the pectoralis major. It arises from the upper borders of the third, fourth, and fifth ribs and from their outer surfaces close to the costal cartilages. The fiber bundles extend diagonally upward to insert on the coracoid process of the scapula. With the scapula fixed, the pectoralis minor elevates the ribs, aiding inspiration, especially a forced inspiration. If the ribs are fixed, it pulls downward and forward on the scapula (Figure 7.3a and c) (Table 7.3).

Anterior muscles of the shoulder girdle and muscles of the thorax

149

TABLE 7.3

Muscles of the pectoral region

Muscle	Origin	Insertion	Action	Nerve
Pectoralis major	Sternal half of clavicle, sternum, costal cartilages, aponeurosis of external abdominal oblique	Crest of greater tubercle of humerus	Flexes, abducts, and medially rotates humerus, draws body upward in climbing	Lateral and medial pectoral
Pectoralis minor	Anterior surface of ribs 3–5	Coracoid process of scapula	Draws scapula down and forward and elevates ribs	Medial pectoral
Subclavius	Cartilage of first rib	Inferior surface of clavicle	Draws clavicle down and forward	Nerve to subclavius

The *serratus anterior* lies beneath the pectoralis major and pectoralis minor. It arises from the first to ninth ribs by slender digitations and is inserted on the inner surface of the scapula at the vertebral border. Its contraction moves the scapula forward and downward, rotating it somewhat as the arm is abducted (Figure 7.3*a* and *b*) (Table 7.4).

The intercostal muscles are located between the ribs and consist of two muscles that fill in each intercostal space. The *external intercostal* is the thicker, outer sheet of muscle, and the *internal intercostal* forms the thinner, inner sheet. There are 44 of these muscles occupying 11 intercostal spaces on each side. The external intercostal muscles extend from the posterior tubercles of the ribs to the anterior costal cartilages. The fibers run diagonally from the lower border of one rib to the upper border of the next adjacent rib. They pull adjacent ribs toward each other and so elevate the ribs, increasing the volume of the thoracic cavity, as in inspiration.

The internal intercostal muscles extend posteriorly from the sternum to the angle of the ribs. The fibers run downward diagonally but in a direction opposite to that of the external intercostal fibers. When the lowest pair of ribs is held firmly by the *quadratus lumborum* muscles (see Figure 7.19), the internal intercostal muscles depress the ribs, decreasing the volume of the thoracic cavity, as in expiration (Figure 7.4).

The *diaphragm* (Figure 7.5) is a dome-shaped muscle that forms a wall between the thoracic and abdominal cavities. It is attached around the lower circumference of the thorax to the inner surfaces of the xiphoid process and the lower six costal cartilages and also by tendinous and muscular slips to the lateral surfaces of the upper lumbar vertebrae. The muscle fibers converge upward and are inserted into an aponeurosis called the *central tendon*. As the muscles contract, the diaphragm loses some of its convexity, and the capacity of the thoracic cavity is increased. Supported by other muscles, which act on the ribs, inspiration is accomplished by the elevation of the ribs and the lowering of the diaphragm. To accomplish a forced expiration, as in coughing, sneezing, laughing, or crying, a deep inspiration must precede the act. Muscles that depress the ribs and others that act to compress the abdomen then support a forcible expulsion. The diaphragm also aids in such acts as vomiting, defecating, micturating and in childbirth.

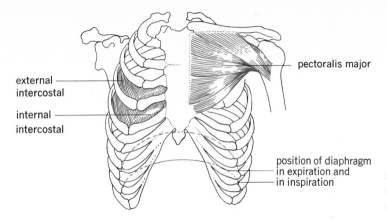

external intercostal

internal intercostal

pectoralis major

position of diaphragm in expiration and in inspiration

FIGURE 7.4
Intercostal muscles and the pectoralis major.

The diaphragm arches over the liver on the right side of the body and over the stomach on the left. It has three large openings: The vena cava passes through one opening in the right half of the central tendon; the esophageal opening located posterior to the central tendon on the left transmits the esophagus and branches of the vagus nerves; and a more posterior opening permits the passage of the aorta and the thoracic duct.

MUSCLES OF THE BACK

The right and left *trapezius muscles* are roughly trapezoid in shape. They arise from a flat aponeurosis extending down the dorsal midline from the occipital bone; they also attach to dorsal vertebral spines as far down as the twelfth thoracic. The upper part of the muscle is inserted in the clavicle; the median and lower parts insert on the acromion process and spine of the scapula. This is a large muscle, placed over other muscles of the shoulder. Contraction of the trapezius pulls the scapula toward the vertebral column and upward. It aids in lifting with the arms or in carrying

Muscles of the back
trapezius
latissimus dorsi
erector spinae
(sacrospinalis)

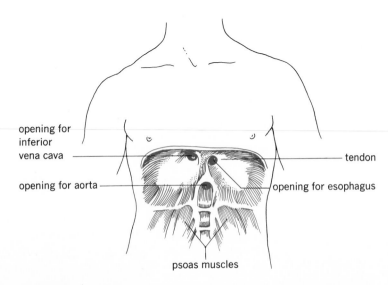

opening for inferior vena cava

tendon

opening for aorta

opening for esophagus

psoas muscles

FIGURE 7.5
Anterior view of the diaphragm.

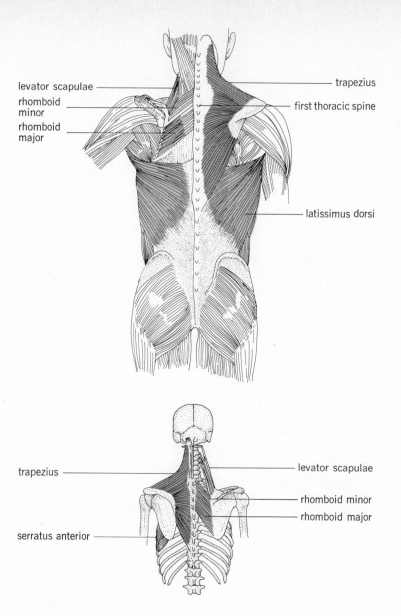

levator scapulae

rhomboid minor

rhomboid major

trapezius

first thoracic spine

latissimus dorsi

trapezius

serratus anterior

levator scapulae

rhomboid minor

rhomboid major

FIGURE 7.6
Superficial muscles of the back.

Skeletal muscles

a load on the shoulders, since it braces the shoulders. It also plays a part in shrugging the shoulders (Figure 7.6 and Plate C).

Other muscles such as the splenial muscles are attached at the back of the head and to the upper vertebrae. They aid in holding up and extending the head.

The *latissimus dorsi muscle* has a very extensive origin—from an aponeurosis attached to the spinous processes of the lower six thoracic vertebrae and all the lumbar vertebrae and from the dorsal surface of the sacrum and the posterior part of the crest of the ilium. There is also some attachment to the lower three or four ribs. From this broad area of origin the muscle tapers upward to a tendinous insertion in the intertubercular groove of the humerus. The action of the latissimus dorsi is to draw the arm downward and backward when the trunk is fixed. It also rotates the arm

TABLE 7.4

Superficial muscles of the back

Muscle	Origin	Insertion	Action	Nerve
Latissimus dorsi	Spinous processes of lower 6 thoracic vertebrae, thoracolumbar fasciae, crest of ilium	Interubercular groove of humerus	Adducts, extends, rotates arm medially, draws shoulder down and backward	Thoracodorsal
Levator scapulae	Transverse processes of cervical 1–4	Medial border above spine of scapula	Draws scapula medially and depresses shoulder	Dorsal scapular
Rhomboideus major	Spinous process of thoracic 2–5 and supraspinous ligament	Medial border below spine of scapula	Adducts and laterally rotates scapula	Dorsal scapular
Rhomboideus minor	Spinous processes of last cervical and 1st thoracic vertebrae	Medial margin of scapula at origin of spine	Adducts and laterally rotates scapula	Dorsal scapular
Serratus anterior	Lateral surface of upper 8 ribs	Ventral surface of medial border of scapula	Draws the scapula forward, rotates scapula as in abduction of arm	Long thoracic
Trapezius	External occipital protuberance, superior nuchal line, nuchal ligament, and spinous process of cervical 7–12	Anterior border of scapular spine, acromion process, lateral third of posterior clavicle	Adducts, rotates, and elevates scapula	Spinal accessory XI and cervical 3, 4

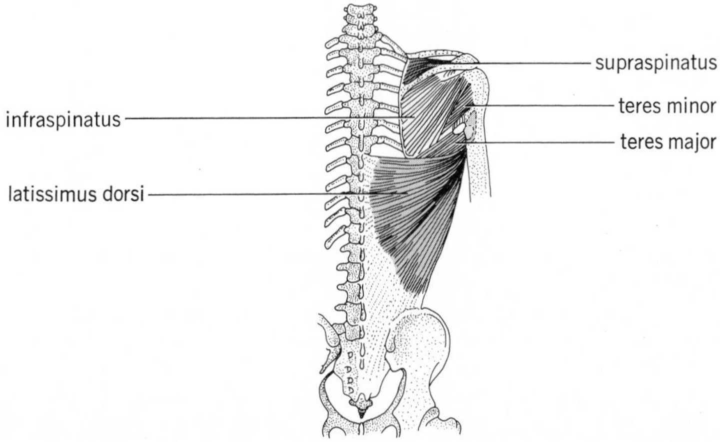

FIGURE 7.7
The latissimus dorsi and deeper muscles of the shoulder.

inward. This is one of the principal muscles involved in a swimming stroke or in bringing the arm forcibly downward, as in striking a blow. When both arms are fixed above the head, the contraction of the latissimus dorsi draws the trunk forward; it is used thus in climbing. This muscle is illustrated in Figure 7.7 (Table 7.4).

The muscles of the back are usually considered as arranged in five layers. The trapezius and the latissimus dorsi are superficial muscles of the first, or outer, layer. We shall consider only one of the muscles from the deeper layers. This is the erector spinae.

The *erector spinae (sacrospinalis)* is a long, deep muscle attached by a strong aponeurosis to the posterior part of the iliac crest, the posterior surface of the sacrum, the spines of all the lumbar vertebrae, and the eleventh and twelfth thoracic vertebrae. Tracing the muscle anteriorly, we find that it divides into several long muscles, which are inserted on the ribs and upper vertebrae at various levels so that there is a consistent arrangement whereby one set of fibers has its origin posterior to the insertion of an adjacent set. With such a system of overlapping fibers, the erector spinae can exert a strong pull dorsad on the vertebral column. This is an important posture muscle, since when both sides act together it serves as an extensor for the spinal column. When acting on one side only, it pulls the spinal column toward that side. In advanced pregnancy the erector spinae aids in balancing the weight by bending the spinal column backward. This is a deep muscle of the back and is not illustrated.

MUSCLES OF THE SHOULDER AND ARM

Muscles of the shoulder and arm
deltoideus
biceps brachii
brachialis
coracobrachialis
brachioradialis
triceps brachii

The *deltoideus* (Figure 7.8b, c) is a short, thick muscle located above the articulation of the arm at the shoulder. It arises from the distal third of the clavicle, the acromion process, and spine of the scapula and is inserted by a stout tendon into the deltoid tuberosity of the humerus. Its contraction raises the arm to a horizontal position or somewhat higher. When the arm is so raised, the contracted muscle can be felt as a hard bunch of muscle above the shoulder joint. There are five other muscles of the shoulder. The reader is cautioned that the muscle selected as an example is not the only muscle of importance in the group.

The *biceps brachii* (Figure 7.9) is a well-known muscle of the anterior part of the arm. It arises from two points of attachment. The short head arises from the coracoid process of the scapula. The long head has its origin on the supraglenoid tuberosity, a bony process of the scapula just above the articulation of the humerus at the shoulder. The tendon of the long head passes over the head of the humerus and into the intertubercular groove; the tendon is protected by the synovial membrane of the joint. The two heads unite about 7 centimeters above the elbow joint and form the body of the biceps brachii. The muscle is inserted on the tuberosity of the radius. Its action is to flex the arm at the elbow and to rotate the forearm outward, a movement called *supination*. The movement of the muscle in supination is most evident when the arm is flexed. When the forearm is fixed, as in climbing or in chinning oneself on a horizontal bar, the body is pulled upward by the flexing of the arm.

The *brachialis* assists the biceps in flexing the arm at the elbow. It arises from the anterior surface of the lower three-fifths of the humerus, passes anteriorly across the elbow joint, and is inserted into the tuberosity of the ulna. It is purely a flexor of the forearm (Figure 7.9).

The *coracobrachialis* has its principal origin on the tip of the coracoid process of the scapula and is inserted on the medial surface of the humerus somewhat proximal to the middle of the shaft. It is a relatively small muscle, aiding in flexing and adducting the humerus (Figure 7.9).

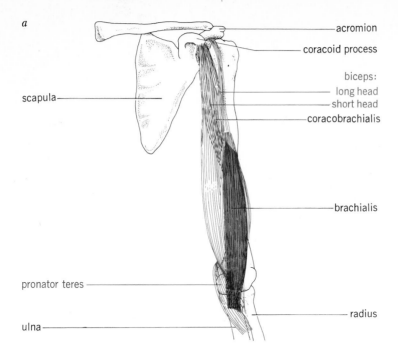

a

acromion
coracoid process
biceps:
long head
short head
coracobrachialis

scapula

brachialis

pronator teres

radius

ulna

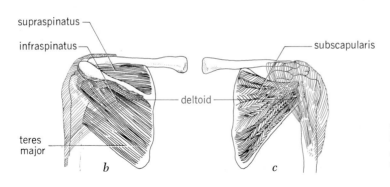

supraspinatus
infraspinatus

subscapularis

deltoid

teres major

b

c

FIGURE 7.8
Muscles of the arm and shoulder:
a and *c* anterior view; *b* posterior
view.

The *brachioradialis* arises from the lateral ridge at the distal third of the humerus and is inserted by a long, flat tendon laterally on the radius at the base of the styloid process. A superficial muscle located above the radius on the forearm, it is primarily a flexor of the forearm but also aids in rotating it (Plate B). Although it is a forearm muscle it is classified here with other arm muscles which are also flexors of the forearm (see Figure 7.11c).

The *triceps brachii* is the principal extensor of the forearm. It arises from three heads; the long head arises from the scapula just below the glenoid fossa, the lateral head from the posterior side of the shaft of the humerus, proximally, and the medial head from a fleshy attachment on the posterior side of the humerus distally from the radial groove. The three heads unite to form a large muscle on the posterior side of the arm. The insertion is by means of a strong tendon on the olecranon process of the ulna. This muscle

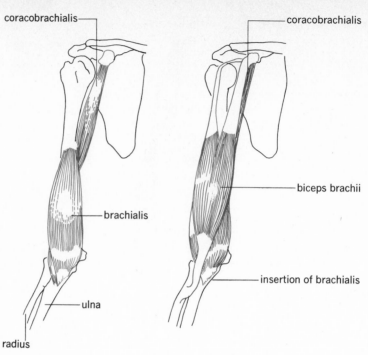

coracobrachialis

coracobrachialis

brachialis

biceps brachii

insertion of brachialis

ulna

radius

FIGURE 7.9
Superficial muscles of the arm
(anterior view).

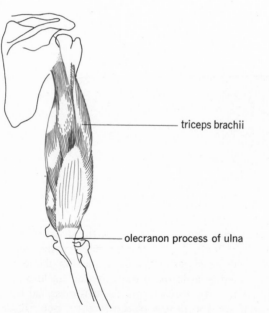

triceps brachii

olecranon process of ulna

FIGURE 7.10
The triceps brachii.
(posterior view)

is the antagonist of the biceps and brachialis and is the extensor of the arm
at the elbow (Figure 7.10) (Table 7.5).

MUSCLES OF THE FOREARM AND HAND In the forearm are a number of muscles
that are classified as flexors and extensors of the wrist and hand. They have
their origin on the humerus and on the radius and ulna. They are inserted
by means of long tendons on bones of the wrist, hand, and fingers. The flexors
are on the anterior side of the arm when viewed in the usual anatomical

position. The extensors are on the posterior side of the arm; some of their tendons are evident over the back of the hand if the fingers are extended to the greatest degree. There are also muscles of the hand that enable one to spread the fingers apart or bring them together. Some of these muscles make it possible to flex and extend the distal joints of the fingers. Other muscles function in moving the thumb (Plates B and C).

ANTERIOR MUSCLES OF THE FOREARM Among the superficial muscles on the anterior aspect of the forearm are the *pronator teres, flexor carpi radialis, palmaris longus,* and the *flexor carpi ulnaris*. This is a group of flexors and

Anterior muscles of the forearm
pronator teres
flexor carpi radialis
palmaris longus
flexor carpi ulnaris

TABLE 7.5
Muscles of the shoulder and arm

Muscle	Origin	Insertion	Action	Nerve
Anconeus	Lateral epicondyle of humerus	Olecranon process, posterior surface of ulna	Weak extensor of arm	Radial
Biceps brachii	Long head, supraglenoid tubercle; short head, coracoid process of scapula	Tuberosity of radius and aponeurosis	Flexes forearm and arm, and supinates hand	Musculocutaneous
Brachialis	Distal two-thirds of front of humerus	Coronoid process of ulna	Flexes forearm	Musculocutaneous and radial
Coracobra-chialis	Coracoid process of scapula	Middle third of humerus	Flexes and adducts arm	Musculocutaneous
Deltoid	Anterior surface, lateral clavicle, acromion process and spine of scapula	Deltoid tubercle of humerus	Abducts arm; aids in flexion, extension, and adduction	Circumflex
Infraspinatus	Medial part of infra-spinous fossa	Middle portion of greater tubercle of humerus	Rotates arm later-ally	Suprascapular
Subscapularis	Medial part of sub-scapular fossa	Lesser tubercle of humerus	Rotates arm medi-ally	Upper and lower subscapularis
Supra-spinatus	Medial part of supra-spinous fossa	Superior portion of greater tubercle of humerus	Abducts arm	Suprascapular
Teres major	Axillary margin of scapula	Intertubular groove of humerus	Adducts, extends, and rotates arm medially	Lower sub-scapular
Teres minor	Axillary margin of scapula	Greater tubercle of humerus	Laterally rotates arm and adducts it	Circumflex
Triceps brachii	Long head, infraglenoid tubercle; lateral head, proximal portion of shaft of humerus; medial head, distal half of shaft of humerus	Olecranon process of ulna	Extends arm and forearm	Radial

Muscles of the shoulder and arm

157

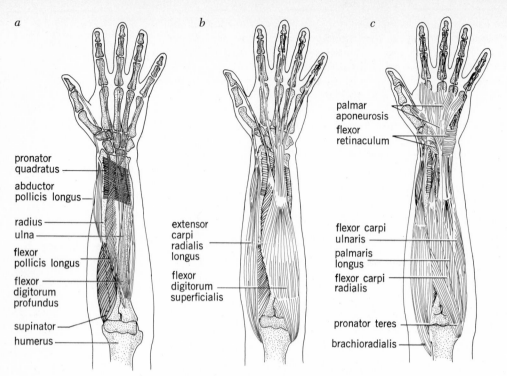

a   b   c

a (labels):
pronator quadratus

abductor pollicis longus

radius
ulna

flexor pollicis longus

flexor digitorum profundus

supinator
humerus

b (labels):
extensor carpi radialis longus

flexor digitorum superficialis

c (labels):
palmar aponeurosis
flexor retinaculum

flexor carpi ulnaris
palmaris longus
flexor carpi radialis

pronator teres

brachioradialis

FIGURE 7.11
Muscles of the anterior forearm.

pronators forming a large part of the musculature at the base of the forearm (Plate B). The pronator teres is a strong pronator of the hand. The flexor carpi radialis is located laterally alongside the pronator teres. It is essentially a flexor of the hand and a weak abductor toward the radius. The palmaris longus is a long slender muscle inserted in the palmar aponeurosis. It is a flexor of the hand and also functions to tighten the palmar aponeurosis. The flexor carpi ulnaris is inserted on the pisiform wrist bone and the palmar fascia. It is a flexor of the hand and an adductor toward the ulna (Figure 7.11a, b, c) (Table 7.6).

Posterior muscles of the forearm

anconeus

extensor digitorum (communis)

extensor digiti minimi

extensor carpi ulnaris

extensor carpi radialis longus

extensor carpi radialis brevis

POSTERIOR MUSCLES OF THE FOREARM There are several superficial muscles located on the posterior aspect of the forearm. These are the *anconeus, extensor digitorum (communis), extensor digiti minimi, extensor carpi ulnaris, extensor carpi radialis longus,* and the *extensor carpi radialis brevis* (Figure 7.12b) (Tables 7.5 and 7.7).

The specialization of the hand requires numerous muscles to provide all the intricate movements of which the hand is capable. As the names indicate, most of the above muscles are primarily extensors. The anconeus is a small triangular muscle located at the proximal end of the forearm (Plate C). It arises on the lateral epicondyle of the humerus and is inserted along the proximal part of the ulna and to the olecranon process of the humerus. It acts with the triceps brachii in extending the forearm.

The extensor digitorum (communis) is primarily an extensor of the fingers. The tendons of this muscle are evident on the back of the hand when the fingers are fully extended. The extensor digiti minimi is a slender muscle with its tendon attached at the base of the fifth digit. It extends the little finger and also aids as an abductor.

Skeletal muscles

158

TABLE 7.6

Muscles of the forearm, anterior view

Muscle	*Origin*	*Insertion*	*Action*	*Nerve*
Flexor carpi radialis	Medial epicondyle of humerus	Base of second metacarpal	Flexes wrist and aids in abducting it	Median
Flexor carpi ulnaris	Medial epicondyle of humerus, olecranon process, and medial border of ulna	Pisiform, hamate, and fifth meta-carpal	Flexes and adducts wrist	Ulna
Flexor digitorum profundus	Anterior and medial aspects of ulna and interosseous membrane	Distal phalanges of fingers	Flexes distal pha-langes and wrist	Median and ulnar
Flexor digitorum superficialis	Medial epicondyle of humerus and coro-noid process of ulna	2d phalanges of fingers	Flexes the phalanx of each finger	Median
Flexor pollicis longus	Middle half of radius, interosseous membrane, coronoid process of ulna	Distal phalanx of thumb	Flexes thumb	Median
Palmaris longus	Medial epicondyle of humerus	Palmar aponeurosis and flexor reti-naculum	Flexes hand	Median
Pronator quad-ratus	Distal portion of ulna	Distal portion of radius	Pronates hand	Median
Pronator teres	Medial epicondyle of humerus and coro-noid process of ulna	Lateral surface of radius	Pronates hand	Median

The extensor carpi ulnaris is located along the ulna in the forearm. It extends the hand at the wrist and aids in drawing the hand laterally toward the ulna. The extensor carpi radialis longus and the extensor carpi radialis brevis have essentially the same function. They extend and abduct the hand (Figure 7.12). (*Carpi* refers to the wrist.) Muscles located entirely within the hand are the lumbricales and the interrossei, which are discussed on pages 161 and 162 (Figure 7.13).

Somewhat deeper muscles of the forearm are the *supinator*, the *abductor pollicis longus*, the *extensor pollicis brevis*, and the *extensor pollicis longus*. The supinator turns the palm upward, as in supination. The abductor pollicis longus is located just beneath the supinator and is inserted on the first metacarpal. Its primary function is to abduct the thumb. The extensor pollicis brevis is closely allied to the abductor pollicis longus but is inserted on the proximal phalanx of the thumb, where it aids in extending this phalanx (Plate C). The extensor pollicis longus is inserted on the distal phalanx and extends the distal phalanx (Figure 7.14) (Table 7.7).

The thumb is well supplied with a number of short muscles which provide flexibility to this important digit. The thenar group forms the thenar eminence, a bulge of short muscles controlling movement of the thumb.

Deeper muscles of the forearm
supinator
abductor pollicis longus
extensor pollicis brevis
extensor pollicis longus

The thenar group
flexor pollicis brevis
abductor pollicis brevis
adductor pollicis
opponens pollicis

159

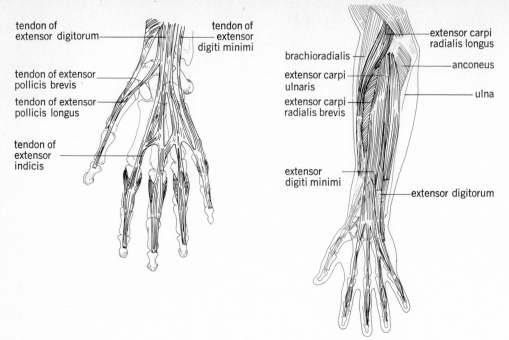

tendon of extensor digitorum

tendon of extensor digiti minimi

tendon of extensor pollicis brevis

tendon of extensor pollicis longus

tendon of extensor indicis

brachioradialis

extensor carpi ulnaris

extensor carpi radialis brevis

extensor carpi radialis longus

anconeus

ulna

extensor digiti minimi

extensor digitorum

FIGURE 7.12
Muscles and tendons of the anterolateral forearm (posterior view).

These muscles are the *flexor pollicis brevis*, the *abductor pollicis brevis*, the *adductor pollicis*, and the *opponens pollicis*. The thenar muscles are inserted on the proximal phalanx of the thumb. The primary function of these muscles is indicated by their names. The opponens pollicis enables the thumb to be brought in opposition to the fingers. When this is being done, the first metacarpal is rotated medially toward the palm (Figure 7.15*a, c*).

The use of the terms *abduction* and *adduction* may be confusing when they are applied to the musculature of the hand. In this case, these terms

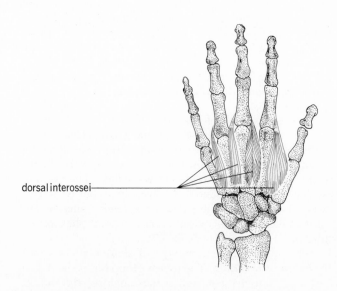

dorsal interossei

FIGURE 7.13
Deep muscles of the hand (posterior aspect); the dorsal interossei.

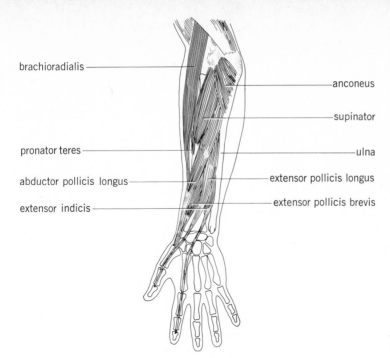

brachioradialis

anconeus

supinator

pronator teres

ulna

abductor pollicis longus

extensor pollicis longus

extensor indicis

extensor pollicis brevis

FIGURE 7.14
Deeper muscles of the forearm.

apply to the median line of the hand rather than to the median plane of the body. The median line is in the third finger.

MUSCLES OF THE POSTERIOR ASPECT OF THE FOREARM WHICH FLEX THE FINGERS An intermediate muscle, the *flexor digitorum superficialis,* and a deep muscle, the *flexor digitorum profundus,* provide strength for flexing the fingers. Arising from attachments on the humerus, radius, and ulna, the tendons of the flexor digitorum superficialis are inserted on the middle phalanges of all four fingers, The flexor digitorum profundus arises from the ulna and the interosseous membrane with its four tendons inserting on the distal phalanges of each finger. Since the position of this muscle is deeper than that of the flexor digitorum superficialis, the tendons of the latter divide to permit the passage of these tendons to the distal phalanges. The primary function is flexing the distal phalanges and aiding the flexing of the other finger joints (Figure 7.15*b*).

MUSCLES OF THE HAND The hypothenar eminence, located on the medial side of the palm proximal to the little finger, is composed of three short muscles. These are the *abductor digiti minimi,* primarily an abductor of the little finger, the *flexor digiti minimi brevis,* a flexor of the proximal phalanx of the fifth digit, and the *opponens digiti minimi,* which enables the little finger to oppose the thumb (Figure 7.15*a, c*).

There are two groups of muscles within the hand itself, the *lumbricales* and the *interosseous* muscles. The lumbricales are deep muscles of the palm composed of four small muscle slips which arise from the tendons of the flexor digitorum profundus muscle and are inserted by slender tendons into the tendons of the extensor digitorum muscle (Figure 7.15*b*). The lumbricales

Muscles of the posterior aspect of the forearm which flex the fingers
flexor digitorum superficialis
flexor digitorum profundus

Muscles of the hand
The hypothenar eminence
abductor digiti minimi
flexor digiti minimi brevis
opponens digiti minimi

lumbricales
interosseous
palmar
dorsal

TABLE 7.7

Muscles of the forearm and hand, posterior aspect

Muscle	Origin	Insertion	Action	Nerve
Abductor pollicis longus	Posterior surface of ulna and radius, and interosseous membrane	First metacarpal	Abducts thumb and wrist	Radial
Brachioradialis	Lateral supracondylar ridge and lateral intermuscular septum	Styloid process of radius	Flexes forearm	Radial
Extensor carpi radialis brevis	Lateral epicondyle of humerus	Third metacarpal	Extends and abducts wrist	Radial
Extensor carpi radialis longus	Lateral supracondylar ridge of humerus and lateral intermuscular septum	Second metacarpal	Extends and abducts hand	Radial
Extensor carpi ulnaris	Lateral epicondyle of humerus and posterior border of ulna	Fifth metacarpal	Extends and adducts hand	Radial
Extensor digiti minimi	Lateral epicondyle of humerus	Extensor expansion of little finger on dorsum of first phalanx	Extends little finger	Radial
Extensor digitorum	Lateral epicondyle of humerus	Into distal phalanx by 4 tendons	Extends fingers and wrist joint	Radial
Extensor indicis	Dorsal surface of ulna and interosseous membrane	Extensor tendon of index finger	Extends index finger	Radial
Extensor pollicis brevis	Middle third of radius and interosseous membrane	Proximal phalanx of thumb	Extends first phalanx of thumb and abducts hand	Radial
Extensor pollicis longus	Middle third of ulna and adjacent interosseous membrane	Distal phalanx of thumb	Extends distal phalanx of thumb and abducts hand	Radial
Interossei, dorsal	Adjacent sides of metacarpal bones	Proximal phalanges of digits 2–4	Abducts fingers from midline	Ulnar
Supinator	Lateral epicondyle of humerus, supinator crest of ulna	Proximal third of radius	Supinates hand	Radial

aid in flexing the fingers at the metacarpophalangeal joints while extending the interphalangeal joints. The interosseous muscles lie between the metacarpal bones. There are two sets, a palmar and a dorsal. The palmar set consists of three small muscles attached to the bases of all the fingers except the middle, or third, finger (Figure 7.16). Their principal action is to adduct the index, fourth, and little fingers toward the middle finger. They also aid in flexing the proximal phalanges of these fingers and in extending the middle and distal joints. The dorsal interosseous set is composed of four small muscles arising between the metacarpal bones. The tendons of these muscles are extended to the second, third, and fourth fingers. The middle finger has two tendons, one on either side. The dorsal interossei abduct the fingers involved, flex the proximal phalanges, and extend the middle and distal joints (Figure 7.13) (Table 7.8).

This discussion has considered some of the principal muscles of the arm and hand. The reader is referred to larger works on human anatomy for complete coverage.

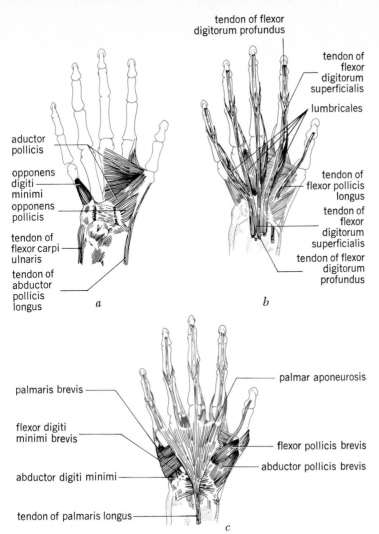

tendon of flexor
digitorum profundus

tendon of
flexor
digitorum
superficialis

lumbricales

aductor
pollicis

opponens
digiti
minimi
opponens
pollicis

tendon of
flexor carpi
ulnaris

tendon of
abductor
pollicis
longus

tendon of
flexor pollicis
longus

tendon of
flexor
digitorum
superficialis

tendon of flexor
digitorum
profundus

a

b

palmaris brevis

flexor digiti
minimi brevis

abductor digiti minimi

tendon of palmaris longus

palmar aponeurosis

flexor pollicis brevis

abductor pollicis brevis

c

FIGURE 7.15

Muscles of the palm of the hand.

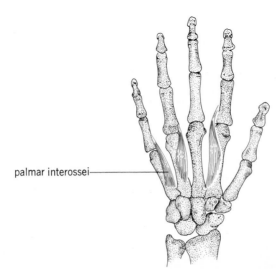

palmar interossei

FIGURE 7.16

The palmar interossei muscles.

TABLE 7.8
Muscles of the hand, palmar aspect

Muscle	Origin	Insertion	Action	Nerve
Abductor digiti minimi	Pisiform and tendon of flexor carpi ulnaris	First phalanx of little finger	Abducts little finger	Ulnar
Abductor pollicis brevis	Flexor retinaculum, scaphoid, and trapezium	First phalanx of thumb	Abducts thumb and aids in flexion	Median
Adductor pollicis	Capitate, second, and third metacarpals	First phalanx of thumb	Adducts thumb and aids in opposition	Ulnar
Flexor digiti minimi brevis	Flexor retinaculum and hook of hamate	First phalanx of little finger	Flexes little finger	Ulnar
Flexor pollicis brevis	Flexor retinaculum and trapezium	First phalanx of thumb	Flexes thumb	Median
Interossei, palmar	Palmar surface of metacarpals of digits 2, 4, and 5	Base of proximal phalanx and expanded extensor tendon of same finger as origin	Adducts digits 2, 4, and 5 toward the midline and aids in extension of fingers	Ulnar
Lumbricales	Tendons of flexor digitorum profundus	Extensor expansion distal to metacarpophalangeal joint	Flex metacarpophalangeal joints and extend interphalangeal joints	Lateral two by median, medial two by ulnar
Opponens digiti minimi	Flexor retinaculum and hook of hamate	Fifth metacarpal	Draws fifth metacarpal toward palm	Ulnar
Opponens pollicis	Flexor retinaculum and trapezium	Lateral border of first metacarpal	Opposes thumb to fingers and abducts, flexes, and rotates first metacarpal	Median
Palmaris brevis	Flexor retinaculum and palmar aponeurosis	Skin over medial border of palm	Deepens hollow of hand	Ulnar

MUSCLES OF THE ABDOMINAL WALL

*Muscles of the
  abdominal wall*
Anterolateral muscles
  external oblique
  internal oblique
  transversus abdominis
  rectus abdominis
Posterior muscles
  iliopsoas
  quadratus lumborum

The *external oblique* (Figure 7.17) is the most superficial of these muscles. It is a broad, thin sheet of muscle arising from the outer surfaces of the lower eight ribs. From these points of origin the fibers pass obliquely across the abdomen, the upper and middle portions of the muscle being inserted in a broad aponeurosis that covers the anterior part of the abdomen. The lower fibers are inserted directly into the crest of the ilium and, by an aponeurosis, with the inguinal ligament (Poupart's ligament). This is not a true ligament; it is a tendinous band extending from the anterior superior iliac spine to the pubic tubercle. It is formed by the folded aponeurosis of the lower part of the external oblique muscle.

There is an opening in the aponeurosis of the external oblique muscle called the *subcutaneous inguinal ring*. It is located just above and laterad to the crest of the pubic bone. This is the external opening of the inguinal canal, which transmits the spermatic cord in the male and the round ligament of the uterus in the female. The opening is somewhat larger in the male than in the female (Figure 7.18).

The *linea alba* is a narrow tendinous sheath extending anteriorly along the middle of the abdomen from the xiphoid process of the sternum to the

Skeletal muscles

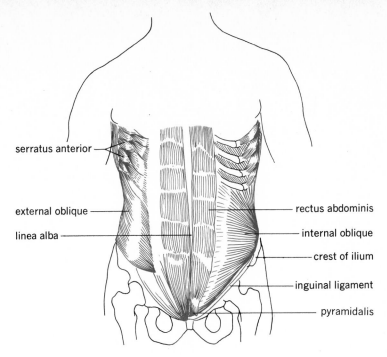

serratus anterior

external oblique

linea alba

rectus abdominis

internal oblique

crest of ilium

inguinal ligament

pyramidalis

FIGURE 7.17
Muscles of the anterior abdominal
wall.

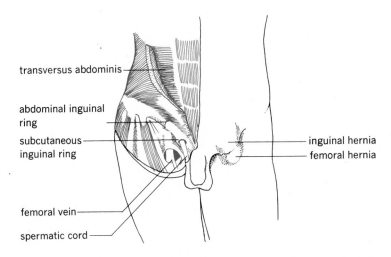

transversus abdominis

abdominal inguinal
ring

subcutaneous
inguinal ring

femoral vein

spermatic cord

inguinal hernia

femoral hernia

FIGURE 7.18
Lower abdominal structures and
the external appearance of
inguinal and femoral hernias.

symphysis of the pubic bones. It is formed by the aponeuroses of the oblique
and transversus muscles of both sides uniting in the midline. The umbilicus
lies in the linea alba a little below the middle. The *pyramidalis muscle* arises
from the pubis and is inserted in the linea alba. Its contraction causes a pull
to be exerted on the linea alba and increases its tension (Figure 7.17).

The *internal oblique muscle* lies directly under the external oblique
muscle, but its fibers run approximately at right angles to those of the muscle
above it. It arises from the inguinal ligament, iliac crest, and deep fascia
in the lumbar region. Spreading upward, the fibers insert on the costal
cartilages of the lower three ribs, into the aponeurosis of the linea alba, and
onto the crest of the pubis.

Muscles of the abdominal wall

165

The *transversus abdominis* (Figure 7.18) is the innermost of the flat muscles composing the abdominal wall. It arises from deep lumbar fasciae of the back and lower ribs, the iliac crest, and the inguinal ligament. Most of the fibers cross the abdomen horizontally and are inserted by an aponeurosis into the linea alba and onto the crest of the pubis. The abdominal wall is greatly strengthened by the three-layer arrangement of the external and internal oblique and the transversus muscles. It is of interest that their fibers run in three directions.

The *rectus abdominis* is a long, flat, paired muscle extending along the anterior side of the body from the xiphoid process and cartilages of the fifth, sixth, and seventh ribs to the crest of the pubis. It lies in a sheath formed by the aponeuroses of the oblique and transversus muscles. The medial border lies adjacent to the linea alba. Each muscle is crossed by three or four tendinous bands. These rectangular divisions of the muscle may be observed in individuals with well-developed musculature (Plate B and Figure 7.17) (Table 7.9).

ACTION OF THE ANTERIOR ABDOMINAL MUSCLES The abdominal muscles act to compress the abdomen, and, assisted by the descent of the diaphragm, they aid in such acts as vomiting, micturating, and defecating and in childbirth. They support the abdominal viscera and aid in expiration. When only one set of muscles contracts, the trunk is flexed toward that side. Acting together, they bend the body forward and so are useful in climbing and jumping.

TABLE 7.9
Muscles of the anterior abdominal wall

Muscle	Origin	Insertion	Action	Nerve
External abdominal oblique	External surface of lower 8 ribs	Anterior half of iliac crest and linea alba	Compresses abdomen, rotates and flexes vertebral column, assists in forced expiration	Intercostals 8–12, iliohypogastric, ilioinguinal
Internal abdominal oblique	Lateral half of inguinal ligament, anterior iliac crest, and lumbodorsal fascia	Lower 4 ribs, linea alba, and by tendinous sheath, to pubis	Compresses abdomen, rotates and flexes vertebral column, assists in forced expiration	Intercostals 8–12, iliohypogastric, ilioinguinal
Pyramidalis	Pubis and anterior pubic ligament	Linea alba	Tenses linea alba	Subcostal, branch of 12th thoracic
Rectus abdominis	Pubic symphysis and crest of pubis	Xiphoid process of sternum and cartilages of ribs 5–7	Tenses abdominal wall and flexes vertebral column	Intercostals 7–12
Transverse abdominis	Lateral third of inguinal ligament, anterior iliac crest, and lumbodorsal fascia	Linea alba, and by tendon to pubis	Compresses abdomen and depresses ribs, assists in forced expiration	Intercostals 7–12, iliohypogastric, ilioinguinal

Skeletal muscles

166

The rectus abdominis muscle gives a powerful contraction to raise the thorax and shoulders when one is lying on his back, or it flexes the pelvis on the trunk when the thorax is fixed. Contraction of the recti muscles protects against body blows and helps to protect the abdominal viscera from injury.

AREAS WHERE RUPTURE COMMONLY OCCURS   The muscular wall of the body is not complete. There are openings in it to permit the passage of arteries, veins, nerves, and other structures. These areas, which must be kept open, constitute weak places in the body wall where a rupture may occur. *Rupture,* or *hernia,* means the protrusion of a part of the viscera through the body wall. *Abdominal hernia* usually means the protrusion of a portion of the intestine, mesentery, or peritoneum through an opening.

The inguinal canal constitutes such an opening. It extends from the internal inguinal ring, an opening in the fascia of the transversus muscle, to the subcutaneous inguinal ring, which lies in the aponeurosis of the external oblique muscle, just above the pubic tubercle. The inguinal canal is about 1½ inches long and lies just above the inguinal ligament and parallel to it. Occasionally the intestine is forced through the inguinal canal to form an *inguinal hernia.* This is a common type of hernia in the male. *Femoral hernia* occurs at the femoral ring, the internal entrance to the femoral canal, located below the inguinal ligament. The femoral sheath covers the femoral artery and vein where they pass from the body cavity to the leg. On the medial side of the femoral vein the sheath also covers the femoral canal, which may permit passage of a pouch of the peritoneum in the case of femoral hernia. Femoral hernia is not a common type, although it is more frequent in women than in men (Figure 7.18).

Umbilical hernia occurs at the umbilical ring (navel). This type is commoner in infants, although it may occur in adults.

Weakening of the muscular wall and lack of tone in muscles are conditions conducive to hernia. Heavy lifting or coughing increases abdominal pressure and so can cause hernia.

POSTERIOR MUSCLES OF THE ABDOMEN AND PELVIC REGION   The *psoas major* and *minor* form a portion of the posterior wall of the abdomen, but the psoas major and the iliacus are of greater interest since they are concerned with movement of the trunk and thigh. They may act to move the trunk when the thigh is fixed, or they may act to move the thigh when the trunk is fixed.

The *iliopsoas* (Figure 7.19) is a muscle composed of two closely related muscles, the *iliacus* and the *psoas major.* The iliacus arises from the crest of the ilium and from the iliac fossa. The fibers pass under the inguinal ligament and converge toward their insertion on the lesser trochanter of the femur. The psoas major arises from the transverse processes, the bodies, and the intervertebral disks of lumbar vertebrae. It is a much longer muscle than the iliacus, but it too passes under the inguinal ligament, and they have a common tendinous attachment on the lesser trochanter of the femur. The iliacus and psoas major act together in flexing the thigh on the pelvis when the pelvis is fixed. They also rotate the femur outward. If the femur is fixed, however, as when one bends forward from a standing position, the iliopsoas muscle aids in flexing the trunk on the thigh. In the standing position its tension helps to prevent the trunk from tilting too far backward on the pelvis. A powerful antagonist is the gluteus maximus muscle on the posterior side of the hip joint.

Muscles of the abdominal wall

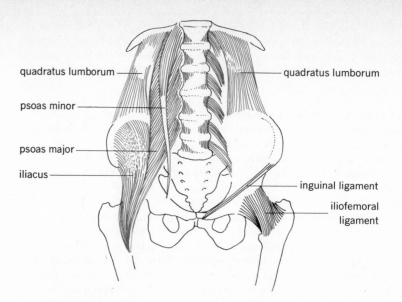

quadratus lumborum

quadratus lumborum

psoas minor

psoas major

iliacus

inguinal ligament

iliofemoral
ligament

FIGURE 7.19
Deep musculature of the hip and
thigh.

The *quadratus lumborum* is a broad muscle forming the greater part
of the posterior wall of the abdomen. Its origin is on the crest of the ilium,
and it extends forward to its insertion on the inferior border of the twelfth
rib. It flexes the vertebral column laterally and fixes the eleventh and twelfth
ribs during forced expiration (Figure 7.19) (Table 7.10).

MUSCLES OF THE PELVIC FLOOR

*Muscles of the pelvic
floor*
levator ani
coccygeus

Two muscles form the pelvic diaphragm; they are the paired *levator ani*
and *coccygeus muscles*. These muscles support the abdominal and pelvic
viscera. The levator ani is a thin sheet of muscle arising from the body of

TABLE 7.10
Muscles of the pelvic region, anterior view

Muscles	Origin	Insertion	Action	Nerve
Iliacus	Iliac fossa and lateral margin of sacrum	Lesser trochanter of femur with psoas major	Flexes and medially rotates thigh	Femoral
Obturator externus	Margins of obturator foramen	Floor of trochanteric fossa	Rotates thigh laterally	Obturator
Psoas major	Transverse processes of lumbar vertebrae	Lesser trochanter of femur, with iliacus	Flexes and medially rotates thigh	2d and 3d lumbar
Psoas minor	Bodies of 12th thoracic and 1st lumbar vertebrae	Pectineal line on ilium and iliac fascia	Flexes lumbar vertebral column	1st lumbar
Quadratus lumborum	Iliolumbar ligament and crest of ilium	Lower border of 12th rib, transverse process of upper 4 lumbar vertebrae	Depresses 12th rib inferiorly and flexes trunk to same side	12th thoracic and first 3 lumbar

Skeletal muscles

168

the pubis and from the obturator fascia back to the spine of the ischium. The fibers extend downward posteriorly and are inserted along the median raphe (seam) with fibers from the muscle on the opposite side. The posterior fibers insert on the coccyx. The anal canal passes through in the midline, and the urethra passes more anteriorly. In the female the vagina penetrates the muscle, anterior to the anal canal. The levator ani resists the downward pressure of the thoracic diaphragm. Acting together, they constrict the lower part of the rectum and pull it forward (Figure 7.20).

The coccygeus may be considered as essentially the posterior part of the levator ani. It arises from the spine of the ischium and is inserted on the sides of the upper coccygeal and lower sacral vertebrae. It aids in supporting the pelvic and abdominal viscera and flexes the coccyx slightly.

Perineum is a term referring to the muscles of the pelvic floor and associated structures of the pelvic outlet. Obstetricians limit the term to the region between the anus and the vulva.

Included among several muscles of the perineum is a voluntary sphincter muscle around the lower part of the anal canal and around the anus. Its contraction closes the anus. There is also a sphincter muscle around the urethra. Its action is to compress the urethra after micturition. In the male the sphincter urethrae surrounds the urethra in the region between the prostate gland and the base of the penis. The female urethra is surrounded by its sphincter as it leaves the pelvis and passes through the perineum.

POSTERIOR MUSCLES OF THE HIP, THE GLUTEAL MUSCLES

The *gluteus maximus* (Figure 7.21) is the largest and most superficial of the gluteal muscles. Its fibers are coarse and covered with a thick layer of fat. Since this muscle is concerned with sitting and with an erect posture, it is much better developed in human beings than in quadrupeds. The gluteus maximus arises from the outer surface of the ilium and sacrum. It is inserted

*Posterior muscles of
the hip*
The gluteal muscles
gluteus maximus
gluteus medius
gluteus minimus

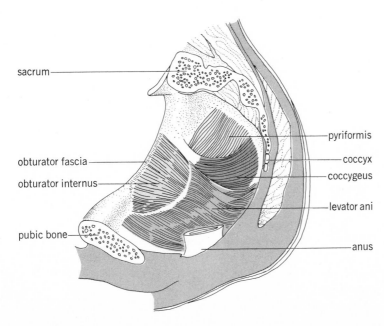

FIGURE 7.20
Muscles of the male pelvic floor, viewed from above.

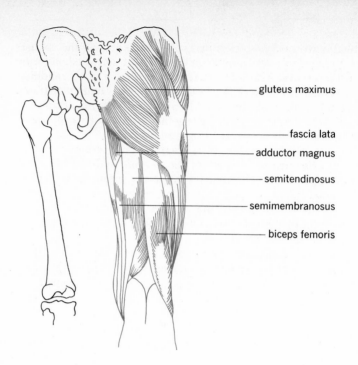

FIGURE 7.21
Posterior muscles of the hip and thigh.

gluteus maximus

fascia lata

adductor magnus

semitendinosus

semimembranosus

biceps femoris

TABLE 7.11
Muscles of the gluteal region, posterior view

Muscle	Origin	Insertion	Action	Nerve
Gemelli, superior and inferior	Ischial spine	Tendon of obturator internus to greater trochanter	Rotates thigh laterally	Branches from sacral plexus
Gluteus maximus	Upper portion of ilium, the sacrum, and coccyx	Gluteal tuberosity and iliotibial tract	Principal extensor and powerful rotator of thigh	Inferior gluteal
Gluteus medius	Middle portion of ilium	Oblique ridge on greater trochanter of femur	Abducts and medially rotates thigh	Superior gluteal
Gluteus minimis	Lower portion of ilium	Greater trochanter and capsule of hip joint	Abducts and medially rotates thigh	Superior gluteal
Obturator internus	Margin of obturator foramen	Medial surface of greater trochanter of femur	Rotates thigh laterally and aids abduction	Branches from lumbar 5 and sacral 1 and 2
Piriformis	Pelvic surface of sacrum	Greater trochanter of femur	Rotates thigh laterally and abducts it	Branches from sacral 1 and 2
Quadratus femoris	Ischial tuberosity	Greater trochanter and shaft of femur	Rotates thigh laterally	Branch from lumbar 4 and 5, and sacral 1

on the gluteal ridge of the femur and into a heavy sheath of connective tissue called the *fascia lata*. The muscles of the thigh are covered by this dense sheath. It extends downward from the hip to the knee joint, where it blends with the capsular ligament around the knee.

The action of the gluteus maximus is complex. When the pelvis is fixed, it extends the thigh. It is used, therefore, in jumping or walking upstairs, but it plays little part in ordinary walking. As an antagonist of the iliopsoas, it aids one in regaining an erect position after bending forward. This action is accomplished with the femur fixed and the pelvis as the movable part. The muscle supports the pelvis and trunk on the femur as well as strengthening the knee joint. Other gluteal muscles are the *gluteus medius* and the *gluteus minimus* (Table 7.11).

MUSCLES OF THE THIGH

The *quadriceps femoris* is a very large muscle of the anterior part of the thigh. It arises by four heads and differentiates into four muscles: the *rectus femoris*, the *vastus lateralis*, the *vastus medialis*, and the *vastus intermedius*. The rectus femoris is the only muscle of the group attached to the pelvis. It arises by two tendons from the ilium. The other three muscles arise along the femur, and all four have a common insertion into the tendon that passes across the knee joint and is attached to the tuberosity of the tibia. The tendon below the knee joint becomes the patellar ligament. The patella develops in it as a sesamoid bone. The quadriceps femoris is a powerful extensor of the leg at the knee; the rectus femoris also flexes the thigh on the pelvis. The quadriceps femoris enables one to rise from a squatting position, as in setting-up exercises. It supplies most of the power in kicking a football or in the leg kick in swimming. It is said to be three times as powerful as the flexors that oppose it (Figure 7.22).

Anterior muscles of the thigh
quadriceps femoris
sartorius
Medial
gracilis
adductor longus
Posterior
biceps femoris
semitendinosus
semimembranosus

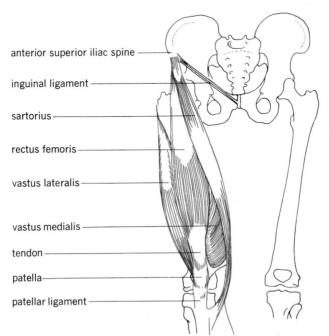

anterior superior iliac spine
inguinal ligament
sartorius
rectus femoris
vastus lateralis
vastus medialis
tendon
patella
patellar ligament

FIGURE 7.22

Anterior muscles of the thigh; the quadriceps femoris muscles and the sartorius.

TABLE 7.12
Superficial anterior muscles of the thigh

Name	Origin	Insertion	Action	Nerve
Quadriceps femoris				
Rectus femoris	Anterior inferior iliac spine and upper margin of acetabulum	Tibial tuberosity	Extends leg and flexes thigh	Femoral
Vastus lateralis	Intertrochanteric line and linea aspera of femur	Tibial tuberosity	Extends leg	Femoral
Vastus medialis	Intertrochanteric line and linea aspera of femur	Tibial tuberosity	Extends leg	Femoral
Vastus intermedius	Upper shaft of femur	Tibial tuberosity	Extends leg	Femoral
Sartorius	Anterior superior iliac spine	Medial margin of tibial tuberosity	Flexes both thigh and leg	Femoral
Tensor fasciae latae	Iliac crest	Iliotibial tract	Tenses fascia lata	Superior gluteal

The *sartorius* is a long, slender muscle running diagonally across the quadriceps femoris from its origin on the anterior, superior spine of the ilium to the medial side of the knee, where it is inserted on the upper part of the tibia near the tuberosity. It takes its name from the Latin *sartor* (tailor) and refers to the action of the muscle in pulling the thigh into a cross-legged position formerly assumed by tailors. This is the longest muscle in the body. Although it is an anterior muscle of the thigh, it can, by virtue of its attachment, flex the leg at the knee. It is the only one of the anterior muscles that flexes the leg (Figure 7.22) (Table 7.12).

MEDIAL MUSCLES On the medial side of the thigh there are several adductor muscles. They arise from the front of the pelvis and attach at different levels to the femur. They draw the thigh inward or pull the legs toward each other. The *gracilis* and the *adductor longus* are two of this group of muscles (Figure 7.23) (Table 7.13). One of the functions of the *tensor fasciae latae* (Plate B) is to assist in abducting the thigh.

POSTERIOR MUSCLES The *biceps femoris*, the *semitendinosus* and the *semimembranosus muscles* are three muscles of the posterior aspect of the thigh. They are commonly referred to as hamstring muscles, because butchers use the tendons to hang up hams. The biceps femoris arises by two heads. The long head arises from the tuberosity of the ischium, and the short head arises from the middle and distal part of the femur. The muscle is inserted on the head of the fibula. The semitendinosus and the semimembranosus arise with the long head of the biceps femoris from the tuberosity of the ischium. The long tendons of these muscles pass back of the knee joint and outline the popliteal space, which lies between them. The tendon of the biceps lies on the outer side. The tendons of the semitendinosus and the semimembranosus muscles lie on the medial side and are inserted on the proximal part of the tibia (Figure 7.24).

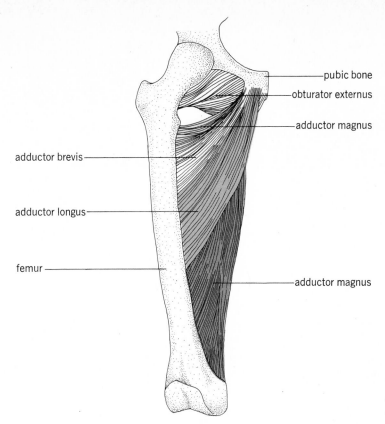

pubic bone

obturator externus

adductor magnus

adductor brevis

adductor longus

femur

adductor magnus

FIGURE 7.23
Adductor muscles of the thigh
(anterior view).

TABLE 7.13
Muscles of the hip and thigh, medioanterior view

Muscle	Origin	Insertion	Action	Nerve
Adductor brevis	Inferior pubic ramus	Upper part of linea aspera and pectineal line	Adducts, flexes, and medially rotates thigh	Obturator
Adductor longus	Between pubic rami near symphysis	Middle third of linea aspera	Adducts, flexes, and medially rotates thigh	Obturator
Adductor magnus	Pubic arch and ischial tuberosity	Linea aspera and adductor tubercle	Adducts, flexes, and laterally and medially rotates thigh	Obturator and sciatic
Gracilis	Inferior pubis near symphysis	Upper portion of tibia	Adducts, aids flexion, and medially rotates leg	Obturator
Pectineus	Pectineal line of pubis	Pectineal line of femur	Adducts thigh, aids in rotating it medially	Obturator and femoral

Muscles of the thigh

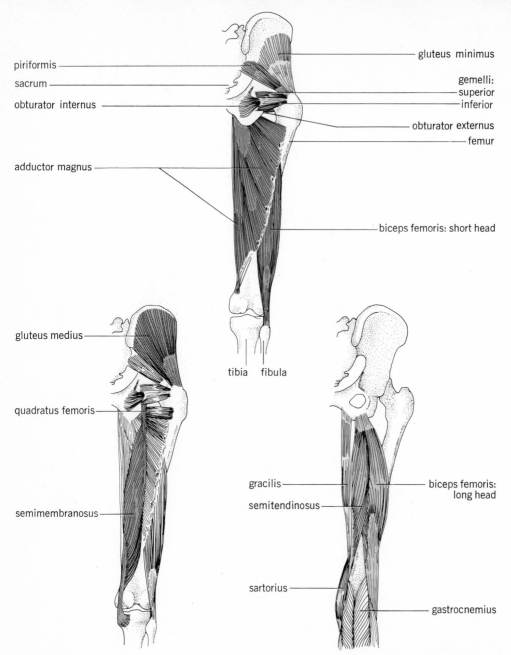

piriformis

sacrum

obturator internus

adductor magnus

gluteus minimus

gemelli:
superior
inferior

obturator externus

femur

biceps femoris: short head

tibia   fibula

gluteus medius

quadratus femoris

semimembranosus

gracilis

semitendinosus

biceps femoris:
long head

sartorius

gastrocnemius

FIGURE 7.24
Posterior muscles of the hip and
thigh.

Skeletal muscles

174

The action of this group of muscles is concerned with flexing the leg
at the knee and extending the thigh at the hip joint. The posterior thigh
muscles extend across two joints, the hip and the knee joints. Their antagonist,
the rectus femoris, also extends across these two joints but on the opposite
side. It should be noted that the hip and knee joints flex in opposite directions
(Table 7.14).

TABLE 7.14

Posterior muscles of the thigh

Muscle	Origin	Insertion	Action	Nerve
Biceps femoris, long head and short head	Long head: ischial tuberosity; short head: linea aspera of femur	Head of fibula and lateral condyle of tibia	Flexes leg and rotates leg laterally; long head extends thigh	Sciatic
Semimembranosus	Ischial tuberosity	Medial condyle of tibia	Extends thigh, flexes and medially rotates leg	Sciatic
Semitendinosus	Ischial tuberosity, fused with long head of biceps	Upper medial part of tibia	Extends thigh and flexes leg	Sciatic

MUSCLES OF THE LEG AND FOOT

Though the word *leg* commonly refers to the entire lower limb, the anatomical term refers only to the portion between the knee and the ankle. Since the foot is normally at right angles to the leg, *flexion* at that point means raising the foot above its normal position. *Extending* the foot means lowering it from its resting position; *inversion* of the foot refers to raising the medial border and turning the sole inward. *Eversion* means turning the foot outward, but this is a more limited movement.

ANTERIOR MUSCLES The anterior muscles of the leg are the *tibialis anterior* (Figure 7.25) and *extensor muscles* concerned with extending the toes. The tibialis anterior arises from the lateral condyle and from the proximal half of the shaft of the tibia. The muscle lies along the outer side of the tibia.

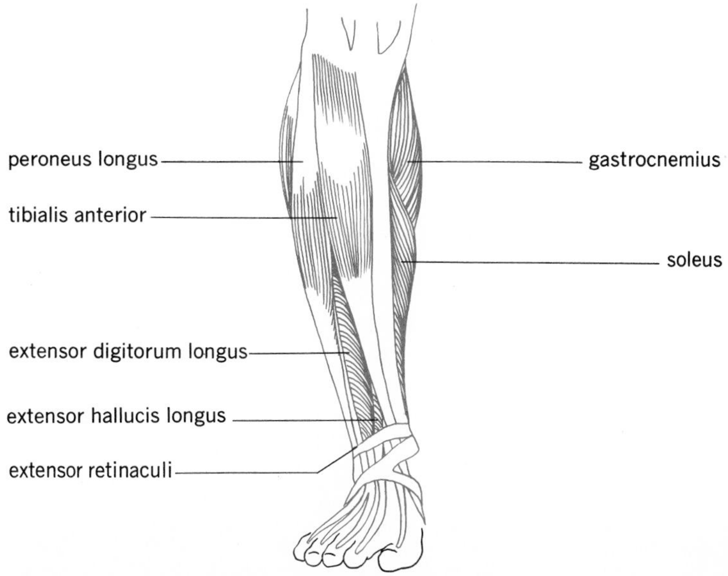

Muscles of the leg and foot
Anterior
  tibialis anterior
  extensor digitorum
    longus
  extensor hallucis longus
Lateral
  peroneus longus
  peroneus brevis
Posterior
  gastrocnemius
  soleus
  flexor digitorum longus
  flexor hallucis longus
  tibialis posterior

peroneus longus

tibialis anterior

gastrocnemius

soleus

extensor digitorum longus

extensor hallucis longus

extensor retinaculi

FIGURE 7.25

Anterolateral muscles of the leg.

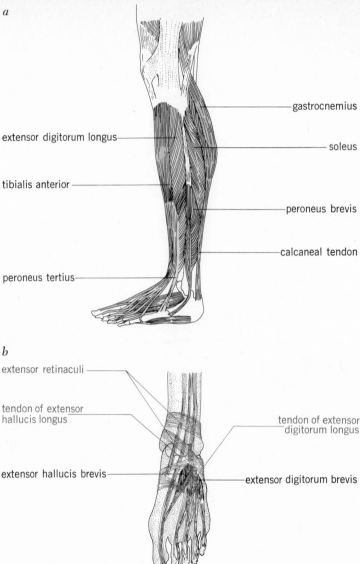

a

extensor digitorum longus —

tibialis anterior —

peroneus tertius —

— gastrocnemius

— soleus

— peroneus brevis

— calcaneal tendon

b

extensor retinaculi —

tendon of extensor hallucis longus —

extensor hallucis brevis —

tendon of extensor digitorum longus

extensor digitorum brevis

FIGURE 7.26
Deeper muscles of the leg: *a* lateral view; *b* extensor tendons of the foot (dorsal aspect).

Its tendon crosses to the inner side of the ankle and is inserted on the base of the first metatarsal bone and on the median cuneiform bones. The tibialis anterior flexes and inverts the foot. The extensor muscles of the anterior portion of the leg are the *extensor digitorum longus* and the *extensor hallucis longus*. (*Hallucis* comes from a Latin stem and refers to the great toe.) The extensor digitorum longus (Figure 7.26) arises from the lateral condyle of the tibia and from the upper anterior surface of the fibula. It lies laterad to the tibialis anterior and terminates in a tendon that divides into four slender tendons, each of which is inserted on the dorsal surface of one of the four lesser toes.

The *extensor hallucis longus* (Figure 7.26*b* and 7.27) arises from the middle anterior portion of the fibula. It is one of the deeper muscles of the

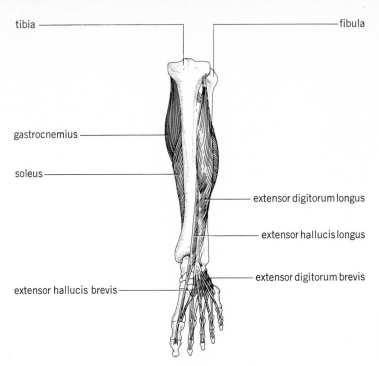

tibia
fibula
gastrocnemius
soleus
extensor digitorum longus
extensor hallucis longus
extensor digitorum brevis
extensor hallucis brevis

FIGURE 7.27

Muscles of the leg and dorsum of the foot (anterior view).

anterior portion of the leg, but its tendon appears near the surface on the dorsal side of the ankle. The tendon here lies between the tendons of the extensor digitorum longus and the tibialis anterior. It then passes across the dorsal portion of the foot to its insertion on the dorsal surface at the base of the distal phalanx of the great toe.

The extensor muscles extend the toes by drawing them up dorsally. They also aid in flexing the foot upon the leg.

LATERAL MUSCLES   The peroneal muscles (Figure 7.26) occupy a lateral position on the leg. The *peroneus longus* arises from the lateral condyle of the tibia and from the upper part of the shaft of the fibula. The muscle terminates in a long tendon at the ankle. The tendon passes behind the lateral malleolus of the fibula enclosed in a common tendon sheath with the tendon of the peroneus brevis. It passes obliquely across the sole of the foot in another tendon sheath and is inserted on the first cuneiform bone and on the base of the first metatarsal bone. The *peroneus brevis* arises from the lower lateral part of the shaft of the fibula. Its tendon passes behind the lateral malleolus and then forward to insert on the dorsal surface of the fifth metatarsal bone. These peroneal muscles extend the foot and evert it. The peroneus longus and the anterior tibialis offer strong support for both the longitudinal and transverse arches of the foot (Table 7.15).

POSTERIOR MUSCLES   The superficial muscles on the posterior side of the leg are the *gastrocnemius* and the *soleus*. Deeper muscles include the *flexor digitorum longus* and the *flexor hallucis longus*. The *tibialis posterior* is the deepest of the posterior muscles.

The *gastrocnemius* and *soleus* (Figure 7.28) are the muscles of the calf of the leg. The gastrocnemius arises by two heads from areas just above the

TABLE 7.15

Muscles of the leg, anterolateral view, and of the foot, dorsal aspect

Muscle	Origin	Insertion	Action	Nerve
Extensor digitorum brevis	Dorsal surface of calcaneus	By 4 tendons into extensor expansion	Extends toes	Deep peroneal
Extensor digitorum longus	Tibia, proximal three-fourths of fibula, and interosseous membrane	Tendons to middle and terminal phalanges of 4 lateral toes by extensor expansion	Extends toes	Deep peroneal
Extensor hallucis brevis	Distal lateral surface of calcaneus	Proximal phalanx of great toe	Extends toes in conjunction with extensor·digitorum brevis	Deep peroneal
Extensor hallucis longus	Middle half of fibula and interosseous membrane	Distal phalanx of great toe	Extends great toe	Deep peroneal
Peroneus brevis	Lower two-thirds of fibula	Fifth metatarsal	Everts and abducts foot and aids in plantar flexion	Superficial peroneal
Peroneus longus	Upper two-thirds of fibula and intermuscular septua	First cuneiform and first metatarsal	Everts and aids in plantar flexion	Superficial peroneal
Peroneus tertius	Distal third of fibula and interosseous membrane	Fifth metatarsal	Dorsally flexes and everts foot	Deep peroneal
Tibialis anterior	Upper half of tibia and interosseous membrane	First cuneiform and first metatarsal	Dorsally flexes and inverts foot	Deep peroneal

medial and lateral condyles of the femur. The soleus arises from the proximal posterior portions of both the tibia and fibula. It lies beneath the gastrocnemius. The calcaneus tendon (Achilles) is a common tendon for both muscles. It is a strong tendon inserted on the calcaneus (heel bone). These muscles are extensors of the foot. They enable one, when standing, to rise up on the toes. The gastrocnemius, since it is attached above the knee joint, also aids in flexing the leg at the knee. The gastrocnemius and the soleus are referred to as composing the *triceps surae* muscle. The two heads of the gastrocnemius and the proximal attachment of the soleus give rise to this name (from *sura*, Latin for the calf of the leg).

The *flexor digitorum longus* (Figure 7.29) has its origin along the posterior side of the shaft of the tibia. It extends downward on the medial and posterior side of the leg. A strong tendon passes behind the medial malleolus of the tibia and passes transversely under the arch of the foot, where it divides into four tendons, each inserted on the ventral surface of one of the lesser toes. The contraction of this muscle flexes the four lesser toes and inverts the foot. It is an aid in supporting the longitudinal arch of the foot.

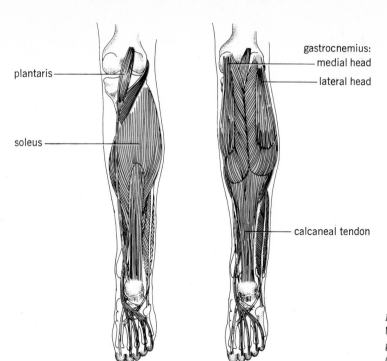

plantaris

soleus

gastrocnemius:
medial head
lateral head

calcaneal tendon

FIGURE 7.28
Muscles of the calf of the leg; the gastrocnemius and soleus (posterior view).

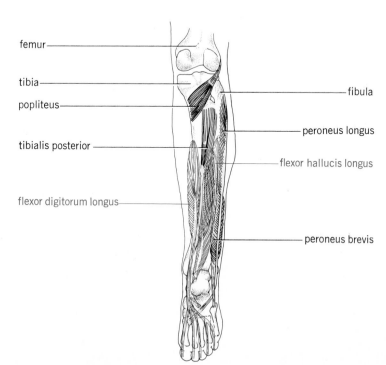

femur

tibia

popliteus

tibialis posterior

flexor digitorum longus

fibula

peroneus longus

flexor hallucis longus

peroneus brevis

FIGURE 7.29
Deeper posterior muscles of the leg.

The *flexor hallucis longus* (Figure 7.29) arises from the lower posterior surface of the shaft of the fibula. The muscle extends downward on the lateral, posterior side of the leg. Its long tendon crosses over to the medial side posteriorly above the heel. It passes under the side of the talus on the medial side of the ankle and under the arch of the foot. It is inserted on the distal phalanx of the big toe. The action of this muscle is to flex the big toe and to invert the foot. It supports the longitudinal arch of the foot, especially when the weight is on the toes.

The *tibialis posterior* arises from the upper part of the shaft of both the tibia and fibula and also from an aponeurosis between these two bones. The muscle extends downward just posterior to the tibia. Its tendon passes obliquely to the medial side just above the ankle (Figure 7.29). The tendon passes under the medial malleolus of the tibia and forward under the arch of the foot, where it is inserted on several tarsal and metatarsal bones. The tibialis posterior acts with the tibialis anterior to invert the foot. It aids in maintaining the arch of the foot. Since the medial malleolus acts as a pulley, the muscle extends the foot and thus acts as an antagonist to the tibialis anterior (Table 7.16).

MUSCLES OF THE FOOT

Intrinsic muscles of the foot
extensor digitorum brevis
flexor digitorum brevis
flexor hallucis brevis
interossei, dorsal and plantar

The foot is specialized to support the body and does not have so much flexibility as the hand. The musculature in general resembles that of the hand. The muscles of the leg that, by their tendons, control movements of the foot

TABLE 7.16
Muscles of the leg, posterior view

Muscle	Origin	Insertion	Action	Nerve
Flexor digitorum longus	Middle half of tibia	By 4 tendons into distal phalanges of lateral 4 toes	Flexes lateral 4 toes	Tibial
Flexor hallucis longus	Distal two-thirds of fibula and intermuscular septum	Distal phalanx of great toe	Flexes great toe	Tibial
Gastrocnemius	Medial and lateral condyles of femur	With soleus into calcaneus via calcaneal tendon	Flexes leg and plantar flexes foot	Tibial
Plantaris	Popliteal surface of femur	Posterior calcaneus	Plantar flexion of foot	Tibial
Popliteus	Lateral condyle of femur	Posterior tibia above soleal line	Flexes leg and rotates femur medially	Tibial
Soleus	Upper third of fibula and soleal line of tibia	With gastrocnemius into calcaneus via calcaneal tendon	Flexes foot	Tibial
Tibialis posterior	Interosseous membrane and tibia and fibula on either side	Navicular, with slips to cuneiform; cuboid; 2, 3, and 4 metatarsals	Adducts and inverts foot, aids in plantar flexion	Tibial

Skeletal muscles

180

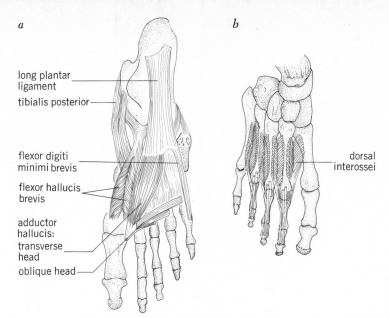

a   *b*

long plantar ligament

tibialis posterior

flexor digiti minimi brevis

flexor hallucis brevis

adductor hallucis: transverse head

oblique head

dorsal interossei

FIGURE 7.30

Intrinsic muscles of the foot: *a* plantar view; *b* the dorsal interossei.

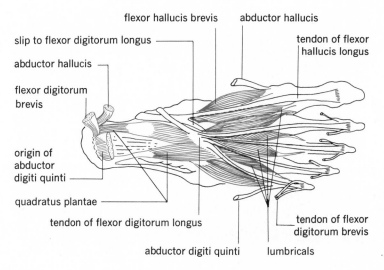

flexor hallucis brevis   abductor hallucis

slip to flexor digitorum longus

abductor hallucis

flexor digitorum brevis

origin of abductor digiti quinti

quadratus plantae

tendon of flexor digitorum longus

abductor digiti quinti   lumbricals

tendon of flexor hallucis longus

tendon of flexor digitorum brevis

FIGURE 7.31

Muscles of the sole of the foot; second layer. (*After Barry J. Anson (ed.), "Morris' Human Anatomy," 12th ed., McGraw-Hill Book Company, New York, 1966*)

are called *extrinsic muscles. Intrinsic muscles* are those located within the foot itself (Figure 7.30). On the dorsum of the foot the *extensor digitorum brevis* aids the long tendons of extrinsic muscles in extending the four medial toes. This muscle is not represented in the hand (Figure 7.27).

The bottom of the foot is the *plantar surface* (Figure 7.31). The plantar aponeurosis is a heavy sheath of connective tissue under the skin in the sole of the foot. It seems to support and strengthen the foot. The *flexor digitorum brevis* lies just above the plantar aponeurosis. It flexes the four lateral toes and helps support the arch of the foot. The *flexor hallucis brevis* flexes the great toe. There are other intrinsic muscles of the foot concerned with various movements and with the support of the arches. Tendons in the ankle and foot are generally covered with synovial sheaths. A number of sesamoid bones

Muscles of the foot

form in the tendons of the foot, and there are numerous bursae for the protection of muscles and joints (Figure 7.32) (Table 7.17).

SUMMARY   Skeletal muscles ordinarily are attached to bones by tendons. Tendons may be flat sheets of tissue and are then called aponeuroses. Fasciae are connec-

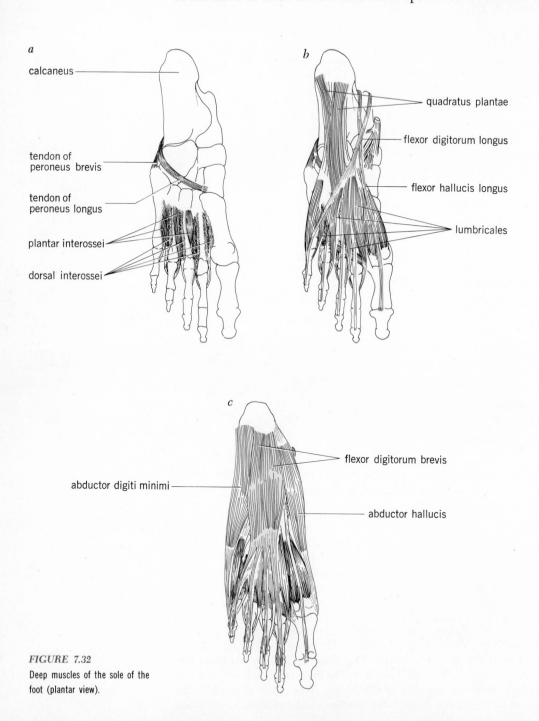

a

calcaneus

tendon of peroneus brevis

tendon of peroneus longus

plantar interossei

dorsal interossei

b

quadratus plantae

flexor digitorum longus

flexor hallucis longus

lumbricales

c

flexor digitorum brevis

abductor digiti minimi

abductor hallucis

FIGURE 7.32
Deep muscles of the sole of the foot (plantar view).

TABLE 7.17
Muscles of the sole of the foot, plantar view

Muscle	Origin	Insertion	Action	Nerve
Abductor digiti minimi	Calcaneus and plantar aponeurosis	Proximal phalanx of little toe	Abducts little toe	Lateral plantar
Abductor hallucis	Calcaneus and plantar aponeurosis	Proximal phalanx of great toe	Abducts great toe	Medial plantar
Adductor hallucis Oblique head	2–4 Metatarsals; sheath of peroneus longus tendon	Proximal phalanx of great toe	Adducts and flexes great toe	Lateral plantar
Transverse head	Metatarsophalangeal ligaments 3, 4, and 5		Adducts great toe, supports transverse arch of foot	
Flexor digiti minimi brevis	Metatarsal 5 and sheath of peroneus longus tendon	Proximal phalanx of little toe	Flexes little toe	Lateral plantar
Flexor digitorum brevis	Calcaneus and plantar aponeurosis	By 4 tendons into middle phalanx of lateral 4 toes	Flexes lateral 4 toes	Medial plantar
Flexor hallucis brevis	Navicular and cuneiform bones, and long plantar ligament	2 tendons to medial and lateral sides of proximal phalanx of great toe	Flexes great toe	Medial and lateral plantar
Interossei, dorsal	Adjacent metatarsal bones	Proximal phalanges of both sides of 2d toe, and lateral side of 3d and 4th toes	Abducts toes	Lateral plantar
Interossei, plantar	Medial side of metatarsals 3, 4, and 5	Proximal phalanges of 3d, 4th, and 5th toes	Adducts lateral 3 toes toward 2d toe	Lateral plantar
Lumbricales	Tendons of flexor digitorum longus	Extensor expansion over lateral 4 toes	Aids in flexion of toes	Medial and lateral plantar
Quadratus plantae	Calcaneus and long plantar ligament	Into tendons of flexor digitorum longus	Supports action of tendon of flexor digitorum longus	Lateral plantar

tive-tissue layers covering muscles and related structures. Synovial bursae are sacs located between tendons or around joints to prevent damage by friction.

There are three classes of levers; the leverage of body movement is mostly of the third class. Usually the more fixed attachment of a muscle is its origin and the more movable attachment, its insertion.

Muscles of the head and face include several muscles concerned with facial expression and mastication. The orbicularis oculi muscles around the eyes and the orbicularis oris muscle around the mouth are voluntary sphincter muscles. The masseter, temporal, and pterygoid muscles are muscles of mastication.

Muscles of the foot

Muscles of the neck include the platysma and the sternocleidomastoideus.

The pectoralis major is the large anterior muscle of the chest; the pectoralis minor lies directly underneath it. The external and internal intercostal muscles occupy the spaces between the ribs. They are concerned with respiratory movements of the thorax.

The deltoideus is located above the articulation of the arm at the shoulder. It raises the arm to a horizontal position. Flexors of the arm include the biceps brachii and the brachialis. The triceps brachii extends the arm at the elbow. Muscles of the forearm are flexors and extensors for the wrist and hand.

Superficial muscles located on the anterior aspect of the forearm are the pronator teres, flexor carpi radialis, palmaris longus, and the flexor carpi ulnaris.

On the posterior aspect of the forearm are the anconeus, extensor digitorum, extensor digiti minimi, extensor carpi ulnaris, extensor carpi radialis longus, and the extensor carpi radialis brevis.

Deeper muscles of the forearm are the supinator, abductor pollicis longus, extensor pollicis brevis, and the extensor pollicis longus.

The thenar muscles are the flexor pollicis brevis, abductor pollicis brevis, adductor pollicis, and the opponens pollicis.

The flexor digitorum superficialis and the flexor digitorum profundus aid in flexing the fingers.

Hypothenar muscles are the abductor digiti minimi, flexor digiti minimi, and opponens digiti minimi.

Muscles within the hand itself are the lumbricales and the interosseous sets.

Two prominent superficial muscles of the back are the trapezius and the latissimus dorsi. A deep muscle of the back is the erector spinae.

The diaphragm is a voluntary skeletal muscle that forms the floor of the thoracic cavity. It is dome-shaped and moves up and down as one breathes. Muscles of the abdominal wall are the external and internal oblique muscles, the transversus abdominis, and the rectus abdominis. The linea alba is a tendinous sheath extending along a midventral line from the xiphoid process of the sternum to the symphysis pubis. The common types of abdominal hernia are inguinal, femoral, and umbilical.

The levator ani and coccygeus muscles form the pelvic diaphragm. Sphincter muscles constrict the anus and the urethra.

The iliopsoas muscle flexes the thigh on the pelvis when the pelvis is fixed; it can flex the trunk on the thigh if the femur is fixed. It is an antagonist of the gluteus maximus, the large muscle of the buttock.

The large muscle of the anterior part of the thigh is the quadriceps femoris. It is an extensor of the leg at the knee. The sartorius muscles enable one to cross the legs. Adductor muscles, such as the gracilis, pull the thigh inward. The tensor fasciae latae assists in abducting the thigh.

The biceps femoris, the semitendinosus and the semimembranosus muscles are three muscles located on the posterior aspect of the thigh. They flex the leg at the knee and extend the thigh at the hip.

The tibialis anterior flexes and inverts the foot. Extensor muscles located anteriorly on the leg are the extensor digitorum longus and extensor hallucis longus. The peroneal muscles occupy a lateral position on the leg. They extend the foot and evert it.

The gastrocnemius and soleus are muscles of the calf of the leg. Deeper muscles include the flexor digitorum longus and the flexor hallucis longus. The tibialis posterior, the deepest of the posterior muscles of the leg, acts with the tibialis anterior to invert the foot and aids in supporting the longitudinal arch.

On the dorsum of the foot, the extensor digitorum brevis joins the tendons of extrinsic muscles to extend the four medial toes. On the plantar surface, the flexor digitorum brevis flexes the four lesser toes; the flexor hallucis flexes the great toe.

QUESTIONS

1 Discuss the mechanics of body movement in relation to force applied to a system of levers.
2 Which muscles of the face are responsible for chewing, smiling, and frowning?
3 Name some of the antagonistic muscles of the appendages and indicate their action.
4 The human hand is extremely versatile. Consider some of the muscles that provide this versatility.
5 What is the function of a pronator muscle? An adductor? An opponens?
6 Indicate on a drawing the location of some of the extensors of the wrist and fingers.
7 Which muscles compose the thenar group?
8 Name two sets of muscles located within the hand itself.
9 What is meant by synergistic muscles? Can you name some?
10 Explain the action of the diaphragm. What structures pass through its openings?
11 Locate three areas in the abdominal body wall where hernia may occur. What is hernia?
12 What muscles form the floor of the pelvis?
13 Explain the action of the iliopsoas and the gluteus maximus muscles when the femur is fixed; when the trunk is fixed.
14 Why is it that the back muscles are usually strong, but the ventral or abdominal muscles are apt to be weak?
15 Discuss the muscles of the leg and foot relative to the support of the longitudinal arch of the foot.
16 How do the abdominal muscles support breathing?

SUGGESTED READING

Several textbooks on human anatomy are listed at the back of the book.

unit three

the nervous system and the sense organs

chapter 8
neural integration

The nervous system has reached its highest degree of development in man. The human body is a large organism of great complexity. While lower animals may have relatively simple nervous systems to coordinate their metabolic activities, the nervous system of man is exceedingly intricate and well adapted to the needs of a highly developed organism. Man has gone far beyond lower animals in the development of the brain. Such complicated functions as those involved in consciousness, abstract thought, memory, and the interpretation of emotions illustrate some of the activities of the human brain.

Nervous tissue is ectodermal in origin. Functionally, it has developed two outstanding characteristics, excitability and conductivity. *Excitability* means that it is capable of reacting to stimuli; *conductivity* refers to the ability to transmit nerve impulses along a nervous pathway.

THE NEURON The structural and functional unit of nervous tissue is the nerve cell, or *neuron*. It is a microscopic grayish cell with fibers radiating out from it. A structural classification includes bipolar cells, which in their development have two processes, and multipolar cells, with numerous processes (Figure 8.1). The cells in the dorsal sensory ganglion of a spinal-nerve root originate as bipolar nerve cells. Multipolar cells are a common type found in the brain and the spinal cord. They include the pyramidal cells of the cerebral cortex and the characteristic flask-shaped Purkinje cells of the cortex of the cerebellum, as well as motor cells of the spinal cord.

A functional classification of neurons includes (1) *motor neurons (motoneurons)*, or *efferent neurons*, which convey impulses to muscles; (2) *sensory* or *afferent neurons*, which transmit stimuli inward from various receptors; and (3) *association neurons*, which are found between sensory and motor units and convey impulses from sensory to motor neurons (Figure 8.2). Such connecting neurons are also referred to as *internuncial, interneurons,* or *central neurons.*

The dendrites and the cell membrane are the receptive parts of the neuron. Specialized peripheral receptors may act as transducers in sensory neurons. The axon conducts impulses away from the receptive areas. The functions of the cell cytoplasm probably should be considered separately from the conducting function of the surface membrane. The specializations of the terminal filaments, or telodendria, provide distribution of the impulse by synaptic transmission or by neurosecretion.

STRUCTURE The *cell body,* or *perikaryon,* of a neuron contains a well-defined nucleus and nucleolus. Within the cytoplasm are small masses of a deep-staining material called *Nissl substance.* This chromidial substance is a

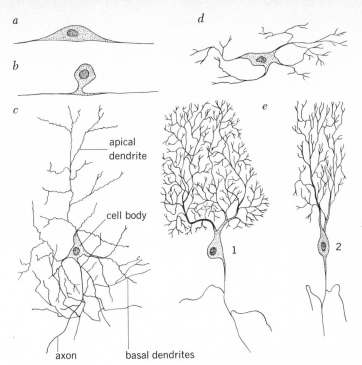

apical
dendrite

cell body

axon   basal dendrites

nucleoprotein, largely RNA, associated with the endoplasmic reticulum, and is concerned with the physiological activity of the cell. The cytoplasm also contains neurofibrils, long thin fibrils found in both the cell body and its processes. Cell bodies vary considerably in size, roughly from 5 to 50 microns in diameter. They are a vital part of the neuron. If, as a result of mechanical injury or disease, the cell body is unable to survive, then the processes also die and are unable to carry nervous impulses. Paralysis of muscles follows if injury to neurons prevents nervous impulses from reaching the muscles.

Groups of nerve-cell bodies appear gray and largely make up the *gray matter* of the brain and spinal cord (Figure 8.3). Nerve cells in an adult are not capable of mitotic division and so are unable to increase in numbers. Centrosomes are seldom demonstrated in mature nerve cells.

Dendrites (Greek *dendron*, tree) are branches of a neuron that conduct a nerve impulse toward the cell body. In spinal motor neurons there are usually several short dendrites with treelike branches. The pyramidal cell (Golgi type I) of the cerebral cortex is characterized by a long apical dendrite with relatively few side branches. There are also basal dendrites. The dendrites have a rough-appearing irregular surface. The Purkinje cells of cerebellar cortex reveal a most intricate type of branching and are somewhat fan-shaped in appearance. The dendrite of a spinal sensory neuron may be 3 feet long and is commonly called the *peripheral process*. Dendrites contain Nissl bodies.

The *axon* is the process that conducts an impulse away from the cell body. The area of the cell from which the axon arises is the axon hillock. The part of the axon just distal to the hillock before the axon becomes covered by the myelin sheath is the inital segment. Axons do not contain Nissl substance, which may also be absent from the axon hillock. The axon

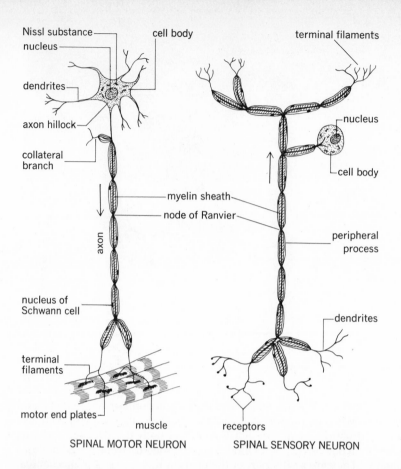

Nissl substance
nucleus
cell body
dendrites
axon hillock
collateral branch
axon
nucleus of Schwann cell
terminal filaments
motor end plates
muscle

terminal filaments
nucleus
cell body
myelin sheath
node of Ranvier
peripheral process
dendrites
receptors

SPINAL MOTOR NEURON   SPINAL SENSORY NEURON

FIGURE 8.2
Two common types of neurons, spinal motor neuron and spinal sensory neuron.

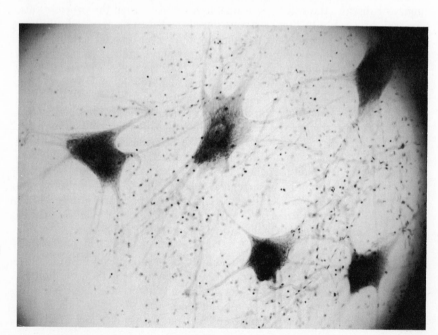

FIGURE 8.3
Motor nerve cells from spinal cord smear. The numerous small cell bodies are neuroglial cells. (*Courtesy of General Biological Supply House, Inc., Chicago.*)

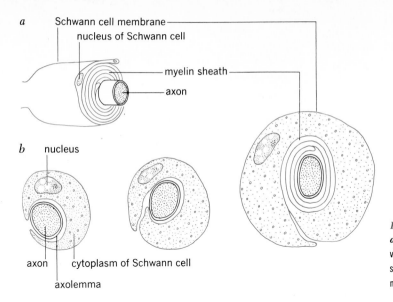

FIGURE 8.4
a Myelinated axon, anterolateral view; *b* development of myelin sheath from Schwann cell membrane.

of a neuron may be very short, as in association neurons (Golgi type II), or it may be 2 or 3 feet long, such as those extending from the gray matter to the spinal cord to the muscles of the fingers or toes. Axons terminate in fine branches called *terminal filaments,* or *telodendria.* The tips of these terminal branches appear to be highly specialized to provide for the transmission of the nerve impulse to succeeding neurons or to a muscle effector. The difference between axon and dendrite is based primarily not on minor differences in structure but on difference in function. The nerve impulse conducted over a series of neurons passes from the axon and terminal filaments of one neuron by synaptic transmission to the dendrites and cell body of the succeeding neuron.

The myelin sheath Peripheral fibers of neurons are covered with a myelin sheath formed by the plasma membrane of Schwann cells. The fiber, enclosed in its own membrane, the *axolemma,* first becomes embedded in the Schwann cell. The membrane of the Schwann cell then becomes tightly wrapped around the fiber in many layers. This wrapping is the result of the growth of the membrane rather than a rotation of the Schwann cell or the axon. The membrane is always double, since the original invagination brings the two sides of the plasma membrane together, leaving a cleft, the mesaxon (Figures 8.4 and 8.5). As in most cell membranes, the molecular structure consists of a layer of protein material on the outer and inner sides, whereas the middle layer is composed of lipid material.

The fiber has a segmented appearance where the successive myelin sheaths are interrupted at the *nodes of Ranvier.* A single Schwann cell forms the myelin sheath between two nodes. The Schwann cell terminates at the node, and the node itself is covered only by a thin basement membrane. The nodes represent relay stations for the propagation of a nerve impulse. Depolarization and repolarization occur at the node area, where there is no myelin to act as an insulating layer. The impulse "jumps" from one node to another in an action described as *saltatory conduction.*

Unmyelinated fibers are also embedded in Schwann cells, but they are not completely enclosed, and there is no coiling of the membranes around

The neuron

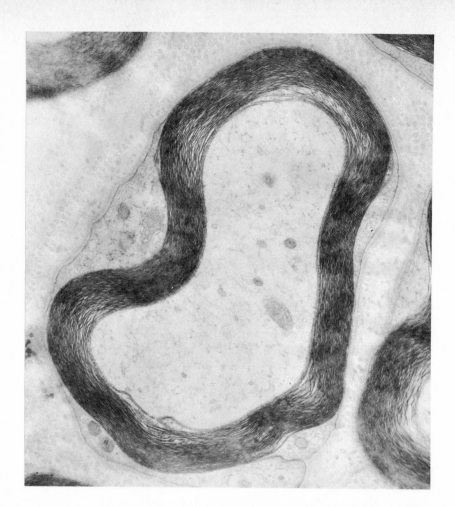

FIGURE 8.5
Electron micrograph of myelin
sheath. (*From James A.
Freeman, "Cellular Fine
Structure," McGraw-Hill
Book Company, New York,
1964.*)

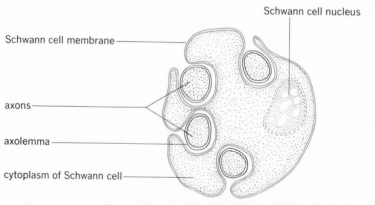

Schwann cell nucleus

Schwann cell membrane

axons

axolemma

cytoplasm of Schwann cell

FIGURE 8.6
Unmyelinated axons embedded in
Schwann cell membrane.

Neural integration

them. Several fibers of this type may be embedded in a single Schwann cell,
but the membranes remain separate (Figure 8.6).

In the central nervous system, myelin is formed around nerve fibers
by neuroglial cells or glial cells of a type called *oligodendroglia* or *oligocytes*.
They are essentially the Schwann cells of the myelinated fibers in the brain

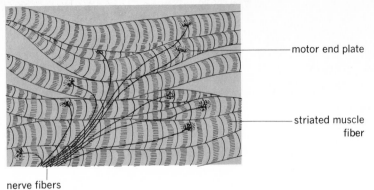

motor end plate

striated muscle fiber

nerve fibers

FIGURE 8.7

Motor end plates on the fibers of skeletal muscle.

a

b

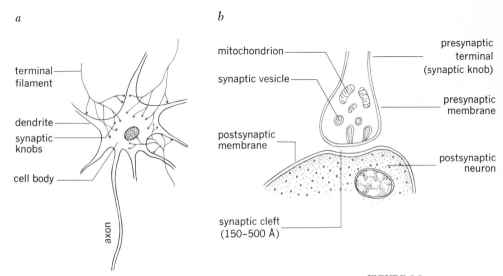

terminal filament

dendrite

synaptic knobs

cell body

axon

mitochondrion

synaptic vesicle

postsynaptic membrane

synaptic cleft (150–500 Å)

presynaptic terminal (synaptic knob)

presynaptic membrane

postsynaptic neuron

FIGURE 8.8

a Synaptic knobs resting on a cell body; *b* diagrammatic representation of the synaptic cleft between the presynaptic terminal and the postsynaptic neuron. Synaptic vesicles are thought to release a transmitter substance.

and spinal cord. The impulse is conducted more rapidly in myelinated fibers than in nonmyelinated ones.

The terminal filaments, or telodendria, of peripheral motor neurons have enlargements at their distal ends, called *motor end plates*. The end plates are closely applied to the fibers and form the myoneural junction (Figures 6.9 and 8.7). These enlargements are essentially the same as those occurring at typical synaptic endings in other types of neurons. The neuromuscular junction is sometimes described as an axosomatic synapse.

The neuron, as a distinct structural unit, must have some method of transmitting a nerve impulse from one neuron to another. The typical *synapse* is an area where the distal ends of terminal filaments come into close morphological and physiological relationship with the dendrites and cell bodies of succeeding neurons. The presynaptic terminal filaments are enlarged into synaptic knobs, or boutons, and closely applied to the cell membrane of the postsynaptic neuron (Figure 8.8*a*). Electron micrographs reveal a space of about 100 to 200 angstroms, a *synaptic cleft*, between the presynaptic and postsynaptic plasma membranes (Figure 8.8*b*). Synaptic knobs contain mitochondria and synaptic vesicles. The vesicles are thought

The neuron

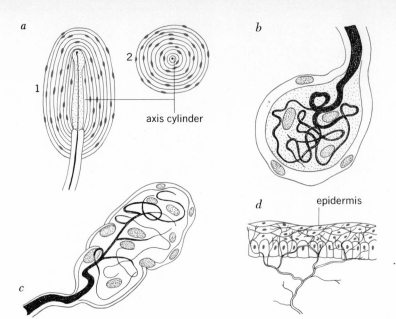

to contain a chemical transmitter substance, since transmission at a synapse is believed to be primarily chemical rather than physical. Several substances may be capable of acting as transmitters. Acetylcholine seems to be generally accepted as being a transmitter substance although positive proof may be lacking. There is some indication that serotonin may be a transmitter substance in the central nervous system, whereas in the sympathetic nervous system noradrenaline apparently functions in this way. Acetylcholine is hydrolyzed and made ineffective in a fraction of a second by the action of the enzyme *acetylcholinesterase.*

In the condition known as *myasthenia gravis,* muscles become weak and unable to function properly, presumably because acetylcholine is deficient or because its effectiveness is destroyed by abnormal cholinesterase activity. A third possibility might be failure of the muscle membrane in some way to react to the transmitter substance. The condition can be treated with drugs that depress cholinesterase activity at the myoneural junction.

Various types of *sensory receptors* or *end organs* are found at the peripheral ends of sensory neurons. Some of these structures are associated with the special senses, to be taken up later. All such structures act as receptors for sensory stimuli conveyed inward toward the central nervous system by afferent neurons (Figure 8.9).

NERVES

Though we commonly think of nerves as tiny white fibers, judging them by the appearance of their terminal branches, many nerves are large, extremely complicated structures, made up of bundles of fibers, the axis fibers of neurons. These bundles are separated and held in place by connective tissue in which there are fat cells. Capillaries of blood vessels are present also. A cross section of nerve is shown in Figure 8.10. Dense fibrous connective tissue, the *perineurium,* surrounds each nerve bundle, or fascicle. Indi-

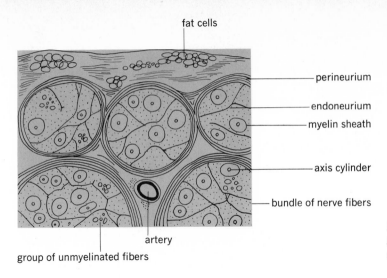

fat cells

perineurium

endoneurium

myelin sheath

axis cylinder

bundle of nerve fibers

artery

group of unmyelinated fibers

FIGURE 8.10
A portion of a nerve in cross
section, greatly enlarged.

vidual nerve fibers are embedded in a loose fibrous connective tissue which contains reticular fibers and is called the *endoneurium.*

Large peripheral nerves contain fibers of varying diameters, which may be both myelinated and unmyelinated. Smaller nerves may be composed entirely of myelinated fibers, or they may be entirely unmyelinated. When a nerve branches, the axis fibers of certain neurons are directed into the nerve branch. Though some nerves may contain only motor neurons or only sensory neurons, most nerves are composed of both motor and sensory neurons and are called *mixed nerves.*

DEGENERATION AND REGENERATION Portions of neurons cannot survive if severed from their cell bodies. Therefore, if a nerve is cut, the axis fibers that are separated from their cell bodies degenerate. The neurilemma surrounding the severed fibers undergoes extensive degenerative changes. If regeneration occurs, the developing axis fiber grows from the cell body into the old neurilemma of the severed portion. The neurilemma appears to be an essential factor in regeneration of nerve fibers. Regeneration does not occur in the nerve tracts of the brain or spinal cord.

Cell bodies sometimes fail to survive if their fiber is injured. Since mature nerve cells are unable to divide by mitosis, neurons cannot be replaced. A fairly large number of brain cells apparently die during the normal lifetime of an individual. In aged persons, probably millions of nerve cells have died, and their loss eventually impairs motor and sensory efficiency.

THE NERVE IMPULSE

The impulse that travels along a nerve was once thought to be an electric current. It was therefore assumed that it would move along a nerve at the same speed that electricity is conducted by a wire. Later research indicated, however, that the nerve impulse travels much more slowly. In myelinated nerves of mammals it travels at about the same speed as a bullet from a revolver, the maximum velocity being about 90 to 100 meters per second. The speed of the nerve impulse is much slower in cold-blooded animals.

The nerve impulse

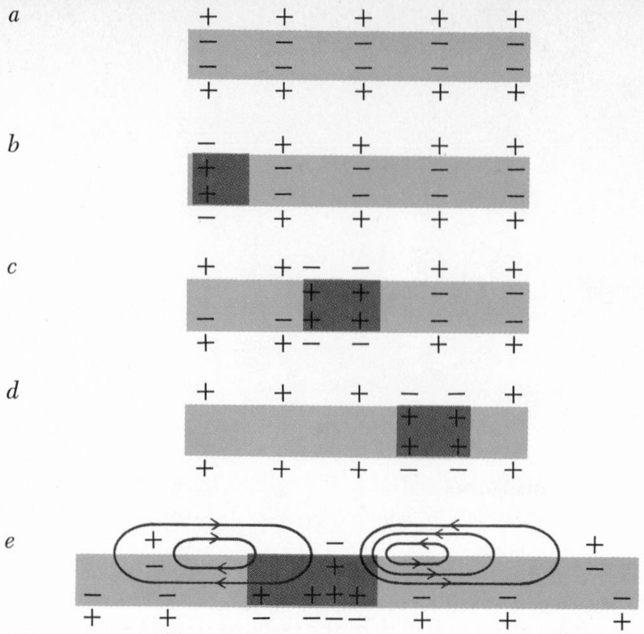

FIGURE 8.11

The progress of the nerve impulse along the nerve fiber: *a* the resting-nerve fiber, with a positive charge on the outside of the membrane and a negative charge within; *b* the initiation of the nerve impulse, at the left; for a fraction of a second the charges on the membrane are reversed; *c* the nerve impulse passes rapidly (left to right) along the nerve fiber; *d* the charge on the membrane returns to normal after a brief refractory period; *e* the flow of local currents as recorded by surface electrodes. Note that the action currents flow away from the active region on the inside of the fiber and toward the active region on the outside in the external medium. The active region here is in the center of the diagram, and the direction of the impulse is toward the right. [(*e*) After Bernard Katz, "Nerve, Muscle, and Synapse," McGraw-Hill Book Company, New York, 1966.]

In the frog, at room temperature, the maximum velocity is about 30 meters per second. Conduction rates vary with the temperature, the kind of animal involved, the diameter of the fiber, and the nature of the fiber covering, that is, whether or not the fiber is myelinated. In general, the largest fibers record the highest conduction rates and the greatest spike potential. Conversely, smaller fibers record slower conduction rates and lower potentials. Conduction in a myelinated fiber is about 10 times as rapid as in a non-myelinated fiber of the same diameter. Even within a single nerve, a cathode-ray oscillograph reveals several spike potentials, for the nerve is composed of fibers of various sizes and the rate of conduction varies with the size of the fiber.

The nerve impulse is an action potential, an electrical change in the nerve-fiber membrane that progresses along the nerve fiber, jumping from one node to another and deriving its energy from metabolic sources within the neuron. The action potential is self-propagating and represents a progressive depolarization of the membrane (Figure 8.11). The nervous excitation (the nerve impulse) is a series of ionic changes resulting in a change of electrical sign, as the membrane in excitation becomes much more permeable to sodium ions.

The membrane covering a neuron is electrically polarized when inactive, the inside of the fiber being about 70 millivolts negative to the outside. This is the membrane potential, which is variable.

In a resting neuron, sodium ions, which are positive, are in excess outside the membrane in a ratio of about 10:1. Potassium ions, also positive, are in excess within the fiber, the ratio being about 30:1. It is obvious that these ratios represent a high degree of ionic imbalance.

Formerly the membrane was considered to be passive but selectively permeable to sodium ions, and the action potential was thought to increase this permeability so that sodium ions rushed into the fiber during the passage of the nerve impulse while potassium ions moved out. After the passage of

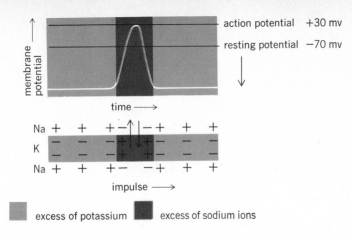

action potential +30 mv
resting potential −70 mv

impulse ⟶

excess of potassium   excess of sodium ions

FIGURE 8.12

Diagram to illustrate the behavior of sodium and potassium ions as the nerve impulse passes along a fiber. In the resting fiber, there is an excess of sodium ions on the outside of the nerve membrane, and an excess of potassium ions, inside the fiber. As the nerve impulse passes, sodium ions rush inside the fiber until they are in excess there, while potassium ions move outside the membrane. For about 0.001 second the inside of the nerve fiber becomes positive and the outside negative as the nerve impulse passes. (*After Bernard Katz, Nerve Impulse, in "The Physics and Chemistry of Life," A Scientific American Book, Simon & Schuster, Inc., New York, 1955.*)

the impulse the ions simply reversed their flow and the resting potential was restored. While it is true that the ions move in this way, their action is believed to involve more than just a matter of selective permeability. More consideration is given to the active transport of ions, especially sodium ions.

It is postulated that in the resting neuron, the so-called "sodium pump" continually forces sodium ions outward through the membrane against an electrochemical gradient. At the same time, potassium ions are able to move into the fiber. To maintain a ratio of 30:1, however, there must be some sort of active transport for potassium ions also. The energy for this active transport must come from the cell's own metabolism.

Chloride ions are able to diffuse through the cell membrane in either direction at equal rates. Actually, chloride ions, being negative, Cl⁻, tend to be repelled by the negative charge inside the fiber and are forced outward through the membrane. The internal negativity also tends to keep them outside. The interstitial fluid on the outside of the axon membrane is composed largely of sodium and chloride ions, which represent about 90 percent of the ionic concentration, whereas such ions represent less than 10 percent of the total concentration inside the fiber.

The nerve impulse has been described as a traveling electrical disturbance. When the nerve fiber receives a stimulus of threshold level, a wave of depolarization sweeps over the fiber. Actually it is more than a depolarization because for about 0.001 second during the passage of the impulse the charges are reversed, the outside becomes about 10 to 20 millivolts negative, and the inside becomes positive (Figure 8.12).

A change in membrane permeability permits sodium ions to rush into the fiber momentarily, and potassium ions to move out. The membrane equilibrium potential is considerably reduced at this time, to perhaps around 45 millivolts. Almost immediately, however, sodium ions are forced out, potassium ions move back into the fiber, and the resting potential is restored.

So far as axons are concerned, fibers exhibit the *all-or-nothing* type of response. A stimulus strong enough to produce a response, that is, a stimulus of threshold level, produces an adequate or maximal response in an individual neuron fiber under a given set of conditions. Unlike a current of electricity, the nerve impulse is self-propagating and does not become progressively weaker as it moves farther away from the point of stimulation. This question may well be considered: Why is it, then, that a strong stimulus produces a greater response? The answer lies in the facts that (1) a strong

The nerve impulse

stimulus may increase the frequency of the impulses passing along a nerve fiber and (2) more fibers probably are involved.

The all-or-nothing principle, however, does not apply to all parts of the neuron. The receptor parts, that is, the dendrites and the cell membrane, react to a specific stimulus, usually a chemical secreted at the neuron endings. (An axon can act as a receiver in some cases, an axoaxon type of synapse.) The chemical transmitter substance reacts at specific sites on the postsynaptic membrane, causing the membrane to become more permeable to the passage of ions. Sodium ions, Na^+, move rapidly inside and as a result there is an increase in the positive charge inside the cell. This ionic disbalance initiates the excitatory postsynaptic potential.

A summation effect is commonly developed at the terminal enlargements of telodendria on the postsynaptic membrane. At the neuromuscular junction the effective postsynaptic potential is necessarily of an excitatory nature, but at synaptic endings between neurons the transmitter substance may develop either an excitatory postsynaptic potential (EPSP), or it may produce an inhibitory postsynaptic potential (IPSP). The EPSP, as we have seen, depolarizes the membrane in forming the action, or spike, potential. The IPSP causes the internal voltage to advance farther to the negative, making depolarization more difficult to attain. If a series of inhibitory impulses drives the interior voltage of the cell 10 millivolts farther toward the negative, it would be more difficult for the membrane potential to reach threshold level and cause the neuron to "fire."

It is not known exactly by what means the inside of the fiber becomes more negative when an IPSP is produced. If the membrane suddenly became more permeable to potassium ions but continued to exclude sodium ions and the potassium ions rushed out, the internal negative equilibrium potential would be increased beyond that of the resting state, perhaps from 70 to 75 or 80 millivolts negative. But it is known that sodium ions rush in during the transmission of the excitatory depolarization wave. It may be determined eventually that in an inhibitory response sodium ions are largely excluded, thereby increasing the negativity inside the membrane.

One hypothesis to explain some of the features of the IPSP postulates specific receptor sites on the postsynaptic membrane (see the articles by J. C. Eccles listed at the end of the chapter). When the transmitter substance reaches these sites, fine channels open in the postsynaptic membrane, permitting rapid passage of ions. An excitatory transmitter substance presumably would open larger channels, so that the larger sodium ions could move rapidly inward through the membrane. Potassium ions, with a lesser potential gradient, would move outward much more slowly. The inhibitory transmitter substance presumably would open smaller channels that would be too small to permit sodium ions to pass. Since potassium ions could flow out through these channels and chloride ions could flow in, the net effect would cause the hyperpolarization characteristic of the IPSP.

The EPSP may not reach threshold level, in which case it declines slowly, and the neuron does not respond. Since the EPSP is capable of summation, a series of two or more subthreshold stimuli may summate and initiate depolarization. The number of synapses on the dendrites and cell membrane and whether or not they are near each other are factors that affect summation. The IPSP also is capable of summation.

TYPES OF SUMMATION Two types of summation are recognized. If two or more presynaptic terminals (end knobs) become activated at the same time on the postsynaptic neuron, they are, of course, separated by a space and their activity may produce an additive effect in the EPSP, referred to as *spatial summation*. If the activity is limited to one terminal end knob but this terminal filament continues to fire several times, the consequent effect on the postsynaptic neuron may afford an example of *temporal summation*. In this case the several stimuli are distributed in a matter of time.

There is a tendency to oversimplify synaptic events. Hundreds of synaptic endings may be involved. If the synaptic activity tends to be excitatory even though it does not reach threshold level, it will be likely to cause some change toward depolarization in the postsynaptic neuron, so that a few additional stimuli (recruitment) will cause the neuron to fire. This is an example of *facilitation*, an important factor in all neural activity. *Convergence* of several presynaptic neurons upon a single postsynaptic neuron leads to facilitation in the neural pathway. Facilitation is considered as the basis for habit formation. After a series of stimuli traverses a given pathway, subsequent stimuli are believed to follow with greater facility, and in this way various activities are performed with greater ease. Facilitation applies to mental habits as well as physical ones.

INHIBITORY CHEMICAL TRANSMITTERS   The existence of an inhibitory chemical transmitting agent is still open to question. The release of a specific inhibitory transmitter substance from certain presynaptic endings would aid in clarifying the mode of operation of the IPSP. Neutral amino acids, such as gamma-aminobutyric acid (GABA) and glycine, have been subjected to extensive research. They do act as inhibitors, but they may act more as regulators within the central nervous system than as transmitters. Further research should establish their precise function more accurately.

PRESYNAPTIC INHIBITION   We have considered postsynaptic inhibition, but we should mention presynaptic inhibition also. Inhibition of the presynaptic neuron is thought to be the result of activity of interneurons which lower the excitability at presynaptic terminals by a slight depolarization. The depolarization is not of sufficient magnitude to initiate an action potential, but it "conditions" the stimuli reaching the postsynaptic neuron.

Theoretically, the propagated nerve impulse could travel in either direction along the axon, but it would not be possible for an impulse to traverse a synapse from the postsynaptic neuron cell body toward the presynaptic terminals. The postsynaptic neuron does not produce any transmitter substance from its cell body or dendrites, to carry the impulse across the synaptic cleft. That the impulse can pass in only one direction at the synapse illustrates the Bell-Magendie law.

REFRACTORY PERIOD   Experimentally a nerve can be stimulated at a very rapid rate by controlled electric stimuli. The nerve responds by initiating impulses at a rate that corresponds to that of the stimulus. There is an upper limit, however, to the rate of conduction of nerve impulses. As the rate of electric stimuli is increased, the rate of nervous response fails to increase. The passage of the nerve impulse uses up the energy of the nerve fiber, and for half a millisecond it is unable to conduct another impulse. This is referred to as the *absolute refractory period*.

The potential of the nerve fiber is gradually restored during a *relative refractory period*. During this period the response of the nerve fiber is below normal. Stronger stimuli are required to initiate an impulse. The impulse may be slow and weak. At the end of the relative refractory period the nerve potential is restored sufficiently to conduct impulses in a normal fashion. Since the metabolic activities of the nerve have been altered by the passage of the impulse, it may take much longer to restore the nerve to a resting condition. A normal threshold is reached in about 0.5 to 2 milliseconds, depending on the size of the fiber.

Although the refractory period places a limit to the frequency at which nerve impulses can be conducted, this frequency is well above normal physiological needs. Nerves and muscles have a refractory period, but there is ample latitude for accelerating their action.

If the nerve impulse is transmitted across the synapse there will be a *synaptic delay* of about half a millisecond. This small time lapse accounts for events occurring at synapse. These include the release of the transmitter substance, its diffusion into the synaptic cleft, and time for its effect to be produced on the postsynaptic neuron. Apparently calcium ions are required to permit the release of the transmitter substance into the synaptic cleft from synaptic vesicles. The breaking-out of the contents of the vesicles is called *exocytosis*.

NERVE METABOLISM Careful measurements have shown that nervous tissue uses oxygen and gives off carbon dioxide, as do all other tissues in the body. The measurements have also shown that nerves give off minute amounts of heat even though the nerve is resting. The amount of heat given off increases when these nerves become active. Even so, the amount of heat given off by nervous activity is so small that it probably does not contribute any significant amount to body heat. It has been estimated that the initial heat of muscular contraction, based on a single contraction of frog muscle, is 30,000 times as great as the initial heat of nerves. However, the greatest use of oxygen and the greatest release of carbon dioxide occur during the recovery period, in repolarization.

The energy for conduction is apparently obtained from the rapid breakdown of creatine phosphate. Since the impulse is self-propagating, energy must come from the neuron itself. Mitochondria as power sources are present in dendrites, the cytoplasm of the cell body, and in the telodendria. The development of electric potential across the nerve-cell membrane requires an outlay of energy which is most likely derived from the adenosine triphosphate system. ATP is also the probable energy source for the active transport of ions, for example, the sodium pump.

Since nervous tissue shows a very rapid recovery after the passage of a nerve impulse, it has been difficult to demonstrate fatigue in nerve fibers. Continued activity reduces the irritability of a nerve fiber, lowers metabolic activity, and has the effect of extending the refractory period. However, it is most likely that the cell body and the synapse are readily susceptible to fatigue, not the nerve fiber. Synaptic fatigue is a much more realistic problem. Continued stimulation of the presynaptic neuron may very well exhaust the transmitter substance, so that the response may be weakened and ineffective.

CONVERGENCE OF NERVE IMPULSES Sensory receptors can initiate sensory impulses, which travel in toward the central nervous system over many individual neurons. Motor impulses arise from various parts of the brain. When these impulses are directed toward a muscle, they must converge and travel over ventral root neurons to cause contraction of muscles. The spinal motor

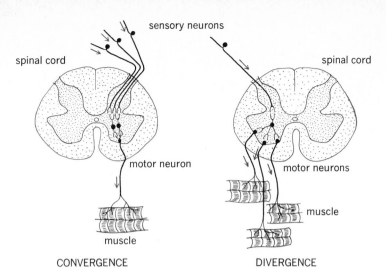

CONVERGENCE   DIVERGENCE

neuron represents the *final common pathway* over which all impulses must pass to reach the muscle effector. Figure 8.13 illustrates the convergence of nerve impulses. Since a strong stimulus evokes a more immediate and more vigorous response than a weak stimulus, the convergence of stimuli from various sources may exert a strong effect on the muscles. An example is the tensing of muscles as a reaction to pain. It may be difficult to visualize a neuron network with thousands of presynaptic terminals converging on a few postsynaptic cell bodies, but this is the common pattern of convergence.

DIFFUSION OF IMPULSES Though motor stimuli are ordinarily directed into a final common path, the number of connections that can be made are almost infinite. A certain degree of inhibition or resistance to the passage of a nervous impulse at the point of synapse may help to channel impulses into relatively few restricted pathways. Habit formation, or facilitation, can induce them to follow a definite course. Under certain conditions widespread diffusion of nerve impulses to muscles may occur, causing convulsions, that is, uncontrolled contractions of numerous muscles. Diffusion of nervous impulses occurs much more readily in children than in adults. It is assumed that in children there is less inhibition to the diffusion of nerve impulses so that a strong pain stimulus, for example, can result in convulsions. Here a strong sensory stimulus produces impulses that spread to various motor effectors (Figure 8.13).

WHITE MATTER AND GRAY MATTER

A cross section of the spinal cord reveals an interior portion of gray matter shaped roughly like the letter H, surrounded by white matter (Figure 8.14). The white matter is composed of the axis fibers of neurons. Nearly all are myelinated, but they have no neurilemma. They are grouped into great columns, which are concerned with the transmission of nerve impulses either up or down the cord. Some columns are ascending pathways; others are descending pathways. A cross section reveals that the cord is oval in outline

White matter and gray matter

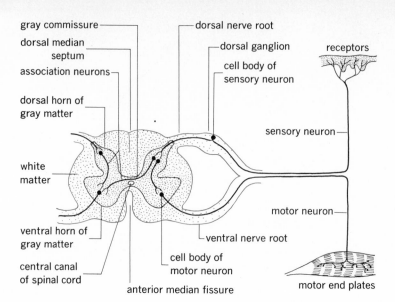

FIGURE 8.14

Cross section of the spinal cord, revealing the essential structures of the reflex arc.

with a dorsal median septum and a ventral median fissure. Although the gray substance of the cord is usually portrayed in cross section, it is important to realize that it consists of several columns composed largely of the cell bodies of neurons and unmedullated fibers. The slender dorsal columns contain numerous fibers of the dorsal spinal nerves. The cell bodies of some of the afferent ascending neurons are located there, their fibers traversing the white matter of the cord.

The cell bodies of neurons forming the anterior roots of spinal nerves lie in the larger ventral columns. Descending neurons in the white matter make synaptic connections with these cell bodies, establishing a pathway by which motor impulses from the brain are transmitted to muscles.

Between the dorsal and ventral columns in the pars intermedia there is an intermediolateral column, which contains the cell bodies of motor neurons of the sympathetic system. These are the cell bodies of preganglionic neurons that extend out to the chain of sympathetic ganglia by way of the white branch (see discussion in Chapter 10).

The gray matter also contains association, or connecting, neurons, which form synaptic connections between the dorsal and ventral roots or between the right and left sides of the cord. Many of these neurons make connections vertically between different levels of the cord.

The central canal is a minute tube extending the length of the cord. It opens anteriorly into the fourth ventricle of the brain and ends posteriorly in the filum terminale of the cord. It contains cerebrospinal fluid. A cross section of the cord shows the central canal within the gray commissure, a group of nerve fibers connecting the two columns of gray matter. The gray commissure lies just above the base of the anterior median fissure of the cord.

STRUCTURE OF THE SPINAL NERVE

The spinal nerve has two roots close to the cord: a dorsal afferent, or sensory, root and a ventral efferent, or motor, root (Figures 8.14 and 8.15). Diagrams usually portray the nerve roots with a cross section of the cord. In such

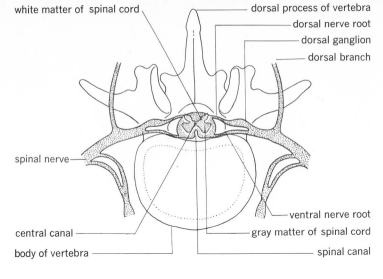

FIGURE 8.15
Cross section of the spinal cord within a thoracic vertebra.

white matter of spinal cord

dorsal process of vertebra

dorsal nerve root

dorsal ganglion

dorsal branch

spinal nerve

ventral nerve root

central canal

gray matter of spinal cord

body of vertebra

spinal canal

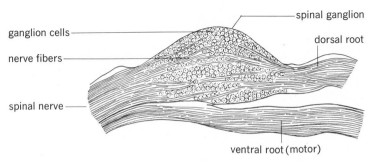

FIGURE 8.16
Longitudinal section through a dorsal-spinal nerve root ganglion.

spinal ganglion

ganglion cells

dorsal root

nerve fibers

spinal nerve

ventral root (motor)

diagrams the motor roots usually are shown in a ventral position. If one considers the diagram as drawn from a horizontal position, the individual is lying on his stomach. If one thinks of the spinal cord as being vertical, then the ventral motor root is anterior.

The dorsal afferent root is composed of sensory neurons, which bring impulses in toward the central nervous system. The cell bodies of these neurons are located outside the cord in the ganglion of the dorsal root (Figure 8.16).

A ganglion consists of a group of nerve cells usually located outside the brain or spinal cord. However, certain basal nuclei of the brain constitute the basal ganglia. Afferent cell bodies arise from embryonic neural-crest cells as bipolar nerve cells. The two processes later become fused at the cell body to form a unipolar or pseudounipolar cell, and the single process divides into two branches. One fiber grows outward to a sensory receptor located, for example, in the skin, in a muscle, or tendon. The other fiber grows inward toward the cord through the dorsal root of the spinal nerve. The fiber entering the cord may not make an immediate connection with an association or motor neuron. Certain of these fibers may extend to upper or lower levels of the cord before making these connections.

The cell bodies of the neurons composing the ventral root are located in the ventral columns of gray matter of the cord. Their fibers extend outward, usually terminating in a muscle or gland.

The sensory and motor roots merge as they pass outward through openings between the vertebrae, forming a *mixed nerve* (one in which there are both sensory and motor fibers). All spinal nerves are mixed nerves.

Spinal nerves divide into two large branches just outside the vertebral column. The dorsal branches innervate the muscles and skin of the posterior portions of the neck and trunk. The ventral branches are distributed to the ventral part of the body wall and to the extremities.

REFLEX ACTIVITY

The spinal nerves and the cord form a pathway for nervous stimuli, permitting reflex movements. The sudden retraction of the hand or foot as the result of a pain stimulus, the protective closure of the eyelids when an object comes close to the eye, or the coordinated muscular movements that enable us to keep our balance are common examples of reflex movements. We are usually more or less aware of such reflex movements, but there are other reflexes, involving smooth muscles, that are not brought to our attention. Sensory stimuli may affect the heart, stomach, blood vessels, or any other organ reflexly so that automatic adjustments to varying conditions may be made constantly.

THE REFLEX ARC   The structural elements of a reflex arc include the sensory receptor, the sensory neuron, usually association neurons within the spinal cord, and the motor neuron. Sensory stimuli are received by specialized structures called *receptors*. These peripheral receptors are of many kinds and include structures in the skin for the reception of stimuli arising from change in temperature, touch (tactile sense), pressure, and pain, as well as receptors for the higher senses located in the eye, ear, and nose and on the tongue. Apparently the fine nerve endings themselves act as receptors for the sense of pain. The sensory neuron carries the nerve impulses inward toward the brain or cord. Association neurons can connect directly with motor neurons on the same side of the cord, or they can make multiple connections, some crossing over to the opposite side. The motor neurons carry impulses to muscles. A diagrammatic representation of the structural elements of a reflex arc is shown in Figure 8.14. Though reflex arcs are commonly described as though only one sensory neuron and one motor neuron were involved, it is doubtful whether any reflex act is as simple as that. Probably there are always a number of neurons taking part in a reflex act.

Simple reflexes are those that do not involve centers in the brain. Usually they are simple nervous circuits through the cord. The reaction to a pinprick, winking the eye when some object comes near it, and the knee jerk from a sudden blow to the patellar tendon are all simple reflexes. The reaction seems to be almost immediate, but actually conduction is slower over the reflex arc than it is over a nerve fiber, because in the reflex pathway the nerve impulse is slowed somewhat in passing over the synapses. The strength of the stimulus may vary. A weak stimulus, such as touching a hair, may elicit only a moderate response, while a severe pain stimulus can cause a pronounced muscular reaction, involving many muscles. Each sensory neuron is potentially in contact with many motor neurons, and if the stimulus is strong enough, there may be considerable spread or diffusion of the stimuli to motor effectors.

The *stretch reflex*, or *myotatic reflex*, is a simple reflex arc, yet a very

Neural integration

important postural act. In skeletal muscles there are specialized muscle fibers which have a proprioceptive function. These are sensory receptors which interpret the degree of stretch in muscles. They are mechanoreceptors and are called *muscle spindles*. The stretch reflex involves only the receptor in the muscle spindle, the afferent or sensory neuron, and the motor neuron which returns the impulse to the same muscle from which it originated. The knee jerk is often cited as a common example. When the patellar tendon is tapped, the *quadriceps femoris* muscle is stretched, the tendon reflex initiates a nerve stimulus, and when the impulse returns to the muscle, contraction occurs, causing the leg to jerk forward.

The smooth operation of postural muscles is much more complex than it at first appears. In a standing position, it may appear that the extensor muscles of the legs are all important, but the flexors are involved also and are in a partial state of contraction. The muscles of the trunk and legs are in a "fixed" state of contraction against the force of gravity. By somewhat different afferent pathways involving interneurons, antagonistic muscles are able to operate smoothly in reciprocal fashion. Reciprocal innervation permits one muscle to contract while its antagonist is inhibited, as seen, for example, in the leg muscles during walking. Inhibition to stretch is of value in preventing undue extension of muscles. The stretch reflex also affords an example of negative feedback, since it represents a balance of control between the strength of the motor stimulus in contraction and afferent sensitivity limiting the degree of contraction, the net result being that the muscle is enabled to maintain a constant length and acquire greater smoothness in its operation.

Reflexes involving the brain become extremely complicated. Many activities can be carried on below the level of consciousness. The respiratory, cardiac, and vasoconstrictor centers are located in the medulla. The cerebellum functions as a coordinator of muscular activity and helps to maintain balance. Midbrain, thalamus, and cerebral cortex function to perfect posture and coordinate movements.

Eventually many stimuli reach the level of consciousness. After we have jerked a finger away from a hot iron, we are aware of pain, and we may consider how badly the finger is burned and whether to put an ointment on it or not. A workman surrounded by machinery may not be able to take a step backward to retreat from danger and so may have to consider which way he will go to reach safety. If one has had to reason his way or consciously direct his muscular movements, he has involved the highest level of the brain, the cerebral cortex.

THE CONDITIONED REFLEX   Many reflexes are conditioned by experience. The early work of Pavlov, the Russian physiologist, demonstrated that the salivary response in dogs can be conditioned in various ways, such as by the ringing of a bell, the flash of a light, or by the sight of objects of different sizes and shapes. If a bell is rung as the dog is fed, the animal soon associates the ringing of the bell with the presence of food. Since salivation is a normal reaction associated with hunger and anticipation of eating, the dog can be conditioned so that its saliva will flow at merely the sound of the bell.

The conditioned reflex may be the basis of habit formation. The voiding of urine by the young is probably controlled as a spinal reflex act. When the urinary bladder becomes distended to a certain degree, a train of nerve impulses develops that results in the emptying of the bladder. As the child develops, however, he learns to control this act; cerebral control is super-

Reflex activity

205

imposed over the simple reflex, and it becomes a conditioned reflex. It appears that simple reflexes may be conditioned by experience; these experiences, good or bad, may be made the basis of habit formation.

SUMMARY   The structural and functional unit of the nervous system is the neuron. Neurons are classified as motor, or efferent; sensory, or afferent; and association neurons. Structural parts of a neuron are the cell body, dendrites, axon, and terminal filaments, or telodendria. The fiber may be myelinated, that is, have a sheath derived from the Schwann-cell membrane coiled around the fiber. The Schwann cell terminates at the node of Ranvier, so that there is a Schwann cell with its myelin sheath between nodes. Telodendria have enlargements at their tips which contain mitochondria and vesicles presumably containing a transmitter substance. The transmitter substance is discharged into the synaptic cleft.

Where the terminal filaments of peripheral motor neurons make contact with muscle fibers, there are enlargements which are the motor end plates at the myoneural junction. The nerve impulse arriving at the motor end plate releases acetylcholine, which initiates an electrochemical change, the action potential, in the muscle-fiber membrane. The enzyme acetylcholinesterase hydrolyzes acetylcholine in a fraction of a second and destroys its effectiveness.

The nerve impulse is an action potential. It is a propagated impulse and moves along the fiber by saltatory conduction from one node to another. It is a progressive depolarization of the membrane.

The membrane of the resting neuron is positively charged on the outside and negatively charged on the inside. For a brief fraction of a second as the nerve impulse passes, the charges are reversed. Sodium ions rush into the fiber, and potassium ions move out. The resting potential is restored almost immediately as sodium ions are forced out and potassium ions move back into the fiber. It is thought that sodium ions are forced out against an electrochemical gradient by active transport, the so-called "sodium pump."

Axons and peripheral processes exhibit an all-or-nothing type of response. If the stimulus reaches threshold level, the impulse is conducted along the fiber, and the response will be adequate without regard to the strength of the stimulus. The nerve impulse, being self-propagating, must derive its energy from metabolic processes within the cell itself.

The synapse is the area where the enlarged ends of terminal filaments come into close contact with the dendrites and cell bodies of succeeding neurons. A synaptic cleft lies between synaptic knobs or presynaptic terminals and the postsynaptic membrane. The presynaptic terminals contain mitochondria and vesicles. The vesicles are thought to contain the chemical transmitter substance. Transmission at the synapse between neurons is similar in most respects to transmission at the myoneural junction.

A summation effect may be developed on the postsynaptic membrane. Also, at the synapse the transmitter substance may develop either an excitatory postsynaptic potential (EPSP) or an inhibitory postsynaptic potential (IPSP). The IPSP causes the internal voltage to move farther to the negative and therefore to make depolarization, or "firing," of the neuron more difficult. The IPSP is a property of the synapse; it would be incorrect to refer to "an inhibitory action potential."

Nerves exhibit a refractory period during which time they are unable to conduct another impulse.

A simple reflex arc includes the sensory receptor; the sensory, or afferent, neuron; association neurons within the spinal cord; and the motor, or efferent, neuron. A conditioned reflex is one that is conditioned by experience. The stretch reflex does not involve interneurons.

1 Experimentally, it should be possible to start nerve impulses at opposite ends of a nerve fiber and to have them travel toward each other. When they meet, they are extinguished. Why?
2 Why is there a ganglion on the dorsal-spinal nerve root but not on the ventral motor root?
3 Distinguish between neuron and nerve.
4 Describe the development of the myelin sheath.
5 Describe the presynaptic terminal and discuss transmission across the synaptic cleft. Why does the impulse move in only one direction?
6 What is meant by sodium pump?
7 If the all-or-nothing type of response applies to axons and peripheral processes, explain why a stronger stimulus produces a greater response.
8 Outine the pathway of a simple reflex arc.
9 A man is slightly injured. He makes an involuntary exclamation and steps back, away from danger. Trace the nerve impulses over various pathways.
10 Explain how a reflex act may be conditioned.
11 Explain how the nerve impulse is propagated.
12 Is there an upper limit to the rate at which nerve impulses can be conducted? Why?
13 Illustrate the principle of convergence of nerve impulses. Illustrate divergence.
14 Discuss physiological mechanisms resulting in an excitatory postsynaptic potential. How is an inhibitory postsynaptic potential developed? What advantage would an IPSP afford?
15 What neurons are involved in a stretch reflex?

SUGGESTED READING

BJÖRKLUND, A., and ULF STENEVI: Nerve Growth Factor: Stimulation of Regenerative Growth of Central Noradrenergic Neurons, *Science*, **175**:1251-1253 (1972).
De ROBERTIS, E. D. P.: Molecular Biology of Synaptic Receptors, *Science*, **171**:963-971 (1971).
ECCLES, J. C.: Ionic Mechanism of Postsynaptic Inhibition, *Science*, **145**:1140-1147 (1964).
————: The Synapse, *Sci. Am.*, **212**:56-66 (1965).
EVERETT, N. B.: "Functional Neuroanatomy," 6th ed., Lea & Febiger, Philadelphia, 1971.
FREEMAN, JAMES A.: "Cellular Fine Structure," McGraw-Hill Book Company, New York, 1964.
HEBB, C.: The Central Nervous System at the Cellular Level: Identity of Transmitter Agents, *Ann. Rev. Physiol.*, **32**:165-192 (1970).
KATZ, B.: "Nerve, Muscle and Synapse," McGraw-Hill Book Company, New York, 1966.
————: Quantal Mechanism of Neural Transmitter Release, *Science*, **173**:123-126 (1971).
O'BRIEN, JOHN S.: Stability of the Myelin Membrane, *Science*, **147**:1099-1107 (1965).
PORTER, KEITH R., and MARY A. BONNEVILLE: "An Introduction to the Fine Structure of Cells and Tissues," 4th ed., Lea & Febiger, Philadelphia, 1973.
ROBERTSON, DAVID J.: The Membrane of the Living Cell, *Sci. Am.*, **206**:65-72 (1962).

chapter 9
the brain and spinal cord

The human brain is an organ of great complexity. It guides our movements, interprets the senses, and, in the greatest achievement of all, enables us to think. It is the seat of consciousness, guides our emotions, and is the keeper of our memories. The brain and spinal cord compose the central nervous system; whereas, the cranial and spinal nerves constitute the peripheral system.

The nervous system is first indicated in the embryo by a median depression in the dorsal ectoderm called the *neural groove*. As the groove deepens, the sides grow up and fuse dorsally, forming a long, hollow tube that will become the brain and spinal cord (Figure 9.1). The walls of this tube are filled with nerve-cell bodies and fibers. While it is forming, there is a differentiation of the dorsal edge of the fold on either side into the neural crests. The development of the neural crests is of interest as these nerve-cell bodies later become the dorsal ganglia of the spinal nerves.

The anterior end of the neural tube enlarges and later is constricted to form the three primary divisions of the brain: the *forebrain, midbrain, and hindbrain* (Figures 9.2 and 9.3). The posterior part of the neural tube becomes the spinal cord. The developing brain has relatively large cavities, the ventricles, which become proportionately smaller as the brain grows larger. The skull forms around the embryonic brain, and the vertebrae grow around the spinal cord, forming a protecting structure of bone around this portion of the central nervous system.

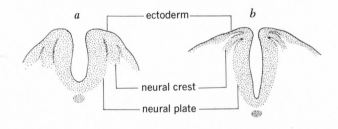

FIGURE 9.1

Development of the neural tube and neural crest (transverse section): *a* neural groove open; *b-d* gradual closure to form the neural tube. (*Redrawn from Bradley M. Patten, "Human Embryology," 3d ed., McGraw-Hill Book Company, New York, 1968.*)

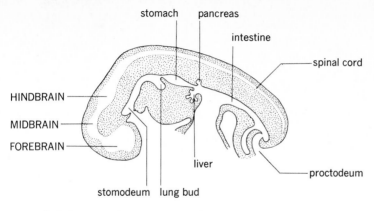

stomach   pancreas

intestine

spinal cord

HINDBRAIN

MIDBRAIN

FOREBRAIN

liver

proctodeum

stomodeum   lung bud

FIGURE 9.2
Diagram illustrating the early development of nervous and digestive systems.

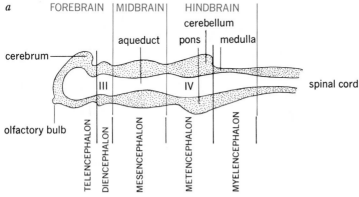

a

FOREBRAIN |MIDBRAIN| HINDBRAIN

cerebellum

aqueduct   pons   medulla

cerebrum

III   IV   spinal cord

olfactory bulb

TELENCEPHALON
DIENCEPHALON
MESENCEPHALON
METENCEPHALON
MYELENCEPHALON

b

FOREBRAIN |MIDBRAIN| HINDBRAIN

lateral
ventricles

III   IV

FIGURE 9.3
Diagrammatic representation of an early embryonic brain: *a* sagittal section; *b* horizontal section.

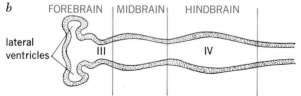

lateral ventricle   cerebral hemisphere

pineal body

superior and inferior
cerebral colliculi

thalamus

hypothalamus

corpus striatum

pituitary gland

optic nerve

olfactory bulb

aqueduct

cerebellum

medulla

I and II

III   IV

pons

TELENCEPHALON
DIENCEPHALON
MESENCEPHALON
METENCEPHALON
MYELENCEPHALON

FIGURE 9.4
Diagram showing the principal structures developed from the three original divisions of the brain.

BASIC DIVISIONS The forebrain may be divided into two parts: an anterior part, the *telen-cephalon,* and a posterior part, the *diencephalon.* The midbrain, or *mesen-cephalon,* becomes a part of the brainstem. The hindbrain is also divided into two parts: an anterior part, the *metencephalon,* and a posterior part, the *myelencephalon.* The structures associated with these various parts of the brain will be taken up in detail later (Figure 9.4).

Basic divisions of the brain and their principal structures are:

Primary divisions	Secondary divisions	Structures
Forebrain	Telencephalon	Olfactory bulbs
		Cerebral cortex
		Lateral ventricles
	Diencephalon	Epithalamus, pineal body
		Thalamus
		Hypothalamus
		Stalk of pituitary gland
		Mammillary bodies
		Greater part of third ventricle
Midbrain	Mesencephalon	Colliculi
		Cerebral aqueduct
		Cerebral peduncles
		Substantia nigra
		Red nuclei
Hindbrain	Metencephalon	Cerebellum
		Pons
		Fourth ventricle
	Myelencephalon	Medulla
		Pyramidal tracts

MENINGES The brain and spinal cord are protected by the meningeal membranes. These are three in number and consist of a tough, fibrous outer layer, the *dura mater;* a delicate intermediate membrane, the *arachnoid;* and an inner vascular layer, the *pia mater* (Figure 9.5).

The dura mater within the cranium is composed of two layers, which adhere very closely except as they are separated by venous sinuses. The outer layer becomes fused with the cranial bones and is the internal periosteum. The inner layer of the dura lines the vertebral canal extending down over the spinal cord. Here, the epidural space intervenes between the dura and the bony canal, a condition not present in the cranial cavity. The epidural space contains fat tissue and numerous fine veins.

Folds of the meningeal dura form the *falx cerebri,* the *falx cerebelli,* the *tentorium cerebelli,* and the *diaphragma sellae.* The falx cerebri is usually described as a sickle-shaped fold located between the two cerebral lobes in the longitudinal cerebral fissure. The falx cerebelli is a small triangular fold formed between the cerebellar hemispheres. It is attached above to the

The brain and spinal cord

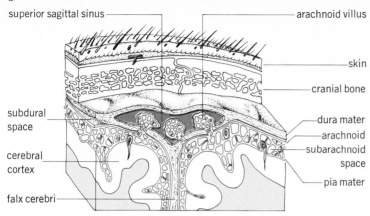

a

superior sagittal sinus — arachnoid villus

— skin

— cranial bone

subdural space

cerebral cortex

falx cerebri —

— dura mater
— arachnoid
— subarachnoid space
— pia mater

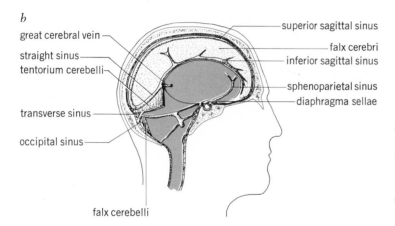

b

great cerebral vein —
straight sinus —
tentorium cerebelli —

transverse sinus —

occipital sinus —

superior sagittal sinus
— falx cerebri
— inferior sagittal sinus
— sphenoparietal sinus
— diaphragma sellae

falx cerebelli

FIGURE 9.5
a A cranial section of the brain showing the meningeal membranes. Note the arachnoid villi; *b* lateral view illustrating the venous sinuses and the folds of the cranial dura mater.

tentorium cerebelli, which forms a roof over it. The tentorium cerebelli covers the cerebellum and lies under the occipital lobes of the cerebrum. It joins the falx cerebri along the median line, and together they form a venous sinus called the *straight sinus*. The diaphragma sellae covers the sella turcica and partially covers the hypophysis (pituitary gland). There is a small circular opening which permits the passage of the stalk of the hypophysis.

The delicate arachnoid membrane lines the dura mater and extends down over the cord. It does not ordinarily follow the sulci closely, as does the pia mater, and so there is a subarachnoid space at each depression between the convolutions of the brain (Figure 9.5). The subarachnoid spaces are filled with cerebrospinal fluid, which protects the brain and cord from mechanical injuries. It is a lymphlike fluid with a few white cells in it. Diseases that attack the central nervous system or the meninges can alter the composition or increase the amount of fluid. Diagnosis and treatment of such diseases can be greatly aided by taking a sample of the fluid for study and analysis. The sample is obtained by a procedure called *lumbar puncture*. Since the spinal cord ends at the first or second lumbar vertebra and since the dura and arachnoid membranes extend below this point, it is possible to insert a needle between the vertebrae and into the subarachnoid

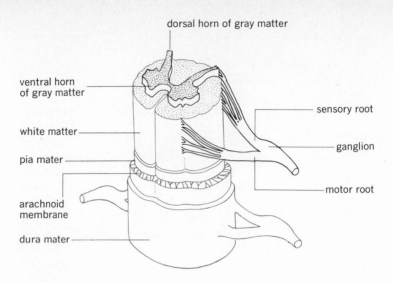

dorsal horn of gray matter

ventral horn
of gray matter

white matter

pia mater

arachnoid
membrane

dura mater

sensory root

ganglion

motor root

FIGURE 9.6
The meninges covering the spinal
cord.

space of this region and obtain cerebrospinal fluid. Spinal anesthetics are introduced into the subarachnoid space in the same region by lumbar puncture. *Meningitis* is an inflammation of the meninges.

The pia mater contains a dense network of blood vessels. It is applied very closely to the brain surface, following down into the sulci and fissures. It is closely applied to the spinal cord also (Figure 9.6).

VENTRICLES

The ventricles, or cavities, of the brain communicate with each other and are continuous with the central canal of the spinal cord. They are lined by a membrane called the *ependyma*. Ependymal cells are derived from embryonic neural-tube epithelium. The ventricles, the central canal of the cord, and the subarachnoid spaces are filled with cerebrospinal fluid. The largest cavities are the lateral ventricles of the cerebral hemispheres. Each lateral ventricle communicates with the third ventricle of the diencephalon by way of an interventricular foramen (foramen of Monro). The third ventricle connects with the fourth through the cerebral aqueduct (aqueduct of Sylvius), which traverses the midbrain. The fourth ventricle is continuous with the central canal of the spinal cord (Figure 9.7).

In each ventricle there is a modification of the ependymal lining, covering a specialized area where a network of capillaries from the pia mater project into the ventricle. These are the *choroid plexuses,* the source of the cerebrospinal fluid.

The *cerebrospinal fluid* filters out through the membrane and circulates slowly through the ventricles. Probably the greater amount of fluid is derived from the choroid plexuses of the lateral ventricles. The fluid flows posteriorly through the interventricular foramens into the third ventricle, then through the cerebral aqueduct and into the fourth ventricle. The fourth ventricle has three openings in its roof that permit the fluid to flow into the subarachnoid spaces around the cerebellum and the medulla. The fluid then flows slowly down through the subarachnoid space, covering the spinal cord, and also enters the central canal of the cord. The direction of flow in the

The brain and spinal cord

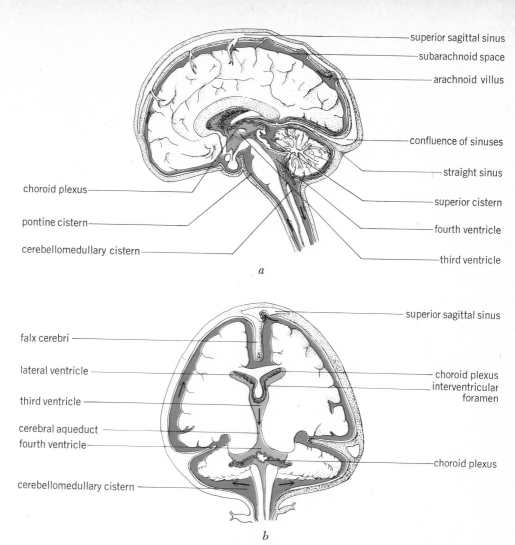

superior sagittal sinus
subarachnoid space
arachnoid villus
confluence of sinuses
straight sinus
superior cistern
fourth ventricle
third ventricle
choroid plexus
pontine cistern
cerebellomedullary cistern

a

superior sagittal sinus
falx cerebri
lateral ventricle
choroid plexus
interventricular foramen
third ventricle
cerebral aqueduct
fourth ventricle
choroid plexus
cerebellomedullary cistern

b

FIGURE 9.7
Cerebral fluid circulation; *a* sagittal section; *b* frontal section.

subarachnoid space of the cord appears to be downward on the posterior or dorsal side and upward on the anterior or ventral side. It flows over the cerebrum and is returned to the blood through arachnoid villi which project into a venous sinus, the *superior sagittal sinus*. A number of factors may influence the circulation of the cerebrospinal fluid: (1) higher blood pressure at the choroid plexuses than at the arachnoid villi, (2) osmotic pressure, (3) variation in venous pressure, and (4) changes in the position of the body.

For a long time neuroglial cells have been regarded as composing the connective tissue of the brain and spinal cord, but there is considerable doubt that they function primarily as supporting tissue. In structure or function they do not closely resemble other types of connective tissue.

There are three principal types of neuroglial cells, namely, *astrocytes*,

NEUROGLIA

213

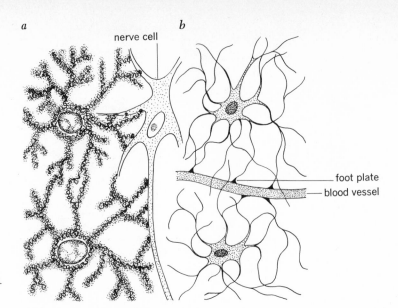

a nerve cell *b*

foot plate
blood vessel

FIGURE 9.8
Two types of neuroglial cells: *a*
protoplasmic astrocytes of gray
matter; *b* fibrous astrocytes, typical of white matter.

oligodendrocytes, and *microgliocytes*. These cells vary in structure and function. Astrocytes and oligodendrocytes arise from neural-tube cells and therefore are ectodermal in origin, but microgliocytes arise from the mesoderm.

Astrocytes are associated with the neurons and blood vessels of the
central nervous system. Around the small blood vessels, especially under the
pia mater, they form a membrane of neuroglial cells and their protoplasmic
processes. There are two kinds of astrocytes, protoplasmic and fibrous. The
protoplasmic astrocytes have larger cytoplasmic processes than the fibrous
astrocytes, are more irregularly branched, and are found predominantly in
the gray matter. Fibrous astrocytes with long thin processes are found for
the most part in white matter (Figure 9.8).

A peculiar feature of fibrous and protoplasmic astrocytes is the development of enlargements, or *foot plates*, along blood vessels and neurons. Glial
cells also form delicate membranes around blood vessels of the brain and
spinal cord, thus separating blood vessels from nervous tissue. This peri-
vascular glial membrane may prove to be the structural element of the
blood-brain barrier, but recent research seems to indicate that the perivas-
cular barrier is not continuous. The capillary endothelium appears to be the
only continuous barrier between the blood and the neurons. Astrocytic glial
cells have been shown to exhibit an action potential with a duration many
times longer than that of nerve cells. When stimulated electrically, they are
capable of slow contraction lasting several minutes.

Fibrous astrocytes form glial scar tissue after injury to the brain or
spinal cord. Neuroglial cells are commonly involved in primary tumors of
the central nervous system. Such tumors are called gliomas.

Oligodendrocytes are found abundantly in the central nervous system
(Figure 9.9*a*). They are smaller than astrocytes, and ordinarily only their
small round nuclei are seen. Cells of this type are commonly clustered around
the large cell bodies of neurons, where they are called *satellite cells*. In the
white matter of the central nervous system they are found in rows along
nerve fibers; in this respect they may be regarded as functioning somewhat
in the same manner as the Schwann cells of the peripheral nervous system,

a   *b*

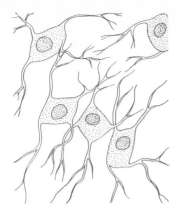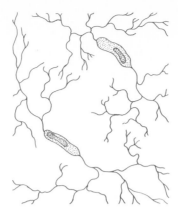

FIGURE 9.9
Neuroglial cells: *a* oligodendrocytes;
b microgliocytes.

by producing myelin. They are also associated with capillaries of the central nervous system.

In general, there are about 10 times as many glial cells as there are neurons, but the physiological functions of neuroglia are not well defined. There are indications that some may have a nourishing function, regulating the neuron's nutrition. They seem to stick to nerve tissue, and this has given rise to the name *glial* (from the Greek for glue).

Microgliocytes have very small irregularly shaped cell bodies with two or more processes (Figure 9.9*b*). The processes are finely branched with short, spinelike projections. They have no foot plates and are active, motile, phagocytic cells somewhat like the reticuloendothelial cells of some organs of the body.

CEREBRUM

The cerebrum of man has had a long developmental history. From meager beginnings in the lower vertebrates, it has developed into a large, dominant part of the mammalian brain (Figure 9.10). Nervous activity within the cerebral cortex results in conscious thought. It provides for the higher intelligence of man and for all that may be inferred under such general terms as reasoning ability, good judgment, memory, and willpower. We become aware of the outside world through the interpretations of the special senses of sight, hearing, taste, smell, and touch. We are capable of initiating a voluntary muscular response, of directing our own body movements through the activity of the cerebral cortex. Certain emotions—feelings of charity, appreciation of beauty, a desire to do right—are functions of the cerebrum; more primitive emotions such as rage may be in part the expression of a lower part of the brain, the hypothalamus. Many acts that are ordinarily reflex may be dominated or controlled by the cerebrum, as when we cough voluntarily or hold the breath for a time.

The cerebrum grows until it completely covers the brainstem dorsally. The two hemispheres are separated medially by the great longitudinal fissure. *Fissures* are deep depressions on the brain surface. The numerous lesser depressions of the surface are called *sulci* and separate the elevations, which are *convolutions*. The surface of the brain has a covering, or *cortex*, of gray

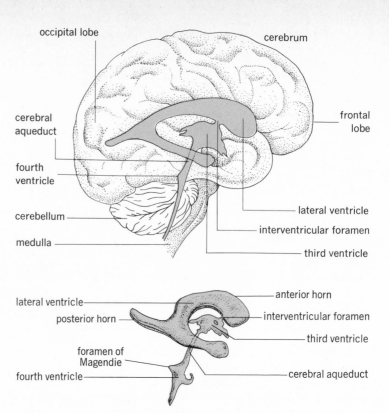

occipital lobe

cerebrum

cerebral aqueduct

fourth ventricle

cerebellum

medulla

frontal lobe

lateral ventricle

interventricular foramen

third ventricle

lateral ventricle

posterior horn

foramen of Magendie

fourth ventricle

anterior horn

interventricular foramen

third ventricle

cerebral aqueduct

FIGURE 9.10
The cerebrum with ventricles of the brain in lateral view.

matter. It is composed of millions of unmyelinated nerve-cell bodies and fibers.

Beneath the cortex the inside of the brain appears white. This is because it is made up largely of myelinated nerve fibers. These fibers usually lie in tracts and may connect the cortex with the spinal cord, or they may extend between different parts of the brain itself. The nerve fibers that compose the white matter are usually classified as *association fibers*, *projection fibers*, and *commissural fibers*. Association fibers connect convolutions, or gyri, within the same hemisphere, whereas projection fibers extend from the cortex to lower structures in the central nervous system. Projection fibers may be efferent to lower motor neurons, or they may be afferent fibers from the brainstem to sensory areas of the cortex. Commissural fibers extend between corresponding parts of the two hemispheres. The largest commissure connecting the right and left hemispheres of the cerebrum is called the *corpus callosum*.

LOBES Fissures divide the cerebrum into anatomical areas called *lobes*. They are the frontal, parietal, temporal, and occipital lobes (Figure 9.11a). The *insula*, formerly called the *island of Reil*, is sometimes considered a lobe. The lobes, with the exception of the insula, lie under cranial bones of the same name. Thus the frontal lobe lies below the frontal bone and is separated from the parietal lobe by the central sulcus (the *Rolandic fissure*). The parietal lobe is therefore posterior to the frontal lobe. The temporal lobe lies on the side of the brain below the lateral cerebral fissure (the *Sylvian fissure*). The occipital lobe is demarcated by the parietooccipital fissure. The insula (the island of Reil) is an area of gray matter located under the lateral

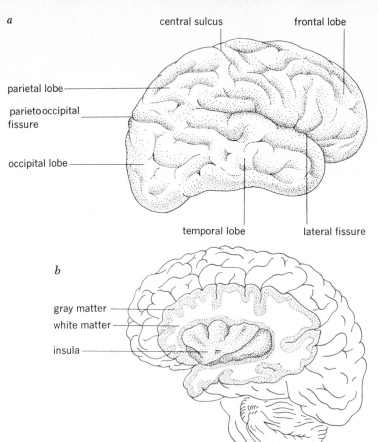

a central sulcus   frontal lobe

parietal lobe

parieto occipital fissure

occipital lobe

temporal lobe   lateral fissure

b

gray matter

white matter

insula

FIGURE 9.11

a The lobes of the cerebrum in lateral view. *b* Section of the left cerebral hemisphere cut away to expose the insula, an area of gray matter covering the corpus striatum.

cerebral fissure and can be seen only when the adjacent portions of the frontal and temporal lobes are raised (Figure 9.11*b*).

CEREBRAL LOCALIZATION It is not possible to localize many of the functions of the cerebrum. No one can point out the area that governs willpower or any one part of the brain that is the seat of music appreciation. Still, a number of cerebral functions originate in localized areas.

The *motor area* lies just anterior to the central sulcus. In the neurons of this area motor impulses arise. These are voluntary impulses that cause contractions of voluntary muscles. Experimentally it can be demonstrated that when certain parts of the motor area are stimulated, certain groups of muscles or individual muscles respond. The area governing the thigh lies at the upper part of the motor area, while the muscles of the face and tongue are controlled from the lower part (Figure 9.12). It is interesting that a greater area proportionately is devoted to governing the muscles of the hand and tongue than to the muscles of the trunk. The motor speech area is concerned with the coordination of all the muscles of the face, tongue, and throat necessary for speaking. It is commonly called *Broca's area*. In right-handed persons the speech area is developed in the left hemisphere.

Not only is the motor area inverted with respect to the body position it controls, but the motor area of the right cerebral hemisphere governs the

Cerebrum

217

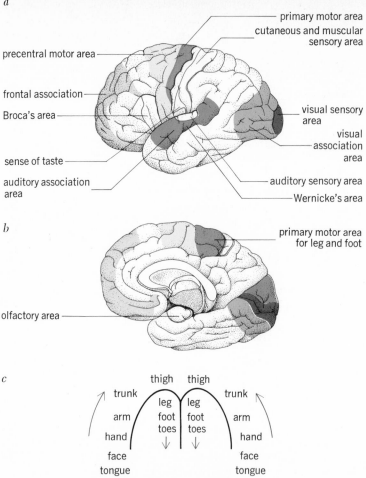

primary motor area

cutaneous and muscular sensory area

precentral motor area

frontal association

Broca's area

visual sensory area

visual association area

sense of taste

auditory association area

auditory sensory area

Wernicke's area

b

primary motor area for leg and foot

olfactory area

FIGURE 9.12

Localized functional areas of the cerebral cortex: *a* left cerebral hemisphere, lateral view; *b* medial view; *c* primary motor area, frontal section. The labels indicate the part of the body governed from motor areas that lie between the cerebral hemispheres and laterally along the central sulcus. The lateral motor area of the brain is inverted with respect to the body.

c

thigh thigh

trunk leg Y leg trunk

arm foot foot arm

hand toes toes hand

face ↓ ↓ face

tongue tongue

primary motor area (frontal section)

movement of muscles of the left side of the body and vice versa. The reason for this is that the nerve tracts from the motor area cross over either at the base of the medulla or at various levels in the spinal cord (Figure 9.13). It would appear that in right-handed persons the coordinating centers of the left cerebral lobe are somewhat better developed.

A knowledge of the location and function of the motor area has been of great value in the diagnosis of many types of brain injuries. Motor paralysis of a given part can result from injury or pressure involving the motor area of the brain. Cerebral hemorrhage, often referred to as a "stroke" or *apoplexy*, while usually occurring deep within the brain in the internal capsule, has the same effect as a surface injury. For example, if a blood clot located below the motor area of the right cerebral hemisphere prevents nervous impulses from reaching the muscles, a motor paralysis of the left side of the body results.

The precentral motor area is not as specific as we are often led to believe. The instrument ordinarily used for experimental exploration is a stimulating electrode. If the upper part of the motor area is stimulated, muscular movements occur in the lower extremity on the opposite side of

The brain and spinal cord

218

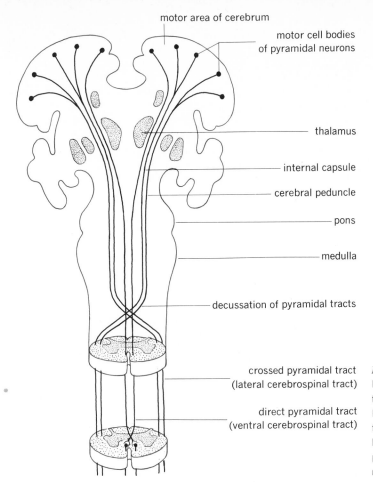

motor area of cerebrum

motor cell bodies of pyramidal neurons

thalamus

internal capsule

cerebral peduncle

pons

medulla

decussation of pyramidal tracts

crossed pyramidal tract (lateral cerebrospinal tract)

direct pyramidal tract (ventral cerebrospinal tract)

FIGURE 9.13
Diagram illustrating the crossing of the descending pyramidal tracts. Neurons of the direct pyramidal tract may cross at any spinal level, where they synapse with appropriate peripheral motor neurons.

the body. But if one stimulates the middle or lower part of the motor area with adequate stimuli, lower extremity movements occur also. It is found that in the area for trunk movement, the muscles of the trunk have the lowest threshold, that is, are most easily stimulated, but, evidently, the various regions are not specific.

It is generally agreed that the cortex is composed of five to seven layers of neurons, but the nonspecific effect apparently is not caused by diffusion of impulses at the cortical level. Electrical stimulation of the surface area can produce very different motor effects from those produced by normal nerve stimuli originating in this area. The sequence of neurons or pattern of distribution may be very different in the normal functioning organism.

Stimulation of the cortex of unanesthetized monkeys with implanted electrodes elicits motor effects in all areas of the cortex. Such experimentation suggests that probably all areas of the cortex are actually *sensorimotor;* that is, they are not exclusively sensory or exclusively motor. The region directly anterior to the central sulcus is predominantly motor, whereas the region posterior to the central sulcus is predominantly sensory, or afferent; but even areas such as the visual cortex contain motor fibers. Motor fibers in localized sensory areas produce a motor response in muscles associated with the activity of the sense organ involved; for example, stimulation in

Cerebrum

the visual area may evoke eye and head movements but only the precentral motor area controls movement of the body as a whole.

Originally, we presented the classical concept that voluntary movement originates in the motor area of the cerebrum. We must now consider that other areas influence the motor area even to the extent of causing some doubt that all direct motor impulses normally originate there. The motor cortex receives sensory stimuli from a wide variety of receptors, from the eyes, ears, and a great many other sources. Since coordination plays such an important part in all bodily movement, lower subcortical coordinating areas may be largely responsible for routine movement. Consider that, in walking, the position of the leg muscles varies at all times. This requires sensory feedback impulses from the muscles to the brain in order to maintain essential muscular coordination. Even zones formerly considered remote, such as the parietal and occipital zones, may have some influence on spatial orientation. The premotor zone neurons coordinate certain muscular activity, and the frontal zone plays some part in programming movements.

Posterior to the central sulcus there lies a sensory area for the interpretation of sensations such as touch, pain, temperature, pressure, and muscle sense. This area is called the *cutaneous sensory area*. It is thought that the subdivisions of this area correspond approximately to those of the motor area just anterior to the central sulcus. The right sensory area interprets sensations received from the left side of the body and vice versa.

The *visual interpreting area* is located at the back of the brain in the occipital lobes. Nervous impulses arising in the retina are conveyed to the interpreting area by nerve tracts. This sensory area enables one to interpret and understand what he sees. A closely associated function is concerned with the ability to understand written words or symbols. Failure of this function is recognized as a form of *aphasia* in which the individual is unable to read his own language. His eyes may be able to see and follow the printed words, but the symbols mean nothing to him.

The *auditory areas* are concerned with the interpretation of the sense of hearing and are located in the temporal lobes. Each area receives nervous impulses from both ears. Injury to closely related areas can result in a type of sensory aphasia in which the individual loses his memory of the meaning of words. Although hearing may remain unimpaired, the words or their meaning is not recognized. An area located in the temporal lobe close to the auditory area is known as *Wernicke's area*. Injuries to this area result in failure to understand the spoken language.

Interpreting areas for the sense of taste and smell have not been definitely located. Apparently a cortical area associated with the sense of taste is located near the Sylvian fissure and at the base of the central sulcus where sensory impulses from the face and tongue are interpreted. Located ventrally below the frontal lobes of the cerebrum are the two olfactory bulbs, where important synaptic connections are made between the fibers of the olfactory nerve and the interpreting area of the brain. The interpreting area for the sense of smell is thought to lie deep within the cerebrum associated with the limbic system. This is part of the primitive brain known as the *rhinencephalon*.

It may be of interest to give a brief consideration to the development of the embryonic cerebrum. The older part, from an evolutionary standpoint, is the *paleocortex*, which includes the *corpus striatum*, the *hippocampus*, the *septum*, and the *pyriform* cortex (Figure 9.14). These four structures form the rhinencephalon of the more primitive vertebrate brains, now called the *limbic system* in man. The corpus striatum and the gray matter above it

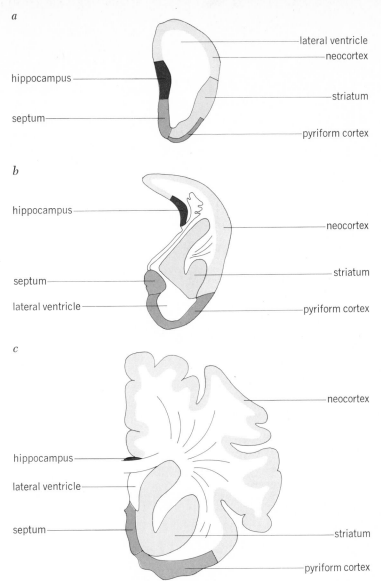

a

lateral ventricle
neocortex

hippocampus

septum

striatum

pyriform cortex

b

hippocampus

neocortex

septum

striatum

lateral ventricle

pyriform cortex

c

neocortex

hippocampus

lateral ventricle

septum

striatum

pyriform cortex

FIGURE 9.14
Early development of the cerebrum. Five main subdivisions are illustrated in coronal sections: *a* 15-mm embryo; *b* 50-mm embryo; *c* adult (human). The hippocampus, septum, striatum, and pyriform cortex form the rhinencephalon (called the limbic system in man), the primitive olfactory part of the brain. The striatum and the cortex of gray matter above it form the insula. The older structures of the rhinencephalon become involved in functions other than olfactory in the human being and now form part of the limbic system. (*Limbic* means border.) (*After W. J. S. Krieg, "Functional Neuro-anatomy," 2d ed., McGraw-Hill Book Company, New York, 1953.*)

form the insula. The hippocampus appears to be concerned with long-term memory. The rhinencephalon was the primitive olfactory brain, but in the human being it has been superseded by the neocortex, although it still retains some olfactory connections.

The newer portion of the brain, phylogenetically, is the *neocortex*. This more highly developed neocortex has grown posteriorly and has covered over the paleocortex. It is now the greatly enlarged cortex which provides the motor area and the sensory association areas, with the exception of the olfactory sense.

The *basal ganglia* consist of areas of gray matter deep within the white matter in the lower part of the brain. They include the *caudate nucleus,* the *putamen,* and the *globus pallidus.* The latter two form the *lentiform* (lens-shaped) *nucleus.* The caudate nucleus has a thickened head portion and a long slender tail, at the end of which is an enlargement called the *amygdala*

Cerebrum

221

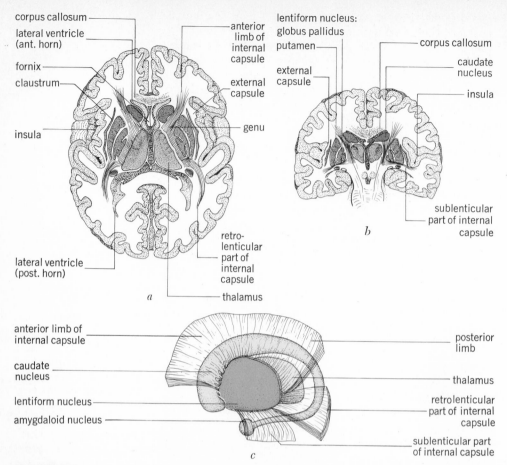

FIGURE 9.15

Internal structure of the brain: *a* horizontal section; *b* coronal section; *c* basal ganglia and thalamus.

The caudate and lentiform nuclei, including the internal capsule, compose the corpus striatum. The internal capsule is an important nerve tract, probably best seen in a horizontal section of the brain. It is V-shaped, with anterior and posterior limbs meeting at the apex which is called the *genu*. The internal capsule is composed of motor and sensory fibers from all parts of the cerebral cortex. Cerebral hemorrhage, often called a "stroke," commonly occurs at the internal capsule.

The brain and spinal cord

222

or amygdaloid body (Figure 9.15*a*, *b*, *c*). Not much is known about the function of the amygdala except that it is closely associated with the hypothalamus. Experimental injury to the amygdala in animals appears to produce an emotional response. The basal ganglia are not easily reached for experimentation without injury to other parts of the brain, so their functions are not clearly understood, but they have elaborate connections with the thalamus, hypothalamus, and reticular system, to be discussed later. The basal ganglia are thought to help coordinate movements of skeletal muscles, as in walking. There is also some indication that they have an inhibitory effect on impulses from the motor cortex to the skeletal muscles. Malfunction of some of the basal ganglia give rise to muscular tremors characteristic of Parkinson's disease.

The *diencephalon* is the posterior part of the forebrain. The lines of demarcation are not distinct, since many structures lie partly in the telencephalon and partly in the diencephalon. Thus the choroid plexus of this region is continuous with that of the telencephalon, since the greater part of the third ventricle is the cavity of the diencephalon and a part of it is the cavity of the telencephalon. The thalamus region may be divided into an upper portion, or epithalamus; an intermediate portion, the thalamus proper; and a lower portion, the hypothalamus. From the epithalamus there arises an outgrowth that becomes the *pineal body*. This structure apparently functioned in prehistoric reptiles as a median, light-sensitive third eye; in man it is usually considered as an endocrine gland, although its function is not well understood. The pineal body contains an enzyme which catalyzes serotonin to melatonin. The most striking example of the action of melatonin is its effect on the melanocytes of the frog, where it causes bleaching of the frog skin. In man, the pineal body appears to influence body rhythms such as "biological clock" activities through indirect response to light (see discussion of the pineal body in Chapter 20 on endocrine glands).

The *thalamus* is the largest of a number of areas of gray matter deep within the lower part of the brain. It is an important relay center for both motor and sensory impulses and has extensive cortical connections. It has been likened to a telephone switchboard where messages are received and relayed. In its development it has retained synaptic connections with the optic nerves and is often referred to as the optic thalamus (Figure 9.15). All the senses, however, except olfactory fibers have connections by way of the thalamus.

The *hypothalamus* is one of several structures located in the basal portion of the diencephalon. It is situated below the thalamus and close to the base of the stalk (infundibulum) of the *hypophysis* (pituitary). Various autonomic functions are controlled from centers in the hypothalamus. The body-temperature regulatory center is located here and also a center concerned with thirst and water metabolism. A feeding and hunger center has also been identified. It is a regulatory center for both the sympathetic and parasympathetic divisions of the autonomic nervous system. "Sham" rage in animals has long been associated with stimulation at the hypothalamic level. Functionally, the hypothalamus is closely associated with various endocrine activities of the hypophysis.

The optic chiasma and the mammillary bodies in addition to the infundibulum and hypophysis are visible in a ventral view of the brain. The optic chiasma is formed by a crossing-over of some of the fibers of the optic nerve. The mammillary bodies do not have clearly defined functions, but since they have synaptic connections with the hypothalamus, they may have related activities.

LEARNING AND MEMORY PROCESSES

Memory is the ability to recall events or information about previous experiences that may have happened only a few minutes ago or that may have occurred many years ago. There seems to be a short-term type of memory, such as remembering a telephone number until one has dialed it, and a long-term memory of facts and events which may last for the lifetime of the individual.

One theory is that of facilitation. This theory attempts to explain memory on the basis that the passage of an impulse over a set of neurons and synapses may make it easier in some way for similar impulses to follow the same path. Eventually, a certain pathway would become facilitated and easily recalled. This hypothesis explains the process of learning by repeating a list of facts or learning the words of a foreign language by repetition. Facilitation is also given as an explanation of habit formation; muscular performance becomes easier and more skilled with repetition.

The mechanism of memory, however, is still largely unknown. If it can be localized, possibly it may be associated with the hippocampus, an area of the *limbic system*, but it is more likely, considering the complexity of neuronal connections, that it is not localized. The hypothesis that short-term memory is produced by neuronal circuits has gained some credence. It is considered that the impulse goes round and round a circuit of neurons and synapses in the process of sustaining memory for a brief period.

Another line of investigation concentrates on the nature of conduction across the synapse. Learning may increase the conductivity of a certain group of synapses, but it is more likely that the postsynaptic membrane may increase its sensitivity to the transmitter substance. The sensitivity might very well increase during the learning experience and decrease during a period of forgetting. Some very interesting experiments performed by Penfield seem to indicate that memories can be stimulated in certain cases by introducing electrodes into an area of the temporal lobe below the auditory area. He calls this the interpretive cortex. These experiments have been performed during surgical operations on the brain under local anesthesia.

With very mild electrical stimulation in the temporal area, patients sometimes recall events that occurred in their past quite vividly and in great detail. Penfield has never obtained these results from stimulating any other part of the brain. The memory record may not be in the interpretive area itself but in some deeper area related to it.

It is not surprising that many investigators have turned to research on ribonucleic acid (RNA) and protein metabolism in their search for the process that involves memory storage. A molecular theory for long-term memory is an attractive hypothesis but very difficult to establish on an experimental basis. It holds that memory is stored according to the configuration of molecules, especially RNA and protein. Learning, in itself, may very well involve changes in the structure of the RNA molecule and its metabolism within brain cells, but this concept awaits further investigation.

It is known that there is a loss of brain cells with aging. This loss in itself would eventually reduce mental ability. The aged person often becomes forgetful or is unable to recall recent events although he may remember the earlier events in his life quite well. Other related factors may include a reduced blood supply to the brain or a reduction in the oxygenation of brain cells.

CHEMICAL COMPOUNDS AFFECTING THE BRAIN

The chemical mechanisms of the brain have assumed a position of great importance in present-day research. Studies of anticonvulsants, chemical transmitters, or mediators of the nerve impulse, chemical inhibitors, tranquilizers, and drugs that produce hallucinations have all been in the forefront of investigation.

The amine *serotonin* (5-hydroxytryptamine) was discovered in 1947 and was found to be present in brain tissue. It is known to stimulate contraction in the smooth muscle of blood vessels and is a strong vasoconstrictor. Blood platelets contain serotonin and release it following an injury. Although its function in the brain is somewhat obscure, it is considered to be a putative transmitter in the central nervous system. The amine is present in considerable amounts in the hypothalamus, together with norepinephrine. Experimentally administered serotonin apparently does not cross the blood-brain barrier, but the amino acid precursor 5-hydroxytryptophan does. This amino acid is converted to serotonin in the tissues.

Another compound, lysergic acid diethylamide (LSD), produces hallu-

The brain and spinal cord

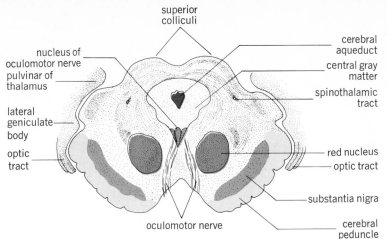

FIGURE 9.16
A section through the midbrain revealing the red nuclei and the substantia nigra. These structures are part of an intrinsic system involving the basal ganglia and the thalamus. They function as an intricate relay system of short neurons linking various basal ganglia and in carrying impulses between the cerebral cortex, basal ganglia, and the spinal cord. [*Redrawn from J. Parsons Schaeffer (ed.), "Morris' Human Anatomy," 12th ed., McGraw-Hill Book Company, New York, 1966.*]

cinations and severe mental disturbances even though administered in minute amounts. When serotonin is used experimentally to induce contraction in smooth muscle, lysergic acid diethylamide acts as an antagonist to serotonin; but such an antagonism in the brain has not been proved.

Several drugs have attained prominence in the treatment of certain types of mental illness. They exert a quieting effect, affording relief from anxiety states and emotional imbalance, and are commonly known as *tranquilizers*. Several are derived from the root of a toxic plant of the genus *Rauwolfia*. Tranquilizing drugs have been used clinically to great advantage in the treatment of mental illness. The alkaloid reserpine is one of these. It is known to release serotonin from brain cells. The released serotonin presumably is metabolized by the enzyme monamine oxidase. One of the actions of reserpine is to block the production of norepirephrine. Reserpine is used also as an antihypertensive agent to reduce blood pressure.

The exact function of tranquilizing drugs is not known, but they seem to exert their effect on subcortical areas, such as the midbrain reticular formation and the hypothalamus, and on certain nuclei associated with the primitive olfactory area, now thought to be concerned with emotion.

The long-term effects of tranquilizing drugs are not known. A long period of experimentation will be necessary to determine their value. The human nervous system is well adapted and quite capable of withstanding ordinary emotional emergencies and anxieties. Tranquilizing drugs should not be used as an easy way to avoid cares and worries that are a part of everyday life.

Another substance considered to be a possible neurotransmitter is *dopamine*. The amino acid tyrosine is acted upon by the enzyme tyrosine hydroxylase and converted to 3,4-dihydroxyphenylalanine, commonly called *dopa*. Dopamine is produced as a result of the conversion of dopa to dopamine by the action of dopa decarboxylase.

Dopamine nerve tracts involve the *substantia nigra*, the caudate nucleus, and the putamen of the *corpus striatum*. The caudate nucleus and putamen are basal ganglia and are included in the extrapyramidal system. The *substantia nigra* is a layer of gray matter embedded in the cerebral peduncles. In a cross section through the midbrain, this dark-colored layer lies anterior or ventral to the *red nucleus* on either side (Figure 9.16). The

Chemical compounds affecting the brain

225

FIGURE 9.17

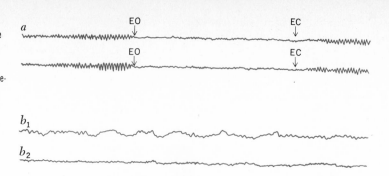

a Normal EEG. This illustrates the effects of eye opening (EO) and eye closing (EC) on the alpha waves which were taken with parietal to occipital lobe linkages on the left and right sides, respectively. *b₁* Abnormal EEG. Note that the sine wave activity seen about equally on left and right sides in *a* is sharply reduced on the left side here and very slow irregular waves are recorded, whereas the waves from the right hemisphere *b₂* are normal. The reduced normal activity and abnormal slow waves resulted from occlusion of the middle cerebral artery on the left side. (*Courtesy of Hartford Hospital, Hartford, Connecticut.*)

substantia nigra contains a dark pigment which is probably some form of melanin formed from dopa or dopamine by polymerization.

The tremors of Parkinson's disease appear to involve the extrapyramidal pathways and are relieved in many cases by the administration of L-dopa. Dopamine does not cross the blood-brain barrier, but L-dopa readily enters the brain. This does not prove that dopamine is a neurotransmitter, but it may indicate a stimulating effect, perhaps in synaptic activity.

Amphetamines are used extensively as energizers, but little is known about their actual effect on the brain. It is known that they depress the appetite and make sleeping more difficult. As energizers, they increase nervous activity, probably at the synapse. One can be sure that these drugs do not create energy; they simply permit the body to go into debt for stored energy by stimulating undue activity and through loss of sleep and rest. This may result in poor health, for the debt must be paid.

BRAIN WAVES

The active brain produces electrical impulses in steady rhythm. These brain waves can be recorded by an instrument called an encephalograph. Electrodes are placed on the scalp and the amplified electrical impulses are recorded as an electroencephalogram (EEG), a series of wavy lines or tracings (Figure 9.17). The resting rate of impulses from the cerebral cortex of a person who is awake but with eyes closed and not unduly excited is around 10 to 12 per second; these are called alpha waves. It is thought that they arise in the reticular formation and then stimulate the cerebral cortex so that one is able to stay awake.

An alert state produces greater cerebral activity, and beta waves are recorded at a frequency that varies considerably but is often in the range of 16 to 30 hertz. The origin of brain waves is not known with certainty, but they may be the result of current flow in the mass of pyramidal cell dendrites of the cerebral cortex. The rhythmic pattern may originate in the thalamus, where a pacemaker has been discovered, but this will require further study.

Waves of very low frequency, fewer than four or five per second, occur during deep sleep. These are delta waves, characteristic of restful sleep. Dreaming, accompanied by eye movements, produces a period of fast waves resembling those of the alert state.

Abnormal brain waves recorded by the EEG aid in detecting mechanical injuries to the brain as well as brain tumors and epilepsy. Two kinds of epilepsy, petit mal and grand mal, produce quite different types of brain waves. Petit mal, in which a person may lose consciousness for a few seconds, is recorded on the EEG as a slow rhythm with a spike and a broader dome-shaped pattern, usually alternating. In grand mal, the person loses consciousness and falls into convulsions. The EEG indicates great neural excitement and a confusion of strong impulses.

HEADACHE

One of the common ailments of mankind is attributed to pain in the head. A headache is usually a symptom of some malfunction elsewhere in the body. It may be the result of unusual fatigue, eyestrain, indigestion, sinus trouble, too much alcohol, or strong emotional tension. However, not much is known about the actual source of the pain. The brain itself is considered to be insensitive to pain. If there is unusual pressure within the blood vessels, though, pain receptors in the walls may respond and produce pain. Pain receptors in the meninges and venous sinuses also may be involved. Migraine headaches may be severe and are of several kinds. Older people are more often affected. There is often an accompanying gastrointestinal disturbance. One type is characterized by wavy lines or flashes of light in the eyes, causing temporary, localized blindness, apparently in the retina but more likely arising from the brain. Persons subject to this type of migraine are often abnormally sensitive to light. The head may feel sore for days following a migraine attack.

THE BRAINSTEM

Great nerve tracts connecting the spinal cord with higher synaptic levels in the cerebrum compose the white matter of the brainstem. The gray matter consists of numerous ganglia and nuclei interwoven into the white matter. Some of these are the ganglia of the cranial nerves, since all the cranial nerves arise from the brainstem except the first pair, the olfactory nerves. Large motor and sensory tracts pass through this region, and many reflex centers are located here. The brainstem includes structures of the midbrain and hindbrain, including the pons and medulla but not including the cerebellum.

The *midbrain* is the upper portion of the brainstem. In lower animals, the optic lobes develop in this region, but in higher animals and in human beings the midbrain becomes covered over by the cerebrum and loses most of its optic-tract connections. The greater part of the midbrain consists of nerve tracts that carry impulses between the cerebrum and the cerebellum, medulla, and spinal cord. The anterior part is composed largely of two great nerve tracts, the cerebral peduncles, or crura cerebri. The *cerebral aqueduct* (aqueduct of Sylvius) extends from the third ventricle to the fourth and traverses the midbrain. Between the aqueduct and the cerebral peduncles lie the *red nuclei* (Figure 9.16), two masses of gray matter connected by nerve tracts with the cortex of the cerebrum, the thalamus, the cerebellum, and the spinal cord. The red nuclei, through their connections, are thought to be concerned with muscle tone and with delicate or skilled movements. The nuclei of the oculomotor and trochlear cranial nerves also are located in the midbrain. Posterior to the cerebral aqueduct is an area called the

The brainstem

227

tectum. Within this dorsal area are four rounded structures called the *superior* and *inferior colliculi,* or the *corpora quadrigemina* (Figure 9.11). They are the optic lobes of the brains of lower vertebrates, but in man they have become subdivided into four structures. The superior colliculi still retain some connections with the optic tract and are a center for certain visual reflexes in most mammals, and possibly in man. The inferior colliculi have become auditory in function and may therefore be considered as an auditory reflex center in mammals, including man.

THE RETICULAR FORMATION   A diffuse mixture of gray matter throughout the white matter of the brainstem is known as the *reticular formation.* The various visceral connections are not well defined from an anatomical standpoint. It has been established that the reticular formation receives fibers from the cortex, from the hypothalamus, and from the nuclei of cranial nerves associated with the brainstem region. It also receives collateral branches from ascending sensory tracts and acts as a reflex center. Nerve impulses are relayed or projected to certain cranial nerves and to descending tracts of the spinal cord.

Some nerve cells in the reticular formation have an inhibitory function as they project to descending spinal neurons. They raise the threshold of motor neurons receiving impulses and inhibit muscular activity. Other nerve cells are excitatory and facilitate muscular contraction.

Considerable interest has developed in regard to the part played by the reticular formation in arousal from sleep and in maintaining a state of consciousness. Stimulation of the reticular formation causes an unanesthetized animal to be aroused from sleep. It is known that ascending spinal sensory tracts give off collateral branches to the reticular system. One interpretation of arousal is that sensory impulses excite the reticular system and then gradually diffuse through subcortical areas. When they finally arrive at the cerebral cortex, awakening or a return to consciousness occurs. Loss of consciousness can be produced by injury or destruction of the midbrain reticular formation in animals. A small lesion at the upper level of the reticular formation and close to the posterior part of the hypothalamus is said to produce a state resembling sleep in animals. Apparently, subcortical areas in addition to cortical areas are involved in sleep and conscious states. An understanding of the function of the reticular formation and its relation to sleep, arousal, and consciousness may prove to be of the greatest importance in human physiology.

The *metencephalon* lies just below the midbrain. Its most obvious structures are the cerebellum, which lies posterior to the brainstem, and a bridge of nerve tracts, called the *pons,* extending across the anterior part of the brainstem. The fourth ventricle is the cavity of this region. All ventricles contain choroid plexuses, which are highly vascular folds and elaborations of the pia mater.

CEREBELLUM   Two cerebellar hemispheres are located below the occipital lobes of the cerebrum. Between the two hemispheres is an area called the *vermis.* The *cerebellum* has a cortex of gray matter, but it differs from that of the cerebrum in certain respects. It is not convoluted in the same manner but appears as a series of layers. Within the cortex are the large cells of Purkinje

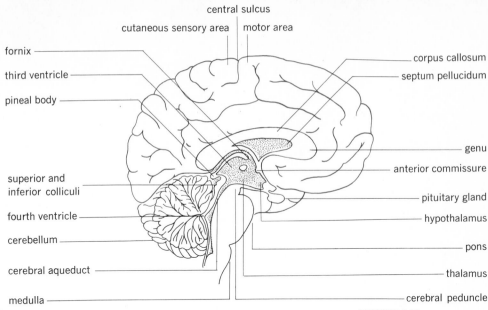

central sulcus

cutaneous sensory area | motor area

fornix

third ventricle

pineal body

corpus callosum

septum pellucidum

genu

anterior commissure

superior and
inferior colliculi

fourth ventricle

cerebellum

cerebral aqueduct

medulla

pituitary gland

hypothalamus

pons

thalamus

cerebral peduncle

FIGURE 9.18

Sagittal section through the brain.

(Figure 8.1), with their remarkable branching dendrites. These efferent cells are found only in the cortex of the cerebellum.

The interior of the cerebellum is largely composed of white matter, although the gray matter of the cortex descends deeply into the white matter and elaborates into an inverted treelike pattern of branching. The branching pattern, as seen in sections of the interior, gave rise to the name *arbor vitae*, or tree of life, as a descriptive name for this area. There are some nuclei of gray matter within the cerebellum, the largest of which are the dentate nuclei. Great nerve tracts, the cerebellar peduncles, connect the cerebellum with the cerebrum, the pons, and the medulla (Figure 9.18).

The *pons* consists essentially of horizontal nerve tracts that serve to connect the two hemispheres of the cerebellum anteriorly, and vertical tracts that connect the cerebrum with the medulla. There are also nuclei of gray matter including the nuclei of the trigeminal (Vth cranial), abducent (VIth cranial), and the motor nucleus of the facial (VIIth cranial) nerves.

The cerebellum has been called the "secretary" to the cerebrum. It does not initiate motor responses but functions to coordinate muscular movements so that the action will be smooth and efficient instead of jerky and uncoordinated. The cerebellum is also concerned with the equilibrium of the body. It is connected by nerve fibers with the semicircular canals of the inner ear, which are also concerned with equilibrium. The cerebellum is able to direct the muscular coordination that tends to keep the body balanced in various positions. It is concerned also in coordinating impulses received from the sense of hearing, the sense of sight, and the tactile sense.

The cerebellum helps to maintain the tone of muscles. Birds from which the cerebellum has been removed allow the wings to droop and are never able to fly. Mammals with the cerebellum removed exhibit a peculiar uncoordinated walking gait. All functions are below the level of conscious activity; sensory impulses received do not produce sensation.

Localization of function in the cerebellum does not appear to be as definite as in the cerebrum. Each cerebellar hemisphere controls the coordi-

Cerebellum

229

nation of movement of the appendages on the same side of the body, whereas the vermis controls the coordination of the trunk musculature. However, an injury to the right hemisphere may affect the right side of the body as well as the appendages of that side. Localization of projection areas has been demonstrated in animals, where the sense of touch and the proprioceptive sensations from muscles seem to be located middorsally on the cortex. Auditory and visual stimuli are received in an area located at about the middle of the dorsal aspect. Equilibrium seems to be controlled from two centers, one in an anterior and one in a posterior cortical area. It seems significant that the tactile area and the proprioceptive areas from muscles coincide and also that the auditory and visual areas overlap. Snider suggests that this association is relevant to behavior in that we draw the hand away from an unexpected touch and we look in the direction of an unexpected sound.

THE MEDULLA

The *medulla* is the base of the brainstem, the myelencephalon (Figure 9.3a). It is continuous with the spinal cord but does not have the same internal structure. Though the nerve tracts are continuous, some are larger and better defined in the medulla. Some of the fibers cross to the opposite side (the decussation of the pyramidal tracts). The continuous gray matter of the cord is broken up into groups of nuclei in the medulla. These include the nuclei of the IXth, Xth, XIth, and XIIth cranial nerves. The central canal of the cord is continuous anteriorly through the medulla, where it opens into the lower part of the fourth ventricle. The medulla, however, is not just the upper part of the spinal cord. It contains vital reflex centers, such as the cardiac inhibitory center, which by way of the vagus nerve acts in slowing the heart rate; the vasoconstrictor center, responsible for the constriction of peripheral blood vessels and the consequent rise of arterial pressure; and a respiratory center, which provides the nervous stimulus for regular respiratory movements. The medulla also controls a number of common reflex activities, such as laughing, coughing, and sneezing, and many of the activities of the digestive tract.

CRANIAL NERVES

I	Olfactory	Sensory
II	Optic	Sensory
III	Oculomotor	Motor
IV	Trochlear	Motor
V	Trigeminal	Mixed
VI	Abducent	Motor
VII	Facial	Mixed
VIII	Vestibulo-cochlear	Sensory
IX	Glosso-pharyngeal	Mixed
X	Vagus	Mixed
XI	Accessory	Motor
XII	Hypoglossal	Motor

The cranial nerves are 12 pairs of nerves arising from the brain within the cranial cavity. They are numbered from I to XII and are also named. The cranial nerves are essentially like spinal nerves, but they are more highly specialized. Though some are mixed nerves, there are also nerves that have lost one branch or the other and are, for the most part, motor or sensory nerves.

I The *olfactory nerve* arises from sensory receptors located in the upper part of the mucous membrane that lines the nasal cavity. The separate fibers grow inward through the cribriform plate of the ethmoid bone to the olfactory bulbs, where they make synaptic connections with the secondary neurons leading inward to the olfactory-interpreting centers of the brain. The nerve is purely sensory and is concerned with carrying nervous impulses that give rise to the sense of smell.

The olfactory nerve is peculiar in that the individual fibers grow inward. There is therefore no sensory ganglion on the olfactory nerve. The fibers, furthermore, are unmyelinated.

II The *optic nerve* is a sensory nerve concerned with the sense of sight. It arises from

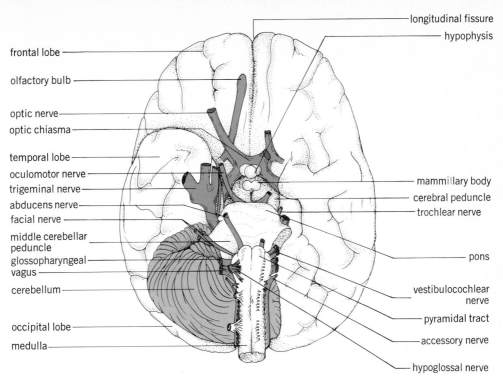

frontal lobe

olfactory bulb

optic nerve

optic chiasma

temporal lobe

oculomotor nerve

trigeminal nerve

abducens nerve

facial nerve

middle cerebellar peduncle

glossopharyngeal

vagus

cerebellum

occipital lobe

medulla

longitudinal fissure

hypophysis

mammillary body

cerebral peduncle

trochlear nerve

pons

vestibulocochlear nerve

pyramidal tract

accessory nerve

hypoglossal nerve

FIGURE 9.19

Ventral view of the brain including the bases of cranial nerves.

ganglion cells located in the retina of the eye, and its fibers form the optic tract which leads back to the lateral geniculate body of the thalamus. From there, sensory impulses are conveyed by secondary neurons to the visual interpreting area in the occipital lobe of the cerebrum. In a ventral view of the brain (Figure 9.19) the optic nerves can be seen to form a structure something like a letter X. This is the optic chiasma, where the fibers from the inner half of the retina of either eye cross over and go to opposite sides of the brain. The crossing of some of the fibers of the optic nerves probably results in better coordination of responses between the eye and the brain.

III The *oculomotor;* IV the *trochlear;* and VI the *abducent nerves* are essentially motor nerves to the muscles that move the eyeball. The oculomotor and trochlear nerves arise from nuclei of gray matter located beneath the cerebral aqueduct in the midbrain. The trochlear-nerve nucleus is posterior to the nucleus of the oculomotor nerve. The nucleus of the abducent nerve is in the lower part of the pons, beneath the fourth ventricle. The oculomotor and trochlear nerves emerge near the anterior border of the pons; the abducent nerve emerges at the lower border of the pons in the fissure between the pons and the medulla. The oculomotor nerve also carries fibers of the parasympathetic system to the circular muscle of the iris and to the ciliary muscle of the eye. There may be a few sensory fibers in nerves to the extrinsic eye muscles, but the presence of sense organs in these eye muscles has not been fully determined.

V The *trigeminal* is a mixed nerve with both motor and sensory nuclei. The motor nucleus and the sensory root are located in the pons laterad and below the fourth ventricle. The sensory nuclei are located in a large semilunar ganglion (gasserian ganglion) outside the brainstem. There are three large branches of the trigeminal nerve: ophthalmic, maxillary, and mandibular.

The ophthalmic branch is a sensory branch and carries impulses originating in the surface of the eye, in the lacrimal gland, and from the nose and forehead.

The maxillary branch is also a sensory branch and has a broad distribution of its

Cranial nerves

nerves. Among the structures supplied are the teeth and gums of the upper jaw, the upper lip, and the cheek.

The mandibular branch is a mixed nerve. It has many small branches. Some of these nerves supply the teeth and gums of the lower jaw, the chin, the lower lip, and the tongue. It is motor to the muscles concerned with mastication and also contains sensory fibers from proprioceptors in muscle tendons (Table 9.1).

VII The *facial nerve* is a mixed nerve. Its motor nucleus lies in the lower part of the pons, and fibers are supplied to the muscles of the face and forehead. The sensory branch is very small. Its fibers arise from the geniculate ganglion located in the temporal bone and are distributed to the anterior two-thirds of the tongue. They are concerned with the sense of taste. The motor branch also carries fibers of the parasympathetic system to the sublingual and submaxillary salivary glands. Parasympathetic fibers stimulate vasodilatation and secretion in these glands. Fibers concerned with taste sensation and parasympathetic fibers pass through the tympanic cavity in the *chorda tympani* branch.

VIII The *vestibulocochlear nerve* (acoustic) is a sensory nerve concerned with the sense of hearing and with equilibrium. It is composed of two nerves of different origin and function. The cochlear nerve carries auditory impulses. Its ganglion lies in the cochlea. From receptors in the spiral organ of Corti the auditory impulse is conveyed inward to the medulla. It crosses over to the opposite side and passes upward through the pons and midbrain over a series of neurons to the auditory-interpreting area in the temporal lobe of the cerebrum. A few efferent fibers are present also.

The vestibular nerve arises in the vestibular ganglion of the portion of the ear associated with the semicircular canals. It enters the medulla but has important connections with the cerebellum. It is concerned with maintaining equilibrium. The functions of the semicircular canals and vestibule are discussed later in connection with a study of the ear.

IX The *glossopharyngeal nerve* arises from the medulla and supplies the tongue and

TABLE 9.1
Cranial nerves 1 through 6

No.	Name	Type	Function
1	Olfactory	Sensory	Sense of smell
2	Optic	Sensory	Sense of vision
3	Oculomotor	Motor	Innervates extrinsic muscles of eyes except lateral rectus and superior oblique. Also innervates the levator palpebrae superioris. Parasympathetic fibers to sphincter of pupil and to ciliary muscle of lens
4	Trochlear	Motor	Superior oblique muscle of eye
5	*Trigeminal*	Mixed	
	Ophthalmic branch	Sensory	Anterior surface of eye; nasal mucous membrane; skin of forehead
	Maxillary branch	Sensory	Teeth of upper jaw; skin area of face above mouth and below eyes
	Mandibular branch	Mixed	Sensory to teeth of lower jaw, skin, lower lip, and tongue. Motor to muscles of mastication. Muscle sense
6	Abducens	Motor	Lateral rectus muscle

pharynx. It is a mixed nerve; the motor fibers are distributed to muscles of the pharynx, while sensory fibers are supplied to the tonsils, mucous membranes of the pharynx, and posterior part of the tongue. Stimuli resulting in the sense of taste originate from receptors located in the large papillae at the back of the tongue. This nerve also carries fibers of the craniosacral system. Secretory and vasodilator fibers are distributed to the parotid salivary gland.

X The *vagus nerve* is the longest cranial nerve. Its pathway lies from the medulla through the neck and thorax to the abdomen. It is a mixed nerve: sensory branches convey impulses from the mucous membranes lining the respiratory and digestive tracts; voluntary motor fibers are distributed to certain muscles of the pharynx and larynx.

The right and left vagus nerves send branches to the cardiac and pulmonary plexuses. Above the stomach they unite to form the esophageal plexus. Branches supplying the abdominal viscera arise below the esophageal plexus and contain involuntary fibers from both vagus nerves.

The vagus nerve is also one of the principal nerves of the craniosacral (parasympathetic) division of the autonomic nervous system. It carries inhibitory fibers to the heart and secretory fibers to the gastric glands and pancreas, as well as vasodilator fibers to the abdominal viscera. Autonomic fibers are also supplied to the bronchial tubes, esophagus, stomach, pancreas, gallbladder, small intestine, and ascending colon.

XI The *accessory nerve*, a motor nerve, is composed of two parts, a cranial and a spinal portion. The cranial part arises from a nucleus in the medulla and emerges from the side of the medulla just below the roots of the vagus. The spinal part arises from the spinal cord in the upper cervical region and ascends, passing upward through the foramen magnum. It then turns and descends beside the vagus. The cranial portion is accessory to the vagus and supplies most of the pharyngeal and laryngeal muscles. The spinal portion innervates the sternomastoid and trapezius muscles.

XII The *hypoglossal nerve* is a motor nerve distributed to the muscles of the tongue. It arises from the medulla. Injury to this nerve causes difficulty in speaking or swallowing.

Sensory fibers have been discovered in many of the cranial nerves formerly thought to contain only motor fibers. The cranial nerves listed as motor should be considered to be only predominantly motor, probably containing relatively few sensory fibers to muscles supplied. Some investigators (e.g. Merton) maintain that we have no sense organs in the extrinsic eye muscles that determine the direction of movement of the eyes (Table 9.2).

SPINAL CORD

The spinal cord is that portion of the central nervous system within the vertebral canal. It is continuous with the base of the brain anteriorly; posteriorly it tapers to a threadlike strand below the second lumbar vertebra. There are cervical and lumbar enlargements in the regions where large nerves to the appendages are given off.

In the early fetus the spinal cord extends the length of the spinal canal, but as the fetus grows, the vertebral column grows in length at a greater rate than the spinal cord. Hence the cord is drawn forward in the vertebral canal, and the roots of lumbar, sacral, and coccygeal nerves travel down the spinal canal to reach their normal places of exit. The canal below the second lumbar vertebra then contains the threadlike strand of the cord surrounded by lumbar, sacral, and coccygeal nerves. This tail-like group of nerves is called by a descriptive name, the *cauda equina* (Figure 9.20).

The cord is suspended rather loosely in the spinal canal. Since its diameter is considerably less than that of the canal, the vertebral column can be moved freely without injury to the cord.

Spinal cord

233

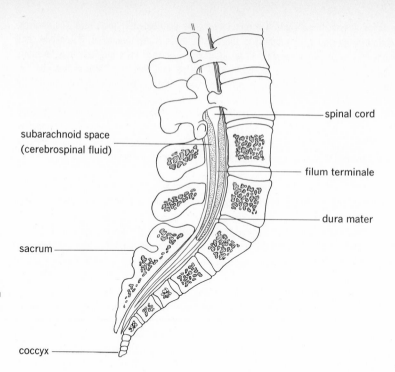

spinal cord

subarachnoid space
(cerebrospinal fluid)

filum terminale

dura mater

sacrum

coccyx

FIGURE 9.20

Longitudinal section through the lower part of the vertebral column and the spinal canal, showing the relationship between the spinal cord and the meninges in this region. (Diagrammatic)

TABLE 9.2
Cranial nerves 7 through 12

No.	Name	Type	Function
7	Facial	**Mixed**	
		Motor	Muscles of the face
		Sensory	Taste, anterior two-thirds of tongue
			Parasympathetic fibers to lacrimal, submandibular, and sublingual glands
8	*Vestibulocochlear*		
	Vestibular branch	Sensory	Hearing
	Cochlear branch	Sensory	Equilibrium
9	Glossopharyngeal	Mixed	Motor to swallowing muscles of pharynx
			Sensory to pressure receptors in carotid sinuses. Parasympathetic fibers to parotid salivary glands
10	Vagus	Mixed	Motor to muscles of pharynx, larynx; thoracic and abdominal viscera
			Sensory to mucous membranes of respiratory and digestive tracts
11	*Accessory*	Motor	
	Cranial portion		Pharyngeal and largyngeal muscles
	Spinal portion		Sternomastoid and trapezius muscles
12	Hypoglossal	Motor	Muscles of the tongue

The brain and spinal cord

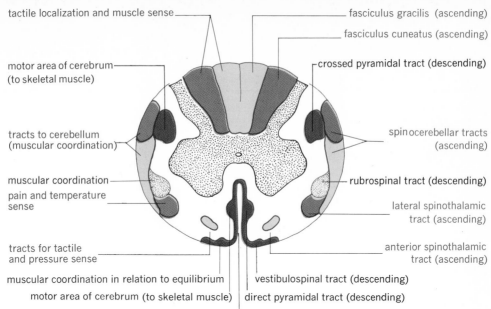

tactile localization and muscle sense — fasciculus gracilis (ascending)

fasciculus cuneatus (ascending)

motor area of cerebrum — crossed pyramidal tract (descending)
(to skeletal muscle)

tracts to cerebellum — spinocerebellar tracts
(muscular coordination) (ascending)

muscular coordination — rubrospinal tract (descending)

pain and temperature sense — lateral spinothalamic tract (ascending)

tracts for tactile and pressure sense — anterior spinothalamic tract (ascending)

muscular coordination in relation to equilibrium — vestibulospinal tract (descending)

motor area of cerebrum (to skeletal muscle) — direct pyramidal tract (descending)

ventral median fissure

FIGURE 9.21

Diagrammatic cross section of the spinal cord, indicating the general location of some of the principal nerve tracts. The motor area of the cerebrum gives rise to both the crossed and the direct pyramidal tracts.

CONDUCTION PATHWAYS OF THE SPINAL CORD *Some ascending tracts* The vertical neurons of the cord are arranged in orderly bundles. Many of these tracts have been carefully investigated and their origin, termination, and function recorded. Two of the large posterior ascending tracts are the *fasciculus gracilis* and the *fasciculus cuneatus* (Figure 9.21). The cell bodies of the neurons composing these tracts lie in the dorsal ganglia of spinal sensory nerves, and their fibers extend upward to the medulla, ending in the *nucleus gracilis* or the *nucleus cuneatus.* (A nucleus is a group of nerve-cell bodies within the central nervous system.) Other neurons connect the nuclei and the thalamus; a third set of neurons conveys impulses from the thalamus to the sensory interpreting areas of the cerebral cortex. This is the pathway by which the position of muscles (voluntary muscle sense) is interpreted and the sensations of touch are received.

Large tracts in the lateral part of the cord are direct cerebellar tracts. They are concerned with muscular coordination. Sensory impulses that may be considered as the unconscious muscle sense are conveyed upward to the cerebellum.

Spinothalamic tracts lie in lateral and ventral portions of the white matter. These pathways convey impulses to the thalamus and then to the cerebral cortex, where they may be interpreted as pain, temperature, pressure, touch, and muscle sense (Table 9.3).

Some descending tracts The neurons composing the pyramidal tracts have their origin in the motor area of the cerebral cortex. Most of them cross over to the opposite side in the medulla and descend in the lateral part of the cord; hence the name *crossed pyramidal tract (lateral corticospinal tract).* The fibers that do not cross in the medulla form two smaller ventral columns called the *direct pyramidal tracts (ventral corticospinal tracts).* The neurons of both tracts make synaptic connections with the motor-nerve roots of spinal nerves at various levels. The neurons of the direct pyramidal tract cross in the anterior white commissure just before they make a synaptic connection

Spinal cord

235

TABLE 9.3
Principal ascending tracts of the spinal cord

Name	Location in spinal cord	Origin	Terminal endings in brain	Functional sensation
Fasciculus gracilis Fasciculus cuneatus	Posterior white columns	In spinal ganglia on same side	Medulla	Touch, pressure, conscious muscle sense, motion
Spinocerebellar (direct cerebellar), dorsal and ventral	Lateral white columns	Neuromuscular receptors	Cerebellum	Unconscious muscle sense, muscular coordination
Lateral spinothalamic	Lateral white columns	Cell bodies in posterior gray columns of opposite side	Thalamus	Pain and temperature sense on opposite side
Anterior spinothalamic	Anterior white columns	As above	Thalamus	Touch and pressure

with a spinal-nerve root. These are the pathways of voluntary motor impulses to the muscles of the trunk, arms, and legs. The direct pyramidal tract supplies muscles of the trunk; the crossed pyramidal tract supplies muscles of the arms and legs. About two-thirds of the descending fibers cross over.

Neurons of the pyramidal tract do not synapse until they descend to a peripheral spinal motor-nerve outlet at a certain spinal level. Their fibers extend from the cerebral cortex to the spinal motor-nerve cell or to short association neurons before they synapse. The *extrapyramidal system* involves a greater area of the cortex, including the supplementary motor area. The neurons of this system may synapse several times at subcortical levels before reaching a spinal motor outlet. The subcortical levels include the basal ganglia, the red nucleus, and the reticular formation. Some corticothalamic and corticohypothalamic fibers are included also. The extrapyramidal system is concerned with complex voluntary movements in support of the pyramidal system. Probably the best explanation of its function is that a balance between the two systems permits a refinement of muscular movement. The tremors associated with Parkinson's disease are usually considered to involve pathology in certain areas of the basal ganglia but may also involve fibers between the cortex and other subcortical areas.

The *rubrospinal tracts* descend through the lateral part of the cord. The cell bodies of these neurons are located in the *red nucleus* of the midbrain. Their fibers cross immediately and descend to various levels of the cord where they make connections with spinal motor-nerve roots. Since the red nucleus has both cerebral and cerebellar connections, much of the voluntary muscular control may be transferred to involuntary muscular coordination over these pathways. Experimentally this pathway has been shown to be concerned with reflexes that aid in righting the body and with the tone of muscles affecting posture.

The vestibulospinal tract originates from the nucleus of the vestibular (VIIIth cranial) nerve in the medulla, and the neurons end at various levels

TABLE 9.4
Principal descending tracts of the spinal cord

Tract	Location in spinal cord	Origin in brain	Terminal endings	Motor function
Pyramidal tracts				
Lateral corticospinal (crossed pyramidal tract)	Lateral white columns	Voluntary motor areas	Anterior gray or anterolateral columns of spinal cord	Voluntary motor impulses, especially to muscles of the arms and legs
Ventral corticospinal (direct pyramidal tract)	Anterior or ventral columns	Voluntary motor areas	Anterior gray or anterolateral columns of spinal cord	Mainly to muscles of the trunk
Extrapyramidal tracts				
Rubrospinal	Lateral white columns	Red nucleus	Anterior gray or anterolateral columns of spinal cord	Muscular coordination, postural control
Vestibulospinal	Anterior or ventral columns	Vestibular nucleus of VIIIth cranial nerve in medulla	Anterior gray or anterolateral columns of spinal cord	Maintaining equilibrium

near the origin of the spinal motor-nerve roots. Since they receive impulses from the semicircular canals of the ear, their function is to adjust muscular coordination in relation to maintaining equilibrium (Table 9.4).

INJURIES TO SPINAL CORD   The cord functions as a pathway for impulses between the body and the brain. The large tracts in the dorsal part of the cord are ascending pathways. Injury to these tracts causes lack of coordination, since spinal reflexes are disturbed. Walking becomes uncoordinated, the movements are jerky, and the individual may find it difficult to keep his balance. Locomotor ataxia (*tabes dorsalis*) is an injury of this sort due to degeneration of some of the large ascending dorsal tracts of the cord.

The nerve fibers of the cord are not capable of regeneration after an injury. Degeneration of fibers proceeds upward, away from the cell bodies, in ascending tracts, and downward in descending motor tracts.

Severing a motor pathway results in muscular paralysis. The paralysis may be of two types, depending on which tracts are severed and where the injury occurs. If the injury occurs in the pyramidal tract or if the motor neurons of a spinal nerve are injured, the paralysis may be complete and then would be of the *flaccid type*. A lack of muscle tone and considerable atrophy of the muscles occurs. *Spastic paralysis* portrays an exaggerated tonicity and uncoordinated reflexes. It is caused by certain extrapyramidal injuries in the brain or cord.

Thirty-one pairs of spinal nerves arise from the cord. They are grouped as follows: cervical, 8 pairs; thoracic, 12 pairs; lumbar, 5 pairs; sacral, 5 pairs; coccygeal, 1 pair (Figure 10.1).

SPINAL NERVES

There are eight pairs of cervical nerves rather than seven, because the first pair arises from the medulla and emerges above the atlas. The remaining spinal nerves arise from the cord and emerge through openings between the vertebrae.

Nerves leaving the spinal cord form complex interlacing networks in the cervical, brachial and the lumbosacral regions. Such networks are called *plexuses*. Motor branches of the first four cervical nerves form the *cervical plexus*. Peripheral branches innervate muscles of the neck and shoulder. The phrenic nerve, which is distributed to the diaphragm, arises from this plexus.

The *brachial plexus* is a large plexus of the neck and shoulder composed of motor branches of the Vth to VIIIth cervical nerves and the Ist thoracic nerve. Among the nerves that supply the arm are the axillary, median, radial, and ulnar nerves.

The *lumbosacral plexus* consists of branches from all the lumbar nerves and the Ist to IIId sacral nerves. This is a very large network, with nerves extending to the lower extremities. Included among several nerves are the femoral nerve and the very large sciatic nerve. Inflammation and injury to the latter nerve may cause a neuralgic condition called *sciatica*.

SUMMARY   The brain in its early embryonic development forms three primary divisions: the forebrain, midbrain, and hindbrain. The forebrain is divided into two secondary divisions, the telencephalon and the diencephalon. The hindbrain also is divided into two secondary divisions, the metencephalon and the myelencephalon. The midbrain, or mesencephalon, is the connecting link between the forebrain and the hindbrain. The principal parts of the brain are the cerebrum, cerebellum, and medulla.

The brain and spinal cord are covered by meningeal membranes: the dura mater, arachnoid, and pia mater.

Neuroglial cells are closely associated with the neurons and blood vessels of the central nervous system. There are three principal types: astrocytes, oligodendrocytes, and microgliocytes.

The cavities of the brain are called ventricles. The lateral ventricles communicate with the third ventricle through the interventricular foramens (of Monro). The third and fourth ventricles are connected by way of the cerebral aqueduct. The fourth ventricle is continuous with the central canal of the spinal cord. The ventricles and the central canal contain cerebrospinal fluid.

The cerebrum is divided into lobes: the frontal, parietal, temporal, and occipital lobes. The central sulcus separates the frontal and parietal lobes. The lateral cerebral fissure demarcates the temporal lobe. The occipital lobe lies posterior to the parietooccipital fissure. The insula (island of Reil) is sometimes considered as a lobe of the cerebrum.

There are certain areas of cerebral localization. The motor area lies just anterior to the central sulcus. It is inverted with respect to the area of the body that any given portion controls. The right motor area governs the left side of the body and vice versa, because the pyramidal tracts, arising on one side of the brain, cross over to the opposite side. The decussation occurs in the medulla or at the spinal level where the spinal nerve emerges. The cutaneous sensory area and the interpreting area for muscle sense lie directly posterior to the central sulcus. The visual cortex is located in the occipital lobe. Auditory impulses are interpreted in the temporal lobe.

The thalamus lies in the lateral walls of the diencephalon surrounding

The brain and spinal cord

238

the third ventricle. The hypothalamus is located in the basal portion of the diencephalon. The thalamus is described as a relay center for nerve impulses. Several autonomic functions are regulated by the hypothalamus. Among them are regulatory centers for the control of body temperature and water metabolism.

The cerebellum functions to coordinate muscular movements and to maintain equilibrium. The large cortical cells with branching dendrites are Purkinje cells. The pons serves to connect the two cerebellar hemispheres anteriorly. There are also vertical tracts that connect the cerebrum with the medulla.

The medulla is the base of the brainstem. It contains a number of vital centers, such as the cardiac inhibitory center, the respiratory center, and the vasoconstrictor center.

The reticular formation, located in the brainstem, is thought to function in arousal from sleep and in maintaining a state of consciousness.

There are 12 pairs of cranial nerves and 31 pairs of spinal nerves. Spinal nerves form the cervical, brachial, and lumbosacral plexuses.

Ascending tracts in the spinal cord carry impulses concerned with touch, pressure, pain, temperature, and muscle sense. Descending tracts, such as the pyramidal tracts, are motor-impulse pathways to muscles of the trunk and the appendages.

1 Trace the flow of the cerebrospinal fluid.
2 List the numerous functions of the cerebrum.
3 What is meant by cerebral localization?
4 An injury affecting the upper part of the motor area on the left side may cause motor paralysis of what part of the body? Why?
5 Locate the sensory-interpreting areas of the brain.
6 Name some of the structures found in the diencephalon.
7 Discuss the functions of the thalamus and hypothalamus.
8 Write a paragraph about the structures and functions of the midbrain.
9 How does the cerebellum function?
10 Describe the different kinds of neuroglial cells and discuss their function.
11 In what ways do cranial nerves differ from spinal nerves?
12 A man is seated at a desk, writing. He hears the door open and arises. Discuss the brain centers and nervous pathways involved.
13 What function is ascribed to Broca's area?
14 Name some of the structures included in the rhinencephalon.
15 Discuss some of the ideas concerning memory.

ANDERSSON, B.: Thirst and Brain Control of Water Balance, *Am. Scientist,* **59**(4):408–413 (1971).

CERASO, J.: The Interference Theory of Forgetting, *Sci. Am.,* **217**:117–124 (1967).

DEUTSCH, J. A.: The Cholinergic Synapse and the Site of Memory, *Science,* **174**:788–794 (1971).

DINGMAN, W., and M. B. SPORN: Molecular Theories of Memory, *Science,* **144**:26–29 (1964).

FERNSTROM, J. D., and R. J. WURTMAN: Brain Serotonin Content: Increase Following Ingestion of Carbohydrate Diet, *Science,* **174**:1023–1025 (1971).

GESCHWIND, N.: The Organization of Language and the Brain, *Science,* **170**:940–944 (1970).

GORDON, B.: The Superior Colliculus of the Brain, *Sci. Am.,* **227**:72–82 (1972).

HEIMER, L.: Pathways in the Brain, *Sci. Am.,* **225**:48–57 (1971).

HUBEL, D. H.: The Visual Cortex of the Brain, *Sci. Am.,* **209**:54–62 (1963).

———— and T. N. WIESEL: Receptive Fields and Functional Architecture of Monkey Striate Cortex, *J. Physiol.,* **195**:215–243 (1968).

KATZ, B.: Quantal Mechanism of Neural Transmitter Release, *Science*, **173:**123–126 (1971).

LASANSKY, A.: Nervous Function at the Cellular Level: Glia, *Ann. Rev. Physiol.*, **33:**241–256 (1971).

LIPPOLD, O.: Physiological Tremor, *Sci. Am.*, **224:**65–73 (1971).

LURIA, A. R.: The Functional Organization of the Brain, *Sci. Am.*, **222:**66–78 (1970).

McGEER, P. L.: The Chemistry of Mind, *Am. Scientist*, **59:**221–229 (1971).

MERTON, P. A.: How We Control the Contraction of Our Muscles, *Sci. Am.*, **226:**30–37 (1972).

PENFIELD, W.: The Interpretive Cortex, *Science*, **129:**1719–1725 (1959).

RAPAPORT, S. I., M. HORI, and I. KLATZO: Reversible Osmotic Opening of the Blood-Brain Barrier, *Science*, **173:**1026–1028 (1971).

SNIDER, R.: The Cerebellum, *Sci. Am.*, **199:**84–90 (1958).

WHITMAN, R.: The Fortification Illusions of Migraines, *Sci. Am.*, **224:**88–96 (1971).

WOOLDRIDGE, D. E.: "The Machinery of the Brain," McGraw-Hill Book Company, New York, 1963.

chapter 10
the autonomic nervous system

The autonomic nervous system provides the key to homeostasis in the body. Through very complex feedback systems and reflex activity it helps to regulate and control the normal balanced functioning of organs and organ systems. Emotional states are supported by very extensive bodily changes. A state of fear can induce the desire to run, but one cannot run far unless physiological adjustments support the effort. Anger can mean that one is prepared to fight, but it will not be an efficient battle unless one's circulatory system makes necessary adjustments to provide strength and endurance for the contest. The situation need not be a highly emotional one; work and exercise are supported by the same physical adjustments.

Regulation of the internal environment of the body with regard to temperature and body fluids is a normal function of the autonomic system. A good example is the adjustment the body makes to a marked change in the temperature of the surrounding medium. As the room temperature rises, the sweat glands are stimulated to secrete and the surface of the skin becomes moist. If the humidity of the air is low enough, evaporation tends to cool the surface. At the same time peripheral vasodilatation of the arterioles and capillaries of the skin permits a greater volume of blood to be brought to the surface. Skeletal muscles tend to relax when the body is exposed to high temperatures.

When the body is exposed to cold, the skin surface is nearly dry and the surface arterioles constrict to keep more blood away from the cool surface and thus conserve heat. Skeletal muscles increase their tone, and shivering may be induced to produce more heat from the contraction of muscles.

Physical adjustments to an emergency are largely controlled by the autonomic nervous system. Preparations to strengthen the body for a critical situation include acceleration and strengthening of the heartbeat, a rise in blood pressure, release of glucose from the liver, and the secretion of a small amount of epinephrine by the adrenal glands. Breathing is made easier by the relaxation of muscles in the bronchial tubes. During an emergency digestion can wait, and so the activity of the digestive system is altered and depressed; the blood supply is largely diverted from the digestive system to the skeletal muscles. These effects are obtained mainly by the stimulation of one division of the autonomic nervous system, the sympathetic, or *thoracolumbar,* portion.

The autonomic nervous system

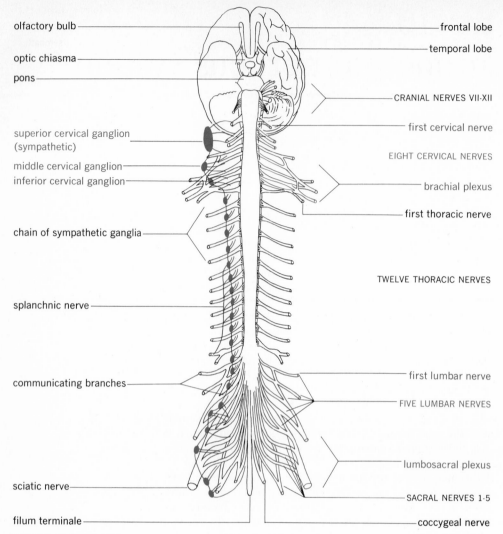

olfactory bulb ——————————————————— frontal lobe

———————————— temporal lobe

optic chiasma ————————————

pons ————————

———— CRANIAL NERVES VII-XII

———— first cervical nerve

superior cervical ganglion
(sympathetic) ————————————

EIGHT CERVICAL NERVES

middle cervical ganglion ————————
inferior cervical ganglion ————————

———— brachial plexus

———— first thoracic nerve

chain of sympathetic ganglia ————————

TWELVE THORACIC NERVES

splanchnic nerve ————————

communicating branches ————————

———— first lumbar nerve

FIVE LUMBAR NERVES

———— lumbosacral plexus

sciatic nerve ————————

———— SACRAL NERVES 1-5

filum terminale ————————— coccygeal nerve

FIGURE 10.1

Ventral view of the brain and spinal cord. Cranial and spinal nerve roots are shown, including the chain of sympathetic ganglia, on one side only. [*Redrawn from J. Parsons Schaeffer (ed.), "Morris' Human Anatomy," 12th ed., McGraw-Hill Book Company, New York, 1966.*]

The autonomic nervous system

242

The autonomic system is divided somewhat artificially into a thoracolumbar, or sympathetic, portion and a craniosacral, or parasympathetic, part. The thoracolumbar division is composed of a chain of ganglia and nerves on either side of the spinal cord, extending from the cervical region through the thoracic and lumbar regions (Figure 10.1). Throughout the thoracic and lumbar regions each ganglion is connected to a spinal nerve by a communicating branch (Figure 10.2). Fibers extend upward to the head from the superior cervical ganglion; they also extend downward from sacral ganglia, thus increasing the distribution of sympathetic fibers (Figure 10.3*a*).

The craniosacral, or parasympathetic, division is associated with certain cranial and sacral nerves and will be discussed later. The terms *thoracolumbar* or *craniosacral* appear well adapted for anatomical considerations; the terms *sympathetic* and *parasympathetic* seem better adapted when referring to the physiology of the autonomic system (Figure 10.3*b*).

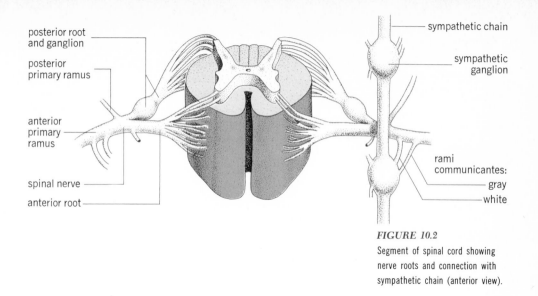

posterior root
and ganglion

posterior
primary ramus

anterior
primary
ramus

spinal nerve

anterior root

sympathetic chain

sympathetic
ganglion

rami
communicantes:
gray
white

FIGURE 10.2
Segment of spinal cord showing
nerve roots and connection with
sympathetic chain (anterior view).

SYMPATHETIC, OR THORACOLUMBAR, DIVISION

Motor impulses from the spinal cord to smooth muscles are conveyed over
two sets of visceral efferent fibers instead of one, as in somatic motor nerves.
A synaptic connection ordinarily is made in a ganglion of the thoracolumbar
chain, although this is not necessarily so. There are *preganglionic neurons,*
with cell bodies located in the intermediolateral column of gray matter of
the spinal cord and with fibers extending, ordinarily, from the cell body
to the autonomic ganglion outside the cord, and a *postganglionic neuron,*
with its cell body located in a ganglion and with its fiber extending to
visceral muscle (Figure 10.4). The preganglionic fiber can extend through
the autonomic ganglion to a collateral ganglion, in which case there is a
short postganglionic fiber to the organ supplied.

PREGANGLIONIC NEURON   The cell body of the preganglionic neuron is smaller
than that of a motor neuron of the central nervous system. The particles
of its Nissl substance are finer and more rounded. The axon emerges from
the spinal cord as a part of the motor root of a spinal nerve but soon leaves
it to enter the autonomic ganglion. The majority of these axons are myelinated. A group of myelinated fibers presents a white appearance, and so
this connection of preganglionic fibers between the spinal nerve and the
sympathetic ganglion is called the *white branch,* or *white ramus communicans.* When the preganglionic neuron enters the sympathetic ganglion, it
makes a synaptic connection with many postganglionic neurons. This arrangement is significant, since it provides for the rapid, widespread response
characteristic of the sympathetic system (Figure 10.5).

The thoracic and the first three lumbar nerves are connected with
the autonomic chain of ganglia by a white ramus; hence the name *thoracolumbar* for this division. Cervical ganglia are supplied by preganglionic
fibers extending upward from the thoracic nerves through the lateral chains
of ganglia. The lower lumbar and sacral ganglia are supplied by fibers
extending downward.

*Sympathetic, or
thoracolumbar, division*

243

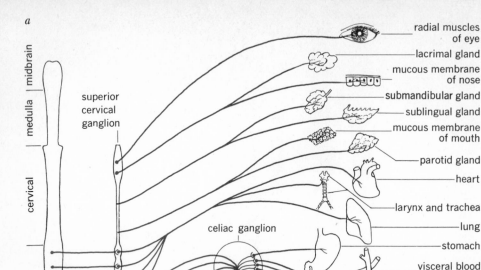

a

medulla | midbrain

superior
cervical
ganglion

cervical

celiac ganglion

thoracic

superior
mesenteric
ganglion

lumbar

inferior
mesenteric
ganglion

sacral

radial muscles
of eye

lacrimal gland

mucous membrane
of nose

submandibular gland

sublingual gland

mucous membrane
of mouth

parotid gland

heart

larynx and trachea

lung

stomach

visceral blood
vessel

liver

pancreas

suprarenal gland

small intestine

colon

kidney

bladder

penis

testis

THORACOLUMBAR (SYMPATHETIC) SYSTEM

FIGURE 10.3

Diagrammatic representation of the autonomic nervous system: *a* thoracolumbar system; *b* craniosacral system. (*Adapted from "Blakiston's New Gould Medical Dictionary," 2d ed., McGraw-Hill Book Company, New York, 1956.*)

The autonomic nervous system

POSTGANGLIONIC NEURON The postganglionic neuron of the thoracolumbar division has its cell body in a lateral chain ganglion or in a collateral ganglion. The fiber extends to involuntary muscle tissue or to glandular cells; thus the cell bodies of the ganglia of the lateral chain are entirely motor (Figure 10.5). Postganglionic fibers may take two courses extending beyond the lateral ganglia. They may proceed inward by way of a visceral branch to terminate in the muscles of the viscera, or they may rejoin the spinal nerve by way of the gray root, usually called the *gray ramus communicans*, and terminate in the involuntary muscles of the peripheral region, such as the muscles in the walls of blood vessels, or in sweat glands of the skin. Since these postganglionic fibers are not myelinated, the nerve appears gray in contrast with the white branch of myelinated preganglionic fibers.

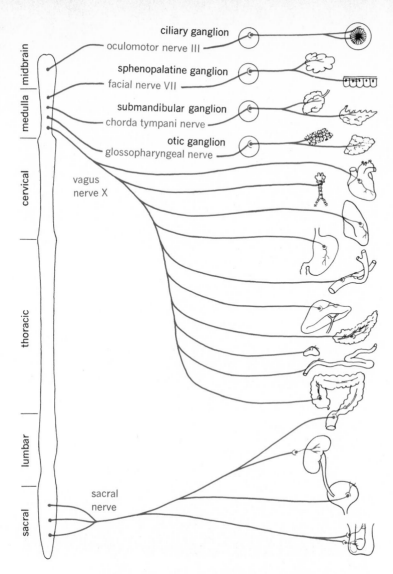

ciliary ganglion

oculomotor nerve III

sphenopalatine ganglion

facial nerve VII

submandibular ganglion

chorda tympani nerve

otic ganglion

glossopharyngeal nerve

vagus
nerve X

midbrain | medulla | cervical | thoracic | lumbar | sacral

sacral
nerve

CRANIOSACRAL (PARASYMPATHETIC) SYSTEM

Both white and gray rami are attached to the thoracic and upper lumbar spinal nerves, but the cervical, lower lumbar, sacral, and coccygeal nerves have no white rami. However, each spinal nerve has a gray branch and therefore receives sympathetic postganglionic fibers.

SYMPATHETIC PLEXUSES The great plexuses of the autonomic system are the *cardiac; celiac,* or *solar;* and *hypogastric plexus.* Though these plexuses are regarded as essentially sympathetic, they also receive fibers from the parasympathetic system.

The cardiac plexus lies under the arch of the aorta just above the heart. It receives branches from the cervical sympathetic ganglia and from the right and left vagal nerves (parasympathetic) and has a regulatory effect on the heart.

*Sympathetic, or
thoracolumbar, division*

245

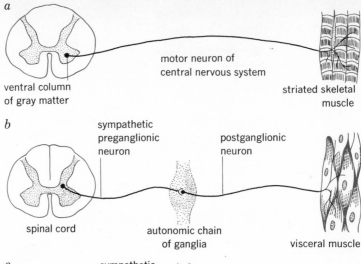

a

ventral column
of gray matter

motor neuron of
central nervous system

striated skeletal
muscle

b

sympathetic
preganglionic
neuron

postganglionic
neuron

spinal cord

autonomic chain
of ganglia

visceral muscle

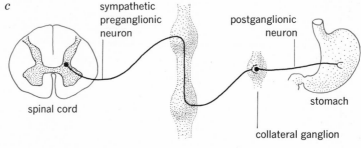

c

sympathetic
preganglionic
neuron

postganglionic
neuron

spinal cord

collateral ganglion

stomach

FIGURE 10.4

Common types of motor neurons: *a* spinal peripheral motor neuron with no synapse between the spinal cord and skeletal muscle; *b* typical sympathetic pathway; *c* the sympathetic preganglionic neuron, in this case, passes through the chain of ganglia to a collateral ganglion; the shorter postganglionic neuron leads to the organ supplied.

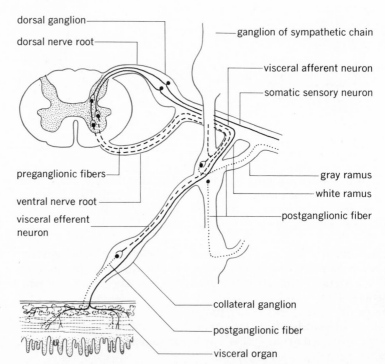

dorsal ganglion

dorsal nerve root

ganglion of sympathetic chain

visceral afferent neuron

somatic sensory neuron

preganglionic fibers

ventral nerve root

visceral efferent neuron

gray ramus

white ramus

postganglionic fiber

collateral ganglion

postganglionic fiber

visceral organ

FIGURE 10.5

Diagram of a cross section of the spinal cord with spinal nerve roots and sympathetic ganglion. Preganglionic fibers are represented by dashes; postganglionic fibers, by dotted lines; afferent fibers, by solid lines. [*Adapted from Barry J. Anson (ed.), "Morris' Human Anatomy," 12th ed., McGraw-Hill Book Company, New York, 1966.*]

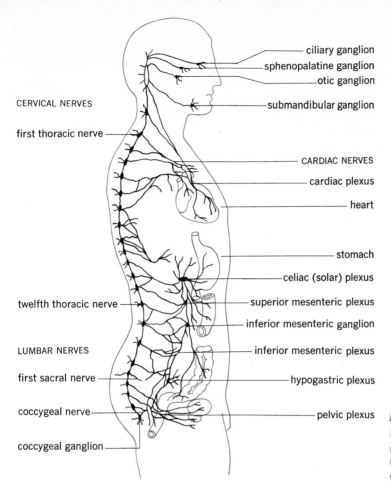

ciliary ganglion
sphenopalatine ganglion
otic ganglion
submandibular ganglion

CERVICAL NERVES

first thoracic nerve

CARDIAC NERVES

cardiac plexus

heart

stomach

celiac (solar) plexus

twelfth thoracic nerve

superior mesenteric plexus

inferior mesenteric ganglion

LUMBAR NERVES

inferior mesenteric plexus

first sacral nerve

hypogastric plexus

coccygeal nerve

pelvic plexus

coccygeal ganglion

FIGURE 10.6

Diagram illustrating the chain of ganglia and the larger plexuses of the sympathetic system (lateral view).

The celiac, or solar, plexus is the largest network of cells and fibers of the autonomic system. It lies behind the stomach and is associated with the aorta and the celiac arteries. The ganglia receive the splanchnic nerves from the sympathetic system and branches of the vagus from the parasympathetic system. A blow to this region may slow the heart, reduce the flow of blood to the head, and depress the breathing mechanism.

The hypogastric plexus forms a connection between the celiac plexus above and the two pelvic plexuses below. It is located in front of the fifth lumbar vertebra and continues downward in front of the sacrum, forming the right and left pelvic plexuses. These plexuses supply the organs and blood vessels of the pelvis (Figure 10.6).

CIRCULATORY EFFECTS Vasoconstriction is a function of the sympathetic system. Although vasodilatation may be a function of the parasympathetic system, experimental results are not conclusive. It appears that sympathetic nerves also can include vasodilator fibers. Other factors may influence the blood vessels, such as hormones circulating in the bloodstream, the CO_2 content of the blood, and temperature.

Sympathetic fibers are conveyed to the blood vessels of the arms and legs by way of the spinal nerves of the peripheral nervous system.

Sympathetic, or thoracolumbar, division

247

Vasoconstriction may be localized or general. In an emergency calling for quick action, general vasoconstriction causes a rise in blood pressure. At the same time vasoconstriction may reduce the flow of blood to the digestive tract in a localized area. Muscular exercise requires an increased flow of blood to the skeletal muscles and, therefore, vasodilatation of the blood vessels supplying them. Coronary arteries supplying the heart muscle are dilated also. The action of the sympathetic system is supported by uptake of epinephrine from the bloodstream.

PARASYMPATHETIC, OR CRANIOSACRAL, DIVISION

The craniosacral, or parasympathetic, division of the autonomic nervous system is associated with certain cranial and sacral nerves in which autonomic fibers are incorporated; hence the name *craniosacral division*. The oculomotor (IIId cranial) nerve, arising in the midbrain, innervates certain voluntary muscles that move the eyeball; in addition, it carries parasympathetic fibers to involuntary muscles within the eyeball. Preganglionic fibers are distributed to the ciliary ganglion located behind the eyeball. Postganglionic fibers arising in the ganglion extend to the ciliary muscles of the eye and to the sphincter of the pupil.

The facial (VIIth cranial), glossopharyngeal (IXth cranial), vagus (Xth cranial), and accessory (XIth cranial) nerves constitute a group of cranial nerves arising from the medulla. Since they also contain parasympathetic fibers, they are a part of the craniosacral division. The vagus supplies the viscera of the thorax and abdomen; this may be why there are no parasympathetic fibers arising from the thoracic or lumbar regions of the cord.

The sacral portion of this system is identified with certain sacral nerves that carry parasympathetic fibers to the pelvic viscera.

PREGANGLIONIC AND POSTGANGLIONIC FIBERS Typically, the parasympathetic preganglionic fiber extends from its nucleus in the brain or sacral region of the spinal cord to the organ supplied. The postganglionic fiber is often a very short fiber located within the organ itself. The preganglionic fiber may end in a collateral ganglion, as in the case of preganglionic fibers extending out to the ciliary ganglion of the eye. The postganglionic fibers in this case are longer than those incorporated within certain organs (Figure 10.7).

The parasympathetic system functions as an antagonist of the sympathetic system if an organ is supplied by both systems. If the sympathetic system is the accelerator system, as in the heart, for example, then the parasympathetic system is the inhibitor. Its function in this case is to slow the accelerated heart and thus restore the normal heart rate. Even though it acts as an inhibitor, it does not ordinarily depress the heart rate below normal unless unduly stimulated, as from the action of drugs or pressure on a nerve.

CRANIAL NERVES THAT CARRY PARASYMPATHETIC FIBERS If the oculomotor nerve is cut experimentally, the pupil dilates. The parasympathetic fibers within the oculomotor nerve carry nervous impulses that cause the pupil to constrict. Cutting the nerve destroys the balance between parasympathetic and sympathetic innervation. The sympathetic nervous impulses then cause the

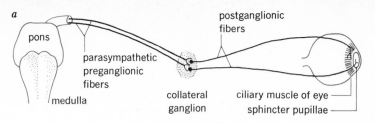

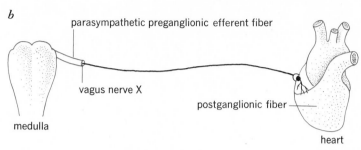

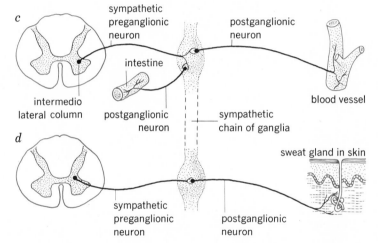

FIGURE 10.7

Various types of autonomic neurons: *a* parasympathetic neurons with a synapse in a collateral ganglion; *b* typical parasympathetic pathway, with a short postganglionic fiber within the organ supplied; *c* sympathetic neuron to visceral muscle, showing a synapse in the sympathetic ganglion; *d* typical synapse of a sympathetic preganglionic neuron; the postganglionic fiber reaches the sweat gland through a peripheral nerve.

pupil to dilate. The "drops" placed in the eye for optical examination apparently act in much the same way by blocking the parasympathetic nerve endings.

It has been mentioned that there are four cranial nerves arising from the medulla that carry autonomic fibers and therefore are a part of the craniosacral system. These nerves are the facial, glossopharyngeal, vagus, and accessory nerves.

The facial nerve includes parasympathetic fibers that are secretory to the lacrimal gland and to the sublingual and submandibular salivary glands. The lacrimal gland is supplied with postganglionic fibers from the sphenopalatine ganglion. The sublingual and submandibular salivary glands receive postganglionic fibers arising in the submandibular ganglion (Figure 10.8).

Preganglionic fibers in the glossopharyngeal nerve extend outward to the otic ganglion. Postganglionic fibers arise in the otic ganglion and supply the parotid salivary gland. These glands, including the lacrimal, have a double innervation. They derive their sympathetic innervation by way of

Parasympathetic, or craniosacral, division

249

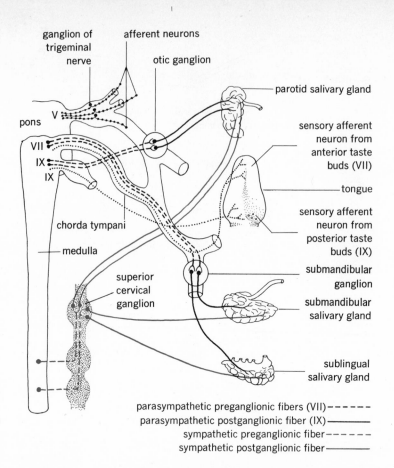

ganglion of trigeminal nerve

afferent neurons

otic ganglion

parotid salivary gland

pons

V

VII

IX

IX

sensory afferent neuron from anterior taste buds (VII)

tongue

chorda tympani

sensory afferent neuron from posterior taste buds (IX)

medulla

submandibular ganglion

superior cervical ganglion

submandibular salivary gland

sublingual salivary gland

parasympathetic preganglionic fibers (VII)------
parasympathetic postganglionic fiber (IX)————
sympathetic preganglionic fiber------
sympathetic postganglionic fiber————

FIGURE 10.8

Diagram illustrating autonomic innervation of the salivary glands. Afferent fibers of cranial nerves V, VII, and IX are shown by dotted lines.

the superior cervical sympathetic ganglion and carotid plexuses. The action of the two sets of nerves is not clear. Apparently they both contain secretory fibers, but the secretory action of the parasympathetic system seems to be dominant.

The vagus nerve contains both motor and visceral afferent fibers. The motor fibers are long preganglionic fibers that extend to the organ supplied. Very short postganglionic fibers are contained within the organ. Motor fibers are supplied to the larynx, trachea, bronchioles, heart, esophagus, stomach, small intestine, and some parts of the large intestine. Stimulation of the vagus acts as an inhibitor to the heart, causing its rate of beating to slow or to stop. To the muscles of the wall of the digestive tract, branches of the vagus act as accelerator nerves. Peristalsis is increased by parasympathetic stimulation. Parasympathetic fibers to the glands of the digestive tract have a regulatory function on secretion, but food content of the stomach or intestine and hormones circulating in the blood can also stimulate secretion (Figure 10.9).

Parasympathetic fibers from both the right and left vagus nerves enter the great plexuses of the sympathetic system. There is, however, a definite parasympathetic nerve supply to such organs as the pancreas, liver, and kidneys. Nervous stimulation of these organs is, for the most part, merely regulatory. Hormones in the blood normally cause the pancreas and liver to secrete, but stimulation of the vagus increases the flow of pancreatic juice

The autonomic nervous system

250

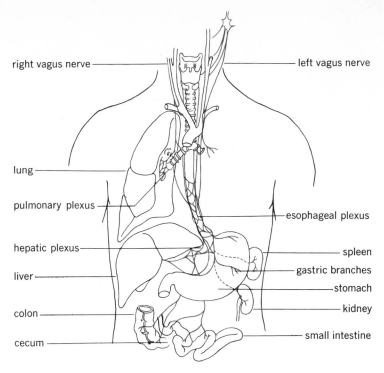

right vagus nerve

left vagus nerve

lung

pulmonary plexus

esophageal plexus

hepatic plexus

spleen

gastric branches

liver

stomach

colon

kidney

cecum

small intestine

FIGURE 10.9

Distribution of parasympathetic fibers to the viscera by the vagus nerves.

and bile. Though sympathetic stimulation of the kidneys by way of the splanchnic nerves results in vasoconstriction and therefore reduced flow of urine, many other physiological factors affect the function of the kidneys.

A part of the accessory nerve contains visceral motor and cardiac inhibitory fibers.

Certain types of allergy offer examples of overstimulation of the parasympathetic system. Epinephrine can be used to counteract these effects, since it is associated with the action of the sympathetic system.

THE SACRAL AUTONOMICS The sacral portion of the craniosacral system is composed of preganglionic fibers incorporated in the second, third, and fourth sacral nerves. The fibers extend to the pelvic plexuses, where they enter into close relationship with fibers of the sympathetic system. Parasympathetic fibers innervate the urogenital organs and the distal part of the colon. Postganglionic fibers are considered to be in the organs supplied or in small ganglia located close by. These parasympathetic fibers are motor to the muscles of the distal two-thirds of the colon, to the rectum, and to the urinary bladder. They carry vasodilator impulses to the penis and clitoris. Inhibitory impulses pass to the internal sphincter muscle of the bladder and to the internal sphincter of the anus (Table 10.1).

PARASYMPATHETIC PLEXUSES *Enteric plexuses* The digestive tube has its own intrinsic nerve supply, consisting of the *myenteric plexus*, located between the longitudinal and circular muscles, and a *submucous plexus*, located under the mucous layer in the submucosa (Figure 10.10b and c). This part of the nervous system extends the entire length of the digestive tube. It may be assumed that parasympathetic fibers entering the wall of the digestive tract

Parasympathetic, or craniosacral, division

251

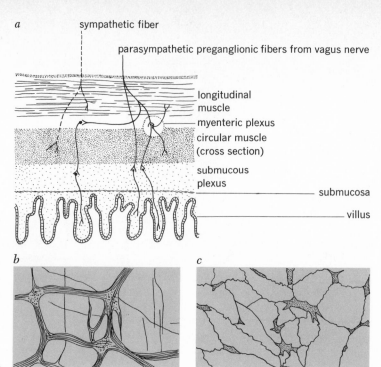

a

sympathetic fiber

parasympathetic preganglionic fibers from vagus nerve

longitudinal
muscle

myenteric plexus

circular muscle
(cross section)

submucous
plexus

submucosa

villus

b

c

FIGURE 10.10

Nerve plexuses: *a* a longitudinal section through the wall of the small intestine, illustrating the location of the myenteric and submucous plexuses (these intestinal plexuses are considered to act as postganglionic neurons of the parasympathetic system and may act in this capacity for the sympathetic system also); *b* surface view of the myenteric plexus; *c* surface view of the submucous plexus.

are preganglionic fibers that make synaptic connections with neurons of the enteric system. Sympathetic fibers entering the muscular wall, however, are postganglionic fibers and until recently have been considered to terminate on the muscles that they supply without making synaptic connections on the intestinal plexuses (Figure 10.10*a*).

Earlier investigations, mentioned above, which indicated that sympathetic postganglionic neurons end directly on intestinal muscle may not be correct in light of new evidence. The interpretation formerly accepted that the inhibitory effects of norepinephrine were directly transmitted to muscle cells is now in doubt. It now appears that sympathetic postganglionic neurons may synapse on the ganglion cells of the submucous and myenteric plexuses along with parasympathetic endings. However, the problem has not been entirely resolved.

The enteric plexuses function in maintaining rhythmic peristaltic movement along the digestive tract. Peristalsis is maintained if both sympathetic and parasympathetic nerve supplies are cut. The nerves of the autonomic system, however, exert a regulatory effect (Table 10.1).

PHYSIOLOGY OF THE AUTONOMIC NERVOUS SYSTEM

SYMPATHETIC AND PARASYMPATHETIC RELATIONSHIPS *Autonomic effects* are usually conditioned by other factors such as the presence of hormones in the bloodstream or by circulatory effects. The secretion of a gland can be depressed by the stimulation of an inhibitor nerve; secretion can also be depressed by vasoconstriction of blood vessels supplying the gland, thus limiting its blood

The autonomic nervous system

TABLE 10.1

Some of the known autonomic effects

Structure	Sympathetic	Parasympathetic
Eye		
Iris	Dilation of the pupil	Contraction of the pupil
Ciliary muscle	Relaxation of ciliary muscles, accommodation for distance vision	Contraction of ciliary muscles, accommodation for close-up vision
Bronchial tubes	Dilatation	Constriction
Heart	Accelerates and strengthens action	Depresses and slows action
Blood vessels		
Coronary arteries	Dilatation	Constriction
Abdominal and pelvic viscera	Constriction	
External genitalia	Constriction of blood vessels	Dilatation of blood vessels Erection
Stomach		
Muscles	Depresses activity	Increases activity
Glands	Alters secretion	Increases secretion
Liver	Stimulates glycogenolysis	
Visceral muscle of intestine	Depresses peristalsis	Increases peristalsis
Adrenal medulla	Secretion of epinephrine	
Adrenergic fibers	Most of the sympathetic postganglionic fibers; among the exceptions are those to the sweat glands	
Cholinergic fibers	All preganglionic fibers; all parasympathetic postganglionic fibers; motor-nerve fibers to voluntary muscle; postganglionic sympathetic fibers to the sweat glands	

supply. Though the sympathetic system may be considered as an accelerator to the heart, the situation is reversed in the case of the action of the autonomic system upon the digestive tract. Here the action of sympathetic nerves depresses peristalsis and the secretion of digestive glands during emotional excitement, while the parasympathetic system, as an accelerator, effects a return to normal.

When we speak of the sympathetic and parasympathetic nerves as being *antagonistic*, we mean this in the sense of antagonistic muscles. The nerves from the sympathetic and parasympathetic systems can produce opposite effects, but they provide a correlated adjustment to meet many physiological conditions. *Autonomic effects* are not always clearly antagonistic. The accommodation reflex of the eye whereby the lens and iris are adjusted to facilitate clear vision appears to be primarily a parasympathetic function so far as the ciliary muscle and the muscles of the iris are concerned. The two sets of muscles of the iris seem to have a synergistic relationship, which causes them to contract or dilate the pupil smoothly in a mild state of opposition to each other. The pupil can also dilate in response to an

Physiology of the autonomic nervous system

253

emotional state such as fear or pain. This is due to stimulation of the sympathetic system.

Man has largely lost the ability to elevate the hair in response to cold or fear, but he still experiences "gooseflesh" or the feeling that the hair is erect. There is a tiny muscle attached to the base of each sloping hair. Under the stimulus of cold or excitement these arrector muscles can contract and give rise to peculiar skin sensations (Figure 4.6).

REGULATORY SYSTEMS The autonomic nervous system affords numerous examples of regulatory control of physiological mechanisms concerned with maintaining a constant internal environment, or homeostasis. The control mechanism also offers examples of the principle of *negative feedback* which governs most of the body functions. The common example is to cite the house or room thermostat as analogous to the function of the hypothalamus in controlling the heat of the body. But homeostatic feedback mechanisms are much more complex. The input may come from various sources, and there may be numerous means of conduction to more than one effector. A feedback system is an example of a reflex act with the following essential elements: a receptor, a conductor, and an effector. In functioning, the sensory receptor is excited by the stimulus, a message is conducted to the reactive site, and the effector reacts. Negative feedback control means that when there is overproduction in the output of the system, various factors will bring about a decrease in the stimulus (input) and effect a return to normal. The normal control of the heart rate is an example. If sympathetic nerve stimuli cause the heart rate to accelerate, control centers in the brain send impulses by way of the vagus nerves to slow it down.

Body-heat regulation The sweat glands are innervated by neurons of the sympathetic system. In the same way that peripheral blood vessels are innervated, sympathetic fibers are carried by nerves of the peripheral nervous system. Though general sweating is the usual response of eccrine sweat glands to high temperature, there is an absence of sweating during a fever. In the latter case, the toxins produced by disease apparently affect the heat-regulatory center in the hypothalamus so that, acting like a thermostat, it is "set" higher than normal. The heat-regulatory mechanism is altered in several ways, causing heat to be conserved and therefore resulting in a rise in temperature. Sweating in response to a strong emotional situation may involve the apocrine sweat glands also and may be most evident on the forehead, in the palms of the hands, the soles of the feet, and in the axillae of the arms. Stimulation of the sweat glands involves the integrated action of the hypothalamus, centers in the brainstem, and, indirectly, hormone secretion.

In general, the hypothalamus acts as a coordinating center for the autonomic nervous system. In this way it aids in regulating the functions of the viscera.

Experimentally it has been shown that stimulation of the hypothalamus in anesthetized animals results in dilation of the pupils, increase in blood pressure, and stimulation of respiration. That it has some control over the emotions is indicated by the "sham" rage animals exhibit when cortical connections are cut and the hypothalamus is no longer under the control of a higher center. Experimentally this is the so-called "thalamus animal," with centers in the midbrain intact.

Adrenergic fibers The terminal filaments of most sympathetic postganglionic neurons produce an epinephrine-like substance (norepinephrine) and are classified as *adrenergic*. Sympathetic fibers to sweat glands, blood vessels of the skin, and the arrectores pilorum muscles are exceptions. These postganglionic fibers enter spinal nerves through the gray rami and reach the skin incorporated in peripheral nerves.

The effects of norepinephrine, in conjunction with epinephrine, can be general and widespread. There is experimental evidence that the chemical substance resulting from excitation of sympathetic postganglionic fibers is carried by the bloodstream and can affect organs remote from the point of origin. It is interesting to note that the sympathetic ganglia and the medullary portion of the adrenal gland have the same embryonic origin. They both arise from neural-crest cells (Figures 9.1 and 19.2).

There has been considerable investigation into the nature of an adrenergic receptor that would mediate the various effects of norepinephrine. If the postulated chemical receptor is verified, it may eventually be determined that it is adenosine 3'5'-monophosphate (cyclic AMP) acting as the so-called "second messenger." It is known that dopamine stimulates adenyl cyclase activity. Adenylate cyclase is the enzyme that catalyzes the conversion of ATP to cyclic AMP. Dopamine is an intermediate substance in the biosynthesis of norepinephrine.

$$\text{ATP} \xrightarrow[\text{cyclase}]{\text{adenylate}} \text{cyclic AMP}$$

$$\text{Dopa} \xrightarrow[\text{decarboxylase}]{\text{dopa}} \text{dopamine} \xrightarrow[\beta\text{-oxidase}]{\text{dopamine}} \text{norepinephrine}$$

Cholinergic fibers Parasympathetic fibers also produce a chemical mediating substance. In this case the substance is acetylcholine, which is promptly converted to choline and acetic acid by the action of an enzyme called *cholinesterase*. Since acetylcholine does not remain in its most active state for any great length of time, it is probable that its effects are entirely local. Unlike norepinephrine, it is probably not carried by the bloodstream.

All preganglionic fibers, whether sympathetic or parasympathetic, have been shown to liberate a cholinergic substance, probably identical with acetylcholine. This means that the transmission of the nervous impulse across the point of synapse between the preganglionic and postganglionic fiber is accomplished by the production of acetylcholine.

As we have indicated, postganglionic sympathetic fibers to the sweat glands and to smooth muscles of the skin are cholinergic. These fibers are carried by peripheral nerves. Voluntary motor nerves to skeletal muscles are also cholinergic. On the basis of chemical transmitter substances it appears that the division of the autonomic system into sympathetic and parasympathetic is somewhat artificial.

Visceral afferent fibers There are afferent fibers arising in the viscera and associated with the autonomic nervous system. These fibers do not give rise to sensation in the ordinary sense. No sense of touch or pain has been developed internally, as it has been developed externally in the skin and

Chemical transmitters

255

body wall. The viscera are relatively insensitive to pain in the way that pain from a cut or burn on the skin or the crushing of a finger is recognized.

The receptors of visceral afferent neurons are located in the viscera. Large, medullated fibers extend through the autonomic ganglion and the white ramus to the cell body located in the ganglion of the dorsal nerve root (Figure 10.5). They do not synapse in the autonomic ganglion. From cell bodies located in the ganglia of dorsal spinal-nerve roots, processes enter the dorsal horn of gray matter of the cord along with the processes of sensory spinal nerves. Visceral afferent impulses also can reach the medulla by way of the vagus nerve. Even though the vagus nerve may not carry recognizable pain impulses, afferent stimuli of a reflex nature cause the pupil of the eye to dilate when the vagus nerve is stimulated in the abdominal region.

The exact way in which visceral pain arises is not well understood. It may be caused by certain kinds of muscular contraction in visceral muscle or by pressure within a hollow organ. Internal pain is not readily localized; it may be referred to some surface area. One explanation of such *referred pain* is that, since the visceral afferent neuron enters the gray matter of the cord in close association with the spinal sensory nerve, the pain is interpreted as arising in the surface area supplied by the spinal nerve; thus pain arising in the gallbladder is felt as coming from the surface area above the organ. Cardiac pain can be felt in the scapular area or, as in the case of angina pectoris, can radiate down the left arm. A reflex arc can be established with spinal motor neurons, which would account for the tensing of muscles over an area of internal pain.

Though referred pain is often described as an acute, sharp, stabbing pain, there is a direct type of pain that is described as a dull pain. It is usually not well localized, but the individual is aware of general discomfort. Afferent visceral neurons carry the impulses that give rise to the direct type of pain; they enter the cord, but little is known about the mechanism by which one becomes aware of such pain.

A new development in the study of the autonomic nervous system may have future significance. Animals subjected to instrument training (reward and punishment) have been able to increase or decrease heart rate, blood pressure, intestinal muscle contractions, and a number of other visceral responses. There has been some success with human subjects also in modifying blood pressure and heart rate. This concept is in the early stages of investigation, but physiologists may have to reconsider the belief that the autonomic nervous system is purely involuntary and not subject to learning.

SUMMARY   The autonomic nervous system is divided into a craniosacral, or parasympathetic, division and a thoracolumbar, or sympathetic, division. The craniosacral division is associated with certain cranial and sacral nerves. The thoracolumbar division is characterized by a chain of ganglia on either side of the spinal cord in the thoracic and lumbar regions. A preganglionic neuron extends from the spinal cord to one of the ganglia of the lateral chain. A postganglionic neuron extends from the autonomic ganglion to the organ supplied.

In the craniosacral division, preganglionic fibers are carried in cranial or sacral nerves. Postganglionic fibers arise in collateral ganglia or within the organ supplied.

The autonomic nervous system

The great plexuses of the autonomic system are the cardiac, celiac, and hypogastric plexuses.

The hypothalamus is a center for autonomic regulation.

Cranial nerves that carry parasympathetic fibers are the oculomotor (IIId cranial), facial (VIIth cranial), glossopharyngeal (IXth cranial), vagus (Xth cranial), and accessory (XIth cranial) nerves.

The enteric plexuses represent the intrinsic nerves of the intestine. The intestine also receives postganglionic sympathetic fibers and preganglionic parasympathetic fibers. The enteric-system neurons are considered to act as postganglionic neurons of the parasympathetic system.

Sympathetic fibers that release an epinephrine-like substance at their terminal filaments are said to be adrenergic. Sympathetic postganglionic fibers are adrenergic except those innervating sweat glands and the smooth muscles of the skin.

Parasympathetic fibers release acetylcholine as a chemical transmitter substance, and, like motor neurons of peripheral nerves, they are cholinergic. Sympathetic preganglionic fibers also are cholinergic.

Visceral afferent fibers carry impulses interpreted as deep-seated pain. Often it is a referred pain, arising in a certain area but interpreted as coming from some other area. The spinal visceral afferent fiber passes through the autonomic ganglion, but its cell body is located in the ganglion of the dorsal spinal-nerve root.

1 List some of the physiological adjustments that take place in climbing stairs.
2 Discuss the effects of worry or a severe fright.
3 Why is it advantageous to be happy and to have pleasant surroundings at mealtime?
4 If the oculomotor nerve is severed, why does the pupil of the eye dilate?
5 Explain how it is possible to stimulate the vagus nerve and yet observe that the heart rate slows or stops altogether.
6 Discuss the role of the autonomic system in the control of body temperature.
7 By what pathway do sympathetic fibers reach peripheral blood vessels and the sweat glands?
8 Why does the hair seem to stand on end when one is frightened?
9 Explain the chemical mediation of nerve impulses at the synapse.
10 What is meant by referred pain?
11 Explain negative feedback.
12 How does the negative-feedback mechanism contribute to homeostasis?

SUGGESTED READING

AXELROD, J.: Noradrenaline: Fate and Control of Its Biosynthesis, *Science*, **173**:598-600 (1971).

AXELSSON, J.: Catecholamine Functions, *Ann. Rev. Physiol.*, **33**:1-30 (1971).

BENZINGER, T. H.: The Human Thermostat, *Sci. Am.*, **204**:134-147 (1961).

CANNON, W. B.: "Bodily Changes in Pain, Hunger, Fear and Rage," Appleton-Century-Crofts, Inc., New York, 1929.

A classical treatise on the function of the autonomic nervous system.

DAVIDOFF, R. A.: Gamma-aminobutyric Acid Antagonism and Presynaptic Inhibition in the Frog Spinal Cord, *Science*, **175**:331-333 (1972).

DICARA, LEO V.: Learning in the Autonomic Nervous System, *Sci. Am.*, **222**(1):30-39 (1970).

HARDMAN, J. G., G. A. ROBISON, and E. W. SUTHERLAND: Cyclic Nucleotides, *Ann. Rev. Physiol.*, **33**:311-336 (1971).

KEBABIAN, J. W., and PAUL GREENGARD: Dopamine-sensitive Adenyl Cyclase: Possible Role in Synaptic Transmission, *Science*, **174**:1346–1349 (1971).

LOEWI, OTTO: On the Humoral Transmission of the Action of Heart Nerves, in M. L. Gabriel and S. Fogel (eds.), "Great Experiments in Biology," Prentice-Hall, Inc., Englewood Cliffs, N.J., 1955.

An original and noteworthy study in experimental physiology.

VON EULER, U. S.: Adrenergic Neurotransmitter Functions, *Science*, **173**:202–205 (1971).

chapter 11
special senses

The sense organs afford an awareness of external stimuli, permitting us to react to forces impinging upon the body or arising within the body. Whatever we are able to learn about the world around us depends upon our perception of various forces arising from an area outside our own bodies. It is an energy relationship, for the forces that give the impression, for example, of warmth or pressure do so because of a release of energy to which sense organs react. It is not by accident that the highly developed sensory organs are located at the anterior end of the body and close to the brain. This has been the course of development in the evolutionary history of all organisms. Ordinarily sense organs are adapted to react to only one kind of stimulus: the eye receives light waves; the ear is stimulated by sound waves. Perception requires that the sensory unit be intact and functioning.

THE SENSORY UNIT

Three essential structures compose a sensory unit. They are the *receptors*, the *neural pathway*, and an *interpreting center* in the brain. Receptors can be of various types so far as their structure is concerned, but their function is to react to certain types of stimuli, which they are adapted to receive. The rod and cone cells of the eye or the specialized nerve endings in the skin are examples of receptors. The nervous pathway consists of sensory neurons that carry nerve impulses inward toward the brain.

PROJECTION OF SENSATIONS The interpretation of sensation occurs in more or less localized areas of the brain, but these sensations are projected to their source. If a finger is cut, the pain is projected to that particular finger. The sense of vision is projected to the object rather than to the area of interpretation in the brain. Sometimes the projection is within the body, as when internal pain is experienced or when the sensations of hunger or thirst are projected to organs of the digestive tract.

CLASSIFICATION OF RECEPTORS

Receptors are commonly classified into three groups: exteroceptors, proprioceptors, and interoceptors. The sense organs that receive stimuli from the exterior, such as the eye, ear, and skin, are *exteroceptors*. They are the receptors of the somatic sensory or somatic afferent system. *Proprioceptors* are located in muscle spindles and tendons and around joints. They are important in the interpretation of the position of any part of the body, in muscular coordination, in postural reflexes, or in determining the degree of stretch in muscle or tendon. They are the receptors for muscle sense. Sensations arising from stimuli to proprioceptors are ordinarily below the conscious

TABLE 11.1
Sensory receptors

Classification	Location	Source of stimulus
Exteroceptors	Skin, tongue, nose, ear, eye	External
Proprioceptors	Muscle spindles, tendons, joints	Stretch reflexes, postural reflexes, position in space
Interoceptors	The viscera	Endings of the visceral afferent nervous system

level but can reach the level of consciousness. *Interoceptors* are the receptors of the visceral afferent system. They are located in the viscera, primarily in the organs of the respiratory, digestive, and reproductive systems. Stimuli arising from these receptors are ordinarily below the level of consciousness, although visceral pain can reach the conscious level. The visceral afferent system is closely interrelated with autonomic functions and was discussed in Chapter 10 (Table 11.1).

CUTANEOUS SENSATIONS

RECEPTORS OF VARIOUS STRUCTURAL TYPES *Simple nerve endings* Invading practically all the tissues, simple nerve endings have been demonstrated in such tissues as oral epithelium, mucous and serous membranes, and skin. Nerve fibers penetrate the basal layers in bundles, breaking up into individual unmedullated fibers in the outer layers as they pass between cells. The endings may be simple or somewhat enlarged. Since similar endings are found in the cornea and in the teeth, there is a strong assumption that naked nerve endings are primarily pain receptors. Free sensory-nerve endings occur also in connective tissue and in muscles and tendons. There is a nerve network around the base of each hair. Nerve endings associated with hair follicles are concerned with the tactile sense.

Various types of stimuli can excite pain receptors. In general, an energy stimulus of extreme degree, one that threatens to cause damage to tissues, such as increasing the amount of heat or pressure, is interpreted as pain.

Encapsulated sensory-nerve endings This type of receptor is characterized by a connective-tissue capsule that surrounds the nerve ending. The capsule varies considerably in shape and thickness in the various types of end organs. They function as receptors for the sense of touch, pressure, and temperature.

While the tactile sense is commonly stimulated by simple nerve endings associated with hair follicles, there are also long, encapsulated structures called *Meissner's corpuscles* that react to tactile stimuli. These receptors have a rather thin covering, and within the capsule there are thin, transverse connective-tissue plates that divide the corpuscle into minute compartments. A nerve fiber enters the capsule and branches into the transverse compartments. Meissner's corpuscles are found in connective-tissue elevations of the skin, called *papillae*. Though present in the skin throughout the body, they are most numerous on the fingertips, the palm of the hand, and the sole of the foot. They are also present in the tip of the tongue, lips, nipples, the glans penis, and clitoris. Recent work seems to indicate that they are

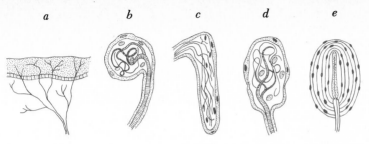

FIGURE 11.1
Various types of receptors: *a* free nerve endings (pain); *b* end bulb of Krause (cold); *c* end organ of Ruffini (warmth); *d* Meissner's corpuscle (tactile); *e* Pacinian corpuscle (pressure). (*After Wendell J. S. Krieg, "Functional Neuroanatomy," McGraw-Hill Book Company, New York, 1953.*)

ordinarily present in groups of two or three. A touch-sensitive spot in the skin ordinarily represents an area above several Meissner's corpuscles (Figure 11.1).

Pacinian corpuscles   Located deeply in subcutaneous tissues and elsewhere throughout the body are oval structures called *Pacinian corpuscles*. They react to pressure stimuli. These structures are laminated; the capsule is composed of layers resembling an onion. A nerve fiber grows into its central cavity. Pacinian corpuscles are found also in the mesentery, in the connective tissue of the pancreas and other organs, and along blood vessels. Those located along blood vessels apparently function in the control of blood pressure. Cholinesterase has been found around the nerve ending, in the core of the Pacinian corpuscle.

Sensory receptors are biological transducers. In a physical sense, they receive energy in various forms (light, sound, heat waves) and convert it into another form of energy, the nerve impulse. Lowenstein has investigated the transducer function of Pacinian corpuscles. Mechanical stimulation of the core in these receptors initiates a weak local electric current, called the *generator potential*. This generator current starts the *action potential*. The nerve ending, in the core is the essential transducer. It is of interest that the generator potential does not travel along the neuron; it only starts the propagated nerve impulse which travels over the fiber. The nerve fiber is myelinated but the core is not. It was determined that the action potential is initiated at the first node, which is located at the beginning of the myelin sheath.

Pressure on the Pacinian corpuscle causes a wave of depolarization to pass along the core membrane. This generator potential is not an all-or-none response and is not self-propagating. It is a graded response, but if it is of sufficient strength when it reaches the low threshold level of the fiber's first node, a propagated action potential is initiated over the afferent neuron.

End bulbs of Krause   Located in the outer portion of the dermis, in the tip of the tongue, and in the cornea of the eye are receptors known as the *end bulbs of Krause*. They are assumed to be receptors for cold, and they underlie cold spots in the skin. Although they vary in shape, the commonest type is an oval capsule with small, branching nerve fibers in its cavity. These receptors ordinarily react to temperatures below the normal temperature of the skin or to temperatures lower than those needed to stimulate receptors for warmth.

End organs of Ruffini   The receptors for warmth are thought to be elongate, encapsulated structures located deep in the dermis and called the *end organs of Ruffini*. The long, cylindrical capsules are filled with fine nerve fibers. They are not as abundant as the end bulbs of Krause, and warm spots on

Cutaneous sensations

the skin are not as abundant as cold spots. Ordinarily they react to temperatures above normal skin temperature.

It has become evident that the various senses of the skin, for the most part, have their own receptors. These receptors are not evenly distributed over the surface of the body. The fingertips are highly specialized for the sense of touch but are not so efficient in indicating temperature. The wrist, for example, gives a better estimate of temperature than can be obtained from the fingertips.

Though there are two receptors for the temperature sense, they actually react only to degrees of heat. The receptors for warmth are stimulated by a degree of heat that is ordinarily greater than normal skin temperature. The cold receptors react to a degree of heat ordinarily less than normal skin temperature or below that required to stimulate the receptors for warmth.

Although much is known about temperature and its regulation, there is a lack of conclusive evidence regarding the identity and function of the receptors concerned with the temperature sense.

The neural pathway   There are several pathways in the spinal cord for the tactile sense. Impulses for both tactile and pressure senses travel in the anterior spinothalamic tract, crossing over directly to the opposite side. From synaptic endings in the thalamus, the impulses travel through the internal capsule to the cutaneous sensory area of the cerebrum, which lies posterior to the central sulcus. Other routes for tactile impulses are in the large posterior tracts of the cord (the fasciculus gracilis and fasciculus cuneatus; see Figure 9.21). These tracts lead to nuclei of the same name in the medulla. The fibers synapse in the medulla, and succeeding fibers cross to opposite sides. They join anterior spinothalamic tracts and pass upward to the thalamus, where again they synapse and a final set of fibers leads to the cutaneous sensory area. Tactile impulses from the head and face are carried by fibers in the trigeminal nerves.

The pathway for pain and temperature impulses from receptors in the skin of the trunk and appendages is in the lateral spinothalamic tract. It is probably entirely a crossed tract. Pain impulses from the head and face are conveyed inward toward the thalamus by fibers in the trigeminal nerves.

THE GUSTATORY SENSE

The senses of taste and smell are considered to be chemical senses, in that they react to stimuli that are chemical in nature. The end organs for the sense of taste are stimulated by chemical substances in solution, and those of the sense of smell by substances in volatile form. Although many taste sensations are closely correlated with the sense of smell, structurally the sense of taste seems to be more closely related to the somatic senses discussed in the preceding pages.

The sense of taste plays an important part in the selection and enjoyment of food. If the nose is closed so that one is unable to benefit from the olfactory sense, one must depend on taste which is limited to only four categories: sweet, sour, salty, and bitter, or some combination of these. The tongue must be moist enough to permit the substances to go into solution. One cannot taste sugar or salt on a perfectly dry tongue.

Certain areas of the tongue react more strongly to one of the four taste sensations than to the others. The posterior part of the tongue reacts to bitter stimuli. The lateral edges respond to sour and salt but are most sensitive to sour stimuli. The tip of the tongue responds to all four categories of taste but is most sensitive to sweet and salt. There appears to be at least a physiological difference in the kinds of receptors in these different regions of the tongue. The middle portion of the surface does not react to any great extent to taste stimuli.

RECEPTORS   The receptors for the sense of taste are located in taste buds. These are minute oval bodies composed of supporting cells and elongated sensory cells. The sensory cells have hairlike processes that converge at the taste pore. Through the opening at the taste pore the sensory cells make contact with chemical stimuli. The sensory cells and their processes within the taste bud are protected from the friction of rough food surfaces against the tongue. The taste buds are located in small, rounded elevations on the tongue, called *papillae,* of which there are various types. The papillae are responsible for the rough appearance of the upper surface of the tongue. The largest are the *vallate papillae,* which form a V-shaped row at the back of the tongue. While taste buds are found primarily on the tongue's upper surface, they are also distributed to a limited extent over the palate, the epiglottis, and the back part of the mouth, especially during embryonic development and in young children. The number of taste buds decreases as the individual becomes older. Elderly persons are assumed to have lost some of the acuteness of the sense of taste exhibited by children. The effects of colds and catarrhal conditions may tend to cover the taste buds and interfere with their normal function (Figure 11.2).

THE NEURAL PATHWAY   The two cranial nerves involved in transmitting impulses from receptors located on the tongue are the facial (VIIth cranial) and the glossopharyngeal (IXth cranial). A branch of the facial nerve supplies receptors on the anterior two-thirds of the tongue, and the glossopharyngeal nerve supplies the posterior one-third. Taste buds in the posterior part of the mouth (pharynx) and around the larynx are supplied by a few fibers of the vagus (Xth cranial) nerve.

The neural pathway for impulses from the tongue leads to gray matter in the medulla, where most of the fibers terminate. The second part of the relay leads from the medulla to a nucleus in the thalamus. Neurons of the third set lead from the thalamus to the lower part of the cutaneous sensory area in the cortex of the cerebrum. The sensory-interpreting area is now thought to be located there in association with the sensory-interpreting area for the face as a whole and close to the sensory area for the tongue.

THE OLFACTORY SENSE

The olfactory sense is closely correlated with the sense of taste, in that the taste of many substances is largely supplemented by their odor. When we taste orange juice with the nose closed, it is merely sweet or acid, but not the taste of orange, for orange is a highly volatile substance, and to classify it accurately the sense of smell must also be used.

The olfactory sense is a chemical sense stimulated by volatile substances. The substances must be soluble in water, at least to some extent, since the olfactory-cell processes project into the mucous covering of the olfactory epithelium. Probably olfactory substances must also be soluble in

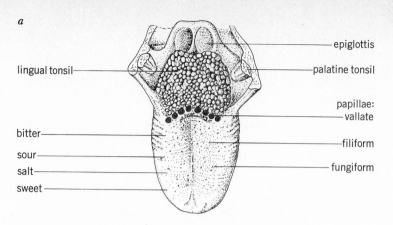

a

lingual tonsil —

bitter —
sour —
salt —
sweet —

— epiglottis
— palatine tonsil

papillae:
— vallate

— filiform

— fungiform

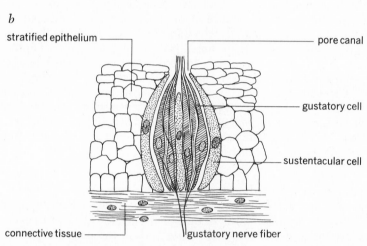

b

stratified epithelium —

connective tissue —

— pore canal

— gustatory cell

— sustentacular cell

gustatory nerve fiber

FIGURE 11.2

a The tongue, illustrating papillae and primary taste centers; *b* vertical section through a taste bud.

lipids or at least be subject to active transport through the lipid layer of the receptor-cell membrane.

There are several theories regarding the mechanism of olfaction. They are generally based on chemical or physical principles. A recent proposal is a stereochemical theory based on the shape of molecules. It hypothesizes an olfactory membrane with receptor sites approximately the same shape as the molecules that it is designed to receive.

The olfactory area occupies about 1 square inch in the upper part of each nostril. It is a rounded yellowish area containing vitamin A, located on the upper part of the superior nasal conchae and the nasal septum just opposite. Since this area is high in the nasal passageway, it is somewhat above the ordinary pathway of air passing through the nose in ordinary respiration. When we wish to smell something in which the odor is not very strong, we sniff at it, thus drawing the air containing the volatile substance higher into the nasal passageway and over the olfactory area (Figure 11.3).

The posterior part of the nasal passageway leads down into the throat by way of the posterior nares. The posterior nares are closed during swallowing, but at other times odors of foods may pass up through them to reach the olfactory area.

The greater part of the nasal passageway is innervated by branches of the trigeminal (Vth cranial) nerve. These sensory branches are concerned

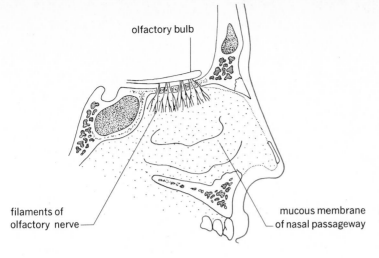

olfactory bulb

filaments of olfactory nerve

mucous membrane of nasal passageway

FIGURE 11.3
The olfactory area in the nasal passageway.

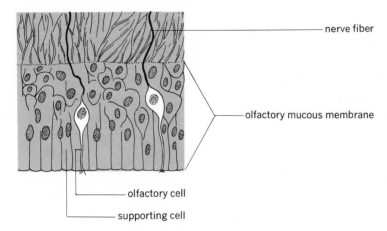

nerve fiber

olfactory mucous membrane

olfactory cell

supporting cell

FIGURE 11.4
Olfactory cells and supporting cells as seen through the olfactory mucous membrane (vertical section).

with pain, temperature, and tactile stimuli within the nostrils. Some substances do not produce an olfactory effect but rather produce an irritating or tactile effect. Substances such as pepper, camphor, phenol, ammonia, ether, and chloroform are not entirely olfactory, since they also stimulate the trigeminal nerve. Chloroform, in addition, gives rise to a sweetish taste as it stimulates the taste buds.

RECEPTORS   The *olfactory receptors* represent a primitive type of cell in which the distal part, located in the nasal mucosa, is the receptor and the basal part is extended inward as a nerve fiber. The bipolar receptor cell is long and slender, with a nucleus at its base. Distally, arising from the peripheral process there is a tuft of hairlike processes, 2 to 12 in number, projecting into the mucus that covers the olfactory membrane. Volatile substances are thought to stimulate the receptor cells by coming in contact with the hairlike processes. The receptor cells lie in columnar epithelium. The yellowish color of the olfactory area is due to pigmentation of the columnar cells (Figure 11.4).

The olfactory sense

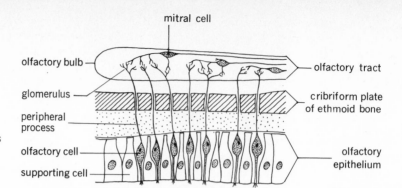

mitral cell

olfactory bulb

glomerulus

peripheral process

olfactory cell

supporting cell

olfactory tract

cribriform plate of ethmoid bone

olfactory epithelium

FIGURE 11.5
Diagram illustrating olfactory cells and their synaptic connections with mitral cells in the olfactory bulb.

THE NEURAL PATHWAY The olfactory nerve is composed of unmyelinated fibers extending inward from receptor cells to neurons within the *olfactory bulbs.* These synaptic relay centers lie beneath the frontal lobes of the cerebrum and on either side of the crista galli of the ethmoid bone. They are part of the brain and contain neurons of a different structural type. Bundles of fibers from the nasal mucosa pass through openings in the cribriform plate of the ethmoid and enter the olfactory bulbs. There, they make contact with the dendrites of *mitral cells,* the impulse first passing through areas of intricate synaptic branching called *glomeruli.* The glomeruli seem to function as relays for impulses directed to various parts of the brain. Neurons of the mitral cells form the olfactory tract (Figure 11.5).

The olfactory tract Neurons of the olfactory bulbs lead inward to form the olfactory tract. Some neurons pass over to the opposite side in the anterior commissure. The olfactory tract passes along the ventral side of each frontal lobe. The posterior connectives of the olfactory tract have not been fully worked out. For this reason the location of the interpreting area for the sense of smell remains in doubt. The tract itself contains some gray matter dorsally, and the interpreting area may be located either in the tract itself or in areas of the brain close to the olfactory bulb and tract. The olfactory sense differs from others in having no known direct connections with the thalamus. There is, however, an indirect connection with the hippocampus, located in the inferior horn of the lateral ventricle. Some of the fibers enter the pyriform cortex, an area in the lower part of the temporal lobe and below the hippocampus in the rhinencephalon. Fibers have been traced also to some of the basal nuclei, such as the amygdaloid and caudate nuclei, as well as to the mammillary bodies. Since the fibrous connections extend to so many parts deep within the brain, it has been difficult to determine an exact interpreting center for the olfactory sense.

The olfactory ability of human beings is not as highly developed as that of most animals. The ability of hunting animals to track their prey by scent requires olfactory ability of a high order. However, man does have the ability to recognize a great variety of odors, both good and bad. The florist may be able to identify many different flowers by their individual odors, as the chemist can recognize a great variety of volatile substances. The variety of recognizable odors is not easily classified and put into definite categories as are the sensations of taste.

The sensation of smell develops quickly, but it also adapts quickly to an individual odor. After one smells a flower awhile, the odor seems to have vanished. This condition is not caused by fatigue of the receptor mechanism, for one can readily smell some other odor. It illustrates the property of *adaptation* and explains why one can so easily become accustomed to prevalent odors. When one walks into a room where someone is smoking, he will probably be quite aware of the odor of burning tobacco. But after remaining in the room for some time, one probably will be accustomed to the odor and be generally unaware of the condition of the air. Persons have been asphyxiated by escaping coal gas while they were asleep, apparently unaware of their danger.

HEARING AND EQUILIBRIUM

Auditory receptors react to physical energy in the form of sound waves. The ear represents an intricate mechanism designed to transmit sound waves by mechanical means inward to the receptors. The ear consists of three primary parts: (1) the external ear, (2) the middle ear, and (3) the inner ear. The auditory pathway consists of the auditory (VIIIth cranial) nerve and the interpreting center, which is located in the temporal lobe of the cerebrum.

EXTERNAL EAR The outer ear is a structure designed to collect sound waves and to direct them into the *external auditory* meatus; it is composed of elastic cartilage, covered with skin. The lobe of the ear contains vascular connective tissue rather than cartilage. The meatus is about an inch in length and leads inward to the tympanic membrane. The meatus is not straight; pulling upward and backward on the outer ear tends to straighten it. The skin covering the ear is extended into the meatus. It becomes thinner and more sensitive as it approaches the eardrum. Close to the external opening there are stiff protective hairs and sebaceous glands, which produce earwax, or *cerumen*. The glands are thought to be modified sweat glands. Both the hair and the glands tend to keep foreign objects out of the meatus. Since the eardrum is a delicate membrane and subject to injury, persons should be advised against cleaning wax out of the ears with hard objects, such as matches, pencils, or hairpins. Wax may be gently washed out, if necessary, with clean warm water, preferably by a physician.

The *tympanic membrane* completely covers the inner extremity of the external auditory meatus and thus separates the outer and middle parts of the ear. It is a very thin fibrous membrane covered with a thin layer of skin on the outer side and with mucous membrane on the inner side. It lies at an angle of about 55° to the lower border of the canal. Most of the membrane is tightly stretched as a vibrating membrane, but there is a small, thin section in the upper part which is not stretched and is called the *flaccid part*. It is in this weaker, flaccid part that injuries from concussion or inflammation are likely to occur (Figure 11.6).

MIDDLE EAR An air space in the petrous portion of the temporal bone houses the middle ear. It is continuous with the spongy bone of the mastoid process. The auditory (eustachian) tube connects the cavity of the middle ear with the nasopharynx and is concerned with the equalization of pressure on the eardrum.

Within the cavity there is a lever system of three very small bones. The bones are the *malleus, incus,* and *stapes.* The malleus is shaped somewhat like a mallet, with the handle portion attached to the inner surface of the

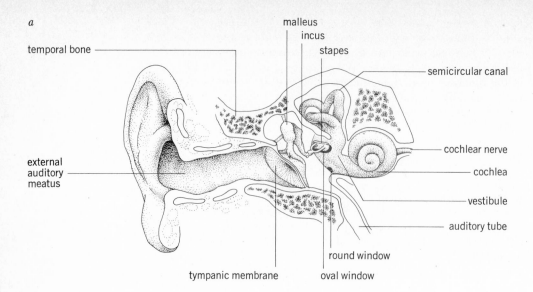

a

temporal bone

malleus
incus
stapes

semicircular canal

cochlear nerve

cochlea

external
auditory
meatus

vestibule

auditory tube

round window

oval window

tympanic membrane

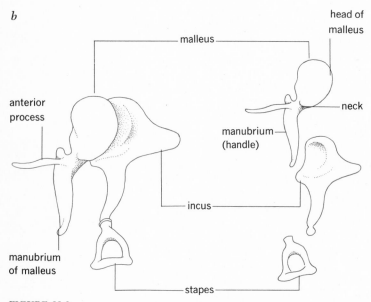

b

head of
malleus

malleus

anterior
process

neck

manubrium
(handle)

incus

manubrium
of malleus

stapes

FIGURE 11.6
a The structure of the ear, including detail of the middle ear; *b*
auditory ossicles.

eardrum. The rounded head fits into a depression in the incus and is held
tightly in place by ligaments so that these two bones move as one. The incus
is the intermediate bone in the series. The stapes is a stirrup-shaped bone
that fits into the oval window of the vestibule. The joint between the incus
and stapes is freely movable, the stapes performing a rocking movement
at the oval window. That these are, indeed, small bones is indicated by their
measurements. The malleus is about 8 millimeters long, the incus is approximately 7 millimeters, and the stapes measures only 4 millimeters.

There are two small muscles in the middle ear. The *tensor tympani*
attaches by a tendon to the manubrium of the malleus and by its contraction
tightens the eardrum. The second muscle is the *stapedius*, a tiny muscle,
the smallest of all the skeletal muscles. Its tendon attaches to the neck of

Special senses

268

the stapes. It opposes the action of the tympanum and the lever system as the stapes is pushed inward. Contractions of both muscles would have a damping effect on the vibrations of the ear bones and so attenuate and protect the delicate inner ear from intense vibrations.

A branch of the facial (VIIth cranial) nerve, the *chorda tympani*, passes through the cavity of the middle ear.

The impact of sound waves causes the stretched membrane to vibrate. The energy of these vibrations is carried across the air space of the middle ear by a lever system formed by the three ear bones. The rocking motion of the stapes sets up waves in the fluid of the inner ear. However, there is no gain in pressure by virtue of the lever system. Since the area of the tympanic membrane is some 20 times larger than that of the base of the stapes, the pressure exerted on the fluid within the cochlea is about 20 times that exerted on the tympanic membrane by the impact of sound waves. It is obvious that a fluid has greater inertia than air, and therefore the greater pressure exerted on the footplate of the stapes is essential to equalize the strength of the vibrations between air and fluid.

The *auditory tube* extends from the middle ear to the pharynx, a distance of about $1\frac{1}{2}$ inches. It is only about $\frac{1}{8}$ inch in diameter at its narrowest point. Its function is to permit air to enter the middle ear behind the eardrum to equalize air pressure on either side. When riding up and down hills, it is a common experience to notice the effect of a change in air pressure on the eardrums. The lower opening of the tube is closed except during such movements as swallowing or yawning. One may aid the equalization of pressure on the ear membrane, therefore, by swallowing or opening the mouth wide. If for any reason the auditory tube is blocked or does not open freely, an inequality of air pressure may cause the eardrum to bulge inward or outward. Under these conditions the membrane cannot vibrate freely, and hearing will be impaired (Figure 11.6).

Although it is essential for the equalization of pressure, the auditory tube may also become a liability, in that it offers a passageway by which infections may invade the middle ear. Infections of the throat can cause inflammation of the tube and middle ear. The tube is shorter and more horizontal in children, providing somewhat more ready access for invading organisms. An earache ordinarily means an active infection behind the eardrum. Such infections should be brought to the attention of a physician without delay. If uncontrolled, the infection may invade the spongy bone of the mastoid process, perhaps making a surgical operation necessary to clean out the infected area in this bony process. The middle ear has a rather thin, bony roof. The infection can spread upward and invade the meninges that lie just above. A common condition is the bulging outward of the eardrum due to the accumulation of material behind it. A physician can pierce the eardrum in order to relieve the pressure and permit drainage through the external auditory canal. If the eardrum is not pierced, it may rupture and thus permit drainage. This is more likely to occur in children, and it produces "running ears." Small ruptured areas usually heal, but larger ruptures may persist as perforated eardrums and cause impaired hearing. Persons with openings through the eardrums should be advised against diving or swimming underwater lest infections reach the middle ear.

INNER EAR   The major structural divisions of the inner ear are the vestibule, the cochlea, and the semicircular canals (Figure 11.7). These structures develop in the harder portions of the temporal bone and make up the osseous

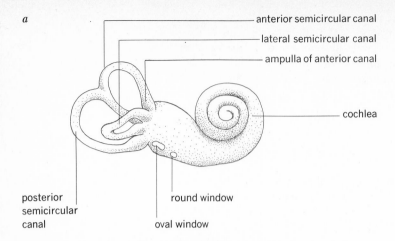

a

anterior semicircular canal

lateral semicircular canal

ampulla of anterior canal

cochlea

posterior semicircular canal

round window

oval window

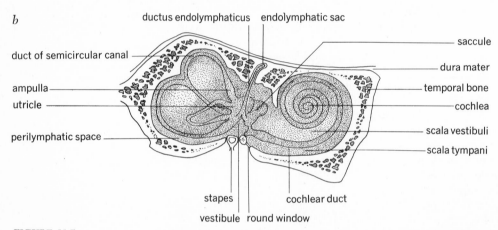

b

ductus endolymphaticus   endolymphatic sac

duct of semicircular canal

ampulla

utricle

perilymphatic space

saccule

dura mater

temporal bone

cochlea

scala vestibuli

scala tympani

stapes   cochlear duct

vestibule round window

FIGURE 11.7

The inner ear, illustrating the membranous and bony labyrinths; *a* right bony labyrinth; *b* membranous labyrinth.

labyrinth. A fluid, the *perilymph,* separates the osseous labyrinth from the smaller membranous labyrinth within. The fluid of the membranous labyrinth is called the *endolymph.*

The *vestibule* is a chamber in the bony labyrinth and is located between the semicircular canals and the cochlea. The stapes rocks in and out at the oval window of the vestibule. Below this opening is the round window. The vestibule also has communications with the semicircular canals and the cochlea. Within the bony vestibule there are two small sacs of the membranous labyrinth. The larger of the two is called the *utricle* and is associated with the membranous semicircular canals; the other, the *saccule,* is associated with the membranous duct of the cochlea.

The *cochlea* consists of a bony and membranous spiral labyrinth. There is a bony core, the *modiolus,* from which projects a thin bony shelf, the *spiral lamina.* Like the threads of a screw, the spiral lamina winds around the central modiolus. The bony canal is therefore partially divided by this shelf of bone projecting out into the cavity. The upper spiral is called the *scala vestibuli,* and the lower one is called the *scala tympani.* The scala vestibuli leads from the oval window to the apex of the cochlea. The scala tympani extends from the round window to the apex, where the two spiral passageways are confluent. The small aperture permitting confluence at the apex is called the *helicotrema.* The fluid of the osseous labyrinth is perilymph (Figure 11.8).

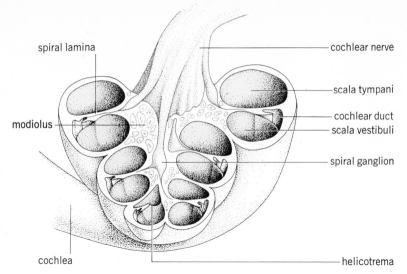

spiral lamina

cochlear nerve

modiolus

scala tympani

cochlear duct

scala vestibuli

spiral ganglion

cochlea

helicotrema

FIGURE 11.8

Section through the cochlea of the inner ear.

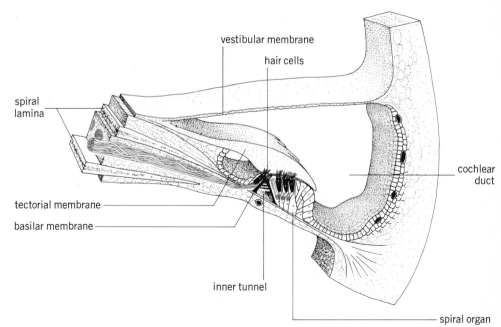

vestibular membrane

hair cells

spiral lamina

cochlear duct

tectorial membrane

basilar membrane

inner tunnel

spiral organ

FIGURE 11.9

The cochlear duct and spiral organ (organ of Corti). The figure represents a cross section of the cochlear duct and spiral organ in one turn of the cochlea. The hair cells are thought to be stimulated by movement of the tectorial membrane, and nerve impulses are transmitted to the spiral ganglion by nerve fibers.

The *cochlear duct (scala media)* is the membranous labyrinth of the cochlea. It is separated from the scala vestibuli above by the thin vestibular membrane (of Reissner). Below, the basilar membrane lies between the cochlear duct and the scala tympani (Figure 11.9). The basilar membrane, therefore, forms the floor of the cochlear duct, extending from the bony shelf of the spiral lamina. The basilar membrane actually becomes wider as the spiral approaches the narrowing apex of the cochlea. The cochlear duct is connected with the saccule in the vestibule and extends upward through the cochlea, where it lies between the scala vestibuli and the scala tympani. It ends as a blind sac at the apex. Its outer wall is composed of fibrous periosteum and epithelial tissue (Figure 11.10*a*).

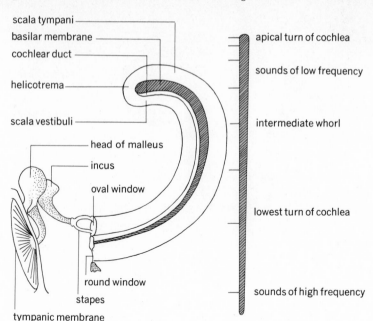

a   *b*

scala tympani

basilar membrane

cochlear duct

helicotrema

scala vestibuli

head of malleus

incus

oval window

round window

stapes

tympanic membrane

apical turn of cochlea

sounds of low frequency

intermediate whorl

lowest turn of cochlea

sounds of high frequency

FIGURE 11.10
Diagrammatic representation of the cochlea and the basilar membrane: *a* the cochlea extended to show relative positions of the scalae; *b* the basilar membrane extended to show areas affected by sound waves of low and high frequency.

When viewed in vertical section, the cochlea reveals a spiral of $2\frac{1}{2}$ turns, each whorl being divided into three ducts: the scala vestibuli, the cochlear duct, and the scala tympani.

The *spiral organ (of Corti)* contains the receptors for the sense of hearing. It lies on the basilar membrane and is composed of supporting cells and hair cells. The supporting cells are tall epithelial cells of four types. They are attached to the basilar membrane, and their free surfaces form a reticular membrane over the spiral organ. The hair cells are neuroepithelial cells, with hairlike processes that extend through openings in the reticular membrane into the endolymph of the cochlear duct. They are arranged in outer and inner rows. Above the spiral organ lies the tectorial membrane, attached to the spiral lamina and extending over the hair cells. Though it usually appears well above the spiral organ in prepared sections, in the living state apparently it rests on the hair cells and transmits stimuli to them (Figure 11.9).

THE NEURAL PATHWAY   The fibers of the cochlear nerve arise from bipolar cell bodies located in the spiral ganglion. The spiral ganglion, in turn, is located in the modiolus. In the early embryo a fiber grows inward from each cell body and terminates about the bases of hair cells, while the other fiber grows in the opposite direction as a part of the acoustic-nerve pathway. The cochlear nerve is a branch of the vestibulocochlear (VIIIth cranial), or auditory, nerve. The cochlear-nerve fibers terminate in nuclei in the medulla. Many fibers cross over to the opposite side in the medulla and then ascend in lateral tracts until the auditory interpreting center in the temporal lobe of the cerebrum is reached. There are many relays of neurons in the auditory pathway. Among the more important nuclei are the inferior colliculi of the midbrain and the medial geniculate bodies of the thalamus.

The spoken word presumably stimulates word-understanding areas in both temporal lobes, but in the majority of individuals, the left area appears to be better developed. There is evidence also that the right side of the brain receives noise and other nonverbal stimuli whereas the left side of the brain receives the greater amount of verbal stimuli (R. Cohn).

AN OUTLINE OF CONDUCTION Let us review in outline form the series of structures set in motion by the initial impact of sound waves upon the ear membrane. The eardrum vibrates and transmits its movement as follows:

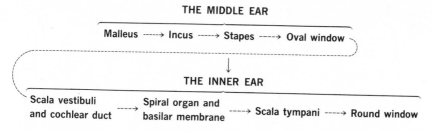

THE MIDDLE EAR

Malleus ----→ Incus ----→ Stapes ----→ Oval window

THE INNER EAR

Scala vestibuli and cochlear duct ----→ Spiral organ and basilar membrane ----→ Scala tympani ----→ Round window

The transmission of impulses includes mechanical movements of the three ear bones in the air space of the middle ear. It also includes transmission by means of pressure waves in the liquid of the inner ear. The movement of the endolymph within the membranous labyrinth stimulates the receptors (hair cells) for the sense of hearing.

SOME PHYSICAL ASPECTS OF SOUND When a shot is fired or when a bell is rung, the atmosphere is disturbed by sound waves radiating away from their source. The succession of sound waves consists of alternate condensations and rarefactions of the air. These waves can be likened to waves in water and are commonly shown diagrammatically as a curved line representing a series of hills and valleys. Such sound waves can vary in frequency (pitch) and amplitude (loudness). When they impinge upon a stretched membrane such as the eardrum, they exert a push-and-pull effect and cause it to vibrate in sympathy with the frequency of the force exerted upon it. However, sound waves are not ordinarily as simple as those described. The sound waves produced by most musical instruments are compound; they are composed of a series of vibrations, which produce overtones. The wave diagram can therefore be complex rather than a regular series of hills and valleys.

If a thundercloud is very distant, we may see the lightning many seconds before we hear the thunder. Sound travels at a relatively slow rate (1,100 feet per second) compared with the speed of light (approximately 186,000 miles per second).

The frequency of sound waves (pitch) may vary from low vibrations of around 16 per second to very high frequencies of 20,000 per second. There are, of course, vibrations both below and above these frequencies which are inaudible to the human ear. An example often cited is that a whip moved gently through the air; the waves created by its movement are inaudible. But if the whip is moved swiftly through the air, a high, piercing sound is heard. Apparently many animals hear frequencies higher than the upper limit for man, and there are supersonic frequencies presumably inaudible to any vertebrate ear.

Intensity of sound is commonly associated with loudness, but they are not necessarily identical. Intensity is simply the physical energy of sound transmitted through the atmosphere. Its loudness is subjective, a matter of

Hearing and equilibrium

auditory perception. Audibility curves show that hearing is good within a range of 1,000 to 3,000 hertz but falls off rapidly for frequencies below or above this range. It has been observed that the range of frequencies common to the human voice falls within the range of optimum reception by the human ear. The intensity of sound can be so great that it is felt rather than heard, as in the case of heavy concussions. Such disturbance close to the ear can give rise to pain.

The ear is able to recognize quality, or timbre, in musical sounds. It is easy to distinguish between the sound of a violin, piano, or horn, even though they may all play the same note. The difference lies in the overtones produced. These are partial vibrations and produce a compound wave.

THE PHYSIOLOGY OF HEARING   The resonance theory of hearing was developed by the great physicist Helmholtz. The fibers of the basilar membrane running crosswise were compared with the strings of a piano. In the basilar membrane the shortest fibers are at the base of the cochlea, the longest at the top, or apex. Helmholtz knew that stretched strings, such as those of the piano, vibrate in sympathy with sound waves that strike them. He reasoned that the shorter fibers of the basilar membrane might vibrate in sympathy with sound waves of high frequency and the longer fibers might respond to sound waves of low frequency. Though subsequent investigation has indicated that this may be true in a general way, the fibers of the basilar membrane are not piano strings, there is little evidence that they are stretched, and they are bound together in the membrane.

Another factor to be considered is the effect of inner-ear fluid on the vibration of the basilar membrane fibers. Waves in the fluid are responsible for the vibrations of the membrane. The longer fibers near the helicotrema are farthest away from the stapes footplate at the oval window where the waves are initiated, and therefore vibrate under a greater fluid "load." The shorter fibers vibrating at high frequency are near the oval and round windows, where the slight "load" favors their rapid vibration.

It is generally accepted that certain areas of the basilar membrane vibrate in sympathy with sound waves of given frequency and that these areas change their position to correspond with a change of frequency. Thus the area of greatest vibration in response to high frequencies would be the shorter fibers at the lower end of the basilar membrane, whereas lower frequencies would affect the upper part (Figure 11.10b). This is, in general, an elementary explanation of the concept of hearing known as the *place theory*. It is assumed that the hair cells in the spiral organ immediately above the vibrating area are stimulated and that they initiate nerve impulses. These impulses ultimately reach a specific cortical area of the temporal lobe, which interprets the impulse as sound of a certain pitch.

The *traveling-wave theory* is essentially a place theory, but traveling waves have been shown to affect the entire membrane. As the frequency increases, the maximum of vibrations moves toward the lower part of the basilar membrane; as the frequency decreases, the maximum moves toward the upper part. The area of maximal vibrations shifts its position along the membrane according to the frequency. The area of maximal vibration determines the pitch.

An important step in the interpretation of the action of the cochlea was taken many years ago when it was discovered that activity of the spiral organ causes changes of electric potential, which can be accurately recorded by electrical devices.

The principle of the telephone is essentially that the transmitter converts sound waves into electric impulses. These impulses have the same frequency as the sound waves received at the transmitter. Electric impulses may travel along wires for great distances, but at the receiving instrument they are converted back into sound waves of the same frequency as the original stimulus.

Wever and Bray placed two electrodes about the inner ear of an anesthetized cat. Using an amplifier wired in with a telephone receiver, they were able to hear sounds through the cochlea of the cat's ear. Sound waves transmitted to the cochlea caused an electric potential to be developed there. The sounds heard at the receiver were of the same frequency as the original sound waves introduced into the cat's ear.

An important contribution to our knowledge of the physiology of hearing is obtained by studying the rate of discharge of the auditory nerve in relation to sounds of given frequency and intensity. Such studies as those of Galambos and Davis indicate that both intensity and frequency of sound stimuli affect the frequency of nerve impulses over individual fibers. These investigators came to the conclusion, however, that the auditory fibers perform in every important respect like all other sensory fibers that have been studied. It was found that, at a constant frequency, an increase in sound intensity causes an increase in the rate of discharge of a single fiber. Another finding is that an increase in intensity causes the nerve fiber to be stimulated over a wider range of frequencies. At the threshold level of intensity the fiber may be activated by a narrow band of frequencies, which becomes broader as the intensity is increased. Probably an inhibitory mechanism functions to suppress the weaker stimuli; in this way, the effect of the stronger stimuli is enhanced.

There are a few motor or efferent fibers in the auditory nerve. They arise in the olivary nuclei of the medulla and have endings in the cochlea. Not much is known about their function, but it is generally considered that they suppress afferent fibers, perhaps performing a protective function against sudden bursts of noise.

Much of the recent research deals with recordings of *cochlear microphonic potentials*. These are alternating receptor potentials, probably produced by a change of position of the hair cells. There is no direct proof that these are generator potentials or that they stimulate auditory nerve fibers, although some physiologists consider that they may release a chemical transmitter capable of initiating a neural response.

One of the hazards of modern times lies in exposing the ears to loud sounds. Continuous exposure to highly amplified music may bring about partial deafness. Men who work around jet engines may become nearly deaf even though their ears are partially protected. The firing of heavy artillery is a liability to the ears.

THE VESTIBULAR APPARATUS   The *semicircular canals* are deeply embedded in the temporal bone. Three osseous canals, about 1 millimeter in diameter, are hollowed out of bone and lined with connective tissue. The three canals occupy a superior-posterior position with reference to the cochlea. They are located in planes that are approximately at right angles to each other. The anterior and posterior canals are nearly vertical; their medial extremities join and enter the vestibule as a single cavity. The lateral canal lies in a horizontal plane. The canals communicate with the vestibule by five openings rather than six because of the fusion of two of the ducts (Figure 11.7).

Hearing and equilibrium

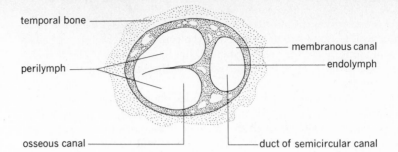

temporal bone

membranous canal

endolymph

perilymph

osseous canal

duct of semicircular canal

FIGURE 11.11

Cross section of a semicircular canal, showing the semicircular duct and adjacent perilymph spaces.

Within the osseous canals are the membranous canals. They follow the general contour of the osseous canals but are much smaller, occupying hardly a third of the space within the bony canals. They are supported by strands of connective tissue and by fluids. There are two fluids: The *perilymph* is the fluid of the osseous canals, and the *endolymph* is the fluid within the membranous canals (Figure 11.11).

Each canal has an enlargement near one end, called the *ampulla*. Within this structure there is a crest of hair cells called the *crista ampullaris*. The hair cells are end organs of the vestibular nerve. Rising above the crista and hair cells is a wedge-shaped gelatinous structure, the *cupula*. Hair cells extend upward into the cupula. Movement of the cupula in response to movement of the endolymph is thought to stimulate the hair cells (Figure 11.12).

The membranous labyrinth of the vestibule does not follow the outline of the osseous structure. The vestibule contains two membranous sacs, the *utricle* and the *saccule*. The utricle is the larger. It lies in the superior-posterior portion of the vestibule and is closely associated with the semicircular canals, which open into it. It also communicates with the saccule by means of a very fine duct. On the inner surface of a small out-pocketing there is a thickened oval area called the *macula*. It contains cells of two types, supporting cells and hair cells. On the surface of the macula is a gelatinous substance that contains minute calcareous concretions called *statoconia*, or *otoconia*. These are similar to the larger "ear stones," or otoliths, of certain invertebrates and fish. The macula is about 3 millimeters long, and since it is oval in shape, its width is a little less than its length (Figure 11.13).

The saccule is a small oval sac located in the lower anterior recess of the vestibule. It contains a macula similar to that located in the utricle. The saccule connects with the cochlear duct by means of a very slender tube. Another tiny duct from the posterior part of the saccule joins with a duct from the utricle to form the *ductus endolymphaticus*. It ends as a blind sac under the dura mater on the inner surface of the temporal bone (Figure 11.7).

The cells of the cristae and maculae are supplied by branches of the vestibular nerve, which in turn is a branch of the vestibulocochlear (VIIIth cranial) nerve. The pathway of the vestibular nerve leads back through the pons to nuclei in the cerebellum.

The utricle is considered to be affected by gravity (Figure 11.14). Therefore a change of position, as when an animal is tilted away from a horizontal plane, would affect the macula in the utricle. Motion sickness of various kinds, including seasickness, is thought to be caused by continuous

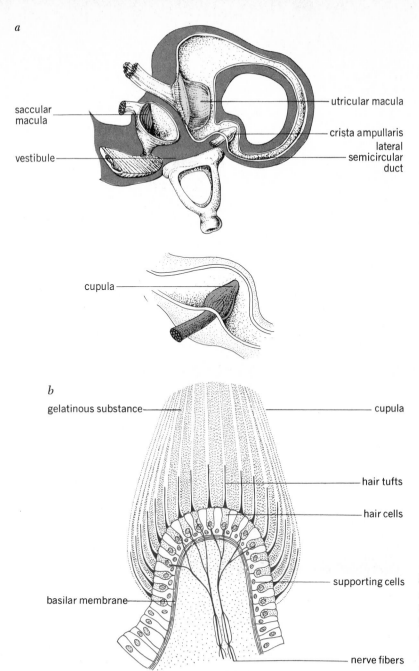

a

saccular
macula

vestibule

utricular macula

crista ampullaris
lateral
semicircular
duct

cupula

b

gelatinous substance

cupula

hair tufts

hair cells

supporting cells

basilar membrane

nerve fibers

FIGURE 11.12

a Vestibular system, showing cristae in the semicircular ducts and the maculae in the utricle and saccule; *b* enlargement of the crista ampullaris and cupula of the semicircular canal.

stimulation of the maculae rather than by a disturbance of receptors in the ampullae of the semicircular canals. The function of the saccule is less well known. It is closely associated with the utricle and may be largely proprioceptive, concerned with helping maintain the normal upright position of the head.

Maintenance of equilibrium Many experiments have indicated that the vestibular apparatus plays an essential part in equilibrium, postural reflexes, and right-

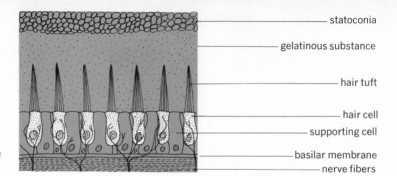

statoconia

gelatinous substance

hair tuft

hair cell

supporting cell

basilar membrane

nerve fibers

FIGURE 11.13
Section through the macula of the utricle.

ing reflexes. The semicircular canals are primarily involved when there is a sudden change of position, either in a direct line or in rotation; a familiar example of response to acceleration in a direct line is the placing reaction of a cat. If a cat is blindfolded and lowered rapidly head down, as in jumping down from some high place, the forelegs extend and the toes spread—a position assumed for making a safe landing. Probably the maculae play a part in this also. As an example of the righting reflex, the cat is able to right itself and land on its feet when dropped from an inverted position. Since the vestibular apparatus is concerned primarily with righting the head, further consideration must be given to proprioceptors of the neck and body which coordinate the orientation of both head and body in maintaining the equilibrium necessary to secure a proper landing.

Acceleration effects from the semicircular canals during rotation may affect muscles of the neck, trunk, and appendages as well as the extrinsic muscles of the eyes. As a person seated in a swivel chair is rotated, his eyes tend to fix on some object until the head is turned away; then the eyes quickly move back and focus on a new object. There is a slow phase of movement, in which the eyes turn in the opposite direction as they attempt to hold their focus on some object as the head turns. The quick phase will be in the direction of the turning. Such movements of the eyes are called *nystagmus*. It has been observed in vertebrate animals from fish to mammals. One of the interesting aspects of nystagmus lies in the fact that it is an acceleration effect and not caused by whirling motion alone. If a person or animal is subjected to rotation at a uniform speed for some period of time, the nystagmus is no longer evident. When the rotation is stopped suddenly, postrotatory nystagmus is again evident; this time the quick movement will be in the opposite direction from that of the previous rotation. If a person is rotated in a vertical position with the head erect, the horizontal canals are affected and the nystagmus is horizontal. If the head is in such a position that the vertical canals are primarily affected, the nystagmus will be vertical; that is, the eyes move up and down. There is a certain amount of inertia in the endolymph; accordingly, in accelerated turning to the right, the endolymph in the right horizontal canal moves toward the ampulla and the crista moves toward the utricle. In other words, the endolymph, because of its inertia, moves in the opposite direction from the turning of the head. In the left horizontal canal, the movement of the endolymph is in the direction away from the ampulla. Movement of the endolymph deflects the hair cells of the crista and initiates impulses in the vestibular nerve.

The nerve pathway leads from the semicircular canals to the medulla, where secondary neurons ascend and make synaptic connections within the

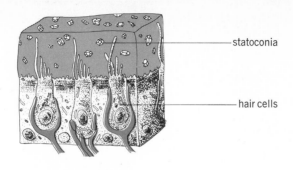

statoconia

utricular macula

hair cells

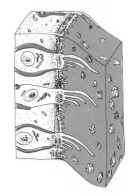

statoconia
fall with
gravity

FIGURE 11.14
The utricular macula, showing how
a change of position affects the
hair cells.

nuclei of the IIId, IVth, and VIth cranial nerves, which are motor nerves to the extrinsic eye muscles. Nystagmus is therefore a vestibular reflex. There are also nerve connections with the cerebellum.

A similar reflex is demonstrated by *past pointing* after rotation. The blindfolded subject is unable to bring his arm straight down in a vertical plane. Instead he points his finger a little to one side, in the direction that he has been rotated previously. This effect is linked up with postrotatory nystagmus, in that one points in the same direction as the slow movement of the eyes. In the case of vestibular reflexes of this type, there are widespread motor stimuli to the muscles of the neck, trunk, and appendages, all concerned with an attempt to maintain balance during rotation or to regain a normal position after rotation.

The effects of whirling, such as dizziness and staggering, are well known. Sometimes such stimulation of the horizontal canals affects the autonomic nervous system, causing nausea, pallor, sweating, and vomiting.

Dohlman, using a technique developed by Steinhausen, was able to make the cupula visible within the ampulla in living fish. This was accomplished by the introduction of dyes, such as india ink, into the ampulla. A drop of oil placed in the semicircular canal is shown to move toward the ampulla during acceleration. At the same time the cupula swings inward toward the utricle as a result of movement of the endolymph within the canal and ampulla. The cupula swings back in the opposite direction when the rotation stops. The movement of the cupula is correlated with observed nystagmus. The evidence indicates that nystagmus is an effect caused by deviation of the cupula. It lasts until the cupula returns to its normal position.

The action of the vestibular apparatus plays an important part in enabling the body to maintain its balance even though man has assumed an erect position. There is a precise coordination between the proprioceptive

Hearing and equilibrium

279

sense organs located in muscles and tendons and the higher centers of the brain. The eyes aid in helping to maintain equilibrium, while the cerebellum is a center for muscular coordination.

SUMMARY   Receptors are classified as exteroceptors, proprioceptors, and interoceptors. Encapsulated sensory-nerve endings are described as Meissner's corpuscles, Pacinian corpuscles, end bulbs of Krause, and end organs of Ruffini.

Taste buds contain chemoreceptors for the sense of taste. The taste buds are located in papillae on the tongue. The facial and glossopharyngeal nerves are concerned with the sense of taste.

The olfactory area is a small area in the upper part of each nostril and contains receptor cells for the olfactory sense. The olfactory nerve is composed of fibers from receptor cells that pass inward through openings in the cribriform plate of the ethmoid bone. The fibers synapse in the olfactory bulb. A second set of neurons leads inward toward the brain, forming the olfactory tract. A great variety of odors can be distinguished, but adaptation is quickly made to specific odors.

The ear consists of three primary parts: the external, middle, and internal ear. The external ear leads inward from the pinna, or auricle, through the external auditory meatus to the tympanic membrane.

The middle ear is a cavity containing a lever system of three ear bones: the malleus, incus, and stapes. The auditory tube extends from the pharynx to the middle ear. It functions in permitting air to enter behind the eardrum to equalize pressure on either side.

The major structures of the inner ear are the vestibule, cochlea, and semicircular canals. The vestibule contains an oval window and a round window. Within the bony vestibule are two small sacs, the utricle and the saccule. The utricle is associated with the semicircular canals; the saccule apparently is more closely associated with the membranous duct of the cochlea.

The cochlea is coiled into $2\frac{1}{2}$ turns and consists of a bony and membranous spiral labyrinth. The modiolus is the bony core of the cochlea. There are cavities, called scalae, in the cochlea. The scala vestibuli is continuous with the vestibule; the scala tympani leads to the round window at its base. These two passageways are confluent at the apex of the cochlea by a small aperture, the helicotrema.

The cochlear duct, or scala media, is the membranous labyrinth of the cochlea. It contains endolymph. The spiral organ lies on the basilar membrane of the scala media. Receptors for the sense of hearing are the hair cells located in the spiral organ. The cochlear nerve is a branch of the vestibulocochlear nerve.

The place theory holds that certain areas of the basilar membrane vibrate in sympathy with sound waves of given frequency. High frequencies affect the narrow, lower part of the basilar membrane, whereas low frequencies affect the wider, upper part of the membrane.

The semicircular canals are three in number. The anterior and posterior canals are nearly vertical, whereas the lateral canal is horizontal. The osseous canals are hollowed out of bone, but within each osseous canal there is a much smaller membranous canal or duct. Each canal has an enlargement at one end, called an ampulla. Within the ampulla is the crista ampullaris, a crest of hair-cell receptors. The cupula rises above the crista. Movement of the endolymph causes movement of the cupula and stimulates the hair cells. The hair cells are the end-organ receptors of the vestibular nerve. The pathway of the vestibular nerve leads to the cerebellum.

The semicircular canals function in maintaining equilibrium principally when there is a sudden change of position, either in a direct line or during rotation. When a person is rotated and then stopped suddenly, a postrotatory nystagmus is observed. The eyes move back and forth, the quick movement being in the opposite direction from the previous rotation. Whirling can cause dizziness, and the autonomic nervous system can become involved, causing nausea and sweating. Motion sickness, such as seasickness, is thought to be caused by continuous stimulation of the maculae rather than by disturbance of receptors in the semicircular canals. Equilibrium involves not only the semicircular canals and maculae but (1) proprioceptive organs in muscles and tendons, (2) the eyes, and (3) the cerebellum as the center for muscular coordination.

QUESTIONS

1 The sensory unit is composed of what structures?

2 How can receptors be classified?

3 List the receptors located in the skin and describe their structure and function.

4 Why is it that if the nasal passageway is closed, the sense of taste seems limited?

5 Where are the receptors located on the tongue?

6 Does the tongue respond to stimuli other than those that stimulate the sense of taste?

7 Just where is the olfactory area located in the nasal passageway?

8 Describe the appearance of olfactory receptors.

9 Trace the pathway of a nerve impulse from a tactile receptor to the interpreting center in the brain.

10 Explain the property of adaptation to odors.

11 Discuss the function of the auditory tube and its relation to the middle ear.

12 List the structures involved in transmitting the energy of sound waves from the ear membrane to the spiral organ.

13 Explain how we hear.

14 Where are the receptors for the sense of hearing located?

15 Is the narrowest part of the basilar membrane at the top or bottom of the cochlea? The shortest fibers may be expected to react to sounds of high or low frequency?

16 What structures compose the osseous labyrinth? The membranous labyrinth? The perilymph is the fluid of which labyrinth?

17 Discuss the structure and function of the utricle and saccule.

18 Explain the reaction of the semicircular canals to acceleration.

SUGGESTED READING

AMOORE, JOHN E., JAMES W. JOHNSTON, Jr., and MARTIN RUBIN: The Stereochemical Theory of Odor. *Sci. Am.*, **210**:42-49 (1964).

BÉKÉSY, GEORG V.: Current Status of Theories of Hearing, *Science*, **123**:779-783 (1956).

    A summary of the four major theories of hearing.

————: The Ear, *Sci. Am.*, **197**:66-78 (1957).

    A popular account; well illustrated.

COHN, R.: Differential Cerebral Processing of Noise and Verbal Stimuli, *Science*, **172**:599-601 (1971).

DOHLMAN, B.: Some Practical and Theoretical Points in Labyrinthology, *Proc. Roy. Soc. Med.*, **28**:1371-1380 (1953).

ELDREDGE, D. H., and J. D. MILLER: Physiology of Hearing, *Ann. Rev. Physiol.*, **33**:281-310 (1971).

GALAMBOS, R., and H. DAVIS: The Response of Single Nerve Fibers to Acoustic Stimulation, *J. Neurophysiol.*, **6**:39-57 (1943).

GREEN, J. D.: The Function of the Hippocampus, *Endeavour*, **22**:80-84 (1963).

LOWENSTEIN, W. R.: Biological Transducers, *Sci. Am.*, **203:**99–108 (1960).

————: Mechano-electric Transduction in the Pacinian Corpuscle. Initiation of Sensory Impulses in Mechanoreceptors, in "Handbook of Sensory Physiology," vol. 1, pp. 269–290, Springer-Verlag, New York, 1971.

McCUTCHEON, N. B., and J. SAUNDERS: Human Taste Papilla Stimulation: Stability of Quality Judgments over Time, *Science*, **175:**214–216 (1972).

MELZACK, RONALD, and PATRICK D. WALL: Pain Mechanisms: A New Theory, *Science*, **150:**971–979 (1965).

ROSENZWEIG, MARK R.: Auditory Localization, *Sci. Am.*, **205:**132–142 (1961).

————: Sensory Receptors, *Cold Spring Harbor Symp. Quant. Biol.*, **30**(1965).

WEVER, E. G., and C. W. BRAY: "The Nature of Acoustic Response: The Relation between Sound Frequency and Frequency of Impulses in the Auditory Nerve," *J. Exp. Psychol.*, **13:**373–387 (1930).

chapter 12
special senses: vision

Visual receptors react to the physical stimulus of light. The eye looks out upon the world and by a very intricate mechanism reports its observations to the brain. The exact means by which receptors convert light radiations into nerve impulses and the interpretation of these impulses by the brain are still among the most perplexing problems in physiology.

ACCESSORY STRUCTURES

Before we take up the action of the eye itself, let us consider the accessory structures surrounding the eye. The eyeball lies in a bony orbit, which protects it on all sides except the anterior. It is cushioned by a padding of fat that lines the bony orbit.

There is a protective bony ridge above the eyes, which with the eyebrows helps to ward off objects falling into the eyes from above. The eyebrows also direct perspiration away and tend to shade the eyes from the direct rays of the sun.

The *eyelids* are movable folds of skin that form a protective curtain over the anterior surface of the eyeball. The upper lid is the larger and is more freely movable. It is pulled up by the levator palpebrae superioris muscle. The sphincter muscles, the orbicularis oculi, close the eyelids. The eyelashes are stiff, curved, protective hairs placed along the margin of each lid. Sebaceous glands are associated with the eyelashes. Occasionally one or more of these glands may become infected. This painful swelling around the base of the eyelash is called a *sty*. Posterior to the bases of the eyelashes are the tarsal, or Meibomian, glands. They secrete an oily fluid that tends to retain the tears and prevents the eyelids from sticking together.

The *conjunctiva* is a thin mucous membrane covering the anterior surface of the eyeball and lining the eyelids. It is transparent, since light must pass through it to reach the receptors of the eye. The portion between the eyeball and the lining of the eyelids is folded and loose to permit free movement of the lids and the eyeball. At the inner angle of the eye there is a fold of the conjunctiva called the *plica semilunaris*. It is crescent shaped and is commonly known as the *third eyelid* or *nictitating membrane*. Deep in the inner angle of the eye is a fleshy mound known as the *caruncle*. In the embryo this structure develops as a separate portion of the lower lid and has glands similar to those found in the eyelid. It may even bear an eyelash or two in a few cases. The caruncle is an island in a space formed at the inner angle of the eye. The eyelids form the boundaries of this conjunctival space, which holds a small amount of tear fluid and is referred to as the *lacrimal lake*. The caruncle forms part of the yellowish secretion that tends to gather in this area, especially overnight.

The conjunctiva is well supplied with blood vessels and nerve endings. Irritation of this membrane can cause the blood vessels to become plainly

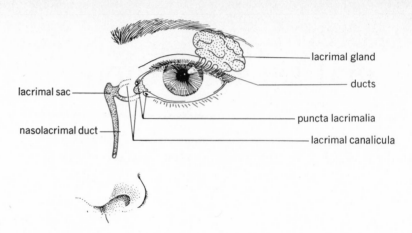

FIGURE 12.1
The lacrimal structures of the eye.

visible. The pain felt when a foreign body gets into the eye indicates that there are abundant sensory-nerve endings in the conjunctiva. Inflammation of the conjunctiva (*conjunctivitis*) may appear in many different forms. Granulated lids is one form; pinkeye is a highly contagious form.

LACRIMAL STRUCTURES

The *lacrimal gland* produces tears. It is an elongate compound gland located in a depression of the frontal bone above the outer angle of the eye. Several ducts carry the lacrimal secretion from the gland to the conjunctiva of the upper eyelid. Ordinarily just enough fluid is secreted to keep the surfaces moist. Much of this secretion is lost by evaporation. The remainder flows toward the inner angle of the eye where it is drained off through two minute openings, called the *puncta lacrimalia*. The puncta are the openings of two lacrimal canals, which carry the secretion to the lacrimal sac. An extension of the lacrimal sac, the nasolacrimal duct, leads the secretion down into the nasal passageway (Figure 12.1).

The lacrimal secretion contains various salts in solution in water and a small amount of mucin. It is slightly antiseptic, since it contains the enzyme *lysozyme* which has a bactericidal action. However, the principal function of tears is to moisten the surface of the eyes. Irritation of the conjunctiva or the nasal membranes increases the flow of tears, as do emotional states. The lacrimal glands are not fully functional at birth. A young baby's eyes should therefore be protected against intense light, dust, and the drying effect of wind.

EXTRINSIC EYE MUSCLES

There are six extrinsic muscles attached to the eyeball. Four of them, the recti muscles, arise from a ring tendon (Zinn's ring) at the back of the orbit. They are about 40 millimeters long and are inserted on the eyeball in positions corresponding to their names. The *superior rectus* is attached to the superior surface of the eyeball and permits the eye to turn upward. The *inferior rectus* is attached underneath the eyeball and enables the eyeball to turn downward. The *lateral rectus* is attached to the outer side; the *medial rectus* is attached on the inner side. The latter two muscles turn the eyeball to one side or the other, as their names indicate. There are also two oblique muscles, which aid in rotating movements of the eyeball. The *inferior oblique muscle* arises anteriorly from the floor of the orbit near the fossa for the lacrimal sac. It passes backward laterally between the inferior rectus tendon and the floor of the orbit, inserting posteriorly on the eyeball under the lateral rectus. This is the shortest of the ocular muscles.

The *superior oblique* is the longest and most slender of the ocular muscles. Arising from the apex of the orbit (the edge of the sphenoid bone in the region of the aperture that transmits the optic nerve), it passes forward on the upper nasal side between the superior and medial rectus muscles. Above the eye on the medial side it passes through a loop of cartilage commonly called the *trochlea*. It becomes tendinous at the trochlea and turns downward and backward, passing under the superior rectus to insert on the posterior aspect of the eyeball. The oblique muscles probably to not act independently but act with the rectus muscles to roll the eyes.

The oculomotor nerve innervates all the rectus muscles with the exception of the lateral rectus, which is innervated by the abducens nerve. The inferior oblique muscle is also innervated by the oculomotor nerve. The trochlear nerve supplies the superior oblique muscle (Figure 12.2*a* and *b*).

The eyeball is suspended in the orbit by the extrinsic muscles and by a complex series of ligaments and other connective tissues. One of the most important of these suspensory structures is the *fascia bulbi,* or Tenon's capsule. This is a serous sac that is attached to the eyeball anteriorly in a region around the cornea but does not cover the cornea. Posteriorly it envelops the eyeball completely so that the eye muscles, nerves of the eye, and blood vessels pass through it. The eyeball rests and moves on these structures, which are further supported by a padding of fat within the orbit (Table 12.1).

TABLE 12.1
Extrinsic muscles of the eye

Muscle	Origin	Insertion	Action	Nerve
Inferior oblique	Anterior orbital floor	Between insertion of superior and lateral recti muscles	Rotates eyeball upward and laterally	Oculomotor
Rectus: inferior, lateral, medial, superior	Common tendinous ring around optic foramen	By tendons into sclera at their respective positions	Medial, rotates eyeball medially; lateral, rotates eyeball laterally; superior, rotates eyeball upward and medially; inferior, rotates eyeball downward and medially	Inferior, medial, and superior by oculomotor, lateral by abducens
Superior oblique	Roof of orbital cavity, medial to optic foramen by a narrow tendon	A narrow tendon; passes through trochlear pully and inserts between rectus superior and rectus lateralis	Rotates eyeball downward and laterally	Trochlear

Extrinsic eye muscles

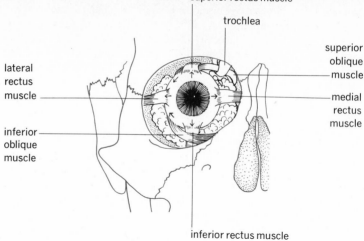

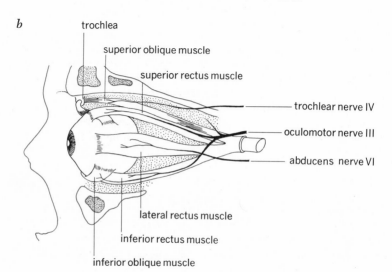

FIGURE 12.2

Extrinsic muscles of the eye: *a* superior view; *b* lateral view.

STRUCTURE OF THE EYEBALL

The eyeball is nearly spherical, but the distance from front to back is a little greater because the cornea, which appears as an arc of smaller dimension, is located on the anterior surface of the larger sphere. The eyeball is remarkably constant in size in either sex, but it is slightly smaller in the female (0.5 millimeter in all diameters). The eye may appear large or small because of its position in the orbitals. If it protrudes somewhat so that more of the white shows than would normally be exposed, the eyeball appears to be abnormally large. The eye also can appear deep-set and small or sunken, as after a long illness, when considerable body fat has been used up.

The wall of the eyeball is composed of three layers. The outer coat is a fibrous tunic consisting of the sclera and cornea. The intermediate layer is a highly vascular, pigmented tunic composed of the choroid, a muscular ciliary body, and the iris. The inner nervous tunic is the retina. The refracting

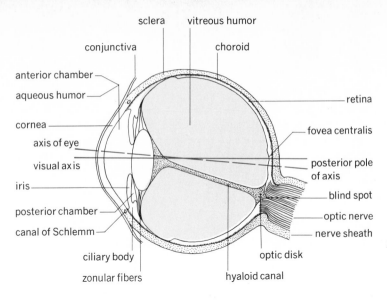

sclera   vitreous humor

conjunctiva   choroid

anterior chamber

aqueous humor

cornea

axis of eye

visual axis

iris

posterior chamber

canal of Schlemm

ciliary body

zonular fibers

retina

fovea centralis

posterior pole
of axis

blind spot

optic nerve

nerve sheath

optic disk

hyaloid canal

FIGURE 12.3
Horizontal section through the eyeball.

media of the eye are the aqueous humor, the lens, and vitreous body (Figure 12.3).

THE FIBROUS TUNIC   The *sclera* is the white coat of the eye. It covers the eyeball except for the cornea and consists of a dense interlacing of white fibrous tissue. This opaque covering helps to maintain the shape of the eyeball and protects the more delicate structures from injury. The anterior surface is covered by the conjunctiva, and small blood vessels can be seen through this layer in living specimens. The sclera itself is not abundantly supplied with blood vessels. However, at the junction of the sclera and cornea there is an important venous sinus, the *sinus venosus* of the sclera (canal of Schlemm), which encircles the base of the cornea. This thin-walled sinus at the filtration angle of the eye provides an outlet for the aqueous humor into the venous system of the eyeball.

The *cornea* is the transparent part of the fibrous tunic. It is a portion of a sphere, superimposed upon the anterior surface of a larger sphere, the eyeball. It represents about one-sixth of the total area. The cornea proper is composed of modified fibrous tissue, which during development becomes clear and presents a fairly homogeneous appearance. Over the outer surface there is an epithelial layer that is continuous with the epithelial layer of the conjunctiva. The embryonic cornea contains capillaries, but they are lost in the functional eye. The cornea contains no blood vessels except around its border; it is nourished by lymph, which diffuses slowly through the tissue. Oxygen is absorbed directly from the atmosphere. The corneal epithelium has an abundant nerve supply of fine unmyelinated nerve endings from the ciliary nerves, making it highly sensitive. Injury to the fibrous portion of the cornea heals by the formation of opaque scar tissue.

Astigmatism is usually caused by imperfect curvature of the cornea. If all the arcs of the cornea were equal, light waves passing through would form a cone of light behind the cornea and would focus at a single point. If there is a variation in the curvature so that, for example, the vertical curvature is greater than the horizontal, there will be two points of focus. The light rays from the greater curvature will focus at a shorter distance

Structure of the eyeball

than those from the lesser curvature. The focus in this case will not be sharp, and images may appear blurred; bright lights may appear to have shafts of light extending outward in something of a star-shaped pattern. Seen through the corrected lenses of a telescope, the stars appear as spheres, as they actually are; but with the unaided eye, the stars usually appear to have shafts of light extending outward, indicating that the eyes are astigmatic to a certain extent.

A test for astigmatism is to have a person look at vertical and horizontal lines that cross; or a person may observe a figure with radiating lines, like the spokes of a wheel, under test conditions. If the vertical lines appear more distinct or darker than the horizontal lines, the inference is that the vertical curvature gives a sharper image than the horizontal. The type of astigmatism in which the vertical curvature is greater than the horizontal is the commonest form. Astigmatism is corrected by glasses in which cylindrical lenses are designed to compensate or equalize the vertical and horizontal curvatures of the cornea; in other words, the greater curvature of the lens compensates for the lesser curvature of the cornea and vice versa.

The *anterior chamber* of the eye lies immediately posterior to the cornea. It is filled with the *aqueous humor*. The *posterior chamber* is a small area located on either side between the lens and the iris.

THE VASCULAR COAT   The intermediate layer of the eyeball is composed of the choroid, the ciliary body, and the iris. The *choroid* is a dark-brown membrane that lines the sclera. It is highly vascular, since it is concerned with maintaining the nutrition of the retina. There are numerous irregular-shaped pigment cells, which with the blood vessels give the membrane its dark color. It therefore darkens the interior of the eye, absorbing light rays and preventing their reflection. The optic nerve passes through the choroid at the back of the eyeball. The membrane extends forward to the ciliary body. Human eyes do not shine in the dark because they lack the reflecting layer (*tapetum*) in the choroid that is a characteristic structure in the eyes of most animals.

The *ciliary body* may be divided into three regions: the ciliary ring, the ciliary processes, and the ciliary muscle. The ciliary ring is a darkened area about 4 millimeters wide and continuous with the choroid. The ciliary processes are folds or ridges arranged as along the meridians of a sphere. There are about 70 of them radiating out from behind the iris and tapering toward the ciliary ring. They afford attachment for zonular fibers (suspensory ligament) of the lens and for the ciliary muscles (Figure 12.4).

The ciliary muscle is composed largely of fibers running in a meridional direction, but there is also a small band of circular muscle. When the meridional fibers contract, they pull forward on the ciliary ring and the ciliary processes, thus slackening the zonular fibers. The lens then becomes more convex as in accommodation for close-up work, such as reading. There is some evidence that the circular muscles play a greater part as antagonists of the meridional fibers than had been indicated formerly.

The *iris* is the most anterior portion of the vascular tunic. It represents the highly colored part of the eye; the eyes of human beings are described as blue, gray, green, or brown, depending on the distribution and the amount of pigment. The albino has no pigment in the iris, and the iris appears pink. The circular opening in the iris is the pupil. It appears black because it opens into the dark recess of the eyeball. The iris is anterior to the lens but posterior to the cornea and the anterior chamber of the eye. There are two layers

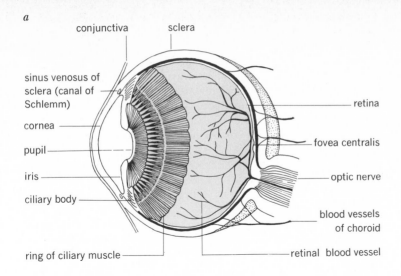

a

conjunctiva

sclera

sinus venosus of sclera (canal of Schlemm)

cornea

pupil

iris

ciliary body

ring of ciliary muscle

retina

fovea centralis

optic nerve

blood vessels of choroid

retinal blood vessel

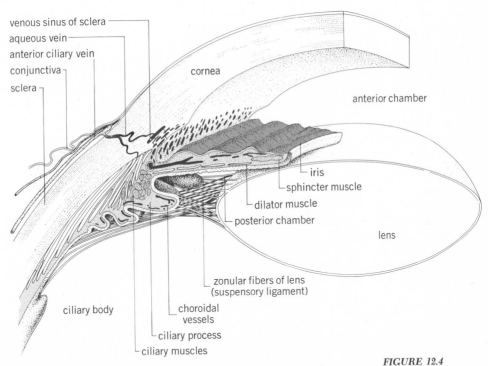

b

venous sinus of sclera

aqueous vein

anterior ciliary vein

conjunctiva

sclera

cornea

anterior chamber

iris

sphincter muscle

dilator muscle

posterior chamber

lens

zonular fibers of lens (suspensory ligament)

ciliary body

choroidal vessels

ciliary process

ciliary muscles

FIGURE 12.4

a Section through eyeball showing ciliary muscle; *b* detailed section through anterior part of the eye (greatly enlarged).

in the iris. Anteriorly there is a colorless layer and posteriorly a layer that is of a blue color. The blue layer is usually present in all eyes, but in those persons with dark eyes, a dark pigment, *melanin,* develops in the anterior layer and obscures the blue color.

  The inheritance of eye color is more complex than can be considered here; but, in general, blue eye color is a recessive trait, and brown eye color is dominant. Albinism is also a recessive character. If both parents have pure

289

blue or gray eyes, their children generally will have eyes of the same color as the parents, since this is a recessive character. Brown eyes may be either homozygous (*BB*) or heterozygous (*Bb*). Parents that are heterozygous for brown eyes may, therefore, have blue-eyed children (*bb*). Persons with blue eyes should be homozygous recessive for this character. There is more variability in eye color than these simple illustrations indicate, and probably more than one gene is involved.

The iris is dark colored on its posterior surface. It contains two antagonistic sets of muscles. The dilator muscles are radial muscles; their contraction dilates the pupil. They derive their innervation from the sympathetic chain of ganglia of the autonomic nervous system. Preganglionic fibers pass upward from the first and second thoracic levels to the superior cervical ganglion. Postganglionic fibers in the ophthalmic branch of the Vth cranial nerve enter the eyeball in the long ciliary nerve to the dilator muscles of the pupil. The pupils of the eyes dilate widely following the use of marijuana. Morphine and opium greatly constrict the pupil into a pinhole type.

A circular muscle, the sphincter pupillae, surrounds the pupil, and its contraction causes the pupil to constrict. Parasympathetic fibers are carried by the oculomotor nerve to the sphincter muscle and act as constrictors of the pupil. Preganglionic fibers arise in the midbrain and extend outward to the ciliary ganglion located behind the eyeball. Postganglionic fibers in the short ciliary nerve enter the eyeball and innervate the ciliary muscle and the sphincter of the pupil.

The *lens* is not a part of the choroid coat, but since it is directly attached to the ciliary muscles by zonular fibers, it will be considered here. The lens is a clear, transparent tissue located posterior to the pupil and iris. It lies between the anterior chamber and the vitreous body and is slightly yellow in older people. The adult lens measures 9 millimeters in diameter and is about 4 millimeters thick. It is a biconvex lens, but somewhat more convex on the posterior side. In the fetal eye, the lens is almost spherical; even in children, the lens is strongly convex; but in adults past middle age, there is usually a gradual and partial loss of convexity and elasticity. The living lens is able to change its degree of convexity during accommodation for near and far vision.

In its embryonic development the lens forms from a thickening of the epidermal ectoderm above the optic vesicle or optic cup. The optic vesicle is a cup-shaped lateral outgrowth of the brain; in the early embryo it marks the beginning of the development of the eye. The lens is held in place by the zonular fibers, which in turn are attached to the ciliary body (Figure 22.28).

The lens is enclosed by a clear, elastic capsule. Just underneath the capsule on the anterior surface only, there is a single layer of cells constituting the lens epithelium. A study of the structure of the lens itself reveals that it is a tissue composed of specialized *lens fibers*. The fibers appear to be long, slender, and flat, but in cross section they are flattened hexagons. The fibers lie along meridional lines, but no fiber is long enough to reach from pole to pole. The lens fibers are arranged in concentric lamellae like layers of onion tissue. *Cataract* is a condition in which the lens becomes opaque or milk-white in appearance. The change often occurs first in the firm nucleus, or central portion, although there are various types of cataract. Since light cannot pass through the clouded lens in adequate amounts, cataract results in blindness as the lens becomes progressively more opaque. A very delicate operation may be performed to remove the lens. Sight may be restored by this means, but, of course, there is no longer any focusing lens. With strong bifocal lenses in the glasses, a person can read by holding the book at the place where he obtains the best focus. Cyclotron-induced radiation cataracts are a new hazard in the modern world.

The *vitreous body* is a transparent, jellylike substance that fills the cavity of the eyeball posterior to the lens. There is a cavity in the vitreous body anteriorly to accommodate the lens. The vitreous body is largely water, and its refractive index is close to that of water. A lymph canal passes through it from the optic disk to the region of the lens. This is called the *hyaloid artery* in the embryonic eye, where it provides for the nourishment of the lens during its development. The vitreous body helps to maintain the shape of the eyeball and supports the retina.

THE RETINA   The retina is the inner nervous tunic of the eye. The optical portion forms the lining of the posterior part of the eyeball, extending forward to the ciliary ring. At the back of the eye the retina is about 0.4 millimeter thick, becoming thinner as it extends forward. Within this thin, delicate membrane the receptors for the sense of sight are located.

This portion of the eye forms originally as a part of the lateral wall of the brain. In the early embryo there is an infolding of the neural tube, followed by a lateral out-pocketing in the region of the eye, which results in the receptor cells being located at the back of the retina next to the pigment layer. They face away from the source of light; this type of retina, found in all vertebrates, is called an *inverted retina*.

The retina, though very thin, contains several layers of neurons, which may be demonstrated by suitable histological methods. Actually, 10 layers may be differentiated, of which we shall mention only a few. At the back of the retina is the pigment layer, composed of low columnar epithelium containing great amounts of pigment. The next layer contains the receptor cells. They are sensitive to light and are called *rod* and *cone cells* because of their shape. An external limiting membrane extends through the receptor cell area at the base of rod- and cone-cell bodies.

The outer nuclear layer contains the nuclei of the receptor cells (outer layers refer to those layers farther away from the center of the eye). The outer nuclear layer consists of a mass of dark-staining nuclei located adjacent to the rod-and-cone layer.

The adjoining layer is the outer synaptic layer, or outer plexiform layer. It consists of fibers and synaptic endings of receptor cells as they connect with dendrites of bipolar cells of the next layer. It may be identified as a lighter-colored area (Figure 12.5).

Bipolar cells form the next large area near the middle of the retina, containing, in addition, some *horizontal cells* and some *amacrine cells*. Bipolar cells have large round nuclei surrounded by rather thin layers of cytoplasm. They are a connecting link between the receptor cells and the inner synaptic layer. The horizontal cells appear to be association cells, relaying impulses to bipolar cells nearby, but their lateral connections do not seem to be extensive. There appears to be no direct connection between horizontal cells and ganglion cells. Horizontal cells have a small oval nucleus and a rather large amount of cytoplasm.

Greater interest is now directed toward investigation of amacrine cells, which lie close to the ganglion cells and synapse in the inner synaptic (plexiform) layer. The function of these cells is not clear. They synapse, at least in part, on ganglion cells. In the primate retina, bipolar axons synapse

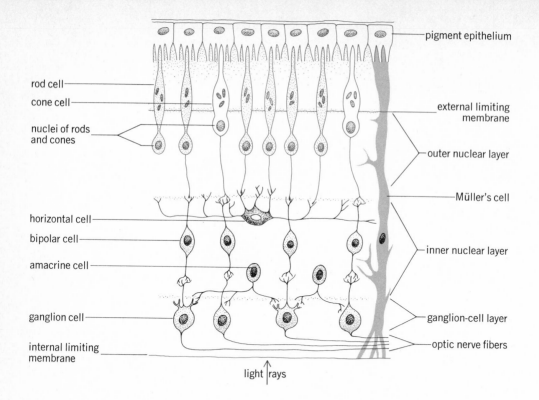

rod cell

cone cell

nuclei of rods
and cones

horizontal cell

bipolar cell

amacrine cell

ganglion cell

internal limiting
membrane

pigment epithelium

external limiting
membrane

outer nuclear layer

Müller's cell

inner nuclear layer

ganglion-cell layer

optic nerve fibers

light rays

FIGURE 12.5

Ganglionic cells of the retina (diagrammatic and simplified).

with both ganglion-cell dendrites and processes of amacrine cells (Michael). Amacrine cells are probably more important in lateral transmission than horizontal cells, since their lateral processes extend much farther. There also is evidence of feedback loops to bipolar neurons. It may be determined later that amacrine cells play some part in the organization of the visual input to ganglion cells.

The nuclei of bipolar, longitudinal, amacrine, and Müller's cells form the inner nuclear layer of the retina. Adjoining is the inner synaptic, or plexiform, layer.

The inner cell layer is composed of large ganglion cells which synapse with bipolar and amacrine cell processes and give rise to axon fibers of the optic nerve.

Müller's cells are considered to be modified neuroglial cells. The fibers extend throughout the retina, curving around ganglion cells and spreading out to form a supporting framework for both the internal and external limiting membranes. They envelop rod and cone cells, their fibers perhaps functioning in the place of myelin as an insulating layer. They may be more than passive supporting cells, since they exhibit a slow change of potential associated with excitation of the rod and cone cells which they surround. Their exact function has not been determined, but they have been shown to contain glycogen and oxidative enzymes.

The rods and cones are receptors of light waves, but action potentials do not originate here. They function as biological transducers and initiate only receptor potentials. It is not clear as yet whether bipolar cells produce action potentials, but certainly their graded potentials are strong enough to initiate action potentials in the ganglion cells which are then transmitted by the optic nerves.

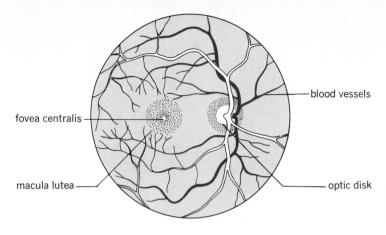

fovea centralis

blood vessels

macula lutea

optic disk

FIGURE 12.6
Posterior portion of the right retina as viewed from the front.

X ●

FIGURE 12.7
Diagram for determining the blind spot of the eye. With the left eye closed or covered, gaze steadily at the X while moving the book from a distance about 12 inches in front of the face slowly toward the face. Note that at a certain distance, the image of the dot becomes invisible. To check the other eye, close or cover the right eye and gaze steadily at the dot, noticing that the X becomes invisible when the book is advanced a certain distance toward the face.

Various transmitter substances have been found in the retina. These include GABA, dopamine, and norepinephrine. Acetylcholine may be present also. It is suggested that presynaptic fibers may be cholinergic. The role of these transmitter substances as they pertain to different kinds of cells and their synapses has not been clarified.

THE BLIND SPOT   Since the fibers that form the optic nerve arise on the anterior side of the retina facing the lens, they must turn and pass back through the retina in order to carry impulses to the brain. The optic nerve is formed by the union of these optic fibers. They all pass through the retina at one spot. Since all receptors are pushed aside to permit the exit of the optic nerve fibers, there is a blind spot in each eye (Figure 12.6).

As an experiment to prove the existence of the blind spot, draw a small circle and a small cross 2 inches apart on a 3- by 5-inch card. Now close the left eye and, staring at the cross, move the card from a position about 6 inches away from the eye to a position a little closer or a few inches farther away. As you maneuver the circle so that its image falls on the blind spot, it will not be seen, since there are no visual receptors there (Figure 12.7).

The blind spot is not located in the exact center of the posterior part of the eye; it is on the inner side of a median line passing through the posterior pole of the sphere. Blood vessels enter the eye at the center of the optic disk or blind spot. The retinal artery then branches into four main branches, which with numerous smaller branches cover the retina. The retinal veins follow the path of the arteries and exit beside the retinal artery in the optic disk. Blood vessels of the retina can be observed under proper illumination. A physician is able to look into the depths of the eye as through a window and make observations regarding the condition of the blood vessels. In ill health and disease the blood vessels change in appearance; they may be notched where one crosses another or even partially obliterated.

Structure of the eyeball

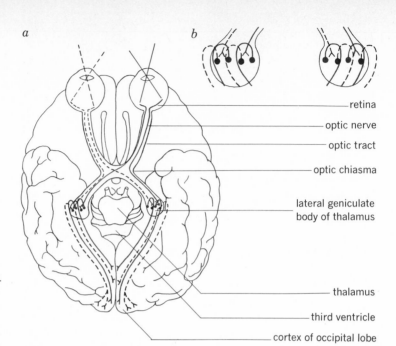

FIGURE 12.8

a Visual pathway, including the optic chiasma and the optic tract. Note that the fibers from the outer half of each retina do not cross in the chiasma; *b* enlargement of lateral geniculate bodies.

retina

optic nerve

optic tract

optic chiasma

lateral geniculate body of thalamus

thalamus

third ventricle

cortex of occipital lobe

THE NEURAL PATHWAY The axons of ganglion cells leave the retina at the optic disk and form the optic nerve, which contains nearly a million of these fibers. As the fibers form the nerve back of the eye, they become myelinated. The nerves from each eye progress toward each other and form a partial crossing, the *optic chiasma*. Only the fibers from the inner, or nasal, half of the retina cross over to the opposite side in the chiasma. The fibers from the outer, or lateral, half of the retina remain on the same side. The fibers now form the optic tracts, passing around the cerebral peduncles to the lateral geniculate bodies of the thalamus. Most of the fibers synapse in the lateral geniculate bodies, and succeeding fibers extend back to the visual areas in the occipital lobes (Figure 12.8).

Some fibers, however, do not synapse in the lateral geniculate bodies but enter the *superior colliculi*. These small structures in the midbrain function as optic reflex centers. Through connections with the oculomotor nerve, the sphincter muscle of the pupil is stimulated and reduces the size of the pupil in response to strong light, the light reflex. Further connections with ciliary nerves to the ciliary muscles cause the lens to become less convex, an accommodation reflex for distant vision. More elaborate nerve connections provide protective reflexes, such as those for closing the eyelids as an object approaches the eyes and for raising the arms to protect the eyes. The superior colliculi also supply impulses to the extrinsic eye muscles concerned with eye movements.

THE RECEPTOR CELLS: RODS AND CONES The photoreceptor cells of the retina are adapted as transducers to receive the energy of light waves and through a photochemical process to initiate nerve impulses that can be interpreted by the brain. Rod cells function to provide vision in the dark or in dim light; cone cells are responsible for color vision in daylight or under adequate lighting conditions (Figure 12.9).

There are roughly 10 times as many rods as cones in the human retina;

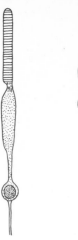

FIGURE 12.9
Receptor cells of the retina, greatly enlarged: rod cell at left; cone cell at right.

the number is estimated to be 1 million cone cells and 10 million rod cells. The structural organization of these receptors is of great interest, since electron microscopy indicates that they are derived from cilia and retain some of their characteristic structures. Rod and cone cells have essentially the same basic structure, but rod cells are typically longer and more slender than cone cells. Rod cells are generally 50 to 60 microns long in the retinas of various animals and about 2 microns wide. Cone cells, being shorter than rod cells, have a somewhat conical appearance and are about three times wider, 6 to 7 microns wide. The basic structure of rod and cone cells consists of an outer segment, a slender connecting stalk, and a basal portion. Rod cells are more numerous toward the periphery of the retina, whereas cone cells are concentrated in the area of most acute vision.

Structural detail of rod cells The rod cell, like other cells, is contained within a thin limiting membrane. The outer segment is characterized by laminated foldings of the endoplasmic reticulum, or ergastoplasm. It is a smooth membrane (without ribosomes), and the visual pigments concerned with the biochemical aspects of vision are believed to be located in or on this membrane.

The central connecting stalk in its embryonic development contains the structural characteristics of a cilium, with nine pairs of filaments and a basal body that resembles a centriole. A second centriole at right angles to the first has been observed in some electron micrographs. The connecting cilium has an off-center appearance. This is caused by the one-sided enlargement of the outer segment during its embryonic development (Figures 12.10 and 12.11).

The basal portion of the rod cell contains numerous elongate mitochondria, the energy source of the cell. The rod-cell nucleus lies just beyond the mitochondrial area.

Rod cells contain a reddish pigment called *rhodopsin,* or visual purple, in the outer segments. It is a colored aldehyde of 11-*cis*-vitamin A plus the protein opsin. The energy of light waves causes the isomerization of rhodopsin, resulting in a rapid change in the physical form of the structure as all-*trans*-retinal is formed. In a brief explanation of *cis-trans isomerism,* we note that when a double bond exists between two carbon atoms, their position is fixed in regard to each other. If additional carbon atoms, such as two

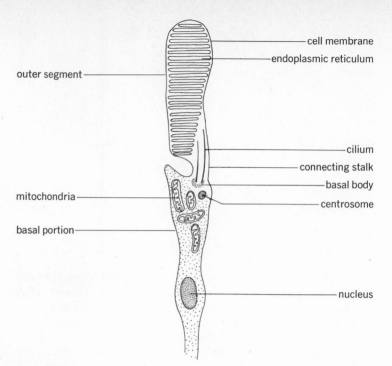

Labels on figure:
- cell membrane
- endoplasmic reticulum
- outer segment
- cilium
- connecting stalk
- basal body
- centrosome
- mitochondria
- basal portion
- nucleus

FIGURE 12.10
Drawing of a rod cell, showing detail based on electron micrographs.

carboxyl groups, are added on the same side of the double bond, this represents a cis position. If located on opposite sides of the double bond, it is a trans position. The cis molecule bends at the numbered position, whereas the all-trans molecule is generally in a more-or-less straight line. This distinction is pertinent here relative to the formation of 11-*cis*-rhodopsin, since only the cis forms of vitamin A and retinal can take part in this reaction. Retinal is a carotenoid pigment closely related chemically to beta carotene. Opsins are colorless proteins chemically bound to a chromophore group (literally color-bearing).

Adenylate cyclase has been identified as an enzyme present in the photoreceptors of the retina. It is inactivated by light, and this inactivation is found to be proportional to the bleaching of rhodopsin (Miller). Evidently the cyclase and cyclic AMP are involved in photoreception, but their exact function has not been determined.

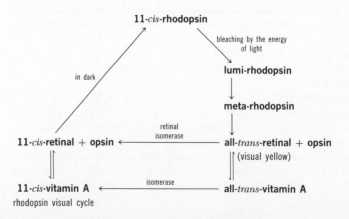

11-*cis*-rhodopsin

bleaching by the energy of light

in dark

lumi-rhodopsin

meta-rhodopsin

11-*cis*-retinal + opsin ← retinal isomerase ← all-*trans*-retinal + opsin (visual yellow)

11-*cis*-vitamin A ← isomerase ← all-*trans*-vitamin A

rhodopsin visual cycle

FIGURE 12.11
Electron micrograph of a rod cell.
(*From James A. Freeman,
"Cellular Fine Structure,"
McGraw-Hill Book Company,
New York, 1964.*)

Energy derived from exposure to light causes rhodopsin to bleach and to be converted to another chemical form called lumi-rhodopsin, but this form is highly unstable and it almost immediately changes into meta-rhodopsin. This compound also is unstable and in a matter of seconds forms all-*trans*-retinal plus opsin. The two kinds of retinal (chemically trans and cis) can each be converted into corresponding types of vitamin A, and the reverse is also true, but these are relatively slow reactions. The cycle is completed when 11-*cis*-retinal plus opsin is converted to rhodopsin in the dark. Rhodopsin is a stable form as long as it is unaffected by exposure to light. The bleaching of rhodopsin by light energy causes excitation of the rod cells, and subsequently the ganglion cells of the optic nerves are stimulated.

Good nutrition requires adequate amounts of vitamin A, which is supplied to the retina by the blood. Occasionally persons are found who are unable to see as well as the average person in a dim light. This visual defect is called *night blindness*, and if it is due to a deficiency of vitamin A, the condition can be improved by adding vitamin A to the diet.

Since the concentration of rods is greater around the periphery rather than in the center of focus for vision with cone cells, there is an advantage in looking a little to one side of an object when trying to locate an indistinct object in the dark.

Dark adaptation is the change in the eye as it accommodates to darkness. The pupil dilates, but there is also a change in the retina, where the rod cells increase in sensitivity. It requires at least half an hour to accomplish dark adaptation. Most of us have had the experience of going into a darkened theater and finding that we were hardly able to locate our seats. Later we find that there seems to be plenty of light available and we can see fairly well.

Light adaptation is accommodation to normal or bright light. It is a somewhat misleading term, since the eyes are ordinarily adapted to daylight vision. It is best observed when we suddenly turn on the lights after being in the dark for some time or upon emerging from a motion-picture theater into bright sunlight. The light appears too bright for our eyes, but after a few minutes the eyes adapt themselves to the new condition. The excess visual purple stored up in the dark-adapted eye is bleached rapidly, the pupil constricts somewhat, and we find that we can see quite well again, even in bright sunlight. Dark adaptation is a slow process, but light adaptation is fairly rapid.

Rod cells are not concerned with color vision. Looking at a flower garden in the deep twilight is much like observing a photographic image on black-and-white film. Light-blue flowers are seen as various shades of gray, and the deep red flowers appear to be black. With adequate illumination, the yellow-red end of the spectrum appears the brightest to the light-adapted eye, but with decreasing illumination, the blue-green end of the spectrum is more readily discerned by the dark-adapted eye. This effect is explained as the result of the ability of the retina to shift from cone vision to rod vision in the process of dark adaptation. It is known as the *Purkinje effect*. After sundown, as daylight decreases, red objects become darker in color; but light-blue objects are still clearly blue, gradually resolving to gray as twilight deepens.

It is often of distinct advantage to have eyes dark adapted, as, for example, under war conditions. One way to accomplish this is to have the men sit in a totally dark room for 30 minutes. There remains the problem of transferring them to the scene of operations without exposure to light. It was discovered that by using goggles with red glass to act as filters it was possible to have good cone vision while permitting the rods to become dark adapted. The red filters screen out the light rays that stimulate the rods but do not greatly affect the light rays stimulating the cones.

Nocturnal animals presumably have good vision in the dark. Some of the nocturnal animals, such as bats, have retinas that are said to contain only rods. The retinas of rats contain a few cone cells among the rods, whereas some owls have a fair distribution of cone cells and are able to see very well in daylight. The eye of man is essentially adapted for vision during the day but functions reasonably well in the dark if there is some weak light available.

Cone cells These cells provide acute detailed vision and the perception of color. It appears to be well established now that there are three light-sensitive pigments, located in three different kinds of cone cells. One pigment exhibits its greatest sensitivity in the wavelength for blue light, one in the spectrum for green light, and one for sensing red light waves. It has become possible experimentally to pass a beam of light through a single retinal cone cell. When the spectra from several different cone cells are plotted, they fall into the three groups indicated above. Cone cells of these groups cannot be distinguished by their appearance, for they have no colored pigment in their

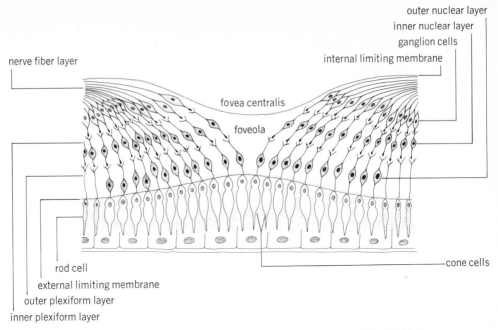

nerve fiber layer

outer nuclear layer
inner nuclear layer
ganglion cells
internal limiting membrane

fovea centralis

foveola

cone cells

rod cell
external limiting membrane
outer plexiform layer
inner plexiform layer

FIGURE 12.12
Horizontal section of the retina
through the macula lutea and
fovea centralis.

structure. A light-sensitive violet-colored pigment was found originally in chicken retinas by Wald. It is considered to be a visual pigment associated with cone cells and is called *iodopsin*. It has since been discovered in a few other vertebrates and is thought to be the red-sensitive pigment of color vision in man. We shall discuss color vision later in connection with color blindness.

The *area centralis* is the region around the visual axis and is located 2 to 2.5 millimeters laterad from the optic disk (Figure 12.6). In man it is a pigmented area and so corresponds in general to the area called the *macula lutea*, or *yellow spot*. Though the area centralis is about 6 millimeters in diameter, the most evident yellow pigment area is only about 3 millimeters in diameter. Structurally, the area is differentiated by an increase in the number of ganglion cells into more than one row. The color of the pigmentation is described as canary yellow or as about the color of grapefruit. It has been extracted and shown to be a carotenoid, related to the xanthophyll of leaves (Wald).

The *fovea centralis* is a bowl-shaped depression in the middle of the central area. It is a tiny pit measuring about 1.5 millimeters from one margin to the other and about 0.24 millimeter deep. In the center of the depression is the *foveola*, an additional concavity measuring only 0.35 millimeter in diameter. The fovea centralis is specialized in many ways to provide the most acute vision possible (Figure 12.12).

The concentration of cone cells increases inward from the periphery until, in the foveola and a surrounding area about 0.5 millimeter in diameter, rod cells are completely excluded. The periphery of the fovea, however, even the rim of the inner slope of the depression, contains both rods and cones. The cone cells of the fovea are longer, much more slender, and more closely packed than elsewhere in the retina. This tends to accentuate acute vision. Larger blood vessels pass around this area, for although there are small blood vessels present in the fovea centralis, even capillaries are not found in the foveola. The retina is thinner in the fovea; actually at the depth of the pit,

Structure of the eyeball

299

in the foveola, most of the layers of neurons either have been pushed aside or are absent altogether. Here the ratio of cone to ganglion cell approaches 1:1. It has been suggested that the slope of the fovea acts to magnify the image projected on the foveal cones. At any rate, everything that we see clearly is projected on this very small area of the fovea centralis. The most detailed observations are made by the projection of a minute image on the foveola. The size of the retinal image varies with the size of the object and its distance from the eye; but since the fovea centralis is only about 1.5 millimeters in diameter and the foveola is only 0.35 millimeter, most retinal images to be clearly seen should be less than 1 millimeter in size. The retinal image may be much more minute; in fact, two points can be distinguished when their retinal images are only 0.003 millimeter apart. The term *retinal image* is in common use, but it would be more accurate to consider it as simply an energy effect on the retina. It is not a visible image.

REFRACTING MEDIA OF THE EYE   In order to reach the retina, light rays must pass through the cornea, the aqueous humor, the lens, and the vitreous body. The refractive index varies for each of these tissues and fluids, and light rays are refracted or bent accordingly in passing through them. Refraction occurs because light rays travel at different speeds as they pass through media that vary in density; for example, a stick lying obliquely and half-submerged in water appears to be bent.

The curved *cornea* is the first refracting tissue. Light waves pass from the air through the cornea and aqueous humor and enter the pupil. The refractive power of the cornea is often underrated. Its refractive power is about twice that of the lens of the eye; rays of light therefore are converged considerably before they reach the crystalline lens. The *aqueous humor* is a clear, watery fluid that fills the area between the cornea and the lens and vitreous body. This space is divided into an anterior and posterior chamber. The anterior chamber is much the larger and lies between the cornea and the lens. The posterior chamber is a small area posterior to the iris. The refractive power of the aqueous humor is about the same as that of water.

When light waves strike the curved surfaces of the lens they are further bent or refracted. The refractive index of the air is 1.00; whereas the refractive index of the lens is 1.40. The refractive indices of other parts of the eye are as follows: cornea, 1.38; aqueous humor, 1.33; vitreous humor, 1.34. Normally light rays are directed to focus on the fovea centralis of the retina.

While the structures through which light must pass have a high degree of transparency, they are not completely transparent. Hecht, using blue-green light under test conditions, found that in order to see a light, one must receive from 54 to 148 quanta at the cornea of the dark-adapted eye. Because of reflection and absorption of the 54 quanta falling on the cornea, only 26 arrive at the retina. Here again only 5 quanta are estimated to be absorbed by the visual purple of the rods, and there are indications that an individual rod cell may be capable of stimulation by 1 quantum of light, the smallest unit of energy. Such results approach the ultimate in sensitivity.

THE FORMATION AND MOVEMENT OF FLUIDS WITHIN THE EYEBALL

The intraocular fluid keeps the eyeball distended and firm and is maintained at an average pressure of 19 mm Hg, the range being about 15 to 25 mm Hg. This fluid includes the aqueous humor and the vitreous body.

The aqueous humor is formed from the inner folds of the ciliary processes. These processes are infoldings of the choroid and are arranged in a circle around the lens. The aqueous humor flows forward through the zonular fibers, through the pupil, and into the anterior chamber of the eye. It also diffuses into the gelatinous consistency of the vitreous body, but since there is little exchange of fluid there, it cannot be said to flow into it.

The outlet for the intraocular fluid is in the angle between the cornea and the iris where the sinus venosus of the sclera (canal of Schlemm) drains the fluid eventually into the veins of the sclera (Figure 12.4). The sinus venosus drains into thin-walled aqueous veins which normally carry aqueous humor rather than blood. Projections from endothelial cells lining the canal of Schlemm have been observed (Worthen). They increase the surface area, and it has been suggested that they may enhance the exchange of fluid in these cells.

Glaucoma is a condition that occurs when the drainage of aqueous humor fails to keep up with the amount produced by the ciliary processes. In this case intraocular pressure may rise to such levels that severe pain is produced and the nerve fibers of the retina may be injured. Damage usually occurs at the optic disk, which is a weak spot in the retina where nerve fibers leave the retina to form the optic nerve. The nerve fibers may be forced against the edge of the disk, and continued pressure on them may cause blindness. The retinal artery enters the eye at the optic disk and may become compressed by the increased pressure within the eyeball. Drugs that constrict the pupil may afford relief in mild cases, because constriction of the pupil tends to enlarge and relax the angle between the iris and the cornea, thus permitting better drainage.

MECHANISM OF VISION

THE LENS The camera has been used many times to illustrate the mechanism of the eye. It is a fair working model. They both have a biconvex lens to bring light rays to a focus. The camera has a diaphragm to control the amount of light admitted; the iris performs this function in the eye. Light rays enter the dark interior of the camera and focus on sensitized material on the film or plate. Light rays also enter the eye and pass through its dark interior, focusing upon the sensitive retina.

A biconvex lens converges light rays, bringing them to a principal focus behind the lens. The principal focal distance varies according to the curvature of the surfaces of the lens and the refractive index of the glass used in its construction. This is the kind of lens used in a magnifying glass; the light rays are brought to a focus as a bright spot of light. Heat rays also are brought to a focus, and the lens is sometimes called a "burning glass."

A convex lens focuses parallel rays at its principal focal distance. For practical purposes, the rays of light from an object 20 feet away are considered to be parallel, since they diverge so little. But the object to be observed or photographed is often closer than 20 feet. In this case the eye and the camera have different ways of focusing the diverging light rays from a nearby object (Figure 12.13).

The eye is capable of changing the convexity of the lens. It becomes more convex to focus strongly divergent rays on the retina. The distance between the lens and the retina remains the same. The glass camera lens cannot change its shape in order to focus on some object only a few feet

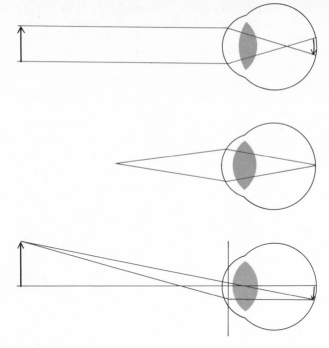

FIGURE 12.13

Accommodation for distant and near vision. Light rays from a distant object are practically parallel and can be focused with the ciliary muscles relaxed and with the normal convexity of the lens. Diverging rays from near objects, however, require greater convexity of the lens to focus sharply on the retina.

FIGURE 12.14

The inversion of the retinal image in the schematic eye.

away. The focusing type of camera moves the lens forward toward the object in this case, in order to increase the distance between the lens and the film.

INVERSION OF THE IMAGE If we look at the back of a camera that focuses on a ground-glass plate, we observe that the image is inverted. This is one of the properties of a convex lens; the rays passing through it are refracted to such an extent that they cross. This means that light rays from the top of the object are projected at the bottom of the image and vice versa. The image is also reversed from left to right. Lantern slides provide another example when placed in the projector upside down in order that the picture projected on the screen may be right side up. The image formed on the retina is inverted by the convex lens of the eye. We do not see things as inverted, partly because we learn by experience to distinguish objects in proper relation to their surroundings. We should keep in mind, however, that the retinal image is not seen; nervous impulses pass back to the visual area of the brain and are then interpreted as an image (Figure 12.14).

The following is an interesting experiment that anyone can perform. Take a card and punch a pinhole in it. Now take the pin and hold it close to the eye, between the eye and the pinhole in the card, while looking at some bright light through the pinhole. When the shadow of the pin falls upon the retina, the image formed by the shadow is seen but the pin appears to be upside down. The brain apparently interprets all retinal images as having been inverted. In this case the shadow is not inverted on the retina, and so the pin is seen upside down.

The visual area is located in the right and left occipital lobes of the cerebrum. Nerve fibers from the right half of each retina carry impulses to the right visual area, whereas fibers from the left half of the retina serve to bring impulses to the left visual area. This means that the fibers arising from the inner half of each retina must cross to the opposite side. The crossing of the fibers occurs back of the eyeballs in the *optic chiasma* (Figure 12.8). The crossing is incomplete, since fibers from the outer half of each retina do not cross. The crossing of the fibers is thought to contribute toward correlation of eye movements and visual perception.

Visual impulses are therefore transmitted to the visual areas of the cortex in the same relative position correlated with the image projected on the retina. Not only are right and left halves correlated, but neurons from the upper half of the retina carry impulses to the upper part of the visual cortex and impulses arising from the lower part of the retina are transmitted to the lower half of the visual area. Evidently the inverted image at the retina is correlated directly with the visual area; impulses transmitted from the inverted retinal image do not arrive right side up at the visual cortex.

ACTION OF THE LENS IN ACCOMMODATION A relatively flat lens may be able to focus parallel light rays and so accommodate for distant vision, but to focus the diverging rays of nearby objects requires a more convex lens. The ciliary muscles are attached anteriorly in the region around the cornea; posteriorly they blend into the choroid coat. The zonular fibers, supporting the lens, are continuous with the capsule of the lens and extend out radially to make connection with the ringlike ciliary body. The ciliary muscles therefore exert tension only indirectly on the zonular fibers. When the ciliary muscles are relaxed, the ciliary ring is thin and the zonular fibers are tightened. Tension on the capsule of the lens causes the lens to become less convex, as in accommodation for distance. In accommodation for near vision, the ciliary muscles contract, pulling the ciliary body and choroid forward toward the cornea. The zonular fibers slacken, and the lens becomes as convex as its elasticity will permit. Looking at some distant object is restful to the eyes, since most of the ciliary muscles are relaxed. Hours of close-up work require these muscles to maintain a more or less constant tension.

ACCOMMODATION REFLEX While looking at a distant object, bring the finger into the field of vision about 12 inches from the eyes. The finger is seen, but the image is blurred. Now a quick change of focus brings the image of the finger into sharp focus, but the distant object is not seen sharply. When the image is not focused sharply on the retina for near vision, sensory impulses pass over the optic nerve to the brain and motor impulses are transmitted by autonomic fibers in the oculomotor nerve to the ciliary muscles, which contract, permitting the lens to become more convex and in sharper focus. This is the mechanism of the *accommodation reflex*. A negative feedback mechanism enables the lens to adjust automatically to secure the best focus. The visual cortex would necessarily be involved in accommodation. The greatest change in the convexity of the lens is at its anterior surface; there is not much change in the posterior surface.

THE LIGHT REFLEX We have noted that the pupil dilates in fear and in pain in response to stimulation by way of the sympathetic system. There is also the response of the pupil to light, the *light reflex*. We have observed that the pupil constricts when exposed to bright light and dilates in the dark or in dim light. The most active part of the light reflex is the swift protective closing of the pupil when the eye is exposed to bright light. Constriction of the pupil is mediated through the parasympathetic system. The pupil also plays an important part in accommodation. The radial muscles of the iris contract somewhat for distant vision, and the pupil therefore dilates to a certain extent. For near vision, the sphincter muscle constricts the pupil. Camera fans know that for best results with a close-up the photographer uses good lighting and reduces the aperture of the lens. The center of the lens is optically the best part of the lens for close work. The action of the iris in constricting the pupil, as when a printed page is brought closer to the eyes, permits light rays to pass through the center of the lens, resulting in a clearer image. These rays can then be brought to a sharp focus at the fovea centralis.

An interesting abnormal condition in regard to the pupillary light reflex is termed the *Argyll Robertson pupil.* It is characterized by loss of the light reflex, the pupil remains small, but the accommodation reflex still responds. This reaction usually indicates syphilis in the central nervous system and so has diagnostic value. It is considered that the contraction of the pupil during accommodation still occurs because the disease has not affected the parasympathetic nerve fibers to the eyes.

CONVERGENCE   The eyes appear to look straight ahead at a distant object, but if an object is brought close to the eyes, they converge markedly until, with the object close to the face, the eyes appear "cross-eyed." Have someone hold his finger vertically at arm's length and then bring it close to the face, keeping the eyes focused on it as long as possible. You will note the extreme convergence of the eyes as the finger approaches the face. Convergence is a necessary corollary of accommodation, since it keeps the retinal image on corresponding points in each eye. It is also essential to binocular vision.

An object may be brought so close to the eye that it is impossible to obtain a sharp focus. The *near point* therefore is defined as the closest position at which an object can be clearly seen, with the eye completely accommodated. The near point is very close in children (8.8 centimeters at ten years of age). As the age increases, the distance of the near point from the eye increases. The increase is gradual in young persons, but between the ages of forty and fifty there is a marked change. An older person finds that he has to hold the newspaper quite a distance from his eyes in order to read it. His near point may be around 50 or even 75 centimeters. If his eyes are normal otherwise, this condition is called *presbyopia,* the farsightedness of the aged. The lens has lost so much of its elasticity that, even though the ciliary muscles contract strongly, the lens does not become convex enough for close-up work, such as reading small print. The necessary additional refraction can be obtained by using the proper convex lenses in glasses.

Parallel rays from a distant object are focused sharply by the normal eye with the ciliary muscles relaxed, or without accommodation. Theoretically the *far point* is at infinity, but practically an object at a distance of 20 feet gives off rays so nearly parallel that the normal eye can focus them without accommodation; this distance is taken as the far point for most practical purposes. In testing the eyes for visual acuity, a properly illuminated test chart is placed at a distance of 20 feet. The chart has rows of letters, each row in a different-sized print. At the end of each row is a number indicating the ability of the normal eye to read the type of that particular size. The rating 20/20 means that the individual tested can read the line indicated for 20 feet and that his eyes are normal. The eyes must be tested one at a time, of course. If the individual being tested is rated 20/40, it means that he can read only the larger letters, marked 40. The normal eye could read those letters at a distance of 40 feet. If the individual has better-than-normal vision, he may be able to read a smaller row of letters—perhaps a row that the normal eye could read only at 15 feet. In this case his rating would be 20/15.

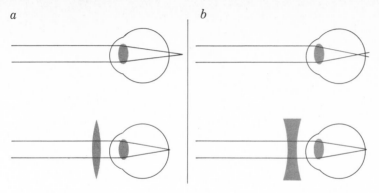

a   *b*

FIGURE 12.15

Diagram illustrating two common eye defects, hyperopia and myopia: *a* hyperopia, a type of farsightedness. If the eyeball is too short, the rays of light focus behind the retina, as shown in the upper illustration. The correction of farsightedness is accomplished by wearing a convex lens before the eye. This lens brings the rays of light into sharp focus on the retina, as illustrated: *b* myopia, or nearsightedness. If the eyeball is too long or if the refractive power of the eye is too great, the rays of light focus at some point in front of the retina. Nearsightedness is corrected by wearing a concave lens before the eye. This type of lens brings the rays to sharp focus on the retina, as illustrated.

An eye that has no optical defects is said to be *emmetropic;* one with optical defects is called *ametropic.* We have discussed astigmatism and presbyopia. There are two other common defects that should also be considered: hyperopia and myopia.

Hyperopia is a type of farsightedness in which the refractive power of the eye is not great enough or the eyeball is too short. It should not be confused with presbyopia, which is a gradual loss of accommodation as one grows older. The light rays from an object in this case are focused behind the retina, resulting in a blurring of vision unless the accommodation is strong enough to secure a sharp focus. Ordinarily objects at a distance may be seen clearly, but the diverging rays from near objects are focused with difficulty, if at all. Reading or other detailed work may cause eyestrain. The condition may be corrected by placing the proper convex lenses in the glasses (Figure 12.15*a*).

Myopia means nearsightedness. In this condition the refractive power of the eye is too great, or the eyeball is too long. Light rays focus somewhere in the vitreous body, a little in front of the retina. The image, when projected back to the retina, will be blurred. To correct this condition, concave lenses are placed in the glasses (Figure 12.15*b*).

Biconcave lenses are thin in the middle and spread the light rays that pass through. A concave lens in glasses will diverge light rays before they pass through the cornea, and they will focus farther back in the eye. The nearsighted person is handicapped in regard to distant objects, but he is capable of doing close work without difficulty. The eyes of those elderly persons who can read fine print without their glasses are almost certainly myopic.

BINOCULAR OR STEREOSCOPIC VISION

The eyes, very fortunately, make every effort to maintain the visual field as the head is moved. It would be most confusing if we shook our head vigorously during conversation and found that the visual field also appeared to move from side to side. As we approach an object, we may keep the object fixed by greater and greater convergence of the eyes. Binocular vision refers to the blending of two images to give one image that looks real and has depth. When the eyes converge upon an object, the right eye sees a little more of the right side of the object while the left eye sees a little more of the left side. The two images are not identical, but when they are superimposed and interpreted by the visual area, we perceive an image that has depth and unity.

Binocular or stereoscopic vision

The stereoscope is an instrument that gives the illusion of depth and solidity when one looks at pictures taken in a certain way. A picture taken by a camera of the ordinary type appears flat in comparison. The two pictures to be viewed through the stereoscope are taken at slightly different angles, equivalent to the angle of convergence of the eyes. Light rays, passing from the two pictures through the lenses of the stereoscope, stimulate corresponding retinal points in each eye, and the image is interpreted as having depth and reality.

Binocular vision aids in judging distance. There are, however, other factors that help in our judgment. The color of distant objects, the relative size of intervening objects, shadows, and perspective all play a part.

Only when corresponding points on the retina are stimulated simultaneously is a single image obtained. Obviously the foveae are corresponding areas. If we fix our eyes on some object across the room and then bring a finger up close to the near point of the eyes, we see two fingers instead of one. If the right and left fingers are held horizontally in front of the eyes at about the distance of the near point while the eyes are fixed on some distant object, the right finger will be seen by the right eye, the left finger by the left eye; but in the middle will be a seemingly detached portion, the part that is superimposed by both eyes.

STRABISMUS In some cases the eye muscles are not properly balanced, and the two eyes do not focus together normally. There may be too much convergence, as in the condition commonly called "cross eyes"; then again one or both eyes may be strongly divergent (divergent strabismus). Surgical and other means can be employed to restore muscular balance. If the two eyes do not focus on the same retinal points, the vision of one eye is usually disregarded and the individual sees only one field.

THE PHYSIOLOGY OF VISION

The processing of visual information has been difficult to determine. There is evidently considerable coding of the visual input by ganglion cells before impulses leave the retina. Synaptic connections in the geniculate bodies continue the sorting and refining process before information reaches the visual cortex. We have discussed some of the functions of the visual cortex, including on-and-off areas, sensitivity to straight lines, contours, and contrasting light effects. The extensive work of Hubel, Wiesel, and others has helped greatly in developing these concepts.

The eye, by a series of rapid movements, scans the object and interprets by a series of retinal images, an estimate of form, size, lines, and color. Rapid eye movements (REM) constitute the *saccade*. The individual is unaware of these movements as the eye fixes on various details. Rapid movements of the eyes during dreaming are well documented. Also, in reading a line of print, the eye jumps along from one point of fixation to another several times in a second.

Returning to rapid eye movements in scanning an object, let us take a green, plastic bucket as an example. By a series of rapid movements, the eye will fix in succession on the contour, the slope of the sides, height, width, and circular top as an indication of size, shape, and recognition. Probably some details will be overlooked, for example, whether or not it has a bail and the nature of its color and composition. Obviously this series of retinal images in a fraction of a second is quite different from a single picture taken with a camera. The series of retinal images enables the viewer to form a concept of the object which he may recognize as a bucket.

The eye is even more sensitive to objects that move. Fortunately, the object viewed does not have to hold still. In fact, there is evidence that in some animals the retina is selective as to the direction of movement; some ganglion cells apparently respond to movement in one direction but give no response when the movement is in the opposite direction (Michael). The element of contrast is quite important in initiating visual stimuli. One has only to think of the difficulty in observing insects, birds, or mammals that exhibit protective coloration and blend in with their background as long as they remain quiet, and of how easy it is to see them when they move.

Perhaps it would be well to repeat here that some ganglion cells send impulses to the superior colliculi which are concerned with eye movements. It is the cortex, of course, which functions to give us the conscious interpretation of the visual image. The retinal image, if we may repeat, is not actually seen, but it enables us to see with its reception of light waves.

An understanding of color vision requires some knowledge of the physical aspects of light waves. Color sensation is produced by different wavelengths of light. It should be remembered that color is the interpretation by the brain of certain wavelengths of light received on the retina. This fact gives rise to the saying "color exists only in the mind."

Sunlight or white light is a combination of the colors of the spectrum. When such a light is passed through a prism, it is separated into its component parts: red, orange, yellow, green, blue, and violet. These are the colors of the spectrum. The longest visible wavelengths, 760 millimicrons (760 millionths of a millimeter), produce the sensation of red, while the shortest visible wavelengths, 390 millimicrons (390 millionths of a millimeter), are interpreted as violet. Intermediate wavelengths produce the other colors of the spectrum. Light waves are commonly measured in angstroms; one ten-millionth of a millimeter equals 1 angstrom.

An object has color because it absorbs certain wavelengths of light and reflects or transmits other wavelengths. The eye receives the reflected or transmitted wavelengths, and they are interpreted as representing the color of the object. Objects that appear white to the eye must reflect all wavelengths. Objects that appear black in sunlight must absorb all wavelengths. Black appears to be a definite sensation; it is not equivalent to seeing nothing as when objects are focused on the blind spot.

Since white light can be separated into the colors of the spectrum, it is also possible to combine these colors to produce white light. However, all the colors are not needed for white light, and all the colors of the spectrum can be formed from three primary colors: red, green, and violet. When two or more wavelengths fall upon the retina at the same time, the result is a color fusion and the sensation is a different color from that of any single wavelength. Thus the wavelengths for red and green give rise to the color sensation of yellow or orange, while the fusion of red and violet is interpreted as purple. The fusion of certain wavelengths of light should not be confused with the mixing of pigments to produce different colors of paint. These

COLOR VISION

Color vision

307

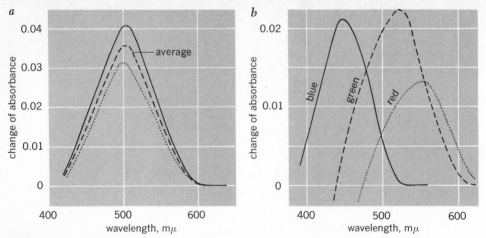

FIGURE 12.16

Spectral sensitivity of rod and cone cells: *a* difference spectra of the visual pigment in a human rod cell in the parafoveal region. The average curve maximum at 505 millimicrons closely resembles the difference spectrum of human rhodopsin. The pigment of the rod consists primarily or entirely of rhodopsin. *b* Difference spectra of visual pigments in single human cone cells in the parafoveal region. [*After Paul K. Brown and George Wald, Visual Pigments in Single Rods and Cones of the Human Retina, Science, 144:45–51 (1964). Copyright, 1964, by the American Association for the Advancement of Science.*]

pigments are not optically pure and so reflect and absorb light waves in quite a different manner. Mixing yellow and blue pigments produces a green color; the fusion of yellow and blue wavelengths of the spectrum in the proper proportion gives rise to a white color sensation.

COMPLEMENTARY COLORS The fusion of two colors or combinations of colors can give white. Depending upon the proportion of color used, a suitable color complement can be found for any given color, and the result of their fusion will be the sensation of white. Color pairs that produce white when fused are complementary colors. In the spectrum the following color pairs are complementary to each other: red and blue-green; orange and blue; yellow and indigo blue. These colors are diametrically opposite each other on a color scale. Closely related colors, upon fusion, produce a more saturated color, that is, one that has less of the sensation of white. Examples are the fusion of red and yellow to form orange or of red and violet to produce purple. The pale or so-called "pastel" shades are examples of unsaturated colors.

It is easy and interesting to perform the following experiment but far more difficult to explain the result. If one places a green filter over one eye and a red filter over the other eye and then looks at a white area, he does not see either color but rather a mixing of the two colors to produce a weak yellow or orange sensation. Obviously the retina of one eye is reacting to rays in the region of the red spectrum, while the other is reacting to green. It would appear that the mixing could not occur in either retina but would more likely occur in the visual cortex.

THE THREE-PIGMENT CONCEPT OF COLOR VISION The investigations of MacNichol and Wald in cooperation with several collaborators leave little doubt that color vision requires three photopigments and that these are incorporated in three types of cone receptors having their individual maximum sensitivities in the blue, green, and yellow regions of the spectrum. Whereas red receptors reach their peak in the yellow area of the spectrum, they extend their sensitivity well into the red wavelengths and so are efficient in sensing red.

The following table of wavelengths indicates the length of light waves, in millimicrons, which are interpreted as a certain color.

Light waves of
435 millimicrons   blue
540 millimicrons   green
570 millimicrons   red

There is a range of color in each case; for example, in the area of light waves shorter than 435 millimicrons are blue-violet and violet. In an intermediate range between green and red there is orange at about 580 millimicrons and yellow at about 550 millimicrons.

The spectral-sensitivity curves as determined by several research workers employing somewhat different procedures vary a little, but in general, blue-sensitive receptors reach their maximum sensitivity at about 435 millimicrons; this is in the blue-violet area of the spectrum. Green-sensitive cones have a maximum sensitivity at 540 millimicrons, which is interpreted as green although their sensing ability extends from blue through green, well into yellow, orange, and red wavelengths. As we have indicated, red-sensitive cones reach their maximum sensitivity at 570 millimicrons in the yellow area of the spectrum but extend through green and well into the red wavelengths (Figure 12.16). In color television, orange-red is a very strong color; that is, it has a dominant color reception.

Blending of colors is accomplished by stimulating more than one type of cone cell at the same time. The stimulation may be essentially equal or it may be unequal. If red and green receptors are stimulated equally, as they are by wavelengths in the yellow area of the spectrum, this is interpreted as yellow. But if the stimulation is unequal and red-receptor cones are stimulated more strongly than the green receptors, then the color is interpreted as orange or orange-red.

Color analysis apparently begins with certain ganglion cells which respond strongly to one of the primary colors but are suppressed by another color. Secondly, some cells in the lateral geniculate bodies respond in a similar manner to impulses initiated by light of different wavelengths. These cells typically operate from fields with on or off centers. Finally, impulses reach the primary visual cortex where the coded messages are interpreted.

THE AFTERIMAGE

Visual sensations develop, last for a certain length of time, and then fade. If the image can be seen for a longer time than the actual exposure, it is called an *afterimage*. It is termed a *positive afterimage* if it has the same color and the same appearance as the object. If it changes from white to black or if it is produced in the complementary color, it is called a *negative afterimage*. It is the persistence of the positive afterimage that enables us to enjoy moving pictures; the afterimage is retained just long enough for the motion-picture machine to substitute the succeeding picture in a series. The picture then appears to have movement. When a bicycle wheel is turned rapidly, the individual images of the spokes soon become blurred and the interior of the wheel appears solid. At the same time, if one were to place his finger lightly on the spokes, he would feel each spoke pass by. Evidently the afterimage of the eye lasts much longer than that for the sense of touch. A positive afterimage may be developed by looking at an ordinary electric light for a few seconds. Upon closing the eyes a bright image will persist for some time and will be in the same color as the original. The persistence of the afterimage appears to be due to nerve impulses that continue to develop in the retina after the initial stimulus has passed.

A negative afterimage is thought to be caused by adaptation of the cone cells for the particular light or color involved. If we stare at a white globe over an electric light and then turn our gaze to a white wall, the afterimage is black, not white. The theory is that white light has adapted all the receptors temporarily where the light globe was focused. When the eyes are turned to the white wall, a second and weaker stimulus of white falling on this area of the retina produces no response. We therefore see a black area surrounded by the white of the wall. Black is the sensation produced by the lack of retinal stimulation.

If we stare at a bright red color for about 30 seconds and then look quickly at a white surface, the afterimage will appear in the complementary color, which is blue-green. It is assumed that the retinal elements sensitive to red have become adapted, and the second stimulus of white light, without the aid of the retinal substance for red, causes the blue-green substance to respond.

COLOR BLINDNESS

The term *color-blind* is usually applied to those who do not have normal color vision. For the most part it is an inappropriate term, since most of these persons see color of one sort or another. There are a very few persons who are truly color-blind; but if so, their vision is much like viewing a photograph on ordinary film. The image is in tones of white and black—a true condition of achromatic vision.

There are many persons who are simply color-weak. They are poor at distinguishing different hues, and their color-perception mechanism adapts readily. While gazing steadily at some small area of color, as in attempting to identify an insect by verifying its color, the color seems to disappear. But if the color area is large and definite, these persons are not aware of any difficulty. Many are quite unaware that they do not have normal color vision.

For some, there appears to be merely a shortening of the spectrum at either end. If the shortening occurs in the longer wavelengths, the individual may not be able to distinguish a red line from the black lines on a chart or graph. All appear black to him. Apparently there is considerable individual variation in ability to distinguish the shorter wavelengths that are interpreted as violet hues.

Color blindness occurs ordinarily when one of the color receptors is deficient. If the red-receptor cones are missing or nonfunctional, the color-blind person is forced to interpret color primarily with green and blue receptors. This is not adequate because true color perception requires both red and green cones to provide contrast. The same is true if green receptors are deficient. The color-blind person is unable to distinguish between various shades of color and is most deficient in the longer wavelengths of green and red; hence the condition is referred to as red-green color blindness. A red-blind person is called a *protanope*, and a green-blind person is a *deuteranope*.

The defective condition called blue color blindness is very rare and differs in its inheritance from the more typical red-green type. It is a dominant trait and is not sex-linked.

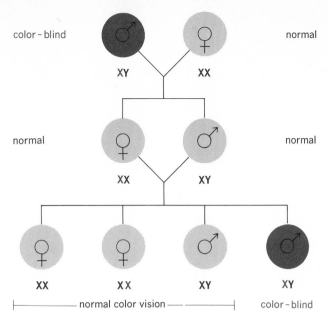

FIGURE 12.17
Schematic design illustrating the inheritance of color blindness. The male is color-blind, and the female has normal color vision. The defect is transmitted through the daughters, but both male and female children have normal color vision. The grandsons who inherit the X-chromosome gene for color blindness will be color-blind; all other grandchildren will have normal color vision.

Color perception has not been completely resolved. Whereas the three-pigment concept has been generally accepted, there is evidence that a two-color on-off system is used in the transmission of impulses from the ganglion cells to the visual interpreting center in the brain. This requires further investigation.

THE INHERITANCE OF COLOR BLINDNESS The genetic form of red-green color blindness is inherited as a sex-linked trait. It is determined by a gene that is incapable of directing the development of full color vision. Since the gene is recessive, the female must possess this gene in both X chromosomes if she is to be color-blind herself. If the gene is present in only one X chromosome in the female (the usual condition, if present at all), the female will not be color-blind but will transmit the condition to some of her children (Figures 12.17 and 12.18). See Chapter 22 for a discussion of the behavior of sex chromosomes.

The gene for color blindness has no normal dominant counterpart in the Y chromosome of the male. The Y chromosome is the one that normally pairs with the X chromosome in the male. The male, therefore, who inherits this gene will be color-blind. Almost without exception the color-blind father will transmit the gene for this trait to his female children, because ordinarily one X chromosome is received from the father and one from the mother.

When the female has one gene for color blindness and marries a normal male, half the male children may be expected to be color-blind and half the female children will be carriers. The other half, both male and female, will inherit the gene for normal vision. Color blindness is far commoner in men, affecting about 4 percent of the male population. It is rare in women, affecting only about 0.5 percent of the female population, according to one estimate.

Red-green color blindness may indicate a deficiency in two different color-sensitive pigments, each presumably controlled by a specific gene. In this case, two loci are thought to be on the X chromosome.

Color blindness

311

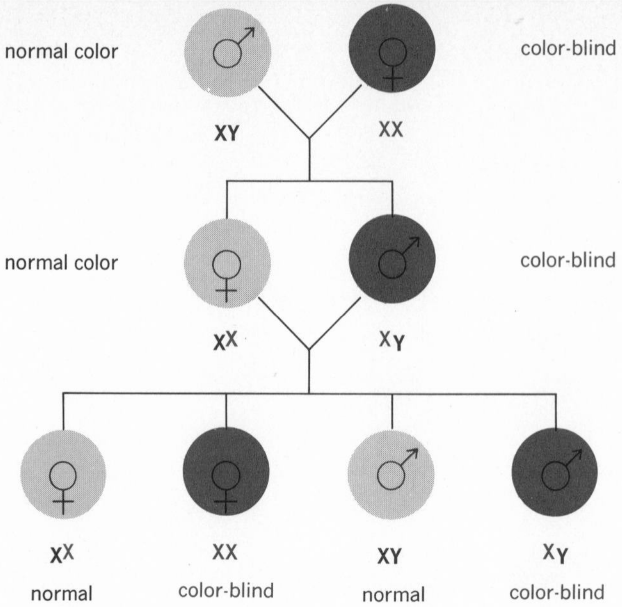

FIGURE 12.18
Schematic design illustrating the inheritance of color blindness. The male has normal color vision, and the female is color-blind. The color-blind mother transmits the defect to both sons and daughters. Only the sons are color-blind, however, because the Y chromosome contains no dominant gene to suppress the expression of the defect. The daughters receive a normal (dominant) gene in the X chromosome from the father.

SUMMARY

The accessory structures of the eye are the eyelids, eyebrows, eyelashes, conjunctiva, and lacrimal apparatus. The conjunctiva is a thin, transparent membrane covering the anterior surface of the eyeball and lining the eyelids. The lacrimal gland produces tears. The lacrimal secretion moistens the anterior surface of the eye, and the excess fluid drains into the nasolacrimal duct.

There are six extrinsic muscles attached to the eyeball. They are the superior, inferior, lateral, and medial rectus muscles plus the superior and inferior oblique muscles.

The wall of the eyeball is composed of three layers: an outer fibrous tunic, consisting of the sclera and cornea; an intermediate vascular layer, called the choroid; and an inner nervous tunic, the retina.

The outer coat, or sclera, is white except for the anterior transparent portion, called the cornea. Astigmatism is a condition usually caused by imperfect curvature of the cornea.

The dark-colored choroid gives rise to the ciliary body and the iris as it extends anteriorly. The muscles of the ciliary body control the degree of convexity of the lens. The iris is the colored portion of the eye. Persons are said to have blue eyes or brown eyes, depending on the color of the iris. The pupil is the circular opening in the iris. It appears black because it opens into the dark recess of the eyeball. The posterior side of the iris is dark colored and contains two antagonistic sets of muscles. Contraction of the radial muscles dilates the pupil, whereas contraction of the sphincter pupillae causes the pupil to constrict.

Parasympathetic fibers of the oculomotor nerve innervate the sphincter muscle, while sympathetic fibers innervate the radial muscles.

The inner nervous coat contains receptors called rod and cone cells. There are several layers of neurons in the retina, the rod and cone cells representing the deepest layer. This is an inverted retina, with the receptor cells turned away from the light source. The optic disk is a blind spot, because the nerve fibers forming the optic nerve are concentrated at this spot and there are no receptors in this area.

Special senses: vision

Rod cells react to low intensities of light, affording vision in dim light. Rhodopsin, or visual purple, is a photosensitive pigment associated with rod cells. Cone cells react to light of high intensity and provide acute, detailed vision. They are also the receptors for the perception of color.

The area centralis, or yellow spot, contains the fovea centralis, which is the area of acute vision. The fovea is a depressed area of the retina. In the center of the depression is the foveola, an area of only 0.35 millimeter in diameter; it contains only cone cells.

Light waves must pass through the cornea, the aqueous humor, the lens, and the vitreous body before reaching the retina. The refractive index varies for each of these tissues and fluids; light waves therefore are refracted or bent in passing through them. The refractive power of the cornea is nearly twice that of the lens.

The lens is biconvex and capable of changing its convexity in accommodation for objects at varying distances. It becomes more convex to accommodate for objects that are close to the eye.

When the radiant energy of light waves is focused on the retina, nerve impulses are initiated in the optic nerve. The nerve pathway includes the optic chiasma, the optic tract, the lateral geniculate body of the thalamus, and, finally, the visual area of the occipital lobes. In the optic chiasma, only the nerve fibers from the inner half of each retina cross over to the opposite side. The fibers from the outer half of each retina remain on the same side as that from which they originated.

The pupils respond to varying light intensities by constricting when subjected to bright light and dilating when exposed for some time to light waves of low intensity. This is the light reflex. The pupil constricts also during the accommodation of the lens for near vision and dilates to a certain extent for distant vision.

Hyperopia is a type of farsightedness in which the refractive power of the eye is not great enough or the eyeball is too short.

In myopia either the refractive power of the eye is too great or the eyeball is too long. An appropriate concave lens in glasses is used to diverge light waves and to offset the refractive power of the eye.

Color sensation is produced by different wavelengths of light stimulating the cone cells of the retina. Visual sensation can last longer than the actual exposure time. If the sensation persists for a time after the exposure has ended, the persisting image is called an afterimage. If it has the same color as the object, it is a positive afterimage; if it is produced in the complementary color, it is a negative afterimage.

Color-blind persons do not have normal color vision. Inability to distinguish red and green colors is the most common type. The genetic type of red-green color blindness is inherited as a sex-linked character. The gene for color blindness is recessive but finds expression in the male (XY) who has inherited the character, because the Y chromosome contains no dominant gene to prevent the development of the defect. If the gene is present in only one X chromosome in the female (XX), she will not be color-blind but can transmit the defect to some of her children. The color-blind female inherits the gene in both of her X chromosomes.

1 Name some of the accessory structures surrounding the eye.
2 Locate the conjunctiva and indicate its function.
3 What are the lacrimal structures?

4 Name and locate the muscles that move the eyeball.

5 Discuss the three layers or coats of the eyeball. Indicate the derivatives of each coat.

6 What is astigmatism?

7 Why is the vertebrate retina inverted?

8 How does it happen that there is a blind spot in the eye?

9 What is visual purple?

10 How do rod cells function?

11 Discuss dark adaptation and night vision.

12 Explain the structure and function of the fovea centralis.

13 What are the refracting media of the eye?

14 What is cataract?

15 Compare the focusing of a camera with that of the eye.

16 Give some of the uses of a convex lens.

17 Explain the action of the lens and ciliary body during accommodation.

18 Discuss the action of the pupil when exposed to light and during accommodation.

19 Why is convergence of the eyes necessary for accommodation?

20 How does presbyopia differ from hyperopia?

21 What structural conditions can cause the eye to be myopic? What kind of lens is used in the glasses to correct the condition?

22 Explain binocular vision.

23 Discuss color vision.

24 What is a positive afterimage? A negative one?

25 Discuss color blindness. How are certain forms inherited?

SUGGESTED READING

BRINDLEY, G. S.: Central Pathways of Vision, *Ann. Rev. Physiol.*, **32**:259-268 (1970).

BROWN, PAUL K., and GEORGE WALD: Visual Pigments in Single Rods and Cones of the Human Retina, *Science*, **144**:45-51 (1964).

COULOMBRE, ALFRED J., and JANE L. COULOMBRE: Lens Development: Fiber Elongation and Lens Orientation, *Science*, **142**:1489-1490 (1963).

De ROBERTIS, E.: Some Observations on the Ultrastructure and Morphogenesis of Photoreceptors, *J. Gen. Physiol.*, **43**:1-6 (1960).

GLICKSTEIN, M.: Organization of Visual Pathways, *Science*, **164**:917-926 (1969).

HECHT, S.: Energy and Vision, *Am. Scientist*, **32**:159-177 (1944).

HENDRICKS, S. B.: How Light Interacts with Living Matter, *Sci. Am.*, **219**:174-186 (1968).

HUBEL, D. H., and T. N. WIESEL: Receptive Fields, Binocular Interaction and Functional Architecture of the Cat's Visual Cortex, *J. Physiol.*, **160**:106-154 (1962).

MacNICHOL, EDWARD F., Jr.: Three Pigment Color Vision, *Sci. Am.*, **211**:48-56 (1964).

MARKS, W. B., W. H. DOBELLE, and EDWARD F. MacNICHOL, Jr.: Visual Pigments of Single Primate Cones, *Science*, **143**:1181-1182 (1964).

MICHAEL, C. R.: Retinal Processing of Visual Images, *Sci. Am.*, **220**:104-114 (1969).

MILLER, W. H., R. E. GORMAN, and M. W. BITENSKY: Cyclic Adenosine Monophosphate: Function in Photoreceptors, *Science*, **174**:295-297 (1971).

NEISSER, U.: The Processes of Vision, *Sci. Am.*, **219**:204-214 (1968).

NODA, H., R. B. FREEMAN, Jr., and O. D. CREUTZFELDT: Neuronal Correlates of Eye Movements in the Visual Cortex of the Cat, *Science*, **175**:661-663 (1972).

NOTON, D., and L. STARK: Eye Movements and Visual Perception, *Sci. Am.*, **224**:35-43 (1971).

POLLEN, D. A., JAMES R. LEE, and J. H. TAYLOR: How Does the Striate Cortex Begin the Reconstruction of the Visual World?, *Science*, **173**:74-77 (1971).

POLYAK, S. L.: "The Retina," University of Chicago Press, Chicago, 1941.

RICHARDS, W.: The Fortification Illusions of Migraines, *Sci. Am.*, **224**:89-96 (1971).

RODIECK, R. W.: Central Nervous System: Afferent Mechanisms, *Ann. Rev. Physiol.*, **33**:203-240 (1971).

THOMAS, E. L.: Movements of the Eye, *Sci. Am.*, **219**:88-95 (1968).

WALD, G.: The Receptors of Human Color Vision, *Science*, **145:**1007–1016 (1964).

———: Molecular Basis of Visual Excitation, *Science*, **162:**230–239 (1968).

WERBLIN, F. S.: The Control of Sensitivity in the Retina, *Sci. Am.*, **228:**71–79 (1973).

WITKOVSKY, P.: Peripheral Mechanisms of Vision, *Ann. Rev. Physiol.*, **33:**257–280 (1971).

WORTHEN, D. M.: Endothelial Projections in Schlemm's Canal, *Science*, **175:**561–562 (1972).

unit four

the circulatory system

chapter 13
the blood

The blood affords a pathway for the transportation of oxygen and nutrient fluids to cells that are often at a considerable distance from the heart, lungs, or digestive tract. It enables tissues to rid themselves of waste materials even though the tissues are located in the feet or hands and relatively far away from the kidneys. The blood is commonly considered to be a liquid tissue, one in which the intercellular structure is liquid rather than composed of fibers or a more or less solid substance. The circulatory system is concerned with conducting the blood through a series of tubes (arteries, veins, capillaries) to the tissues. The heart acts as a pump to supply the motive force.

FUNCTIONS   Most persons are aware that the blood carries oxygen, carbon dioxide, nutritive elements, and waste materials, but they are commonly unaware that the blood also has many other functions. Following is a brief listing of some of the more evident functions.

1 It is the medium by which oxygen is transported from the lungs to the tissues.
2 Carbon dioxide, a product of the metabolism of cells, is transported from the tissues to the lungs.
3 Nutrient materials are absorbed from the intestine and carried to the tissues.
4 Many organic substances that represent breakdown products of metabolism (urea, uric acid, creatinine, purine wastes) are carried by the blood to the kidneys for excretion.
5 Hormones, the secretions of ductless or endocrine glands, are distributed throughout the body by the bloodstream.
6 Like a hot-water heating system in a house, the blood flows from the deeper and warmer parts of the body to the extremities and tends to distribute heat more evenly to all parts of the body. Surface blood vessels in the skin can be dilated so that more blood can come to the surface, thus losing heat more readily, or surface vessels can be constricted to keep more blood away from a cold exterior and thus reduce heat loss. The blood, therefore, is important in the regulation of body temperature.
7 The blood plays an important part in maintaining the acid-base balance of the tissues. Most of the tissues, including the blood, are slightly alkaline in their reaction. The pH of arterial blood is between 7.35 and 7.45.

While the metabolism of the body constantly produces numerous acids and acid substances, the tissues themselves and the body fluids remain remarkably constant at a pH that is a little on the alkaline side. The principal reason for this chemical stability is the fact that the blood contains a number of alkaline substances; the chief of these is sodium bicarbonate. Weak acids produced by metabolic processes are constantly buffered by alkaline substances in the blood and in the tissues, while excess alkalinity is buffered by acids. A buffer solution contains substances that afford a reserve of alkalinity and acidity. If a weak acid or base is added to the solution, either substance is buffered by the appropriate reserve substance and a state of chemical equilibrium is maintained (see Chapter 1, section on buffers).

Metabolic processes commonly form acid products; hence there is a tendency to emphasize the buffering action of alkaline substances against these acid products. Carbonic acid is a weak acid, which is readily neutralized by the bicarbonate of the blood. Carbonic acid breaks down into carbon dioxide and water.

Since CO_2 constantly formed in metabolism is regularly eliminated through the lungs, its influence on the acidity of the blood is greatly reduced. The pH of the blood changes very little as the result of acids taken into the bloodstream. Sodium bicarbonate acts as an *alkaline reserve* to protect the body from the acids produced by its own metabolic processes. Another factor in maintaining the chemical acid-base balance of the blood is the fact that acid substances buffered in the blood are constantly removed by the kidneys.

8 There is a constant relationship between blood volume and the fluid content of the tissues. The capillary wall acts as a selectively permeable membrane, permitting a constant filtration into the tissues of water molecules and other substances in solution. Small molecules, such as those of oxygen, glucose, or amino acids, pass through the capillary wall readily, but larger protein molecules pass through very slowly, if at all. Filtration, in this case the movement of water and dissolved substances out of the bloodstream, is aided by capillary blood pressure.

The blood also contains a number of proteins in colloidal state that tend to attract fluid from the tissues into the bloodstream and hold it there. Food proteins in the process of digestion are broken down to amino acids and absorbed in this form. Amino acids enter into the formation of plasma protein, including albumin and globulin. Plasma protein plays an important part in building up osmotic pressure if osmosis is interpreted as the movement of water through the capillary wall toward the protein. Abnormal conditions such as increased permeability, increased capillary pressure, or decreased plasma-protein content of the blood may permit excessive filtration of fluid into the tissues. The tissues swell and literally become water-logged, a condition known as *edema*. After severe loss of blood, water moves from the tissues into the bloodstream, and the volume of the blood may be quickly restored in this manner. The blood in this case is able to draw on a water reserve normally held in the tissues.

9 The ability of the blood to form a clot and so reduce bleeding has been of survival value to animals and man. The mechanism of clot formation will be discussed later in the chapter.

10 The blood plays an important part in protecting the body from bacteria and other organisms that can cause disease or other abnormal conditions. Some kinds of white blood cells afford protection by ingesting bacteria or other foreign matter appearing in the bloodstream. Another phase of protection is acquired resistance to infections, or acquired immunity. It is well known that in many types of infections the body develops defensive mechanisms that overcome bacteria or neutralize their toxic effects. Vaccinations are a type of protection obtained when dead or attenuated etiological agents of disease, or their products, are introduced into the body. After an incubation period the body builds up an immunity, much the same as if the infectious organisms themselves had entered the body.

Specific antibodies also can be produced in the blood to protect against specific infections. Foreign proteins introduced into the body are called *antigens*. They cause the formation of protective substances called *antibodies*. The protein fraction of the blood called *gamma globulin* contains antibodies. Antibodies operate in several distinct ways: They may establish immunity to a disease through immunological reactions. *Antitoxins* are antibodies that tend to neutralize the toxins produced by certain types of disease organisms, for example, diphtheria and tetanus bacilli. *Agglutinins* cause a clumping of certain kinds of bacilli. They may also cause the clumping of red cells of the blood if in a transfusion, the blood type of the donor is not compatible with that of the recipient. *Lysins* are dissolving agents. If the blood serum of one species of animal is introduced into the blood of another species, the recipient's red cells may be destroyed. The lysin in this case is a *hemolysin*. Opsonins are globulin molecules that combine with foreign materials making it possible for phagocytes to adhere to the surface, thus facilitating phagocytosis. The chemistry of the blood functions in many ways to form a protective mechanism for the welfare of the body as a whole.

The protein fraction of the blood plasma that has become well known through its use in combating disease is *gamma globulin*. It is the fraction of the blood that contains most of the known antibodies. Its removal from blood plasma does not materially reduce the effectiveness of plasma in the treatment of injuries resulting from wounds or in the treatment of shock. The source of supply, therefore, comes largely from blood donated for emergency uses rather than for combating disease.

It is known that, in the general population, many persons have had various diseases and have developed antibodies against them. The chemical fractionation of pooled blood removes gamma globulin containing these antibodies. When injected into a susceptible person, a specific antitoxin or antiserum can give passive immunity for a few weeks. This procedure can be used to protect children especially, during local epidemics.

The reaction to *allergens* is responsible for various allergies, such as asthma, hay fever, urticaria (hives), and eczema. Many persons become sensitized to a great variety of substances, usually proteins, such as pollen, certain food substances, animal fur, feathers, insect bites and stings, serums, molds, and dusts. The tissues most commonly affected are those of the respiratory tract, the digestive tract, and the skin.

INJURIOUS SUBSTANCES DISTRIBUTED BY THE BLOOD The blood does not always carry substances beneficial to the body. It also can be a pathway for foreign substances that can have a deleterious effect. Alcohol and other drugs, some venoms, the metastases of cancer, certain parasites, some forms of bacteria, and toxic substances in general are distributed by way of the bloodstream.

QUANTITY, OR VOLUME, OF BLOOD

The amount of blood in the body has been measured in various ways. Naturally the volume of blood may be expected to vary with the size of the body. It has been estimated at $\frac{1}{20}$ to $\frac{1}{13}$ of the body weight. The blood volume of a man of average size is about 5 quarts.

An outline of the principal constituents of normal human blood is as follows:

Plasma
   Water, 92 percent
   Solids, 8 percent
   Inorganic chemicals: sodium, calcium, potassium, magnesium, chloride, bicarbonate, phosphate, sulfate
   Organic chemicals
       Proteins: serum albumin, serum globulin, fibrinogen
       Nonprotein nitrogenous substances: urea, uric acid, creatine, creatinine, ammonium salts, amino acids
       Nonnitrogenous substances: glucose, fats, cholesterol
   Hormones
   Gases: oxygen, carbon dioxide, nitrogen
Cells:
   Erythrocytes, or red cells; leukocytes, or white cells; blood platelets, or thrombocytes

PLASMA The liquid portion of circulating blood is called the *plasma*. It is a straw-colored fluid, very complex chemically, containing a wide variety of substances. The red cells, white cells, and blood platelets float in this liquid medium. In this respect the blood may be regarded as a liquid tissue; it contains cells, but the intercellular substance is a changing liquid rather than some more substantial building material.

Most of the functions of the blood previously mentioned affect the plasma directly. Even though the blood is continuously engaged in transporting absorbed food products and receiving the waste products of cell metabolism, its chemical content is fairly constant. The plasma is about 92 percent water; the remaining 8 percent of materials in solution make blood thicker than water, as the saying goes. Its specific gravity is greater than 1, more nearly 1.025 as an average. This chemical mechanism is largely self-adjusting, much the same as the maintenance of water balance between the blood and the tissues. As some of the chemical elements of the blood are used by cells in metabolism, more of these same elements are absorbed into the blood from food sources and the chemical balance is maintained. We have seen that waste products of metabolism are steadily absorbed by the blood and are just as regularly removed by excretory organs. We have mentioned the buffering action of the blood, which helps to preserve its chemical balance.

The blood and tissue fluids have been called the *internal environment*. The concentration of inorganic salts in the internal environment resembles that found in seawater, which constitutes the external environment for a great many animals. It is considered by many that life arose in the sea and that a great deal of development took place there. The salt ions of the blood are mostly chloride, bicarbonate, phosphate, and sulfate of sodium, calcium, potassium, and magnesium. In order to maintain a proper acid-base equilibrium, these salts must be present and maintained in proportion to one another. Physiologists have known since the experiments of Sydney Ringer were published, in the period around 1885, that there is a salt balance in the blood. Physiological salt solutions are used to maintain the internal environment of experimental animals during demonstrations, operations on animals, and various other laboratory procedures.

Ringer's solution has many modifications, but it is essentially as follows:

NaCl	0.65 gram	$NaHCo_3$	0.02 gram
KCl	0.014 gram	NaH_2PO_4	0.001 gram
$CaCl_2$	0.012 gram	Water	to 100 milliliters

This solution, or a modification of it, is used more often as a physiological solution for invertebrates and for some vertebrates such as amphibians.

A more recent modification is the Ringer-Locke solution, used especially in mammalian physiology. Since the total salt concentration is higher in the blood of mammals than in lower classes of vertebrates, the Ringer-Locke solution has a higher salt concentration more nearly equal to that of mammalian blood. Through their buffering action, the salts of the blood aid in maintaining an acid-base balance between the blood and the tissues; they are also concerned in maintaining water balance in order that blood cells and tissue cells can carry on their physiological processes in a normal manner.

Organic substances in the blood plasma include such proteins as serum albumin, serum globulin, and fibrinogen. These proteins are not food proteins in the sense that they are directly absorbed from food sources or that they are food proteins being transported to the tissues as such. Fibrinogen is formed in the liver, and most other plasma proteins are thought to be formed there also. The blood proteins, present in a colloidal sol state, exert considerable osmotic pressure, up to 25 to 30 mm Hg. This osmotic pressure is a considerable factor in maintaining the water balance between the blood and the tissues and in regulating the volume of the blood. The proteins give viscosity to the blood, a factor in the maintenance and regulation of blood pressure. Serum globulin is concerned with antibody formation—the reaction of the blood to toxins formed by bacteria or to foreign proteins introduced into the blood. Fibrinogen is essential to the clotting mechanism.

Nonprotein nitrogenous substances found in the blood include urea, uric acid, creatine, creatinine, and ammonium salts. These substances represent breakdown products of protein metabolism and are carried by the blood to the organs of excretion.

Protein foods are reduced to amino acids during the process of digestion and are absorbed as such. Amino acids, the building blocks for all proteins found in the body, are therefore present in the blood plasma.

Glucose (or blood sugar), fats, and cholesterol are nonnitrogenous substances present in the blood. Glucose is a simple sugar derived by digestion from more complex sugars. It is absorbed from the intestine and transported to the liver, where much of it is stored as a complex polysaccharide called *glycogen*. A considerable amount of glucose is absorbed from the blood and stored as glycogen in muscle tissue also.

Though the role of glucose in nutrition is well recognized, it also acts as a physiological constant in the blood. The sugar level of the blood is fairly constant at an average concentration of about 0.1 percent (80 to 120 milligrams per 100 milliliters). A reduction in the blood-sugar level may cause weakness, fainting, or more serious consequences. The kidneys excrete sugar if the sugar level of the blood becomes too high.

Fats are carried by the blood, as well as several fatlike substances, such as cholesterol, and the phospholipids. Fats are absorbed largely by way of the lymphatic system. They break down during digestion into glycerol and fatty acids. Cholesterol is distributed in tissues throughout the body but is found in considerable concentration in nerve tissue, adrenal glands, and skin. It is excreted in the bile.

The blood plasma contains hormones, the secretions of ductless glands. It also contains the chemical substances concerned with the clotting of the blood.

While dissolved gases are transported by the blood, oxygen and carbon dioxide are more closely related to the hemoglobin of red cells, as we shall see later. Only a small amount of carbon dioxide is carried in solution as carbon dioxide, even though it is continuously produced as a waste product of metabolism and constantly absorbed by the blood. After forming carbonic acid, it is buffered by hemoglobin and salts such as sodium phosphate, which remove carbonic acid as such by entering into chemical combination with it. Hemoglobin is one of the chief substances concerned in the transportation of both oxygen and carbon dioxide. Nitrogen is carried in the plasma as an inert gas.

RED CELLS: ERYTHROCYTES

Carried along in the fluid portion of the blood are tiny red cells so numerous that the blood itself appears red. These cells are called *erythrocytes*. As they arise in the red marrow of bone, they are nucleated and are called *erythroblasts*, but shortly before entering the bloodstream, they usually lose the nucleus and become highly specialized cells, that is, red corpuscles. Their

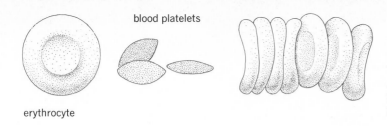

blood platelets

erythrocyte

FIGURE 13.1
Red blood cells and blood platelets.

size is fairly constant at about 7.5 to 8.6 microns in diameter and about 2 microns thick (1 micron equals $\frac{1}{1,000}$ of a millimeter, or $\frac{1}{25,000}$ of an inch). The red cell is a biconcave disk, thinner in the middle than at the edge. When observed singly under a microscope by transmitted light, the cells are only faintly colored. Individually they appear slightly yellowish in color (Figure 13.1).

A chemical factor called *erythropoietin* controls the production of erythrocytes. It is a glycoprotein and generally thought to be produced by the kidneys, although the liver and perhaps other tissues may produce small amounts. It stimulates the production of erythroblasts in the early stages (stem cells) as they develop in red bone marrow.

The *number of erythrocytes* per cubic millimeter of blood is usually stated as an average of 5 million red cells per cubic millimeter for men and 4,500,000 for women. It is well known, however, that active young men frequently have more than 5 million red cells per cubic millimeter. A more accurate estimate, therefore, would be 5,450,000 red cells per cubic millimeter for men and 4,750,000 for women. The blood-cell count is also higher in a newborn infant than it is in older children or adults. Ascending to high altitudes increases the number of red cells in the bloodstream. Other factors causing an increase in the number of red cells are concerned with muscular exercise and with a rise in the environmental temperature. Under these conditions additional red cells can be discharged from the spleen.

HEMOGLOBIN Erythrocytes derive their color from a complex protein called *hemoglobin*. This substance is composed of a pigment, *heme*, containing iron, and the protein *globin*. Hemoglobin has the power to attract oxygen molecules and to hold them in a loose chemical combination known as *oxyhemoglobin*. It is said, therefore, to have a chemical affinity for oxygen. The structure of the hemoglobin molecule has been successfully analyzed by x-ray diffraction and chemical methods. It consists of four folded polypeptide chains of amino acid units. The four chains form the globin, or protein, part of the molecule. In addition there are four atoms of iron, each associated with a pigment, or heme, group of atoms. The heme group provides the red color of the blood and also its oxygen-combining ability. The iron atoms are bivalent or in the ferrous state. It has been estimated that one erythrocyte contains approximately 280 million molecules of hemoglobin (Perutz).

As the blood passes through a capillary network in the thin air sacs of the lungs, oxygen enters into a loose chemical combination with hemoglobin (oxyhemoglobin) and is carried to the tissues. There, as the blood passes through tissue capillaries, the hemoglobin loses oxygen to the tissues and is then referred to as *deoxyhemoglobin*. Arterial blood, after passing through the lungs, is a somewhat brighter red than that found in the veins, but venous blood is never blue. The blue color of veins close to the surface is due to the absorption of red and yellow rays of light and the reflection of blue and

Red cells: erythrocytes

323

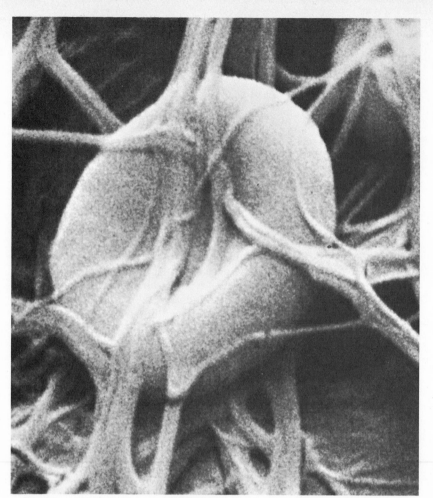

FIGURE 13.2
Scanning electron micrograph of
an erythrocyte enmeshed in fibrin.
Original magnification, ×10,000.
[*Courtesy Emil Bernstein and
Eila Kairinen, Gillette Com-
pany Research Institute, Rock-
ville, Maryland. Cover photo,
Science, 173, Aug. 27, 1971.
Copyright, 1971, by the
American Association for the
Advancement of Science.*]

green light. Erythrocytes not only function to carry oxygen to the tissues, but indirectly they also function in carrying carbon dioxide away from the tissues (Figure 13.2).

The number of red cells in the circulation remains fairly constant. The average life of a red cell is about 120 days but they are constantly replaced. The maintenance of this constancy affords another example of homeostasis. If there is a lack of oxygen in the tissues (hypoxia), erythropoietin is produced and carried to the red bone marrow by the blood. The red bone marrow is then stimulated to produce more erythrocytes, more carriers of hemoglobin. This is another example of a negative-feedback system regulating the supply of oxygen to tissues as oxygen is utilized and becomes deficient. Other factors may influence the production of red cells, such as some endocrine secretions and vitamin B_{12}.

Anemia A condition of the blood in which there is a reduction of the number of red cells or reduced hemoglobin is known as *anemia*. Often both the red-cell count and the percentage of hemoglobin are reduced. The typical patient with anemia appears pale and weak and has a loss of energy. The oxygen-carrying capacity of the blood is reduced by the loss of hemoglobin.

The blood

a
homozygous

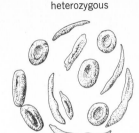

b
heterozygous

There are many kinds of anemia. Basically anemia may be caused by inability of the body to manufacture enough hemoglobin or failure of the red marrow of bone in the production of erythrocytes. In the first type there is a lack of an adequate amount of iron in the diet or its absorption and utilization in the production of hemoglobin. Iron absorbed from the intestine combines with a protein to form a compound known as *ferritin*. It is then found in the liver, bone marrow, and in minute amounts in other tissues. In some kinds of anemia the number of red cells may be normal, but the percentage of hemoglobin is greatly reduced. Occasionally anemia is caused by an abnormal rate of destruction of red cells from chronic bleeding or from some hemolytic substance in the blood.

The failure of the red marrow of bone in the production of erythrocytes results in a very low red-blood-cell count, but individual cells may be normal or large and contain normal or even above normal amounts of hemoglobin. *Pernicious anemia* is essentially of this type. Great progress has been made in understanding and treating anemia. The discovery of an antianemic factor in liver has proved of great benefit to victims of pernicious anemia and similar forms. The understanding of the relation between the erythrocyte maturation factor and vitamin B_{12} has helped greatly in interpreting the factors involved. The reader is referred to discussions of folic acid and vitamin B_{12} in Chapter 18.

Sickle-cell anemia   We have seen that normal adult hemoglobin A (Hb^A) is composed of four polypeptide chains and also contains four heme groups. The four polypeptide chains consist of two alpha chains and two beta chains. The number and location of the amino acids composing each chain have been determined; each alpha chain contains 140 amino acids, and each beta chain consists of 146.

In 1949 Linus Pauling and his coworkers discovered an abnormal hemoglobin (Hb^S) associated with an inherited defect. When the defect is present as a homozygous recessive trait, many red cells of the blood exhibit a peculiar thickened, elongate appearance called *sickling*. The abnormal cells are not strictly half-moon–shaped. They are often misshapen red cells with strands protruding that tangle other blood cells and tend to block small blood vessels in the tissues of affected persons (Figure 13.3). The abnormal cells are rapidly destroyed, giving rise to a condition called *sickle-cell anemia*. The blood of heterozygous individuals ($H^A H^S$) may show some sickling, but these individuals are essentially normal and are not subject to sickle-cell anemia. They do, however, carry the *sickle-cell trait*. Their hemoglobin is about 60 percent normal Hb^A and 40 percent Hb^S. Sickled cells are not very efficient oxygen carriers although individuals with sickle-cell trait do not

Red cells: erythrocytes

325

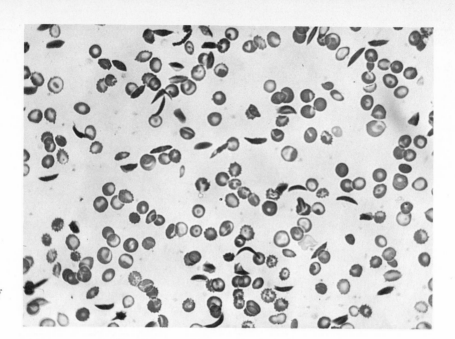

suffer any great deficiency (Figure 13.4). The abnormal hemoglobin of sickle-cell anemia can be identified by electrophoresis. A clinical test for sickle-cell trait is also available. It has been discovered, in addition, that sickled cells regain their normal appearance, temporarily, under hypobaric pressures of 200 to 300 atmospheres.

It is of considerable interest to geneticists, physiologists, and biochemists that the only chemical difference between the hemoglobin of normal adults and that of individuals with sickling trait is the substitution of the amino acid valine for glutamic acid at position 6 in the beta chain. In the heterozygous individual the substitution produces sickle-cell trait, and in the homozygous individual it produces the very serious condition of sickle-cell anemia.

Sickling of hemoglobin occurs almost exclusively in Negroes of African descent. Various surveys indicate that American Negroes, as a group, have about two-thirds African ancestry and one-third Caucasian ancestry. It has been estimated that the sickle-cell trait is present in around 9 percent of American Negroes. If this estimate is correct, the homozygous condition would be comparatively rare, perhaps as low as 0.2 percent.

The incidence of sickle-cell trait in the population of certain areas in Africa is relatively high, perhaps as great as 20 percent. It was determined that in areas where subtertian malaria is endemic there is a much higher incidence of sickle-cell trait than in regions free of malaria. It appears that the sickle-cell trait affords a selective advantage to those individuals carrying the trait over those with normal hemoglobin relative to their resistance or susceptibility to malaria. The mechanism affording this advantage is not known. In areas free of malaria there is, of course, no selective advantage, and the frequency of sickle-cell trait should gradually decline.

Polycythemia The blood condition known as *polycythemia* is characterized by the production of a greater-than-normal number of red cells. The erythrocyte-forming areas become overactive. As a result there is an increased viscosity of the blood and a rise in blood pressure. The heart may become overloaded, and there is also a danger of blood clots.

THE HEMOCYTOMETER It is possible to make a fairly accurate estimate of the number of red cells per cubic millimeter by actually counting a limited number of cells as they are spread out on a ruled microscope slide called a *hemocytometer*. This is a glass slide ruled so that in the center there are squares that measure $\frac{1}{20}$ by $\frac{1}{20}$ of a millimeter. The counting area has a depth of $\frac{1}{10}$ of a millimeter.

If undiluted blood were spread over the counting area, the red cells would be too numerous to count accurately and the blood would clot, accompanied by a breakdown of red cells. The blood is therefore diluted with a suitable physiological saline solution, which will overcome these difficulties. The dilution commonly used is 1:200.

With the diluted blood spread over the counting area, the number of cells in several squares can easily be counted. The cubic measurement of a single square is $\frac{1}{20} \times \frac{1}{20} \times \frac{1}{10} = 1/4,000$ of a cubic millimeter. The problem is essentially this: If there are a certain number of red cells in one square, which equals 1/4,000 of a cubic millimeter, how many are there in the whole cubic millimeter? Of course, the dilution must be considered, and it is customary to count the number of cells in several squares and take the average for any one square.

White cells also can be counted in essentially the same manner. A lower dilution is used, and the squares of the counting chamber are larger, since there are not nearly so many white cells as red cells in a cubic millimeter of blood. A reader interested in this technique will find detailed instructions in laboratory manuals and in clinical textbooks. The average well-informed citizen, however, may be interested in understanding what his doctor means when he refers to a *blood count*. Automation has provided modern electronic blood-counting instruments that are capable of estimating both red and white cells rapidly and accurately.

Hematocrit is the percentage of red cells per 100 milliliters of whole blood by volume after it is centrifuged. The reading is taken in a special graduated tube.

WHITE CELLS: LEUKOCYTES

White cells are nucleated and somewhat variable in size and shape. They are far less numerous than red cells, numbering 5,000 to 9,000 per cubic millimeter. If 7,000 white cells per cubic millimeter is taken as an average, red cells outnumber the white by about 700 to 1. Some kinds of leukocytes, especially neutrophils and monocytes, exhibit ameboid movement and are actively phagocytic. Their lifespan is probably only a few days.

There are two major groups of white cells; the first group includes those cells that have granules in the cytoplasm and possess a nucleus of two or three lobes. They are called *granular*, or *polymorphonuclear*, *leukocytes*. According to some workers, this is the only group properly referred to as leukocytes (Table 13.1).

The second group includes those cells that do not have granules in the cytoplasm and in which the nucleus is more or less spherical in shape. These are the *nongranular white cells;* the grouping includes *lymphocytes* and *monocytes*.

White cells: leukocytes

FIGURE 13.5
(See facing page.) Cells from smear preparation of normal human blood, Wright's stain. In the center, adult red cells and a few blood platelets around a polymorphonuclear neutrophil. Upper left, two basophils (purple granules) and two eosinophils (red granules). Upper right, large and small lymphocytes. Lower right, a group of six monocytes. Lower left, a group of six neutrophils. (From Roy O. Greep, "Histology," McGraw-Hill Book Company, New York, 1973.)

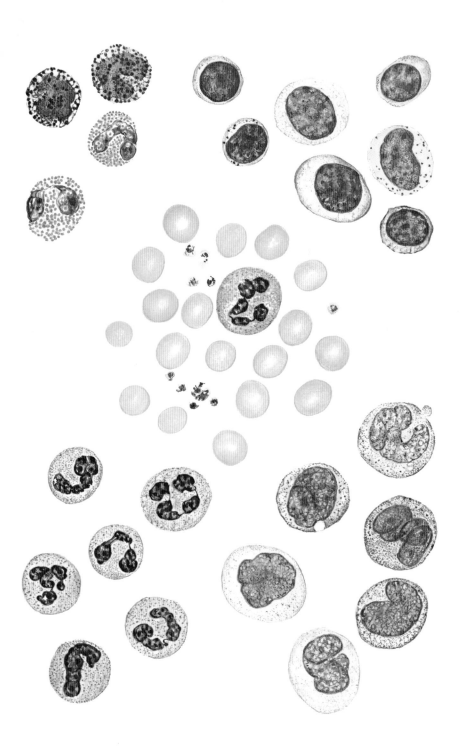

TABLE 13.1
Classification of blood cells

Name	Normal no.	Diameter, microns	Description
Erythrocytes	4.75–5.5 million per cu mm	7.5–8.6	Biconcave disk, nonnucleated
Leukocytes	5,000–9,000 per cu mm		
Granular			
Neutrophils	Approx. 65% of total no. of white cells	10–12	Fine granules in cytoplasm usually stain to lavender; two- or three-lobed nucleus
Eosinophils	2–4 of total no.	12	Large cytoplasmic granules usually stain bright eosin red; nucleus usually bilobed
Basophils	0.2–0.5% of total no.	10	Large cytoplasmic granules usually stain deep, dull blue or purple; thick bilobed nucleus
Nongranular			
Lymphocytes	20–25% of total no.		
Large		12	A thin rim of nongranular cytoplasm, usually stains light blue; large globular nucleus stains dark blue or purple
Small		9	
Monocytes	5–7% of total no.	15–20	Resemble large lymphocytes, but with relatively more cytoplasm staining light blue; nucleus large and deeply indented

Granular leukocytes arise from bone-marrow cells called *myeloblasts*. Lymphocytes in the fetus are thought to arise first in the thymus. Later they are found in lymph nodes, spleen, and other lymphoid tissues as well as in bone marrow.

GRANULAR LEUKOCYTES *Neutrophils* The most abundant type of white cell is the neutrophil. These cells constitute 65 to 70 percent of the total number of white cells. Averaging about 10 microns in diameter, they are somewhat larger than erythrocytes. The nucleus appears to vary with age. The younger mature cells have a bilobed or trilobed nucleus with thick connections between the lobes, whereas in older cells there are more lobes in the nucleus and the connections between the lobes are thin. The granules of the cytoplasm are very fine, and when stained with Wright's or some similar stain, they take both acid and basic stains and appear as a lavender or lilac color (Figure 13.5).

Neutrophils are active ameboid cells and are also phagocytic. They are capable of pushing their way between cells in the wall of the smallest

White cells: leukocytes

blood vessels and moving about through the tissues. The movement of leukocytes through the capillary wall is called *diapedesis. Phagocytosis* refers to the ability of these cells to ingest bacteria or other foreign bodies. They contain protein-digesting enzymes that enable them to digest most of the materials they engulf. Since they are capable of ingesting other cells, they are called *phagocytes.* Their phagocytic activity is important in helping to rid the body of injurious bacteria. Much of the protective function of antibodies is their ability to react with infectious particles, slowing them down or immobilizing them, making phagocytosis more effective.

Eosinophils Resembling neutrophils, eosinophils are slightly larger, and the nucleus is usually bilobed. The granules in the cytoplasm are larger and stain a bright red with acid dyes such as eosin. There cells make up only 2 to 4 percent of the total number of white cells in the blood, but in the tissues they can congregate in considerable numbers. According to some workers, eosinophils are considered to be less active and not so highly phagocytic as neutrophils.

There is a rise in the number of eosinophils in some cases of allergy, possibly in response to toxic substances released by the allergic reaction. It is well known that the eosinophil count rises sharply in reaction to blood parasites such as *Trichinella.* The condition is called *trichinosis* and results when a minute encysted roundworm is eaten in raw or undercooked pork.

Basophils Basophils have a bilobed nucleus, and the large cytoplasmic granules stain a deep blue with basic stains such as methylene blue or with Wright's stain, which contains methylene blue. Basophils constitute only 0.5 percent of the white cells of the blood. Not much is known about the motility of phagocytic activity of basophils, but they are considered to be less active than eosinophils. Basophils in the blood are said to contain histamine and a heparin-like substance. Histamine dilates capillaries and often permits fluid to move through the capillary wall into the tissues; heparin is an anticoagulant of the blood. Apparently tissue basophils become the mast cells of the tissues. The large granules of mast cells are thought to store enzymes.

NONGRANULAR LEUKOCYTES *Lymphocytes* Lymphocytes are considered to be of two distinct sizes, small and large. The small lymphocytes are the smallest of the white cells and constitute 20 to 25 percent of the total number of all leukocytes. The nucleus is comparatively large and spherical and is surrounded with a thin covering of cytoplasm. The cytoplasm is nongranular and stains a light-blue color, while the nucleus is dyed a much deeper blue or purple with Wright's stain. The large lymphocytes represent only about 3 percent of the total of white cells; they contain more cytoplasm, and their nucleus is large, oval, and indented. As we have indicated, lymphocytes are produced in lymphoid tissues such as the thymus, spleen, and lymph nodes. They also arise from bone marrow.

Small lymphocytes formerly were thought to be the major group of cells capable of producing antibodies and so were called the immunologically competent cells. Now it appears that *plasma cells* are very active in synthesizing antibodies of the circulating blood. It is thought that plasma cells arise from lymphoid cells or possibly from medium-sized lymphocytes in lymph nodes or lymphoid tissues. Small lymphocytes have only a thin covering of cytoplasm over the large nucleus, but in the development of the plasma cell, the plasmoblast has more cytoplasm than a lymphocyte and the mature plasma cell has considerable cytoplasm, containing a well-developed endoplasmic reticulum studded with ribosomes. The ribosomes are thought to produce immunoglobin in much the same manner that the ribosomes of other kinds of cells synthesize protein.

It seems likely that fixed macrophages of the reticuloendothelial system do play some part in absorbing antigens, although their precise function is not clear. The antigen in some way stimulates the rapid development of plasma cells and lymphocytes. It appears that in the fetus the thymus is the essential structure capable of establishing an immunologic response in lymphoid precursor cells or in thymic lymphocytes. There may be a thymic humoral factor, possibly a hormone, that influences the development of lymphocytes in the spleen and lymph nodes. These lymphocytes then become capable of multiplying and producing antibodies if an antigen is introduced.

Experiments with mice indicate that the thymus produces a hormone capable of giving lymphoid cells the ability to become competent immunologically. Morphological studies indicate that the first lymphocytes that can be clearly recognized arise from the epithelium of the thymus. The thymus is active in the fetus and by birth is producing lymphocytes that seem to be capable of reacting selectively with antigens. When such lymphocytes react with a specific antigen they are said to be committed to this particular antigen and become capable of multiplication.

Some investigators do not recognize a thymic humoral factor and believe that immunologically competent thymic cells simply migrate to the spleen where they multiply. Studies along this line indicate that the spleen does not differentiate its own lymphocytes, as the thymus does, but depends on migration from other sources for immunocompetent cells. There is an indication (Auerbach) that cells may migrate from the thymus to bone marrow and then to the spleen.

Lymphocytes are known to be able to transform into large phagocytic cells called *macrophages*. The macrophage may be simply a lymphocyte that has ingested other cells as well as antigen and has become greatly enlarged in the process.

It is evident that the role of the thymus in producing immunologically competent cells is not thoroughly understood, but considerable progress has been made. Once the cells in the spleen and lymph nodes are activated, the thymus has performed its principal function in constituting and regulating the immunity mechanism and it begins to regress.

Monocytes Closely resembling large lymphocytes in appearance, monocytes are about 15 microns in diameter, with a large, deeply indented nucleus. The nucleus stains deep blue or purple with Wright's stain. There is relatively more cytoplasm than in the large lymphocyte. There are comparatively few monocytes in the blood—about 5 percent of the total white-cell count. Monocytes are actively motile and phagocytic. It is thought that they function in contributing to the repair and reorganization of tissues.

Monocytes and macrophages are capable of engulfing old, worn-out neutrophils, mast cells, antigens, and particles of tissue in the process of cleaning up an area of inflammation or infection after the initial stages have been passed and recovery is in progress.

THE FUNCTION OF LEUKOCYTES IN INFLAMMATION AND DISEASE The white cells of the bloodstream are usually in transit from their place of origin to their destination in the tissues. Unlike the red cells, which must remain within a closed circulatory system, the white cells are able to pass through the

White cells: leukocytes

capillary wall and move about through the tissues. They are most important in protecting the body against the invasion of bacteria, since by virtue of their phagocytic ability they are able to ingest great numbers of bacteria or other particles of foreign matter.

The protective action of white cells can be illustrated by following the sequence of the events when a splinter becomes lodged under the skin. There is some initial injury to surrounding tissues, and probably a minor infection from bacteria introduced on the splinter. Toxic products released by bacteria tend to destroy tissues locally, and blood vessels in the immediate vicinity are affected. They tend to dilate, thus bringing more blood to the affected area. Therefore the area reddens somewhat and feels warm, since the temperature of the blood is warmer than the normal skin surface.

White cells pass through capillary walls and congregate in great numbers in the infected area. The white cells are actively phagocytic and rapidly ingest bacteria, as well as particles of tissue cells. Some of the phagocytes are destroyed by toxic substances liberated by the bacteria. The breakdown of phagocytic cells liberates their digestive enzymes.

The festering that often occurs around the base of a splinter is called *suppuration*, or the formation of pus. The exudate from blood vessels, dead tissue cells, bacteria, and living and dead leukocytes form a more or less liquid material called *pus*. In the normal course of events the splinter becomes loosened and can be easily withdrawn. The white cells clean up the infected area, and regeneration of the skin and tissue cells restores the area to normal.

The biological events associated with inflammation are very complicated. Phagocytic cells have been observed to stick to the inner surface of the capillary wall at the site of inflammation. Soon they begin to move through the wall by diapedesis. Outside the wall in the exudate they are attracted to bacteria by a process called *positive chemotaxis*. Presumably, some chemical attractant is released in the area, but the nature of the chemical is not known. The polymorphonuclear cells in the inflamed area are predominantly neutrophils. Lymphocytes and monocytes are present also in great numbers. Tissue macrophages increase in number and move about through the injured area.

Sometimes the inflammation is deep-seated, as in the case of an infected appendix. The white-cell count may rise remarkably as the protective elements of the body attempt to control the infection. In the case of an abscess around the root of a tooth, tonsillitis, or appendicitis, and in many infectious diseases, the white-cell count may rise to 13,000, 18,000, or even higher than 30,000 white cells per cubic millimeter. This condition is called *leukocytosis*. A rise in the white-cell count affords excellent corroborative evidence to a physician that there is an infection within the body. Often the severity of the infection is indicated by the number of leukocytes present in the blood.

A rise in the white-cell count does not mean ordinarily that the various kinds of white cells retain their normal numerical relation to each other. Inflammation and certain infectious diseases, especially those involving the round or coccus forms of bacteria, cause a rise in the number of neutrophilic cells. Some chronic infections cause an increase in the number of lymphocytes, and, as we have indicated earlier, some cases of asthma, certain skin diseases, and some cases of infection by parasitic roundworms (*Trichinella*) cause an increase in the number of eosinophils (eosinophilia).

A *differential count of white cells* indicates the proportion of each kind

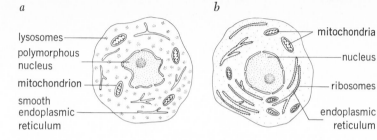

a
b

lysosomes

polymorphous
nucleus

mitochondrion

smooth
endoplasmic
reticulum

mitochondria

nucleus

ribosomes

endoplasmic
reticulum

FIGURE 13.6

a A megakaryocyte, diagrammatic;
b a plasma cell, diagrammatic.

present. Ordinarily, the number of each kind of white cell in 100 cells is counted. After counting 100 or 200 cells, the percentage of each type is determined.

Some diseases, especially pneumonia and typhoid fever, cause a reduction in the white-cell count. Certain drugs appear to reduce the number of white cells. Destruction or degeneration of red bone marrow also causes a reduction of blood cells. *Leukopenia* is the term used to indicate a reduction of white cells below normal. Reduction in the number of white cells indicates lowered resistance to infection.

The normal rise in the number of white cells as the result of infection, as we have seen, is properly called *leukocytosis*. There is, however, an abnormal condition in which white cells are produced in tremendous numbers but not in response to any known infection. This condition is called *leukemia*. The red bone marrow, for some unknown reason, produces white cells at an uncontrolled rate. In another type of leukemia, lymphoid tissues are overactive in the production of lymphocytes. A deficiency in the number of red cells (erythropenia) is frequently associated with leukemia.

Infectious mononucleosis is a disease affecting, in the majority of cases, young persons, in the age group of sixteen to twenty-five years. It is generally considered to be a viral disease, characterized by fever, enlargement of cervical lymph nodes, and lymphocytosis with atypical lymphocytes. In many cases the upper eyelids become puffy and sag downward. The spleen is usually enlarged. Normally healthy young persons can expect to recover and resume their usual activities in 3 or 4 weeks. It is not contagious by exposure in the sense that a disease such as measles is contagious.

BLOOD PLATELETS

Minute, granular, disk-shaped objects in the blood are called *blood platelets* (thrombocytes). They arise from fragmentation of giant cells in the red marrow of bone. These giant cells are derived from cells of the reticulo-endothelial system and are called *megakaryocytes* (Figure 13.6). Even though they arise by fragmentation of a large cell, they are not irregular in shape; platelets are normally round disks. They are much smaller than erythrocytes, are nonnucleated, and average around 250,000 to 400,000 per cubic millimeter. Blood platelets contain a number of chemical substances; probably the most important are the phospholipid cephalin, a prothrombin activator similar to tissue thromboplastin but weaker, and serotonin, a vasoconstrictor substance. Blood platelets tend to stick together if there is a vascular injury and so are able to plug small breaks in capillaries. Platelets contain no nuclei and their life-span is short, probably 5 to 10 days.

Blood platelets

The ability of the blood to form a clot protects the individual from excessive bleeding from minor wounds; it undoubtedly has played an important part in survival. Clotting is essentially a chemical and physical process, although many of the details are still unknown. Apparently there is a thromboplastic substance, *tissue thromboplastin*, that exudes from injured tissues and causes clotting when it interacts with prothrombin of the blood. The breakdown of blood platelets releases phospholipids (cephalins), which, interacting with serum proteins, form plasma thromboplastin. Thromboplastin is not a single chemical substance; it is a name given to a group of substances capable of initiating an interaction that causes prothrombin to release active thrombin in the clotting process.

Blood can be drawn carefully so that it does not touch injured tissues. If it is placed in a container such as a test tube or glass beaker, the blood platelets break down and clotting occurs promptly. If the walls of the container are paraffined so they are not wettable by an aqueous solution such as blood, the blood platelets disintegrate slowly and clotting is delayed. This seems to be the expression of a physical principle that platelets tend to break down on rough surfaces or on surfaces that can be wet by aqueous solutions. Prompt refrigeration will prevent blood from clotting, and the platelets remain intact.

The amount of thromboplastin that can be produced from the disintegration of blood platelets appears to be remarkably small compared with the volume of blood plasma. It is considered, however, that a globulin in the blood, called the *antihemophilic factor* (AHF), factor VIII, enhances the breakdown of blood platelets, causing them to disintegrate rapidly and completely once the clotting process is set in motion.

Another factor in the clotting process is the presence of free calcium ions. When sodium oxalate is added to a sample of blood, it goes into solution and reacts with the calcium to form insoluble calcium oxalate, which precipitates. Fluorides or citrates can also be used as anticoagulants. They also unite with the calcium but form soluble compounds.

In uniting with the calcium of the blood, these substances remove free calcium ions and in this way prevent calcium from taking any part in the clotting reaction. Blood treated with these substances will remain unclotted; if ionizable calcium salt is added later, a clot will form.

Accelerator globulin, also called labile factor or factor V, which is formed in the liver, is an essential factor in the clotting process. It activates the change of prothrombin to thrombin.

Several mechanisms have been proposed to explain the function of various factors involved in the clotting process. The factors are usually arranged in stepwise sequence. The concept of arranging the various clotting factors in pairs, so that one factor acts as an enzyme and the other appears as the substrate, seems of special importance. Davie and Ratnoff view this sequence as a waterfall or cascade in which a series of factors are converted to enzymes, one after another, until prothrombin is converted to thrombin.

The clotting process is actually much more complicated than the following outline would indicate. Several proposed factors are not mentioned, and the complete interaction of all the chemical entities in the clotting process is not well understood.

Thrombin is an enzyme that plays an active part in the clotting reaction. It is present in the blood in an inactive form called *prothrombin*. The phases of the clotting process may be outlined as follows:

Phase 1 Blood-platelet breakdown (or injured-tissue substance) produces thromboplastin.

Phase 2 Thromboplastin, interacting with free calcium ions, accelerator globulin, and several other factors, converts prothrombin to active thrombin.

Phase 3 Thrombin interacts with fibrinogen. Fibrinogen comes out of solution as threads of fibrin. Fibrin plus blood cells and blood platelets forms a clot.

FIBRINOGEN AND FIBRIN Fibrinogen, a protein in solution in the blood plasma, is formed in the liver. As a protein it can be coagulated by heat or precipitated by certain salts, but its normal converting agent is the enzyme thrombin. When thrombin is present in its active state, fibrinogen comes out of solution as strands of a material called *fibrin*, which forms the framework of the clot. Blood cells and blood platelets are held in this framework as the clot develops.

Defibrination can be accomplished mechanically by whipping blood and causing the fibrin threads to collect on the brush or other object with which it is being whipped. When fibrin is removed, the blood no longer has the ability to clot.

There is evidence to suggest two systems for blood clotting, an intrinsic system and an extrinsic system. The two systems vary somewhat, particularly in the use of certain cofactors. The intrinsic system involves a series of reactions in which the blood platelets in cooperation with several cofactors, plus calcium ions, initiate the clotting sequence. The platelets appear to release a substance similar to thromboplastin, but this platelet substance apparently requires the aid of several cofactors in order to induce clotting.

In the extrinsic system, tissue factor or thromboplastin, aided by several cofactors plus calcium ions, initiates the clotting process. In the final stages, from prothrombin to fibrinogen and fibrin, both systems are essentially alike. A simplified diagram follows.

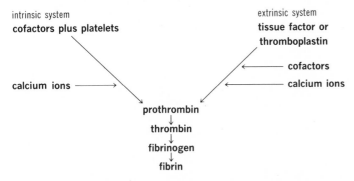

HEPARIN Blood does not ordinarily clot internally in the body. It is assumed that, unless there is access to injured surfaces, not enough thromboplastic substances are liberated to convert prothrombin into thrombin and thus start the series of chemical reactions that results in clotting. Even so, additional safeguards are present in antiprothrombic substances such as *heparin*. This substance was originally found in the liver but is now thought to be produced by large basophilic cells (mast cells) in tissues of various organs. Heparin reduces the ability of the blood to clot by blocking the change of prothrombin to thrombin. It can also be used to aid in reducing clots when internal clotting has already occurred. In either case it acts in conjunction with a plasma cofactor. Internal clotting is called *thrombosis*. The clot, or *thrombus*, may form in some blood vessel of the arm or leg and do comparatively little harm, but if it should block the blood supply to the brain, the heart (coronary

The clotting mechanism

335

thrombosis) or lungs, it can be very serious. An *embolus* is a clot that has become dislodged from its place of origin. The condition is known as *embolism*.

Blood clots are sometimes dissolved by a plasma protein called *fibrinolysin* or *plasmin*. This substance is present in the inactive form in the plasma, where it is known as *profibrinolysin* or *plasminogen*. The inactive lysin must be activated before it can act as a digestive enzyme to dissolve a clot.

HEMOPHILIA

In the peculiar condition known as *hemophilia*, the blood clots very slowly, and internal bleeding is especially difficult to control. The defect is inherited as a sex-linked, recessive character in the same way that color blindness is inherited. Women are not ordinarily affected. If a normal woman marries a man who is a hemophiliac, their sons will not be affected, but their daughters will transmit the condition to some of their own sons. Thus the defect skips a generation so far as its active expression in males is concerned (see Figure 12.17).

It is generally accepted that an unidentified substance called the *antihemophilic globulin factor* (factor VIII) is not present in the blood of hemophiliacs. When normal human plasma globulin is added to hemophilic blood, clotting proceeds normally. Transfusion of normal blood, normal plasma, or purified globulin into a hemophiliac can serve to avert a bleeding crisis. However, this is only a temporary expedient, lasting at best for a few days. There is no cure or treatment with lasting effectiveness.

The mechanism of blood clotting varies in different types of hemophilia. In one type, factor IX is missing, and in another rare type it is factor XI. As we have seen, in the usual type of hemophilia, factor VIII is missing.

SERUM

When a sample of blood is drawn into a container and allowed to stand, a clot forms and the cells and clot are separated from a yellowish fluid, which is blood *serum*. It differs from plasma in that the chemical substances concerned with clotting have interacted; that is, fibrinogen has been precipitated as fibrin, and a clot has been formed (Figure 13.7).

STORAGE OF WHOLE BLOOD AND PLASMA

Blood transfusions were used on a large scale in World War I. It was found that sodium citrate was a safe anticoagulant. Prompt refrigeration then made it possible to keep whole blood for 5 to 7 days. Blood banks were established at various bases and later in many hospitals. A better whole-blood preservative was used in 1944. This consisted of sodium citrate, citric acid, and dextrose. Refrigerated whole blood could then be preserved and kept in good condition for a period of 21 days. More efficient preservation has continued to lengthen the time that blood can be kept in usable condition.

During the greater part of World War II blood plasma was used very extensively as a substitute for whole blood in transfusions. It could be preserved, either frozen or dried, and had far less bulk than whole blood.

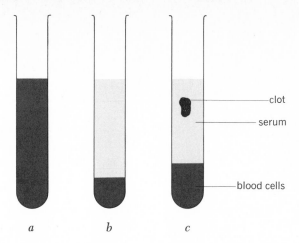

FIGURE 13.7
Diagram illustrating the formation of a clot when blood is drawn into a container and allowed to stand: *a* whole blood; *b* the formation of fibrin threads (not visible) and some settling of blood cells; *c* the separation of the serum, the clot, and the blood cells.

Dried lyophilized plasma remains effective for years and can be made ready for use simply by redissolving it in distilled water. Since a great part of the effectiveness of plasma in combating shock is due to the serum albumin that it contains, several agencies concentrated on fractionating the various plasma proteins, with particular emphasis on obtaining serum albumin. Although there were definite advantages in reduced shipping bulk for these newer preparations, there arose at about the same time a definite conviction that the only completely satisfactory material for transfusion is whole blood. Actually, other fractions of the blood plasma became of more scientific interest than did the serum-albumin portion.

BLOOD GROUPING

The transfusion of whole blood brings up the problem of blood grouping. It was learned many years ago that it was not safe to transfuse bloods of different species and that the blood of two human beings often was found to be incompatible. When the blood is incompatible upon mixing, the red blood cells clump together or agglutinate.

Little was known as to why some bloods were compatible and others incompatible until the research work of Karl Landsteiner, around 1900, revealed that in the human species there are four blood groups. The groups are A, B, AB, and O. These blood groups are definitely inherited according to mendelian laws.

Agglutination results when the antigen of a certain group of red cells interacts with its specific antibody in the serum. Obviously, both antigen and antibody cannot be present in the same blood, for the red cells would then be agglutinated. The letters indicating the four blood groups also indicate the antigen on the red cells. The red cells of Group A have antigen A; Group O has no antigens.

The plasma or serum contains antibodies, which will cause agglutination of red cells if they come in contact with the antigen. These antibodies normally occur in blood and are designated alpha and beta, or anti-A and anti-B. They are also called *isoagglutinins*. The plasma of Group A contains anti-B substances, while the plasma of Group B contains anti-A substances (Table 13.2). Group AB plasma does not contain any antibodies or agglutinins;

Blood grouping

337

TABLE 13.2
Blood groups and corresponding antigens and antibodies

Blood group or phenotype	Antigens of red cells	Antibody in plasma or serum	Genotype
A	A	Anti-B	AA or Ao
B	B	Anti-A	BB or Bo
AB	AB	None	AB
O	None	Anti-A and anti-B	OO

the serum of Group O contains both anti-A and anti-B agglutinins (Table 13.2).

One can see the absolute necessity of determining the blood type before a transfusion is made. It is always desirable to use blood from a donor of the same blood type as the recipient, but if such blood is not available, other types can be used, as long as the red cells of the donor are not agglutinated by the plasma of the recipient. Practically, the plasma of the donor is so diluted in mixing with the total blood volume of the recipient that the antibodies that it contains do not ordinarily cause any reaction. For this reason, blood plasma, which is usually pooled, can be safely used without typing. The donor's red cells, however, receive the full effect of the antibodies in the recipient's plasma; therefore they must be carefully typed before transfusion.

To determine the blood group of an individual, his red cells need to be tested only with the serum of Group A and of Group B, as illustrated in Figure 13.8.

Figure 13.9 illustrates the reaction of the blood when red cells of the donor are mixed with plasma or serum of the recipient; the four blood groups

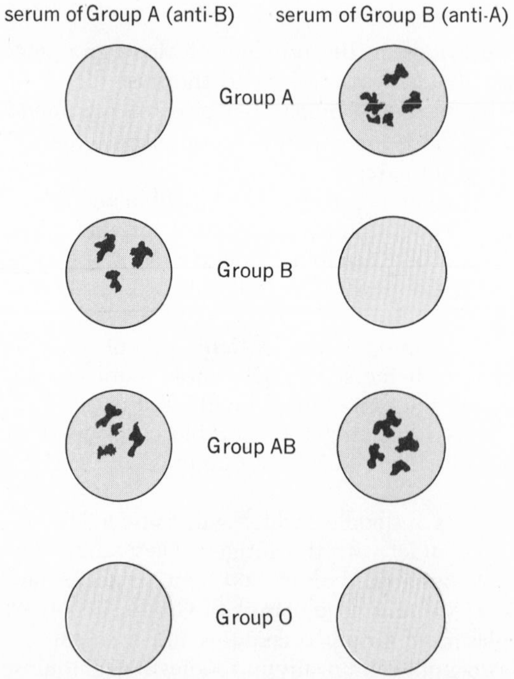

FIGURE 13.8
Determination of the four blood groups. The serums of Group A and Group B are tested for reaction against the unknown cells of the four groups: A, B, AB, and O. Four large drops of Group A serum and four large drops of Group B serum are placed on microscope slides. A small drop of blood of the unknown type is added to each drop of serum. Agglutination occurs only when the antigens of the red cells react with the corresponding antibodies in the serum. Refer to Table 13.2.

serum of Group A (anti-B)   serum of Group B (anti-A)

Group A

Group B

Group AB

Group O

The blood

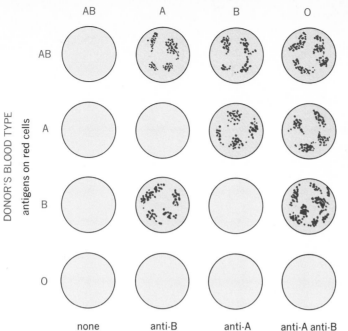

DONOR'S BLOOD TYPE
antigens on red cells

AB

A

B

O

none   anti-B   anti-A   anti-A anti-B

FIGURE 13.9

A scheme designed to show the reaction occurring when antigens on red cells of the donor's blood type are introduced into the plasma of the recipient's blood type. Agglutination indicates a positive reaction. Persons with type O blood have been called universal donors, whereas persons with type AB blood are termed universal recipients. However, all blood should be carefully typed and tested before transfusion.

of donors are tested against the four groups of recipients. All tests should be observed with the aid of a microscope to determine agglutination, since extensive rouleaux formation, that is, red cells stacked together, might otherwise be interpreted as agglutination.

In the early stages of our knowledge of blood typing and whole-blood transfusions, type O blood was in greatest demand. In the general white population, about 45 percent have type O blood. Individuals with type A blood represent about 41 percent of the population, while type B individuals total about 10 percent and type AB only 4 percent. In the American Negro population the distribution of blood types is: O—48 percent, A—26 percent, B—23 percent, AB—3 percent.

THE Rh FACTOR

In addition to the blood-group antigens, several other known antigens are associated with red cells. One of these is commonly called the *Rh factor*, because the antibody was originally developed in rabbits and guinea pigs by injecting the red cells of *Macacus rhesus*, the rhesus monkey. There are many genotypes of the Rh factor, but this discussion will consider only the two phenotypes: Rhesus-positive (Rh+) and Rhesus-negative (Rh−) individuals.

It has been discovered that about 85 percent of the white population is Rh-positive and only 15 percent is Rh-negative. The dark-skinned races are almost 100 percent Rh-positive, American Negroes being about 93 percent Rh-positive. The red cells of an Rh-positive individual will be agglutinated by the anti-Rh substance found in the plasma of rhesus monkeys—a positive reaction between the antigen present in red cells and the antibody present in the plasma of the monkey. Since human serum does not

The Rh factor

339

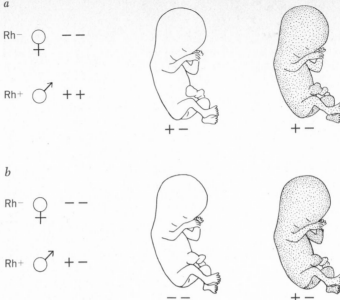

a

Rh− ♀ − −

Rh+ ♂ + +

+ −   + −

b

Rh− ♀ − −

Rh+ ♂ + −

− −   + −

FIGURE 13.10

Diagram illustrating possible effects of the Rh factor on the fetus: *a* Even though the father in this case is Rh + and the mother Rh −, the first child probably will not be affected with erythroblastosis fetalis. The second child is more likely to be affected. *b* If the father is heterozygous for the Rh factor, the fetus may be Rh −, in which case there could be no reaction with the Rh − blood of the mother. An Rh + fetus, however, increases the possibility of an antibody reaction.

ordinarily contain this specific antibody, test serum is obtained from animals in which the antibody has been carefully developed. The red cells of about 15 percent of the white population are not agglutinated by the anti-Rh substance; the reaction, therefore, is Rh-negative. These persons do not have the Rh antigen in their red cells. It is important to realize that if the antigen of an Rh + individual is introduced into the blood of an Rh − individual by transfusion or otherwise, the Rh − individual's blood is capable of forming antibodies.

Before these reactions were understood it was known that some transfusions between compatible types of blood or even between bloods of the same type would occasionally cause agglutination. The first transfusion should be safe in appropriate blood groups, regardless of the Rh factor. Only when a second or third transfusion or a series of transfusions from Rh + into Rh − blood occurs can agglutination result. After the first of such transfusions, antibodies form in the recipient's blood and react against any further transfusions of Rh + blood. Often two, or perhaps even three, transfusions of Rh + into Rh − blood can be accomplished before the antibody becomes strong enough to give a severe reaction, but obviously, with our present knowledge, such transfusions should not be practiced.

A second noteworthy example of the action of the Rh factor occurs in cases of pregnancy in which the father is Rh + and the mother Rh −. The factor is definitely inherited, and Rh + is dominant over Rh −. Of course, the male parent can be heterozygous for this factor, in which case the fetus could be negative; but the commoner occurrence is that the fetus is Rh + (Figure 13.10).

Since the fetus develops its own blood and circulatory system, even this circumstance of developing an Rh + fetus in the Rh − mother should not cause any difficulty; frequently it does not, especially with the firstborn child. But in numerous cases the second or third child develops an abnormal blood condition with agglutination, anemia, and jaundice, a condition given the medical name *erythroblastosis fetalis.*

It is presumed that there can be an occasional break in the placental barrier that ordinarily separates the circulatory system of the mother and that of the fetus. If the antigen in the red cells of the Rh+ fetus should be introduced into the blood of the Rh− mother, the mother's blood would form antibodies, just as it would in a direct transfusion with Rh+ blood. The antibody formed during pregnancy tends to react more strongly against the fetal red cells during the later months of pregnancy, frequently resulting in a stillbirth or a serious blood disturbance in the newborn infant. The antibodies do not affect the Rh− mother, since there is no Rh antigen in her red cells; but if she should receive a transfusion of Rh+ blood, the well-developed antibody might cause serious agglutination.

In the present state of our knowledge it is highly desirable, in case of pregnancy, to determine the Rh characteristics of both parents. If the mother is Rh− and the father is Rh+, periodic tests of the mother's plasma should be made to determine if there has been any development of Rh antibodies as the result of carrying an Rh-positive fetus. Such knowledge will enable a physician to anticipate erythroblastosis and to take appropriate measures to combat any unfavorable development. The Rh− negative mother can now be given an anti-Rh antibody to inactivate any Rh-positive blood cells that may have entered her bloodstream from the fetal circulation. In this case, the mother's blood in the second pregnancy is in the same condition as it was in the first pregnancy and the second child's blood should be normal.

Severe erythroblastosis in the newborn infant is often treated by replacing the infant's Rh-positive blood with Rh-negative blood in an exchange transfusion. This procedure stops the destruction of red cells, and the symptoms of jaundice and anemia tend to disappear. The infant's blood-forming tissues will then begin developing blood to replace gradually the transfused Rh-negative blood. The antigen is not present in the Rh-negative blood, and the antibodies are largely removed during the replacement transfusion. The antibodies remaining in circulation will gradually disappear as the infant develops its own Rh-positive blood.

OTHER FACTORS

There are many antigens on human red cells in addition to those already mentioned. The M, N, and MN factors are well known. Blood factors, with the exception of the ABO groups, are detected by immunization and by the production of specific antibodies against the agglutinogens. The M and N factors are detected by using rabbit plasma from animals immunized against human red blood cells.

The ABO blood groups, subgroups, factors M and N, and the Rh factor are often used in the field of legal medicine in attempts to prove parental relationships. For example, neither antigen A nor B could be present in a child's blood unless it were present in the blood of one or both of the parents. Neither would it be possible for type AB parents to have type O children. The MN group is inherited as MM, NN, or MN. Blood groups have their limitations in the legal field, but they are often useful in establishing an unchangeable means of identification.

RADIATION HAZARDS AFFECTING THE BLOOD

One of the hazards of modern living is the possibility of exposure to atomic radiation and fallout. The blood-forming tissues are very susceptible to radiant energy but show considerable ability to recover. Soon after exposure to radiation, leukocytes decrease in number. The clotting mechanism of the blood is adversely affected, causing a tendency toward bleeding. Later, anemias occur as a result of injury to the tissues that form erythrocytes.

A radioactive-fallout product, strontium 90, is of especial concern, since it is absorbed like calcium and deposited in bone. Radioactive fallout can be expected to contaminate the ground and to enter into the food chain through being taken up by plants. When individuals eat plant foods directly or drink milk, strontium 90 will become incorporated in the human organism. It may then affect the blood-forming tissues or cause cancer to develop. Strontium 90 is capable of emitting beta rays for a very long time. It has a physical half-life of 28 years. (Half-life is the period of time required for a given quantity of an element to lose half its radioactivity. A half-life of 28 years means that the element will lose half its radioactivity by the end of that time.)

Most experimental evidence regarding injurious effects of radiation has been obtained from experimentation on animals. Shielding a portion of the body from excessive radiation affords the animal a better chance for recovery. Shielding the spleen, in particular, has been shown to produce a favorable effect toward recovery.

Bone-marrow transplants have proved to be successful in animals, and eventually this technique may be used to save the lives of persons who have been exposed to excessive amounts of radiation.

THE SPLEEN AND BLOOD SUPPLY

The spleen, which is an important accessory of the circulatory system, is a highly vascular organ located to the left of the stomach and somewhat behind it. It lies below the diaphragm and directly above the left kidney. It is covered by the peritoneum, a membrane lining the abdominal cavity. Though it is capable of changing its size, it is ordinarily 5 or 6 inches long and about 4 inches wide. It is a soft, pliable organ of dark purplish color (see Plate G) (Figure 13.11).

The internal structure of the spleen is rather complex. Just beneath the serous coat is a connective-tissue capsule, from which a fibrous framework (the trabeculae) projects into the interior. The capsule and the internal fibrous structure contain smooth muscle, which accounts for the ability of the organ to contract. The areas between the fibrous framework are filled with splenic pulp, composed of loose reticular tissue supporting small arteries and veins and containing red and white blood cells in great numbers. In the splenic pulp there is a complicated network of blood spaces, the splenic sinuses. Here also are lymphatic nodules composed of lymphoid tissue and concerned with the development of lymphocytes. Large phagocytic white cells of the reticuloendothelial system (macrophages) are found in great numbers in the splenic pulp. They function in the removal from the blood of old or agglutinated red cells, pieces of red cells, and foreign matter.

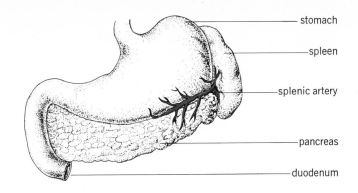

a

stomach

spleen

splenic artery

pancreas

duodenum

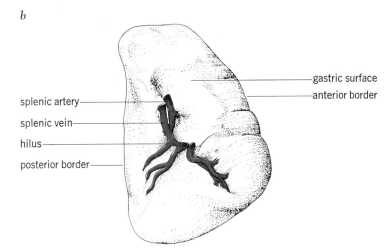

b

gastric surface

anterior border

splenic artery

splenic vein

hilus

posterior border

FIGURE 13.11

a The spleen, anterior view of stomach, pancreas, and spleen; *b* the spleen, visceral surface.

FUNCTIONS The spleen has had numerous functions ascribed to it, but three are of primary importance: (1) the spleen is capable of holding a reserve blood supply; (2) it is concerned with the destruction of old or agglutinated red cells and platelets; (3) it acts in the formation of lymphocytes.

As an accessory organ of the circulatory system, the spleen is capable of directing a considerable amount of reserve blood into the arteries in response to an emergency, thus increasing the oxygen-carrying capacity of the blood. Stimulated by impulses from the sympathetic nervous system and by epinephrine from the adrenal glands, the spleen is strongly activated in emotional states and in times of physical stress. When there is a loss of blood, the spleen aids not only in making up the volume of the blood but also in supplying additional red cells.

Although the spleen is not essential to the life and well-being of man and animals, it enables them better to meet emergencies. Exercise and an increase in external temperature causes the spleen to decrease its volume of blood. Animals with spleens removed appear to have less resistance to the toxic effects of disease.

The large phagocytic cells of the spleen, liver, and other tissues apparently do not ingest normal, healthy red cells; they are most useful in taking

The spleen and blood supply

old red cells, fragments, and agglutinated masses of cells out of circulation. The spleen has been called the "graveyard" for old blood platelets.

In the embryo the spleen forms both red cells and lymphocytes. After birth, erythrocytes are formed in the red marrow of bone. The spleen continues to be an important organ for the manufacture of lymphocytes and monocytes throughout life.

SUMMARY   The blood has many functions such as its ability to transport oxygen, carbon dioxide, food materials, waste materials, and hormones. The blood plays an important part also in regulating body temperature and in maintaining both the acid-base balance and the water balance of the tissues. Its protective functions include its clotting ability, the phagocytic activity of leukocytes, and the production of antibodies.

Erythrocytes are the red cells of the blood. They are biconcave disks, averaging 5 million cells per cubic millimeter of blood. They arise as nucleated erythroblasts but ordinarily lose their nuclei by the time they enter the bloodstream. They derive their color from the oxygen-carrying compound called hemoglobin.

White cells, or leukocytes, are classified as granular and nongranular. The granular leukocytes are neutrophils, eosinophils, and basophils. Nongranular leukocytes are lymphocytes and monocytes. White cells number about 7,000 per cubic millimeter of blood.

Blood platelets are not cells but fragments of large cells, called megakaryocytes. They number about 250,000 to 400,000 per cubic millimeter. When they break down, thromboplastin is formed; it initiates the clotting process.

After thromboplastin has been released, the clotting process proceeds as follows: thromboplastin, calcium ions, and accelerator globulin interact with prothrombin, causing the release of thrombin. Thrombin, in turn, interacts with fibrinogen to produce threads of fibrin. Fibrin threads form the framework of the clot. Heparin reduces the clotting ability of the blood. Hemophilia is a condition in which an antihemophilic factor is absent in the blood and the clotting ability is therefore greatly reduced.

Plasma is the fluid portion of the blood. It is very complex chemically and contains the clotting components. After the blood has been permitted to clot, the amber-colored fluid is called blood serum.

There are four blood groups: A, B, AB, and O. These letters represent the antigens, if present, on the red cells. The red cells of Group A have antigen A, but the red cells of Group O have no antigens. The plasma or serum contains antibodies. The plasma of Group A contains anti-B substances and the plasma of Group B contains anti-A substances. Group AB plasma has no antibodies, whereas the serum of Group O contains both anti-A and anti-B agglutinins. Persons with Group O blood have been termed universal donors; persons with type AB blood, universal recipients.

The Rh antigen, if introduced by transfusion or otherwise into the blood of a person who is Rh-negative, will cause the formation of antibodies. About 85 percent of the white population are Rh-positive; about 15 percent, Rh-negative.

Erythroblastosis fetalis can result from pregnancies in which the father is Rh-positive and the mother Rh-negative. The mother's blood may develop antibodies against the Rh-positive blood of the fetus.

The spleen is an important organ concerned with reserve blood supply.

The blood

344

Old erythrocytes and blood platelets are taken out of circulation by the spleen, and lymphocytes are formed in this organ as well as in lymph nodes. The blood-forming tissues are especially sensitive to atomic radiation.

1 List and explain the functions of the blood.
2 Differentiate between plasma and serum.
3 Why has the blood and tissue fluid been called the internal environment?
4 What part do erythrocytes play in the functioning of the blood?
5 Is it possible to count the red cells? How?
6 List the different kinds of leukocytes and give their functions.
7 What is the function of plasma cells?
8 Follow the sequence of events that results in the formation of a clot.
9 Why does not internal clotting occur more readily?
10 Name the four ABO blood groups, and discuss blood typing. Why would some transfusions cause agglutination of the recipient's blood?
11 Discuss the Rh factor.
12 What are some of the functions of the spleen?

SUGGESTED READING

AUERBACH, R.: "Experimental Analysis of Lymphoid Differentiation in the Mammalian Thymus and Spleen," pp. 539–557, Organogenesis, Holt, Rinehart and Winston, Inc., New York, 1965.
CLARKE, C. A.: The Prevention of "Rhesus" Babies, *Sci. Am.*, **219:**46-52 (1968).
DAVIE, EARL W., and O. D. RATNOFF: Waterfall Sequence for Intrinsic Blood Clotting, *Science*, **145:**1310-1312 (1964).
FRIEDMAN, T.: Prenatal Diagnosis of Genetic Disease, *Sci. Am.*, **225:**34-42 (1971).
GATTI, R. A., O. STUTMAN, and R. A. GOOD.: The Lymphoid System, *Ann. Rev. Physiol.*, **32:**529-546 (1970).
KABAT, D.: Gene Selection in Hemoglobin and in Antibody-synthesizing Cells, *Science*, **175:**134-140 (1972).
PERUTZ, M. F.: The Hemoglobin Molecule, *Sci. Am.*, **211:**64-76 (1964).
ROBBINS, JAY H.: Tissue Culture Studies of the Human Lymphocyte, *Science*, **146:**1648-1654 (1964).
SPEIRS, ROBERT S.: How Cells Attack Antigens, *Sci. Am.*, **210:**58-64 (1964).
ZUCKER, MARJORIE: Blood Platelets, *Sci. Am.*, **204:**58-64 (1961).
ZUCKERKANDL, EMILE: The Evolution of Hemoglobin, *Sci. Am.*, **212:**110-118 (1965).

chapter 14
the heart and general circulation

The heart is a muscular organ the contractions of which force the blood to circulate through a closed system of arteries, arterioles, capillaries, and veins. The blood is conducted away from the heart through arteries, which divide into smaller and smaller branches. Finally, at the periphery, the blood passes through very fine capillaries and starts its return flow into small venules. Larger and larger veins bring the blood back to the heart.

THE HEART The heart lies in a double-walled pericardial sac located in the interpleural space, or *mediastinum*, a space bounded laterally by the lungs and extending from the backbone to the sternum dorsoventrally. In addition to the heart, many blood vessels closely associated with the heart lie in this area. The diaphragm forms the floor of the mediastinum as well as of the entire thoracic cavity. Although the heart is centrally located, its tip (or apex) is inclined downward and toward the left. As the heart fills with blood and starts to contract, its beat is felt against the wall of the chest between the fifth and sixth ribs and about 3 inches to the left of the midventral line (Plates F and G).

The heart varies in size, but it is about 5 inches long and $3\frac{1}{2}$ inches wide in the adult male. In general, a large person can be expected to have a proportionately larger heart; it is often described as being about the size of the closed fist.

The human heart contains four chambers. At the upper end are the right and left *atria*. The atria receive blood from veins. Their muscular wall is thin and is usually observed in a collapsed condition when hearts are preserved for study. The right and left *ventricles* form the lower part of the heart. They have thick muscular walls, which enable them to force the blood out into arteries. In its actual position, the right ventricle is largely anterior to the left (Figure 14.1).

PATH OF BLOOD THROUGH THE HEART Large caval veins bring the blood back from the body and empty into the right atrium. The blood flows down into the right ventricle, even before the atrial walls begin to contract. Between the right atrium and the right ventricle there is a cylindrical valve, which permits the blood to pass from atrium to ventricle but closes as the ventricle starts to contract. This is the *tricuspid valve*, so called because it is composed of three cusps. Valves increase the efficiency of the heart action by preventing the blood from moving backward into the space it has just occupied. The atrioventricular valves are supported by tendinous strands below, which attach to small mounds of muscle projecting from the ventricular wall. The strands are *chordae tendineae*, and they attach to papillary muscles. The

The heart and general circulation

346

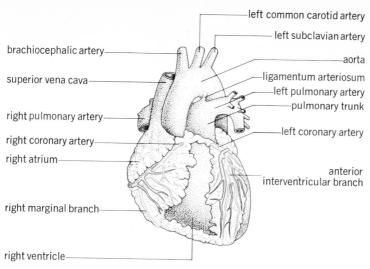

Labels on figure:
- left common carotid artery
- left subclavian artery
- brachiocephalic artery
- aorta
- superior vena cava
- ligamentum arteriosum
- left pulmonary artery
- pulmonary trunk
- right pulmonary artery
- right coronary artery
- left coronary artery
- right atrium
- anterior interventricular branch
- right marginal branch
- right ventricle

FIGURE 14.1
The heart and associated arteries and veins in anterior view.

chordae tendineae prevent the valve from being forced back into the atrium by the pressure of the blood as the muscular ventricular wall contracts. Since the tricuspid valve prevents the blood from being forced back into the atrium, the only exit for the blood is through the *pulmonary artery* to the lungs (Figure 14.2).

At the base of the pulmonary artery there are three valves, somewhat half-moon-shaped, called *pulmonary semilunar valves*. Their concave surfaces are directed upward, like cups, to hold the column of blood in the pulmonary artery. When contraction of muscles in the ventricular wall forces the blood upward, the valves collapse and so offer little resistance to the passage of the blood into the pulmonary artery. As soon as this forward flow has ceased, the valves close and prevent the blood from flowing backward.

The pulmonary artery leads to the lungs, where it divides into many small branches. Arterioles finally subdivide into a capillary network traversing the walls of air sacs in the lungs. Here, in the heart-lung cycle, carbon dioxide is given off by the blood, to be exhaled, and oxygen is taken up.

The lung capillaries give rise to small veins, or *venules*, which, in turn, give rise to larger veins. Eventually, four *pulmonary veins* carry freshly oxygenated blood back to the left atrium, two veins from each lung. The blood passes through the left atrium and goes on down into the left ventricle.

The thin atrial walls contract, and this action is followed by a strong contraction of the ventricles. Between the left atrium and the left ventricle lies the *mitral, or bicuspid, valve*. The valves located between atria and ventricles are also referred to as right and left atrioventricular, or AV, valves. The mitral valve closes as the left ventricle starts to contract. The blood is forced up through the only exit available, the large aortic artery. At the base of the aorta there are three *aortic semilunar valves*, which prevent backflow of a column of blood in the aorta into the left ventricle as it relaxes. Branches of the aorta distribute the blood all over the body (Figure 14.3).

The heart repeats this action 60 to 80 times per minute for all the years of one's life.

PERICARDIUM The heart and the bases of the great blood vessels are enclosed in a two-layered membrane called the *pericardium*. If one side of a hollow

semi-lunar prevent backflow

bicuspid-left
tri - Right

The heart

347

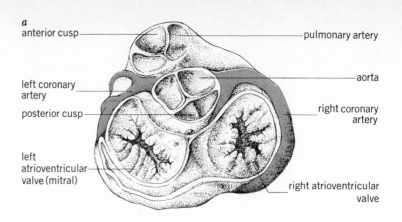

a

anterior cusp —————————————————————— pulmonary artery

left coronary artery

posterior cusp ————— aorta

right coronary artery

left atrioventricular valve (mitral)

right atrioventricular valve

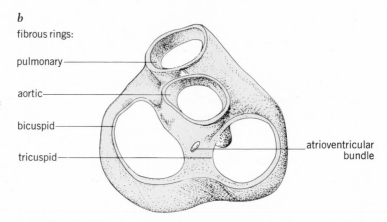

b

fibrous rings:

pulmonary ————

aortic————

bicuspid————

tricuspid————

atrioventricular bundle

FIGURE 14.2
Structure of the heart: *a* valves of the heart viewed from above; *b* fibrous framework (cardiac skeleton); this is the framework of dense collagenous and connective-tissue fibers that separates the musculature of the atria from that of the ventricles.

The heart and general circulation

348

rubber ball were forced in against the other side, it would be a fair example of this double-layered membrane. The outer layer is composed of fibrous tissue, and the more delicate, inner layer is a serous membrane. The serous membrane is closely applied to the heart, but it also lines the fibrous pericardium. The serous portion forms a closed sac, the *pericardial cavity*. The heart is not in a cavity; it is merely covered by two layers of membrane. The pericardial cavity is only a potential cavity under normal conditions; the outer serous lining is in contact with the inner serous membrane, with only a very small amount of pericardial fluid to prevent friction as the heart beats.

ENDOCARDIUM The cavities of the heart are lined by a serous membrane called the *endocardium*. This layer is continued over the valves and chordae tendineae and fuses with the membrane lining the large blood vessels of the heart. An inflammation of the lining membrane is termed *endocarditis*. The inflammation is considered by some investigators to be a secondary infection commonly accompanying such diseases as scarlet fever and rheumatic fever. Since the endocardium covers the valves of the heart, as well as lining the openings guarded by the valves, the inflammation can alter the shape of either so that they no longer close tightly. In this case a small amount of blood may leak past the valves, causing a low sound in a stethoscope, called a *murmur*. If the valvular opening is constricted as a result of inflammation, the condition is referred to as *stenosis*. The same valve can be affected by both types of injury.

Fortunately, the heart ordinarily will compensate for considerable loss of efficiency from valves that do not close tightly. The heart with valvular deficiency often will perform well enough to permit ordinary activity, but it will reach its peak load before a normal heart would. Persons with well-defined heart murmurs are therefore advised not to engage in strenuous activities that might overtax their hearts.

MYOCARDIUM The musculature of the heart is referred to as the *myocardium*. It is composed of cardiac-muscle tissue, which is involuntary, finely striated, and has branching fibers. The reader may wish to review the discussion of muscle tissue in Chapter 6. The muscles of the heart are arranged in complex patterns of irregular whorls, the heart contracting with a twisting or wringing motion. It is an all-or-nothing type of contraction. The atrial walls are relatively thin, while the ventricular walls are thick and strong. This is especially true of the left ventricle, which forces blood out over the body. Its walls are much thicker than those of the right ventricle, which only moves the blood through the lungs.

NERVE SUPPLY The nerve supply to the heart is considered to be largely regulatory. The heart receives motor impulses through branches of the right and left vagus nerves. Stimulating the vagus nerve depresses heart action. Accelerator nerves are the cardiac branches of spinal nerves, which are associated with the thoracolumbar autonomic system. The heart also has afferent nerves by which pressor or depressor effects can be secured indirectly.

There is, however, a very high degree of automatism about the beating of the heart. The heart of cold-blooded and warm-blooded animals, if properly prepared, can be completely removed from the body and continue to beat for some time with no nerve connections whatever. The distention of the muscular wall as the chambers of the heart fill with blood provides one sort of stimulus. Certain salts (electrolytes) in the blood provide a chemical stimulus. There is also an inherent rhythmic contraction that is characteristic of cardiac tissue.

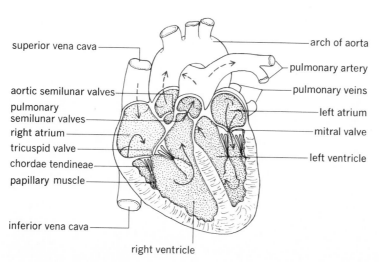

superior vena cava

arch of aorta

pulmonary artery

aortic semilunar valves

pulmonary veins

pulmonary semilunar valves

left atrium

right atrium

mitral valve

tricuspid valve

left ventricle

chordae tendineae

papillary muscle

inferior vena cava

right ventricle

FIGURE 14.3

Chambers and valves of the heart, showing direction of blood flow, anterior view.

The heart

349

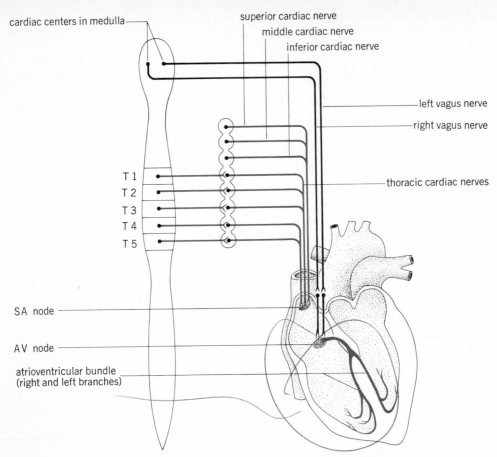

cardiac centers in medulla

superior cardiac nerve
middle cardiac nerve
inferior cardiac nerve

left vagus nerve
right vagus nerve

T 1
T 2
T 3
T 4
T 5

thoracic cardiac nerves

SA node

AV node

atrioventricular bundle
(right and left branches)

FIGURE 14.4
Autonomic innervation of the
heart.

SINOATRIAL NODE The wave of contraction that spreads over the atria originates in a small area of specialized tissue called the *sinoatrial node* (SA node). It is located in the wall of the right atrium near the entrance of the superior vena cava in the mammalian heart. The heartbeat originates there, and the SA node is considered to be the pacemaker (Figure 14.4). Depolarization of the SA node creates an action potential, and an excitation wave spreads over the atria from one muscle fiber to another as the atria contract.

THE ATRIOVENTRICULAR NODE The musculature of the atria, however, is separate and distinct from that of the ventricles. A band of connective tissue around the crown of the heart separates the two sets of muscles. A strand of differentiated muscular tissue in the *septum* (or wall), which completely separates the right and left sides of the heart in an adult, affords a pathway by which excitation can reach the ventricles (see Figure 14.5). In the septum between the atria is located another area of specialized tissue called the *atrioventricular node* (AV node). The neuromuscular bundle (the bundle of His) extends downward from this node. It divides into right and left branches and conducts impulses to the muscles of the right and left ventricles, respectively. The terminal branches produce a dense network of minute filaments, the *Purkinje network*, located underneath the endocardium of the ventricles. Since the muscles of the atria are the first to be stimulated, the atria are the first to

The heart and general circulation

350

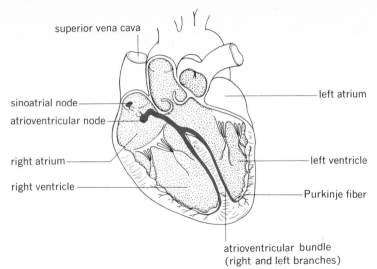

superior vena cava

sinoatrial node

atrioventricular node

right atrium

right ventricle

left atrium

left ventricle

Purkinje fiber

atrioventricular bundle
(right and left branches)

FIGURE 14.5
Cardiac conduction: sinoventricular conduction system. The heartbeat originates at the sinoatrial node, the pacemaker of the heart. It spreads rapidly over all the atrial musculature and converges on the atrioventricular node. Impulses pass down the atrioventricular bundle at about 500 centimeters per second; the conduction rate then slows to around 50 centimeters per second as impulses pass through the ventricular muscle enabling both ventricles to contract at the same time.

contract. The wave of contraction spreads through normal cardiac tissue from SA node to AV node. But the rate of conduction through the neuromuscular bundle is about 10 times as fast as it is through unspecialized cardiac muscle. The impulse therefore descends rapidly through the neuromuscular bundle and is distributed to all parts of the ventricular musculature by way of the Purkinje network. The muscular wall of the ventricles then contracts almost immediately after the atria contract. The conduction time from atria to ventricles is around 0.12 to 0.2 second in human beings. Normally, the atrioventricular bundle is the only pathway by which impulses can be transmitted to the ventricles. It is therefore a very important passageway. In experiments with animals, if transmission across the atrioventricular bundle is interrupted by cutting or tying, the ventricles lose pace with the atria. The ventricles can be inhibited temporarily and stop contracting, but ordinarily they resume beating at a much slower rate than that maintained by the atria. Various conditions can affect conduction in the human heart. There may be a partial or complete blocking of conduction in the atrioventricular bundle. If the blocking is complete, the ventricular contraction rate drops to around 35 per minute, while the atrial rate remains around 70 per minute.

THE CARDIAC CYCLE If we are permitted to look at the beating heart of an animal, we observe first that there is a regularity or rhythm to its movement. Ordinarily it is possible to observe that the atria contract first and that this movement is followed by contraction of the ventricles. The contraction phase, during which blood is forced out of the chambers of the heart, is referred to as *systole*. One may refer to either atrial systole or ventricular systole, but when used alone the word refers to the contraction of the ventricles. *Diastole* refers to the dilating or relaxing of chambers of the heart so that they may fill with blood. Unless atrial diastole is indicated, the term refers to the relaxing of the ventricles. The action of the heart is so arranged that the atria are already entering diastole by the time the ventricles are beginning the systolic phase. The ventricles then fill with blood, and there is a short period of diastasis during which the heart muscle is inactive but

The heart

351

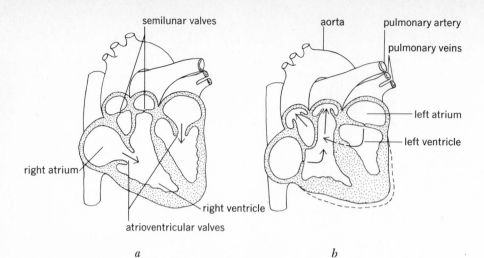

semilunar valves   aorta   pulmonary artery

pulmonary veins

left atrium

left ventricle

right atrium

right ventricle

atrioventricular valves

a   *b*

FIGURE 14.6

Diagram illustrating the heart cycle: *a* relaxation phase, with the ventricles in diastole. The atrioventricular valves open as the ventricles fill with blood. The pulmonary and aortic semilunar valves are closed; *b* contracting phase, showing the ventricles in systole. The atrioventricular valves are closed, and the semilunar valves are open.

filling continues. This is a brief period of rest for the heart musculature; it is almost immediately ended by atrial systole as the cardiac cycle is repeated. The reader may recall that cardiac tissue has a long refractory period and is not subject to summation or tetany.

During systole the pressure within the ventricles is high. Since it becomes higher than the pressure in the great arteries leaving the heart, blood rushes out into these arteries as the semilunar valves are forced open. The atria are then already in their diastolic phase, the pressure within is low, and blood flows into them from the great veins. As the ventricles complete their relaxation phase, the pressure within these chambers becomes lower than the atrial pressure. The atrioventricular valves then open, and the ventricles begin to fill with blood even though the atria are still in a resting stage. Atrial systole produces a final and positive surge of blood through the atrioventricular valves, which then close as the ventricles start to contract (Figure 14.6).

HEART SOUNDS If one listens to the heart by placing his ear against the chest over the heart or by means of a stethoscope, two well-defined sounds may be heard. The first sound is lower and of longer duration than the second sound. Though the sound can be caused by vibrations arising from various sources, it is primarily associated with the closure of the atrioventricular valves. The first and second heart sounds are commonly expressed by the syllables lub-dup.

The second sound is associated with the closure of the semilunar valves as the ventricles enter their diastolic phase. This sound is louder, sharper, and of higher pitch than the first sound. The sounds are altered in hearts with defective valves.

CARDIAC OUTPUT OR STROKE VOLUME The amount of blood moved by each contraction of a single ventricle has been estimated by several methods. The resting heart moves about 60 milliliters of blood from each ventricle at each contraction. The amount of blood ejected by a ventricle beating 70 times per minute totals 4,200 milliliters, or 4.2 liters. During exercise each ventricle is capable of moving several times this amount. It is further estimated that all the blood in the body passes through the heart in about 60 seconds, even in the resting state.

The heart and general circulation

352

HEART RATE So many factors influence the heart rate that statements concerning rate are only relative to various conditions. To say that the average heart beats about 70 times per minute most likely refers to studies on the heart rate of young men, seated, and in a postabsorptive state. The heart rate for young women is somewhat faster, on an average, than that of young men. One method of judging the sex of the fetus during the latter part of pregnancy is to count the number of heartbeats per minute. The heart rate of the male fetus usually is around 130 to 135 beats per minute, while that of the female is ordinarily 140 to 145.

The heart rate of a person in a recumbent position commonly is 58 to 68 times per minute. A typical rate in the standing position would be around 80. The digestion of a meal increases the strength and rate, partly because of activity of the muscles of the stomach and intestine. Muscular exercise increases the heart rate; riding a bicycle may increase the rate to 120 to 130 beats per minute. There is some variation in the figures concerning heart rates obtained by different investigators. The average for young male college students with some athletic training is perhaps as follows:

Reclining position: 58 beats per minute
Standing position: 78 beats per minute
Immediately after light exercise: 90 beats per minute

A. H. Turner reported the following average heart rates for a group of healthy young women: reclining, 64.7; seated, 71.8; standing, 85.1 beats per minute.

Emotional excitement affects the heart. Many have felt the pounding of the heart as a result of being afraid; others have felt a depressing effect from overwhelming fear.

Age and size influence the heart rate. A newborn child begins postnatal life with a heart rate of about 140 per minute. As the child becomes older and larger, the heart rate becomes slower. Typical figures are as follows:

Three years old: around 100 beats per minute
Ten to twelve years old: around 90 beats per minute
Young male adult: around 70 beats per minute
The aged: around 75 to 80 beats per minute

The heart rate is affected by the secretions of ductless, or endocrine, glands. The secretions of the thyroid and adrenal glands have a marked effect. Thyroxine, the secretion of the thyroid gland, increases the heart rate. Epinephrine, the secretion of the adrenal glands, increases both the rate and strength of the heartbeat.

The heart is also under the control of nerves from the sympathetic and parasympathetic systems. Sympathetic fibers reach the heart by way of the cardiac nerves from the cervical ganglia and the upper four or five thoracic ganglia of the sympathetic chain. The adrenergic substance released at the nerve endings increases the rate of firing at the SA node and so increases the heart rate. It also decreases the refractory period of the atrioventricular node and neuromuscular bundle, thus increasing the rate of conduction.

Fibers of the parasympathetic system reach the heart by way of the vagus nerves. Acetylcholine released at the nerve endings slows the heart rate. It depresses the rate of firing in the SA node and increases the refractory period for conduction along the neuromuscular bundle of His.

The electrocardiograph is an instrument that amplifies and records the small voltages produced by the beating heart. It is essentially a string galvanometer with a recording device. The electrocardiogram (ECG) is a record of differences in electric potential initiated by the fibers of heart

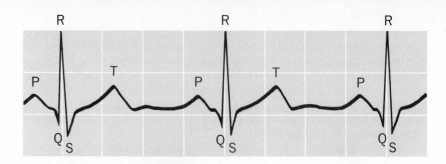

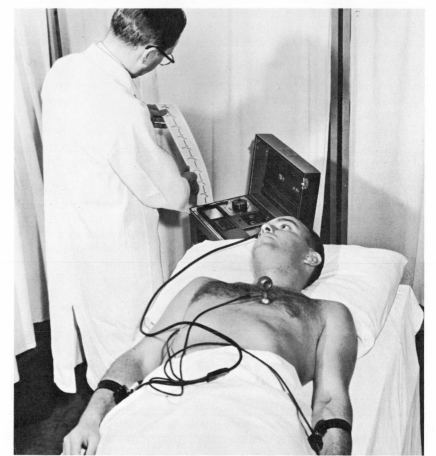

FIGURE 14.7

The drawing is taken from an electrocardiogram recording the action of a normal heart. The P wave indicates the contraction of the atria; the QRS period represents the contraction of the ventricles, whereas the T wave corresponds to the relaxation of the ventricles. The P, QRS, and T waves represent one heart cycle. Irregularities in these waves may indicate an abnormal heart condition. (*Courtesy of the American Heart Association.*)

muscle as they contract. Electric currents spread over the heart and into the tissues around the heart. Some of these electric currents spread over the surface of the body and can be recorded from electrodes placed on the right and left wrists and the right arm and the left leg or the left arm and the left leg. These are the standard positions, but other leads may be used.

The typical ECG records P, QRS, and T waves. The P wave represents excitation of the atria; QRS waves indicate excitation of the ventricles, whereas the T wave indicates repolarization of the ventricles (Figure 14.7).

The ECG provides an accurate record of heart action, which can be used in the study and diagnosis of various heart conditions.

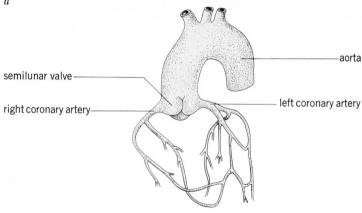

a

semilunar valve
right coronary artery

aorta
left coronary artery

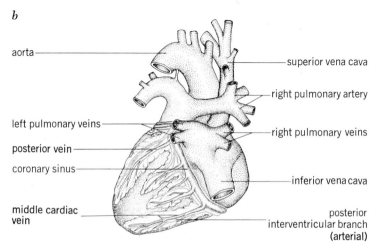

b

aorta
left pulmonary veins
posterior vein
coronary sinus
middle cardiac vein

superior vena cava
right pulmonary artery
right pulmonary veins
inferior vena cava
posterior interventricular branch (arterial)

FIGURE 14.8

Large arteries and veins associated with the heart, plus the coronary circulation: *a* the arch of the aorta and its branches. At the base of the aorta are the coronary arteries arising from aortic sinuses just above the semilunar valves. *b* The heart, posterior view, illustrating the distribution of the coronary veins. The blood is returned to the right atrium by the posterior cardiac veins, which open into the coronary sinus, but several anterior cardiac veins open directly into the right atrium.

CORONARY BLOOD VESSELS   The heart muscle does not receive its oxygen and nutrition from the blood that passes through the chambers of the heart. Cardiac tissue has its own blood supply, which comes to it by way of the right and left *coronary arteries*. Any blocking of the coronary arteries is serious, therefore, because of the inability of heart muscle to continue its work without proper nourishment. The distribution of blood vessels throughout cardiac tissue is so complete that no other tissue is better nourished. After passing through a dense capillary network, the blood passes into a venous system that eventually empties into the coronary sinus and so into the right atrium (Figure 14.8).

ARTERIES, VEINS, AND CAPILLARIES

The circulatory system includes a series of tubes through which blood is pumped to all parts of the body. The heart is the motive force; arteries, veins, and capillaries are the tubes. Arteries are responsible for carrying the blood from the heart to the tissues through even smaller branches. The occasionally used term *aterial tree* refers to the arterial trunk and its treelike branching. The trunk arteries are often very large; the aorta is about an

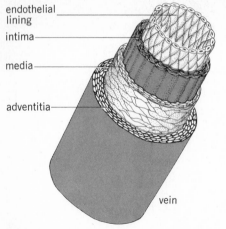

a

endothelial
lining

intima

media

adventitia

vein

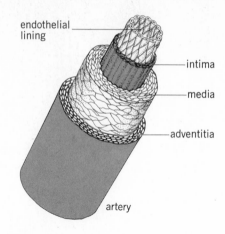

b

endothelial
lining

intima

media

adventitia

artery

FIGURE 14.9

The structure of blood vessels: *a* Veins: Veins have thinner walls than arteries, largely because the media is thinner; *b* artery: Note the thickness of the media, a mixture of elastic connective tissue and smooth muscle fibers.

inch in diameter at its base in an adult. Successive branches have increasingly smaller diameters; the smallest arteries may have diameters of only 0.3 millimeter. The total diameter of all the branches, however, is much greater than the diameter of the trunk artery, and so the blood flows with decreasing pressure and velocity through the smaller arteries and arterioles toward the capillaries.

The venous system is concerned with gathering the blood from the tissues and returning it to the heart. From the smallest venules, the succeeding branches become larger and larger as one vein joins another. The veins bringing blood into the heart are large trunk blood vessels often lying side by side with arteries of equal size.

The capillaries are minute tubes that lie between the smallest arteries (arterioles) and the venules. Smaller than the diameter of a hair, many capillaries are so tiny that even the red cells of the blood must pass through in single file. Capillaries vary in diameter from 5 to 20 microns and are ordinarily about 0.5 millimeter in length. The capillary wall consists of flat, endothelial cells that are only one cell layer thick. Through the capillary wall oxygen and food materials pass to the tissues, while carbon dioxide and other breakdown products of metabolism enter the bloodstream and are carried away.

STRUCTURE OF ARTERIES AND VEINS   The conducting vessels of the circulatory system, including the heart, develop from endothelial tubes, which persist as the endothelial linings of these structures. The endothelial intima affords a smooth surface over which the blood moves. The arterial wall consists of three layers: the endothelial lining, which contains some elastic tissue; an intermediate layer of smooth muscle and elastic tissue; and an outer layer of loose connective tissue, composed of both collagenous and elastic fibers. In large arteries such as the aorta there is little muscular tissue; the wall is composed largely of elastic tissue. This elastic tissue provides a tough yet resilient wall; it has extensibility and elasticity.

Veins have the same three layers in their walls as arteries, but since the middle layer is poorly developed, the walls of veins are thin and contain

The heart and general circulation

little muscle or elastic tissue. Walls of veins tend to collapse after death, while the thicker-walled arteries retain their shape (Figure 14.9*a*). Some veins, especially those of the extremities, have flaplike valves, which prevent backflow of the blood.

PULMONARY AND SYSTEMIC ARTERIES   There are two main divisions of the vascular system: the *pulmonary* and the *systemic.* The pulmonary circulation consists of the pulmonary artery and its branches, the capillary network in the lungs, and the pulmonary veins. The pulmonary artery, unlike other arteries, carries venous blood from the right ventricle to the lungs. It divides into the right and left pulmonary arteries, which supply the right and left lungs, respectively. The blood, after passing through a capillary network within the lungs, flows into venules and small veins forming the pulmonary veins. There are four pulmonary veins, two from each lung. They are short veins without valves; they return oxygenated blood from the lungs to the left atrium of the heart.

The aorta and its branches supplying blood to the tissues of the body, as well as the capillaries and the veins returning blood to the heart, compose the systemic circulatory system.

PRINCIPAL SYSTEMIC ARTERIES

THE AORTA   The largest trunk artery is called the *aorta.* It arises from the left ventricle, ascends to a position above the heart, arches to the left, and descends behind the heart. It is referred to by different descriptive names in different locations as the ascending aorta, the arch of the aorta, and the descending aorta. The descending aorta is the thoracic aorta above the diaphragm and the abdominal aorta below it (Plate D and Figure 14.10).

The principal branches of the aorta are as follows:

Ascending aorta	*Thoracic aorta*	*Abdominal aorta*
coronary	pericardial	celiac
brachiocephalic	bronchial	superior mesenteric
right common carotid	esophageal	renal
right subclavian	mediastinal	spermatic or ovarian
left common carotid	intercostal	inferior mesenteric
left subclavian		inferior phrenic
		lumbar
		middle sacral

RIGHT AND LEFT CORONARY ARTERIES   The first branches to arise from the ascending aorta are the coronary arteries. They arise from dilatations at the base of the aorta just above the semilunar valves. The dilatations are called *aortic sinuses.* The coronary arteries grow around the crown of the heart, giving off branches to the muscles of the atria and ventricles. The coronary arteries are of great importance, since they supply the heart musculature (Figure 14.8).

BRANCHES OF THE ARCH OF THE AORTA   *Brachiocephalic artery*   The first branch off the arch of the aorta is the brachiocephalic artery. A large artery in diameter but only 4 to 5 centimeters long, it gives rise to the right common carotid and to the right subclavian arteries (Figure 14.10).

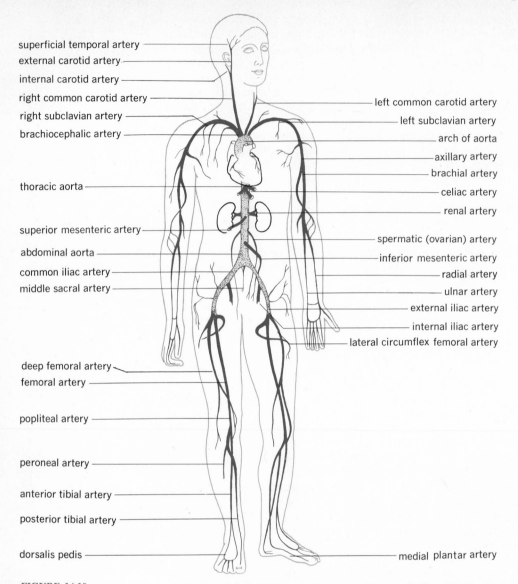

superficial temporal artery
external carotid artery
internal carotid artery
right common carotid artery
right subclavian artery
brachiocephalic artery

thoracic aorta

superior mesenteric artery
abdominal aorta
common iliac artery
middle sacral artery

deep femoral artery
femoral artery

popliteal artery

peroneal artery

anterior tibial artery

posterior tibial artery

dorsalis pedis

left common carotid artery
left subclavian artery
arch of aorta
axillary artery
brachial artery
celiac artery
renal artery

spermatic (ovarian) artery
inferior mesenteric artery
radial artery
ulnar artery
external iliac artery
internal iliac artery
lateral circumflex femoral artery

medial plantar artery

FIGURE 14.10

The principal arteries.

The heart and general circulation

Carotid arteries   The left common carotid artery branches directly from the arch of the aorta. Both right and left common carotid arteries then pass obliquely upward, covered by several muscles, especially the sternomastoid muscles. At the upper level of the larynx each artery divides into external and internal carotid arteries. The external carotid in its course up the side of the head gives off many branches and rapidly decreases in size. The internal carotid passes upward from the bifurcation of the common carotid and then turns inward and enters the cranium. It supplies the anterior part of the brain, the forehead, nose, and the orbit of the eye. The ophthalmic artery is the branch entering the orbit alongside the optic nerve. Orbital branches supply the extrinsic muscles of the eyeball and accessory structures of the eye; ocular branches supply the eyeball itself. The central artery of the retina and the ciliary arteries are branches of the ocular group. Branches

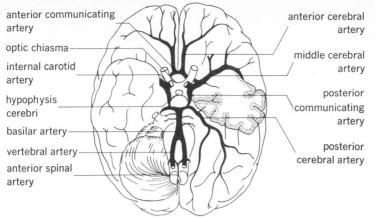

anterior communicating artery

optic chiasma

internal carotid artery

hypophysis cerebri

basilar artery

vertebral artery

anterior spinal artery

anterior cerebral artery

middle cerebral artery

posterior communicating artery

posterior cerebral artery

FIGURE 14.11

Arteries at the base of the brain. The arterial circle (circle of Willis) is formed by the two anterior cerebral arteries (which are branches of the right and left internal carotid arteries) and the anterior communicating artery; the two posterior cerebral arteries and the posterior communicating arteries.

of the internal carotid arteries (the anterior cerebral arteries), with the aid of other arteries supplying the brain, form an interesting circle at the base of the brain called the *arterial circle* (*circle of Willis*). It is a remarkable anastomosis of arteries aiding distribution of blood to the brain (Figure 14.11).

Carotid sinus and carotid body　There is a slight enlargement at the base of the internal carotid artery called the *carotid sinus*. In the region between the bases of the external and internal carotid arteries there lies a small structure composed of epithelioid tissue termed the *carotid body*. There are similar structures in the arch of the aorta, called the *aortic bodies*. Sensory receptors of these regions are of two types, chemoreceptors and pressoreceptors. The receptors play an important part in the reflex regulation of circulation and respiration (Figure 14.12).

Subclavian arteries　Although the right subclavian artery arises as a branch of the brachiocephalic artery, the left arises directly as a branch off the arch of the aorta. It arises a little above the clavicle and then passes below it into the shoulder area. The subclavian is a good example of the custom of giving the same artery different names as it passes through different regions. The *subclavian* becomes the *axillary* as it passes through the axilla of the arm. In the area above the elbow, it is the *brachial* artery. At the elbow it divides into the *ulnar* and *radial arteries*. The ulnar is the larger and passes down the ulnar side of the forearm. The radial, looking like a direct continuation of the brachial, but much smaller, passes down the radial side of the forearm toward the thumb. The pulse is commonly taken from the radial artery at the wrist. Branches of the two arteries anastomose in the palm through the volar arches (Figure 14.10).

BRANCHES OF THE THORACIC AORTA　The arteries of the thoracic region are numerous but small. Included in this group are the pericardial, bronchial, esophageal, mediastinal, and intercostal arteries. Briefly, the pericardial arteries are small arteries on the posterior surface of the pericardium. Branches of the bronchial arteries supply the walls of the bronchial tubes and the tissues of the lungs. The esophageal branches (usually four or five small branches) anastomose with other small arteries to form a network along the esophagus. The mediastinal arteries are numerous small branches in the posterior part of the mediastinum. They supply the lymph nodes and areolar tissue of this region. Nine pairs of intercostal arteries arise from the dorsal

Principal systemic arteries

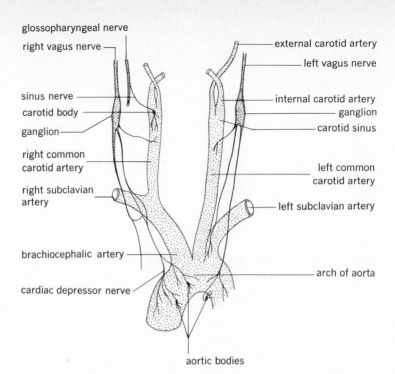

glossopharyngeal nerve

right vagus nerve

external carotid artery

left vagus nerve

sinus nerve

carotid body

ganglion

internal carotid artery

ganglion

carotid sinus

right common
carotid artery

left common
carotid artery

right subclavian
artery

left subclavian artery

brachiocephalic artery

arch of aorta

cardiac depressor nerve

aortic bodies

FIGURE 14.12

Diagram indicating the branches off the arch of the aorta and the location of the carotid and aortic bodies.

side of the thoracic aorta and supply the muscles and skin of the back as well as the intercostal muscles.

BRANCHES OF THE ABDOMINAL AORTA The abdominal aorta diminishes considerably in size as it progresses posteriorly, since it gives off a number of large branches. Its branches with a few exceptions supply the walls and viscera of the abdominal cavity.

Celiac artery This is the first large branch of the aorta below the diaphragm (Figure 14.13). The celiac has three branches: the *left gastric, splenic,* and *hepatic.* The left gastric is a small artery passing from left to right along the lesser curvature of the stomach. Toward the right it forms an anastomosis with the right gastric branch of the hepatic artery. The gastric arteries give off branches to both the anterior and posterior surfaces of the stomach. The splenic is the largest branch of the celiac artery. It extends behind the stomach and above the pancreas to the spleen, where its branches enter at the hilus to supply the splenic tissue. Small branches are given off to the pancreas. Close to the spleen there is a large branch, the left gastroepiploic, which passes around the lower or greater curvature of the stomach and anastomoses with the right gastroepiploic. The epiploic arteries give off numerous branches to the stomach and greater omentum. The hepatic artery supplies the tissues of the liver. Among its several branches is the cystic artery, which branches over the gallbladder.

Superior mesenteric artery The superior mesenteric is a large artery that sends branches to the greater part of the entire intestinal tract. Exceptions are the upper part of the duodenum and the distal part of the colon. Its branches are the mesenteric tissues for support in order to reach the intestine. It has its origin about $\frac{1}{2}$ inch below the origin of the celiac artery.

The heart and general circulation

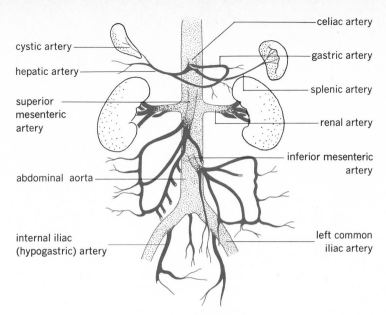

cystic artery

hepatic artery

superior mesenteric artery

abdominal aorta

internal iliac (hypogastric) artery

celiac artery

gastric artery

splenic artery

renal artery

inferior mesenteric artery

left common iliac artery

FIGURE 14.13
Principal branches of the abdominal aorta.

Renal arteries The renal arteries arise laterally from the aorta and are distributed to the kidneys. They are large arteries, each breaking up into several branches before entering the hilus of the kidney.

Spermatic arteries Arising from the anterior part of the aorta below the renal arteries, the spermatics are long, slender arteries that course downward and pass through the inguinal canals. They extend into the scrotum, and their branches terminate in the testes (Figure 14.10 and Plate D).

Ovarian arteries The female ovarian arteries are homologous with the spermatic arteries of the male. They arise at a similar location and descend into the pelvis, where each is distributed to an ovary. One branch forms an anastomosis with the uterine artery.

Inferior mesenteric artery Branching from the aorta not far above the place where the aorta itself divides to form the two common iliacs, the inferior mesenteric supplies the latter part of the colon and partially supplies the rectum. It is smaller than the superior mesenteric artery.

Phrenic arteries Small arteries supplying the diaphragm are the phrenic arteries.

Lumbar arteries Similar to the intercostal arteries, the lumbar arteries arise from the posterior surface of the aorta in the region of the first four lumbar vertebrae. There are usually four pairs of lumbar arteries. They supply the muscles and skin of the lumbar region of the back.

Middle sacral artery A small artery, the middle sacral artery, originates from the posterior surface of the aorta about $\frac{1}{2}$ inch above its bifurcation into the common iliac arteries. It descends medially along the anterior surface of the fourth and fifth vertebrae, the sacrum, and the coccyx. Terminal branches supply nodules on the anterior surface of the coccyx, which compose the coccygeal gland.

Principal systemic arteries

COMMON ILIAC ARTERIES   The aorta divides at the level of the fourth lumbar vertebra and gives rise to the common iliac arteries. Each artery passes downward for about 2 inches, diverges, and gives rise to two branches: the internal iliac, or hypogastric, and the external iliac.

INTERNAL ILIAC, OR HYPOGASTRIC, ARTERIES   The internal iliac artery has numerous branches supplying the pelvic muscles and viscera. Branches extend also to the gluteal muscles, to the medial area of each thigh, and to the external genitalia. The uterine artery in the female is an important branch. It becomes the largest branch of the internal iliac during pregnancy. Pelvic viscera supplied by branches of the internal iliac include the urinary bladder, the rectum, the prostate gland in the male, and the uterus and vagina in the female.

Internal iliac, or hypogastric, arteries in fetal circulation   The internal iliac arteries are large, vital arteries in the fetal circulatory scheme. They represent a direct extension of the common iliacs in their fetal development. Each internal iliac gives off branches to the pelvic viscera and then extends up the anterior wall of the abdomen as an umbilical artery. Reaching the umbilicus (navel), each umbilical artery passes out through the umbilical cord to the placenta (see Figure 14.16). After the umbilical cord is severed at birth, the greater part of the abdominal umbilical artery atrophies and leaves in its place a strand of fibrous tissue called the *lateral umbilical ligament.* The proximal part remains as the umbilical artery of postnatal life. The umbilical artery gives rise to the vesical arteries, which supply the urinary bladder.

EXTERNAL ILIAC ARTERY   Extending from the base of the internal iliac about 4 inches downward, the external iliac artery diverges through the pelvis to the point where it passes underneath the inguinal ligament; here it enters the thigh and becomes the femoral artery. Like the subclavian artery, the external iliac is called by different names in different parts of its course.

FEMORAL ARTERY   After passing below the inguinal ligament, the femoral artery lies in a compartment of the femoral sheath, with the femoral vein on its medial side and the femoral nerve on its lateral side. The blood vessels and nerve are close to the anterior surface of the thigh at this point. They are covered by some fatty tissue. There are lymph nodes and lymphatics in this region also. The femoral artery sends small branches to the wall of the abdomen, to the external genitalia, and to the muscles of the thigh.

The *femoral artery* becomes the *popliteal artery* as it passes through the popliteal space at the back of the knee joint. Below the knee it gives rise to the *posterior tibial* and *anterior tibial arteries.* The posterior tibial gives off a large branch, the *peroneal artery.* Both arteries descend beneath the muscles of the calf of the leg. The posterior tibial arteries form the *plantar arteries* of the sole of the foot. The anterior tibial artery passes forward between the tibia and the neck of the fibula and extends downward close to the tibia. On the anterior surface of the ankle bones it becomes the *dorsalis pedis artery* and branches over the top of the foot.

The heart and general circulation

The venous system is responsible for returning blood to the heart. The veins originate in the tiny venules that receive blood from the capillary network. Through a system of coalescence of ever-larger vessels, the large trunk veins are formed. The systemic veins are the pathway for the return flow from the systemic arteries. They return the blood from all over the body to the heart. Since the flow of blood is from smaller vessels to larger ones, it is advantageous to study the veins in this manner—not by working out from the large vessels at the heart, as in the study of the arterial system. The following are some of the principal veins.

PULMONARY VEINS The blood returns to the heart by way of the pulmonary veins. Originating in the capillary network of the alveoli, successively larger branches finally form two pulmonary veins emerging from each lung. They empty separately into the left atrium of the heart. There are no valves in these veins.

VEINS OF THE HEART The return flow from the coronary arteries is taken up by the cardiac veins, most of which empty into a large sinus, the *coronary sinus*, which empties into the right atrium of the heart. Its opening is partially protected by a valve, consisting of a single semicircular fold of membrane. The valve of the coronary sinus prevents backflow of blood into the sinus during contraction of the right atrium (Figure 14.8).

LARGER VEINS OF THE HEAD AND NECK *External jugular veins* Superficial veins of the scalp and the deeper veins of the face flow into the external jugular vein. The external jugular descends over the sternomastoid muscle, but underneath the platysma, and empties into the subclavian vein on either side (Figure 14.14 and Plate E).
The brain receives a very abundant blood supply from the internal carotid arteries, and numerous veins on the surface and within the brain are concerned with gathering the blood and returning it by way of the internal jugular veins. Branches of external or surface veins can be seen, for the most part, in the sulci or gyri of the cortex. Deep within the substance of the brain are many veins that drain into large sinuses. These veins and sinuses have very thin walls and possess no valves. Veins within the brain lack a muscular coat.

Internal jugular veins The internal jugular vein arises as a continuation of the transverse sinus. It is a large vein descending laterally beside the common carotid artery. At its base it joins the subclavian vein; together they form the brachiocephalic vein. In addition to its principal function of returning the blood from the brain, the internal jugular receives tributaries from the face and from the neck. The right vein is usually larger than the left, and both are much larger than the external jugular veins.

Brachiocephalic (innominate) veins The right and left brachiocephalic veins vary in length and position. The right brachiocephalic is almost vertical in position and appears to be an extension of the right internal jugular vein. It is only about 1 inch in length. The left brachiocephalic is almost horizontal in position as it passes above the heart to join the right brachiocephalic and form the superior vena cava. It is $2\frac{1}{2}$ to 3 inches in length. The brachio-

The venous system

363

cephalic veins receive the return flow from the internal jugular and sub-clavian veins, but in addition, the vertebral, deep cervical, inferior thyroid, and internal thoracic veins are tributaries.

Superior vena cava Formed by the confluence of the right and left brachio-cephalic veins, the superior vena cava extends downward about 3 inches and empties into the right atrium of the heart. It receives the azygos vein as a tributary. The superior vena cava contains no valves.

Azygos vein The azygos vein originates along the dorsal wall of the abdomen as the ascending lumbar vein. It ascends through the thorax along the right side of the vertebral column and empties into the superior vena cava just above the pericardium. From the standpoint of comparative anatomy, the azygos has always emptied into the veins above the heart, not into the inferior vena cava (Figure 15.10).

Hemiazygos vein Arising from the lumbar vein and occasionally from the left renal vein, the hemiazygos vein ascends to the left of the vertebral column as high as the eighth or ninth thoracic vertebra, where it turns to the right and enters the azygos vein. There is also a small accessory hemiazygos vein. The hemiazygos varies in regard to its size, position, and connections (Figure 15.10).

VEINS OF THE UPPER EXTREMITY There are two sets of veins in the forelimb: deep and superficial. The deep veins are associated with arteries and are known as the radial, ulnar, brachial, axillary, and subclavian veins. The superficial veins form a great network of anastomoses with each other and with the deep veins. There are valves in both the superficial and deep veins. Both groups finally open into the axillary and subclavian veins.

Cephalic vein The cephalic vein arises in the radial network of superficial anastomoses. Above the elbow it passes along the lateral side of the biceps muscle. At the shoulder it penetrates deeply and empties into the axillary vein.

Basilic vein The basilic vein has its origin in a venous network on the ulnar (little-finger) side of the back of the hand. It curves around to the anterior surface before it reaches the elbow and continues upward along the medial side of the biceps brachii muscle. At the shoulder it unites with the brachial vein to form the axillary vein. The axillary becomes the subclavian vein in the area along the clavicle. The thoracic duct of the lymphatic system flows into the left subclavian vein at its junction with the internal jugular. On the right, the right lymphatic duct enters the right subclavian vein at the same junction.

VEINS OF THE LOWER EXTREMITY The veins of the lower extremity are divided into superficial and deep veins, as are the veins of the upper extremity. The superficial veins are formed of a great network of anastomoses close to the surface, while the deep veins are covered deeply with muscles and follow the large arterial trunks.

Superficial veins The great *saphenous vein* is the longest vein in the body. It receives branches from the sole and dorsal aspect of the foot and crosses the ankle on the medial side, just anterior to the medial malleolus. It ascends along the medial side of the leg and thigh and enters the femoral vein an inch or so below the inguinal ligament. There is also a small saphenous vein, which passes up the back of the leg and empties into the popliteal vein at the knee. The saphenous veins have valves, but not as many as the deep veins of the lower extremity. The long saphenous vein is especially likely to enlarge under conditions of prolonged standing because of back pressure of the blood. Such enlarged veins are called *varicose veins*.

Deep veins   Below the knee there are two veins placed beside each artery. The names of the veins correspond to the names of the arteries that they accompany. The *anterior* and *posterior tibial veins* arise from veins of the foot. They ascend through the leg and drain into the *popliteal vein* at the back of the knee. The popliteal becomes the *femoral vein* as it passes through the upper two-thirds of the thigh. Just above the inguinal ligament it becomes the *external iliac vein*.

VEINS OF THE PELVIS AND ABDOMEN   The *internal iliac veins (hypogastric)* join the external iliac veins to form the *common iliac veins*. Tributaries to the internal iliac veins are numerous and carry the return flow from the internal iliac arteries. The *uterine veins* of the female are important tributaries. The common iliac veins unite to form the inferior vena cava at the level of the fifth lumbar vertebra.

The *inferior vena cava* is the largest vein returning blood from the lower parts of the body. Formed by the union of the common iliac veins, it extends upward through the abdomen and thorax to the right atrium of the heart. It lies deeply protected by the viscera just anterior to the vertebral column and to the right of the aorta. Numerous small tributaries enter the inferior vena cava. For the most part these carry the return flow from smaller branches of the abdominal aorta, and their names correspond to names given to the arteries. The largest tributaries are the short renal veins, which enter at nearly right angles.

The *renal veins* return the blood from the kidneys. The left renal vein is a little higher, since the left kidney is higher than the right. It is also longer, crossing over the aorta. The left spermatic vein in the male, or the left ovarian in the female, and the left suprarenal are tributaries (Figure 14.14). The renal circulation provides for the elimination of waste products from the blood (see Chapter 19).

The *hepatic veins* arise in the liver and flow into the inferior vena cava. There are usually two or three large veins and several small veins. They carry the return flow from the hepatic artery and the blood that comes to the liver through the portal system.

HEPATIC PORTAL SYSTEM   Veins ordinarily transmit blood directly back to the heart; there are certain pathways, however, in which the blood passes through a capillary network before returning to the heart. Such a venous pathway has capillaries at either end and is called a *portal system*. The portal system in human beings is a hepatic portal system; that is, blood returning from the spleen, stomach, and intestine enters the portal vein and passes through capillaries and sinusoids in the liver before returning to the heart. The *splenic vein* returns blood from the spleen, stomach, and pancreas. The *inferior mesenteric vein*, returning blood from the left colon, is a tributary of the splenic vein. The *superior mesenteric vein* joins the splenic to form the *portal vein*. The mesenteric blood vessels lie on a supporting tissue called *mesentery* (Figure 14.15).

The *splenic vein* arises from the union of several branches at the spleen.

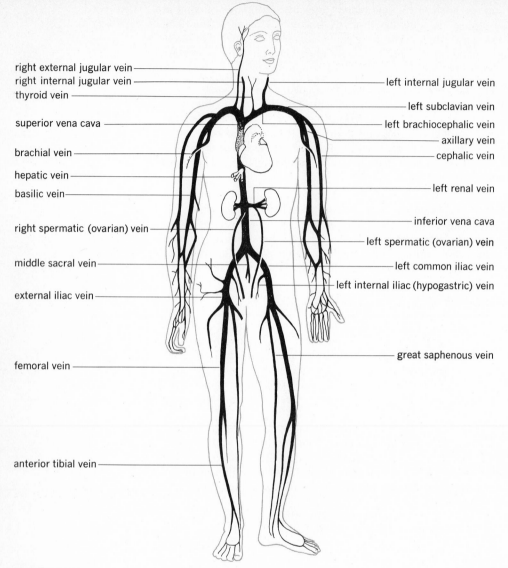

right external jugular vein

right internal jugular vein

thyroid vein

superior vena cava

brachial vein

hepatic vein

basilic vein

right spermatic (ovarian) vein

middle sacral vein

external iliac vein

femoral vein

anterior tibial vein

left internal jugular vein

left subclavian vein

left brachiocephalic vein

axillary vein

cephalic vein

left renal vein

inferior vena cava

left spermatic (ovarian) vein

left common iliac vein

left internal iliac (hypogastric) vein

great saphenous vein

FIGURE 14.14

The venous system. Only the major veins are shown.

The heart and general circulation

It passes behind the pancreas and receives several small pancreatic veins. The left gastroepiploic vein from the greater curvature of the stomach is a tributary. The largest vein joining the splenic is the inferior mesenteric. This vein transmits blood from the rectum and the descending colon.

The *superior mesenteric vein* is the returning pathway for blood from the small intestine and the ascending and transverse colon. It is the larger of the two mesenteric veins. Many of its tributaries correspond to branches of the superior mesenteric artery. It joins the splenic vein at the level of the pancreas, and together they form the portal vein.

The *portal vein* is a short but large vein. It extends upward about 3 inches from its origin, near the neck of the pancreas, to the liver, where it divides into a right and left branch as it enters the liver. The left gastric vein from the lesser curvature of the stomach is a tributary, and at the liver

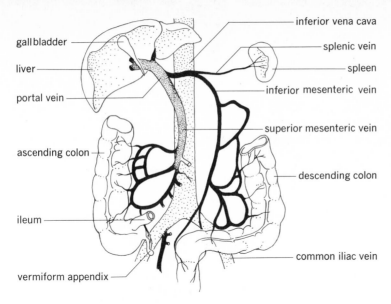

gallbladder

liver

portal vein

ascending colon

ileum

vermiform appendix

inferior vena cava

splenic vein

spleen

inferior mesenteric vein

superior mesenteric vein

descending colon

common iliac vein

FIGURE 14.15
The hepatic portal system.

the right branch usually receives the cystic vein from the gallbladder. Within the liver, the branches of the portal vein break up into enlarged capillary-like blood vessels called *sinusoids*. The blood leaves the liver through the sinusoids that unite to form small branches of the hepatic veins, by which blood enters the inferior vena cava (Figure 14.14).

The portal system is concerned with the transportation of absorbed food products from the intestine to the liver. Glucose, in particular, is stored in the liver and in other tissues in the form of glycogen. The portal system acts also as a reservoir for the storage of blood. The spleen has great capacity for holding blood in reserve and may be considered an important part of the portal storage system.

FETAL CIRCULATION

The circulatory plan of the fetus is somewhat different from that of the adult, largely because the lungs are not functional in the fetus. The *placenta* acts as an organ of respiration by providing for the exchange of O_2 and CO_2 between the maternal and fetal bloodstreams. The placenta provides for passage of food products into the fetal circulation and also is concerned with excretion of waste products. The fetal portion of the placenta is a flat disk of highly vascular tissue attached to the lining of the uterus.

The *umbilical cord* extends from the placenta to the umbilicus, or navel, of the fetus. It contains the umbilical vein and two umbilical arteries. There is also a mucoid connective tissue in the umbilical cord, called *Wharton's jelly*.

The *umbilical vein* within the abdomen of the fetus passes upward to the undersurface of the liver. A considerable portion of the bloodstream passes through the liver. The umbilical vein gives off two or three branches that enter the liver directly. The portal vein joins a second large branch before it enters the liver. A smaller branch, the *ductus venosus*, bypasses the liver and enters the inferior vena cava. The blood that passes through the liver enters the inferior vena cava through the hepatic veins.

Fetal circulation

367

Upon entering the inferior vena cava the blood from the hepatic veins becomes mixed with blood returning from the lower extremities of the fetus. It enters the right atrium, and much of it passes through an opening in the wall between the two atria directly into the left atrium. The opening is called the *foramen ovale*. Some blood from the right atrium passes into the right ventricle and out to the lungs through the pulmonary artery, but the tissues of the lungs do not require a great amount of blood so long as the lungs are nonfunctional. Even so, much of the blood that passes out over the left branch of the pulmonary artery flows directly into the dorsal aorta through a connecting blood vessel, called the *ductus arteriosus*.

Blood from the left side of the heart is forced out into the aorta, as in the adult. The descending aorta supplies the abdominal viscera and the lower extremities, but a considerable amount of this blood is returned to the placenta by the two umbilical arteries. The umbilical arteries in the fetus are almost direct extensions of the internal iliac arteries.

A diagram of fetal circulation (Figure 14.16) shows that the umbilical vein carries the most highly oxygenated blood. Much of the food-laden blood passes through the liver, and this organ is proportionately large in the fetus. Since much of the blood from the umbilical vein goes to the left side of the heart directly, the blood passing over the arch of the aorta and to the brain is fairly rich in oxygen and food products.

MODIFICATIONS IN THE VASCULAR SYSTEM AT BIRTH The lungs become functional at birth, and more blood passes through the pulmonary arteries to the lungs. When the placental supply of oxygen is cut off and the fetus also loses its ability to eliminate CO_2 through the placenta, the CO_2 content of the blood rises, and the respiratory center is stimulated reflexly. The infant, therefore, gasps for breath, and respiration is established. A secondary valvular fold covers the foramen ovale, incompletely at first, but it usually adheres to the septum and closes the opening completely toward the end of the first year after birth.

The ductus arteriosus degenerates and is modified into a ligament connecting the left pulmonary artery and the aorta. The umbilical vein and the ductus venosus also persist only as ligaments. Only the basal or proximal parts of the umbilical arteries remain functional. The distal parts from the urinary bladder to the umbilicus degenerate within a few days after birth. In the adult, fibrous cords persist, representing the obliterated branches of these arteries.

The changes in circulation that occur at birth, while pronounced, are not revolutionary. As the fetal heart and lungs mature, the amount of blood passing through the foramen ovale and the ductus arteriosus is regulated to meet the needs of the developing body. In the early stages of development the greater portion of the blood passes directly across the atrial septum to the left side of the heart. As the lungs mature, the foramen ovale is more occluded and more blood is directed to the lung tissues. At birth practically all the blood entering the right side of the heart passes over the pulmonary artery to the lungs, but the change-over is not as great as the analogy of suddenly putting a lid over a wide-open valve. The foramen ovale may be considered as balancing the work of the atria. In the same way, the ductus arteriosus serves to balance the workload of the ventricles. At birth the ductus arteriosus is constricted, probably by the action of its smooth muscle. The constriction is made permanent by replacement of its cavity by connective tissue. Occasionally a congenital condition develops owing to failure

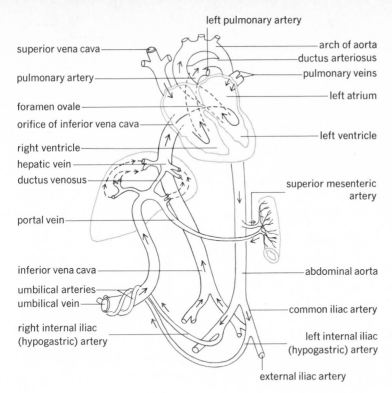

left pulmonary artery

superior vena cava

pulmonary artery

foramen ovale

orifice of inferior vena cava

right ventricle

hepatic vein

ductus venosus

portal vein

inferior vena cava

umbilical arteries

umbilical vein

right internal iliac
(hypogastric) artery

arch of aorta

ductus arteriosus

pulmonary veins

left atrium

left ventricle

superior mesenteric
artery

abdominal aorta

common iliac artery

left internal iliac
(hypogastric) artery

external iliac artery

FIGURE 14.16
Plan of the fetal circulation in a
mature fetus. Arrows indicate the
direction of blood flow.

of the foramen ovale or the ductus arteriosus to close properly. The pulmonary blood supply is inadequate, the blood is insufficiently oxygenated, and a condition of *cyanosis* develops. Babies with such circulatory defects are commonly called *blue babies*.

Heart defects ordinarily are not as simple as described above. The entrance to the pulmonary arteries may be too narrow, or the aorta may be displaced to the right, thus constricting both the aorta and the pulmonary artery, or the septum between the ventricles may not be complete, leaving an opening in the wall between the two lower chambers.

SUMMARY

The heart, located in the interpleural space, or mediastinum, has four chambers: two atria and two ventricles. Blood enters the right atrium, passes through the tricuspid valve, and fills the right ventricle. As the ventricle contracts, the blood is forced up into the pulmonary artery, which carries it to the lungs. Backflow is controlled by the closing of the pulmonary semilunar valves. The blood returns to the left atrium by way of the pulmonary veins. It passes from the left atrium to the left ventricle through the mitral valve. When the left ventricle contracts, the blood is forced up into the aorta and out over the body. Backflow in the aorta is controlled by the closure of the aortic semilunar valves.

The heart is enclosed in a two-layered membrane called the pericardium. The endocardium is a serous membrane lining the cavities of the heart. The myocardium refers to the musculature of the heart. Cardiac tissue is an involuntary, striated type with transverse bars.

The wave of contraction that spreads over the atria originates at the sinoatrial (SA) node. When the excitation reaches the atrioventricular node, it is conveyed to the ventricular musculature by way of the neuromuscular bundle of His. The atrioventricular bundle is located in the median septum of the heart.

Heart sounds are caused by the closing of valves. The first sound is caused primarily by the closure of the atrioventricular valves, whereas the second sound indicates the closing of the semilunar valves.

Systole refers to the contracting phase of the heart, usually the contraction of the ventricles. Diastole occurs when the heart muscles are relaxed and the chambers fill with blood.

The heart rate varies, depending upon a number of factors. The average heart rate for a young adult male, seated, is 70 beats per minute. Heart rates for women are a little faster than those recorded for men. The heart rate of the female fetus is around 140 to 145 beats per minute, whereas the average male fetus is 130 to 135.

Heart muscle derives its oxygen and nutrition from the coronary arteries. Blocking of the coronary arteries by a clot (thrombosis) or by deposition of fatty materials can lead to serious heart conditions. The coronary arteries are the first branches off the ascending aorta.

There is an unequal branching of arteries off the arch of the aorta. On the right, the brachiocephalic artery gives rise to the right common carotid and the right subclavian; on the left, the left common carotid and left subclavian arise separately and directly off the arch.

The principal branches of the abdominal aorta are the celiac, superior mesenteric, renal, spermatic or ovarian, inferior mesenteric, lumbar, and middle sacral.

An enlargement at the base of the internal carotid artery is called the carotid sinus. The carotid body lies between the bases of the external and internal carotid arteries. Around the arch of the aorta lie the aortic bodies. These structures contain chemoreceptors and pressoreceptors for the reflex regulation of circulation.

The venous system functions to return the blood to the heart. The pulmonary veins return freshly oxygenated blood from the lungs to the heart. The return flow from the head and brain enters the superior vena cava by way of the external and internal jugular veins, which, in turn, flow into the brachiocephalic veins.

The largest veins of the arm are the cephalic and basilic. They carry venous blood to the subclavian veins.

The azygos vein originates along the dorsal wall of the abdomen as the ascending lumbar vein. It traverses the thorax and empties into the superior vena cava. One of its tributaries is the hemiazygos vein.

The largest superficial vein of the lower extremity is the great saphenous vein. The anterior and posterior tibial veins flow into the popliteal vein back of the knee joint. The popliteal vein becomes the femoral vein, a deep vein of the thigh. The femoral vein becomes the external iliac vein of the pelvic region.

The internal iliac veins (hypogastric), with the external iliac veins, form the common iliac veins. The uterine veins of the female are important tributaries to the internal iliac veins. The common iliac veins form the inferior vena cava.

The inferior vena cava is a large vein of the trunk. It lies beside the dorsal aorta. It has numerous small tributaries, but relatively few large veins enter it. The two largest tributaries are the renal veins and the hepatic vein.

The hepatic portal system consists of the return flow of blood by way of the splenic, inferior mesenteric, and superior mesenteric veins. These veins form the portal vein, which enters the liver. The portal system returns food-laden blood from the intestine to the liver.

Fetal circulation differs from postnatal circulation in several respects. In the embryo and fetus, the umbilical cord extends from the placenta to the umbilicus of the fetus. It contains the umbilical vein and two umbilical arteries. The umbilical vein carries oxygen and food-laden blood from the placenta. The umbilical vein is joined by the portal vein before it enters the liver. A more or less direct channel through the liver is called the ductus venosus.

Within the fetal heart, the blood may pass directly from the right atrium to the left atrium through the foramen ovale, or it may enter the right ventricle and pass out over the pulmonary artery. Since the lungs are not functional in the fetus, much of this blood passes directly into the aorta by way of a connection between the left pulmonary artery and the aorta, called the ductus arteriosus.

1 Locate the heart.
2 Describe its gross structure.
3 Tell what you know about the microscopic structure of heart muscle.
4 Explain the function of the heart valves.
5 Give an explanation of the factors causing heart sounds.
6 Indicate some of the factors influencing the heart rate.
7 Discuss the function of capillaries.
8 Trace the blood from the liver until it returns to that organ.
9 How do veins differ from arteries?
10 Name and locate some of the larger veins.
11 Trace the blood from the heart to the small intestine and back to the heart.
12 Describe fetal circulation.

ADOLPH, E. F.: The Heart's Pacemaker, *Sci. Am.*, **216**:32–37 (1967).
GUYTON, A. C., T. G. COLEMAN, and H. J. GRANGER: Circulation: Overall Regulation, *Ann. Rev. Physiol.*, **34**:13–90 (1972).
LEVY, M. N., and R. M. BERNE: Heart, *Ann. Rev. Physiol.*, **32**:373–414 (1970).
NETTER, F.: "The Heart," CIBA Medical Illustrations, vol. 5, CIBA Corp., Summit, N.J., 1969.
ROSS, G.: The Regional Circulation, *Ann. Rev. Physiol.*, **33**:445–478 (1971).
TURNER, ABBY H.: The Circulatory Minute Volumes of Healthy Young Women in Reclining, Sitting and Standing Positions, *Am. J. Physiol.*, **80**:601–630 (1927).

Some standard anatomy textbooks are listed at the back of this book. These books may be consulted for detailed descriptions and illustrations of the circulatory system.

chapter 15
physiology of circulation

The blood flows rapidly from the deeper parts of the body to the extremities and returns to the heart in a matter of about 60 seconds. It can take various routes, and some are longer than others. For example, the heart-lung cycle is a much shorter circuit than the round trip between the heart and the foot.

The blood also travels at different rates in various parts of the circulatory system. The mean velocity in the larger arteries is about 20 centimeters per second. The rate of flow in arteries is not uniform, being somewhat faster in the larger than in the smaller ones. Furthermore, the flow is faster at the crest of the systolic wave than it is during diastole.

RATE OF FLOW THROUGH CAPILLARIES   The velocity of the blood slows to about 0.5 millimeter per second as it passes through the capillaries. Though the diameter of a single capillary is exceedingly small, the total diameter of all the capillaries exceeds the total diameter of the arteries. Therefore, like a river flowing into a great open area, the arterial blood flows slowly through the great capillary bed (Figure 15.1).

Some of the capillaries provide a direct connection between the smallest arterioles and the venules; these are called *a-v capillaries* for convenience. There is always a small volume of blood flowing through these capillaries even during vasoconstriction. The a-v capillaries provide an important pathway for circulation between the peripheral ends of arteries and veins.

Vasoconstriction in the smallest arterioles (*metarterioles*) limits the flow through the capillaries. These small tubes have cells of smooth muscle in their walls and so are capable of constriction or relaxation. The typical capillary branches off the smaller arteriole, often at nearly right angles to it. The flow of blood into a typical capillary is controlled by a band of smooth muscle called the *precapillary sphincter,* at the place where it branches off the arteriole. It can close to such an extent that no blood passes through the capillary (Figure 15.2). In an area of tissue under observation, vasodilatation brings to view hundreds of capillaries that were not visible under conditions of normal vasoconstriction. The skin may appear pale, since surface blood vessels constrict when the body is exposed to chilling. Constriction of surface blood vessels tends to keep most of the blood away from the cool surface and so reduces heat loss. Such vessels may dilate locally when the skin is exposed to cold, as seen, for example, in the pink cheeks of healthy children. Surface vessels commonly expand and protect a tissue from injury by extreme cold. On the other hand, surface blood vessels dilate after exposure to heat, or after exercise in a warm room, and a greater volume of blood is brought to the cooler skin area. Sweat glands pour perspiration onto the surface of the skin. As sweat evaporates, the surface is cooled and the blood loses heat to the cooler area. Vasoconstriction and vasodilatation of blood vessels located in the skin play an important part in temperature regulation.

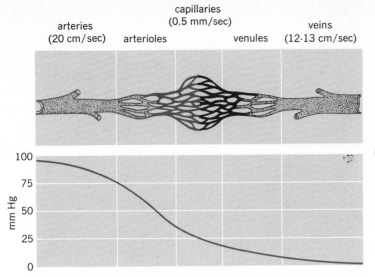

arteries
(20 cm/sec)

arterioles

capillaries
(0.5 mm/sec)

venules

veins
(12-13 cm/sec)

FIGURE 15.1

Diagram illustrating the velocity of blood flow and blood pressure in various parts of the circulatory system. The total capillary bed is much greater than it appears to be in the diagram. Mean arterial pressure is usually a little less than the average of systolic plus diastolic pressures, (120 + 80)/2.

BLOOD FLOW IN VEINS The venous system collects blood from the great capillary bed, and the channel narrows down again as the blood is returned to the heart. The velocity of the blood in the veins may increase, therefore, to more than 250 times the velocity in the capillaries. Since the venous bed is greater than that of the arteries, one would not expect the velocity of blood in the veins to be as great as in the arteries.

Many conditions influence the rate of flow. As in arteries, the velocity is greater in the large vessels than in the small ones. The velocity increases during exercise and is directly proportional to the blood volume of any given vein. The beating of the heart helps to move the blood in the veins, but other factors supplement the heart action. The massaging action of skeletal muscles is of great value. The veins of the lower extremities especially have numerous valves that prevent backflow. The muscle action of exercise promotes the return flow through the veins, but standing quietly for long periods places unusual pressure on the veins of the legs. Elevation of a limb above the level of the heart permits the blood to flow toward the heart, but the effect is not as great as might be expected. Veins tend to collapse

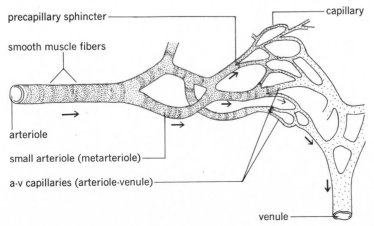

precapillary sphincter

smooth muscle fibers

capillary

arteriole

small arteriole (metarteriole)

a-v capillaries (arteriole-venule)

venule

FIGURE 15.2
Capillary pathways.

Physiology of circulation

373

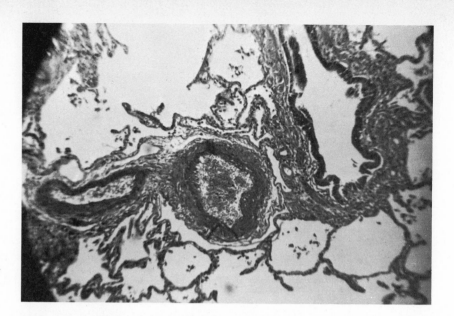

under these conditions and thus increase the resistance to the movement of blood within their walls (Figure 15.3).

Inspiratory movements assist the flow of blood into the atria from the great veins. As the volume of the thorax is increased, air rushes into the lungs. The heart lies within the thoracic cavity, and the decrease in pressure within the cavity also aids in filling the atria of the heart with blood.

Blood vessels, in general, are distensible but veins are much more so than arteries. The veins of the skin, like other surface blood vessels, react to temperature; they dilate when the skin surface is warm and contract as a reaction to chilling. All the veins act as blood reservoirs, but the deep ones especially are capable of holding a considerable quantity of blood. More than half the total blood volume is contained in the veins alone.

THE PULSE   As the heart beats, the ventricles regularly force blood out into the aorta. The force of contraction in the ventricles initiates a wave of increased pressure, which starts at the heart and travels along arteries until it is lost in the capillaries. This wave of increased pressure is the pulse. It can be felt in any of the arteries that are close to the surface. It is common practice to take the pulse from the radial artery at the wrist. The pulse rate is, of course, the same as the heart rate, around 70 per minute in the resting state. The pulse should not be confused with the movement of the blood itself. The pressure wave travels 5 to 8 meters per second, depending upon the age of the subject. As the arteries lose their elasticity with age, the velocity of the pulse wave increases. The mean velocity of the blood has been estimated to be only about 20 centimeters per second. It may travel as fast as 50 centimeters per second in the aorta, but it moves much more slowly in the smaller arteries. It is evident that the pulse wave felt in the artery could not be caused by the passage of the blood itself.

The elasticity of the arterial wall causes the pressure wave to travel more slowly than it would through unyielding pipes. It also permits the blood to flow in a continuous stream through the capillaries. If a larger artery is cut, the blood escapes in spurts; but normally, in the intact circulatory system, much of the force of ejection from the heart is absorbed by the elasticity of the arterial wall. The blood, therefore, moves steadily forward even during the diastolic phase of the heart. If a circulatory scheme were constructed of glass tubing, which is inelastic, the blood would move by spurts in all parts of the apparatus.

A physician or trained observer can learn a great deal about the circulatory system by simply taking the pulse. Not only is the heart rate observed, but also its strength and regularity. Even the blood pressure can be judged by noting whether the pulse is soft or hard. A soft pulse means that arterial tension is low and the artery has retained its elasticity. A hard pulse indicates high arterial tension and that the wall of the artery is hard and inelastic.

BLOOD PRESSURE

The heart pumps blood against the frictional resistance of blood vessels. Cardiac output is always an important factor in blood pressure. If the heart rate or the stroke volume is increased, the blood pressure is elevated. Conversely, a slowing of the heart rate or weakening of its beat causes the blood pressure to fall. Pressure is highest in the large arteries near the heart and diminishes as the blood passes into smaller vessels farther away from the source of pressure.

The viscosity of the blood is a factor in resistance to flow; it increases as the concentration of corpuscles increases. A greater concentration of plasma proteins can increase the viscosity also. Blood of high viscosity offers greater resistance to flow than blood of low viscosity.

The quantity of blood to be moved is a factor influencing blood pressure. Though the body can withstand relatively small losses of blood without an accompanying loss of pressure, the loss of a large amount of blood causes the pressure to fall.

The elasticity of arterial walls affects blood pressure. As the ventricles contract, blood is forced into the aorta under high pressure and the artery is distended somewhat. When the ventricles have completed their contraction and have entered diastole, the blood pressure falls to the diastolic level. The elastic rebound of the arterial wall helps to maintain the diastolic pressure and keeps the blood moving along. Thus a steady flow is delivered to the capillaries.

VASOMOTOR CONTROL OF BLOOD VESSELS Any change in peripheral resistance to the flow of blood is largely a matter of change in the diameter of arterioles. The smooth muscles in the walls of these small blood vessels are supplied with vasomotor nerves. If the muscle fibers are stimulated to contract, the diameter of the blood vessel is reduced; this process is known as *vasoconstriction.* If muscle fibers relax, permitting the diameter of blood vessels to increase, the process is known as *vasodilatation.*

The vasoconstrictor center is located in the medulla, in the floor of the fourth ventricle; the nervous outflow to arterioles is through the postganglionic fibers of the sympathetic system. These are nonmyelinated fibers from cell bodies located in the chain of sympathetic ganglia. The vasoconstrictor center, through vasoconstrictor nerves, maintains a constant flow of impulses to muscles of arterioles, keeping them in a state of tonic contraction. Severing the spinal cord below the medulla, or severing the great splanchnic nerves to the abdominal viscera, causes vasodilatation and a fall in blood pressure. Stimulating the splanchnic nerve causes increased tonic contraction in the arterioles of the abdominal viscera and a consequent rise in the blood pressure. In general, the sympathetic system is responsible for the rise in blood pressure that accompanies emotional states, such as anger and fear, and the physical adjustments attending muscular exercise.

The nervous control of vasodilatation is more obscure. In some instances vasodilatation seems to be caused largely by the inhibition of vasoconstriction. A vasodilator center is said to be located in the medulla, posterior to the vasoconstrictor center. It is assumed that there is a reciprocal inhibition between the two centers; that is, when the vasodilator center is stimulated, the vasoconstrictor center is reflexly inhibited. It is usually difficult to prove experimentally whether a fall in blood pressure is due to stimulation of the vasodilator center or to reflex inhibition of the vasoconstrictor center.

Even though some effects of vasodilatation may be controlled from a center in the medulla, many aspects of vasodilatation affect only a localized area. So far as nervous control is concerned, sympathetic vasoconstriction appears to be far more important than vasodilatation. Oxygen need rather than nervous control appears to be the most effective means of regulating blood flow to the tissues generally. Lack of oxygen is at least one factor in effecting local arteriole dilatation. Blushing caused by rapid vasodilatation of blood vessels in the skin and, conversely, turning pale with fear as a result of peripheral vasoconstriction are probably the best examples of direct control from vasomotor centers in the medulla.

The anterior part of the hypothalamus contains an area concerned with the control of body temperature and maintaining body heat. As the blood temperature lowers, heat production is increased and heat is preserved by peripheral vasoconstriction. In case of a rise in body temperature, the mechanism for increased loss of body heat includes vasodilatation of blood vessels in the skin and perspiration. It is of interest that these are both cholinergic functions.

Parasympathetic vasodilator fibers are found in the chorda tympani branch of the facial nerve supplying the sublingual and submandibular salivary glands and in the glossopharyngeal nerve to the parotid salivary gland. The sacral portion of the parasympathetic or craniosacral division is represented by fibers in the pelvic nerve supplying the penis or clitoris. Vasodilatation causes the erectile tissue in these organs to fill with blood. The walls of blood sinuses in the erectile tissue become turgid as the cavities fill with blood, and the whole organ becomes larger, hard, and erect.

Sympathetic nerves may also contain dilator fibers. The cervical sympathetic outflow supplies vasodilator fibers to the mucous membrane of the mouth, the nostrils, and to surface areas of the face innervated by the trigeminal nerve. Sympathetic vasodilators to the coronary arterioles apparently have been demonstrated in dogs.

THE PRESSORECEPTOR MECHANISM The walls of the carotid sinuses and the base of the aortic arch contain pressoreceptors, or baroreceptors, which are stimulated by a rise in blood pressure. As the blood pressure rises, the walls of the artery and the sinuses are distended slightly, causing impulses from the pressoreceptors to travel to the vasomotor center in the medulla. The vasomotor center is depressed, and the arterial pressure is reduced. The nerves involved include a branch of the vagus to the aortic arch and a branch of the glossopharyngeal to the carotid sinus (Figure 14.12).

Pressoreceptor stimuli at the medulla are followed by a slowing of the heart rate and vasodilatation, especially in the abdominal region supplied by the splanchnic nerves. The blood pressure consequently is lowered.

It is well known that some persons are especially sensitive to pressure over the carotid sinus. Pressure applied to the sinuses is followed by an immediate lowering of blood pressure, which may possibly cause giddiness or fainting.

When arterial blood pressure is lowered, vasoconstriction occurs and the heart rate increases. These pressoreceptor mechanisms illustrate *Marey's law* of the heart. This law refers to the inverse relationship between blood pressure and the heart rate.

An increase in venous pressure close to the heart stimulates pressoreceptors located in the walls of the caval veins and in the right atrium. Afferent neurons in the vagus nerves then carry inhibitory impulses to cardiac control centers in the medulla. As parasympathetic efferent impulses in the vagus nerves are depressed (vagus nerves carry inhibitory impulses to the heart), sympathetic (accelerator) nerves are stimulated and the heart rate increases. This effect is called the *Bainbridge reflex*. The result is an increase in stroke volume and heart rate to relieve venous pressure at the heart.

Experimentally, it has been shown that the heart, with its nerves severed, is still able to respond to an increased amount of blood in the caval veins by increasing its stroke volume and thereby moving an increased amount of blood into the arterial system. Within certain limits, the strength of contraction is determined by the lengthening of ventricular muscle fibers and therefore increasing the volume or capacity of the ventricles. This has been called *the law of the heart*, originally based on work by Starling. The experiment is designed to demonstrate that the heart muscle has the ability to adapt to increased venous volume and also to increased arterial resistance and still maintain an adequate circulation. However, when nerves to the heart are intact, the reflex mechanism ensures the ability of the heart to adjust to sudden changes in blood pressure rather than just maintaining an adequate circulation.

THE CHEMORECEPTOR MECHANISM   Chemoreceptors located in the carotid and aortic bodies react to changes in the chemical composition of the blood. The *carotid body* is a reddish-brown, oval structure about 5 millimeters long and 2.5 millimeters wide, located posterior to the bifurcation of the external and internal carotid arteries on either side. The *aortic bodies*, which are similar to the carotid bodies in structure and function, are located at the base of the right subclavian artery and in the arch of the aorta. These bodies are composed of epithelioid cells and contain *chemoreceptors*. The carotid bodies are innervated by a branch of the glossopharyngeal nerve; a branch of the vagus nerve supplies the aortic bodies. The carotid-body chemoreceptors appear to function primarily in the regulation of respiration, but they have some circulatory effects as well. (Respiratory effects are considered in Chapter 16.) The aortic bodies may have some respiratory effects, but they are primarily concerned with the rise and fall of blood pressure. The chemoreceptors located within these bodies react to lowering of oxygen pressure in the blood or to a rise of carbon dioxide pressure. They also react to a change in acidity of the blood. As the oxygen content of the blood is lowered or as the carbon dioxide content rises, the chemoreceptors are stimulated. Impulses to the vasoconstrictor center in the medulla stimulate vasoconstriction, and the blood pressure rises.

Blood pressure

377

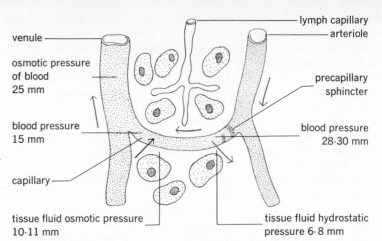

labels on figure:
venule — lymph capillary — arteriole
osmotic pressure of blood 25 mm
precapillary sphincter
blood pressure 15 mm
blood pressure 28-30 mm
capillary
tissue fluid osmotic pressure 10-11 mm
tissue fluid hydrostatic pressure 6-8 mm

FIGURE 15.4

Diagrammatic representation indicating pressures that determine the flow of fluid between the capillary and the tissue spaces. The larger red arrows indicate direction of blood flow.

OTHER REGULATORY FACTORS Epinephrine, a hormone of the adrenal medulla, causes vasoconstriction of the arterioles in the skin, mucous membranes, and the great splanchnic area. The effect is not entirely one of vasoconstriction, since in animal experimentation epinephrine has been shown to dilate arterioles of the coronary arteries. The effects produced by epinephrine closely parallel the action of norepinephrine released at the nerve endings of postganglionic fibers of the sympathetic system. Hormones are discussed in Chapter 20.

Many years ago, Starling, a British physiologist, recognized the importance of plasma-protein osmotic pressure in the exchange of fluids between the blood plasma and the tissue spaces. The capillary wall is the membrane separating the blood plasma from the interstitial fluid. The hydrostatic pressure, or blood pressure, in the capillary and the plasma-protein osmotic pressure represent two opposing forces which regulate the exchange of fluid between the blood and the tissues.

If the hydrostatic (blood) pressure in the capillaries is 28 to 30 mm Hg and the hydrostatic pressure of the tissue fluid is 6 to 8 mm Hg, there is an *effective hydrostatic pressure* of 22 mm Hg, which would tend to force fluid into the tissue spaces (Figure 15.4). Opposing the hydrostatic pressure is a plasma-protein osmotic pressure of approximately 25 mm Hg which would attract tissue fluid toward the blood. However, there is a variable amount of protein in the tissue fluid; thus if the tissue-fluid protein osmotic pressure is as much as 11 mm Hg, this would provide an *effective osmotic pressure* of 14 mm Hg.

MOVEMENT OF FLUIDS THROUGH THE CAPILLARY MEMBRANE Subtracting the effective osmotic pressure from the effective hydrostatic pressure leaves 8 mm Hg as a *filtration pressure* for the movement of fluid into the tissue spaces. This figure represents the balance of forces at the arteriole end of the capillary. At the venous end the blood pressure is much lower, approximately 15 mm Hg, and therefore favors the movement of fluid into the capillary. The *absorption pressure* at the venous end of the capillary is also about 8 mm Hg. The filtration and absorption pressures are generally about equal so that under normal conditions there is a balance between the amount of fluid filtering out of the capillary and the amount being absorbed. A small amount of tissue fluid is absorbed by lymphatic capillaries. In edema, the tissues become filled with fluid, and a much greater amount of fluid passes out of the capillaries than is reabsorbed.

It is difficult to obtain accurate capillary and tissue pressures by direct methods. The hydrostatic capillary pressure at the arteriole end is normally 28 to 35 mm Hg and at the venous end approximately 15 mm Hg. Plasma-protein pressure is fairly steady at 25 to 28 mm Hg. The pressure in the tissue interstitial spaces may be a negative pressure of about −6 mm Hg, according to some experimental data.

Some of the more important aspects of vasomotor regulation have been discussed. The control of arterial pressure is exceedingly complex, involving nervous as well as chemical mechanisms. It is evident also that physical factors must be taken into consideration. However, this introduction should provide a better understanding of what is commonly termed *blood pressure.*

THE MEASUREMENT OF BLOOD PRESSURE   The first measurement of blood pressure is commonly credited to an English clergyman named Stephen Hales. Around 1731–1733 he is said to have measured the blood pressure of a horse by permitting its blood to flow into a vertical glass tube several feet long. In this experiment, he measured the pressure of the blood against the weight of a column of blood in the vertical glass tube.

The use of mercury in the vertical glass tube enables one to read blood pressures in millimeters instead of feet and inches. The weight of mercury, of course, is many times the weight of blood or water. In the *mercury manometer* the blood pressure is measured against the weight of mercury in a U-shaped glass tube, causing the column of mercury to rise in one side of the tube.

The modern device for measuring blood pressure, called a *sphygmo-manometer,* is used in several forms. In the mercury sphygmomanometer, mercury rises in a vertical glass tube as a result of the force of air pressure transmitted from a rubber sac fitted like a tourniquet around the arm. This is obviously an indirect method, but it is the method commonly used for obtaining blood-pressure readings on human subjects (Figure 15.5).

If the rubber bag around the arm is inflated until the air pressure causes the walls of the artery underneath to collapse, the air pressure in the sac is greater than the blood pressure. Holding the fingers on the radial artery at the wrist will indicate when the artery is collapsed, because then the pulse can no longer be felt at this point. If the bag is now deflated very gradually, a faint pulse can be felt at the wrist. This means that the systolic pressure of the heart is just great enough to pry the walls of the artery apart. At this time the air pressure in the rubber sac is approximately equal to the systolic blood pressure, and the blood-pressure reading is taken from the height of the mercury on the millimeter scale. A blood-pressure reading of 118 means that the pressure of the blood is great enough to raise a column of mercury 118 millimeters. This is the systolic blood pressure, the highest pressure developed by the heart during the ventricular systole.

The lower diastolic pressure is the pressure maintained even though the heart is at rest or in the diastolic phase. In taking the diastolic pressure, a stethoscope is placed over the artery just below the rubber sac or cuff. The sound of blood coming through the constricted portion of the artery and making contact with the more or less stationary blood below can be heard through the stethoscope. As the sac is gradually deflated, more blood gets through and a very evident sound is heard. As the diastolic level is reached, around 80 mm Hg, the constriction on the artery has been released to such

FIGURE 15.5
The mercurial sphygmomanometer (blood pressure apparatus) used for measuring arterial blood pressure. (*Courtesy of the Taylor Instrument Company.*)

an extent that the sound disappears. This is the time that the diastolic reading is taken. The diastolic pressure is important because it indicates the degree of arterial tension at the lower level.

Blood-pressure values, like heart rates, vary with physiological conditions. There is no standard blood pressure. A child after the first month should have a systolic pressure of around 70 to 80 mm. By the time the child is twelve years old, the blood-pressure average is around 105 mm. Blood pressure in both sexes is low until puberty, when there is usually a sharp rise. Young adult males, in a resting position, have blood pressures of $\frac{120}{80}$ on an average. Blood-pressure readings for young women are commonly 8 to 10 mm less than for young men of equivalent age. There is a gradual increase in blood pressure with advancing age after twenty years. The difference between the systolic and diastolic pressures is termed the *pulse pressure*. It is usually around 40 mm but varies with changing physiological conditions.

FACTORS AFFECTING BLOOD PRESSURE   Rest, restricted diet, and relaxation have a favorable effect of reducing high blood pressure in many cases. The systolic pressure when one is asleep may be around 25 mm less than during daytime activity. Worry, fear, excitement, and the tenseness with which many persons conduct the day's business are emotional conditions that tend to raise blood pressure. The effect is mediated largely through the sympathetic nervous system. Diseases of the kidneys or renal occlusion often are associated with high blood pressure.

The effects of muscular exercise The blood pressure rises considerably during physical exertion. There is some increase in diastolic pressure, but the greatest rise is in the systolic pressure. Exercises such as weight lifting, which require a maximum effort of strength, raise the blood pressure rapidly. Systolic pressure of 180 or even 200 mm are not unusual with this type of effort. The pulse does not accelerate greatly, and the blood pressure returns to normal quickly after exertion has ceased.

In exercises requiring speed, such as running a 100-yard dash, the heart rate might be expected to increase from 70 beats per minute to around 120 and the systolic blood pressure to rise to around 150 mm Hg. Physical training should be taken into consideration, but in general a greater effort produces an increased heart rate and higher blood pressure. The blood pressures of runners quickly return to normal and frequently drop below normal after a race.

Endurance races produce a more moderate rise in blood pressure, but the recovery phase requires a longer time. It is difficult to obtain blood-pressure readings during exercise, but such readings are highly desirable. Readings taken after exercise has ceased may not mean very much, since blood pressure returns to normal quickly in many individuals. The blood pressure may fall below normal after exercise. Apparently this is because of the slowing of the heart and continued vascular dilatation in the blood vessels supplying skeletal muscles recently active. Also, a rise in body temperature would be accompanied by vasodilatation of blood vessels located in the skin, to provide for heat loss.

The effect of gravity Probably a great many persons have had the experience of having the telephone ring while they are lying down. They jump to their feet, only to find that things black out momentarily; this is associated with some degree of light-headedness or dizziness. In older persons especially, it takes a few seconds to secure a proper circulatory adjustment adequate to supply the brain properly while they are in a vertical position. Circulatory adjustments are rapid in children. They can turn somersaults without coming up dizzy. Loss of elasticity in the circulatory system and greater body size make this more difficult for older adults.

The blood is pumped to all tissues of the body under pressure so that all parts of the body can receive an adequate supply no matter in what position they may be. Thus the brain ordinarily receives an adequate blood supply even though the body is in a vertical position. If the brain does not receive an adequate amount of blood, fainting occurs. The body falls to a horizontal position, thus making it easier for the heart to force blood to the brain. Fainting, in this respect, is a protective device to prevent serious damage to brain cells through lack of blood supply. A person who has fainted should not be lifted to a sitting or standing position.

The blood pressure in leg arteries is 50 to 60 mm higher than in the brachial artery when the individual is standing. Aside from convenience in using the brachial artery for blood-pressure measurements, there is the advantage that it is at heart level. Blood pressures taken below heart level show the effect of gravity on a column of blood from the heart level to the point where the reading is taken.

Venous pressure also shows the effect of gravity. If the arm is allowed to hang idly beside the body, the surface veins can be seen to enlarge. Holding the hand above the head for some time causes the veins to become partially collapsed. The veins of the leg often enlarge. This may be due to a variety of causes but is influenced by the force of gravity. The pressure on the walls of veins of the lower extremities is greater than in the upper part of the body.

VENOUS PRESSURE The venous pump, so called, consists of the compressing action of muscles. Muscular movement compresses the veins, and since valves in the veins of the lower extremities prevent backflow, the blood moves forward or upward. Venous pressure varies greatly in different veins, depending upon their location. When one has been standing still for some time, pressure in the neck veins approaches zero. The subclavian and brachial veins have a low pressure of 6 to 8 mm Hg. Venous pressures are quite relative, but with the hand hanging at the side, the wrist veins may have a pressure of 35 mm Hg. If the hand is held over the head for some time, the pressure drops to a low level. When the vena cava is in a recumbent position, close to the sucking action of the heart, pressure has been estimated to be only 0 to −2 mm Hg. The venous pressure at the ankle in a standing position is 90 to 100 mm Hg, because of the effect of gravity. If a person is seated, the pressure at the ankle is only about half that in the standing position. Normally the venous muscular pumping action is so efficient that venous pressure in the leg veins while walking is only 15 to 30 mm Hg.

HYPERTENSION

HIGH BLOOD PRESSURE AND ARTERIOSCLEROSIS Many older persons are subject to hardening of the arteries and high blood pressure. Arterial walls harden and thicken, losing their elasticity. Their diameter is reduced, and they cannot properly provide for the nutrition of the tissues they supply. Since arteriosclerosis is ordinarily associated with high blood pressure, there is always the danger that some delicate blood vessel may break. Many older persons suffer a cerebral hemorrhage, or "stroke." In this case there is a rupture of a blood vessel in the brain, and the released blood can seriously damage delicate brain tissue. Such brain injuries often result in paralysis.

Atherosclerosis is a common form of arteriosclerosis (hardening of the arteries). Atherosclerosis is characterized by abnormal cholesterol-containing deposits on the inner layer (intima) of the blood vessels. In older persons, there also may be considerable calcification and hardening of the arterial walls. The deposits of cholesterol and other lipids tend to occlude the blood vessels. If this occurs in the arteries that supply the heart muscle, it is conducive to coronary conditions. It is thought that a diet low in fats is advisable for those persons who have atherosclerosis, but there is a difference of opinion concerning the value of low-cholesterol diets.

Hypertension affects a great number of persons. It may have many causes, but basically a constriction of arterioles is the major factor. Vasoconstriction is a sympathetic nervous system effect, but there may be other contributing factors. Renal hypertension involves the renin-angiotensin system, to be considered later in Chapter 19. Many older persons have what is called *essential hypertension*, which refers to high blood pressure of unknown origin. A blood pressure of $\frac{160}{90}$ would indicate hypertension. The kidneys are thought to be involved, but in exactly what way seems uncertain. Perhaps the aging kidneys require a higher pressure to perform excretion at a normal rate. Excessive salt intake increases the retention of water in the tissues and in some way seems to be concerned with vasoconstriction.

Physiology of circulation

Reduction of salt intake frequently lowers the blood pressure in essential hypertension.

The heart may gradually weaken under the strain of continually pumping against high arterial resistance. Since blood pressure rises in emotional situations and with physical activity, older persons with high blood pressure would be well advised to live moderately.

THE LYMPHATIC SYSTEM

The spaces between cells are filled with tissue fluid. This tissue fluid is derived from blood plasma and filters through the walls of capillaries. A system of lymphatic vessels drains the fluid from extracellular spaces. When within the lymphatic vessels, the fluid is called *lymph*. Since tissue fluid and lymph are essentially the same, this definition of lymph is not strictly adhered to.

Lymph is a clear fluid containing a low count of granular leukocytes and a varying number of lymphocytes. It may contain a few erythrocytes. The protein content is less than that of the blood, and the total phosphorus and calcium content is lower. The lymph contains enzymes such as amylase, maltase, protease, and lipase. It exhibits some ability to clot, but the process is slow and the clot soft.

Since the blood passing through the tissues is confined to capillaries, it never normally comes in direct contact with the cells. The lymph, on the other hand, is outside the capillary wall, bathes the tissue cells, and acts as a medium of exchange between the blood and the tissues. Nutrient materials can pass through the capillary wall and be carried by the tissue fluid to the cells. Waste materials of cell metabolism pass from the cell into the tissue fluid and then are absorbed into the bloodstream to be carried away.

Lymph flows very slowly. Its flow is aided by muscular contractions and the pressure built up by filtration of fluid from the capillaries. Even in the larger lymphatic vessels the rate of flow is slow, and the pressure is low. Various estimates place the amount of flow in the thoracic duct of man as averaging around 2 or 3 liters in 24 hours. The pressure in lymphatic capillaries has been estimated at 1.5 to 3 mm Hg, whereas pressure in blood capillaries commonly averages 15 to 30 mm Hg.

GENERAL ANATOMY *Lymphatic capillaries* The lymphatic system is a one-way, or collecting, system. It is concerned with gathering the tissue fluid, not with distributing it. The larger lymphatic vessels eventually drain into veins; the lymph becomes a part of the blood and is distributed by the arterial system. Lymphatic capillaries are the smallest vessels of the lymphatic system. They drain tissue fluid from extracellular spaces.

The lymphatic capillary resembles a blood capillary in structure. Both consist of a single layer of endothelial tissue probably derived embryologically from venous endothelium. The terminal ends of lymphatic capillaries are closed; lymph is absorbed from tissue spaces through the delicate endothelial membrane. Lymphatic capillaries are wider and more irregular than blood capillaries. They anastomose readily and form elaborate plexuses (Figure 15.6).

While lymphatic capillaries are distributed to most of the tissues and organs, no lymphatic vessels have been demonstrated in cartilage, bone

The lymphatic system

383

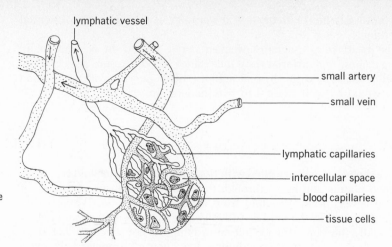

lymphatic vessel

small artery

small vein

lymphatic capillaries

intercellular space

blood capillaries

tissue cells

FIGURE 15.6

Diagrammatic representation of the relationship between the blood and lymphatic capillaries in the intercellular spaces.

marrow, the central nervous system, or in peripheral nerves. The periosteum of bone, however, is well supplied with lymphatic capillaries.

Larger lymphatic vessels drain the capillary network. The walls of these vessels resemble the walls of veins in structure. There are three layers: the inner layer is composed of endothelial cells; the middle layer is largely muscular, with some fine elastic fibers; the third, or outer, coat is largely connective tissue, with some smooth-muscle tissue. The muscle fibers in both the middle and outer coats are longitudinal and oblique. These vessels therefore are contractible, but lymphatic capillaries are not.

The larger lymphatics have a constricted or beaded appearance that is due to the presence of valves on the inner surface which prevent backflow (Figure 15.7). The valves are bicuspid or tricuspid and are placed at shorter intervals than those of veins. Lymphatic capillaries do not contain valves.

The larger lymphatics are white and pursue an irregular course, with frequent anastomoses. Unlike veins, they do not unite to form larger and larger vessels but tend to pursue individual courses. The deep lymphatics frequently form a plexus around blood vessels. The larger lymphatics are collecting vessels and flow from the capillary network toward lymph nodes. They are the afferent vessels of lymph nodes.

Lymph nodes Distributed along the course of lymphatic vessels are small bodies of lymphoid tissue called *lymph nodes* (Figure 15.8). They are usually ovoid or bean-shaped, although the shape can vary considerably. The larger nodes exhibit a depressed area called a *hilus*. Afferent lymphatics can enter the node at various places, but efferent lymphatics emerge at the hilus. Blood vessels enter and also emerge at the hilus. Internally, the lymph node contains a supporting framework of connective tissue that provides spaces called *lymph sinuses*. The lymph stream widens very greatly as it passes through the node; therefore the rate of flow is greatly reduced. The lymph filters through a maze of passageways lined with phagocytic cells. Such cells engulf bacteria or other foreign products from the lymph stream. There may also be a mechanical filtering process. Lymph nodes of the respiratory tract are often black from the filtering of carbon particles. Lymph nodes commonly become swollen and inflamed during severe bacterial infections.

The lymph nodes also afford protection against the invasion of foreign substances. Experimental evidence indicates that lymphocytes of lymph

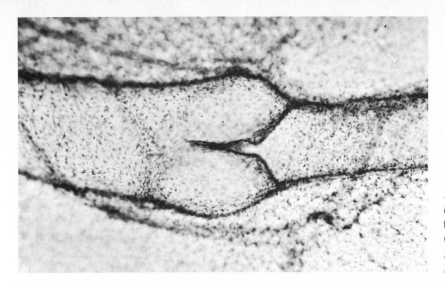

FIGURE 15.7
A portion of a lymphatic vessel from cat mesentery, showing a valve. (*Courtesy of General Biological Supply House, Inc., Chicago.*)

nodes and spleen become immunologically competent cells capable of acting against antigens introduced into the body.

Lymphoid tissues such as lymph nodes, spleen, and thymus give rise to lymphocytes. It has been observed that the lymphocyte count is higher after the lymph has passed through lymph nodes than it was before it entered. Lymph in the thoracic duct, having passed through many lymph nodes, can have a lymphocyte count of 40,000 cells per cubic millimeter, whereas lymph from peripheral lymph spaces may have only a few hundred lymphocytes per cubic millimeter. Lymph nodes are so distributed that most, if not all, of the lymph passes through at least one lymph node before it returns to the bloodstream.

Though lymph nodes are distributed throughout the body along lymphatic pathways, there are great collections of them in certain parts of the body. They consist of deep and superficial lymph nodes, just as there are deep and superficial lymphatic vessels. Groups of large nodes are located in the neck, the axillae, and the groin. Six or seven small lymph nodes are

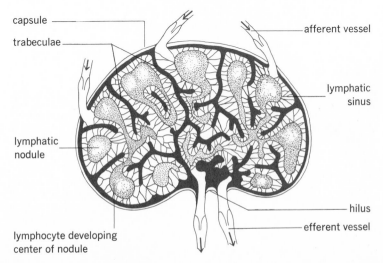

capsule

trabeculae

afferent vessel

lymphatic sinus

lymphatic nodule

hilus

efferent vessel

lymphocyte developing center of nodule

FIGURE 15.8
A lymph node. Lymphocytes develop and proliferate in germinal centers which are located in the central portions of follicles. Lymph sinuses are not blood sinuses.

The lymphatic system

385

embedded in the fat of the popliteal space behind the knee. The mesentery also affords support for numerous nodes (Figure 15.9).

The lymphatics in the region of the small intestine have a special function in the absorption of digested fat from the intestine. The inner wall of the small intestine is lined with small processes called *villi*. Each villus contains a central lymph channel, or *lacteal*. The lacteals are continuous with the lymphatic vessels of the intestinal wall and the mesentery (Figure 17.16). Lymph carrying absorbed fat has a milk-white appearance and is called *chyle*. Intestinal villi and abdominal lymphatics may also appear white when filled with chyle. The chyle also passes through lymph nodes distributed along intestinal blood vessels and in the mesentery.

The *cisterna chyli*, or *chyle cistern*, is a dilatation of the base of the thoracic duct concerned with the collection of lymph. It is located just

FIGURE 15.9
The distribution of major lymph nodes and plexuses.

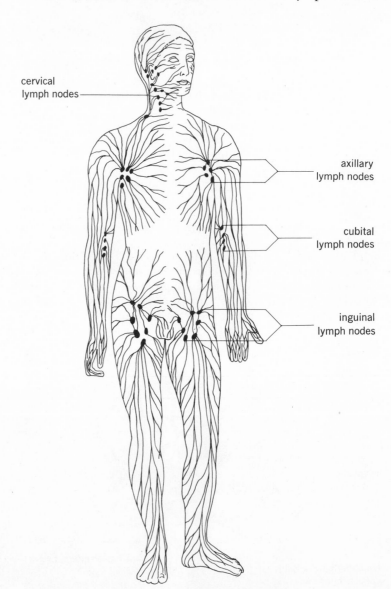

cervical lymph nodes

axillary lymph nodes

cubital lymph nodes

inguinal lymph nodes

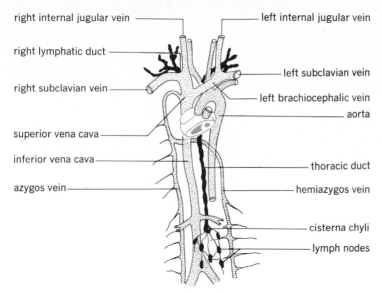

right internal jugular vein

left internal jugular vein

right lymphatic duct

left subclavian vein

right subclavian vein

left brachiocephalic vein

aorta

superior vena cava

inferior vena cava

thoracic duct

azygos vein

hemiazygos vein

cisterna chyli

lymph nodes

FIGURE 15.10

The thoracic and right lymphatic ducts. The azygos and hemiazygos veins are illustrated also.

anterior to the body of the second lumbar vertebra. It receives several tributaries bringing lymph from the lower limbs, the intestine, the pelvic viscera, and the kidneys (Figure 15.10).

The *thoracic duct* arises in the cisterna chyli. It passes upward through the diaphragm and mediastinum anterior to the vertebral column. It curves to the left, passing behind the left internal jugular vein, and opens into the left subclavian vein in the angle between the left subclavian and left internal jugular veins. The thoracic duct is about 18 inches long and is provided with paired valves. It is the left lymphatic duct in man. Its tributaries drain all parts of the body except the upper right quadrant (Figure 15.11).

The *right lymphatic duct* is very much reduced in man. The collecting trunks that drain the upper right quadrant of the body commonly fail to unite, each entering the subclavian vein separately or in some combination. When the collecting trunks of this region do unite, they form a short right lymphatic duct, a little over 1 centimeter long. It enters the right subclavian vein at its junction with the right internal jugular vein.

Lymphagogues   Many substances are capable of acting as capillary poisons, causing an increased infiltration through damaged capillary walls with accompanying loss of protein. Some examples are the following: extracts of strawberries, extracts of crayfish muscle, extracts of leech heads, and various proteins—histamine and peptone. These have been called *lymphagogues of the first class. Lymphagogues of the second class* are such substances as hypertonic glucose and hypertonic NaCl. These substances, when injected into the blood, pass readily through the capillary wall into tissue spaces. An accumulation of such a substance in the tissue spaces attracts fluid from the capillaries by osmosis. The lymph flow, draining the tissue spaces, is greatly increased.

Edema   An abnormal accumulation of fluid in the tissue spaces is called *edema*. The accumulation may be great enough to cause a swelling of the affected part. Pressing with the finger leaves a depression, which is slow in filling up. Edema may result from many causes, such as increased capillary pressure,

The lymphatic system

387

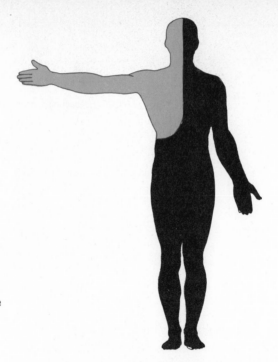

FIGURE 15.11
Areas of the body drained by the right and left lymphatic systems. The colored area is drained by the right lymphatic duct. The rest of the body is drained by the thoracic duct.

increased capillary permeability, reduction of plasma proteins, and lymphatic obstruction. Edema is often associated with heart disease and kidney disease.

Lymphoid tissues   Aside from lymph nodes, there are a number of lymphoid tissues in the body. They consist essentially of a framework of reticular tissue enclosing groups of cells, which are largely lymphocytes. Organs that are lymphoid in structure are the spleen, the thymus gland, the palatine tonsils, the pharyngeal tonsils (adenoids), and the solitary and aggregated lymph follicles. Lymphoid infiltrations are common in the mucous membrane of the intestinal tract and in the lungs. Solitary nodules are common in the intestine, and in the lower part of the small intestine these nodules form groups of patches of lymphoid tissue in the lining. These are *aggregated nodules,* or *Peyer's patches.* The solitary nodules are small, about the size of a pinhead, but the aggregated nodules may be found in patches 2 or 3 inches long. They produce lymphocytes which may later become immunologically competent cells. Lymphoid tissues such as tonsils are generally larger in children than in adults.

SUMMARY   The blood makes a complete cycle around the body in about 60 seconds. The mean velocity of blood in the larger arteries is about 20 centimeters per second. Since the total diameter of all the capillaries is greater than the total diameter of all the arteries, the blood slows to 0.5 millimeter per second as it passes through capillaries. The velocity increases to around 130 millimeters per second as it returns through the veins.

The pulse is a pressure wave, which travels along the arteries at a rate of 5 to 8 meters per second. This is obviously much faster than the velocity of the blood itself. The pulse is lost as the blood filters through the capillaries.

Blood pressure is maintained by the force of the heart beating against the frictional resistance of blood vessels. Several factors affect blood pressure:

(1) cardiac output, which in itself is dependent upon heart rate and stroke volume; (2) viscosity of the blood; (3) quantity of the blood to be moved; (4) elasticity of the arterial walls.

Vasoconstriction is a decrease in the diameter of blood vessels caused by contraction of smooth muscles in the walls. The greatest effect is in the arterioles. Relaxing of these muscles causes vasodilatation. Vasomotor control centers are located in the medulla.

Pressoreceptors are found in the walls of the carotid sinus, around the base of the aortic arch, and in the great caval veins near the heart. As the blood pressure rises, afferent stimuli travel to vasomotor centers in the medulla and the heart rate is depressed. There is also a peripheral dilatation, aiding in the control of blood pressure.

Marey's law of the heart refers to the inverse relationship between blood pressure and heart rate.

Increased venous pressure stimulates pressoreceptors located in the caval veins close to the heart. Afferent impulses in the vagus nerves depress the parasympathetic innervation of the heart, causing an increase in the heart rate. This mechanism is called the Bainbridge reflex. The heart responds to increased venous pressure by an increase in stroke volume and heart rate.

Chemoreceptors located in the carotid and aortic bodies react to changes in the chemical composition of the blood. They are especially sensitive to any lowering of oxygen pressure in the blood or to a change in the acidity of the blood. Chemoreceptors function in the following way: as the oxygen content of the blood falls, the chemoreceptors are stimulated; impulses to vasoconstrictor centers in the medulla stimulate vasoconstriction, and the blood pressure rises.

The heart with its nerves cut responds to an increased amount of blood by increasing its work load. Within limits, the ventricles enlarge, stretching the muscle fibers, and the fibers respond by moving a greater amount of blood. This has been called the law of the heart.

The hydrostatic or blood pressure forces fluids out of capillaries, whereas plasma-protein osmotic pressure attracts fluids toward the bloodstream. Thus two opposing forces regulate the exchange of fluids between the blood and the tissues. The capillary wall is the membrane between the blood and the tissue fluid.

The effective hydrostatic pressure is the difference between the capillary blood pressure and the hydrostatic pressure of the tissue fluid. The effective osmotic pressure is the difference between the osmotic pressure exerted by the plasma proteins and the osmotic pressure exerted by the proteins of the tissue fluid.

The difference between the two effective pressures, about 8 mm Hg, represents the filtration pressure which moves fluid into the tissue spaces. The blood pressure is much lower at the venous end of the capillary, about 15 mm Hg, thus favoring absorption of fluid.

The instrument for measuring blood pressure is called a sphygmomanometer. Young men, seated and resting, commonly have systolic blood pressures around 118 to 120 mm Hg. Young women, under the same conditions, commonly have systolic readings 8 to 10 mm less than those recorded for young men. The diastolic reading is about 40 mm less than the systolic reading. The difference between the two readings is called the pulse pressure.

Lymph is the medium of exchange between the blood and the tissues. The lymphatic system consists of lymphatic capillaries, larger lymphatic vessels, the thoracic duct, the right lymphatic duct, and lymph nodes. Lymph capillaries in the villi of the intestine are called lacteals. Lymph carrying absorbed fat is called chyle. Chyle from the intestinal lymphatic vessels drains into the cisterna chyli. It then moves slowly up through the thoracic duct and empties into the left subclavian vein. The left lymphatic system drains all portions of the body except the upper right quadrant. This portion is drained by the right lymphatic duct, which is very much reduced in size.

Lymph nodes are generally distributed over the body and around organs. Superficial lymph nodes are well distributed, but there are large groups under the jaw, in the axillae of the arms, and in the groin. Lymph nodes function (1) to protect the body against the invasion of bacteria and other foreign substances and (2) to produce lymphocytes, which have an immunological function.

QUESTIONS

1 Why does the blood flow more slowly through the capillaries than through the veins?
2 What is the pulse?
3 Indicate some of the factors influencing blood pressure.
4 Describe the function of the vasoconstrictor center in the medulla.
5 Discuss the pressoreceptor mechanism.
6 Where are the chemoreceptors located? What is their function?
7 Explain the meaning of filtration pressure in regard to the movement of fluids through the capillary wall.
8 Discuss the action of the heart under various conditions of exercise.
9 How is blood pressure affected by exercise?
10 What are some of the liabilities of high blood pressure?
11 In what way does the lymph serve as a medium of exchange?
12 Discuss the structural nature of the lymphatic system.
13 What are the functions of lymph nodes?
14 Explain the function of the lymphatic system in the absorption of fat.
15 Name some lymphoid tissues.

SUGGESTED READING

BAINBRIDGE, F. A.: The Influence of Venous Filling upon the Rate of the Heart, *J. Physiol.*, **50:**65-84 (1915).
BRONK, D. W., and G. STELLA: Afferent Impulses in the Carotid Sinus Nerve, *J. Cellular Comp. Physiol.*, **1:**113-130 (1932).
CARRIER, OLIVER, Jr.: The Local Control of Blood Flow: An Illustration of Homeostasis, *Bioscience*, **15:**665-668 (1965).
CLARK, E. R., and E. L. CLARK: Observations on Living Mammalian Lymphatic Capillaries: Their Relation to Blood Vessels, *Am. J. Anat.*, **60:**253-298 (1937).
COMROE, J. H.: The Location and Function of the Chemoreceptors of the Aorta, *Am. J. Physiol.*, **127:**176-191 (1939).
GATTI, R. A., O. STATMAN, and R. A. GOOD: The Lymphoid System, *Ann. Rev. Physiol.*, **32:**529-546 (1970).
JOHNSON, E. A., and M. LIEBERMAN: Heart: Excitation and Contraction, *Ann. Rev. Physiol.*, **33:**479-532 (1971).
MAYERSON, H. S.: The Lymphatic System, *Sci. Am.*, **208:**80-90 (1963).
VAN BOXEL, J. A., J. D. STOBA, W. E. PAUL, and I. GREEN: Antibody-dependent Lymphoid Cell-mediated Cytotoxicity: No Requirement for Thymus-derived Lymphocytes, *Science*, **175:**194-195 (1972).
WOOD, J. E.: The Venous System, *Sci. Am.*, **218:**86-96 (1968).
ZWEIFACH, B. W.: The Microcirculation of the Blood, *Sci. Am.*, **200:**54-60 (1959).

unit five

the respiratory system and internal respiration

chapter 16
external and internal respiration

The meaning of the term *respiration* is often misunderstood. Basically, it involves providing the tissues with oxygen and removing carbon dioxide. The exchange of oxygen and carbon dioxide in the lungs is essentially included in the process of breathing or pulmonary ventilation and is commonly considered as external respiration. Internal respiration involves the complex physical and biochemical processes of cellular metabolism, the ultimate utilization of oxygen in energy metabolism, and the production of carbon dioxide and water. Some authorities believe that the term respiration should be confined to these cellular processes.

The common pathway for air entering or leaving the lungs is through the nose. There the air is filtered, moistened, and warmed as it passes over warm, moist membranes lining the nasal passageway. The nasal cavity opens posteriorly into the nasopharynx; the air passes through the oropharynx, or posterior part of the mouth (Figure 16.1).

At the entrance to the *trachea*, the tube leading to the lungs, is a cartilaginous structure, the *larynx*. It contains the vocal folds and is protected from above by a movable cartilaginous lid called the *epiglottis*, which opens as one breathes and closes when one swallows. Below the epiglottis and revealed by a narrow fissure in the larynx is the *glottis*, across which are the *vocal folds*. The opening between them is V-shaped when the individual is resting. When one sings a high note, the vocal folds are stretched and their inner edges lie close together. Singing a low note, or taking a deep breath, causes the vocal folds to be pulled apart, and the aperture between them, the opening into the glottis, becomes oval (Figure 16.2).

The vocal folds are responsible for the voice. The sounds of the voice are caused by air being forced over the folds and causing them to vibrate; their vibrations, in turn, affect a column of air above. When the vocal cords become inflamed, as in *laryngitis*, they cannot vibrate freely, often causing a temporary loss of voice. It is also a common experience to find that, when choking from food or liquid in the larynx, one is unable to speak while the vocal folds are so irritated. Raucous singing or shouting, if prolonged, may damage the vocal folds.

The difference between the male and female voice is due to the influence of sex hormones. In childhood there is little difference between a boy's voice and a girl's voice; but after puberty the male larynx enlarges, and the vocal folds become longer and thicker, giving rise to the deeper voice of the adult male.

Pressure can be built up by strong expiratory action of muscles concerned with breathing. With the glottis closed and the diaphragm held firmly, pressure can be exerted downward to promote defecation or to exert pressure upon the urinary-bladder. Pressure may be released upward through the throat as the result of irritation to produce a cough. If the air is directed forcefully through the nose, the action is called *sneezing*.

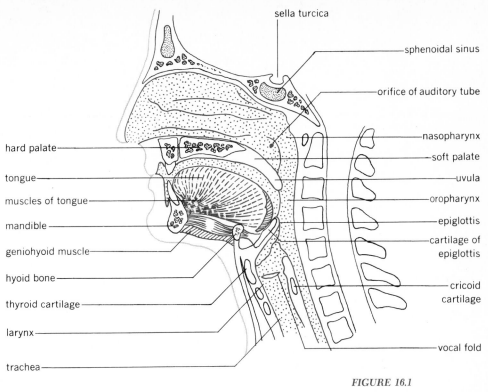

sella turcica

sphenoidal sinus

orifice of auditory tube

hard palate

tongue

muscles of tongue

mandible

geniohyoid muscle

hyoid bone

thyroid cartilage

larynx

trachea

nasopharynx

soft palate

uvula

oropharynx

epiglottis

cartilage of epiglottis

cricoid cartilage

vocal fold

FIGURE 16.1
A midsagittal section through the anterior portion of the head and neck.

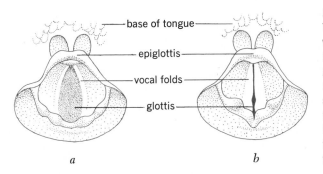

base of tongue

epiglottis

vocal folds

glottis

a　　　　*b*

FIGURE 16.2
The interior of the larynx as viewed from above: *a* glottis dilated as during inspiration. The glottis and vocal folds are also more relaxed in the production of sounds of low frequency; *b* during vocalization, to produce sounds of higher frequency. The aperature of the glottis is greatly reduced, and the vocal folds are stretched.

The glottis leads into the trachea, a long membranous tube supported by rings of cartilage and fibrous connective tissue (Figure 16.3). The rings of cartilage are not quite complete posteriorly but resemble the letter C. The cartilage rings and connective tissue permit great flexibility in the trachea and yet are strong enough to resist compression. The tracheal tube is lined with a mucous membrane and ciliated epithelium. Particles of foreign matter

THE TRACHEA

The trachea

393

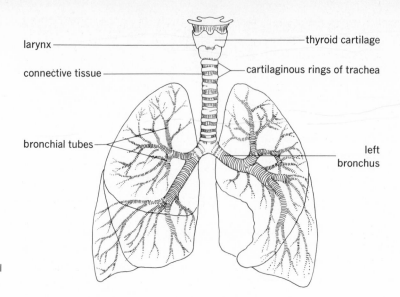

larynx —————— ————— thyroid cartilage

connective tissue ————— ————— cartilaginous rings of trachea

bronchial tubes ————

left
bronchus

FIGURE 16.3
The larynx, trachea and bronchial tubes (anterior view).

inhaled in the air become engulfed in mucus. Constant beating of cilia tend to wave the mucus upward toward the pharynx.

THE LUNGS

INTERNAL STRUCTURE OF LUNGS The trachea divides into two *bronchi*, which enter the right and left lung, respectively. Within the lungs the bronchi subdivide into smaller branches, called *bronchial tubes*, and finally into *bronchioles* (Figures 16.4 and 16.5). The familiar condition of bronchitis is an inflammation located primarily in the bronchial tubes.

Each bronchiole leads into a thin-walled sac with numerous pouches. The pouches are called *alveoli*, and they resemble a bunch of grapes clustered around the stem. The wall of the alveolus is very thin and highly vascular.

FIGURE 16.4

Bronchioles and air spaces. A capillary network is shown around the alveolar sacs. The atria and alveoli have an epithelial lining which is very extensive. The total surface area for gas exchange is approximately 40 square feet. During normal breathing only about 20 percent of this area is utilized.

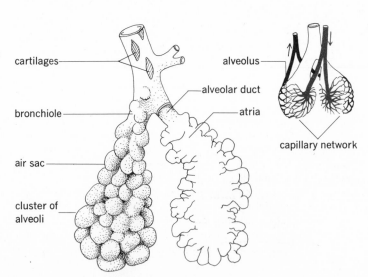

cartilages ————

alveolus ——

alveolar duct ——

bronchiole ————

atria ——

air sac ————

capillary network

cluster of alveoli ————

External and internal respiration

394

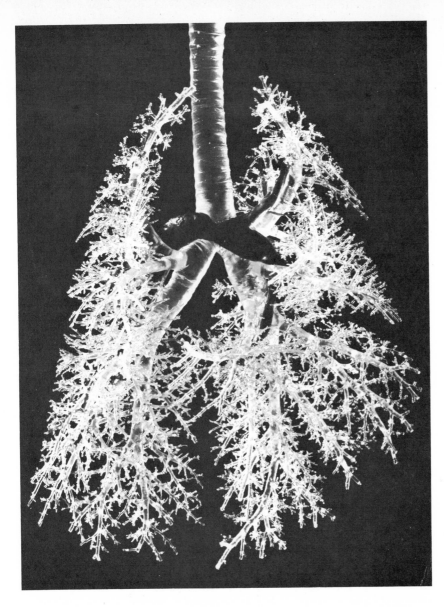

FIGURE 16.5
The bronchiotracheal tree of a dog. (*Courtesy of Ward's Natural Science Establishment, Inc.*)

It is lined with squamous epithelium which is covered with tissue fluid. Through the walls of capillaries in the alveolus the exchange of gases between the blood and the air takes place. In addition to the elaborate branching of the bronchial tubes, the lungs contain a very abundant circulatory tree. The pulmonary artery divides into smaller and smaller branches until finally it resolves itself in capillary networks around bronchioles and alveoli. Pulmonary capillaries anastomose to form tiny venules as the blood leaves the alveoli and give rise to the pulmonary veins, which carry freshly oxygenated blood back to the left atrium of the heart.

The tissue of the lungs is supplied by bronchial arteries directly from the aorta. Most of this blood is returned to the azygos veins by bronchial veins.

The lungs

The inner surface of the alveolar wall presents an interface between the air and a liquid. Each alveolus is lined by a thin layer of fluid. Liquids display surface tension which would tend to constrict or collapse the alveoli. If the liquid were simply pure water, the surface tension would make breathing very difficult and would require considerable effort. However, the alveolar air surface liquid is covered by a lipoprotein called *surfactant* (surface active agent), which lowers the surface tension and makes breathing easier. The "respiratory distress syndrome of the newborn," especially likely to occur in premature infants, is caused largely by a lack of adequate amounts of surfactant. In these cases, the alveoli may collapse, causing breathing to become extremely difficult and exhausting.

Even though the movement of oxygen and carbon dioxide takes place through the alveolar and capillary walls, the respiratory membranes are very thin, averaging less than 1 micron in thickness. The capillaries are so minute that red cells touch the capillary wall as they pass through, thus aiding the diffusion of respiratory gases.

EXTERNAL APPEARANCE OF LUNGS   The lungs occupy the thoracic cavity. The heart and other structures within the mediastinum lie between the lungs. There is, therefore, a concave cardiac impression on the medial surface of each lung. The upper surfaces, or apices, are more or less conical as they rise a little above the level of the clavicles; the lower surfaces are flattened and concave where they rest on the diaphragm.

The two lungs vary in shape and size. The right lung is the larger, consisting of three lobes. Its width is greater than that of the left lung, but it is about an inch shorter. The left lung is divided into two lobes and is somewhat narrower and longer (Figure 16.6).

A thin serous membrane covers each lung and continues over the thoracic wall, the diaphragm, and the lateral aspects of the mediastinum.

FIGURE 16.6

Anterior view of the lungs in the thoracic cavity.

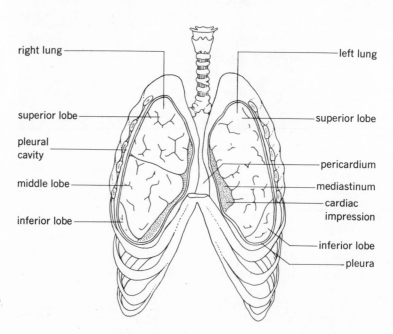

right lung

left lung

superior lobe

superior lobe

pleural cavity

pericardium

middle lobe

mediastinum

cardiac impression

inferior lobe

inferior lobe

pleura

External and internal respiration

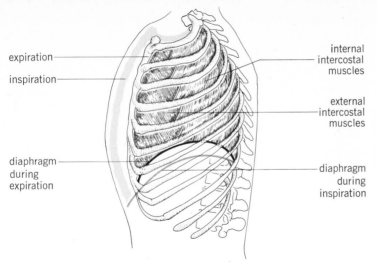

FIGURE 16.7
Changes in the thoracic cavity during inspiration and expiration: lateral view of thoracic cage. During inspiration the diaphragm descends and the ribs are elevated, thus increasing the size of the thoracic cavity. In expiration the ribs fall back into place and the diaphragm becomes more dome-shaped.

This membrane is called the *pleura* and consists of two layers. The *pulmonary pleura* covers the lung; the *parietal pleura* lines the wall of the chest and covers the diaphragm. Since the lungs fill the thoracic cavity, the two layers are in contact and the pleural cavity between the two layers is only a potential one. The serous membrane normally secretes only enough serum to moisten the surfaces that move upon each other with each respiratory movement. *Pleurisy* is a condition in which there is an inflammation of the pleural membranes. In one type considerable fluid is secreted into the pleural cavity.

Puncture of the chest wall, which permits air to enter between the two layers of the pleura, will cause the lung to collapse. Artificial pneumothorax (opening the pleural cavity) is used to rest an infected lung in certain types of tuberculosis.

RESPIRATORY MOVEMENTS

There are two respiratory movements in breathing: One is accomplished by drawing air into the lungs and is called *inspiration;* the other movement forces air out of the lungs and is called *expiration.* Breathing occurs rhythmically at an average rate of about 16 times per minute. The lungs do not take any active part in respiratory movements. Since the thorax is a closed cavity filled by the lungs, any change in the volume of the cavity will affect the volume of air in the lungs. During inspiration the volume of the chest cavity is increased by the descent of the diaphragm and the elevation of the ribs (Figure 16.7). The diaphragm is always dome-shaped but is depressed during inspiration. The ribs normally slope downward, but during inspiration they are brought upward to a more horizontal position by muscular action. Throwing out the chest is accomplished by elevating the ribs. Since the movements resulting in inspiration require that the ribs and diaphragm move from their normal position, inspiration is said to be active, whereas expira-

Respiratory movements

397

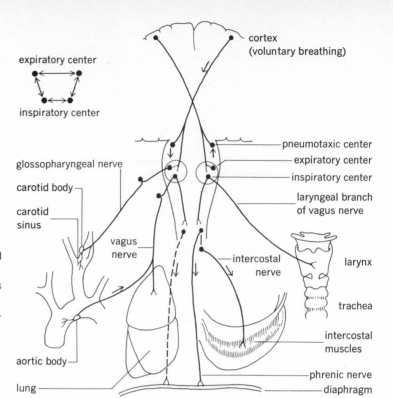

expiratory center

inspiratory center

cortex
(voluntary breathing)

pneumotaxic center

expiratory center

inspiratory center

laryngeal branch
of vagus nerve

glossopharyngeal nerve

carotid body

carotid
sinus

vagus
nerve

intercostal
nerve

larynx

trachea

intercostal
muscles

aortic body

phrenic nerve

lung

diaphragm

FIGURE 16.8

Diagram illustrating the centers
and nerves that exercise control
over breathing. The inspiratory and
expiratory centers of the medulla
are shown in the inset. The arrows
indicate the close interrelation of
nerve impulses between these centers. (*After Frank R. Winton
and L. F. Bayliss, "Human
Physiology," McGraw-Hill
Book Company, New York,
1949.*)

tion, in which these structures return to their normal position, is passive.
Forced expiration, however, is active.

FACTORS CONTROLLING RESPIRATORY MOVEMENTS   Several factors influence the
respiratory rate and depth of breathing. There are two control centers in
the medulla: an inspiratory center and an expiratory center. Both centers
lie beneath the posterior part of the floor of the fourth ventricle; the expiratory center lies immediately below this area and the inspiratory area somewhat deeper and more posterior. Although these respiratory areas have been
determined in experimental animals, there is also evidence that the bulbar
type of poliomyelitis affects similar areas in human beings.

Motor tracts from the inspiratory center descend through the spinal
cord to the third, fourth, and fifth cervical levels, where they synapse with
motor neurons to form the two phrenic nerves, which innervate the diaphragm. Stimulation of the phrenic nerves causes muscles of the diaphragm
to contract, thus lowering the diaphragm and increasing the volume of the
thoracic cavity during inspiration. Cutting the phrenic nerves stops the
movement of the diaphragm (Figure 16.8).

Another set of neurons descending from the medulla synapses with
thoracic motor neurons supplying the external intercostal muscles. The
external intercostals raise the ribs and so enlarge the thoracic cavity during
inspiration.

The internal intercostal muscles are innervated by a similar set of
neurons, but their function is expiratory. They aid in lowering the ribs during
a deep or forced expiration.

External and internal respiration

398

The inspiratory and expiratory centers in the medulla are believed to exhibit reciprocal inhibition of each other. In normal respiration the inspiratory center is dominant, because it has the lower threshold and is therefore more readily stimulated.

The vagus nerves are an important pathway by which afferent impulses arising from sensory receptors within the lungs can provide rhythmical expiration. As inspiration proceeds, elastic lung tissues are stretched and sensory receptors in these tissues are stimulated, producing afferent impulses. As a result of this type of stimulation, which apparently resembles the stretch reflex in muscles and tendons, an ever-increasing flow of nerve impulses passes over the vagus nerves to the expiratory control center. Since the expiratory center has a high threshold, the lungs continue to inflate until the control center is finally stimulated. The inspiratory center is then reciprocally inhibited and expiration follows. It would appear that this mechanism is one means by which the lungs may be protected from overinflation. The lung reflex is an important afferent mechanism and is called the *Hering-Breuer vagal reflex*, after the two physiologists who studied it and published an account of their findings, in 1868.

When both vagus nerves are cut, breathing becomes slower and deeper but still retains its rhythm. The change in breathing pattern may be interpreted as due to the loss of vagal afferent impulses, which normally stimulate the expiratory center and inhibit the inspiratory center.

Delay in stimulating the expiratory center permits the lungs to fill deeply, and the rhythm of breathing is slow. The fact that there is still rhythm would seem to indicate that some other center must influence respiration.

THE PNEUMOTAXIC CENTER   Located in the upper part of the pons is a third control area capable of substituting as an inspiratory inhibitory mechanism when vagal reflexes are eliminated. The evidence for the existence and functioning of a pneumotaxic center is the observation that in an animal whose vagal nerves have been severed, deep breathing continues, as though other controls, such as the pneumotaxic center, had taken over to substitute for the Hering-Breuer stretch reflex. If a section is then made across the brainstem between the pons and the medulla, thus severing connections between the pneumotaxic center and the inspiratory-expiratory centers, there is an immediate cessation of rhythmic breathing. Since all inspiratory inhibiting mechanisms are apparently eliminated, the animal may maintain a deep inspiration until death or deep inspirations may be interspersed with short expirations, but there is no longer any rhythm to respiratory movement. In another animal experiment in which pneumotaxic connections with respiratory centers in the medulla are severed but vagal connections remain intact, rhythmic breathing continues until both vagus nerves are cut (Figure 16.8).

It is probable that in ordinary breathing, afferent impulses from the lungs are the more important in maintaining the rhythm of breathing, but with increased respiratory activity, stimulated by a rise in body temperature, the pneumotaxic center becomes more dominant.

CHEMICAL REGULATION OF RESPIRATION   Aside from neural mechanisms which exert control over breathing, there are chemical mechanisms that enable the body to adapt respiration to changes in metabolism. Chemical mechanisms

Respiratory centers

can affect the respiratory center in the medulla, or they can affect receptors located in the carotid and aortic bodies. Chemoreceptors also play a part in the regulation of blood pressure. This action was discussed in Chapter 15. A rise in carbon dioxide concentration, increased acidity of arterial blood, or a decrease in oxygen content stimulates the chemoreceptors and increases respiration reflexly.

Chemoreceptors Ingenious experiments to isolate the carotid and aortic chemoreceptors have been performed. It was found that respiration can be increased reflexly in experimental animals by increasing the CO_2 pressure and increasing the acidity of a fluid perfusing the isolated chemoreceptors in these arteries. However, the variations in the perfusion fluids used in these experiments were greater than any variation in the chemical composition of the blood under normal conditions.

In the following experiment a normal dog was permitted to inhale nitrogen for 30 seconds. The pulmonary ventilation was approximately doubled. The chemoreceptors in the carotid areas were then denervated. There was no increase in ventilation after inhaling nitrogen for 60 seconds. Actually ventilation was reduced, since lack of oxygen depresses the respiratory center. In a supplementary experiment it was shown that the respiratory response to inhalation of 5 percent carbon dioxide in the normal and carotid-denervated animal was essentially the same. Both showed an increase in respiration, but the increase was due to the direct effect of CO_2 on the respiratory center in the medulla.

 It is probably that in ordinary breathing the chemoreceptors exert only a small regulatory effect, but when there is a lowering of the oxygen content of the arterial blood, they become a more important source of stimulation for respiration. They also control the deep, powerful respiratory movements observed when a person or animal is under the influence of ether anesthesia.

Practical applications include the use of oxygen in hospitals for illness involving respiratory deficiency and in high-altitude flying.

Pressoreceptors and their influence on respiration It is well known that a sudden, substantial rise in arterial blood pressure depresses the respiratory rate. A rise in blood pressure stretches the walls of the aortic arch and the carotid sinus areas where the pressoreceptors are located (Figure 16.9). The pressoreceptors are stimulated, and nerve impulses arise, which inhibit the respiratory center reflexly (Figure 16.10). Conversely, a considerable drop in arterial blood pressure causes an increase in breathing rate above the normal (hyperpnea). This may be merely the result of loss of inhibitory impulses. Though

FIGURE 16.9

Diagrammatic representation indicating the location of carotid and aortic chemoreceptors and pressoreceptors.

External and internal respiration

a

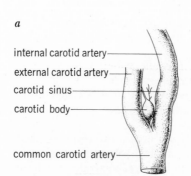

internal carotid artery
external carotid artery
carotid sinus
carotid body

common carotid artery

b

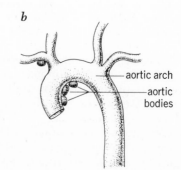

aortic arch
aortic bodies

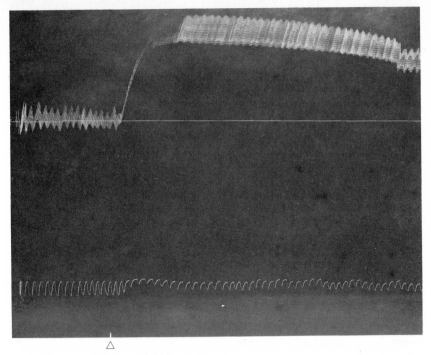

FIGURE 16.10
Effect of a rise in blood pressure
on the rate and depth of breathing in a rabbit. The rise in blood
pressure (upper tracing) was stimulated by an injection of epinephrine (Adrenalin). Note that breathing (lower tracing) is depressed.
The point at which epinephrine
was injected is indicated by the
triangle at bottom. (*Courtesy of
E. G. Boettiger.*)

pressoreceptors probably play their part in respiration, they do not appear
to be as essential as chemoreceptors.

FACTORS DIRECTLY AFFECTING THE RESPIRATORY CENTER

Among the factors that affect the neurons of the respiratory center directly
are changes in carbon dioxide concentration, variations in the amount of
carbonic acid, temperature of the blood at the medulla, and rate of blood
flow through medullary tissue.

Only in carefully controlled experiments can one factor be demonstrated to the exclusion of other factors. Ordinarily all chemical and nervous
mechanisms are coordinated in the control of respiration. If any chemical
or physical regulators are to be singled out for special consideration, changes
in CO_2 concentration and changes in acidity of the blood are probably the
most important in the control of breathing.

Simple experiments serve to illustrate the problems introduced when
the oxygen or carbon dioxide in the air is permitted to vary from normal
amounts. If an animal's breathing is directed so that all the air inspired comes
from a small container and the expired CO_2 is removed, the O_2 in the
container will gradually diminish. Under these conditions of marked lack
of oxygen, the respiratory rate is not increased to any great extent.

Now, if the experiment is varied so that the CO_2 exhaled is returned
to the breathing chamber, the CO_2 content of the air in the chamber will
rise as the O_2 is diminished. Under these circumstances the respiratory rate
increases markedly, and breathing becomes deep and vigorous (Figure 16.11).

Adding even a little CO_2 to the air to be inhaled will accelerate
respiration. This effect is not due entirely to the direct action of CO_2 on

*Factors directly affecting
the respiratory center*

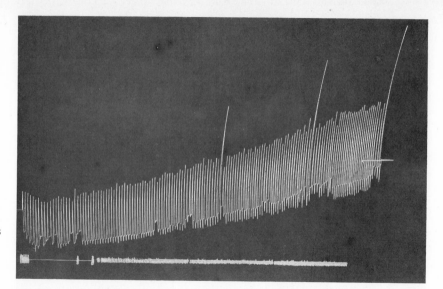

Effect of carbon dioxide on rat respiration. The rat breathes in a closed circuit. As the CO_2 content of the air in the breathing device rises, the respiratory rate increases and the breathing becomes stronger and deeper. (*Courtesy of E. G. Boettiger.*)

the respiratory center, since there is at the same time an increase in the acidity of the blood, which also elevates the respiratory rate. An increase in the acidity of the blood (carbonic acid) affects the respiratory center directly, but it also acts reflexly on chemoreceptors in the carotid and aortic sensory areas. Impulses arising from the chemoreceptors reach the respiratory center in the medulla by way of afferent fibers in the glossopharyngeal and vagus nerves.

The CO_2 concentration in arterial blood can be reduced by breathing deeply and rapidly for a minute or two, a procedure called *voluntary hyperventilation*. After such forced breathing, there follows a period up to 2 minutes when breathing is inhibited (apnea). When breathing is resumed, it is irregular or periodic until oxygen and carbon dioxide pressures of arterial blood again return to normal.

Though it is possible to hold the breath for 30 seconds or so through voluntary nervous control from the cerebral cortex, chemical stimuli soon overpower all voluntary inhibition and one is forced to take a breath. Ordinarily, since breathing is controlled from centers in the medulla, one is not conscious of its rate or depth. Under conditions of rest, breathing varies considerably in different individuals; the average rate is about 16 times per minute in an adult. Like the heart rate, the respiratory rate varies with age, physical activity, and many other factors. The rate is much faster in little children than in adults. Exercise increases the respiratory rate.

The lungs of the human fetus are not functional until birth. The fetal blood is well supplied with oxygen through the placenta, and carbon dioxide is efficiently removed. At birth the placental circulation ceases to function, and it is assumed that as the CO_2 pressure rises in the blood, the newborn child is stimulated to gasp for air and so takes his first breath.

EFFECT OF EXERCISE ON RESPIRATION Vigorous exercise increases respiration markedly. All the factors that stimulate respiration are brought into action. These factors include increased acidity of the blood, a rise in body temperature, and increased secretion of epinephrine into the blood. Nerve impulses from the cerebral motor area, from the heart and lungs, and from muscles and joints affect the respiratory center.

SOME ABNORMAL CONDITIONS *Dyspnea* refers to difficulty in breathing or a feeling of being unable to obtain enough air to breathe. The causes are numerous, but dyspnea may be due to cardiac failure, pulmonary edema, or emphysema.

In *pulmonary edema,* fluid collects in the alveoli and interstitial spaces of the lungs. It may be associated with pneumonia or with cardiac failure, especially of the left side of the heart. If, as a result of congestion, the capillary blood pressure rises above that of the plasma-protein osmotic pressure, fluid begins to fill the alveolar spaces. Ordinarily, the lungs remain "dry" because the capillary blood pressure is low and fluid is absorbed out of the lungs under the influence of the higher osmotic pressure exerted by the plasma proteins. The blood pressure in the pulmonary artery is only about 25 mm systolic (8 mm diastolic), and the resistance afforded by the lung capillaries is remarkably low.

Emphysema is a condition of increasing importance in an aging population. The alveolar walls break down progressively, thus reducing the area available for the exchange of gases. Many alveoli become confluent, leaving open spaces surrounded by fibrous tissue. As the alveoli break down, their capillaries also degenerate, tending to raise the pulmonary blood pressure. There seems to be little doubt that smoking increases the likelihood of developing emphysema; it also increases the incidence of lung cancer.

COMPOSITION OF AIR The air we breathe contains oxygen, nitrogen, water, and a trace of carbon dioxide. There are some rare gases, such as argon and krypton, but they are usually disregarded in a physiological gas analysis or considered as inert gases along with nitrogen. The significance of external respiration lies in the exchange of oxygen and carbon dioxide between the air and the blood. The inspired air loses O_2 to the blood and takes up CO_2. The interchange takes place through the thin capillary walls in the alveoli of the lungs.

The following table gives the average composition of dry inspired and expired air with the human subject at rest.

	O_2, vol. %	CO_2, vol. %	N_2, vol. %
Inspired air	20.9	0.04	79
Expired air	16.3	4.5	79.2
Difference	-4.6	$+4.46$	$+0.2$

It is apparent that the amount of oxygen used is greater than the amount of carbon dioxide given off. This inequality is due to the fact that some of the oxygen used in metabolic processes appears in H_2O and not as CO_2. There appears to be an increase in N_2 in the expired air, but this is due to the lack of balance between the amount of O_2 inspired and the amount of CO_2 expired. Nitrogen is an inert gas and plays no part in metabolism.

It may be of interest to consider the average amounts of oxygen, carbon dioxide, and nitrogen present in arterial and venous blood. Actual values vary with the activity of individuals or various organs or with the rate of metabolism of tissues. Even with an individual in a recumbent position there is still considerable variation.

Average amounts of oxygen, carbon dioxide, and nitrogen present in arterial and venous blood are given in the following table.

	O_2, vol. %	CO_2, vol. %	N_2, vol. %
Arterial blood	20	48	1.7
Venous blood	13	55	1.7

THE EXCHANGE OF AIR IN THE LUNGS   The lungs fill with air as the newborn child takes its first breath; they are never completely devoid of air thereafter. Even after a forced expiration in an adult, there still remains a *residual volume* of air equal to 1,000 to 1,200 milliliters.

When a person is breathing quietly, there is a constant flow of *tidal air* in and out of the lungs. The volume of this tidal air for the average person is about 500 milliliters (1 pint).

After a normal inspiration of tidal air, an additional 3,000 milliliters can be inhaled by taking as deep a breath as possible. The additional amount is called the *inspiratory capacity*.

The additional amount of air that can be exhaled forcibly after a quiet expiration of tidal air is the *expiratory reserve volume*. It amounts to about 1,100 milliliters in the adult.

If a person inhales the maximum amount of air and then exhales as much as possible, the total exchange is his *vital capacity*. It measures approximately 4,500 milliliters for the average adult. Vital capacity varies with the size of the individual, among other factors. The average young woman may have a vital capacity of 4,000 milliliters, whereas the average young man may have a vital capacity of 5,000 milliliters. Vital-capacity measurements are of clinical importance, since vital capacity diminishes with certain diseases affecting chest muscles and lungs.

TRANSPORT OF RESPIRATORY GASES   Oxygen is absorbed by the tissues from tissue fluid, which, in turn, is supplied by capillary blood rich in oxygen. It passes through the wall of capillaries, the tissues taking up only enough to meet their metabolic needs. The oxygen content of the blood is not increased as the demand for oxygen grows greater in muscular activity; because of the dilatation of capillaries, a greater amount of blood flows through the tissues. Carbon dioxide continuously produced in the cells of tissues reaches a higher partial pressure there. It diffuses through the cell membrane into the fluid of lymph spaces, through the capillary wall into the blood plasma, and into red cells to be carried back to the heart and lungs in the venous blood. Small amounts of nitrogen are found in the blood, but since it is an inert gas it takes no part in metabolism.

We have considered the transportation of oxygen as it is carried in loose chemical combination with hemoglobin. In this state it is called *oxyhemoglobin*. When oxyhemoglobin reaches the tissues, the oxygen is given up readily. Diffusion of oxygen into the tissues is aided by the lower oxygen pressure in the tissues.

Carbon dioxide also enters into a loose chemical combination with

External and internal respiration

404

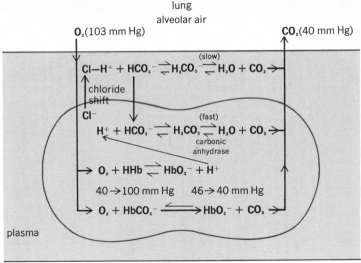

lung
alveolar air

O_2 (103 mm Hg)　　　　　　　　　　　　CO_2 (40 mm Hg)

(slow)
$Cl—H^+ + HCO_3^- \rightleftharpoons H_2CO_3 \rightleftharpoons H_2O + CO_2 \rightarrow$

chloride
shift

Cl^-　　　　　　　　　　　(fast)
$H^+ + HCO_3^- \rightleftharpoons H_2CO_3 \rightleftharpoons H_2O + CO_2 \rightarrow$
carbonic
anhydrase

$O_2 + HHb \rightleftharpoons HbO_2^- + H^+$

$40 \rightarrow 100$ mm Hg　　$46 \rightarrow 40$ mm Hg

$O_2 + HbCO_2^- \longrightarrow HbO_2^- + CO_2 \rightarrow$

plasma

pulmonary capillary

FIGURE 16.12

The movement of respiratory gases in the lungs. The red blood cell is in a lung capillary, such as in the wall of the alveolus. The breakdown of carbonic acid is rapid in the lungs under the influence of the enzyme carbonic anhydrase. As oxygen comes into the blood cell under higher pressure, CO_2 filters out into the alveolar air. This diagram illustrates the chloride shift also. (*After L. L. Langley, "Review of Physiology," 3d ed., McGraw-Hill Book Company, New York, 1971.*)

hemoglobin. About 20 percent of the CO_2 in the blood is transported as $HHbCO_2$. It combines with an NH_2 group in the hemoglobin molecule to become carbaminohemoglobin, $HbNH \cdot COOH$.

A small amount of CO_2 is carried by the plasma in a dissolved state or in combination with the water of the plasma as carbonic acid, H_2CO_3. The total amount is only around 5 percent.

Carbon dioxide enters the blood plasma, where it slowly forms carbonic acid; the carbonic acid dissociates into bicarbonate ions, HCO_3^-, and H^+ ions. Since the reaction of CO_2 with water is a relatively slow process, only a small amount of bicarbonate is formed in the plasma. However, the same reaction within the red cell occurs in a fraction of a second. The reaction within the cell is catalyzed by the enzyme *carbonic anhydrase* (Figure 16.12).

$$CO_2 + H_2O \xrightleftharpoons[\text{anhydrase}]{\text{carbonic}} H_2CO_3 \rightleftharpoons HCO_3^- + H^+$$

About 75 percent of the CO_2 in the blood is transported as bicarbonate ions, HCO_3^-.

The reaction of CO_2 and water is a very important one within the red cell but relatively unimportant in the blood plasma. The excess hydrogen-ion concentration is quickly buffered by the hemoglobin of the red cell, preventing any marked lowering of the pH of the blood.

As the bicarbonate-ion concentration rises within the red cell, it soon becomes much higher than the concentration outside in the plasma. Bicarbonate ions, therefore, diffuse outward through the red-cell membrane. Potassium ions, which are present also within the red cell, do not move readily through the membrane. As the negative bicarbonate ions move outward in the blood plasma, negative chloride ions diffuse in and restore the ion balance within the red cell. This is called the *chloride shift* (Figure 16.13).

When the blood reaches the lungs, the CO_2 pressure is higher in the venous blood than in the alveolar air, resulting in the diffusion of CO_2 from the blood into the alveoli. The chemical reactions in the lung capillaries are the reverse of those occurring in the tissue capillaries.

Factors directly affecting the respiratory center

405

FIGURE 16.13

The movement of respiratory gases from a red blood cell into a tissue cell. In the red cell, carbon dioxide and water unite rapidly to form carbonic acid under the influence of the enzyme carbonic anhydrase. Bicarbonate ions, therefore, are formed more rapidly within the cell than in the plasma. As the excess bicarbonate ions move out of the red cell, chloride ions diffuse in; this is referred to as the chloride shift. Cl^-, chloride ion; Hb, hemoglobin; HbO_2, oxyhemoglobin; HCO_3^-, bicarbonate ion; H_2CO_3, carbonic acid; HHb, reduced hemoglobin. (*After L. L. Langley "Review of Physiology," 3d ed., McGraw-Hill Book Company, New York, 1971.*)

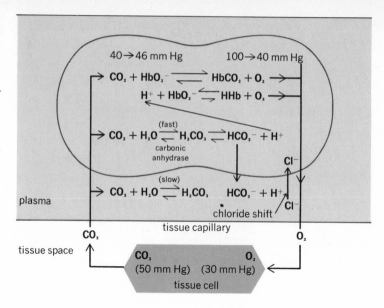

INTERNAL RESPIRATION AND CELLULAR METABOLISM

From time to time we have referred to various aspects of cellular metabolism. In Chapter 2 we considered the functions of enzymes with reference to glycolysis and the electron-transport system. In Chapter 3 we discussed cellular respiration in relation to the functions of mitochondria. We shall now consider the overall aspects of cellular metabolism and respiration.

Cellular metabolism is concerned with the utilization and release of energy from foods. In a general way, carbohydrate metabolism is the reverse of photosynthesis in that glucose is degraded to carbon dioxide and water. The carbon dioxide is given off as a by-product and is exhaled; some of the water is excreted, but much of it is used over and over again.

In this discussion of cellular metabolism we start with the process of *glycolysis* (Figure 16.14). The glycolytic pathway was outlined as a result of extensive work of chemists; it is also referred to as the Embden-Meyerhof pathway. In simplified form, glycolysis includes the conversion of glycogen to glucose 1-phosphate and the conversion of glucose to glucose 6-phosphate. Each step is catalyzed by enzymatic action. The glucose phosphates are then degraded through a series of steps to pyruvic acid.

THE TRICARBOXYLIC ACID CYCLE *Aerobic respiration* refers to the degradation of organic substances, terminating in the release of CO_2 and the formation of H_2O as hydrogen combines with molecular oxygen. It involves the tricarboxylic acid cycle and the electron-transport system. These reactions take place in the mitochondria. However, this phase of aerobic respiration is closely related to glycolysis, the degradation of glucose to pyruvic acid.

The pyruvic acid enters the tricarboxylic acid cycle as it is oxidized to acetyl coenzyme A. At this point, fatty acids and amino acids also enter the energy cycle. The vitamin thiamine acts as a coenzyme in pyruvic acid

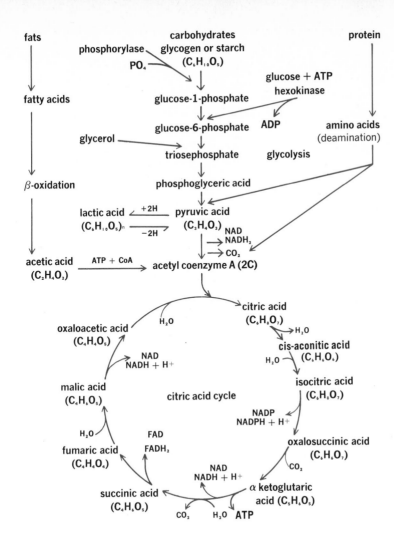

FIGURE 16.14
Outline of glycolysis and the tricarboxylic acid cycle. One molecule of glucose produces two molecules of pyruvic acid; therefore two turns of the cycle are necessary to accomplish the breakdown of one molecule of glucose. Two turns of the cycle produce 16H, $4CO_2$, and 2ATP molecules.

metabolism, and the vitamin pantothenic acid forms part of the molecule of coenzyme A.

The energy of acetyl coenzyme A takes part in the condensation of oxaloacetic acid to form citric acid. An enzyme provisionally called a condensing enzyme catalyzes the reaction. Citric acid, a six-carbon compound with three carboxyl groups (COOH groups), is the familiar acid of citrous fruits.

In the reaction that follows, the citric acid molecular structure is changed slightly by the loss of water to form *cis*-aconitate and then by the addition of water to form isocitrate. As the reactions of the citric acid cycle progress, isocitrate is oxidized to form oxalosuccinic acid. The enzyme concerned in this reaction is isocitrate dehydrogenase, and the coenzyme is nicotinamide adenine dinucleotide (NAD). Here NAD is reduced to $NADH_2$. The subsequent decarboxylation of oxalosuccinic acid produces ketoglutaric acid, with CO_2 given off.

A series of complicated reactions follow. Ketoglutaric acid undergoes oxidative decarboxylation, and CO_2 is given off in the process. Coenzyme

Internal respiration and
cellular metabolism

A is again involved in this reaction, and the first step produces succinyl coenzyme A. NAD is reduced to $NADH_2$, with a subsequent release of energy. Coenzyme A is removed by enzymatic action to form succinic acid, and the energy produced is used in the formation of ATP.

Succinic acid then undergoes dehydrogenation to form fumaric acid. Flavin adenine dinucleotide (FAD) is reduced in this reaction, with a subsequent production of energy.

Fumaric acid plus H_2O, catalyzed by the enzyme fumarase, forms malic acid. Malic acid is then oxidized, and oxaloacetic acid is formed again. This is not the same oxaloacetic acid molecule with which the process started, but now that the molecule is restored, the cycle is reconstituted and ready to start over again.

The electron-transport system In discussing the tricarboxylic acid cycle we have mentioned from time to time that a substance is reduced, with a subsequent release of energy. This refers to the hydrogen atoms (protons and electrons) produced in the cycle. We shall now discuss the disposition of these electrons. Since these electrons are never "free," they are readily captured by electron acceptors. The hydrogen atom ionizes into a hydrogen ion H^+ and an electron e^-. In the electron-transport system, hydrogen ions seem to be held in abeyance until the end of the series, when they finally unite with molecular oxygen to form water. Electrons, on the other hand, are passed along a series of acceptors or carriers in what has been likened to a bucket brigade or a waterfall sequence. As each electron carrier accepts an electron, the carrier is reduced and subsequently oxidized to release electrons to the next acceptor in the series. At certain places in the series the energy obtained by these oxidations is converted to ATP (see Figure 16.15).

Electrons are transferred in packets of two from the tricarboxylic acid cycle to an electron acceptor. In some cases the first acceptor is nicotinamide adenine dinucleotide (NAD), while other dehydrogenases are associated with nicotinamide adenine dinucleotide phosphate as an acceptor (NADP). Note the use of the vitamin substance nicotinamide as a part of the coenzyme. Some of the energy produced at this point goes into the formation of a molecule of ATP.

In the next step, flavin adenine dinucleotide (FAD) accepts electrons and in turn is reduced to $FADH_2$. Here again, a vitamin, riboflavin, is involved as part of the coenzyme. Electrons from succinic acid or succinate are passed directly to FAD by the action of succinic dehydrogenase.

The reduced flavoproteins then pass electrons along to a quinone, ubiquinone (coenzyme Q), which in turn passes them along to the cytochromes. Ubiquinone, as one of the quinones, is chemically a relative of vitamin K.

The cytochrome series The cytochromes are cellular pigments (hemoproteins) containing a heme molecule, which, in turn, contains an iron atom. The iron atom is capable of changing from the ferrous to the ferric form, or the reverse, so the cytochromes are readily oxidized and reduced. The cytochromes are commonly designated as cytochrome a, b, or c and cytochrome oxidase. In this series the cytochromes a, b, and c are all electron carriers; but cytochrome oxidase is truly an oxidase, since it transfers electrons to oxygen. The dehydrogenases transfer electrons from one organic molecule to another but not to molecular oxygen. Cytochrome oxidase, as we have seen, transfers electrons to molecular oxygen in the formation of water.

Referring to Figure 16.15, we see that two more molecules of ATP

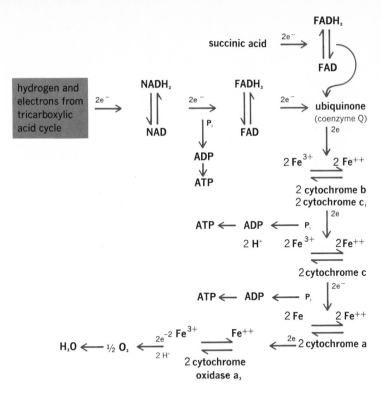

FIGURE 16.15

The electron-transport system. Note that electrons are transferred in packets of two and that three molecules of ATP are formed in this process of oxidative phosphorylation.

are produced during the passage of electrons through cytochromes b and c_1 to c, and c to a, each being alternately reduced and oxidized as the electrons are passed along. As we have indicated, the cytochrome molecules exist in either the ferric form, Fe^{3+}, or the ferrous form, Fe^{2+}. The iron atom can change from the ferric to the ferrous form or from the ferrous to the ferric form. This represents a change in valence between the two forms.

In the electron-transport series there is a phosphorylation at three sites where ATP is produced. Thus, for every atom of oxygen reduced to form one molecule of water, three molecules of ATP are produced. It is estimated that electron-transport phosphorylation is the source of over 90 percent of all the ATP synthesized from ADP and phosphate.

This introduction to cellular respiration will give some indication of the manner in which foods are oxidized with a release of energy. The intermediate steps involve the use of a great many specific enzymes. Energy is given off at various steps in the process of degrading carbohydrates, proteins, and fats. Part of this energy is used in resynthesis in order to keep the energy wheel revolving around and around.

Some idea of the energy released may be gained by considering that each molecule of glucose completely broken down to carbon dioxide and water yields 38 ATP molecules. One mole of glucose completely oxidized releases 686,000 calories of energy. The breakdown of ATP to ADP represents an energy equivalent of 8000 calories. The oxidation of one mole of glucose then, provides for 304,000 calories stored as ATP. The efficiency of energy transfer here is around 44 percent. The remaining 56 percent is released as heat and is therefore not available for cellular metabolism.

Internal respiration and
cellular metabolism

EFFECTS OF RAPID DECOMPRESSION

Persons who work under conditions of increased air pressure, such as divers and workers in caissons, are subject to a different set of conditions from those who live and work in areas of normal or low atmospheric pressures. One does not become acclimated, as does the person living in high mountain areas. Exposure to increased atmospheric pressure does not give rise to symptoms of discomfort; only when the subject is returned to normal atmospheric pressure too quickly do dangerous conditions develop.

The symptoms caused by rapid decompression are those of *caisson disease* and are commonly called the *bends.* They include pain in muscles and joints and disturbances of the central nervous system, involving both motor and sensory nerves. Under conditions of high atmospheric pressure the nitrogen of the blood diffuses slowly through membranes and is absorbed into tissues, especially fatty tissues. Rapid decompression is analogous to suddenly opening a bottle of carbonated beverage. Carbon dioxide has been dissolved in the beverage under pressure; when the cork is removed, it comes out of solution as a gas and causes the drink to foam. Similarly, in rapid decompression excess nitrogen dissolved in the blood during exposure to high pressure comes out of solution and causes bubbles of the gas to form in the blood and tissues. Gas bubbles collect in the lung capillaries and other areas, causing severe symptoms.

A similar situation occurs when a person is rapidly decompressed by ascent in an airplane from sea level to a high altitude. In this case, the change is from a normal pressure to a low pressure instead of from high pressure to normal. Since nitrogen is an inert gas, it causes no damage even if absorbed in increased amounts, so long as it stays in solution. Rapid decompression in aviators, causing nitrogen to come out of solution, is called *aeroembolism.* The symptoms are identical with those of caisson disease. Descending to a lower altitude should bring about recompression and afford relief from distress.

ARTIFICIAL RESPIRATION

Though there are several mechanical devices, such as pulmotors and resuscitators, for establishing artificial respiration, these machines are often not available in an emergency. If the victim of an accident has stopped breathing, it is unlikely that there will be time enough to obtain an inhalator machine or to remove the victim to a hospital. It would be advantageous, therefore, if a high percentage of the public were able to give artificial respiration manually. Many persons are acquainted with the Schafer (prone pressure) method. The subject is placed on the ground in a prone position with arms outstretched or one arm flexed to cushion the head. The head is turned to one side and the tongue pulled forward. A heavy, folded garment may be placed underneath the abdominal area. The operator kneels astride the prone form, facing forward, and places his hands on the lowest ribs. He then leans forward, letting his weight gently compress the subject's chest and abdomen, thus causing air and perhaps water to be expelled from the respiratory tract. The pressure is maintained for about 2 seconds, and then the operator leans back for about 2 seconds, leaving his hands in position. Relaxing the pressure permits the thorax to expand to its normal position and draw air into the

External and internal respiration

lungs. Rhythmical movements are established at 12 to 16 times per minute to simulate the rate of normal respiration. Artificial respiration may have to be continued for a half-hour or more before natural breathing is established.

A more positive intake of air is accomplished by lifting the hips a few inches off the ground. The hips can be raised by taking hold of the belt or clothing or by putting a towel under the hips to aid in lifting. Raising the hips, alternated with pressure on the lower ribs, as in the Schafer method, provides both "a pull and a push" to accentuate the movement of air in and out of the lungs. This method quickly fatigues the operator. Two persons lifting can perform the operation much more easily. If the operator is alone, he can grasp the prone person by the clothing at the side of the hip, roll that side upward, and let it fall back in place. He can then move forward to apply pressure on the chest to induce expiration. A team of operators can apply this method efficiently, and it is superior to the Schafer method.

THE BACK-PRESSURE–ARM-LIFT METHOD A fairly adequate method of artificial respiration is the back-pressure–arm-lift (Holger Nielsen's) method, because it provides a positive push-and-pull movement with the subject in a favorable position. The subject is placed face down in the prone position. The elbows are bent, and one hand is placed on top of the other. The head is turned to one side, the cheek resting upon the hands.

The operator kneels at the head of the prone person. He places his hands on the subject's back above the ribs. He then rocks forward until the arms are nearly vertical and his weight exerts pressure on the subject's chest, forcing air out of the lungs.

The operator then rocks slowly backward, grasping the subject's arms just above the elbows. Just enough pull and lift is applied to feel the resistance and tension in the prone person's shoulders. The operator keeps his own elbows straight in drawing the subject's arms toward him. Completing the pull or expansion phase, the subject's arms are dropped as the push-pull cycle is completed. The cycle is repeated at a steady rate of 12 times per minute, allowing about equal time for the compression and expansion phases and a minimum rest period. The operator should adapt his movements as nearly as possible to the rhythm of normal breathing.

THE MOUTH-TO-MOUTH RESPIRATION TECHNIQUE (FIGURE 16.16) The mouth-to-mouth respiration technique is now considered to be superior to any of the manual methods of artificial respiration. Tilting the head back, the operator blows into the subject's mouth, watching his chest rise. The subject's nose is pinched shut with the operator's free hand. When working to resuscitate a child, the breath required to inflate the lungs is not as great as in an adult and the breathing can be a little more rapid, at a rate around 20 per minute. The resting respiratory rate for adults is around 16 per minute.

Foreign matter may occlude the mouth or throat. The mouth can be cleared manually. Children can be held briefly by the ankles, in the case of drowning, to let water run out. Sometimes a slap on the back will dislodge obstructions. If the teeth are clenched, it may still be possible to blow through the teeth or through the nose.

The volume of CO_2 in the operator's breath does not appear to be a serious problem, since there is still a considerable amount of oxygen in the exhaled air. A special mouthpiece has been devised to aid the operator, but it is not necessary in an emergency.

a   b   c

FIGURE 16.16

Mouth-to-mouth method of artificial respiration: *a* push the subject's head back so that the chin points upward; *b* pull the jaw forward into a jutting-out position (this helps to move the base of the tongue away from the back of the throat and tends to keep the tongue from blocking the air passageway); *c* place your mouth over the subject's open mouth and pinch the subject's nostrils to close them and prevent air leakage. Blow into the subject's mouth until you see his chest rise. Breathe into the mouth regularly about 12 times per minute for an adult or about 20 times per minute for a child.

The American Red Cross has adopted the mouth-to-mouth technique and has issued a manual for instruction in this method.

Inhalation of carbon monoxide may require quick application of artificial respiration. Hemoglobin has a much greater attraction for CO than it does for oxygen. The blood is depleted of its oxyhemoglobin while another compound, carboxyhemoglobin, takes its place. Carbon monoxide is an extremely dangerous gas, and since it has no odor, it is difficult to detect.

Carbon monoxide is a component of the exhaust in gasoline engines, in the combustion of charcoal and other fuels, and in the fumes from leaking burners, stoves, furnaces, or flues. One should be extremely careful to ventilate rooms if there are any exhaust fumes or leaking of fumes. Garage doors should always be open if the automobile engine is running.

When people are crowded into inadequate space they may feel uncomfortable and believe that there is not enough oxygen available. Normally in houses, theaters, and meeting places, enough oxygen is available but unless there is adequate ventilation, the temperature and humidity may rise under crowded conditions. In such cases, evaporation from the skin becomes more difficult and this causes discomfort. If fans are available to circulate the air, evaporation is enhanced and people become more comfortable. For the most part it is the same air that is being circulated, but easier evaporation makes the difference.

SUMMARY

The structures through which air must pass in order to reach the alveoli of the lungs are as follows: the nasal passageway, nasopharynx, pharynx, larynx, trachea, bronchi, bronchioles, and alveolar ducts. The epiglottis is a movable lid that closes above the glottis during the act of swallowing and opens during breathing. The vocal folds are located in the larynx.

The alveoli are thin-walled and very vascular. It is here that the exchange of oxygen and carbon dioxide between the blood and the air takes place.

The lungs are located in the thoracic cavity. The heart and related structures lie in a cavity between the lungs, called the mediastinum. The right lung is the larger and consists of three lobes; the left lung has only two lobes. Each lung is covered with a thin serous membrane, the pleura. The pleura consists of two layers; the pulmonary pleura covers the lungs, whereas the parietal pleura lines the thoracic wall and covers the diaphragm.

External and internal respiration

412

There is a potential pleural cavity between these two layers, but normally it contains only a little serous fluid.

During inspiration, the chest cavity is enlarged by muscular elevation of the ribs and lowering of the diaphragm, permitting air to flow into the lungs. Inspiration, since it requires muscular effort, is said to be active. Expiration, in which the ribs fall back into place and the diaphragm returns to its normal position, is said to be passive.

External respiration refers to the exchange of oxygen and carbon dioxide within the lungs; internal respiration is the term applied to the exchange of these gases in the tissues.

Respiratory control centers are located in the medulla. There is an inspiratory center and an expiratory center. In addition, there exists a pneumotaxic center in the pons, which appears to exert a regulatory control over rhythmic breathing.

Aside from nervous control of respiration, chemoreceptors are stimulated by a rise in carbon dioxide concentration, increased acidity of arterial blood, or a decrease in oxygen pressure to cause an increase in breathing rate.

Oxygen is transported as oxyhemoglobin. Carbon dioxide also enters into loose chemical combination with hemoglobin. Approximately 20 percent of the CO_2 in the blood is carried as $HHbCO_2$.

A small amount of CO_2, about 5 percent, is carried dissolved in the blood plasma or as carbonic acid, H_2CO_3. Carbon dioxide reacts slowly with the water of the blood plasma to form carbonic acid, which dissociates into bicarbonate ions, HCO_3^-, and hydrogen ions, H^+. However, this reaction within the red cell occurs very rapidly, catalyzed by the enzyme carbonic anhydrase. About 75 percent of the CO_2 of the blood is transported as bicarbonate ions, HCO_3^-.

The bicarbonate-ion concentration within the red cell soon becomes much higher than in the plasma outside the cell. Bicarbonate ions therefore diffuse out through the red cell membrane. Negative chloride ions move inward to take the place of the bicarbonate ions to restore the electric balance within the red cell. This is the chloride shift.

Cellular respiration is a process concerned with the oxidation of food materials and with a consequent release of energy. The energy released by the breakdown of nutrient materials is not directly available but is transferred to the ADP–ATP system in the form of high-energy phosphate bonds. Hydrogen ions and electrons produced in the tricarboxylic acid cycle are a principal source of energy. Electron acceptors are necessary to transfer electrons to the cytochrome system. Nicotinamide adenine dinucleotide (NAD), flavin adenine dinucleotide (FAD), and ubiquinone are electron acceptors. The cytochromes are hemoproteins. Electrons are passed along the cytochrome series to cytochrome oxidase. Cytochrome oxidase is a respiratory enzyme that acts as a catalyst, activating molecular oxygen in the formation of water. The series of reactions involving the tricarboxylic acid cycle and electron transport occurs as a function of mitochondria.

The flow of tidal air in and out of the lungs measures about 500 milliliters. Taking a deep breath indicates the inspiratory reserve capacity. It averages around 3,000 milliliters. The expiratory reserve volume amounts to approximately 1,100 milliliters. Vital capacity averages 4,500 milliliters. The residual volume is about 1,000 to 1,200 milliliters.

There are several manual methods of artificial respiration. The most effective technique is the mouth-to-mouth method.

1 Consider some of the ways in which the respiratory system is closely related to the circulatory system.

2 Trace the passage of air from the nose to the alveoli of the lungs.

3 How do the vocal folds produce the voice?

4 Explain how respiratory movements draw air in and out of the lungs.

5 Discuss the nerve centers and their control of respiratory movement.

6 Discuss the chemical regulation of breathing.

7 How do chemoreceptors and pressoreceptors function?

8 Explain the effect of exercise on breathing.

9 Indicate the differences in the composition of inspired and expired air. Is there a corresponding difference between arterial and venous blood?

10 What are the effects of rapid decompression?

11 Discuss cellular respiration.

12 The chemical reactions of the tricarboxylic acid cycle and the electron transport system occur in what structures?

SUGGESTED READING

CLEMENTS, JOHN A.: Surface Tension in the Lungs, *Sci. Am.*, **207:**120-130 (1962).

COMROE, JULIUS H., Jr.: The Lung, *Sci. Am.*, **214:**57-68 (1966).

ENNS, T.: Facilitation by Carbonic Anhydrase of Carbon Dioxide Transport, *Science*, **155:**44-47 (1967).

GOLDSBY, R. A.: "Cells and Energy," The Macmillan Company, New York, 1967.

LEHNINGER, A. L.: Energy Transformation in the Cell, *Sci. Am.*, **202:**102-114 (1960).

McELROY, W. D.: "Cell Physiology and Biochemistry," 3d ed., Prentice-Hall, Inc., Englewood Cliffs, N.J., 1971.

MITCHELL, R. A.: Respiration, *Ann. Rev. Physiol.*, **32:**415-438 (1970).

REDDING, R. A., W. H. J. DOUGLAS, and M. STEIN: Thyroid Hormone Influence upon Lung Surfactant Metabolism, *Science*, 175:994-996 (1972).

WEIBEL, E. R., and D. M. GOMEZ: Architecture of the Human Lung, *Science*, **137:**577-585 (1962).

WEST, J. B.: Respiration, *Ann. Rev. Physiol.*, **34:**91-116 (1972).

WINTER, P. M., and E. LOWENSTEIN: Acute Respiratory Failure, *Sci. Am.*, **221:**23-29 (1969).

unit six

digestion

chapter 17
the digestive system

The animal body must take in food, provide for its digestion and absorption, and utilize the energy derived therefrom. The principal part of the digestive system consists of a long tube extending from mouth to anus. It is greatly coiled in the abdominal region. Within this tube and its derivatives, food materials are prepared for digestion and absorption. The tube, in different regions, is known as the *mouth, pharynx, esophagus, stomach, small intestine,* and *large intestine* (Figure 17.1). Large glandular organs such as the liver and pancreas contribute to the process of digestion. The mucous lining of the stomach and intestine invaginates numerous small glands whose cells are concerned with secreting digestive enzymes and other fluids.

The musculature of the digestive tract seems to be adapted for the special functions it is to perform. Striated muscles of the mouth and pharynx provide the voluntary motive force for chewing and swallowing. Involuntary smooth muscles propel the food materials along the tube and provide churning movements in the stomach and intestine.

Digestive enzymes, produced by glands of the digestive system, are protein in nature and act as organic catalysts to help break down the complex chemical structure of food materials into simpler compounds. When digestion has been completed, these simpler compounds are absorbed through the wall of the intestine and carried away by the blood. Considerable water is reabsorbed in the large intestine, and finally the waste products are eliminated as fecal matter.

MOUTH

The digestive tract begins at the oral opening, or mouth. The oral cavity is lined with mucous and submucous layers and contains, among other structures, the *tongue* and *teeth.* The roof of the mouth is the *hard palate,* formed by the maxillary and palatine bones and covered by mucous membrane. The *soft palate* is a posterior continuation of membrane covering muscle and nerve fibers. A tip of membrane hangs down into the throat cavity at the back of the mouth. This is the *uvula.* It is drawn back out of the way during swallowing or when we open the mouth and say "ah."

TONSILS

At the back of the mouth on either side of the uvula are the palatine arches. Tucked in between the arches on either side are reddish masses of lymphoid tissue commonly known as the *tonsils.* These are the *palatine tonsils.* Small mounds of lymphoid tissue below the tongue are called *lingual tonsils.* There are also *pharyngeal tonsils,* which are often referred to as *adenoids.* The tonsils are specialized masses of lymphoid tissue. The surface is marked with pits and depressions surrounded by lymphoid tissue. Lymphocytes are produced and pass to the surface through the epithelium. They can be observed

The digestive system

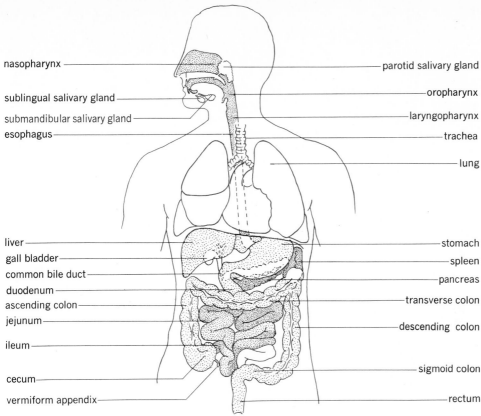

nasopharynx

sublingual salivary gland

submandibular salivary gland

esophagus

parotid salivary gland

oropharynx

laryngopharynx

trachea

lung

liver

gall bladder

common bile duct

duodenum

ascending colon

jejunum

ileum

cecum

vermiform appendix

stomach

spleen

pancreas

transverse colon

descending colon

sigmoid colon

rectum

FIGURE 17.1
General topography of the digestive system. The alimentary canal,
anterior view.

in saliva as salivary corpuscles. The pits, however, occasionally become inflamed by the action of bacteria harbored there. Infected tonsils are commoner in children than in adults. Adenoids, especially in children, can become so much enlarged that they interfere with nasal breathing; they can also compress the opening of the auditory tubes.

TONGUE

The tongue is a muscular structure of many functions. It is known primarily as a special organ for the sense of taste, but the sense of touch and the temperature sense are also well developed on the tongue. It is capable of mashing the softer particles of food and guides other particles between the upper and lower teeth to be crushed. In the act of swallowing, the tongue propels a ball (or bolus) of solid food and liquids back into the pharynx.

TEETH

Hard, enameled teeth located in sockets of the jaw bones provide for cutting, tearing, and crushing food. Human beings have two sets of teeth: a deciduous, or milk, set of 20 teeth to fit the jaw of a child and an adult set of 32 teeth for the jaw of an adult. The deciduous teeth appear gradually and are shed gradually. This means that a child of eight or ten years can have some deciduous teeth and some adult teeth in use at the same time (Figure 17.2).

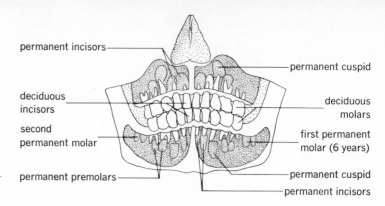

permanent incisors

permanent cuspid

deciduous incisors

deciduous molars

second permanent molar

first permanent molar (6 years)

permanent premolars

permanent cuspid

permanent incisors

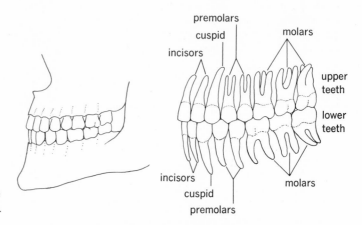

premolars

cuspid

molars

incisors

upper teeth

lower teeth

incisors

molars

cuspid

premolars

FIGURE 17.3

The teeth: permanent teeth of the upper and lower jaw (lateral view).

Not all mammalian teeth are alike; they are differentiated into incisors; canines, or cuspids; premolars; and molars. *Incisors* are the front teeth. They are flat, chisel-like teeth for cutting. There are eight of them, four in each jaw. The *cuspids*, or *canines*, are long and rounded for tearing food. They are located on either side of the lateral incisors, two in each jaw. These are the fangs of carnivores. Premolars, or bicuspids, are found only in the adult, or permanent, set. There are four in each jaw. There are commonly two cusps to a tooth; since they are located just anterior to the molar teeth, they are called *bicuspids* or *premolars*. The *molar teeth* are the large, flat teeth at the back of the jaw. The crushing surface is provided with four or five tubercles, which aid in mastication. There are eight molar teeth plus four *third molars*, or *wisdom teeth*, in the adult (Figure 17.3).

A newborn baby does not ordinarily have any erupted teeth, but they are forming in the jaw bones. Before the end of the first year the child probably will have cut its lower central incisors and upper incisors. At the age of 2 to $2\frac{1}{2}$ years the child may be expected to have his complete set of deciduous teeth. The second molars are usually the last of the deciduous teeth. There are no premolars in the deciduous set (Figure 17.2).

The deciduous teeth may be indicated as in the following formula for one-half of each jaw.

The digestive system

$$\text{Incisors } \frac{2}{2} \quad \text{Cuspids } \frac{1}{1} \quad \text{Molars } \frac{2}{2} = \frac{5 \times 2}{5 \times 2} = 20$$

The deciduous teeth are smaller than adult teeth, and the roots are smaller. The incisors are the first to be shed, around the seventh and eighth years. Children of this age present a peculiar toothless appearance before the adult incisors take their place. The second molars are the last of the baby teeth to be lost, usually when a child is around eleven or twelve years of age.

SIXTH-YEAR MOLARS Children of six or seven years actually possess two sets of teeth. The adult set is forming in sockets in the jaw bones, with their enamel caps still below the gums. By the time a child is six or seven years old, the jaw has lengthened considerably; there is room for the first adult molar to appear immediately behind the second baby molar. This is the so-called sixth-year molar, the first molar of the adult set.

The dental formula for the human adult is as follows:

$$\text{Incisors } \frac{2}{2} \quad \text{Cuspids } \frac{1}{1} \quad \text{Premolars } \frac{2}{2} \quad \text{Molars } \frac{3}{3} = \frac{8 \times 2}{8 \times 2} = 32$$

The deciduous teeth are gradually replaced by the permanent teeth, with the exception of the third molar, during the period between six and thirteen years of age. The second molars usually appear during the twelfth or thirteenth year. The third molars, or wisdom teeth, are very late in making their appearance, if they appear at all, usually between the seventeenth and twenty-fifth years, often at around eighteen or nineteen years. The jaws are usually long enough to accommodate them by this time, but sometimes they arise at an angle, grow into the second molar, and turn sideways. This is an *impacted* wisdom tooth, which usually has to be removed by a dentist. Wisdom teeth are sometimes formed with a poor coating of enamel, in which case dentists commonly advise their removal.

Proper occlusion of the teeth is important for their chewing function and in the configuration of the face. A receding chin or out-jutting lower jaw may be due to faulty occlusion of the teeth. In extreme old age, after the loss of all the teeth, the alveolar processes of the jaw bones wear down, and the chin is drawn up close to the nose when the jaws are closed.

STRUCTURE A tooth may be divided into three regions: The visible part above the gums is the *crown;* the *neck* is a constricted portion between the crown and the root; the *root* is the part embedded in a socket in the jaw bone.

The crown of a tooth is composed of enamel, the hardest substance in the body. The developing tooth is covered with an ectodermal epithelium (the enamel organ) that forms enamel, beginning with the top of the crown. The enamel is thickest on the wearing surface. Histologically, enamel consists of hexagonal rods called *enamel prisms.* Their arrangement accounts for the radial lines and striations characteristic of this substance. The enamel covering becomes thin as it terminates in the neck region of the tooth.

The chewing surfaces of premolar and molar teeth, especially, wear down with use, and the cusp patterns change with age. Teeth develop in sockets, or alveoli. The *gums, or gingiva,* are composed of dense connective tissue covered by mucous membrane of stratified squamous epithelium. The developing tooth and socket are at first covered by this connective tissue. When the tooth is "cut," it emerges through the gums. The alveolus is lined with periosteum, or periodontal membrane, which has several functions. It aids in attaching the tooth to its socket, acts as a nourishing membrane,

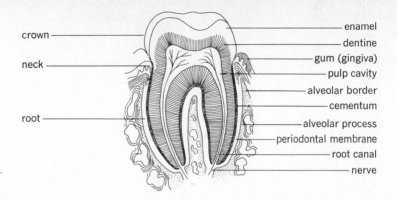

crown

neck

root

enamel
dentine
gum (gingiva)
pulp cavity
alveolar border
cementum
alveolar process
periodontal membrane
root canal
nerve

FIGURE 17.4

Section of a tooth *in situ*, showing its structure.

and is capable of forming the bonelike substance called *cement* over the root of the tooth (Figure 17.4).

The cement is a thinner and softer layer than enamel. It covers the tooth from the termination of the enamel to the tip of the root. Cement is described as bonelike because it contains some of the lacunae, canaliculi, and haversian canals characteristic of bone.

Dentine The greater bulk of the tooth is formed of a bonelike structure called *dentine*. In section the dentine has a striated appearance that is due to fine parallel tubes called *dental canaliculi*. They radiate outward from dentine-forming cells (odontoblasts) located along the border of the pulp cavity. Unlike bone, the dentine contains no cells. Long protoplasmic fibrils from odontoblast cells extend through the tiny canals. The dentine is a yellowish color. It is sometimes called the *sensitive dentine*, because of pain reactions caused by the dentist's drill. Since there are apparently no nerve fibers in the dentine, one explanation is that the protoplasmic strands within the canaliculi act as nerve fibers. Dentine forms slowly throughout the life of the tooth. Injury accelerates the formation of new, or secondary, dentine. Normally in older persons the pulp cavity is reduced by the continued formation of dentine.

Pulp cavity and pulp The pulp cavity lies in the central portion of the tooth, extending down into the root. The pulp is composed of loose connective tissue. It is highly vascular and contains a great many nerves. Odontoblast cells are located around the periphery. Blood vessels and sensory nerves enter through a tiny orifice at the tip of the root and pass through the root canal to the pulp cavity proper.

Probably the most important factor in producing good teeth is a proper diet in early childhood. Proper diet for the mother during pregnancy also contributes to good teeth in the offspring. One should not expect too much of toothpastes and toothbrushes. They clean the exposed surfaces of teeth but cannot be expected to rid the teeth of harmful bacteria. Mouthwashes strong enough to kill bacteria in the recesses between the teeth would be likely to injure delicate membranes of the gums and mouth. The development of dental caries is not an inflammatory process but rather a dissolving process. The enamel and other parts of the tooth are slowly softened and dissolved by the products of acid-forming bacteria or by acid foods and drinks. The more serious diseases of the teeth and gums, such as pyorrhea, should be referred to a dentist promptly.

The digestive system

420

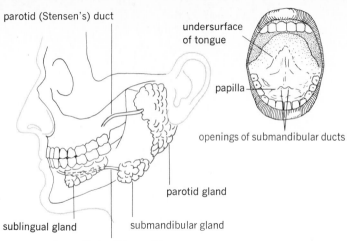

parotid (Stensen's) duct

undersurface of tongue

papilla

openings of submandibular ducts

parotid gland

sublingual gland

submandibular gland

submandibular (Wharton's) duct

FIGURE 17.5

The location of the salivary glands and their ducts, lateral view.

SALIVARY GLANDS

As mastication proceeds, the salivary glands pour their secretion, called *saliva*, over the food. It is a watery secretion, which moistens the food, holds particles together, and aids in the digestion of starch. The secretion of the three pairs of large salivary glands, called the *sublingual, submandibular,* and *parotid glands* (Figure 17.5), is supplemented by the secretion of numerous minute glands located in the mucous membrane lining the mouth. These are the *buccal glands,* the secretions of which are considered to be essentially the same as those produced by the main salivary glands.

The sublingual gland is the smallest of the salivary glands and is located below the anterior portion of the tongue and beneath the mucous membrane in the floor of the mouth. A varying number of small ducts open separately into the mouth below the tongue. The larger sublingual duct, the duct of Bartholin, may join the submandibular duct or may open separately into the mouth below the anterior part of the tongue.

The submandibular (submaxillary) gland is located posterior to the sublingual gland and deeper in the floor of the mouth. Its position, roughly speaking, is inside and a little below the mandible, close to the insertion of the masseter muscle. A long submandibular duct, Wharton's duct, opens through the floor of the mouth beneath the anterior portion of the tongue. The duct is thin-walled and about 5 centimeters long. The papilla, through which it opens, can be seen if the tongue is raised.

The parotid gland, the largest of the salivary glands, lies at the angle of the jaw and in front of the ear. The external carotid artery and the posterior facial vein pass through it. The duct of the parotid gland (Stensen's duct) passes across the outer surface of the masseter muscle and turns inward to open through the mucous lining of the cheek at a small orifice opposite the crown of the second upper molar tooth.

The parotid gland is involved in the virus disease known as *mumps.* The gland becomes greatly swollen and inflamed. All the salivary glands may become affected as well as other glandular tissues such as the testes. Serious inflammation and swelling of the testes may result in sterility, usually unilateral.

FUNCTION   The salivary glands are innervated by two sets of nerves from the autonomic nervous system. The facial (VIIth cranial) nerve carries parasympathetic fibers by way of the chorda tympani nerve to the sublingual and submandibular glands. The parotid gland is innervated by a branch of the glossopharyngeal (IXth cranial) nerve. The sympathetic innervation to the three pairs of salivary glands is through the superior cervical ganglia at the cephalic end of the chain of sympathetic ganglia (Figure 10.8).

Experimental stimulation of various sets of nerves to the salivary glands has determined some generally accepted concepts concerning secretion. Interpretation of effects is complicated by the fact that autonomic nerves carry both "secretory" fibers and fibers that regulate the circulation of blood through the glands. Within the glands themselves are two kinds of cells: serous and mucus-secreting. The mucus-secreting cells appear to be mainly in the sublingual and submandibular glands. These glands lend themselves well to experimentation. If the chorda tympani nerve supplying these glands is stimulated electrically, they are at once stimulated to secrete an abundant supply of thin, watery saliva. The stimulation also produces vasodilatation. It would be simple to conclude that vasodilatation goes hand in hand with increased secretion, but when the secretory function of the gland is paralyzed by atropine, similar stimulation of the nerve produces vasodilatation but depresses secretion.

Stimulation of the sympathetic nerves also produces secretion, but it is a very limited secretion of thick, mucous saliva. Stimulation of sympathetic nerves also causes vasoconstriction. Investigators must take into consideration the fact that salivation differs considerably in different species.

Experimental stimulation of the salivary glands produces results that are not necessarily comparable to the normal response. Thinking of the taste of sour food, such as lemons or vinegar, will usually cause a flow of saliva. Salivation at the thought or sight of food is a good example of a conditioned reflex. Placing substances in the mouth to stimulate the taste buds elicits salivation as a direct reflex act. Pain impulses, such as those invoked by the dentist's drill, cause a salivation of the ropy type because of the inclusion of considerable mucus. Salivation can be inhibited by fear and by embarrassment.

Experimental stimulation of salivary glands compared with observations of normal behavior leads to some similar conclusions. The parasympathetic system promotes a normal flow of saliva in the process of digestion. There is no question that stimulation of the parasympathetic nerves induces vasodilatation in the arterioles supplying the salivary glands and that stimulation of sympathetic nerves causes vasoconstriction. Conditions that would be unfavorable to digestion, such as pain, fear, or unpleasant associations, also activate the sympathetic system.

SALIVARY SECRETION   Saliva is a watery fluid with a specific gravity only a little greater than that of water. Under normal conditions various samples show a range from slightly acid to slightly alkaline. The average amount secreted per day is between 1,000 and 1,500 milliliters. The chemical composition consists of several inorganic salts such as sodium and potassium chloride, sodium bicarbonate, calcium carbonate, and phosphates of sodium and calcium. Some oxygen and carbon dioxide are included as gases. Organic substances include some of the blood proteins, urea, mucin, and the enzyme *salivary amylase,* or *ptyalin.* Mucin is secreted largely by the sublingual and submandibular glands. The composition of saliva varies considerably, depending on the type and strength of the stimulus. Blood cells in the saliva are usually lymphocytes, the so-called salivary corpuscles. The crust that commonly collects on the teeth is called *tartar*. It is a deposit of calcium carbonate and phosphate plus some organic materials from the saliva.

THE DIGESTIVE FUNCTION OF SALIVA   The salivary enzyme that acts upon starches was originally called *ptyalin*. Chemical nomenclature now requires that enzymes have the ending *-ase*, and so this enzyme is now called *salivary amylase*. *Amyl* is derived from the Greek work for starch. Since the food does not remain long in the mouth before being swallowed, much of the starch digestion is carried on in the stomach until the enzyme is inhibited by the acid of the stomach. This period is probably 15 to 30 minutes. Cooked starch is much more easily digested, since the uncooked starch granule has a cellulose covering. The salivary digestion of starch is not fully understood, but the larger starch molecule breaks down through dextrins to maltose and glucose.

It appears that human salivary amylase, acting in vitro, is capable of converting starch to glucose if the relative amount of the enzyme is great enough. Salivary amylase does not normally convert maltose to glucose, however, and there appears to be no maltose in human saliva. The saliva of many animals, including the dog, seems to have no digestive action at all.

EXCRETORY FUNCTION OF SALIVA   One should not overlook the fact that many substances can be excreted in the saliva. Mercury and lead can be excreted when introduced into the body. In lead poisoning there is a deposition of lead salts on the gums close to the teeth, known as the *blue line*. Cyanides are found as trace elements in our food. They are very toxic but are readily converted into thiocyanates, which are nontoxic. Thiocyanates are excreted by the salivary glands as well as by the kidneys. Urea is excreted by the salivary glands, especially in certain kidney diseases. Diabetics sometimes notice a sweet taste in the mouth caused by the excretion of sugar into the mouth. Viruses are sometimes excreted in the saliva. A well-known example is the virus of rabies.

PHARYNX

The pharynx is the aperture back of the mouth. It is lined by a continuation of the mucous membrane that lines the mouth and nasal passageways. The mouth opens into the pharynx through an isthmus called the *fauces*. The opening is bounded by the palatine arches on either side, the uvula hangs down from the soft palate above, and the base of the tongue lies below. The nasopharynx communicates with the nasal passageways and continues posteriorly. The internal openings of the eustachian tubes are nearby. The pharynx communicates below with the larynx and esophagus.

ACT OF SWALLOWING

Swallowing consists of the muscular propulsion of food from the mouth through the pharynx and into the esophagus, the tube leading to the stomach. The bolus, or mass of food, has usually been chewed and mixed with saliva. Mucin from the saliva aids in holding the particles together. The muscular action of the tongue propels the food backward through the fauces into the pharynx. The constrictor muscles of the pharynx contract and force the food into the esophagus. During the act of swallowing, the soft palate is raised to prevent food from entering the nasal passageways. At the same time, the larynx is raised to a position under the base of the tongue and covered by the epiglottis. One can readily observe the elevation of the larynx during swallowing.

There can be no intake of air through the trachea during swallowing; the respiratory center is inhibited momentarily. There is a brief inspiration just preceding the act, but the epiglottis is then closed during the passage of food materials. Sometimes when persons are talking, laughing, and eating as well, a mistake in timing occurs and solids or liquids are directed into the larynx. A violent coughing reflex ensues.

ESOPHAGUS

The muscular tube extending from the pharynx to the stomach is the esophagus. It is about 10 inches long and about an inch in width. The wall of the tube is composed of four coats. The outer coat is thin, fibrous connective tissue. The muscular portion contains an outer layer of longitudinal fibers and an inner layer of circular muscles. The musculature of the upper part of the tube is thick, striated muscle like that of the pharynx, while that in the lower part of the tube is smooth, visceral muscle. The muscular coat is lined with an areolar submucous layer supporting blood vessels; finally a mucous layer lines the tube (Figure 17.6).

Liquids and solids are propelled through the esophagus by wavelike (or peristaltic) contractions of the muscular wall. Liquids are moved very rapidly, with gravity playing some part in their descent to the stomach in man. In animals that drink with the head down, peristalsis evidently overcomes the force of gravity. The food is passed through the upper striated portion of the tube more rapidly than through the lower smooth-muscle portion. Some animals, such as the dog, have striated muscle throughout the length of the esophagus and can swallow very quickly. As the bolus of food approaches the cardiac sphincter muscles, which guard the entrance to the stomach, these muscles relax and permit the food to enter the stomach. The cardiac sphincter may not be a true sphincter muscle in a strict sense, but muscle fibers in this region act as sphincters in opening and closing the upper orifice of the stomach.

The first stage of swallowing is voluntary, but when the food reaches the pharynx, the second and third stages are purely reflex and involuntary. Motor nerves concerned with the act are branches of the hypoglossal, trigeminal, glossopharyngeal, and vagus.

The act of vomiting forces the stomach contents upward through the esophagus and mouth. Vomiting is a reflex act controlled from a vomiting center in the medulla. Stimuli of many different kinds may cause vomiting. Among the better-known stimuli are disturbances of the stomach and intestine, stimulation or irritation of the back of the mouth or pharynx, and disturbances of the semicircular canals, as in motion sickness. The act is preceded by nausea; the cardiac spincter and esophagus are relaxed, while the diaphragm and muscles of the abdomen supply the propelling force. Physical signs are pallor, perspiration, a feeling of weakness, and excessive salivation.

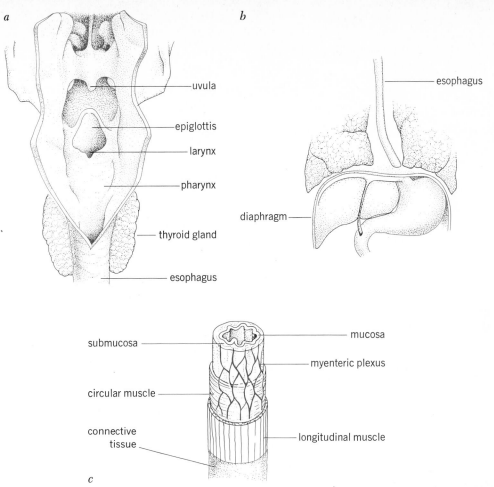

a

uvula

epiglottis

larynx

pharynx

thyroid gland

esophagus

b

esophagus

diaphragm

submucosa

mucosa

myenteric plexus

circular muscle

connective tissue

longitudinal muscle

c

FIGURE 17.6

The esophagus, *a* superior opening; *b* inferior portion; *c* structure.

The abdominal cavity is the largest body cavity. It extends from the dome-shaped diaphragm above to the lesser pelvis below. Posteriorly are the vertebrae and the deep muscles of the back. Abdominal muscles form the sides and the front walls. Contained within the abdomen are the stomach, liver, pancreas, intestine, spleen, and kidneys. The abdominal wall has several openings that transmit various structures. The umbilical opening transmits the umbilical blood vessels in the fetus. The esophagus penetrates the diaphragm to enter the stomach. The vena cava and the aorta also pass through the diaphragm. The thoracic duct and the azygos vein accompany the aorta through the diaphragm. The floor of the abdominal cavity has two openings on either side; one transmits the femoral artery and vein, and the other transmits the spermatic cord in the male and the round ligament in the female.

ABDOMEN

Abdomen

The abdominal cavity is lined with a serous membrane called the *peritoneum*. It is folded and reflected over abdominal organs. The intestine is held in place by folds of peritoneum called the *mesentery*. The mesentery is a double layer; between the two layers it supports mesenteric blood vessels, lymphatic vessels, lymph nodes, and nerves. A large fold connected to the greater curvature of the stomach and to the transverse colon covers the intestine anteriorly like an apron and is called the *greater omentum*. It is usually well supplied with fat, and it functions as a protective and insulating layer. The *lesser omentum* is a double layer of peritoneum attached between the stomach and liver. Inflammation of the peritoneum is called *peritonitis*.

STOMACH

The *stomach* is an enlarged portion of the digestive tract located in the abdominal cavity directly below the diaphragm on the left side. It is continuous with the esophagus and the cardiac sphincter, which surrounds its upper orifice. The shape of the stomach varies. When empty, it is more like a tube; when filled, it is more saclike. Certain areas of the stomach can be readily identified. The *fundus* is the rounded, upper portion extending above the cardiac sphincter (Figure 17.7). At the lower end is the *pyloric antrum*, a slight enlargement above the pyloric orifice, which opens into the upper part of the intestine, or *duodenum*. The pyloric opening is guarded by the pyloric sphincter muscle. Between the pylorus and the fundus is the body of the stomach. The lesser curvature of the stomach is concave toward the right and faces the liver. The greater curvature is convex on the opposite side and is several times longer.

The *stomach wall* is composed of four layers typical of the digestive tract. The outer serous layer is formed from the peritoneum. It is reflected over the organ in such a way that the two layers come together at the lesser

FIGURE 17.7
The stomach (anterior view).

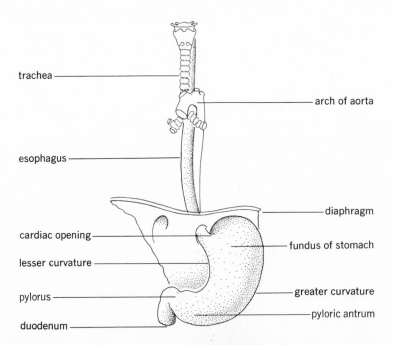

trachea

arch of aorta

esophagus

diaphragm

cardiac opening

fundus of stomach

lesser curvature

pylorus

greater curvature

pyloric antrum

duodenum

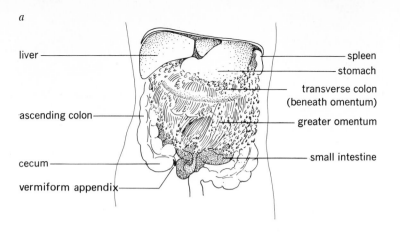

liver — — spleen
— stomach
— transverse colon
(beneath omentum)
ascending colon — — greater omentum
cecum — — small intestine
vermiform appendix —

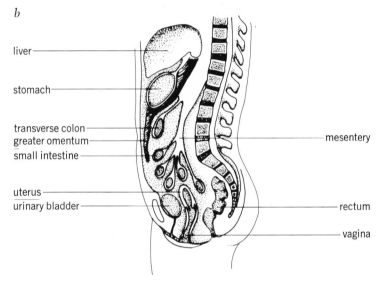

liver —
stomach —
transverse colon —
greater omentum —
small intestine —
uterus —
urinary bladder —
— mesentery
— rectum
— vagina

FIGURE 17.8
a The greater omentum covering the abdominal viscera (anterior view); *b* the trunk in sagittal section showing the location of the mesentery and greater omentum (female).

curvature and extend to the liver as the *lesser omentum*. The two layers join along the greater curvature and extend downward as the *greater omentum* (Figure 17.8).

The thick muscular coat consists of three layers of visceral muscle: longitudinal, circular, and oblique. The position of these muscle layers enables the stomach to perform churning and peristaltic motions in preparing the food for further digestion. The longitudinal layer is the outermost layer. At the cardiac opening it is continuous with the longitudinal muscles of the esophagus. The circular layer is the intermediate layer and is continuous with the circular layer of the esophagus. At the lower end of the stomach the fibers form a strong ring of muscular tissue, the pyloric valve. The internal oblique layer is composed of fibers radiating from the cardiac area. The submucous coat consists of loose, areolar tissue. It connects the muscular and mucous layers. The mucous layer is thick and glandular. Typically, if the stomach is not distended, the mucous lining is in folds, mostly longitudinal. The surface is pitted with the openings of glands, which secrete the

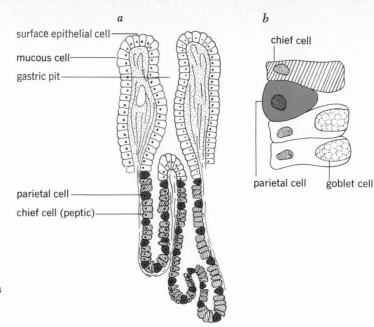

surface epithelial cell
mucous cell
gastric pit

parietal cell
chief cell (peptic)

a *b*

chief cell
parietal cell   goblet cell

FIGURE 17.9

Section of stomach wall showing glands. The cells shown enlarged are surface epithelial cells, mucous cells, parietal cells, and zymogen or chief cells.

gastric fluid. The inner surface of the mucous layer is covered by columnar epithelium containing goblet cells (Figure 17.9).

There are three kinds of cells in the small, tubular principal glands of the stomach. First, there are the *mucous cells*, located in the neck region of the gland or in the lower part of the pit. They secrete *mucin*. Secondly, there are the *chief cells*, or peptic cells, which line the tubule and secrete *pepsinogen*. Zymogen granules, thought to be the source of the pepsinogen, can be demonstrated at the bases of these cells. Microanalysis indicates that the chief cells also contain dipeptidase. This is usually considered to be an intestinal enzyme acting on dipeptides. The third type is the parietal cell.

Parietal cells are large spheroid cells distributed at intervals along the outer side of the chief cells. These cells produce a substance that is converted to hydrochloric acid when it reaches the cavity of the gastric pit. The cells themselves do not contain HCl. Secretory canaliculi can be demonstrated in these cells. Parietal cells are not present in the glands located at the pyloric end of the stomach.

FUNCTION The saclike stomach enables one to eat a considerable quantity of food at one time. The food is moved back and forth by muscular contractions and mixed with the gastric juice. Through the action of digestive enzymes and hydrochloric acid, the food becomes partially liquefied and is called *chyme*. When it reaches the proper consistency, it is forced out through the pylorus and into the duodenum, a little at a time.

Pepsin is the principal digestive enzyme found in the stomach. It is a protease, breaking down the complex protein molecule through cleavage stages known as *proteoses* and *peptones*. It does not complete the work of breaking down proteins to their end products, amino acids. Proteases of the pancreas and intestine accomplish this final breakdown before the protein is absorbed into the blood. When produced by the chief cells, pepsin is in an inactive form called *pepsinogen*. It is activated upon contact with hydrochloric acid (Table 17.1).

Hydrochloric acid activates pepsinogen and provides the proper acid medium for digestion by pepsin. The concentration of HCl in the gastric juice as it appears in the glands is around 0.5 percent. The acidity of the stomach contents is seldom as great as this, since the hydrochloric acid is diluted by saliva, food, and mucin. In addition, the secretion of glands in the pyloric portion of the stomach is slightly alkaline and has some neutralizing effect. Even so, the acidity of the gastric juice is usually found to be about pH 1.6 to 2.6. When gastric contents are regurgitated into the throat, it is the hydrochloric acid that causes the burning sensation.

In addition to activating pepsinogen and providing an acid medium

TABLE 17.1
Digestive fluids and principal digestive enzymes

Source	Fluids	Enzymes	Substrate	End products
Mouth	Saliva	Amylase (ptyalin)	Starch is hydrolyzed	Dextrins and maltose
Stomach	Gastric juice (acidified by HCl)	Pepsin	Proteins are hydrolyzed	Proteoses Peptones Polypeptides
		Lipase	Emulsified fats (as in cream, egg yolks)	Glycerol Fatty acids
		Rennin (in infants)	Caseinogen	Paracasein (precipitated by calcium to form curd)
Pancreas	Pancreatic fluid	Trypsin (in a slightly acid, neutral, or alkaline medium)	Undigested proteins and proteoses	Peptones Polypeptides Amino acids
		Amylase (amylopsin)	Starch, dextrins	Disaccharides maltose
		Lipase (steapsin)	Fats are hydrolyzed	Glycerol Fatty acids
Small intestine Intestinal glands		Peptidases	Polypeptides Dipeptides	Amino acids
		Lipase	Neutral fats	Glycerol Fatty acids
		Amylase	Carbohydrates	Disaccharides

TABLE 17.1 (Continued)

Source	Fluids	Enzymes	Substrate	End products
Brush border of villi and breaking down of epithelial cells		Disaccharidases		
		Lactase*	Lactose is hydrolyzed	Glucose (dextrose) Galactose
		Maltase*	Maltose is hydrolyzed	Glucose (dextrose)
		Sucrase*	Maltose is hydrolyzed	Glucose (dextrose)
		Sucrase* (invertase) (several of lesser importance)	Sucrose (cane sugar) is hydrolyzed	Glucose and fructose
	Intestinal fluid	Enterokinase	Activates pancreatic trypsinogen	
	Bile	(Contains no enzymes)	Fats are split by lipase alone, but they are not absorbed unless emulsified by bile salts.	

*Lactase, maltase, and sucrase are inverting enzymes.

for pepsin, hydrochloric acid is capable of swelling and softening proteins such as meat fibers; it renders gastric juice highly antiseptic; it is capable of initiating the inversion of cane sugar and of curdling milk.

Rennin is an enzyme that acts upon milk and is present in the gastric juice of infants. It is probably produced in an inactive form called prorennin. Rennin or a similar synthetic product has been used in cheese making for many years. It coagulates milk, producing a white mass called *curd* and separating it from the fluid material called *whey*. Rennin acts upon the soluble protein *casein*, converting it into *paracasein*, which combines with calcium ions to form the curd. The action of rennin is to convert milk into a form suitable for gastric digestion. Rennin does not seem to be effective in the stomach of the adult, in which the acid content of the gastric juice is high. In the adult, milk is curdled by direct action of HCl. In the infant, however, the acid concentration in the stomach is much lower and thus is more favorable for the action of rennin. Rennin is thought to be produced by chief cells.

Though some investigators have reported some fat digestion in the stomach, the presence of gastric lipase remains in doubt. It is possible that some pancreatic lipase is introduced into the stomach from the duodenum.

It has been amply demonstrated that the taste of food or even the sight of food will cause gastric-fluid secretion. The latter is a good example of a conditioned reflex, and the response is primarily from a nerve stimulus. If the vagus nerve is severed, the response is largely lost; but the possibility of a hormonal response should not be overlooked. The response has value in that the stomach is thereby better prepared to receive and digest food. It is commonly referred to as the *psychic phase* of gastric digestion.

The flow of digestive juices can be reduced or inhibited, on the other hand, by unpleasant sights, tastes, or smells. This effect is mediated through the sympathetic system. A strong emotional reaction is also unfavorable to good digestion.

Pleasant surroundings, good food, and peace of mind favor good digestion. The phrase "laugh and grow fat" has some basis in fact. The rush of present-day living is often not conducive to good digestion.

When food reaches the stomach, distention of the stomach wall, plus the influence of *secretagogues* in the food itself, initiates the release of the hormone *gastrin*. Secretagogues are chemical substances in certain foods which are released as the food is digested. Meat and some other proteins have a high secretagogue content. Gastrin is absorbed by the blood, and when it returns to the gastric mucosa, it stimulates the parietal cells to secrete HCl. Histamine, when introduced into the body, has a similar effect, but gastrin released normally is much more effective. The secretion of gastric juice after food has entered the stomach is known as the *gastric phase* of digestion.

There is a third phase of gastric digestion commonly called the *intestinal phase*. Experimentally, when food that has never been in the stomach is introduced into the upper part of the small intestine of an animal, the food so introduced causes the gastric glands to secrete. The effect appears to be hormonal, but no hormone has been isolated.

When unusual concentrations of food substances, notably fats and carbohydrates, are passed into the duodenum too rapidly, a hormonal factor is liberated to inhibit the stomach. The hormonal factor is *enterogastrone*, secreted by glands in the mucous lining of the duodenum. Since all hormones are absorbed by the blood, enterogastrone is absorbed and carried to the stomach. Inhibition of the stomach appears to be a protective device to prevent overloading of the duodenum, but it also functions in reducing gastric acidity.

Hyperacidity of the stomach is usually a temporary condition resulting from overeating or perhaps from eating when one is under the influence of some unfavorable situation. Constant hyperacidity may indicate a more serious condition. *Gastric ulcers*, sore spots in the stomach mucosa and in many respects resembling canker sores in the mouth, are commonly associated with hyperacidity, usually of long standing. Normally, the mucous lining of the stomach protects the deeper layers from the action of pepsin and HCl. The open sores of ulcers lack this protection and are irritated by these secretions.

The question is often asked: Why doesn't the stomach digest itself? The question is a difficult one, and the complete answer is not known. Pepsin is considered to be in an inactive form when it is produced by the gland. After it is activated, copious amounts of mucin should help to protect the stomach wall. It is possible that the mucosa produces a protective antienzyme to inactivate pepsin, but no such substance has been isolated. It is well known that after death the digestive enzymes do attack the stomach.

Partially digested food leaving the stomach enters the upper part of the small intestine, or *duodenum*. The small intestine extends from the pylorus to the ileocecal valve, where it enters the large intestine. The duodenum is the **DUODENUM** widest portion, extending to the *jejunum*, or middle portion. The *ileum* is the terminal part. There is no strict morphological demarcation of these different parts, only a gradual change in size and other morphological characteristics. The length of the small intestine in the human adult is commonly given as 23 feet, but there are indications that the living intestine is less than half this length. It is sometimes divided into the duodenum and mesenteric intestine for convenience, the duodenum not being supported by the mesentery and only partially covered by peritoneum. The mesenteric intestine is the jejunum and ileum (Figure 17.1 and 17.10).

The duodenum is said to have derived its name from the practice of estimating its length by the width of 12 fingers, about 25 centimeters. This shortest and widest part of the small intestine makes a loop downward from a little beyond the pylorus, the distal end of the loop rising almost as high as the proximal end. The greater part of the pancreas lies in the loop of the duodenum. The pancreatic and bile ducts enter the duodenum about 10 centimeters beyond the pylorus. The reaction within the duodenum, unlike that in the stomach, is alkaline. Just beyond the pyloris are numerous mucus-secreting glands called *Brunner's glands*. The mucus secreted by these glands affords protection from the strongly acidic gastric juice emerging from the stomach.

PANCREAS

The pancreas is a slender gland extending from the loop of the duodenum upward behind the stomach to a length of 12 to 15 centimeters. It is a compound gland, similar in many respects to the salivary glands, and is somewhat pink in color. Digestive enzymes are secreted; they enter the duodenum through the pancreatic duct. The main pancreatic duct extends the length of the gland, receiving many tributary branches. Near its base it is commonly joined by a small accessory pancreatic duct. Close to the duodenum, the pancreatic duct lies beside the common bile duct, the two joining and opening into the duodenum by a common orifice (Figure 17.10 and 17.11*a*).

The pancreas secretes three major enzymes, one for each of the three kinds of food: *trypsin* (or *tryptase*) is a proteolytic enzyme; *amylase* acts on starches; and *lipase* acts upon fats. There are other proteolytic enzymes, such as *chymotrypsin*, which is similar to trypsin, and the *peptidases*, which act upon reduced proteins called *peptides*. All enzymes exhibit a high degree of specificity for a certain type of substrate upon which they act; that is, protein-digesting enzymes do not digest carbohydrates or fats. Some are limited to acting only on certain stages in the hydrolysis of a molecule.

The control over the secretion of pancreatic fluid is of two types: nervous and hormonal. Stimulation of the vagus nerve causes the pancreas to secrete a fluid rich in digestive enzymes. The results of experimental stimulation of the sympathetic splanchnic nerves, on the other hand, have been given various interpretations. Some have obtained secretion similar to that obtained when the vagus (parasympathetic) nerve was stimulated. Stimulation of the sympathetic nerves brings about vasomotor changes, which complicate interpretation of the results. Hormonal control is due largely to the action of the hormones secretin and pancreozymin.

TRYPSIN The protein-digesting enzyme of the pancreas, trypsin, is secreted by glandular cells in an inactive form called *trypsinogen*. This material, when collected from the pancreatic duct, has little ability to digest proteins.

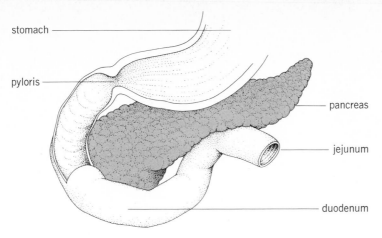

stomach

pyloris

pancreas

jejunum

duodenum

FIGURE 17.10
A portion of the small intestine, the duodenum and proximal part of the jejunum.

However, if permitted to mix with intestinal juice, it becomes active trypsin. It is activated by the coenzyme *enterokinase*, produced by cells of the intestinal mucosa. Trypsinogen also may be activated by cells of the intestinal mucosa. Trypsinogen also may be activated by calcium. Trypsin digests proteins in an alkaline medium. The optimum pH is near 8.

Trypsin acts upon proteoses and peptones of gastric digestion, breaking them down to peptides. Proteins that have not been acted upon by pepsin are also broken down. The degree of digestion is determined by optimum conditions and the length of time allowed. Apparently trypsin does not ordinarily complete the breakdown of proteins to amino acids. This is accomplished by the peptidases of the small intestine.

AMYLASE The starch-digesting enzyme of the pancreas, amylase, is similar to salivary amylase but somewhat more active. In an alkaline to neutral medium, it is capable of hydrolyzing starch to various dextrins and to maltose. In the light of recent work on salivary amylase, it is possible that glucose also is formed.

LIPASE There is only one enzyme concerned with the digestion of fats. Lipase is the lipolytic enzyme; that is, it acts upon lipids or fats. Lipase splits fat molecules into fatty acids and glycerol. The action is aided by bile salts from the liver. Bile salts exhibit a detergent action on fat particles which reduces their surface tension. The process is called emulsification. Bile salts also aid in the absorption of fatty acids. Emulsified fats present a greater surface area for the action of the digestive enzyme. The fatty acids involved are organic acids and highly insoluble in water, although they are soluble in bile-salt solutions. It has been shown that emulsification of fats with high melting points is difficult to accomplish. Such fats are not easily digested. Fats with low melting points appear to be more easily digested. Glycerol does not present any special problem, since it is readily soluble in body fluids.

SECRETIN Hormonal control over pancreatic secretion involves the hormone *secretin*. When the acid contents of the stomach enter the normal alkaline condition of the duodenum, a hormone is released from the duodenal mucous lining. It is absorbed by the blood and carried to the pancreas, causing this organ to secrete a fluid which is largely water and bicarbonate. The secretion is thin, and low in enzyme content, but is more alkaline than that obtained

Pancreas

by stimulation of the vagus nerve. The hormonal mechanism stimulates the pancreas to secrete at the time food is present to be acted upon. The study of endocrinology has developed greatly in the past few years, but the discovery of secretin goes back to the work of Bayliss and Starling, in 1902. When bile is introduced into the duodenum, secretin is absorbed and the pancreas is stimulated to secrete. Finally, there is evidence that another hormone is involved. This hormone, called *pancreozymin*, stimulates the flow of digestive enzymes into the pancreatic fluid.

INSULIN   The pancreas not only produces digestive enzymes; it also gives rise to the hormone *insulin*, which is concerned with carbohydrate metabolism. The hormone is not produced by the large glandular cells that secrete enzymes but by groups of very small cells, called the *islet cells of Langerhans*, of which there are two kinds: *alpha* and *beta* (Figure 17.11c). The beta cell secretes insulin. Alpha cells secrete another hormone called *glucagon* which will be considered later (page 443). Insulin is not dispersed by way of the pancreatic duct; it is absorbed and distributed by the bloodstream, as are all hormones. The chemical structure of insulin has been found to be a protein consisting of two polypeptide chains held together by two disulfide linkages. One chain is composed of 21 amino acids, and the other contains 30 amino acids.

REGULATION OF CARBOHYDRATE METABOLISM

When the pancreas fails to produce enough insulin, a serious derangement of carbohydrate metabolism occurs; this condition is known as *diabetes mellitus*. In the absence of insulin, tissues are unable to utilize glucose, and the blood sugar level rises from a normal 0.1 to 0.2 percent or more. As expressed in milligrams per 100 milliliters, the blood glucose level in diabetics can rise from a normal of 80 to 120 milligrams per 100 milliliters to 200 to 300 milligrams per 100 milliliters. It is generally considered that the blood "threshold," or the level at which glucose begins to appear in the urine, is around 180 milligrams per 100 milliliters. High blood sugar is called *hyperglycemia*. As the sugar level of the blood rises above normal, the kidneys begin to excrete this substance. Since the kidneys do not ordinarily excrete sugar, a chemical test for sugar in the urine is routine procedure in the diagnosis of diabetes. The kidneys do excrete sugar after a high carbohydrate intake, but this is a temporary *glycosuria*.

The sugar reserves of the body are rapidly depleted in diabetes. Proteins are then utilized, and there is also a disturbance of fat metabolism. The oxidation of fats is increased, resulting in the accumulation of certain organic acids (ketone substances) in the blood and urine. One of the breakdown products of fatty acids is acetoacetic acid. This is a toxic substance and may be responsible for the development of diabetic coma in extreme untreated cases.

Early workers removed the pancreas from dogs and discovered that there was a great increase in the production of urine, that the urine contained sugar, and that these animals were unable to live more than a month without the pancreas. They determined also that it is not the digestive portion of the pancreas that is essential to life. Investigators tied off the ducts of the pancreas, causing the glandular cells producing digestive enzymes to degenerate and leaving only the islands of Langerhans still active. The dogs lived

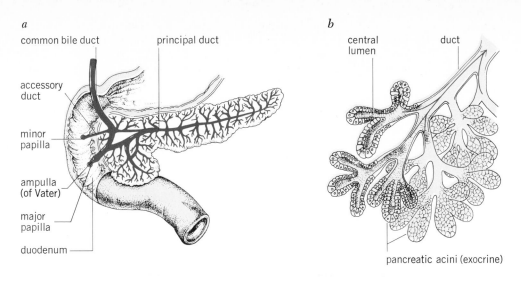

a

common bile duct   principal duct

accessory duct

minor papilla

ampulla (of Vater)

major papilla

duodenum

b

central lumen   duct

pancreatic acini (exocrine)

c

islet of Langerhans (endocrine):

beta cell

alpha cell

sinusoids

reticular tissue

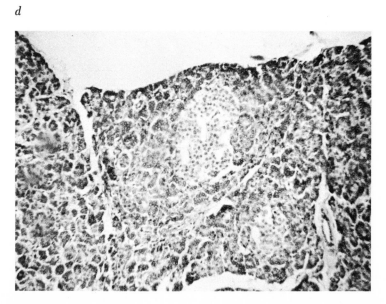

d

FIGURE 17.11

The pancreas: *a* system of ducts; *b* exocrine portion (these are the acinar cells that produce the pancreatic digestive enzymes with an output of about 1,200 milliliters per day); *c* the endocrine portion. Beta cells of the islets of Langerhans produce insulin; alpha cells produce glucagon. These hormones are absorbed directly into the bloodstream; *d* photomicrograph of islet cells, which are located in the lighter areas. (*Courtesy of L. Kelemes M.D.*)

Regulation of carbohydrate metabolism

435

and were not diabetic. There were many attempts to extract a hormone from the pancreas, but these attempts met with failure until 1922. In that year Banting and Best prepared an extract of pancreatic tissue from islet cells that was free of digestive enzymes; when injected into depancreatized dogs, it reduced their diabetic symptoms. This outstanding work was done with the collaboration of J. J. R. Macleod and others at Toronto, Canada.

Insulin is a complex protein. It can be prepared in crystalline form with salts of zinc, nickel, cadmium, or cobalt. Most forms must be injected hypodermically, since, if taken orally, insulin is readily digested before it can be absorbed. Some types of insulin have been developed for oral administration, and apparently these are sometimes useful in mild cases of diabetes. Highly soluble insulin is injected subcutaneously two or three times daily by the diabetic. It is absorbed by the blood rather rapidly, causing the available concentration in the blood to fluctuate and the blood sugar level of the diabetic to vary. The normal secretion of islet cells is apparently produced steadily and in minute amounts.

A search was made for less-soluble forms of insulin, which would prolong their absorption by the blood. It was found that insulin precipitated with certain protein derivatives is absorbed slowly and therefore remains effective over a longer period of time. Protamine zinc insulin is one of the long-acting forms. Protamine is a fish-sperm protein; the addition of zinc renders the solution more stable. NPH (Neutral Protamine Hagedorn) insulin is another long-acting form developed more recently. Since these insulin preparations are not highly soluble, a single injection in the morning may be sufficient for 24 hours if the diabetic condition is mild.

When insulin is injected, there is an almost immediate drop in the blood sugar level. There may be some question as to the disposition of the blood sugar. Part of the glucose is converted to glycogen and stored in the liver and muscles; part is oxidized, with a consequent release of energy; and part undergoes synthesis into fat. Insulin does not cure diabetes, but it does regulate carbohydrate metabolism to such an extent that diabetics can live essentially normal lives. With a lowering of the blood sugar level, glycosuria tends to disappear; the metabolism of fats approaches normal, and therefore abnormal acids in the blood and urine are reduced or disappear. Diabetics usually have increased thirst, which is allayed only by drinking considerable amounts of water. A large amount of water is necessary to provide for the filtration of the quantity of glucose and salts excreted. Insulin, in reducing the glycosuria, also reduces diuresis.

More recent experimental work indicates that there is a difference in tolerance to diabetes between the carnivorous and herbivorous animals investigated. Almost all the early investigators used dogs or cats as experimental animals and found that they could not live without the pancreas for more than a few weeks. On the other hand, herbivorous animals, such as rabbits, sheep, goats, and monkeys, can survive for several months. These animals gradually lose weight and eventually die from causes directly or indirectly related to loss of insulin; but the inference here is that insulin, like other hormones, is merely a regulator of physiological activities. Without the regulator, these physiological activities may vary in different animals, but they do not necessarily cease to function.

Hyperinsulinism is a condition in which the islet cells produce too great a secretion of insulin. There is an accompanying hypoglycemia (low blood sugar), and the central nervous system seems to be primarily affected, probably from the effects of glucose starvation. Tumors of the islet cells appear to be a common cause of excess secretion but hypoglycemia is not necessarily associated with hyperinsulinism.

Insulin Shock Repeated injections of insulin resulting in extreme hypoglycemia to the extent of causing convulsive seizures have been used in treating certain types of mental cases. Considerable mental improvement has been reported in many cases, although the treatment is not without physical danger. Various stages of insulin shock can usually be terminated by the administration of glucose.

TRANSPORT OF GLUCOSE THROUGH THE CELL MEMBRANE   In the discussion of theories of transport across cell membranes in Chapter 2, it was indicated that cell membranes are not readily permeable to glucose. Since most cells utilize considerable amounts of glucose in their metabolic activities, it is generally considered that there must be an active transport of glucose through the cell membrane. The exact mechanism of transport is not known, but it is generally thought that glucose diffuses through the cell membrane with the aid of some sort of carrier mechanism.

The precise way in which insulin functions is not thoroughly understood, but apparently it accelerates the transport of glucose through the cell membrane of certain kinds of cells. Muscle cells especially (or their membranes) require insulin to accelerate the passage of glucose into the cells. However, a number of other kinds of cells, notably intestinal epithelial cells, liver cells, neurons, and erythrocytes, do not require insulin to permit the passage of glucose through their cell membranes. The cells of the kidney tubules and the brain apparently are also in this category. The absorption rate may not be as rapid in these cells as the rate of absorption required to meet the needs of muscle cells, but insulin does not seem to be a factor.

ROLE OF OTHER GLANDS   The pancreas is not the only gland concerned in the maintenance of normal carbohydrate metabolism. The thyroid, adrenal, and pituitary glands are involved also.

Thyroid imbalance produces profound changes in metabolism in general. When linked with diabetic conditions, the disturbance in carbohydrate metabolism is marked. Hyperthyroidism increases general metabolic activities; when it accompanies diabetes, the effects of this condition are much more severe.

Among the effects noted upon injection of epinephrine from the adrenal glands are hyperglycemia and glycosuria. The conversion of liver and muscle glycogen is accelerated. The hyperglycemia that follows injection of epinephrine is a direct countereffect to that following the injection of insulin. While epinephrine has a part in the normal regulation of blood sugar, it should not be inferred that it has any direct relation to the basic causes of diabetes.

A striking effect is obtained by removing both the pancreas and the pituitary gland from an experimental animal. In this case the blood sugar level remains within normal limits, with glycosuria much less than in animals with just the pancreas removed. The lives of these animals are greatly prolonged. The removal of the pituitary alone would ordinarily reduce the blood sugar level; the removal of the pancreas alone would cause the blood sugar level to rise. Apparently the removal of both glands causes a balance to be effected, and carbohydrate metabolism is maintained. There is no protective regulation of carbohydrate metabolism, however, and the metabolism is easily disbalanced by variations in diet, the effects of disease, and many other conditions.

In another series of experiments, a condition similar to diabetes has been induced by injecting anterior pituitary extracts into dogs and cats over a period of several days. The condition disappears within a few days after termination of injections. However, if injections are continued for several weeks, a condition of permanent diabetes may result. The islet cells of the pancreas are injured extensively by this treatment. The diabetogenic factor of the pituitary is evidently involved in causing these changes.

LIVER

The liver is the largest glandular structure in the body (Plate F). It is located just under the diaphragm in the upper right portion of the abdominal cavity. Somewhat wedge-shaped, the thicker portion is on the right, with the thin part of the wedge to the left, lying over the stomach (Figure 17.12). It is also in contact with the right kidney, the duodenum, and the right colon at its flexure. The surface is largely covered by peritoneum. The color is dark reddish-brown. The gland is a soft tissue, easily torn. It has an abundant blood supply, receiving around 1,500 milliliters of blood per minute. The liver is proportionately very large in the embryo and fetus.

The liver is an organ of many complex functions: (1) It secretes bile, (2) it has a glycogenic function, (3) it plays an important part in protein and fat metabolism, (4) it exercises a protective function, and (5) it gives rise to serum albumin, serum globulin, fibrinogen, and heparin. This list is by no means a complete summary of its functions, however.

HISTOLOGICAL STRUCTURE   The substance of the liver consists of *lobules* 1 to 2 millimeters in diameter. The lobule is considered to be the unit of gross structure. It consists of cords of cells radiating from a center. The columns of cells anastomose freely, and there is considerable irregularity. Between the rows of cells is a more or less open capillary network, the *liver sinusoids*. The lobules are held together by interlobular connective tissue, which also

FIGURE 17.12
Liver and gallbladder (anterior view).

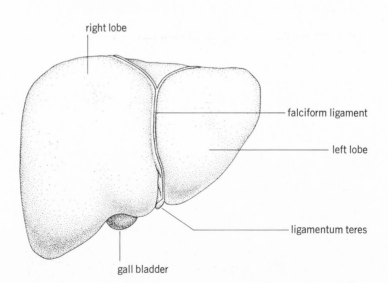

right lobe

falciform ligament

left lobe

ligamentum teres

gall bladder

The digestive system

438

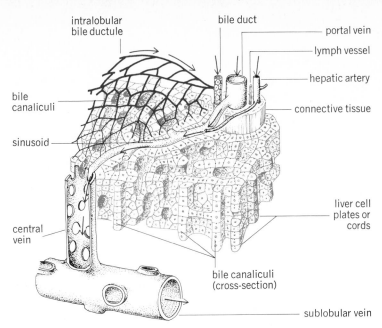

intralobular bile ductule

bile duct

portal vein

lymph vessel

hepatic artery

connective tissue

bile canaliculi

sinusoid

liver cell plates or cords

central vein

bile canaliculi (cross-section)

FIGURE 17.13
A portion of a liver lobule: detailed structure.

sublobular vein

supports the blood vessels, lymphatics, and nerves. Over all, there is a serous membrane, derived from the peritoneum, and a thin fibrous coat.

This description and the illustration may be regarded as an example of a part of a liver lobule considered to be typical. Actually, it is too simplified and stylized. The lobules, especially in the human being, are not well defined, and the cords of cells do not necessarily lie in a flat plane. A more realistic description would represent the microscopic anatomy as resembling somewhat the three-dimensional architecture of a sponge. However, the simplified drawing is useful for teaching purposes.

The liver is supplied with blood from the portal vein and the hepatic artery. Within the liver the portal vein and the much smaller hepatic artery give off numerous branches. Their terminal branches are the interlobular veins and interlobular arteries, respectively, which form a network of blood vessels around the lobules. From these peripheral blood vessels, the blood filters through the liver sinusoids, which are capillary-like vessels with incomplete walls. The sinusoids permit the blood to flow toward the center of the lobule and into the central veins (Figure 17.13). From the central veins, the blood flows into the sublobular veins and so into the hepatic veins. The hepatic veins empty into the inferior vena cava.

Located in the sinusoids are the large, stellate Kupffer cells. They are phagocytic macrophages of the reticuloendothelial system, capable of ingesting bacteria or other foreign material from the blood. They play an important part in filtering the blood that flows through the hepatic sinusoids. It has been demonstrated that the macrophages are capable of storing India-ink carbon particles until these cells are practically black in appearance (Figure 17.13).

GALLBLADDER AND BILE DUCTS The gallbladder is a large, conspicuous green sac located on the undersurface of the liver, with its distal extremity close

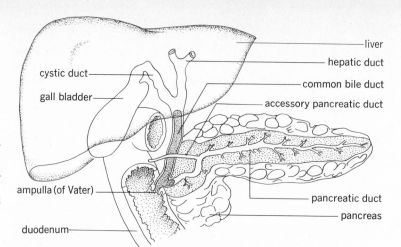

to the anterior border. It commonly is 7 to 10 centimeters long and 2 or 2.5 centimeters wide and holds about 30 milliliters of bile. The gallbladder may be regarded as an enlargement of the *cystic duct*, which drains it. The cystic duct and the neck of the gallbladder have a peculiar twisted appearance because of the attachment of a spiral valve within (Figure 17.14).

The *hepatic duct* arises in the liver. Interlobular bile ducts drain into larger ducts and finally give rise to two large tributaries, one from the right portion of the liver and one from the left. These large ducts unite to form the hepatic duct. The bile descends from the liver and through the hepatic duct. It may pass up the cystic duct to the gallbladder or down the common bile duct.

The *common bile duct* arises from the union of the hepatic and cystic ducts. The duct is about 7.5 centimeters long; it passes downward and enters the duodenum from the left, along with the pancreatic duct, 7 to 10 centimeters below the pylorus. The lower portion of the duct is embedded in pancreatic tissue, and in a majority of individuals it unites with the pancreatic duct at its base. The short, dilated, common passageway is known as the *ampulla* (*of Vater*). A sphincter muscle at the base of the bile duct and in the ampulla is capable of closing the duct. Closure of the sphincter causes the bile to back up into the gallbladder.

Cholecystokinin The smooth muscle of the gallbladder is innervated by a branch of the vagus nerve as well as by fibers of the sympathetic system. Though some nervous control exists, hormonal control appears to be of greater importance. The gallbladder can be caused to empty experimentally by the introduction of fat or fat-digestion products into the intestine, even though all nervous connections have been severed. The hormone involved is secreted by the intestine and has been called *cholecystokinin*. Injections of the hormone cause the contraction of the gallbladder. Normally, as the contents of the stomach enter the duodenum, fats and fat-digestion products stimulate the formation of cholecystokinin. The hormone is absorbed by the blood and, when carried to the gallbladder, causes that structure to contract.

BILE The liver secretes a yellowish fluid called *bile*. It is an alkaline watery fluid containing bile pigments, bile salts, cholesterol, mucin, and other constituents, both inorganic and organic. The bile is not as concentrated as it leaves the liver as it is after it has been held in the gallbladder. This indicates that the gallbladder is capable of absorbing a considerable amount of water and some salts while the bile is stored there. The inner lining also is capable of secreting mucus into the bile.

The *bile pigments* are derived from the breakdown of hemoglobin in old red cells. The disintegration of red cells, which is a continuous process in the bloodstream, releases the hemoglobin. It is broken down in various tissues such as the spleen, liver, and connective tissues, where it is converted into bilirubin, the principal pigment in human bile. Biliverdin is formed as an oxidation product of bilirubin. These pigments are carried by the blood to the liver, where they become bile pigments. Bilirubin is an orange color, and biliverdin is green. A reduction product of bilirubin in the intestine is urobilinogen. It is responsible for the brown color of fecal material. Most of the bile pigment is excreted with the feces, but a small amount is absorbed through the capillary network in the lining of the intestine. This pigment is returned to the liver by way of the portal system to be used again.

BILE SALTS   Sodium glycocholate and sodium taurocholate are bile salts that play an important part in the digestion of fats. These salts are formed in the liver by the chemical combination of cholic acid with the amino acid glycine, in one case, and with the amino acid taurine, in the other. Bile contains no enzyme, but fat digestion by pancreatic lipase is increased when bile is present. Although fat digestion and absorption are poorly understood, it appears that bile salts act as detergents and aid in the emulsification of fats. The fat is broken down into small globules, with decreased surface tension, that provide a much greater surface area for the digestive action of lipase. Bile salts also play an important part in the absorption of glycerol and the insoluble fatty acids. The bile salts, too, are absorbed and used over again by the liver.

Cholesterol is a substance present in tissues and fluids of the body. Chemically it is a sterol, a complex secondary alcohol. It is related chemically (that is, it has a similar molecular structure) to vitamin D, the sex hormones, and hormones of the adrenal cortex. It is fairly soluble in bile, though highly insoluble in water. It is usually considered as an excretory product in the bile. The concretions which sometimes form in the gallbladder (gallstones) are largely composed of cholesterol. Small stones passing down the bile ducts often cause severe pain. Larger stones can block the flow of bile through the ducts and cause obstructive jaundice. The position of the gallbladder is such that it can be removed, if necessary, without serious consequences. Obviously, bile can no longer be stored if the gallbladder is removed, but bile continues to flow down the hepatic and common bile ducts. The digestion of fats is not as efficient after removal of the gallbladder because the same quantity of bile cannot be released at just the time that fats enter the duodenum. By reducing the amount of fat in the diet, however, many persons are able to lead normal lives even though the gallbladder has been removed.

JAUNDICE   The normal outlet for the excretion of bile is through the intestine. If the liver is unable to function properly, bile pigments may be absorbed by the blood in excessive amounts. In some cases the liver may be functioning properly, but the bile may be unable to pass readily into the intestine. This condition occurs commonly when there is a stoppage of the bile ducts because of the presence of gallstones in the passageway.

The absorption of abnormal amounts of bilirubin by the blood causes the tissues to take up unusual amounts of this pigment. A yellow coloration is especially noticeable in the skin and white of the eye. *Jaundice* is the name given to such a condition. There are various types of jaundice, such as obstructive, infectious, and hemolytic jaundice. In the obstructive type the fecal matter is light-colored because the bile is blocked from its normal outlet through the intestine. Unusual amounts of bilirubin are usually excreted by the kidneys, and the urine is dark-colored.

Infectious jaundice usually indicates the presence of a toxic substance, which reduces the ability of the liver to excrete bile pigment. Bilirubin is therefore absorbed by the blood plasma in excessive amounts.

Hemolytic jaundice may occur when there is an abnormal breakdown of red cells of the blood because of various diseases such as malaria or to toxic substances. The release of unusual amounts of hemoglobin causes the bilirubin content of the blood to rise considerably above normal limits.

GLYCOGENIC AND GLYCOGENOLYTIC FUNCTIONS OF THE LIVER Absorbed glucose is carried by the portal system to the liver, where it is stored in the form of glycogen (animal starch). The process of converting glucose to glycogen is called *glycogenesis*. As the tissues use glucose in their metabolism, the liver converts some of its stored glycogen back to glucose, thereby helping to maintain a constant blood-sugar level. The process of breaking down glycogen to glucose is called *glycogenolysis*.

Glycogenesis The blood carries a small amount of glucose, and the liver contains some, but there is no storage of glucose in the body. Glucose is converted to glycogen and stored largely in the liver, but a small amount is stored in skeletal muscle. The conversion of glucose to glycogen occurs in a series of steps. The first step involves the phosphorylation of glucose by a hexokinase (glucokinase) in which phosphoric acid is attached at carbon 6 to form glucose 6-phosphate. Magnesium ions and ATP as a phosphate donor take part in this reaction.

$$\text{Glucose} \xrightarrow{\text{glucokinase, Mg}^{2+},\text{ ATP}} \text{glucose 6-phosphate} + \text{ADP}$$

Another enzyme, phosphoglucomutase, transfers the phosphate group to carbon 1.

$$\text{Glucose 6-phosphate} \underset{\text{Mg}^{2+}}{\xrightleftharpoons{\text{phosphoglucomutase}}} \text{glucose 1-phosphate}$$

Glucose 1-phosphate is then converted to glycogen by way of a uridine pathway. Uridine triphosphate (UTP) is required to form uridine diphosphate glucose (UDP glucose) as the next intermediate step.

$$\text{Glucose 1-phosphate} \xrightleftharpoons{\text{UTP}} \text{UDP-glucose}$$

The final step transforms UDP-glucose to glycogen. This reaction is catalyzed by the enzyme glycogen synthetase (UDP-glycogen-transglucosylase) and involves also a branching enzyme.

The digestive system

$$\text{UDP-glucose} \xrightarrow{\text{glycogen synthetase}} \text{glycogen}$$

Glycogenolysis As the body requires more glucose, glycogen is broken down step by step by action of the enzyme *phosphorylase*, found predominantly in the liver and in muscle. It is inactive in a resting state and must be activated before it is effective. Two hormones, epinephrine and glucagon, are the principal activators. In addition to phosphorylase, there is also a debranching enzyme. The complete breakdown of glycogen by phosphorylation produces glucose 1-phosphate, which may be converted to glucose 6-phosphate, a reversible reaction catalyzed by the enzyme phosphoglucomutase. Glucose 6-phosphate is converted to glucose and phosphoric acid by the enzyme glucose 6-phosphatase.

$$\text{Glucose 6-phosphate} \xrightarrow{\text{glucose 6-phosphatase}} \text{glucose} + PO_4$$

This reaction occurs only in the liver and is responsible for a rise in blood-sugar level. Glycogenolysis occurs in muscle also, producing glucose 1-phosphate and glucose 6-phosphate, but the enzyme phosphatase is not present in muscle, and therefore muscles are not able to produce free glucose available to be absorbed by the blood and to cause an increase in blood sugar.

Hormone control over phosphorylase in the liver is provided by the action of epinephrine and glucagon. When the blood-sugar level falls below normal, these hormones activate the enzyme *adenyl cyclase*, which is present in liver cells. The enzyme catalyzes the conversion of ATP to cyclic AMP. It is the cyclic AMP that activates liver phosphorylase, as we have seen.

The liver is also capable of forming glucose from certain amino acids of food proteins. During starvation, glucose is derived from protein in the tissues themselves. It can also be obtained from glycerol that is split off the fat molecule during digestion. The ability of the liver to form glucose from noncarbohydrate sources is called *gluconeogenesis*.

The liver plays an important role in maintaining the blood sugar at its normal level. This level is around 0.1 percent, or 80 to 120 milligrams per 100 milliliters, but it varies considerably with various bodily conditions. Any great reduction in blood sugar (hypoglycemia), as in the case of insulin shock, gives rise to serious disturbances leading to convulsions and coma. A rise in blood-sugar level (hyperglycemia), as in diabetes mellitus, gives rise to no discomfort. If the blood sugar rises above a certain level (the renal threshold for glucose), the kidneys excrete the excess sugar.

The regulatory action of glucagon A hormone called *glucagon*, which is produced by the alpha cells of the pancreatic islet cells, and epinephrine from the adrenal glands regulate the blood-sugar level. A low blood-sugar level causes an increased secretion of glucagon, whereas a high blood-sugar level depresses secretion. Fasting, therefore, stimulates glucagon secretion, which causes a rise in the blood-sugar level by accelerating the conversion of liver glycogen to glucose (glycogenolysis). Glucose then enters the bloodstream. Experimentally, the blood-sugar level rises rapidly following the injection of glucagon. There is also an increase in gluconeogenesis, although this effect may be caused by a depletion of glycogen.

OTHER FUNCTIONS OF THE LIVER Proteins are not stored in the body as such, but amino acids are found in the blood, liver, and in tissues generally. In the liver, amino acids undergo a process of deamination, in which the nitrogen, $-NH_2$, group is split off the molecule. The enzyme *deaminase* catalyzes the reaction. The nitrogen is used in the formation of urea, $CO(NH_2)_2$, which is mainly formed in the liver and excreted by the kidneys. Some of the amino acids that have undergone deamination are converted into glucose. Glucose, as we have seen, can be utilized or stored in the liver as glycogen. Deamination of glycogenic amino acids can lead also to the formation of fat. The liver has an important function in breaking down the waste products of protein metabolism into substances that will be nontoxic and capable of being excreted by the kidneys.

The liver plays an important role in the metabolism of fats. Among the enzymes concerned is acetyl coenzyme A, which functions not only in the breakdown of fatty-acid chains but also in the synthesis of lipids. The hepatic enzymes and coenzymes present are in the mitochondria of the liver.

Fats, of course, are a good source of energy. When they are completely oxidized, they produce approximately 9 kilogram calories per gram as compared with 4 kilogram calories for carbohydrates or proteins.

After eating a meal containing a large amount of fat, there is an increase of fat in the liver as a result of the steady flow of absorbed fat to this organ. In the process of breaking down fats, some intermediate products of fatty-acid metabolism are formed. These substances are called *ketone bodies* and are composed of acetoacetic acid and beta-hydroxybutyric acid. Acetone is a breakdown product of either of these acids. Ketone bodies in small amounts are normal breakdown products of fatty acids. They are oxidized in cellular metabolism by way of the tricarboxylic acid cycle. Energy is released, and the end products are CO_2 and H_2O.

Ketosis occurs as a result of overproduction of ketone bodies by the liver. It occurs especially in conditions of diabetes and starvation.

Ordinarily in the process of metabolism energy is derived from the three kinds of food—carbohydrates, proteins, and fats—and there is no undue metabolism of fats at the expense of carbohydrates and proteins. In diabetes, there is a loss in carbohydrate metabolism; the body therefore utilizes more fats and proteins to meet its energy requirements. Excess ketone bodies are then formed from fats and some of the amino acids. The body tissues cannot oxidize all the excess ketone bodies that are formed by the liver; as they accumulate in the blood, they are excreted by the kidneys. Since acetone is a breakdown product, this volatile substance may give the urine, or even the breath, a characteristic odor.

During starvation, stores of carbohydrates are soon depleted, and the metabolic needs are met by utilizing fats and proteins for energy. As in diabetes, ketosis results from inability of the tissues to utilize all the ketone bodies in excess of their normal metabolic needs. Their accumulation causes a rise in the acidity of the blood by depletion of the alkaline reserve.

The protective function of the liver is associated with its ability to detoxify products of catabolism, which might accumulate in dangerous proportions. These products are changed chemically into substances that can be excreted by the kidneys or through the intestinal tract. We have noted that macrophages present in liver sinusoids aid in filtering foreign matter from the blood.

The blood proteins, serum albumin and serum globulin, are formed in the liver. Some of the substances concerned with the clotting mechanism of the blood, such as fibrinogen and heparin, are formed there too. Liver tissue also contains an antianemia factor. The discovery of this factor gave rise to the practice of feeding liver and liver extracts in the treatment of anemia. The antianemia factor in liver was later determined to be the same

TABLE 17.2

A brief summary of liver functions

1 *Intermediary metabolism*
  Glycogenesis, conversion of glucose to glycogen
  Glycogenolysis, the process of breaking down glycogen to glucose
  Gluconeogenesis, formation of glucose from noncarbohydrate sources
  Deamination, of amino and nucleic acids, removal of an amino group, by a deaminase (produces a portion of body heat)
  Lipogenesis, formation of fats from glucose, glycogen, or amino acids (enzymes present are in liver mitochondria)
  Lipolysis, breakdown of triglycerides into fatty acids and glycerol (acetyl coenzyme A), formation of ketone bodies
  Cholesterol metabolism, synthesis, and breakdown, from ingested fats

2 *Storage*
  Glucose, stored as glycogen (release of blood sugar involves cyclic AMP)
  Bile, stored in gallbladder
  Vitamins: fat-soluble, A, D, E, K; water-soluble, thiamine, riboflavin, folic acid, and B_{12}
  Fats and fatty acids, lipids such as phospholipids and cholesterol
  Amino acids, in proteins such as albumin, beta globulin, and in liver protoplasm
  Water, contributes to blood volume
  Minerals: iron and copper; most of the iron is stored as ferritin (iron combined with protein)
  Transfer of blood from hepatic portal system to systemic circulation

3 *Secretion and synthesis*
  Secretion of bile salts into bile
  Formation of blood-clotting factors, prothrombin, fibrinogen
  Blood proteins, serum albumin, serum globulin
  Hemopoietin, antianemic substance
  Elaborates heparin, an agent which acts in the prevention of blood clotting
  Erythrocytes, produced only in the embryo or before birth

4 *Excretory functions*
  Breakdown of hemoglobin in old erythrocytes to form bilirubin and biliverdin
  Ingestion of bacteria, old red cells, or other particles by reticuloendothelial cells (Kupffer cells).
  Detoxification of harmful acids and drugs and excretion of their breakdown products
  Breakdown of sterols such as sex hormones
  Formation of urea from amino acids, and uric acid from nucleic acids

as the extrinsic factor found in food, or vitamin B_{12} (see section on vitamins in Chapter 18) (Table 17.2).

JEJUNUM AND ILEUM

The *small intestine* is the area where most of the absorption of food materials occurs. We have mentioned the duodenum as the anterior portion of the small intestine. The jejunum forms the intermediate portion and the ileum is the distal part. Actually, there is little difference in structure between these parts. There is no strict dividing line between the jejunum and the ileum, but the ileum terminates at the colic, or ileocecal, valve. The ileum enters the side of the colon (or large intestine); the aperture by which it enters is guarded by a sphincter muscle, forming the ileocecal valve.

FIGURE 17.15

Diagram illustrating peristalsis in the intestine.

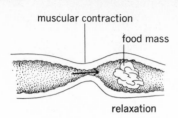

muscular contraction

food mass

relaxation

Like the stomach, the small intestine is covered by a serous membrane and lined by a mucous membrane. The muscular wall is composed of a thin, outer layer of longitudinal fibers and a thicker, inner layer of circular muscle. Rhythmic muscular waves pass over the stomach and intestine, helping to propel the food along. This action is called *peristalsis* (Figure 17.15).

Peristaltic movement of the intestine is largely independent of extrinsic nerve connections. Although stimulation of the vagus increases peristalsis, and stimulation of the sympathetic innervation tends to inhibit peristaltic waves, still, with all extrinsic innervation severed, peristaltic waves continue to pass along the intestine, though in somewhat altered form. A network of neurons within the intestinal wall, known as the *enteric system*, is probably responsible for the independent muscular activity of the intestine, the autonomic innervation being largely regulatory.

A layer of areolar connective tissue lines the muscular layers. This is the submucous layer and is covered with a mucous layer, which lines the intestine. The submucous and mucous layers exhibit circular folds, which project into the cavity of the intestine. These folds increase the absorptive area, and they can serve also to hold the food and prevent it from passing through too rapidly.

VILLI The lining of the small intestine is filled with fine processes projecting into the lumen. These processes of the mucosa, called *villi*, increase the absorptive area of the intestine very greatly. They are freely movable and are concerned with the absorption of food. The fact that they wave back and forth probably makes them more efficient in absorption (Figure 17.16).

Within a villus there is a capillary network of the terminal branches of the mesenteric arteries and veins. There is also a central lymph vessel called a *lacteal*. Carbohydrates, proteins, and some fats are absorbed through the capillary network of the villus. The greater part of fat absorption is through the lacteal, or lymph channel.

Columnar epithelial cells, including goblet cells, cover the surface of the villi. The absorptive surface is further augmented by microvilli that constitute the brush borders (Figures 4.4 and 4.5). At the bases of villi are numerous tubular glands that are called the *crypts of Lieberkühn*. They secrete a portion of the intestinal fluid. It was thought formerly that the disaccharidases and proteases found in the intestinal fluid were secreted as the digestive part of the fluid. It appears now that the only enzymes found in pure succus entericus, uncontaminated by cell fragments, are enterokinase and a small amount of amylase. The other enzymes reported come from broken-down epithelial cells cast off from the surface of the villi and from brush-border cells. Disaccharidases have been shown to be present and functional in the brush borders of the columnar epithelium of the villi. Mucus-secreting goblet cells appear to be merocrine rather than apocrine

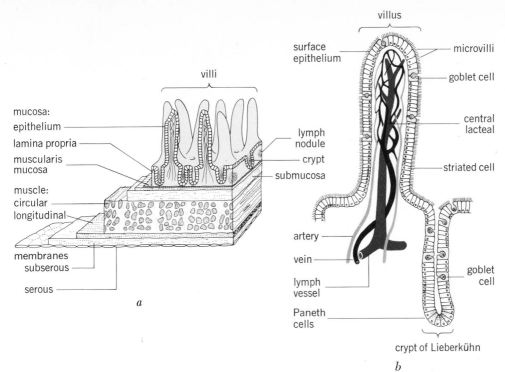

villus

surface epithelium

microvilli

goblet cell

central lacteal

villi

mucosa:
epithelium

lamina propria

muscularis mucosa

lymph nodule

crypt

submucosa

striated cell

muscle:
circular

longitudinal

membranes
subserous

serous

artery

vein

lymph vessel

Paneth cells

goblet cell

a

crypt of Lieberkühn

b

FIGURE 17.16

Detailed structure of the small intestine: *a* layers of the intestinal wall and villi; *b* longitudinal section of a villus and a crypt of Lieberkühn. Paneth cells contain large eosinophilic granules in their cytoplasm. Their function is not well understood; but they may be concerned in the manufacture of digestive enzymes and their secretion.

in function. *Paneth cells* are thought to function in the synthesis of digestive enzymes.

The cells at the bases of the villi have a high rate of mitosis. They appear to divide and start moving up the sides of the villi. The older cells are sloughed off at the top continuously. Some of the enzymes reported as present in the intestinal fluid come from these cast-off disintegrating cells. The brush-border or striated cells of villi have been found to contain disaccharidases. An interesting concept holds that the intestinal fluid produced in quantity by the crypts of Lieberkühn is constantly being circulated until it is absorbed by the villi and functions to hold food substances in suspension in order to enhance enzymatic digestion. The small intestine should not be considered as an undifferentiated membrane, since differences in secretory and absorptive abilities of the crypts and villi in the jejunum and ileum are becoming recognized.

The hormone *enterocrinin*, isolated from extracts of the intestine, stimulates the small intestine to produce *succus entericus* in greater volume (Table 17.3).

Food as taken into the mouth is not immediately available for use by the tissues. For the most part, the molecules of undigested foods are too large and complex to be absorbed from the intestine. From a functional point of view, even food materials in the digestive tube have not really entered the body until they have been absorbed. The process of digestion reduces food materials to a condition suitable for absorption. Large complex molecules are reduced to smaller and simpler molecules by the action of digestive enzymes. Finally these end products of digestion must be absorbed through

Jejunum and ileum

447

TABLE 17.3

Hormones of the digestive tract

Hormone	Origin	Action
Gastrin	Mucosa of pyloric antrum	Causes an increase in the secretion and acidity of gastric juice
Enterogastrone	Mucosa of duodenum and jejunum	Inhibits gastric motility and secretion
Secretin	Duodenal mucosa	Stimulates pancreatic secretion which is weak and watery and low in enzyme content. High concentration of bicarbonate. Stimulates the volume of bile secretion by the liver
Pancreozymin	Duodenal mucosa	Causes the pancreas to secrete pancreatic juice containing enzymes. Stimulates contraction of gallbladder
Cholecystokinin*	Duodenal mucosa	Causes gallbladder to contract
Enterocrinin	Mucosa of small intestine	Controls the secretion of the intestinal juice
Villikinin	Mucosa of small intestine	Stimulates movement of villi
Duocrinin	Mucosa of small intestine	Causes secretion of Brunner's glands

*May be identical with pancreozymin.

the mucosa of the intestine and into the bloodstream to make them available to the tissues.

The process of absorption is more complicated than can be explained by simple diffusion through the mucosa. The absorption rates of various simple monosaccharide sugars have been shown to vary considerably. Glucose and galactose are absorbed readily, but fructose and mannose are absorbed slowly. If the process of absorption through a membrane were the only problem involved, the monosaccharide sugars might be expected to be absorbed at equal rates. Inverting enzymes break down sugars to monosaccharides before they can be absorbed in the intestine. The absorption of hexose sugars appears to involve active transport, although the mechanism is not clear.

The large protein molecules must be converted to amino acids by digestion before they can be absorbed. Proteins, as they are found in food substances, are not water-soluble. Even the breakdown products such as proteoses and peptones are not absorbed. Amino acids, however, are readily absorbed, after which they become the building blocks by which new protein materials can be constructed.

Fats are broken down into glycerol and fatty acids by the action of lipase and bile salts in the process of digestion before being absorbed. Glycerol is water-soluble, but fatty acids are highly insoluble. Bile salts are largely responsible for breaking free fatty acids into minute water-soluble droplets known as *micelles*. Fatty acid from micelles enters the membrane of the villi. In passing through the epithelial cells of the villi, fatty acids are resynthesized to triglycerides (neutral fat); they enter the bloodstream

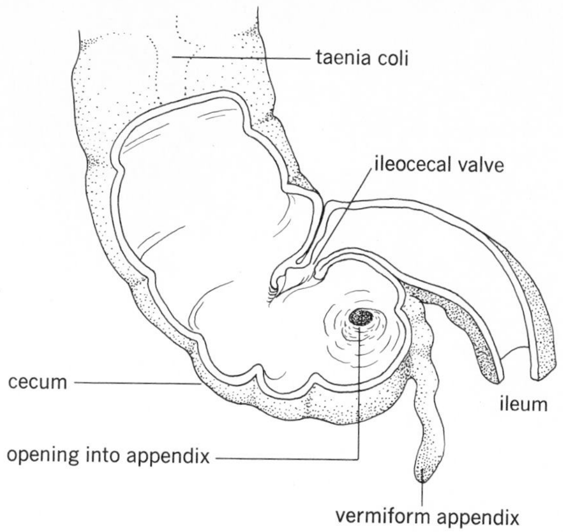

ascending colon

taenia coli

ileocecal valve

cecum

ileum

opening into appendix

vermiform appendix

FIGURE 17.17

The large intestine (anterior view); cutaway to show interior of cecum.

in this form. Triglycerides also enter the lacteals of the villi as minute fat globules called *chylomicrons*. Absorbed fat in the form of chylomicrons gives the lymph a milky appearance.

LARGE INTESTINE

The ileum enters the large intestine laterally. Its entrance is protected by the ileocecal valve. Two horizontal folds of the valve project into the large intestine. The distal end of the ileum is guarded by circular muscles. The valve prevents fecal material from backing up into the ileum; probably it also controls the rate of flow from the ileum into the large intestine, preventing the ileum from discharging its liquid contents too rapidly (Figure 17.17).

The large intestine extends from the ileum to the anus. In succeeding areas it is called the *cecum, ascending colon, transverse colon, descending colon, sigmoid colon, rectum,* and *anal canal.* The various parts of the colon form a sort of frame around the abdominal cavity—the ascending colon on the right, the transverse colon along the top, and the descending colon on the left.

The large intestine has a much greater diameter than the small intestine, but it is only about 5 feet long. It is not greatly convoluted, and in some respects its structure is quite different from that of the small intestine. The longitudinal muscles, for example, are gathered into three bands of fibers, the *taenia coli.* Contractions of the longitudinal muscles cause the colon to be puckered into a series of pouches or sacculations. There are no villi in the large intestine.

The cecum is a blind saclike portion below the entrance of the ileum. It is located in the lower right area of the abdomen just above the right iliopsoas muscle. Occasionally it becomes distended with accumulated waste material. A small prolongation of the cecum is called the *vermiform appendix.*

The appendix, or *vermiform process*, is a blind sac, a process off the lower part of the cecum. The appendix is therefore located in the lower right portion of the abdomen. It is an extension of the intestinal wall and is essentially nonfunctional in man. The appendix is subject to infection and inflammation, giving rise to the condition known as *appendicitis*. The wall of the appendix can be weakened by inflammation and swelling, which can cause the appendix to rupture its bacteria-laden contents into the abdominal cavity. The mesentery and peritoneum are then likely to become inflamed in a much more severe condition called *peritonitis*. So long as harmful intestinal bacteria are confined to the closed sac of the appendix, the infection is limited. Surgical removal of the appendix before rupture is the obvious procedure. Recalling that the appendix is a part of the intestine, one should refrain from using cathartics, which can act unfavorably on an inflamed appendix.

Some cathartics stimulate peristaltic movements of the intestine, some stimulate the mucous glands to secrete, and others prevent the absorption of water or attract water into the intestine. Some salts, such as the chlorides, are readily absorbed if their concentration in the intestine is higher than that of the blood. The tartrates, citrates, and sulfates are commonly used as cathartic salts because they are not absorbed in water-soluble form. Furthermore, in hypertonic solutions these salts cause more water to enter the intestine, thus increasing the bulk and fluidity of the contents.

FUNCTIONS   The material that enters the large intestine rapidly assumes the characteristics of waste matter. One of the principal functions of the colon is the absorption of water. The removal of water changes the intestinal contents from a liquid state to the more solid consistency of fecal matter. Mucus is secreted in considerable quantity. It tends to hold particles in a solid mass. Mucus also protects the lining of the intestine and acts as a lubricant.

The contents of the colon are made up not only of food residues but also of materials excreted from the blood and the digestive glands. The waste materials are subject to bacterial decomposition, and the bacteria themselves make up a small percentage of the waste product.

The large intestine is not responsible for any great amount of digestion. Some digestive enzymes are found in the contents, but they probably enter with food materials from the ileum. The digestion and absorption of food are remarkably efficient, since most of the available food products are removed before they enter the large intestine. Plant foods contain cellulose, which is not digested in human beings and therefore contributes to the bulk of the fecal matter.

Bacterial decomposition gives rise to a number of gases, acids, and amines characteristic of fecal matter. Some of the gases are carbon dioxide, ammonia, hydrogen sulfide, hydrogen, and methane. Acetic, lactic, and butyric acids are commonly present. Amines are breakdown products of amino acids. *Indole* and *skatole* are two such products. They contribute to the odor of the feces.

Some food materials in the diet affect the color of the fecal matter. Plant pigments, such as those derived from beets and spinach, are examples of such foods.

While bacterial decomposition produces toxic substances, it is generally considered that these substances, under normal conditions, are detoxified in the liver and excreted by the kidneys. It has also been shown that intestinal organisms are capable of synthesizing some of the vitamins of the B complex, and vitamin K. This function is regarded as a significant contribution to the nutrition of the individual.

The rectum is usually empty until just before *defecation,* when the feces enter this region of the intestine. The anal canal is closed by an internal and an external sphincter muscle. Feces entering the rectum stimulate sensory-nerve endings and bring about an awareness of impending elimination. Muscular pressure exerted by the diaphragm, the abdominal muscles, and strong peristaltic movements of the colon accomplish the act.

It is usually considered that regular, daily bowel movements are conducive to good health and that it is good hygienic practice to have a regular time for elimination. Adequate amounts of fruits and vegetables in the diet aid in avoiding constipation. Drinking a reasonable amount of water is beneficial. Daily exercise, adapted to the needs of the individual, promotes physical well-being. There is no reason to believe that it is good practice to take laxative medicines regularly. Every effort should be made to secure proper elimination habits by training, proper diet, and good health habits.

SUMMARY

The mouth is the anterior opening of the digestive tract. The tongue is a muscular organ that propels the food back into the pharynx. It functions also as an organ for the sense of taste, touch, and temperature. There are two sets of teeth in mammals, a deciduous set and an adult set. The deciduous teeth number 20, and the set is composed of 4 incisors, 2 cuspids, and 4 molars in each jaw. The adult set consists of 4 incisors, 2 cuspids, 4 premolars, and 6 molars in each jaw, 32 teeth in all.

The structure of the tooth includes the enamel-covered crown, a neck region, and a root covered with cementum. Under the enamel and cementum is a sensitive dentine. The pulp cavity contains blood capillaries and the nerve. The teeth lie in alveolar sockets attached by the periodontal membrane.

There are three pairs of salivary glands: the sublingual, submandibular, and parotid. The sublingual and submandibular glands lie below the tongue in the floor of the oral cavity. The parotid gland is located at the angle of the jaw. Saliva contains mucin and the enzyme amylase, among other constituents. Salivary amylase initiates the digestion of starch.

The esophagus is the muscular tube extending from the pharynx to the stomach. The smooth muscles propel solids and liquids to the stomach by peristalsis.

The abdominal cavity is lined by a serous membrane, the peritoneum. Folds of the peritoneum hold the intestine like an arm in a sling. These folds are the mesentery. A large fold called the greater omentum covers the intestine anteriorly.

The stomach is a muscular, saclike organ located below the diaphragm to the left of the liver. The upper, or esophageal, opening is the cardiac orifice; the lower opening is the pyloris, guarded by the pyloric sphincter. The stomach is lined by a mucous membrane. The principal glands are composed of chief cells, which secrete pepsinogen, and parietal cells, which initiate the formation of hydrochloric acid.

The duodenum is the anterior part of the small intestine beyond the pyloris. It is only about 25 centimeters long. The pancreas lies in the loop of the duodenum and extends upward behind the stomach. It secretes the enzymes trypsin, amylase, and lipase. Trypsin is a protease, breaking down proteins to proteases, peptones, and amino acids. Amylase is a starch-digesting enzyme, hydrolyzing starch to the sugar maltose. Lipase acts upon fats, breaking them down to fatty acids and glycerol. The hormone insulin is also produced in the pancreas and is concerned with the ability of cells to absorb glucose. If cells are unable to absorb glucose, a condition called diabetes develops. The blood-sugar level rises, and the kidneys excrete glucose. A chemical test for sugar in the urine is a routine method for detecting diabetes.

Secretin is a hormone of the duodenum. When acid food of the stomach enters the duodenum, prosecretin is changed to active secretin. Secretin, absorbed by the blood, stimulates the pancreas to secrete a thin watery secretion low in enzyme content.

The liver is an organ of many functions. It secretes bile, has glycogenic and glycogenolytic functions, and is capable of gluconeogenesis. It is also concerned with the chemical breakdown of proteins and fats. It is one of the principal blood filters and also detoxifies products of catabolism, which might otherwise accumulate and become injurious. The liver gives rise to serum albumin and serum globulin as well as fibrinogen and heparin.

The small intestine is divided into three parts: the duodenum, jejunum, and ileum. Most of the absorption of food materials takes place in the small intestine. The fluid of the small intestine is the succus entericus.

The villi are minute processes projecting into the cavity of the intestine. They are concerned with the absorption of food materials and greatly increase the absorptive surface. Each villus contains a capillary network and a lymph vessel called a lacteal.

The ileum enters the large intestine laterally. Below the point of entrance are the cecum and the vermiform appendix. The entrance is guarded by the ileocecal valve. The ascending colon is on the right side of the abdomen; the transverse colon lies just below the stomach; the descending colon is on the left side of the abdomen. Other parts of the large intestine are the sigmoid colon, the rectum, and the anal canal. One of the principal functions of the large intestine is the absorption of water. Food residues are subject to bacterial action, and the material in the large intestine assumes the characteristic of fecal matter.

QUESTIONS

1 Trace the digestion of a piece of bread and butter as it passes through the digestive tract. What will the end products of digestion be?

2 Name and locate the larger glands of the digestive tract. Give the functions of each.

3 What is the relation between the portal system and the digestive system?

4 Show how the deciduous teeth are gradually replaced by the permanent set.

5 Why is it possible for a horse to drink with the head held lower than the stomach?

6 Explain the functions of the stomach.

7 What part does the pancreas play in digestion?

8 Discuss the function of the pancreas as a gland of internal secretion.

9 Explain the role of insulin in carbohydrate metabolism.

10 List some of the numerous functions of the liver.

11 When food is present in the duodenum, the pancreas is stimulated to secrete and the gallbladder is stimulated to contract. Explain the mechanism by which this is accomplished.

12 Indicate the arrangement of bile ducts that permits the removal of the gallbladder without stopping the flow of bile from the liver to the intestine.

13 What is the source of bile pigments?

14 The bile salts have what function?

The digestive system

15 Discuss the causes of jaundice.

16 Explain the glycogenic and glycogenolytic functions of the liver.

17 What is peristalsis?

18 The villi have what function?

19 Describe the function of the small intestine in the absorption of food materials.

20 List the functions of the large intestine.

BANTING, F. G., and C. H. BEST: The Internal Secretion of the Pancreas, *J. Lab. Clin. Med.*, **7:**251-266 (1922).

CARLSON, A. J.: Chemistry of Normal Gastric Juice, *Am. J. Physiol.*, **38:**248-268 (1915).

CORI, C. F.: The Fate of Sugar in the Animal Body, I, The Rate of Absorption of Hexoses and Pentoses from the Intestinal Tract, *J. Biol. Chem.*, **66:**691-715 (1925).

DAVENPORT, H. W.: Why the Stomach Does Not Digest Itself, *Sci. Am.*, **226:**87-93 (1972).

DICKERSON, R. E.: The Structure and History of an Ancient Protein, *Sci. Am.*, **226:**58-72 (1972).

HENDRIX, T. R., and T. M. BAYLESS: Digestion: Intestinal Secretion, *Ann. Rev. Physiol.*, **32:**139-164 (1970).

LELOIR, L. F.: Two Decades of Research on the Biosynthesis of Saccharides, *Science*, **172:**1299-1303 (1971).

LIPMANN, F.: Evolution of Peptide Biosynthesis, *Science*, **173:**875-884 (1971).

MERRIFIELD, R. B.: The Automatic Synthesis of Proteins, *Sci. Am.*, **218:**56-74 (1968).

NETTER, F.: "Digestive System" (in parts), CIBA Collection of Medical Illustrations, CIBA Corp., Summit, N.J. 1957, 1959, 1962.

ORCI, L., K. H. GABBAY, and W. J. MALAISSE: Pancreatic Beta-cell Web: Its Possible Role in Insulin Secretion, *Science*, **175:**1128-1130 (1972).

SUGGESTED READING

chapter 18
foods, nutrition, and metabolism

Foodstuffs such as bread, butter, and eggs are ordinarily broken down into simpler chemical components by enzymatic action in the process of digestion. It is difficult for the average person to regard common foods as chemical substances capable of releasing stored energy. But, as we have seen in the preceding chapter, the end products of digestion are absorbed and carried to the cells, where a final breakdown of their molecular structure releases energy, the principal driving force of metabolism. The release of energy within the cell also is catalyzed by enzymes.

The sun is the ultimate source of energy for life on the earth. Irradiation of plant life by the sun is responsible for the energy resources of coal, oil, and wood. These rich energy sources are not available directly as foods, largely because human beings have no enzyme systems capable of breaking down their molecular structure and releasing the energy. Cellulose, a polysaccharide found in the cell walls of plants, is perhaps a little closer to being a food material, but we are unable to digest it and use its energy. Small insects called *termites* are able to eat and digest woody substances, but only because their digestive tract contains certain kinds of protozoa which produce the enzymes that digest the tissue of the wood.

Green plants possess a remarkable green pigment called *chlorophyll*. This substance is of the utmost importance, because, of all living things, only the chlorophyll-bearing plants can utilize the energy of sunlight directly to form the simple sugar glucose. The process is called *photosynthesis*. It is a chemical and physical process commonly expressed as follows:

$$6CO_2 + 12H_2O + energy \xrightarrow{\text{chlorophyll}} C_6H_{12}O_6 + 6O_2 + 6H_2O$$

A simple equation, however, does not indicate the complexity of the process. Photosynthesis enables the plant to utilize light energy, incorporating this energy into substances of value as foods for both plants and animals. Since the animal body is not adapted to manufacture its own food in this way, it must depend, directly or indirectly, upon energy stored by plants. An animal can feed upon the tissues of another animal, but initially the energy source is derived from plant life.

The chemical elements contained in foods are building blocks for the growth of new tissue and the maintenance and repair of older tissues. In metabolic processes chemical elements are constantly used up and constantly replaced. Food is necessary to provide the materials for replacement.

Upon considering which substances containing stored energy might be available for food, we find that the choice narrows down to three major groups: carbohydrates, proteins, and fats. There are also other food substances essential to life, such as vitamins and certain minerals, which will be considered later. Water is an important constituent of foods. Fruits and vegetables contain a high percentage of water. Body tissues of higher animals are around 75 to 90 percent water. Metabolic processes utilize water, and it can be produced as a by-product of metabolism.

Though only plants have the ability to synthesize glucose, both plants and animals use glucose in their metabolism and derive energy from breaking down its molecular structure. Both plants and animals have the ability to convert glucose into a more complex substance for storage. Thus plants convert simple sugar into starch, and animals convert it into glycogen, which is often called "animal starch." Glucose also can be converted into fats by both plants and animals, or it may form the basic substance from which proteins are derived.

Sugars and starches are examples of carbohydrates. The group is commonly defined as comprising substances containing carbon, hydrogen, and oxygen, with the last two elements in the same proportion as in water, that is, H_2O. Although this definition is true for the commoner sugars, it does not hold for all carbohydrates. The general definition is useful, however, in understanding the chemical structure of the carbohydrates, in which we are at present interested.

MONOSACCHARIDES   Simple sugars such as glucose (or dextrose), fructose (or levulose), and galactose all have the empirical formula $C_6H_{12}O_6$. Therefore they all are hexoses and belong to a class of sugars called monosaccharides. A hexose is a monosaccharide that contains six carbon atoms in the molecule. Although the hexoses all have the same empirical formula, their structural formulas differ; that is, the positions of the atoms within the molecule vary with the type of sugar.

Glucose, as we have seen, is formed in plants by photosynthesis. It can be rapidly changed to starch and stored by the plant, as in dry corn grains or potatoes. Glucose and fructose are found together in fruits and honey. Galactose is derived by hydrolysis from the disaccharide lactose (or milk sugar). Glucose is also found in the blood and tissues of animals. It is stored in the liver and skeletal muscles of animals as a more complex sugar called *glycogen.*

The pentose sugars, or five-carbon sugars, are also monosaccharides. They are important components of nucleic acids and nucleotides. Ribose and deoxyribose are examples.

DISACCHARIDES   Table sugar (sucrose), milk sugar (lactose), and malt sugar (maltose) are disaccharides. They have the common empirical formula $C_{12}H_{22}O_{11}$. Within plant cells, sucrose can be formed by the combination of two monosaccharide molecules and the removal of one molecule of water, as in the following equation:

$$2C_6H_{12}O_6 \rightarrow C_{12}H_{22}O_{11} - H_2O$$

Sucrose is found widely in fruits and vegetables. Lactose is present in cow's milk, representing an average of about 4.9 percent concentration; in human milk it constitutes nearly 6 percent. Elephant's milk is said to contain 8.8 percent lactose, while that of the rabbit contains only 1.95 percent. Maltose is an intermediate sugar. It is commonly formed when the starch molecule is subjected to the action of the enzyme amylase.

When a disaccharide sugar is used by an animal as a food, it is changed by hydrolysis into two monosaccharide molecules in the process of digestion. *Hydrolysis* refers to the decomposition of molecules by the action of water.

Large food molecules are commonly broken up into smaller molecules in this way, the process being catalyzed by enzymatic action. Sucrose, therefore, in the process of digestion, is acted upon by the enzyme sucrase and forms two monosaccharide molecules by hydrolysis, as illustrated in this equation:

$$C_{12}H_{22}O_{11} + H_2O \rightarrow C_6H_{12}O_6 + C_6H_{12}O_6$$

The two monosaccharide sugars formed in the above reaction are not identical. One is glucose and one is fructose. Their structural formulas are slightly different, but both can be absorbed and utilized as a food by the animal body. In the same way, lactose is hydrolyzed into a molecule of glucose and a molecule of galactose, whereas maltose forms two molecules of glucose.

POLYSACCHARIDES Polysaccharides are complex carbohydrates derived from monosaccharides by the removal of one molecule of water from each monosaccharide molecule. The formation of starch from glucose is illustrated by the following equation:

$$nC_6H_{12}O_6 \rightarrow (C_6H_{10}O_5)_n + (n-1)H_2O$$

Since polysaccharides are composed of groups of $C_6H_{10}O_5$ molecules, and since the number of groups varies in different polysaccharides, the empirical formula is written $(C_6H_{10}O_5)_n$. Common polysaccharides are starch, glycogen, dextrins, and cellulose.

Starch, in the process of digestion, is hydrolyzed by the enzyme amylase. The large starch molecule breaks down, forming intermediate products called *dextrins* and *maltose*. The disaccharide sugar maltose yields glucose by hydrolysis. This action is catalyzed by the enzyme maltase.

Cellulose is a plant product. It is found with other substances in woody tissues and in the cell wall of plants. It is not available as a food because, as we have pointed out, the animal body has no enzyme capable of breaking down its molecular structure. The woody parts of plants may have some value, however, in lending bulk to the diet.

FATS AND RELATED COMPOUNDS

Fats are composed of the same basic elements as carbohydrates, but the proportion of these elements is different. The amount of oxygen is less in the fat molecule. When completely oxidized, both fats and carbohydrates yield energy and break down to carbon dioxide and water. More energy can be derived from the combustion of a given amount of fat than from a similar amount of carbohydrate.

Fats or fatlike compounds are referred to as *lipids*. True fats, or neutral fats, are simple lipids. They result from a combination of glycerol with three fatty-acid groups (triglycerides). Glycerol, $C_3H_5(OH)_3$ (glycerine), is not a fat; it is an alcohol. It is soluble in water, whereas fatty acids are highly insoluble in water. Fatty acids are soluble in ether or alcohol.

The commonest fatty acids of plant and animal metabolism are palmitic, $C_{16}H_{32}O_2$; stearic, $C_{18}H_{36}O_2$; and oleic, $C_{18}H_{34}O_2$, acids. Note the relatively high concentration of carbon and hydrogen and the relatively low amount of oxygen in these formulas. Neutral fats derived from these acids are found as a mixture deposited in animal tissues.

There seems to be good evidence that some fatty acids are essential but others are not. Rats on a fat-free diet fail to grow normally and show the effects of dietary deficiency. This condition does not improve upon administration of fat-soluble vitamins, but rats recover when given any one of several unsaturated fatty acids, such as linoleic acid or linolenic acid. These conditions would not be likely to develop from the average human diet, because adequate amounts of fat are usually present.

After digestion, fats are absorbed and deposited in various tissues throughout the body. This fat constitutes a reserve source of energy available for use when needed. The greatest amount is stored in subcutaneous connective tissue. Considerable amounts can be stored around the kidneys, on the mesentery, and on the omentum.

— subcutaneous tissue (kidney etc)

It is well established that in the chemistry of the body, carbohydrates can be converted to fat. The evidence for the conversion of protein to fat in the diet is not so clear. Some of the deaminated amino acids can be converted into fat. However, the amount of fat derived from protein probably is small in normal metabolism.

LIPOID SUBSTANCES   There are a number of compound lipids that are important in physiological processes. Among them are the *phospholipids,* which contain phosphorus. Common phospholipids are *lecithin* and *cephalin.* These two are closely related and very widely distributed in the blood and tissues. When the chemistry of the absorption of fat is more completely understood, it may be found that these phospholipids are necessary intermediate stages in fat metabolism. A knowledge of the action of these substances in relation to body fluids may help to explain how fats are utilized in metabolism. The reaction of lecithin to water is quite different from that of the highly insoluble fatty acids. Lecithin upon hydrolysis yields a nitrogenous substance called *choline.* Choline is probably best known in the form of acetylcholine, considered to be the substance concerned with the chemical phases of the transmission of the nerve impulse. Cephalin is one of the substances involved in the clotting of the blood. Lipids are found also in the membranes covering cells.

STEROLS   These substances are not fats but complex secondary alcohols. They are often combined with fatty acids as esters. (An ester is a compound formed by the condensing of an acid and an alcohol, with elimination of H_2O.) *Cholesterol* is found in tissues throughout the body. It can be present in its free state or as an ester. The cholesterol of the bile is in the free state, as is the cholesterol found in erythrocytes. However, much of the cholesterol of the blood plasma is present as the ester. Cholesterol is related chemically to other substances of great biological significance such as bile acids, the sex hormones, adrenocortical hormones, and vitamin D.

cholestrol

PROTEINS

Protein substances occur in all living matter; they are essential to the living protoplasm of every cell. The protein molecule is often large and extremely complex. It commonly contains a great number of carbon atoms to which are attached atoms of oxygen, hydrogen, and nitrogen. Other elements associated with proteins, in small amounts, include sulfur, phosphorus, iron, copper, and iodine. For example, iron is included in the hemoglobin of blood, and iodine forms a part of thyroid-gland protein. Compared with the formulas of common carbohydrates, the protein molecule appears to be larger and more intricate. It commonly contains more carbon and relatively less oxygen. Protein molecules often represent an unusually great molecular weight, but there are also relatively simple proteins. The structural formulas of proteins vary widely because of the number and position of carbon atoms and the elements attached to them. For this reason, proteins differing only slightly in their molecular formulas exist in almost endless variety.

AMINO ACIDS Amino acids have been called the building blocks of proteins. Though many amino acids have been described, food proteins are usually considered to be derived from 23 amino acids in various combinations. One notices a similarity in the structural formulas of amino acids. In relatively simple organic acids, such as propionic acid or succinic acid, a certain hydrogen atom is replaced by an amino, NH_2, group. One of the simplest amino acids is aminoacetic acid, or glycine. Acetic acid has the formula

$$\begin{array}{l} CH_3 \\ | \\ COOH \end{array}$$

The formula for glycine is written

$$\begin{array}{l} CH_2-NH_2 \\ | \\ COOH \end{array}$$

We note that in aminoacetic acid one hydrogen atom has been replaced by a basic, NH_2, group, but the acid or carboxyl, COOH, group remains the same.

Taking a more complicated example, we find that cysteine, containing sulfur, has the same basic formula.

$$\begin{array}{l} CH_2-S-H \\ | \\ CH-NH_2 \\ | \\ COOH \end{array}$$

It should not be necessary to memorize a list of amino acids or their chemical formulas. However, recognition of their names is essential to an understanding of some phases of physiology. The following is a reference list of the more common amino acids found in proteins:

1 Glycine
2 Alanine
3 Serine
4 Threonine
5 Valine
6 Leucine
7 Isoleucine
8 Cysteine
9 Methionine
10 Aspartic acid
11 Asparagine
12 Glutamic acid
13 Glutamine
14 Arginine
15 Lysine
16 Phenylalanine
17 Tyrosine

18 Tryptophan
19 Histidine
20 Proline

Thyroxine is an amino acid, but it is not found in foods. It is present in the secretion of the thyroid gland.

The great variety of proteins may be classified into three groups: *simple proteins; compound proteins*, or those containing a nonprotein group; and *derived proteins*, or those that have undergone hydrolysis because of the action of acids, alkalies, heat, or proteolytic enzymes.

Simple proteins
  Albumins
    egg albumin, the white of egg
    serum albumin of blood plasma
    lactalbumin from milk
  Globulins
    serum globulin and fibrinogen from blood plasma
    legumin from peas and lentils
    tuberin from potatoes
  Glutelins
    glutenin of wheat
  Prolamines
    gliadin of wheat
    zein of maize
  Scleroproteins
    keratin, a constituent of hair, feathers, and horny structures
    elastin and collagen from connective tissues
  Histones
    globin of hemoglobin
  Protamines, found in combination with nucleic acid in heads of spermatozoa of fish
Compound proteins
  Nucleoproteins: nucleic acid combined with a protein. The protein is usually one of the histones or protamines. Nucleoproteins occur most abundantly in cell nuclei.
  Chromoproteins: protein combined with a pigment, as in hemoglobin
  Glycoproteins; for example: mucin
  Phosphoproteins, for example, casein of milk, vitellin of egg yolk
Derived proteins
  Examples: metaproteins, proteoses, peptones, peptides

The above outline is an abbreviated classification of proteins.

Protein foods are necessary for growth and repair of tissues, but they vary in the number and kind of amino acids of which they are composed. Not all protein foods have the same nutritional value. Some amino acids are absolutely essential for growth; others can be manufactured by the body. Proteins of eggs (ovalbumin), milk and cheese (lactalbumin and casein), meat, and glutenin of wheat contain all the essential amino acids and are called *complete proteins.*

The proteins of corn and gelatin and many others are called *incomplete*, because they do not contain all the essential amino acids. One does not need to be greatly concerned as to whether the various proteins in a normal diet are complete or incomplete. Ordinarily there is a wide range of amino acids in the composition of proteins, and some contain more of the essential amino acids than others.

Essential amino acids are those that the body is unable to synthesize. These amino acids must therefore be available in foods. If the 10 essential amino acids are provided in adequate amounts, the body will be able to synthesize the others. Experimental evidence along these lines has been derived mostly from nutritional studies on rats. There are indications that various animals and man have somewhat different requirements. Following is a list of essential amino acids:

Threonine	Arginine
Valine	Lysine
Leucine	Phenylalanine
Isoleucine	Tryptophan
Methionine	Histidine

Unlike most carbohydrates and fats, the proteins of an animal or plant species are likely to be specific for that species. This is generally true of the blood proteins but may be true of other tissues as well. It is well known that blood cannot be transfused between different species of animals or from animal to man because the introduction of foreign proteins causes coagulation of the blood. This reaction can result even from the transfusion of incompatible types of human blood. Organs or pieces of tissue ordinarily cannot be transplanted from one species to another and be made to live and grow in the new host. The structural formulas of proteins indicate the possibility of almost unlimited combinations of molecules. This may partially explain the wide variety of proteins found in tissues and why it is possible for them to be specific for a given species.

Amino acids are absorbed from the intestine and carried by the blood to various organs and tissues. They may combine with other substances within the cell. The colloid nature of proteins permits them to be retained within the cell membrane. Those amino acids that are not utilized directly may undergo a process of deamination, in which NH_2 groups are removed. The nitrogen is combined with carbon dioxide to form urea, which is excreted in the urine. Though deamination can occur in the kidneys and other tissues, the liver is the principal organ concerned in this process.

The molecule after deamination can be oxidized to carbon dioxide and water with a consequent release of energy, or the chain of carbon, hydrogen, and oxygen atoms can form the framework for conversion into fat or carbohydrate. Conversion of protein to fat enables the organism to store the energy from excess amino acids not utilized directly by the tissues. Conversion of protein to glucose is especially interesting in diabetic patients, who continue to excrete glucose although carbohydrates are excluded from their diet. However, not all amino acids form glucose or glycogen after deamination. Over half the amino acids are glycogenic—some only to a limited degree.

INORGANIC REQUIREMENTS IN THE DIET

The inorganic composition of tissues can be determined by burning the tissue and analyzing its ash constituents. Carbon, hydrogen, oxygen, and nitrogen are volatilized and driven off by this process, but the ash should include calcium, sodium, potassium, magnesium, phosphorus, iron, sulfur, and traces of several other elements. Apparently the organism needs these elements in order to build tissue. Most of them are present in common foods, and adequate amounts are obtained from a normal diet. Sodium, potassium, magnesium, and phosphorus fall in this category. Civilized man includes considerable quantities of sodium chloride in his diet, perhaps more than is healthful. Salt-free diets are sometimes recommended for those who have high blood pressure. However, sodium chloride has an important function in helping to maintain the osmotic pressure of the blood. Salt is lost by excretion in urine and perspiration. Laborers who work under conditions of unusual exposure to high temperatures and who lose a considerable amount of salt by sweating are often given salt tablets to help make up the loss. Adequate amounts of potassium and magnesium are usually obtained from plant foods in the diet. Phosphorus and sulfur are largely obtained from protein foods. Sulfur is a necessary element in the formation of two amino acids, methionine and cystine. If adequate amounts of protein are included in the diet, there should be enough phosphorus and sulfur for physiological needs.

Calcium is essential to the development of bones and teeth, clotting of the blood, contraction of muscle, and nerve excitability. Though calcium is a common element in food, many foods contain only minute amounts, and it apparently is not readily absorbed from some foods. Milk and milk products are good sources of calcium. Growing children need an abundant supply of calcium for the development of good skeletal structures. Calcium is present in the body mostly as carbonate or phosphate salts. Calcium, potassium, and sodium salts are maintained at a nearly constant level in the bloodstream and tissues. The diet is more apt to be deficient in calcium than in the other elements. Vitamin D is essential to proper absorption and utilization of mineral salts deposited in bone.

Iron is a necessary element in the formation of hemoglobin, and it is present in other tissues. When old red cells are broken down in the liver, the iron of the hematin is largely conserved to be used over again. Many foods contain iron in small amounts. Eggs, meat, and some cereals are good iron sources.

An adult in good health and eating an average diet should be able to conserve enough iron so that there should be no iron deficiency. When there is considerable or persistent loss of blood, however, an iron deficiency may develop. Iron is commonly added to the diet in anemia in order to ensure an adequate amount of this mineral for hemoglobin formation.

Iodine is needed only in minute amounts, but it is essential to the normal functioning of the thyroid gland. The hormone of this gland, thyroxine, is an amino acid containing iodine. In iodine deficiency the thyroid gland can enlarge and cause simple goiter (see Chapter 20). Traces of iodine occur in drinking water and in vegetables grown where there is iodine available in the soil. Sea foods are good sources of iodine. The use of iodized salt should supply adequate amounts of this element.

There are also minute amounts of other minerals in cells; these have been called *trace elements*. Some of them appear to be essential for certain enzymatic reactions. Copper, manganese, and zinc are examples. They are not to be confused with radioactive substances commonly called *tracers*. Magnesium is an essential element in muscle metabolism, and fluorides produce hard, cavity-resisting enamel in teeth.

VITAMINS At the beginning of the twentieth century it was generally assumed that the proper proportion of carbohydrates, proteins, fats, and minerals should provide all the essentials of a good diet. It was becoming more evident around 1911 that something more was needed to provide for normal growth and nutrition. There were contributions toward a knowledge of food chemistry before 1911, but the period since that date has been especially noteworthy in the discovery of many essential chemical substances known as

Inorganic requirements in the diet

461

vitamins. The literature on the subject of vitamins has become exceedingly abundant. Many good books and special articles are available.

The word *vitamin* is not strictly a correct term, since not all vitamins are amines and not all are vital for the maintenance of life. They are not necessarily closely related chemically, nor are their physiological effects necessarily closely related. Before the chemical nature of vitamins was determined, they were designated by letters of the alphabet, as vitamins A, B, C, etc. This practice is still followed, but now that the chemical structure of vitamins is better understood, it is considered good usage to refer to them by descriptive names such as niacin, thiamine, or ascorbic acid. It has not been possible to break away from the alphabetical classification completely, and it is often convenient to use. Vitamin A affords a good example. The structural formula is known, and the chemistry of vitamin A substances has been studied extensively, but no chemical name for this vitamin has been generally accepted.

Vitamin A This important chemical substance has the formula $C_{20}H_{30}O$. It is formed from precursors, or provitamins, of which four are well known. They are alpha, beta, and gamma carotene and cryptoxanthin. These substances are hydrolyzed by the animal organism to produce vitamin A. Beta carotene is the best source among the carotenes for conversion to vitamin A. By hydrolysis it yields two molecules of vitamin A, whereas the others yield only one molecule.

$$C_{40}H_{56} + 2H_2O \rightarrow 2C_{20}H_{30}O$$

Strictly speaking, plants do not contain vitamin A but only the substances from which an animal organism can manufacture the vitamin. Therefore certain plant foods may have vitamin A value, even though they do not actually contain any vitamin A. In general, the carotenes are found in yellow vegetables and fruits and in the green leaves of vegetables in which the yellow color is covered over by the green of chlorophyll. Cryptoxanthin is found in orange peel. A very common yellow pigment, xanthophyll, which shows so brilliantly in yellow leaves of trees in the fall, is not a precursor of vitamin A.

Animals take the yellow precursor substances from plants and transform them into colorless vitamin A. We may consider that the hen, the cow, and fish are important animal organisms converting plant provitamins into the form known as vitamin A, for egg yolk, cream, butter, and fish-liver oils are especially good sources of this vitamin. The yellow color of animal products does not necessarily indicate that they contain a greater amount of vitamin A than some colorless substance. White whole milk does not necessarily contain less vitamin A than milk that contains more yellow coloring matter. Animal products may contain both carotenoid substances and vitamin A. Furthermore, the yellow color of some foods is due to carotenoid substances that are not precursors of vitamin A.

Vitamin A and the carotenes are fat-soluble. Their absorption and metabolism within the body are similar to that of fats. Not all fats are good sources of the vitamin, however. Most of the vegetable oils used for cooking, as well as animal fat, such as beef and lard, contain very little. Since animals have the ability to store vitamin A in the liver, this organ may contain relatively large amounts. Fish, after consuming quantities of plant food (or larger fish feeding upon smaller plankton-feeding fish in the food chain), are able to store considerable concentrations of vitamin A in their livers. Cod-liver oil has become an important commercial product containing the two fat-soluble vitamins A and D. Oil from halibut livers is a somewhat better source. Recently it has been shown that the vitamin obtained from the livers of saltwater fish is slightly different in its chemical structure from that found in the livers of fresh-water fish. The predominant form in saltwater fish is referred to as Vitamin A_1, and that from fresh-water fish is called A_2. For nutritional purposes one does not ordinarily distinguish between the two.

Experimental studies by McCollum and Davis, made concurrently with studies by Osborne and Mendel, showed that young animals would thrive on a normal diet including butterfat but would not thrive when lard was substituted for butterfat in the same diet. Soon it was determined that egg fat and cod-liver oil could take the place of butterfat and that most of the commercial fats and oils used in cooking resembled lard in their effect on growth and nutrition. It is now known that the growth-promoting substances found in butterfat and largely deficient in lard are vitamins A and D. It is doubtful that vitamin A alone has any specific growth-promoting factor.

Deficiency conditions express themselves in several different ways. An outstanding deficiency concerns the rod cells in the retina. Vitamin A is essential to the formation of visual purple, or rhodopsin, the dark red pigment associated with rod cells. Vitamin A deficiency may lead to a condition called *night blindness*, for although the synthesis of visual purple from visual yellow, or retinene, is apparently reversible, there is always some loss in this reaction. It is thought that, when light falls upon the rod cells, visual purple is bleached to visual yellow and the energy released in the process initiates a nerve impulse from the retina. Only minute amounts of vitamin A would be necessary to make up the loss from this reaction, but it is essential that enough of the vitamin be available to avoid depletion. Not all cases of night blindness can be traced to vitamin A deficiency, however. Wald and others have made extensive investigations into the relationship of vitamin A and visual purple. Investigation indicates that vitamin A may also be a factor in the development of visual violet, or iodopsin, a visual pigment associated with cone cells.

11-*cis*-rhodopsin

11-*cis*-retinal + opsin ⇌ all-*trans*-retinal + opsin (protein)

dark / ATP — bleaching by light energy

(visual yellow)

NADH ‖ NAD

11-*cis*-vitamin A   **all-*trans*-vitamin A**

Vitamin A deficiency also affects the epithelial tissues in various parts of the body. Perhaps the most striking are the changes that occur in the cornea and conjunctiva of the eye, causing "dry-eye disease," or xerophthalmia. Inflammatory conditions of the eye can be produced in a high percentage of rats fed on a diet lacking vitamin A. Epithelial tissues of the skin and respiratory, digestive, and urinogenital systems are also affected. There is a tendency toward squamous hyperplasia, a change toward stratified squamous epithelium. In the skin the sweat glands and sebaceous glands become inflamed and cause eruptions or sores. Early experimentation showed that rats fed on a vitamin A–free diet failed to gain weight. There was also malformation of bone.

Sources of vitamin A or the provitamin have already been mentioned. The best sources are the green leafy vegetables, yellow fruits and vegetables, and animal products such as egg yolks, butter, cream, or whole milk. Fish-liver oils are excellent sources.

Vitamin D *Rickets* refers to a vitamin-deficiency condition resulting in a malformation of growing bones. It is a condition found in children and young animals that have not had adequate amounts of fat-soluble vitamin D and sunlight. Calcium and phosphorus metabolism is disturbed, producing poor calcification of bone. The leg bones do not support the weight of the body well, and they bend, giving rise to knock-knees or bowlegs (Figure 18.1). Other skeletal malformations ascribed to rickets are narrow chests, scoliosis, and malformations of the skull and pelvis. In young growing bones there is a very evident defect in calcification at the junction of the epiphysis and the shaft of the bone. In adults a deficiency condition occurs in which calcium and phosphorus salts are removed from bone in greater quantity than the amount being deposited. Loss of strength in the bones can lead to deformity. The condition is essentially adult rickets.

Vitamin D increases absorption of calcium from the intestine and

FIGURE 18.1

Rickets in a young child. The bone curvatures and bowlegs are due to a deficiency of vitamin D. (*From "The Vitamin Manual," published by the Upjohn Company: courtesy of Rosa Lee Nemir, M.D.*)

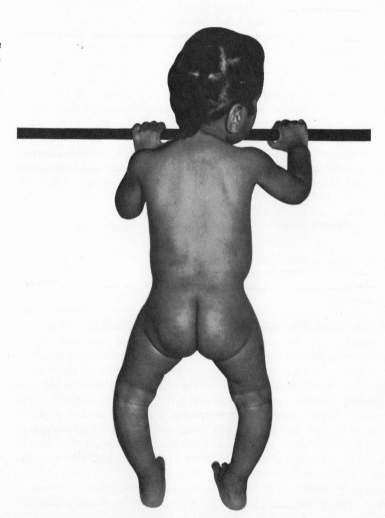

Foods, nutrition, and metabolism

influences the deposition of calcium in bones. It may help also in maintaining the calcium level of the blood.

There are several substances having antirachitic properties. The two of greatest importance are *ergosterol* and *7-dehydrocholesterol*. They are activated by irradiation with ultraviolet light. Ergosterol is a sterol originating in plants. Activated ergosterol is known as vitamin D$_2$ or ergocalciferol. Commercial forms are called *viosterol* and *calciferol*. Yeast is a good source for commercial preparations of ergosterol. Activated 7-dehydrocholesterol produces cholecalciferol or vitamin D$_3$. It is found in animal fats and is the natural vitamin of fish-liver oils. This is also the substance in the skin of mammals that is activated by ultraviolet light. There is no vitamin D$_1$.

Provitamin D substances are activated by wavelengths of light that they are capable of absorbing. These wavelengths have a restricted range within the ultraviolet. Their penetrating power is poor, so that they are not effective through clothing or even through ordinary window glass. Apparently they penetrate the skin to a depth of only about half a millimeter. The wavelengths of light that have antirachitic value are also the wavelengths that cause sunburn and tanning of the skin.

Most abundant sources of vitamin D are the liver oils of bony fish, such as halibut, cod, tuna, and others of the percomorph group. The average diet can include fresh and canned fish, milk, eggs, and butter as good sources. Green vegetables contain small amounts. Though mammalian liver stores adequate amounts of vitamins A and D, liver is not a good source of vitamin D, although it is an excellent source of vitamin A. These two fat-soluble vitamins are commonly found in the same food sources but not necessarily in equal amounts. Since vitamin D is concerned with the utilization of calcium and phosphorus salts, it is essential that the diet contain these salts in adequate amounts. Coupled with sunlight and adequate amounts of cholesterol in the skin, the average diet probably contains enough vitamin D for an adult. It is doubtful whether or not it is adequate for growing children. Infants receiving a quart of whole milk per day should be protected from rickets, but the diet during infancy is commonly supplemented with additional vitamin D from a commercial source. If children are growing rapidly, their need for calcium and vitamin D is increased. Perhaps not all growing children need to have vitamin D added to their diet, but for many it would be helpful. During pregnancy and lactation the vitamin D intake should be increased. A normal amount of this vitamin is stored in the liver.

THIAMINE The search for the antineuritic vitamin was long and arduous. Investigations centered around a cure for a condition of polyneuritis that had been known for centuries to affect populations where the most important item in the diet is polished rice. The vitamin-deficiency condition is called *beriberi*. It is a condition involving inflammation of the peripheral nerves and resulting eventually in paralysis of the appendages. There are also circulatory and sensory effects. The condition also occurs in animals fed on restricted diets (Figure 18.2). Birds are especially susceptible.

When the nature of the vitamin was determined, it was called *vitamin B*, a water-soluble substance that would cure or prevent beriberi. Later experimentation showed that this water extract of food substances contained not just one vitamin but a series of vitamins. The original vitamin, concerned with beriberi, then became B$_1$, and there followed other vitamins called B$_2$, B$_6$, and B$_{12}$, now termed the *vitamin B complex*, which includes a number

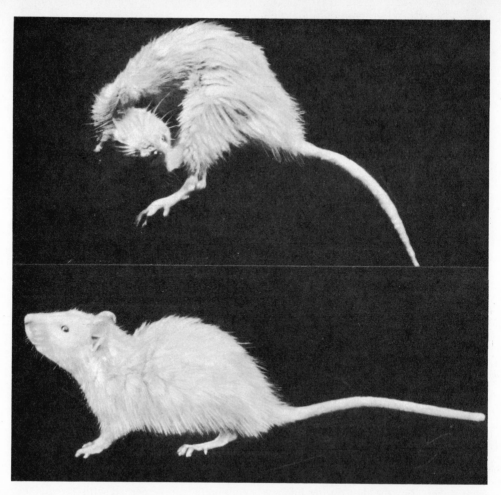

FIGURE 18.2

Rat, twenty-four weeks old, showing thiamine deficiency. The rat in the upper photo has been fed a diet containing practically no thiamine and has lost the ability to coordinate its muscular movements. The lower photo shows the same rat 24 hours later, after receiving a food rich in thiamine. (*Courtesy of U. S. Department of Agriculture, Bureau of Human Nutrition and Home Economics, Washington, D. C.*)

of chemical substances. Now that the chemical nature of these substances is known, there is a generally accepted policy of using chemical names instead of alphabetical symbols. Vitamin B_1 now is known as *thiamine*. Thiamine plays an important part in the proper utilization of carbohydrates. The breakdown of pyruvic acid requires the enzyme carboxylase. Thiamine unites with phosphoric acid to form a pyrophosphate. Thiamine pyrophosphate then acts as a coenzyme with carboxylase in the oxidation of pyruvic acid and is referred to as *cocarboxylase*. It is thought that the accumulation of pyruvic acid due to the failure of an intermediary metabolic process is one of the factors producing symptoms of polyneuritis. This condition is relieved by adding thiamine to the diet.

Thiamine deficiency results in a loss of appetite, which, in turn, reduces the intake of food. This sequence complicates studies of vitamin deficiency, since some of the observed effects may be due to malnutrition rather than to the deficiency of any particular vitamin. Vitamin B_1 has a favorable influence on growth, but, again, an increased intake of food including other vitamins complicates the picture. Thiamine, as well as other vitamins of the B group, may become deficient during pregnancy and lactation. Thiamine is stored by the liver to a limited extent.

Vitamins of the B complex and vitamin K are produced synthetically by bacterial action in the large intestine. In most instances the amount synthesized is not adequate to meet the nutritional needs of the individual, but it does tend to offset vitamin deficiencies in the diet.

The amount of thiamine required by an individual varies slightly with the person's sex, age, and physical condition and the amount of work done. In general, the requirement is 1 or 2 milligrams per day. This amount can readily be obtained from food sources. Thiamine is produced in plants by synthesis but is present in most fruits and vegetables only in minute amounts. There is usually a greater concentration in seeds, and so whole grains, beans, peas, and nuts are good sources. The best animal sources include eggs and lean meat, especially pork muscle. In general, the more highly refined the flours and cereals, the less thiamine is retained in them. Therefore whole grain bread and breakfast cereals contain more thiamine unless white bread and cereals have been enriched by adding a commercially prepared product. Although a good diet should provide enough thiamine, there is a tendency to use too much highly refined flour, breakfast cereals, sugar, and fats that are low in content for this essential vitamin.

Riboflavin The discovery of this vitamin is somewhat unique, for it first became known as a respiratory coenzyme in an oxidation-reduction reaction. This was the "yellow enzyme" described by Warburg and Christian, in 1932. Originally considered to be the heat-stable fraction of vitamin B, it was called B_2 or G. Riboflavin was at first called *lactoflavin*, because it was isolated from milk. Other yellow fluorescent pigments called *flavins* were discovered in various food sources. When it was determined that these flavins were identical and contained the sugar ribose, or a derivative, the vitamin was called *riboflavin*. Riboflavin is an orange-yellow color. In aqueous solution it shows a greenish-yellow fluorescence. Elongate orange-yellow crystals can be obtained.

When riboflavin is phosphorylated, it forms two coenzymes, flavin mononucleotide (FMN) and flavin adenine dinucleotide (FAD). Both may be active as hydrogen carriers in oxidation-reduction reactions. FAD acts as an electron acceptor from coenzymes NAD and NADP in the electron-transport system. As the flavoprotein is oxidized or dehydrogenated, it passes electrons along to the cytochrome system, with the release of hydrogen atoms (Figure 16.15).

The effects of riboflavin deficiency are not as definite as those of some other vitamin deficiencies. Animals do not grow properly, and there is a loss of hair in rats (Figure 18.3). Vascular changes in the cornea, conjunctivitis, and cataract have been described for various animals. Human deficiency conditions include cracks and sores at the corners of the mouth and ocular changes involving the cornea and the conjunctiva. Several investigators have reported dry and scaly skin as a riboflavin deficiency.

Since riboflavin is present in such a wide variety of foods, it is quite likely that noticeable deficiency effects do not commonly occur. Bacteria in the intestine are able to produce the vitamin synthetically and so add to the amount available, but persons who live on inadequate diets may be subject to manifestations of deficiency. Good food sources are leafy vegetables, fruits, yeasts, milk, liver, muscle, and egg white.

Niacin (nicotinic acid) and nicotinic acid amide *Pellagra* denotes a condition characterized by red lesions of the skin, especially on the backs of the hands and on the forearms, legs, and feet. The tongue assumes a bright red color,

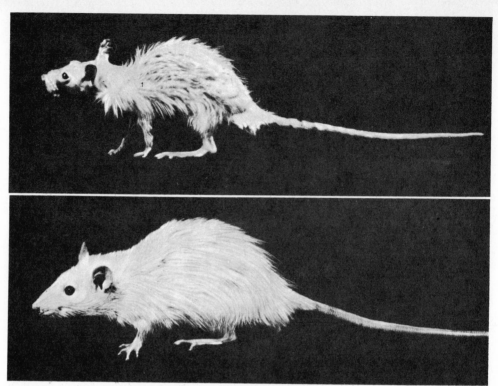

FIGURE 18.3

Riboflavin deficiency. The rat in the upper photo is twenty-eight weeks old and has had no riboflavin. Note the loss of hair, especially around the head. Weight is 63 grams. The lower photo shows the same rat 6 weeks later, after receiving food rich in riboflavin. It has recovered its fine fur and now weighs 169 grams. (*Courtesy of U. S. Department of Agriculture, Bureau of Human Nutrition and Home Economics, Washington, D. C.*)

Foods, nutrition, and metabolism

and eventually there are symptoms of depression or even dementia, indicating derangement of the nervous system. It is found predominantly among persons of poor economic status whose diet is confined largely to cornmeal, white flour, polished rice, and sugar. It is not a condition that can be cured by the administration of riboflavin or thiamine alone.

It was found, around 1926, that yeast contains a pellagra-preventive factor (PP factor) even after being heated in an autoclave to destroy its antineuritic value. Riboflavin is still present after yeast is heated, but it was soon found to have no curative value for pellagra. Actually a good all-round diet is pellagra-preventive, and suitable diets were recommended for those persons located in areas where pellagra was common. After years of study regarding the curative effects of liver extracts on dogs with an experimentally induced condition called *blacktongue*, Elvehjem and his associates, in 1937, discovered that nicotinic acid would cure canine blacktongue. They also isolated nicotinic acid amide from liver extract that had been found to have curative value. Blacktongue in dogs closely resembles pellagra in human beings, and in a short time several investigators reported that pellagra could be cured or its symptoms relieved by the administration of nicotinic acid.

Nicotinic acid had been known for at least 70 years before the discovery of its value as a vitamin. It was obtained from nicotine by chemists in the latter half of the nineteenth century. In the search for a cure for beriberi, nicotinic acid was isolated from rice polishings as early as 1911; but it was discarded when it failed to cure this condition. Thus nicotinic acid takes its place as the first vitamin to be isolated in pure chemical form, even though it was not known as a vitamin until many years later. Nicotinic acid should not be confused with nicotine. The body does not derive the vitamin from nicotine absorbed by smoking tobacco. Although prepared chemically through oxidation of nicotine, nicotinic acid has very low toxicity compared with that of nicotine.

In tissues, nicotinic acid occurs in the form of its amide. The carboxyl group, COOH, of the acid is changed to include an amino group, NH_2, and the molecule then becomes nicotinic acid amide, or nicotinamide. Since tissues or foods contain only the amide, strictly speaking, this substance is the vitamin. The chemistry of the body, of course, has the ability to transform nicotinic acid into nicotinamide. Nicotinamide adenine dinucleotide (NAD) acts as a coenzyme in cellular respiration and in the breakdown of glucose. Some animals, at least, are able to manufacture the vitamin synthetically as a result of bacterial action in the intestine.

Niacin is a water-soluble vitamin and relatively heat-stable. There is always the chance that some of the water-soluble vitamins will be lost in the water used for cooking. The vitamin is found in a wide variety of vegetables, whole wheat, milk, eggs, and lean meat. Among the best sources are liver, kidney, salmon, brewer's yeast, wheat germ, soy beans, and peanuts.

The amino acid tryptophan is a source of niacin, and the amount and kind of protein in the diet affect the amount of niacin available. The amount of tryptophan in animal protein is about 1.4 percent; in vegetable protein it is about 1 percent.

Though pellagra appears to be primarily the result of niacin deficiency, it should be stressed that pellagrins often have other vitamin deficiencies. Treatment of pellagra, therefore, should not be limited to the administration of niacin alone but should include other vitamins. The interrelations of vitamin deficiencies are not so well known that they can be effectively treated by individual vitamins.

Other vitamins of or closely related to the B group are pyridoxine (B_6), pantothenic acid, inositol, biotin, folic acid, B_{12}, p-aminobenzoic acid, and choline.

Pyridoxine (vitamin B_6) Pyridoxal phosphate is essential as a coenzyme in many phases of protein metabolism and is involved in the prevention of a form of dermatitis in experimental animals. Human deficiency effects have not been demonstrated with certainty. This may be largely because of the wide distribution of the vitamin in common foods. It is water-soluble and not stable in light. In animal and plant tissues it is mostly found combined with protein or starch rather than in the free form. Seeds, legumes, wheat germ, liver, kidney, meat, and fish are good sources of pyridoxine. Since this is one of the more recently discovered vitamins, further study will no doubt afford a much better understanding of the part it plays in human and animal nutrition. Pyridoxine is essential in the metabolism of certain amino acids and probably in their transport across membranes. It also is involved in the conversion of proteins to fats and in the metabolism of some of the fatty acids.

Pantothenic acid Reports on the significance of this substance in nutrition began to appear around 1940. It is concerned with preventing a type of dermatitis in chicks and apparently is necessary for proper growth in all animals. Wide interest followed the report that it prevented fur of black rats from turning gray, but there is no evidence that it will prevent gray hair in human beings. Though the vitamin plays some part in human nutrition, clear-cut cases of human deficiency have not been demonstrated. This may be largely because

Inorganic requirements in the diet

469

there are adequate amounts of the vitamin in the food commonly eaten by man.

Pantothenic acid constitutes a part of coenzyme A. Coenzyme A takes part in several known acetylation reactions; however, to be effective, it must be in the form of acetylcoenzyme A. Coenzyme A, in the presence of acetate and ATP and catalyzed by enzymatic reaction, can be converted to acetylcoenzyme A. As acetylcoenzyme A, it is able to accomplish the acetylation of choline to form acetylcholine. In carbohydrate metabolism, pyruvic acid is acted upon by acetylcoenzyme A as it enters the tricarboxylic acid cycle. It then combines with oxaloacetic acid to form citric acid.

There seems to be some interrelation between the pigments of skin and hair, the adrenal cortex, and coenzyme A, but the exact nature of this relationship has not been established. There is need for further research to establish clearly the function of pantothenic acid in nutrition.

Inositol Several substances seem to play some part in nutrition, although their function is not clearly understood. Inositol is one of them. If not a true vitamin, it is at least vitamin-like. It was originally one of the bios factors and was recognized also as part of the phytin complex. Deficiency conditions in the young mouse result in a loss of hair around the trunk and top of the head. The rat develops a dermatitis around the eyes. Inositol aids in preventing the accumulation of fat in the liver. It may also have some function in preventing hardening of the arteries. It can be extracted from muscle tissue and may be found to play some part in muscle metabolism.

Inositol is present in plant and animal tissues. In plants it is commonly present in a form called *phytic acid;* in animal tissue it is found as a more complex compound most abundant in liver, kidney, heart, and skeletal muscle. The part that it plays in human nutrition is still to be determined.

Biotin (vitamin H) As early as 1901, it was discovered that yeast required certain substances for proper growth. These were called *bios factors;* inositol was one of them, as mentioned above. After many years of study another factor was isolated, in 1936, and named *biotin.* Not only yeast but certain bacteria and fungi also were found to require biotin for proper growth.

An interesting relationship between biotin and a protein in raw and dried egg white was investigated. It had been known for some time that there was some factor in egg white that was harmful to experimental animals. Rats fed on such a diet gradually lost their hair and developed a severe dermatitis. There was a characteristic dermatitis around the mouth and eyes. The protein in egg white is called *avidin.* It apparently enters into a chemical combination with biotin in the intestine and prevents the vitamin from being absorbed and utilized. If the egg white is cooked, there is no interference with the action of the vitamin. Though biotin appears to be essential in human nutrition, deficiency conditions are produced only when a very large part of the diet (nearly one-third of the total caloric intake) is raw or dried egg white. The intestinal flora are capable of producing biotin synthetically, though apparently not in large enough quantity. Animals vary in their ability to synthesize the vitamin; chickens apparently require more biotin than do rats.

Biotin takes part in the function of several enzyme systems. It functions also in deamination processes and is closely associated with the vitamins folic acid and pantothenic acid. It has been shown to function in the synthesis of fatty acids and in several carboxylation reactions.

Biotin is found in small amounts in many foods. It is present in fairly large amounts in yeast, milk, liver, kidney, and raw potatoes. Other sources are egg yolk and common garden vegetables, such as carrots and tomatoes.

Folic acid (pteroylglutamic acid) This vitamin was originally isolated from the green leaves of plants such as spinach, peas, and clover. Since its source was the foliage of plants, it was called folic acid. It is found in yeast concentrates and in liver and is essential for the growth of certain kinds of bacteria and protozoa. In vertebrate animals and man it functions in the normal development of red blood cells. It takes part in the synthesis of purines and pyrimidines in the formation of nucleic acids and nucleoproteins. For example, it is involved in the formation of thymidine which, in turn, is a component of DNA.

Folic acid is effective in treating some types of anemia but has no effect in preventing damage to the central nervous system in pernicious anemia. Various studies indicate that there is a physiological relationship between this vitamin and the long-sought erythrocyte maturation factor, vitamin B_{12}. Some animals have the ability to synthesize folic acid through bacterial action in the intestine, and others apparently do not.

Vitamin B_{12} (cyanocobalamine) A red crystalline substance has been isolated as a liver fraction that exerts a powerful effect in promoting the maturation of red cells and also is capable of relieving nervous and digestive disturbances associated with pernicious anemia. This factor is vitamin B_{12} and is apparently the erythrocyte maturation factor (extrinsic food factor) that has been sought for many years. It is a cobalt compound; cobalt itself, when added to the food of rats in small amounts, has been shown to increase the number of red cells and the amount of hemoglobin. Cobalt alone has not proved of value in treating pernicious anemia. Vitamin B_{12} is effective in minute amounts and is many times more effective than folic acid in promoting the production of red blood cells. It is used also in a general way, to promote appetite and growth.

The effectiveness of B_{12} appears to depend on the presence of an enzyme (the intrinsic factor) secreted by the stomach. This enzyme, a mucoprotein, is necessary to facilitate the absorption of B_{12} through the mucosal membrane of the intestine. If B_{12} cannot be absorbed in adequate amounts, pernicious anemia develops. Subcutaneous injections of B_{12} produces an effective response in cases of pernicious anemia. The vitamin taken by mouth is ineffective if the intrinsic factor is inadequate.

In the blood, B_{12} is bound to alpha globulins and is stored mainly in the liver. The vitamin probably functions in the metabolism of all cells but is especially effective in stimulating the production of red cells in the bone marrow. It also is essential in the normal metabolism of nerve cells and the cells of the digestive tract.

Vitamin B_{12} has a high physiological value so that 1 to 1.5 micrograms absorbed per day is adequate to satisfy the needs of normal persons. B_{12} is found in small amounts in meat, milk, and eggs but plant foods are deficient.

Choline As a chemical compound, choline has been known for many years. It was first isolated from bile in 1894. Only recently has investigation shown that it plays an important part in nutrition. There is no known human deficiency, but deficiencies can be demonstrated in experimental animals. Choline occurs in both plant and animal tissues largely as a constituent of phospholipids. It prevents excessive accumulation of fat in the liver of experimental animals when they are fed diets high in fat content. It also prevents fatty livers in depancreatized dogs. Another effect of considerable importance is the development of a hemorrhagic condition in the kidneys of rats fed on a diet extremely low in choline. Though choline can be synthesized within the body, the synthesis is evidently not adequate under extreme dietary conditions.

Choline has a part in various phases of metabolism. Acetylcholine, a derivative, is the chemical substance that provides for the transmission of the nerve impulse across the synapse and at the neuromuscular junction. Choline is the basic substance in the phospholipid, lecithin. As a lipotropic substance, it is able to reduce the fat content of the liver. It accelerates the formation of phospholipids from the fats present in the liver.

Ascorbic acid, or vitamin C A chemical substance found in fresh fruits and vegetables will prevent the development of a deficiency condition called *scurvy*. The chemical substance with antiscorbutic properties has been named, appropriately, *ascorbic acid, or vitamin C*. The chemical name of this substance is long and difficult, and so the coined name, ascorbic acid, is probably a fortunate choice. The name vitamin C is so well established that it is still in common use, and we shall use both names interchangeably.

Scurvy is characterized by weakness and lassitude. There are marked tenderness and swelling of the joints. The gums are red and swollen, and the teeth become infected around their bases, causing them to loosen. There is a change in the capillaries, permitting hemorrhagic conditions beneath the skin, in mucous membranes, and under the periosteum.

Scurvy was very common in Europe in the fifteenth century. It usually broke out in armies during a war. It was the scourge of mariners and explorers when they were forced to live for long periods on dried meats, grains, and beans, which could be conveniently stored. It has been stated that when potatoes were introduced into Europe from the New World scurvy became less common. Potatoes do not represent an abundant source of ascorbic acid, but, like many other vegetables, if they form a regular and substantial part of the diet, they can be very important in warding off a deficiency condition. Fresh meats are in the same category.

Early experimental work proved that limes and lemons contain a substance capable of preventing scurvy. The list of foods containing anti-scorbutic properties was finally enlarged to include all fresh fruits and vegetables in general, but citrus fruits remain one of the best sources. Sprouting grain and bean sprouts contain vitamin C, but it is lacking in dried grain and legumes. The ascorbic acid content of foods is rapidly reduced by oxidation. Prolonged cooking or even prolonged standing and drying of fruits and vegetables reduces their vitamin C potency. In general, commercial canning retains the vitamin better than home-canning methods, although home-canned tomatoes retain the vitamin fairly well. Milk is low in vitamin C; hence the recommended practice of feeding fruit juice or vegetable juice to infants in order to supplement their diet.

Vitamin C apparently takes part in numerous oxidation-reduction reactions. Also, it seems to regulate the formation of intercellular substances such as collagen. Most animals are able to synthesize vitamin C, but the primates and guinea pigs are known to be exceptions. The recommended daily intake for adults is 70 milligrams. For infants and growing children it is 30 to 80 milligrams. One-third to one-sixth of this amount should provide protection from the more evident signs of scurvy. Although considerable amounts of vitamin C occur in the adrenal cortex, it is not stored in the body.

Vitamin E (the tocopherols)   The search for another fat-soluble vitamin evolved around the observation that rats reared upon certain restricted diets often are sterile. Adding green vegetables, wheat germ, or vegetable oils to the diet corrects the deficiency. Investigation finally revealed vitamin E substances, included under the general name *tocopherols*, as the antisterility vitamin for rats.

Little is known about the effect of these substances in the metabolism of human beings, but it appears to be an antioxidant, preventing or inhibiting the oxidation of unsaturated fats. Vitamin E is widespread in plant food materials, and there is little chance of deficiency in the average diet. It is found in green leafy vegetables and also in peas and beans. Wheat-germ oil and vegetable oils are regarded as excellent sources. Animal tissues contain only small amounts.

Vitamin E deficiency in the male rat is manifested by degeneration of the germinal epithelium of the testes. The male becomes incapable of producing living, motile spermatozoa. A female on the same diet is capable of mating and fertilization, but the young embryos die and disintegrate. Their substance is then reabsorbed into the tissues of the mother. The addition of tocopherols to the diet enables the female to produce normal living young. The progress of damage to the testes in the male is not reversible upon adding tocopherols to the diet.

Rats and other animals show muscular weakness and paralysis as a result of vitamin E deficiency extending over many months. It has been observed also that vitamin E exerts a "sparing" effect on the use of vitamin A by the animal body and that more vitamin A is stored in the liver of these animals. Attempts to alleviate muscular weakness or to improve reproductive functions in human beings by administering vitamin E substances have not been successful.

Vitamin K   A condition in chicks kept upon a restricted fat-free diet was studied by Dr. Henrick Dam, of the University of Copenhagen. The condition is one of hemorrhage and anemia; the clotting ability of the blood is very greatly reduced. Since the condition was not relieved by any of the known vitamins, Dam postulated that an unknown vitamin was involved. He called it the *Koagulations vitamin*, and it has become known as *vitamin K*.

There are two natural compounds of vitamin K. The first was isolated from alfalfa, and it is now known as K_1. Vitamin K_2 was isolated from fish meal subjected to bacterial putrefaction. It had been known for some time that bacteria were able to synthesize this vitamin not only in bacterial preparations of various sorts but also in the living intestine. The empirical formula for vitamin K_1 is $C_{31}H_{46}O_2$; that for K_2 is $C_{41}H_{56}O_2$. There are, in addition to K_1 and K_2, several compounds that have vitamin K properties. One of the best known of these synthetic compounds is 2-methyl-1,4-naphthoquinone. It has had considerable use as a substitute for the more expensive K vitamins. The K vitamins are chemically related to ubiquinone (coenzyme Q).

Vitamin K_1 is found chiefly in green plants, in which there may be some relation between the vitamin and chlorophyll. Vitamin K_2 is synthesized by bacteria. The bacteria of the intestinal tract appear to be of great importance in this respect. As with other fat-soluble vitamins, bile promotes absorption from the intestine. Most laboratory animals synthesize and absorb the vitamin readily. Most laboratory animals, therefore, do not show vitamin K deficiency, even though they are fed a deficient diet. Chicks, however, do not absorb K_2 readily; when K_1 is withheld from the diet, they become hemorrhagic as a manifestation of vitamin K deficiency.

Vitamin K is able to raise the prothrombin level of the blood only if the liver is functioning, since prothrombin is formed in the liver. The action of vitamin K is to stimulate the synthesis of prothrombin and blood-clotting factor VII. The prothrombin level of the blood is affected also by any condition that prevents the normal flow of bile into the intestine. In the absence of bile from the intestine, very little vitamin K is absorbed into the bloodstream. The administration of bile salts plus vitamin K has proved helpful in such cases.

In the newborn infant the prothrombin level of the blood is likely to be low. The clotting ability of the blood, therefore, may be impaired. The administration of vitamin K to the mother before birth of the child or to the infant at birth tends to increase the clotting ability of the blood (Table 18.1).

METABOLISM

Life processes, or those processes performed by all living forms in order to maintain life, are grouped under the general heading of *metabolism*. In a very broad sense, the energy relations that result in the assimilation and utilization of food, and in the promotion of growth, are considered to be building-up processes and are often referred to as *anabolism*. *Catabolism*, on the other hand, is taken to mean the breakdown of stored reserves, which leaves the organism with a reduced store of energy. Very often it is difficult to classify some reactions accurately under these headings. For example, is the breakdown of a food substance into simpler substances, so that these, in turn, can be converted into fat and stored, an example of anabolism, catabolism, or both?

Metabolism is essentially an expression of energy relationships. Green plants are able to use the radiant energy of the sun and by the process of photosynthesis incorporate this energy in the synthesis of food material. Animals are unable to manufacture their food directly and so are dependent upon energy stored in the molecules of plants that can be used for food. If energy is utilized in building up the molecules of a food substance, then that energy can be released, provided the animal organism has the ability to break down the molecular structure. Enzyme systems catalyze a series of chemical reactions that enable the animal to release this energy in the case of food substances. Wood and coal, of course, contain abundant sources of energy, but the higher organisms have no enzymes capable of releasing this energy. Wood and coal, therefore, are not available directly as foodstuffs.

The attempt to demonstrate the energy derived from foods presents several well-known problems. In the first place, the law of the conservation of energy indicates that while the total energy of the universe remains constant, one form of energy may be converted into another. Energy in the body can be used in various ways. It can be used, for example, to synthesize chemical compounds; some of these compounds, such as glycogen and fat, are a stored source of energy. Energy can also be used to do work, such as that performed by the contraction of muscles. In this case only a small

TABLE 18.1

The principal vitamins

Vitamin	Major sources	Function	Deficiency
Fat-soluble:			
A Beta carotene	Yellow vegetables, green leafy vegetables, egg yolk, butter, fish-liver oils	Maintenance of epithelial tissues; essential for formation of rhodopsin	Xerophthalmia, night blindness
D_2 Calciferol D_3 Cholecalciferol	Fish-liver oils Egg yolk, milk, butter	Increase absorption of calcium and phosphorus from intestine and their utilization in bones and teeth	Rickets in children
E Tocopherol	Green leafy vegetables, wheat germ	Increases the stability of membranes(?) Acts as an antioxidant in fat metabolism	Degeneration of the germinal epithelium of testes in white rats
K Naphthoquinone	Leafy vegetables	Promotes synthesis of prothrombin and blood factor VII in liver	Blood clotting is impaired
Water-soluble:			
B complex: B_1 Thiamine	Whole grain, cereals, nuts, eggs, pork	Acts as cocarboxylase in oxidation of pyruvic acid	Beriberi, some forms of neuritis
B_2 Riboflavin	Liver, meat, milk, eggs, whole grains	Flavoproteins, as coenzymes in oxidative phosphorylation (electron transport)	Some forms of dermatitis
Niacin	Liver, wheat germ, soy beans, peanuts	Acts as a coenzyme in cellular respiration (NAD, NADP)	Pellagra
B_6 Pyridoxine	Seeds, liver, meat, fish	Pyridoxal phosphate, coenzyme in amino acid and fatty acid metabolism	Depressed hemoglobin synthesis
B_{12} Cyanocobalamin	Liver, meat, milk, eggs	Essential to production and maturation of erythrocytes	Pernicious anemia
Folic acid (pteroylglutamic acid)	Green leafy vegetables, yeast, liver	Essential to synthesis of RNA and DNA, also for normal erythrocyte maturation	Degenerative changes in bone marrow
Pantothenic acid	Vegetables, liver, eggs	Synthesis of coenzyme A	Metabolic functions, especially in carbohydrates and fats
Biotin	Yeast, milk, liver, vegetables (synthesized by bacteria in intestine)	Takes part in fatty acid synthesis and in carboxylation reactions	No definite deficiency in man
Choline	Plant and animal tissues as a constituent of phospholipids	Metabolism of fats, oxidation	Prevents fatty livers in animals
C Ascorbic acid	Citrus fruits, tomatoes, cabbage, most fresh fruits and vegetables	Regulates formation of collagen and other intercellular substances	Scurvy, in severe cases

amount of energy is used in work; a far greater amount is given off as heat. Other cells utilize energy and perform work in various ways. The production of a secretion by secretory cells represents work and requires an outlay of energy. The best way to demonstrate and measure the energy derived from foods is to measure it in the form of heat.

The energy derived from the breakdown of foodstuffs is not applied directly to metabolic functions but, instead, is stored as high-energy phosphate bonds in the ADP–ATP system. Such chemical energy can be released by the aid of specific enzymes to catalyze all sorts of chemical reactions that come under the general heading of metabolism.

Not only carbohydrates but the breakdown products of proteins and fats enter the tricarboxylic acid energy cycle at one stage or another. All these classes of foods find a common pathway and are metabolized by way of the tricarboxylic acid cycle (see Figure 16.14).

At this point it is advisable to review and summarize various aspects of energy metabolism leading to a consideration of basal metabolism and to review the discussion of cellular respiration.

The degradation of carbohydrates proceeds by steps, each step catalyzed by an enzyme. There is some release of energy during glycolysis although only about 10 percent is released in this way. The other 90 percent, approximately, is derived from the oxidation of pyruvic acid by way of the tricarboxylic acid cycle.

Glycolysis proceeds from the phosphorylation of glycogen or glucose to glucose 6-phosphate. ATP is the source of the phosphate and the energy, as it is degraded to ADP. Hexokinase and phosphorylase are the enzymes involved.

$$\text{glycogen} + \text{PO}_4 \xrightarrow{\text{phosphorylase}} \text{glucose 1-phosphate}$$

$$\text{glucose} + \text{ATP} \xrightarrow{\text{hexokinase}} \text{glucose 6-phosphate}$$

The glucose 6-phosphate molecule is further degraded, through a series of steps, to two pyruvic acid molecules, each representing a three-carbon unit. This is a simplification of a series of enzymatic steps worked out by Embden and Meyerhof and often referred to as the Embden-Meyerhof scheme.

When fats are digested, triglycerides first break down into glycerol and fatty acids. In the process of absorption, triglycerides are again formed, and if used for energy, they must again be broken down. Glycerol is then converted to glyceraldehyde 3-phosphate, which can be converted to glucose and used for energy or may enter the glycolytic pathway.

In cellular metabolism, fatty acids are broken down by a beta-oxidation process, a step-by-step removal of two-carbon segments until acetyl Co-A is formed. Acetyl Co-A enters the citric acid cycle.

The breakdown of cellular protein is complicated largely because there are differences in the kinds of amino acids produced. Amino acids are first deaminated; that is, they lose their $-\text{NH}_2$ groups. Some amino acids are said to be *glycogenic;* they are capable of forming glucose or glucose derivatives and so can break down to pyruvic acid. Others are *ketogenic;* that is, they are more nearly related to fatty acids and eventually break down to acetic acid. One of the amino acids, glutamic acid, can be converted directly to alpha-ketoglutaric acid and so enter the tricarboxylic acid cycle at the ketoglutarate level.

As two-carbon units enter the citric acid cycle, they combine with four-carbon oxaloacetic to form the six-carbon compound, citric acid. Citric acid, by a series of four steps, breaks down to the five-carbon alpha-ketoglutaric acid, with a consequent release of CO_2, hydrogen, and energy.

The next step consists of degrading alpha-ketoglutaric acid to the four-carbon compound succinic acid. This reaction yields more CO_2, more hydrogen, and a further release of energy. The energy is incorporated in high-energy phosphate bonds in the reaction ADP–ATP. Hydrogen in this case is passed on to NAD.

Succinic acid by a series of reactions gives rise to the four-carbon starting point of the cycle, namely, oxaloacetic acid. In the process there are two dehydrogenations, and more energy is released. Oxaloacetic acid immediately combines with more two-carbon acetic acid fragments to keep the cycle in constant repetition with a constant production of CO_2, hydrogen, and energy.

Hydrogen and electrons, as we have seen, are passed along to the electron-transport series with a consequent release of energy. These energy reactions take place in the mitochondria.

DIRECT CALORIMETRY Measured amounts of food substances can be burned in an apparatus called a *calorimeter,* and the heat given off can be determined with a high degree of accuracy. Calorimeters have various designs. The bomb calorimeter consists of a strong steel cylinder immersed in a weighed amount of water. The cylinder contains the food sample and is filled with oxygen to obtain complete and immediate oxidation. A platinum wire completes an electric circuit within the cylinder. When an electric current is introduced, foods are oxidized to carbon dioxide and water, except proteins, in which case oxides of nitrogen, sulfur, and phosphorus also remain. The heat liberated causes the temperature of the water to rise. The temperature change of a known amount of water can then be calculated in terms of calories.

A *calorie* is an energy unit defined in terms of heat. The *gram calorie* is the amount of heat required to raise the temperature of 1 gram of water 1°C. More specifically it must raise the temperature of water from 14 to 15°C. The unit of measurement commonly used in physiology is the kilogram calorie, which is written in abbreviated form kcal or Cal. This is the large calorie. It is 1,000 times larger than the small calorie and is the amount of heat required to raise the temperature of 1 kilogram of water 1°C.

The kilogram calories produced by complete oxidation of different food substances as measured in a bomb-type calorimeter are averages based on a knowledge of the relative amounts of these substances in foods. It is known that glucose, when completely oxidized, produces only 3.75 kilogram calories per gram, whereas starch and glycogen produce 4.22 kilogram calories. The average for the oxidation of carbohydrates in the bomb calorimeter is 4.1 kilogram calories per gram. The average for a similar oxidation of fats is 9.5 kilogram calories and that for proteins is 5.6 kilogram calories.

It is of interest to find by experimentation that the oxidation of carbohydrates and fats within the body is as complete as the oxidation of these foods by actual burning. These two kinds of food are oxidized to carbon dioxide and water in the calorimeter or in the body; therefore the heat given off is the same in either case. Proteins are not completely oxidized in the body. The nitrogenous by-products are discarded mainly as urea and ammonia. These end products contain a certain amount of energy, which is released by further oxidation in the calorimeter but not in the body. The body in its metabolism, therefore, derives only 4.3 kilogram calories per gram from protein.

Not all food materials are completely digestible. Plant foods, for example, contain cellulose, which is indigestible. Carbohydrate foods are estimated to be about 98 percent digestible, fats 95 percent, and proteins 92 percent. The figures often quoted for the energy value of foods per gram are carbohydrates, 4.1 kilogram calories; fats, 9.3 kilogram calories; and proteins, 4.1 kilogram calories. These figures were originally determined from experimentation with dogs that were fed selected items of diet. The figures, therefore, may not be entirely accurate for man, eating the average mixed diet.

Figures for practical use in human nutrition can be simplified as follows:

Food	Kilogram calories per gram
Carbohydrates	4
Fats	9
Proteins	4

The methods of direct calorimetry can be applied to an animal or man placed in a large chamber designed to measure directly the heat given off by the body. There are several types of rooms designed for this purpose. They range in size from small compartments for the accommodation of experimental animals to rooms large enough to hold a man on a cot, at a worktable, or on an ergometer bicycle. Most of the designs have in common double-insulated walls to prevent loss of heat and copper tubing or pipes through which water circulates. There is a meter to measure the amount of water circulating, as well as appropriate devices to measure the temperature of the water before it enters the chamber and as it leaves. A rise in the temperature of the water indicates that heat has been absorbed by the water in passing through the tubing. An appreciable amount of heat will be given off as heat of vaporization, for there is always some water given off the body as perspiration even though one is resting quietly on a cot. This is the so-called insensible perspiration. It amounts to about one-fourth the heat loss under standard conditions. Vapor can be collected from circulating air and the heat of vaporization determined. Careful control of all sources of energy causes direct calorimetry to be the most accurate means of measuring metabolism, but the equipment is large and expensive. Few laboratories have the space, staff, or funds to operate this type of calorimeter. However, this type of calorimetry has certain advantages: The metabolism of the individual can be measured at the basal level while he is lying quietly on a cot in a postabsorptive state, or he can be doing work such as operating a typewriter or riding an ergometer bicycle. The observations can be carried on for several days if the experiment requires that length of time. Direct calorimetry now is used mostly to check and verify the results arrived at by indirect calorimetry.

INDIRECT CALORIMETRY One of the problems of direct calorimetry involves the determination of the exact value in calories for each of the three kinds of food. Even if a person is fed only carbohydrate food while in the calorimeter, body fats and proteins are utilized also. We have already observed that most of the nitrogen from the breakdown of protein foods is eliminated in the urine. Nitrogen composes about 16 percent of the protein molecule, or 1 part in 6.25. In the person who is fasting or resting in a postabsorptive state, all the nitrogen recovered from the urine will be a by-product of the breakdown of protein. The weight of the urinary nitrogen in grams multiplied by 6.25 gives the weight of the protein utilized. Since the oxidation of protein in a calorimeter produces approximately 4.3 kilogram calories of heat per gram, the weight of the protein multiplied by 4.3 gives the caloric value of the protein utilized. The caloric value of carbohydrates and fats during the experimental period also can be determined for use in indirect calorimetry. A study of the respiratory quotient explains how this can be accomplished.

RESPIRATORY QUOTIENT The fact that there is a definite ratio between the volume of oxygen used and the volume of carbon dioxide given off in the case of carbohydrates and fats affords an opportunity to calculate the caloric value of these foods. The equation for the production of glucose by photosynthesis is reversible. It may be written

$$C_6H_{12}O_6 + 6O_2 + \text{catalysts} \rightleftharpoons 6H_2O + 6CO_2 + \text{energy}$$

In this balanced equation it is evident that just as much oxygen is used as the carbon dioxide given off. Therefore $\dfrac{6 \text{ (volumes of } CO_2)}{6 \text{ (volumes of } O_2)}$ gives a respiratory quotient of 1 for carbohydrate.

The respiratory quotient for fats is less than that required for the oxidation of carbohydrates. The proportion of oxygen to hydrogen and carbon is much smaller in the molecule of fat. This necessitates the utilization of a greater proportion of hydrogen and carbon to produce a greater amount of energy (9 kilogram calories per gram). When fats are oxidized, oxygen combines with the hydrogen to form water, and more oxygen is needed to complete the oxidation. Since a greater volume of oxygen is used than the volume of carbon dioxide eliminated, the respiratory quotient for fats is less than 1, or about 0.7.

The calculation of the respiratory quotient for protein is more difficult because of the varied and complex nature of these food substances. It lies between the figure given for carbohydrate and that given for fats and is usually considered to be 0.8. When, in addition to protein, equal quantities of carbohydrate and fats are being oxidized, as in a mixed diet, the respiratory quotient is approximately 0.85. The respiratory quotient in the postabsorptive state after fasting for 12 hours is 0.82.

The nonprotein respiratory quotient may be expressed in terms of kilogram calories per liter of oxygen consumed. The burning of a given amount of nonprotein food requires the consumption of a certain amount of oxygen. If this ratio is constant, the heat value for any respiratory quotient can be determined by measuring the volume of oxygen consumed. Taking the respiratory quotient at 0.82, as in fasting, we find that 4.825 kilogram calories of heat are produced per liter of oxygen consumed. Protein metabolism can be determined by calculation from the amount of nitrogen in the urine, as we have stated. For the most exact calculations of metabolism, this is done; but for practical considerations, the amount of error is slight, and calculations do not include this phase of protein metabolism. The measurement of oxygen consumption based on the respiratory quotient leads to a consideration of indirect calorimetry rather than the direct measurement of the heat given off by the body.

BASAL METABOLISM   The study of metabolism has led to attempts to establish a standard test of metabolism that would apply to all individuals. Upon first consideration there seems to be a number of factors that would vary considerably from one individual to another. For example, one person might have eaten a heavy meal just before the test and another eaten little or nothing. This variable factor is solved by requiring all persons to fast for 12 to 16 hours before taking the test. They would all be tested, then, while in a postabsorptive state. Ordinarily this can be accomplished by having the person's metabolism tested in the morning before he has eaten breakfast. Fasting permits the food from the last meal to be digested and absorbed and eliminates food energy as a variable factor. In the second place, some persons take more exercise than others; therefore some individuals would produce more heat as a result of muscular activity. This problem can be reduced to a minimum by having the subject lie quietly on a cot for 15 to 30 minutes before submitting to the test. Of course, there is always some muscular activity in the heart, breathing muscles, and other viscera, but this factor does not change from the beginning to the end of the test period. A basal-metabolism test measures the energy of the resting body in terms of the amount of heat given off. It does not represent the lowest energy output, because that occurs during sleep. Suggestions that basal metabolism be called *standard metabolism* have not met with general acceptance.

Various kinds of respiration apparatus have been devised in order to study basal metabolism. Modern recording spirometers in America commonly represent modifications and improvements on the apparatus developed by Benedict and Roth. Essential to all such equipment is a spirometer chamber into which oxygen is released. The older forms have a cylindrical chamber, called the *spirometer bell,* which is water-sealed. Some modern types have substituted a rubber bellows (Figure 18.4). The spirometer chamber is delicately counterbalanced. It rises and falls as the subject breathes. The up-and-down movement is traced on a revolving kymograph drum. The oxygen in the spirometer chamber is gradually withdrawn as the subject breathes. The spirometer slowly falls, and the absorption of oxygen is indicated by the slope of the tracings on the kymograph drum.

The subject lies on a cot with his nostrils closed by a nose clamp. A rubber mouthpiece and inhalation tube enable him to breathe oxygen from the spirometer chamber. The exhaled air passes through an exhalation tube and is then passed through a compartment containing soda lime to absorb carbon dioxide. It also absorbs some of the water. Flutter valves control the movement of the air. The modern apparatus usually contains a circulating fan, which makes breathing easier.

The person taking the test inhales oxygen from the breathing chamber for a recorded length of time, usually 6 to 15 minutes. At the end of the test period the subject's basal metabolism can be calculated either from the volume of oxygen used or from the carbon dioxide exhaled and absorbed by the soda lime. Ordinarily the computation is based on the volume of oxygen used. Basal metabolism, or the *basal metabolic rate* (BMR), is usually calculated in terms of calories per square meter of body surface per hour.

The person who consumes the greater volume of oxygen should have the greater rate of metabolism, but a large person should be expected to utilize more oxygen than a small person. A large individual having a greater

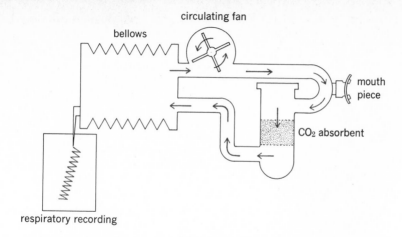

bellows · circulating fan · mouth piece · CO_2 absorbent · respiratory recording

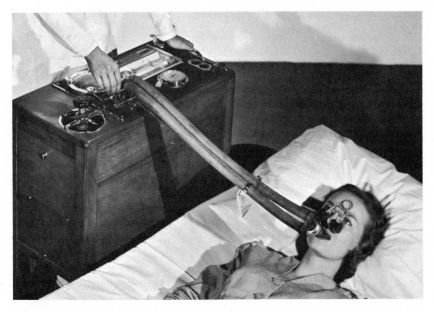

FIGURE 18.4
Basal metabolism apparatus. The diagram indicates the circulation through the apparatus. (*Courtesy of the Sanborn Company.*)

body surface also should give off more heat than a small person, just as a large radiator gives off more heat than a small one. Obviously body size has to be taken into consideration in making a determination of the basal metabolism.

An increase in weight means an increase in the cubic measurement as well as in the surface measurement, but these two do not increase proportionately. It was soon found that an increase in basal metabolism is proportional to an increase in the body surface rather than the weight. The surface of the body can be accurately measured, but for purposes of metabolism it is estimated from a chart showing the surface in square meters in relation to height and weight. A person weighing 137 pounds and standing 5 feet 6 inches tall would have a body surface of 1.7 square meters. Let us suppose that this person took a basal-metabolism test and had an oxygen consumption of 2,400 liters over a test period of 10 minutes. The nonprotein respiratory quotient during fasting is 0.82. The number of kilogram calories per liter of oxygen for this quotient is 4.825. The total heat per hour can be calculated as follows:

$$2.400 \times 4.825 \times \frac{60}{10} = 69.48 \text{ kilogram calories}$$

To find the number of kilogram calories per square meter of body surface per hour, we divide 69.48 by 1.7, which gives 40.87 kilogram calories. The BMR is usually close to 40 kilogram calories per square meter of body surface per hour in young men. It is often expressed on the basis of a 24-hour day, although this is hardly accurate, since one does not remain at the basal level for that length of time. However, it is useful for comparison with the calories needed for a 24-hour day that would include the usual daily work. The calculation on the basis of a 24-hour day for the individual mentioned above who had a heat production of 40.87 kilogram calories per square meter of body surface and 1.7 square meters of body surface is as follows:

$24 \times 40.87 \times 1.7 = 1667.5$ kilogram calories per day at the basal level

Women of eighteen or nineteen years can be expected to average around 37 kilogram calories per square meter per hour. A person 5 feet 2 inches tall and weighing 115 pounds would have 1.5 square meters of body surface. Women of eighteen to nineteen years commonly have basal rates of 1200 to 1400 kilogram calories per 24-hour day.

$24 \times 37 \times 1.5 = 1332$ kilogram calories per day at the basal level

Basal rates are considerably higher in children. The BMR decreases sharply from childhood until around the age of twenty, when it tends to level off. It then decreases gradually with age (Table 18.2).

Physicians commonly express BMR in terms of percentage above or below normal standards of kilogram calories per square meter per hour, worked out for different sex and age groups. According to this system +5 means that the total heat production is 5 percent above the average for individuals of that sex and age group. Since ±15 percent is considered to be a normal limit of deviation, +5 is not significant. The system is open to criticism in that it may not indicate the increase above normal in heat production of the person who has become overweight. The BMR for the overweight person by this method may be considered as normal in relation to body surface, but his actual heat production in kilogram calories per hour may be greatly above normal.

The secretions of some of the endocrine or ductless glands, notably the thyroid, have a marked effect upon metabolism. When the thyroid is overactive, the metabolic rate is commonly high. This subject will be further discussed in Chapter 20.

Foods, as we have seen, are a source of energy; not only do they provide for immediate needs, but they enable the body to conserve its energy resources and store a reserve, as in the formation of stored fat. One of the interesting effects of the utilization of food is a rise in total heat production, which is not accountable to the energy required in the digestion or storage of food. This effect raises the metabolism above the basal level and is called the *specific dynamic action* of food. Carbohydrate and fats raise the metabolism a small percentage above the basal level, but when protein is fed to a fasting individual a very considerable rise in total heat production occurs. The rise is equal to 25 to 30 percent of the energy value of the protein, and

TABLE 18.2

Basal metabolic rate standards, kilocalories per hour per square meter

Age, year	Males Fleisch	Males Robertson and Reid	Males Boothby et al.	Males Aub and Du Bois	Females Fleisch	Females Robertson and Reid	Females Boothby et al.	Females Aub and Du Bois
6	48.3	54.2	53.0		47.0	51.8	50.5	
7	47.3	52.1	52.4		45.4	50.2	48.5	
8	46.3	50.1	51.5		43.6	48.4	46.7	
9	45.2	48.2	49.9		42.8	46.4	46.1	
10	44.0	46.6	48.0		42.5	44.3	45.7	
11	43.0	45.1	47.2		42.0	42.4	45.1	
12	42.5	43.8	46.8		41.3	40.6	43.9	
13	42.3	42.7	46.5		40.3	39.1	42.5	
14	42.1	41.8	46.4		39.2	37.8	41.1	
15	41.8	41.0	46.1		37.9	36.8	39.7	
16	41.4	40.3	45.5	46.0	36.9	36.0	38.6	43.0
17	40.8	39.7	44.4		36.3	35.3	37.6	
18	40.0	39.2	42.9	43.0	35.9	34.9	37.0	40.0
19	39.2	38.8	42.2		35.5	34.5	36.6	
20	38.6	38.4	41.6	41.0	35.3	34.3	36.3	38.0
25	37.5	37.1	40.3		35.2	34.2	36.0	
30	36.8	36.4	39.6	39.5	35.1	34.1	35.8	37.0
35	36.5	35.9	38.9		35.0	33.5	35.7	
40	36.3	35.5	38.3	39.5	34.9	32.6	35.5	36.5
45	36.2	34.1	37.6		34.5	32.2	35.3	
50	35.8	33.8	37.0	38.5	33.9	31.9	34.4	36.0
55	35.4	33.4	36.3		33.3	31.6	33.4	
60	34.9	33.1	35.7	37.5	32.7	31.3	32.8	35.0
65	34.4	32.7			32.2	31.0		
70	33.8			36.5	31.7	30.7		34.0
75	33.2				31.3			
80	33.0			35.5	30.9			33.0

SOURCE: Courtesy of Consolazio, Johnson, and Pecora, "Physiological Measurements of Metabolic Functions in Man," McGraw-Hill Book Company, New York, 1963.

the effect lasts for several hours. Carbohydrate causes a rise in heat production of about 6 percent and fats of about 4 percent when fed separately.

If the total heat production rises 25 percent above the energy value of the protein fed, the body must use its own energy reserve in the utilization of this food substance. Furthermore, if loss of weight is to be prevented, this energy loss has to be taken into consideration. For practical purposes in calculating the total energy requirement, the specific dynamic effect accounts for only 6 to 10 percent of the calories when the nonfasting individual is fed a mixed diet.

The exact source of the heat given off by the specific dynamic action is not entirely understood. It is not due primarily to digestion or to the action of muscles of the digestive tract. So far as proteins are concerned, most of

Metabolism

the effect appears to be in the liver, apparently because of energy released during the deamination of amino acids and in the formation of urea. The heat derived in this way does not contribute to the energy requirements of the tissues, but it does add to the total heat production of the body.

Mental work, strangely enough, requires only a very small outlay of energy. F. G. Benedict and C. G. Benedict found that the extra calories required for 1 hour of intense mental effort could be obtained by eating one oyster cracker or by consuming one-half of a salted peanut.

Muscular work has a pronounced effect in raising the metabolism and therefore the need for more food energy. A man sitting quietly has a total metabolism of about 100 kilogram calories per hour. Moderate exercise can increase the rate of metabolism to around 300 kilogram calories per hour. Heat is produced as a result of muscular activity; some of this heat is used to keep the body warm, and some is lost from the body surface. Muscular exercise on a hot day produces more heat than is needed, and the heat-regulating mechanism attempts to adjust the body to provide greater heat loss through evaporation of sweat, for example. In a very cold environment one may exercise just to keep warm. Muscular activity is the most important single factor affecting the expenditure of energy and thus in determining the amount of food required to maintain something of a balance in metabolism.

Total calories for a 24-hour day   One could not be active at the basal rate of metabolism; neither should a person work continuously for a 24-hour day. Obviously the average 24-hour day is divided into a period for sleeping, the working day, and a rest period. For many the working day may be subdivided into periods of light exercise and periods of moderately heavy muscular work.

Considering the probable food requirement for a man of average size (70 kilograms, or 154 pounds), we form the following estimate:

Activity	Kilogram calories
8 hr sleep at 65 kcal per hour	520
2 hr preparation for the day's work at 150 kcal	300
8 hr moderate work at 200 kcal	1600
2 hr evening chores at 170 kcal	340
4 hr sitting at rest at 100 kcal	400
	3160

Some men in sedentary occupations may need less than 3000 kilogram calories per 24-hour day; many who perform heavy muscular work need more. Women engaged in moderately active work need about 2100 kilogram calories. Growing girls from fifteen to eighteen years require about 2300 kilogram calories. Boys in the same age group require about 3000 kilogram calories (see Table 18.3).

Consuming more food calories than are expended in energy during the day should result in an increase in weight, while working harder and eating less should cause one to lose weight. Even though this is basically true, losing weight by working off fat is usually a difficult way to reduce. Statistics show that a brisk walk for an hour burns only 250 kilogram calories. It is interesting that walking upstairs requires $2\frac{1}{2}$ times more energy than is expended in walking downstairs (see Tables 18.4 and 18.5). After exercise a person usually finds that he has developed an increased appetite and is likely to eat more than his normal amount of food. The best way to reduce is to limit the intake of food calories to a level somewhat below the energy requirement. This will enable one to lose weight slowly, at the rate of a pound or so per week. Severe dieting can be harmful to one's health. Even moderate dieting probably should be undertaken only upon the advice of a physician, for one cannot simply stop eating. The body can supply some reserve fats for fuel, but it cannot supply enough protein, vitamins, and minerals essential to health.

The proper balance between the three kinds of food eaten is desirable whether one is dieting or not. Carbohydrates represent a readily available source of energy and constitute 40 to 50 percent of the diet. It is common for Americans to include a greater than optimum percentage of starch and sugar in their diets. Fats are the best foods for conversion into heat. In addition, they supply certain essential acids (linoleic, linolenic). Various sources recommend that fats compose 35 to 45 percent of the daily diet. Protein is used for the growth and repair of tissues. The normal diet should provide an assortment of protein foods that will contain all the essential amino acids. The recommended allowance for protein in the diet is 10 to 15 percent. Protein foods supply the nitrogen that is essential to the growth and maintenance of protoplasm.

An average diet may be expected to contain the three kinds of food in proper proportion and an adequate supply of vitamins and minerals. A reducing diet should contain the same essentials, but with the amount of carbohydrates and fats reduced. Minerals such as sodium, potassium, and phosphorus are usually present in foods in adequate amounts. Care should be taken that the reducing diet is not deficient in calcium, iron, and iodine. Other considerations include spacing of meals to avoid a feeling of weakness or fatigue and eating enough plant or vegetable material to keep up the bulk of food material along the digestive tract.

WATER BALANCE   Water is not a food, but it plays an important part in the chemical reactions involved in metabolism. Water balance must be considered in relation to any study of a proper diet. The average man has an intake of about 2,500 milliliters of water per day; his water loss is about the same. Only 1,000 to 1,200 milliliters of this amount is acquired by drinking. Food accounts for a similar amount. A potato, lean meat, or a tomato may appear as more or less solid food, but all contain water in considerable quantity. If the diet contains soup and fruits, for example, the necessary intake by drinking may be somewhat less. Finally there is the water of oxidation, produced as a result of metabolic processes through the oxidation of hydrogen in the food or tissues. The water of oxidation amounts to approximately 300 milliliters.

The greatest water loss at average temperature and humidity is through the urine. This loss is 1,400 to 1,500 milliliters per day. About 350 milliliters is lost as vapor in expired air. Insensible perspiration at average room temperature accounts for about 500 milliliters. There is a loss of 150 to 200 milliliters in fecal matter. The proper functioning of the tissues, including the kidneys, enables the body to maintain a normal water balance.

A GENETIC BLOCK IN THE METABOLISM OF PHENYLALANINE   *Phenylketonuria* is a condition resulting from inability of an individual to oxidize excess phenylalanine to tyrosine. Phenylalanine is an essential amino acid and is normally used in the synthesis of proteins. It must be obtained from foods, since the body is unable to synthesize it. Normally, most of the phenylalanine that

TABLE 18.3
Recommended daily dietary allowances,* revised 1968

	Age† Years From Up to	Weight Kg (lbs)		Height cm (in)		K calories	Protein gm	Fat Soluble Vitamins		
								Vitamin A Activity I.U.	Vitamin D I.U.	Vitamin E Activity I.U.
Infants	0-1/6	4	9	55	22	kg x 120	kg x 2.2‡	1500	400	5
	1/6-1/2	7	15	63	25	kg x 110	kg x 2.0‡	1500	400	5
	1/2-1	9	20	72	28	kg x 100	kg x 1.8‡	1500	400	5
Children	1-2	12	26	81	32	1100	25	2000	400	10
	2-3	14	31	91	36	1250	25	2000	400	10
	3-4	16	35	100	39	1400	30	2500	400	10
	4-6	19	42	110	43	1600	30	2500	400	10
	6-8	23	51	121	48	2000	35	3500	400	15
	8-10	28	62	131	52	2200	40	3500	400	15
Males	10-12	35	77	140	55	2500	45	4500	400	20
	12-14	43	95	151	59	2700	50	5000	400	20
	14-18	59	130	170	67	3000	60	5000	400	25
	18-22	67	147	175	69	2800	60	5000	400	30
	22-35	70	154	175	69	2800	65	5000	—	30
	35-55	70	154	173	68	2600	65	5000	—	30
	55-75+	70	154	171	67	2400	65	5000	—	30
Females	10-12	35	77	142	56	2250	50	4500	400	20
	12-14	44	97	154	61	2300	50	5000	400	20
	14-16	52	114	157	62	2400	55	5000	400	25
	16-18	54	119	160	63	2300	55	5000	400	25
	18-22	58	128	163	64	2000	55	5000	400	25
	22-35	58	128	163	64	2000	55	5000	—	25
	35-55	58	128	160	63	1850	55	5000	—	25
	55-75+	58	128	157	62	1700	55	5000	—	25
Pregnancy						+200	65	6000	400	30
Lactation						+1000	75	8000	400	30

*The allowance levels are intended to cover individual variations among most normal persons as they live in the United States under usual environmental stresses. The recommended allowances can be attained with a variety of common foods, providing other nutrients for which human requirements have been less well defined. See text for more detailed discussion of allowances and of nutrients not tabulated.

†Entries on lines for age range 22-35 years represent the reference man and woman at age 22. All other entries represent allowances for the midpoint of the specified age range.

is not used for building protein will be converted to tyrosine, but in individuals suffering from phenylketonuria, a metabolic block prevents this conversion. These individuals are homozygous recessives. The recessive allelic gene is unable to direct the production of the enzyme phenylalanine hydroxylase which is essential for the conversion of phenylalanine to tyrosine.

Under these conditions an abnormal amount of phenylpyruvic acid is produced and is excreted in the urine. If infants are subject to this comparatively rare defect in metabolism, it is imperative that the condition be

Foods, nutrition, and metabolism

Ascorbic Acid mg	Folacin§ mg	Niacin mg equiv.¶	Riboflavin mg	Thiamine mg	Vitamin B_6 mg	Vitamin B_{12} µg	Calcium gm	Phosphorus gm	Iodine µg	Iron mg	Magnesium mg
35	0.05	5	0.4	0.2	0.2	1.0	0.4	0.2	25	6	40
35	0.05	7	0.5	0.4	0.3	1.5	0.5	0.4	40	10	60
35	0.1	8	0.6	0.5	0.4	2.0	0.6	0.5	45	15	70
40	0.1	8	0.6	0.6	0.5	2.0	0.7	0.7	55	15	100
40	0.2	8	0.7	0.6	0.6	2.5	0.8	0.8	60	15	150
40	0.2	9	0.8	0.7	0.7	3	0.8	0.8	70	10	200
40	0.2	11	0.9	0.8	0,9	4	0.8	0.8	80	10	200
40	0.2	13	1.1	1.0	1.0	4	0.9	0.9	100	10	250
40	0.3	15	1.2	1.1	1.2	5	1.0	1.0	110	10	250
40	0.4	17	1.3	1.3	1.4	5	1.2	1.2	125	10	300
45	0.4	18	1.4	1.4	1.6	5	1.4	1.4	135	18	350
55	0.4	20	1.5	1.5	1.8	5	1.4	1.4	150	18	400
60	0.4	18	1.6	1.4	2.0	5	0.8	0.8	140	10	400
60	0.4	18	1.7	1.4	2.0	5	0.8	0.8	140	10	350
60	0.4	17	1.7	1.3	2.0	5	0.8	0.8	125	10	350
60	0.4	14	1.7	1.2	2.0	6	0.8	0.8	110	10	350
40	0.4	15	1.3	1.1	1.4	5	1.2	1.2	110	18	300
45	0.4	15	1.4	1.2	1.6	5	1.3	1.3	115	18	350
50	0.4	16	1.4	1.2	1.8	5	1.3	1.3	120	18	350
50	0.4	15	1.5	1.2	2.0	5	1.3	1.3	115	18	350
55	0.4	13	1.5	1.0	2.0	5	0.8	0.8	100	18	350
55	0.4	13	1.5	1.0	2.0	5	0.8	0.8	100	18	300
55	0.4	13	1.5	1.0	2.0	5	0.8	0.8	90	18	300
55	0.4	13	1.5	1.0	2.0	6	0.8	0.8	80	10	300
60	0.8	15	1.8	+0.1	2.5	8	+0.4	+0.4	125	18	450
60	0.5	20	2.0	+0.5	2.5	6	+0.5	+0.5	150	18	450

‡Assumes protein equivalent to human milk. For proteins not 100 percent utilized factors should be increased proportionately.

§The folacin allowances refer to dietary sources as determined by **Lactobacillus cosei** assay. Pure forms of folacin may be effective in doses less than ¼ of the RDA.

¶Niacin equivalents include dietary sources of the vitamin itself plus 1 mg equivalent for each 60 mg of dietary tryptophan.

detected as soon as possible since serious injury to the brain results from this defect. A simple diaper test consists of putting a drop or two of 10 percent ferric chloride solution on a diaper wet with the infant's urine. If the treated spot retains an orange color this indicates that the urine is normal, but if phenylpyruvic acid is present, a blue-green color appears, indicating that the ferric chloride is reduced by the phenylpyruvic acid.

    This test does not apply to the newborn infant since phenylpyruvic acid would not be excreted until phenylalanine in the blood reaches a high level. The time required for this to take place varies from a few days to a few weeks in infants subject to phenylketonuria.

TABLE 18.4

Metabolic cost of manual-labor activities

Activity	Body wt, kg	kcal/min	kcal/kg/10 min
Metal working	68.1	3.50	0.514
House painting	68.1	3.50	0.514
Carpentry	68.1	3.84	0.564
Farming chores	68.1	3.84	0.564
Plastering walls		4.10	
Truck and automobile repair	68.1	4.17	0.612
Farming, planting, hoeing, raking	68.1	4.67	0.686
Mixing cement		4.70	
Repaving roads	68.1	5.00	0.734
Gardening, weeding	65.0	5.60	0.862
Stacking lumber	68.1	5.83	0.856
Stone, masonry	68.1	6.33	0.930
Pick-and-shovel work	68.1	6.67	0.979
Farming, haying, plowing with horse	68.1	6.67	0.979
Shoveling (miners)		6.80	
Hewing with a pick (miners)		7.00	
Chopping wood	68.1	7.50	1.101
Gardening, digging	63.0	8.60	1.365

SOURCE: Courtesy of Consolazio, Johnson, and Pecora, "Physiological Measurements of Metabolic Functions in Man," McGraw-Hill Book Company, New York, 1963.

A clinical fluorimetric test for the determination of phenylalanine in serum can be performed in a laboratory equipped with a photoflurometer, using proper reagents.

As we have seen, phenylalanine must be obtained from food, but tyrosine may either be obtained from food or formed by oxidation of phenylalanine of the diet. In phenylketonuria, phenylalanine is blocked from being converted to tyrosine, and phenylpyruvic acid accumulates in abnormal amounts. To alleviate this undesirable effect, a special diet is prepared containing a low level of phenylalanine but enough to enable the body to manufacture proteins for growth. On this delicately balanced diet, the child is able to grow and develop normally. The relatively new diet, if used early enough, greatly lessens the danger of brain damage, whereas a normal diet in these cases may very well lead to severe mental defects, even to idiocy.

SUMMARY

There are three kinds of food: carbohydrates, fats, and proteins. Carbohydrates may be classified as monosaccharides, disaccharides, and polysaccharides. Monosaccharides are simple sugars, such as glucose; table sugar, or sucrose, is an example of a disaccharide; polysaccharides are complex carbohydrates, such as starch or cellulose.

Neutral fats result from the chemical combination of glycerol with fatty-acid groups. Examples of compound lipids or lipoid substances are the phospholipids, lecithin and cephalin. Sterols are secondary alcohols. However, they often are combined with fatty acids as esters; cholesterol is an example.

Proteins are used for building new tissue and for tissue repair. The

TABLE 18.5

Metabolic cost of various activities

Activity	Body wt, kg	kcal/min	kcal/kg/10 min
Sleeping	68.1	1.17	0.172
Resting in bed	73.2	1.26	0.174
Sitting, normally	73.2	1.29	0.176
Sitting, reading	73.2	1.29	0.176
Lying, quietly	68.1	1.33	0.195
Sitting, eating	73.2	1.49	0.204
Sitting, playing cards	73.2	1.53	0.210
Standing, normally	73.2	1.50	0.206
Classwork, lecture	68.1	1.67	0.245
Conversing	68.1	1.83	0.269
Personal toilet	73.2	2.02	0.278
Sitting, writing	82.0	2.20	0.268
Standing, light activity	73.2	2.60	0.356
Washing and dressing	68.0	2.60	0.382
Washing and shaving	62.0	2.60	0.419
Driving a car	64.0	2.80	0.438
Washing clothes		3.13	
Walking indoors		3.11	
Shining shoes	73.2	3.20	0.437
Making bed	59.1	3.38	0.572
Dressing	73.2	3.40	0.466
Showering	73.2	3.40	0.466
Driving motorcycle	64.0	3.40	0.531
Cleaning windows	61.0	3.70	0.607
Sweeping floors	73.2	3.91	0.535
Ironing clothes	67.0	4.20	0.627
Mopping floors	73.2	4.86	0.665
Walking downstairs	73.2	7.14	0.976
Walking upstairs	73.2	18.58	2.540

SOURCE: Courtesy of Consolazio, Johnson, and Pecora, "Physiological Measurements of Metabolic Functions in Man," McGraw-Hill Book Company, New York, 1963.

building blocks are amino acids. The protein molecule is usually large and complex, exhibiting almost endless variety.

Vitamins are food substances, present in minute amounts but essential to health. Many have been demonstrated to be coenzymes essential to vital metabolic processes. Vitamins A, D, and E are fat-soluble. Vitamin A is necessary for the formation of retinal and rhodopsin. Vitamin D is concerned with calcium and phosphorus metabolism. A deficiency condition is called rickets.

Thiamine is one of the B vitamins. It plays a part in cellular metabolism. The deficiency condition is beriberi.

Riboflavin is a respiratory coenzyme. It is included in a group of flavoproteins that act as electron acceptors from the coenzymes NAD and NADP. Flavoproteins pass electrons along to the cytochrome system.

A deficiency of niacin is responsible for pellagra. Nicotinamide acts as a coenzyme with NAD and NADP in cellular respiration.

Other vitamins of the B group are pyridoxine (B_6), pantothenic acid, inositol, biotin, folic acid, B_{12}, and choline. Pantothenic acid is the precursor of coenzyme A (see carboxylic acid cycle). Vitamin B_{12} is an erythrocyte maturation factor, effective in the treatment of pernicious anemia.

Ascorbic acid, or vitamin C, prevents the development of scurvy. It is present in fresh fruits and vegetables.

The tocopherols, or vitamin E substances, have been shown to act as antisterility substances for white rats.

Vitamin K stimulates the production of prothrombin by the liver and so increases the clotting ability of the blood.

The energy from the breakdown of foodstuffs is stored as high-energy bonds of the ADP–ATP system. This is largely accomplished by way of the tricarboxylic acid cycle, which is a common pathway in the metabolism of all three kinds of foods.

The amount of heat given off in the utilization of food can be measured by either direct or indirect calorimetry. By either method the approximate energy values are as follows: carbohydrates, 4 kilogram calories per gram; fats, 9 kilogram calories; proteins, 4 kilogram calories.

Basal-metabolism tests attempt to determine the energy of the resting body in a postabsorptive state in terms of the amount of heat given off. Basal metabolic rates for young women are commonly in the range of 1200 to 1400 kilogram calories per 24-hour day. The BMR for young men is more likely to be around 1600 kilogram calories. Calorie needs vary with a number of factors such as age, sex, and weight. The active young man probably needs 3000 to 3200 kilogram calories per 24-hour day, whereas young women, moderately active, probably need only 2300. Boys in the thirteen to eighteen age group need more food and therefore greater caloric intake than the average; older persons need less.

The average American diet consists of carbohydrates, 40 to 50 percent; fats, 35 to 45 percent; and proteins, 10 to 15 percent.

The average intake of water is around 2,500 milliliters per day. About 1,200 milliliters of this amount comes from drinking; the rest is acquired from food and the water of oxidation.

Water loss includes 1,400 to 1,500 milliliters in urine, 350 milliliters as water vapor in expired air, 500 milliliters as perspiration, and 150 to 200 milliliters in fecal matter.

A defect in metabolism resulting from inability to oxidize phenylalanine to tyrosine causes phenylketonuria.

QUESTIONS

1 Which animal produces the most body heat, a mouse or a horse?
2 Which animal produces the most heat per unit of body weight, a mouse or a horse?
3 What is accomplished by the tricarboxylic acid cycle?
4 Of what value is the cytochrome system?
5 Is glucose or a related substance stored by animal organisms? If so where, and in what form?
6 What is the energy source for green plants? What is the energy source for animal organisms?
7 Discuss the chemical relationship of various food carbohydrates with which you are acquainted.
8 What are fats? Name some of the different kinds of lipids. Of what use are fats in metabolism?
9 Discuss the kinds and chemical composition of proteins.
10 Why are minerals necessary in an adequate diet?

Foods, nutrition, and metabolism

11 Write an essay on vitamins or prepare a table indicating the essential facts regarding sources and deficiencies.

12 Prepare a report on vitamins of the B group to be given orally before the class. Consider what questions may be asked and how you will answer them.

13 What is meant by direct calorimetry?

14 Is it direct or indirect calorimetry when a man is placed inside a calorimeter in order to measure the heat given off from the body?

15 Explain indirect calorimetry.

16 What use is made of the respiratory quotient?

17 Discuss basal metabolism.

18 Under what conditions can one determine the basal metabolic rate?

19 What do you understand the term ***specific dynamic action*** to mean?

20 Considering your daily activities, estimate your food requirement in kilogram calories for a 24-hour day.

SUGGESTED READING

BENEDICT, F. G., and C. G. BENEDICT: The Energy Requirements of Intense Mental Effort, *Science*, **71**:567 (1950).

CORRADINO, R. A., and R. H. WASSERMAN: Vitamin D_3: Induction of Calcium-binding Protein in Embryonic Chick Intestine in Vitro, *Science*, **172**:731-733 (1971).

ELVEHJEM, C. A.: Relation of Nicotinic Acid to Pellagra, *Physiol. Rev.*, **20**:249-271 (1940).

Food and Nutrition Board Report: Recommended Dietary Allowances, *Natl. Acad. Sci.—Natl. Res. Council Publ.* 1146, 1964.

FRIEDEN, E.: The Biochemistry of Copper, *Sci. Am.*, **218**:103-114 (1968).

GREGORY, R. A.: Secretory Mechanisms of the Digestive Tract, *Ann. Rev. Physiol.*, **27**:395-414 (1965).

HILL, R. E., and IAN D. SPENSER: Biosynthesis of Vitamin B_6: Incorporation of Three-carbon Units, *Science*, **169**:773-775 (1968).

HOEBEL, B. G.: Feeding: Neural Control of Intake, *Ann. Rev. Physiol.*, **33**:533-559 (1971).

JAENICKE, L.: Vitamin and Coenzyme Function: Vitamin B_{12} and Folic Acid, *Ann. Rev. Biochem.*, **33**:287-312 (1964).

NYQUIST, S. E., F. L. CRANE, and D. J. MORRÉ: Vitamin A: Concentration in the Rat Liver Golgi Apparatus, *Science*, **173**:939-940 (1971).

PATTON, S: Milk, *Sci. Am.*, **221**:59-68 (1969).

WALD, G., and D. STEVEN: An Experiment in Human Vitamin A Deficiency, *Proc. Nat. Acad. Sci. U.S.*, **25**:344-349 (1939).

WHITTAM, R., and K. P. WHEELER: Transport across Cell Membranes, *Ann. Rev. Physiol.*, **32**:21-60 (1970).

YOUNG, V. R., and N. S. SCRIMSHAW: The Physiology of Starvation, *Sci. Am.*, **225**:14-21 (1971).

unit seven

excretion

chapter 19
the kidneys

The kidney's work is not fully appreciated if its function is considered only in relation to the excretion of waste materials. The kidney is, in addition, the chief regulator of the internal environment, since it is largely responsible for maintaining the composition of the blood. These regulatory functions may be summarized under three headings. First, the kidney aids in regulating the acid-base balance of the blood, and indirectly that of the tissues, by excreting excess acid or alkaline substances. Secondly, it helps to maintain the water balance of the tissues by excreting excess water or by conserving water, as the situation demands. It conserves water and many other substances by reducing excretion if there is a reduced intake of liquid or unusual loss as in profuse sweating. Finally, it regulates osmotic equilibrium and ionic balance in the blood and tissues by controlling the excretion of various inorganic substances such as the salts of sodium, potassium, calcium, and magnesium.

The kidneys play an important part in eliminating the by-products of catabolism. Some of these products are toxic and would become highly injurious if allowed to accumulate. Waste products arise in every cell, but through absorption into the bloodstream and continuous excretion by the kidneys they fail to reach injurious concentrations. The constancy of the internal environment is largely the result of this continuous exchange of materials between the blood and the tissues. It depends upon a balance between the processes of anabolism and catabolism and is commonly referred to as resulting in a *dynamic equilibrium*. The organs of excretion play a very important part in helping to maintain such an equilibrium.

EMBRYONIC ORIGIN OF THE URINARY SYSTEM

The urinary system arises in the vertebrate embryo from a crest of mesoderm called the *urogenital ridge*. Looking down on the early embryo, one sees segmentally arranged masses of tissue on either side of the spinal cord. These are *somites*, which give rise to muscles. The nephric region lies just lateral to the somites (Figure 19.1). Laterally, beyond the nephric region, the mesoderm divides into outer and inner layers, which form the walls around the *coelom*, or primitive body cavity. This is most clearly seen in a cross section of an early chick embryo. Structurally, the urinary and reproductive systems are closely related in their method of origin and in the use of common ducts. The latter is well illustrated in the male urethra, which becomes a duct for the passage of urine but which also transmits the spermatozoa and secretions of the reproductive system.

Three regions of nephric tissue eventually become differentiated into three types of kidneys appearing successively in different body areas. These are called the *pronephros, mesonephros,* and *metanephros*. The nephric region from segments 2 to 26 becomes divided into narrow masses of tissue called

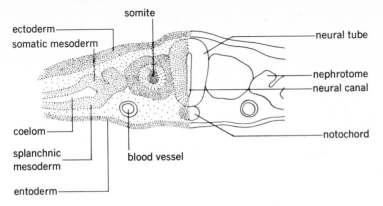

somite

ectoderm

somatic mesoderm

neural tube

nephrotome

neural canal

coelom

splanchnic
mesoderm

blood vessel

notochord

entoderm

FIGURE 19.1

Diagram representing a cross section of a chick embryo (48 hours), showing the position of the nephrotome.

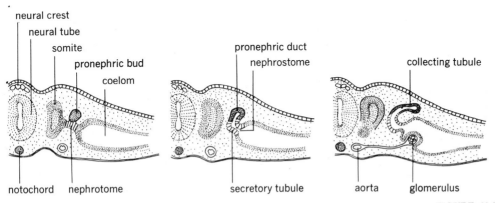

neural crest

neural tube

somite

pronephric bud

coelom

pronephric duct

nephrostome

collecting tubule

notochord   nephrotome

secretory tubule

aorta   glomerulus

FIGURE 19.2

Semidiagrammatic representation of the development of the pronephric tubule in human embryos (transverse sections). (*After Leslie B. Arey, "Developmental Anatomy," W. B. Saunders Company, Philadelphia, 1965.*)

nephrotomes. From each anterior nephrotome a pronephric bud develops into a collecting tubule, which opens into the coelom (Figure 19.2). The pronephros forms so far cephalad (that is, toward the head) that it is sometimes called the *head kidney,* although it is not located in the head. It consists of a few funnel-shaped tubules in the very early human embryo of about 4 millimeters in length. The tubules are transitory and degenerate as the anterior mesonephric tubules begin to appear. The pronephric duct grows posteriad; in segments 14 to 28 it serves as the mesonephric duct.

The mesonephros arises from the same tissue that gives rise to all kidneys. The number of tubules that may be present in the embryonic mesonephros is about 80 for each kidney, but the anterior tubules degenerate as posterior ones develop. In the human embryo of 4 weeks (4 millimeters) the anterior tubules start to degenerate, and the breakdown of tubules progresses posteriorly during the next 4 or 5 weeks. The mesonephros is also a transient kidney in the human embryo but is better developed than the pronephros; there are more tubules, and the end of each tubule surrounds a small capillary network called a *glomerulus.* It is the functional kidney of fish and amphibians, where it is more efficient than the pronephros. The mesonephric duct is retained and becomes a duct of the reproductive system in the male (Figure 19.3).

The metanephros is the bean-shaped functional kidney of human beings and of all amniote animals (reptiles, birds, and mammals). It arises from two sources: the collecting ducts, renal pelvis, and ureter develop as an outgrowth

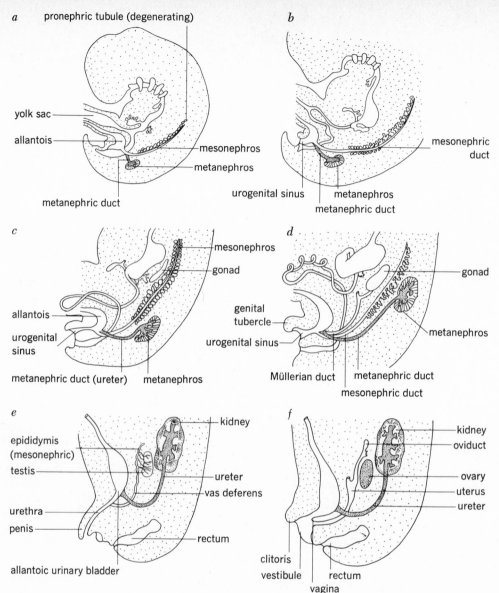

a — pronephric tubule (degenerating)

b

yolk sac

allantois

mesonephros

metanephros

metanephric duct

mesonephric duct

urogenital sinus — metanephros

metanephric duct

c

mesonephros

gonad

allantois

urogenital sinus

metanephric duct (ureter) — metanephros

d

gonad

genital tubercle

urogenital sinus

metanephros

Müllerian duct — metanephric duct

mesonephric duct

e

kidney

epididymis (mesonephric)

testis

ureter

vas deferens

urethra

penis

rectum

allantoic urinary bladder

f

kidney

oviduct

ovary

uterus

ureter

clitoris

vestibule — rectum

vagina

FIGURE 19.3

Diagrams showing the development of mesonephric and metanephric kidneys in the human embryo: *a* early in fifth week as in 5- to 6-millimeter embryos; *b* early in sixth week; *c* seventh week, 14- to 16-millimeter embryo; *d* eighth week, 23- to 25-millimeter embryo; *e* male at about 3 months; *f* female at about 3 months. (*Redrawn from Bradley M. Patten, "Human Embryology," McGraw-Hill Book Company, New York, 1968.*)

of the mesonephric duct; the excretory units arise from the posterior part of the ridge of nephrogenic tissue. The ureter extends from the kidney to the urinary bladder. The metanephric kidney begins to develop while degeneration of the mesonephros is proceeding. At 5 weeks, when the embryo is only 8 millimeters in length, there is evidence of the early formation of the metanephros. The metanephric kidney is large compared with the earlier types and can contain a million tubules. The fetal kidney is externally lobulated; this condition is present at birth and during early infancy. Within the first few years the lobes become indistinct, and the kidney presents a smooth appearance as in the adult.

The human kidney is capable of excretion in the third month of fetal

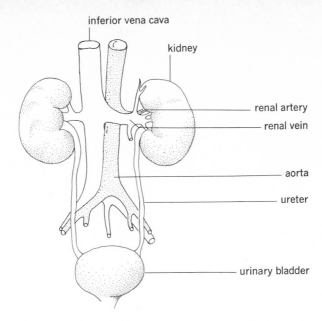

inferior vena cava

kidney

renal artery

renal vein

aorta

ureter

urinary bladder

FIGURE 19.4
The urinary system (anterior view).

life. The urinary bladder fills slowly, and some urine apparently is present in the amniotic fluid (for discussion of fetal membranes, see Chapter 22). The excretory system of the mother is responsible for eliminating the metabolic waste products of the fetus.

KIDNEYS

EXTERNAL APPEARANCE The kidneys lie in the dorsal part of the abdominal cavity, covered by peritoneum and embedded in fat. Each kidney is about 4 inches long and about 2 inches wide. They are somewhat bean-shaped and brown in color. The right kidney, located below the liver, is usually somewhat lower than the left. The concave border of each kidney faces the median plane of the body. The depression near the middle of the concave border is the hilus. Blood vessels and nerves make contact with the kidney at the hilus, and the ureter passes downward to the bladder from this region (Figure 19.4).

INTERNAL STRUCTURE When a fresh kidney is cut in half lengthwise, certain areas and structures may be distinguished. The outer portion is the *cortex*. It is reddish brown and contains renal corpuscles, convoluted tubules, and blood vessels. Masses of cortical tissue fill in between the pyramids of medullary tissue. The medullary tissue is somewhat lighter in color and contains the *renal pyramids* (Figure 19.5). These are wedge-shaped structures containing collecting tubules. The broad base is toward the cortex. The narrow apex forms a soft red papilla, which projects into a small collecting area, the *calyx*. Each papilla contains about 20 small openings through which urine passes into the calyx. The calyx, in turn, empties into a large cavity called the *renal pelvis*. The ureter drains the renal pelvis (Figure 19.6).

Nephron The functional unit of the kidney, the *nephron*, consists of a capsule, a capillary network within the capsule, and the renal tubules. The capsule is called a *glomerular capsule* or *Bowman's capsule*. Within the capsule there

Kidneys

497

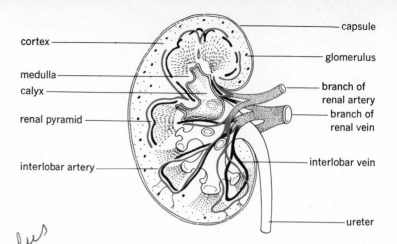

glomerulus

is a microscopic mass of coiled capillaries, the *glomerulus*. The afferent arteriole, which leads blood into the glomerulus, is of slightly greater diameter than the efferent arteriole, which leads the blood away. Renal corpuscles (the capsule plus the glomerulus) are located in the cortical portion of the kidney. A glomerular capsule is equivalent to the greatly enlarged end of the proximal convoluted tubule. As when the finger of a glove is pushed in at the end, there is an inner membrane continuous with the outer membrane of the capsule. The thin inner membrane immediately surrounds the glomerulus (Figure 19.7).

The wall of the proximal convoluted tubule is composed of columnar epithelial cells with peculiar striations on their inner border. This is the so-called "brush border," characteristic of the lining of the tubule, between the capsule and the thin portion. This type of tissue is not present in the thick portion of the distal convoluted tubule.

The proximal convoluted tubule follows an irregular course but eventually becomes straight and forms the loop of Henle (Figure 19.8). The loop of Henle lies deeper in the tissue of the kidney than the glomerular capsule.

FIGURE 19.6

Internal structure of the kidney in longitudinal section.

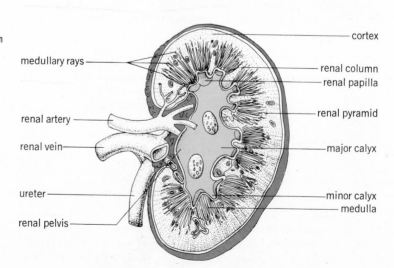

The kidneys

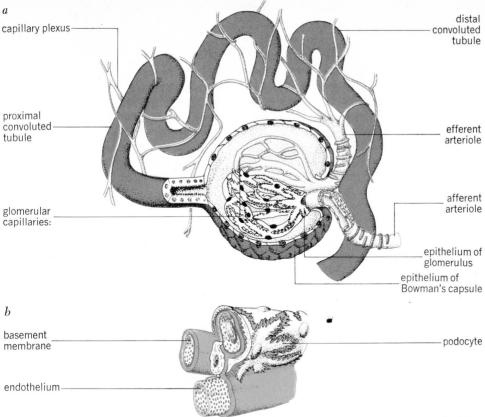

a

capillary plexus

distal convoluted tubule

proximal convoluted tubule

efferent arteriole

glomerular capillaries:

afferent arteriole

epithelium of glomerulus

epithelium of Bowman's capsule

b

basement membrane

podocyte

endothelium

FIGURE 19.7

Renal corpuscle: *a* section through a renal corpuscle; *b* glomerular capillaries greatly enlarged. Podocytes are highly specialized epithelial cells.

The wall of the descending limb abruptly changes from columnar epithelium to flat squamous cells, forming the thin portion of the descending loop. It extends part way up the ascending limb, where columnar cells abruptly reappear and the wall regains its normal thickness. Beyond Henle's loop the ascending tubule follows a tortuous path again and is designated as the *distal convoluted tubule.* It then enters a collecting duct, which receives many such tubules. Approximately 20 collecting ducts are found at the apex of each pyramid, and their contents pass through corresponding openings in the renal papillae and on into the calyces, the renal pelvis, and the ureter.

The proximal tubule and part of the distal tubule are surrounded by a capillary network, which is an extension of the efferent arteriole of the glomerulus. Thus this area is supplied with the same blood that has just passed through the glomerulus. Henle's loop apparently is supplied with capillaries from other efferent arterioles.

The kidneys receive an abundant blood supply through the large but short renal arteries. The renal arteries carry the blood at high pressure from the aorta. Branching readily into smaller arteries, the blood is soon carried by afferent arterioles to the glomeruli (Figure 19.9). Since the diameter of the afferent arteriole is somewhat greater than that of the efferent, the pressure within the glomerulus remains high. The blood circulates at a much lower pressure in the capillary network around the tubules after it has passed through the glomerulus. The volume of blood passing through the glomeruli

Kidneys

499

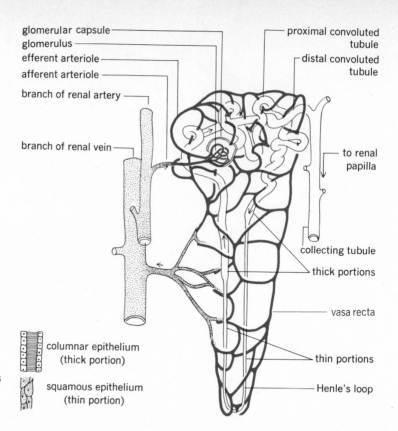

glomerular capsule

glomerulus

efferent arteriole

afferent arteriole

branch of renal artery

branch of renal vein

proximal convoluted tubule

distal convoluted tubule

to renal papilla

collecting tubule

thick portions

vasa recta

thin portions

Henle's loop

columnar epithelium (thick portion)

squamous epithelium (thin portion)

FIGURE 19.8
The nephron unit consisting of a Bowman's capsule, the glomerulus within the capsule, and the tubules.

of both kidneys is around 1,000 milliliters per minute. It varies in the normal adult from 750 to 1,200 milliliters per minute. The total blood volume passes through the kidneys in 4 or 5 minutes. The blood leaves the glomeruli by way of efferent arterioles. These efferent vessels then form a capillary network around the tubules which is called the *vasa recta*. The blood is collected by a number of smaller veins and leaves the kidneys by way of the renal veins, which empty into the inferior vena cava.

The kidneys receive an abundant nerve supply from the autonomic nervous system. The sympathetic nerves reach the kidneys from some of the great plexuses of the abdomen and by way of the splanchnic nerves. Parasympathetic fibers are carried by the vagus nerve. The nerves are largely vasomotor and are concerned with regulating the blood flow and blood pressure within the kidney. The parasympathetic fibers have not been shown to have any effect on the blood flow.

HOW THE KIDNEY FUNCTIONS Each kidney contains approximately 1 million nephrons. The capillary surface available for filtration in 2 million glomeruli is very great. It has been estimated to be over 1 square meter. The volume of fluid filtered per minute in an adult is usually considered to be around 125 milliliters, of which about 99 percent is reabsorbed in the tubules. Stated another way, the glomerular filtration rate is 125 to 130 milliliters per minute.

Many substances in solution in the blood are filtered through the glomerulus and the capsule. The blood proteins are large colloidal molecules

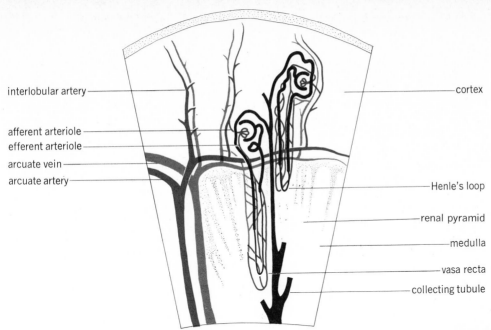

interlobular artery

cortex

afferent arteriole

efferent arteriole

arcuate vein

arcuate artery

Henle's loop

renal pyramid

medulla

vasa recta

collecting tubule

FIGURE 19.9
A renal lobe containing a renal pyramid, nephrons and vascular system.

and do not ordinarily pass into the glomerular filtrate, but crystalloid substances, such as salts and sugars, pass through readily. The glomerular filtrate is actually very similar to protein-free blood plasma or extracellular fluid (see discussion of dialysis in Chapter 2).

If all the valuable materials in the glomerular filtrate were lost in the urine, the organism could not long withstand such depletion. Fortunately, there is very extensive reabsorption into the capillary network surrounding the tubules. As the fluid passes through the tubule, sugar is ordinarily completely reabsorbed, as is most of the sodium and chloride. Some of the urea is reabsorbed and other substances also, in varying amounts. Such extensive reabsorption tends to concentrate excretory products into the urine, which is quite a different substance from glomerular filtrate. Even so, the urine can carry substances of great scientific interest. Vitamins, hormones, and other chemical substances are often recovered from urine, where they can reach a greater concentration than in the blood itself.

In addition to glomerular filtration and active reabsorption in the tubules, a third process involves the ability of the cells in the tubules to pass materials into the urine. The process is commonly called *secretion,* but the term is not appropriate if one applies a strict definition of secretion, that is, the origin within the cell of the substance to be secreted. Other terms used are accumulation and augmentation. Test substances such as dyes have been observed to enter the urine in this way. Creatinine, a nitrogenous compound, will pass through the cells of the tubules when its concentration in the blood is raised experimentally, but this action appears to be limited to human beings and other primates. A group of fish that includes the toadfish, *Opsanus tau,* has no glomeruli; yet their kidneys are able to produce urine. Evidently the tubules in aglomerular fish are able to pass all the materials that make up the normal constituents of urine for fish of this group. This does not prove, of course, that the tubules of glomerular kidneys act in the same way.

COUNTERCURRENT CONCEPT It was thought for many years that a steady concentration of filtrate developed in the tubules from the time the fluid entered the proximal tubule until it reached the excretory duct. More recent investigations indicate that this is not true. Earlier theories required active transport of water and a very high gradient of cellular osmolarity. These two conditions are no longer thought to be valid. (Osmolarity refers to the effective movement by osmosis of an ion in a biological fluid.)

The countercurrent concept involves much more than just the movement of fluid through the nephric tubules. It includes the effect of a high gradient of sodium in the fluid and in the interstitial tissues, the adsorptive function of the blood in the capillary network surrounding the tubules (the vasa recta), and the countercurrent flow in the tubules and capillaries. Sodium ions play a principal role in the osmotic activity of the tubular fluid.

Consideration should be given to the facts that the descending, ascending, and collecting tubules are located close together, that they are embedded in the interstitial tissue of the medullary portion of the kidney, and that the flow is in the opposite direction (a countercurrent) in the three kinds of tubules (Figure 19.10). Recall that there is a sharp turn at Henle's loop between the descending and ascending tubules. Starting with an isotonic filtrate in the proximal convoluted tubule, as the fluid passes through the thin part of the descending tubule, there is a diffusion of sodium ions inward and of water outward, thus increasing the concentration of the fluid. This movement of sodium ions and water is passive. The fluid reaches a maximum concentration at the loop of Henle.

The osmotic activity between the fluid in the ascending tubule and the interstitial tissue is most important. The greater part of the ascending tubule is impermeable to water but possesses a carrier system for sodium ions. As stated earlier, sodium-ion transport plays a principal part in establishing an osmotic gradient. Sodium ions move out from the tubular fluid and into the interstitial fluid by active transport. The filtrate therefore becomes less and less concentrated until it is again isotonic as it enters the collecting ducts.

The interstitial fluid concentration permits sodium ions to enter the descending tubules and in this way to increase the fluid concentration at the loop of Henle. The movement of sodium ions into the descending tubules is passive diffusion.

It is obvious that the high concentration in the interstitial fluid must be maintained if the countercurrent system is to function. The blood circulation through the capillary network (the vasa recta) also acts as a countercurrent. As it flows through the high sodium concentration of the medullary interstitial tissue, water moves toward the area of concentration and sodium moves away, entering the capillaries. The blood flows from the medullary region into the cortical interstitial region, where there is little concentration, and in this case sodium moves out while water moves into the capillaries. In this way, water is removed by the blood, but there is actually little effect on the sodium concentration. Though other ions and other substances are involved in the concentration of the tubular filtrate, we have referred only to the movement of sodium ions and water, because they play the major part and also because this makes for simplification and clarity.

a

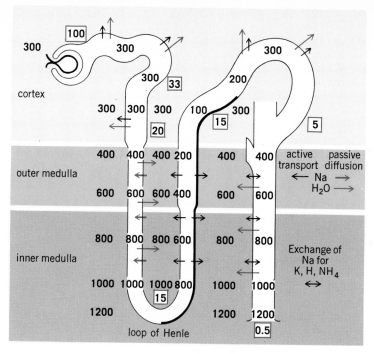

cortex

outer medulla

active passive
transport diffusion
← Na →
H₂O →

inner medulla

Exchange of
Na for
K, H, NH₄
↔

loop of Henle

b

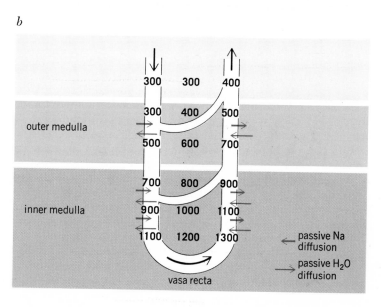

outer medulla

inner medulla

← passive Na
diffusion

→ passive H₂O
diffusion

vasa recta

FIGURE 19.10

a The countercurrent concept of urine concentration. Numbers refer to the concentration of the tubular and interstitial fluid in terms of milliosmols per liter. Boxed numbers indicate the estimated percent of glomerular filtrate remaining within the tubules at various levels. (*From Physiology of the Kidney and Body Fluids, 2d. ed., by R. F. Pitts. Copyright © 1968 by Year Book Medical Publishers, Inc. Used by permission.*); *b* concentration gradients in the capillary network (vasa recta) and interstitial fluid of the medulla. Water diffuses into the blood and is carried away, whereas most of the sodium diffuses out into the interstitial fluid.

KIDNEY ACTION WITH REGARD TO SPECIFIC SUBSTANCES Blood pressure at the glomerulus influences the rate of filtration. The blood pressure there is estimated to be normally about 70 to 75 mm Hg. This is a considerably greater pressure than is found in most capillaries. Capillaries throughout the body commonly function at a blood pressure of 25 to 30 mm Hg. The net

Kidneys

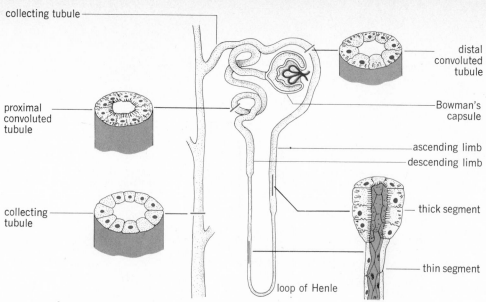

collecting tubule

distal convoluted tubule

proximal convoluted tubule

Bowman's capsule

ascending limb
descending limb

thick segment

collecting tubule

thin segment

loop of Henle

FIGURE 19.11

Uriniferous tubule, showing sections taken at various levels.

effective filtration pressure, however, is greatly reduced by forces opposing it. These forces are mainly the result of the osmotic pressure of the plasma proteins (30 mm Hg), the interstitial-fluid pressure (extracellular fluid) of about 10 mm Hg, and a tubular pressure of about 10 mm Hg. If the opposing forces of approximately 50 mm Hg are added and the sum subtracted from the hydrostatic (blood) pressure of 75 mm Hg at the glomerulus, it is apparent that the effective filtration pressure is around 25 mm Hg. It is this pressure that forces water and substances in solution through the glomerulus (Figures 19.11 and 19.12).

Blood cells are much too large to be filtered through the glomerulus; even the large molecules of the plasma proteins do not pass through ordinarily. The presence of any considerable amount of protein in the urine (proteinuria) is usually considered to be an indication of renal damage, as in certain types of kidney disease. It may accompany congestive heart failure and some conditions of high blood pressure. It is well known, however, that some persons show a functional proteinuria when they assume an erect posture, and a temporary proteinuria is fairly common after severe exercise. Plasma protein in minute amounts, however, is known to pass through the glomerulus; most of it is actively reabsorbed through the wall of the tubule in the normal kidney.

Glucose filters readily through the glomerulus, and normally all of it is reabsorbed in the tubules. But if the blood sugar rises to such a level that not all the glucose filtered by the glomerulus can be reabsorbed, then sugar appears in the urine, a condition known as *glycosuria*. The blood concentration at which glucose begins to appear in the urine is the threshold level for sugar. In diabetes mellitus there is an abundant flow of urine containing glucose. There is also increased thirst. More glucose is present in the blood plasma than can be reabsorbed after filtration. The very fact that the filtrate contains an abnormal amount of sugar in solution tends to raise its osmotic pressure and to decrease the rate of water reabsorption. In other words, more water is used in washing out the increased amount of sugar in the filtrate.

The Kidneys

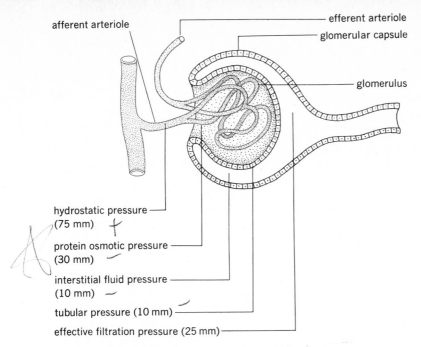

afferent arteriole

efferent arteriole

glomerular capsule

glomerulus

hydrostatic pressure
(75 mm)

protein osmotic pressure
(30 mm)

interstitial fluid pressure
(10 mm)

tubular pressure (10 mm)

effective filtration pressure (25 mm)

FIGURE 19.12

Illustration of effective glomerular filtration pressure (in millimeters of mercury).

Eating salty foods makes one thirsty, and a greater proportion of water is used by the kidneys in the excretion of the excess salt. In another instance, both salt and water are lost by perspiration. Again, there is an increased desire for water to make up the loss; it may be advisable to add salt in extreme cases. It is most interesting that the kidney has the ability to separate salt from water and ordinarily produces urine that has a greater salt concentration than the body fluids.

Drinking large amounts of water or beer produces an abundant flow of urine of low concentration. Low intake of water, or excessive loss by perspiration or hemorrhage, results in the production of a small amount of urine of much higher concentration. The reabsorption of water is accomplished in several ways. Considerable water is reabsorbed as the solvent for substances reabsorbed. For example, glucose is very actively reabsorbed as a solute in water. On the other hand, a considerable increase of dissolved solids in the tubules raises the osmotic pressure and decreases the reabsorption of water. To a certain extent this mechanism enables the kidney to adjust automatically to perform its work, for a greater accumulation of dissolved waste products tends to produce a greater flow of urine. This is the mechanism in diabetes mellitus that produces the diuresis associated with glycosuria; an abnormal amount of sugar that cannot be reabsorbed prevents the normal reabsorption of water.

It can be demonstrated that an antidiuretic hormone (ADH) from the posterior lobe of the pituitary gland exercises a very important control over reabsorption of water in the tubules. Experimental destruction of the posterior lobe or centers in the hypothalamus connected with it produces a very severe diuresis because the reabsorption of water is suppressed. The rare disease called *diabetes insipidus* is caused by the failure of this mechanism. There is an abundant flow of urine of low concentration, accompanied by extreme thirst.

Diuresis means an increase in the excretion of urine; a *diuretic* is an agent that causes the increase. Several salts are known to act as diuretics; caffeine is of special interest as a diuretic, since it or a derivative is present in several of our common beverages. It is thought that the diuresis caused by caffeine is due largely to renal vasodilatation.

The average amount of urine excreted by a person is 1,200 to 1,500 milliliters per day. Its specific gravity is 1.01 to 1.03. The yellow color of urine is due to a pigment called *urochrome*. It is formed from urobilin, which, in turn, is formed from bile pigment. Urine is about 95 percent water and contains inorganic salts and organic wastes.

The inorganic salts include chlorides, sulfates, and phosphates of sodium, potassium, calcium, and magnesium. Of these, sodium chloride is by far the most abundant. Some of the sulfates and phosphates may be in combination with organic substances.

Normal urine is usually slightly acid, with an average pH of approximately 6. Its normal range is around pH 5 to 7. The urine becomes alkaline for a period after a meal, the so-called "alkaline tide." Hydrogen ions are used in the formation of HCl of the gastric juice; for a time during gastric digestion, the kidneys excrete a greater amount of basic materials and thus help the blood to maintain its normal acid-base balance.

Organic wastes include urea, uric acid, and creatinine. *Urea*, $CO(NH_2)_2$, is a product of the deamination of amino acids. The amount varies with the amount of protein in the diet. When urine is allowed to stand, urea breaks down to ammonia and carbon dioxide through the action of bacteria. The enzyme *urease*, which occurs in many bacteria, is responsible for this reaction. *Uric acid* is a breakdown product of nucleoproteins, the compound proteins found especially in the nuclei but also in the cytoplasm of cells. *Creatine*, $C_4H_9N_3O_2$, is present in various tissues but especially in muscle, where it is formed from phosphocreatine during muscle metabolism. (Phosphocreatine breaks down into creatine and phosphoric acid.) It apparently is resynthesized, and the amount excreted as creatinine in the urine remains fairly constant. A rise in the amount of creatine excreted is called *creatinuria*. Fevers, prolonged starvation, diabetes, or other disturbances of carbohydrate metabolism may be expected to create a condition favorable to creatinuria. Actually in healthy individuals very little creatine is excreted as such. It is excreted in the form of its anhydride creatinine, $C_4H_7N_3O$.

RENAL CLEARANCE   Renal physiologists find the study of clearance values of various substances in the blood to be a useful tool. *Clearance* refers to the removal of a substance by the kidneys from a certain volume of blood passing through the glomeruli in 1 minute. The rate of clearance is expressed in terms of the least volume of blood containing a given substance, in relation to the amount of this substance appearing in 1 minute's urine. Clearance varies with the concentration of a given substance in the blood; there is also a corresponding change of concentration in the urine.

When making a test for clearance it is necessary to take samples of both blood and urine. Chemical analysis indicates the concentration of the substance being tested per milliliter of blood and also the concentration of the substance in the urine per minute. If the amount of test substance passing into the urine in 1 minute is divided by the amount found to be present in each milliliter of plasma, the clearance value of the substance can be determined. A formula commonly used is the following:

$$\text{Clearance} = \frac{UV}{P}$$

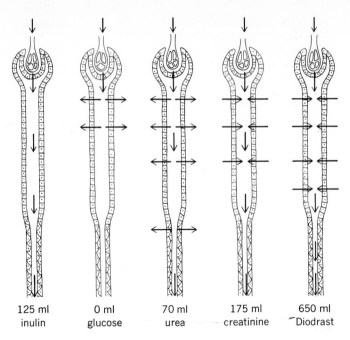

| 125 ml | 0 ml | 70 ml | 175 ml | 650 ml |
| inulin | glucose | urea | creatinine | Diodrast |

clearance value

FIGURE 19.13

Diagram illustrating clearance values for various substances. (*After Houssay.*)

where *U* represents the concentration of the test substance in the urine, *P* represents the concentration of the test substance found in the plasma, and *V* is the volume of urine in milliliters excreted per minute. Normally *V* is about 2 milliliters.

A relatively high clearance value of 75 milliliters for urea is obtained when the rate of urine excretion is greater than 2 milliliters per minute. A common clearance value for urea is 60 milliliters. Some urea is reabsorbed, but selection of a test substance such as inulin, which is not reabsorbed, provides a clear indication of the glomerular filtration rate.

Renal clearance becomes most valuable when the rates of clearance of several filerable substances are compared. The clearance value of glucose is zero, because all the glucose that is filtered through the glomeruli is reabsorbed through the tubules and returned to the bloodstream. Glucose, therefore, is not normally present in the urine. The clearance value for urea is low, since a considerable amount is reabsorbed (Figure 19.13).

The polysaccharide *inulin* is an interesting and valuable test substance. It resembles starch and has to be injected rather than given by mouth, because digestive enzymes convert it by hydrolysis into the sugar fructose. Inulin filters through the glomeruli readily but, because of the size and shape of the molecule, it does not pass through the walls of the tubules by diffusion. The clearance value of inulin is around 125 milliliters. This is also the glomerular filtration rate, since inulin is neither reabsorbed nor diffused into the urine through the wall of the tubule. The glomerular filtration rate of inulin indicates the filtering capacity of the kidneys. A similar substance, *mannitol*, may be used in essentially the same way.

For purposes of comparison, if the ratio of clearance of a given substance is less than the clearance rate for inulin, some of the given substance is evidently being reabsorbed. Thus, if the clearance rate for urea is 75 milliliters and the clearance rate for inulin is 125 milliliters, there is a clear

Kidneys

507

indication that some of the urea is reabsorbed. If the clearance rate of a given substance is greater than the clearance rate for inulin, it may be inferred that the given substance is not only being filtered through the glomeruli but is being augmented by passing into the urine through the walls of the tubules.

An iodine-containing compound, iodopyracet (Diodrast), has a high clearance value, some 650 milliliters per minute. It is not only filtered through the glomeruli, but passes actively through the wall of the tubule into the filtrate as well. The clearance value can be used also to calculate the maximum or total blood flow through the kidneys. Iodopyracet or a similar substance is used in the visual examination of the kidneys and urinary ducts by x-ray. Since its clearance value is much higher than that of inulin, the passage of this substance through the walls of the tubules into the urine must be very great. Much can be learned about the functioning of the tubules in this way. More recently another preparation, *p*-aminohippuric acid (PAH), has come into use. Its passage through the walls of the tubules is similar to that of iodopyracet.

HYPERTENSION OF RENAL ORIGIN The kidneys are becoming recognized as the source of a number of substances that are capable of affecting the vascular system and producing a condition of hypertension. The best known of these substances is *renin* (not to be confused with the digestive enzyme rennin). Renin (pronounced reenin) acts only indirectly through substances in the blood to raise the blood pressure. Under conditions of stress or disease, the kidneys release renin into the bloodstream. Renin, a proteinase, liberates an active substance called *angiotensin* from its inactive precursor, *angiotensinogen*. Angiotensin is a vasoconstrictor substance and causes peripheral vasoconstriction as well as vasoconstriction of arterioles in the great capillary bed, with a consequent rise in blood pressure. The enzyme *angiotensinase* inactivates angiotensin and therefore tends to reduce and regulate the blood pressure.

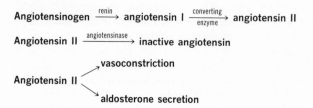

Angiotensinogen $\xrightarrow{\text{renin}}$ angiotensin I $\xrightarrow[\text{enzyme}]{\text{converting}}$ angiotensin II

Angiotensin II $\xrightarrow{\text{angiotensinase}}$ inactive angiotensin

Angiotensin II $<$ vasoconstriction / aldosterone secretion

Angiotensin also stimulates the adrenal cortex to secrete the hormone *aldosterone*. Aldosterone causes increased absorption of sodium by the renal tubules. The kidneys, then, decrease their excretion of salt and water, and body fluids build up. As the blood volume increases, the blood pressure rises. A better understanding of these mechanisms should lead to a more thorough knowledge of the physiological conditions that result in hypertension of renal origin.

KIDNEY DISEASE Like any other organ, the kidney is subject to disease. *Nephritis* is a form of bacterial disease in which the glomeruli are inflamed and their function is impaired. Some substances such as urea are not properly filtered and reach a high concentration in the blood. Proteins, which are not normally found in the urine, then pass through the glomeruli. Severe loss of protein from the blood plasma reduces the osmotic pressure of the plasma and its ability to hold water within the capillaries. Water is poorly excreted and tends to collect in the tissues, a condition called edema. The condition is especially noticeable in the swelling of the ankles. The pressure of the thumb on such tissues leaves a depression that persists for a much longer time than it would in normal tissue.

Abnormal conditions can cause some of the mineral salts such as calcium phosphate to precipitate from the urine and form the basis of renal calculi (or kidney stones) in the tubules or ducts. If such stones obstruct the flow of urine, a serious and painful condition arises. The stones can be removed by surgical procedure.

An artificial apparatus is far from perfect, but when kidney tissue can no longer remove toxic wastes, an artificial kidney is invaluable for maintaining life. The artificial kidney is essentially a dialyzing device which permits substances in solution to pass out of the blood into the dialyzing fluid. The blood can rid itself of waste products and at the same time take up nutritive substances, such as glucose, if necessary. Ions or ionizable substances in the fluid have nearly the same concentration as those in the blood plasma and can interchange freely.

The difficulties encountered in using an artificial kidney include preventing the blood from clotting and the fact that a great amount of blood must be used. The blood flows from a large artery, usually in the arm, through the dialyzer and back into a large vein in the arm or leg. The artificial kidney cannot be used constantly or even every day; its use is limited to a few hours at a time.

URETERS AND URINARY BLADDER

Each kidney is drained by a tube that extends from the kidney to the urinary bladder. These tubes are the ureters. They pass anteriorly near the psoas major muscles and, descending into the pelvis, enter the bladder obliquely on the posterior side. The wall of the ureter is composed of three coats: an outer connective-tissue layer, a muscular coat, and an inner mucous membrane. The smooth muscular layer propels the urine into the bladder by peristaltic movements (Figure 19.4).

The urinary bladder is a muscular structure that varies in size according to the amount of urine it contains. It is capable of holding about 500 milliliters of urine in an adult without overdistention. When full, it is nearly spherical in shape and rises into the abdominal cavity. It is located posteriorly to the pubic symphysis when empty but can rise well above the level of the pubic bones (Figure 19.14).

The muscular coat is composed of three layers and makes up the greater portion of the wall of the urinary bladder. There are an inner and an outer layer of longitudinal muscles separated by a median layer of circular muscles. The internal mucous membrane is a soft reddish lining, which is thrown into folds when the bladder is empty.

A transitional type of stratified epithelium lines most of the urinary tract. It is especially evident in the urinary bladder, which undergoes considerable expansion and contraction. The tissue itself then undergoes a change in appearance. When the tissue is contracted, the cells consist of several layers and are more or less rounded or cuboidal in shape. In the extended tissue, the cells of the deeper layers appear compressed, but the cells of the

Ureters and urinary bladder

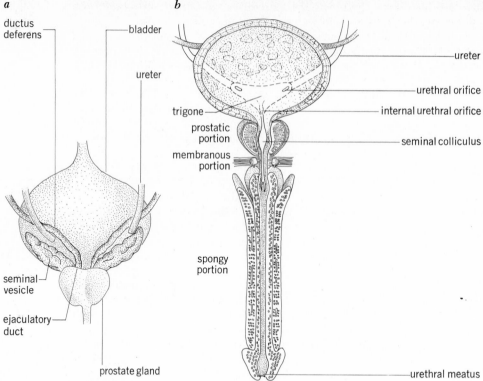

a   *b*

ductus deferens

bladder

ureter

trigone

prostatic portion

membranous portion

spongy portion

seminal vesicle

ejaculatory duct

prostate gland

ureter

urethral orifice

internal urethral orifice

seminal colliculus

urethral meatus

FIGURE 19.14
Urinary excretory pathways: *a* bladder and ureters (male, posterior view); *b* bladder and urethra (male, frontal section).

surface layer become elongated and flattened, resembling stratified squamous epithelium. Some histologists do not regard transitional tissue as a true or valid type of epithelium (Figure 4.1).

The ureters enter the urinary bladder through narrow slits that remain closed except when a peristaltic muscular movement forces urine through the openings. The internal urethral orifice lies in an anterior and medial position to the openings of the ureters, forming a triangular area known as the *trigone*.

The *urethra* is the duct that leads from the urinary bladder to the exterior. It is considerably longer in the male, since it traverses the penis; it can reach a length of 20 centimeters. As it leaves the bladder, it extends through the prostate gland for about 3 centimeters. The internal orifice at the urinary bladder is surrounded by circular smooth-muscle fibers forming the internal vesicle sphincter. The external urethral orifice is a vertical slit at the distal end of the penis.

The female urethra extends from the internal orifice in the urinary bladder along the anterior wall of the vagina to the external orifice located between the clitoris and the vaginal opening. It is about 4 centimeters long.

MICTURITION The act of voiding urine, called *micturition*, is an autonomic reflex act upon which voluntary control is superimposed. The baby voids urine reflexly whenever the pressure of accumulating urine in the bladder becomes great enough to provide an adequate nerve stimulus. Gradually the young child learns to control these impulses, at least when he is not asleep. Control over micturition even during sleep is usually achieved by the three-

The kidneys

year-old child. Inability to establish such control at night is called *nocturnal enuresis,* or *bed-wetting.*

URINALYSIS A routine urinalysis usually consists of a test for specific gravity and chemical tests for the presence of glucose and protein. Just as analysis of ashes tells a great deal about what was burned in a furnace, so a chemical and microscopic examination of urine may afford considerable information concerning the state of metabolism in both normal and pathological conditions.

A simple test indicates acidity of the urine. Microscopic examination may reveal blood cells, epithelial cells, pus cells, casts, calculi, bacteria, or parasites. Chemical analysis indicates the presence of bile pigments, bile salts, hemoglobin, or albumin. Traces of indican (a potassium salt) and ketone bodies may be present in normal urine, but if they are found in greater than trace amounts their presence indicates an abnormal state of metabolism.

OTHER EXCRETORY ORGANS

It is well to recall that there are other organs of excretion, such as the lungs, the skin, and the alimentary tract, including the liver. These organs are discussed under their proper headings.

WATER LOSS THROUGH LUNGS AND SKIN The expired air is nearly saturated with moisture. If the breath comes in contact with a cold surface such as a mirror, the water condenses and becomes visible. The moisture in the breath also becomes visible by condensation in very cold air. The amount of water eliminated in the breath averages 1 pint daily.

Perspiration, or sweat, produced by the sweat glands of the skin, is 98 to 99 percent water. Though it contains essentially the same inorganic constituents as the blood, only traces of these substances are present. Sodium chloride is excreted in significant amounts, especially under conditions of copious and continued sweating. Minute amounts of carbon dioxide and nitrogenous wastes are excreted by way of the sweat glands. The nitrogenous compounds, such as urea, uric acid, and creatine, are also typical constituents of urine. Urea may be excreted in somewhat greater amounts during excessive sweating or when the kidneys are not functioning properly.

The function of the sweat glands in the control of body temperature has been discussed. Their excretory function appears to be less important. Although sweat glands provide considerable moisture in the axillae of the arms, the palms of the hands, and the soles of the feet, there is evidence that much of the normal moisture of the skin is due to movement of water through the skin by osmosis and vaporization rather than perspiration from sweat glands. Such movement of moisture is retarded by the surface layer of the skin (Table 19.1).

SUMMARY

The urinary system arises from the urogenital ridge in the vertebrate embryo. Three different types of kidneys appear successively during embryonic development. They are pronephros, mesonephros, and metanephros kidneys. The metanephric type is the functional postembryonic kidney of human beings and all amniote animals.

The kidneys are brown, bean-shaped organs located in the dorsal part

511

TABLE 19.1

Summary of excretory organs

Organ	Substances excreted
Kidney	**In urine:** Organic wastes: urea, uric acid, creatinine, ammonia Inorganic salts: chlorides (NaCl), phosphates, and sulfates of sodium, calcium, potassium, and magnesium Water; toxic substances; breakdown products of sex hormones
Skin (sweat glands)	**In sweat:** Water Sodium chloride Minute amounts of CO_2 and nitrogenous wastes (urea)
Lungs	Carbon dioxide Water
Intestine, large	Solid wastes of alimentary tract Water Mucus Bile pigments, urobilin Microorganisms Amines, skatole, indole Gases, carbon dioxide, ammonia, hydrogen sulfide, hydrogen, and methane

of the abdominal cavity. They are embedded in fat, covered by the peritoneum, and supplied by the renal arteries.

The internal structure of the kidney exhibits a cortical and a medullary portion. The cortex contains renal corpuscles, convoluted tubules, and blood vessels. The medullary portion consists of wedge-shaped renal pyramids containing collecting tubules. The apex of the papilla projects into a collecting area called the calyx. The calyx empties into a larger cavity, the renal pelvis, which is drained by the ureter.

The nephron is the functional unit of the kidney. It consists of a glomerular capsule, a glomerulus, and a convoluted tubule leading away from the capsule. The proximal convoluted tubule forms the loop of Henle. The distal convoluted tubule is the portion extending beyond the loop of Henle to the collecting duct. Collecting ducts empty at the base of each renal pyramid.

The tubules and the loop of Henle are surrounded by a capillary network, which functions to reabsorb materials from the tubules. Many substances, such as glucose, filter through the glomerulus but are reabsorbed by the capillary network surrounding the tubules.

The renal clearance value for glucose is zero, because normally all the glucose filtered through the glomeruli is reabsorbed. Since a considerable amount of urea is reabsorbed, it has a low clearance value of around 75. The polysaccharide inulin is of interest because it is not reabsorbed or secreted through the wall of the tubule. Its clearance value of 125 milliliters, therefore, represents the glomerular filtration rate. It also indicates the filtering capacity of the kidneys. Substances such as iodopyracet (Diodrast) or *p*-aminohippuric acid (PAH) have a high clearance value, because they are not only filtered through the glomeruli but are secreted through the wall of the tubule as well.

The blood pressure at the glomerulus is normally around 75 mm Hg. Forces opposing this pressure reduce the effective filtration pressure to about 25 mm Hg. Plasma proteins do not ordinarily appear in the urine. If a condition of proteinuria exists, it probably is an indication of some sort of renal damage, as in kidney disease.

Diuresis refers to an increase in the amount of urine produced. The average amount is 1,200 to 1,500 milliliters per 24-hour day.

Hypertension of renal origin can be traced to the action of renin upon renin substrate, one of the alpha globulins of the blood. Inactive angiotensinogen is formed. A converting enzyme changes angiotensinogen to the active form, angiotensin. This substance causes constriction of arterioles in the capillary beds, resulting in a rise in the blood pressure.

The tubes extending from the kidneys to the urinary bladder are the ureters. The urethra is the duct leading from the bladder to the exterior.

Micturition is the act of voiding urine.

The kidneys are not the only excretory organs. The lungs, skin, and alimentary tract also function as organs of excretion.

1 How does the kidney help to regulate the internal environment?
2 Describe the different kinds of kidneys found in the embryo.
3 Describe the internal structure of the kidney.
4 Explain how the kidney functions.
5 What is the meaning of renal clearance?
6 Why is there a glycosuria in diabetes?
7 Discuss the composition of urine.
8 Trace the flow of urine from the kidney to the external urethral orifice.
9 What is angiotensin?
10 Name some organs other than the kidneys that have excretory functions.

SUGGESTED READING

ANDERSSON, B.: Thirst- and Brain Control of Water Balance, *Am. Scientist*, **59**:408–415 (1971).

BRUNNER, HANS R., et al: Hypertension of Renal Origin: Evidence of Two Different Mechanisms, *Science*, **174**:1344–1346 (1971).

GANTEN, D., et al.: Angiotensin-forming Enzyme in Brain Tissue, *Science*, **173**:64–65 (1971).

MARX, S. J., C. J. WOODARD, and G. D. AUERBACH: Calcitonin Receptors of Kidney and Bone, *Science*, **178**:999–1000 (1972).

MERRILL, J. P.: The Artificial Kidney, *Sci. Am.*, **205**:56–64 (1961).

ORLOFF, J., and M. BURG: Kidney, *Ann. Rev. Physiol.*, **33**:83–130 (1971).

ROBERTSON, A. L., Jr., and P. A. KHAIRALLAH: Angiotensin II: Rapid Localization in Nuclei of Smooth and Cardiac Muscle, *Science*, **172**:1138–1139 (1971).

RYAN, J. W., U. SMITH, and R. S. NIEMEYER: Angiotensin I: Metabolism by Plasma Membrane of Lung, *Science*, **176**:64–66 (1972).

unit eight

glands of internal secretion

chapter 20
the endocrine system

Most of the glands we have discussed have been exocrine glands. The salivary glands, for example, are exocrine glands in the sense that their secretions are carried to definite areas by means of ducts. In general, the digestive glands, sweat glands, and mammary glands are exocrine glands, or glands of external secretion.

The endocrine glands, or glands of internal secretion, have no ducts; their products are absorbed and carried all over the body by the bloodstream. The primary effect is often produced in some organ of the body remote from the site of origin of the secretion. The secretions are produced in minute amounts and are constantly broken down and excreted so that large concentrations do not occur. Glands such as the thyroid, pituitary, and adrenals are called *ductless glands, endocrine glands,* or *glands of internal secretion.* Some glands such as the pancreas, testes, and ovaries produce both exocrine and endocrine products (Figure 20.1).

The secretions of glands of the endocrine system are called *hormones.* They act as chemical regulators for many important functions. Hormones are organic substances, but they do not belong to any one class of chemicals; some are proteins, but others are steroids or amines. Even though a hormone may be produced in minute amounts, a deficiency or an overproduction can cause striking changes in the physiology and morphology of an individual. The hormones of invertebrate animals are of a different nature from those found in the vertebrates; vertebrate hormones have little, if any, effect on invertebrate animals.

METHODS OF INVESTIGATION   The methods of study of glands employed by research workers have become varied and complicated. Ideally the investigator should be able to demonstrate that the hormone is present in the blood leaving the gland; that when the gland is removed, certain changes occur; and that administration of the hormone restores the animal to normal. Finally, the active chemical principle should be isolated in pure form. Some of the hormones have not as yet been so isolated, and not all meet these conditions in every respect. Present methods of investigation are largely biochemical.

FUNCTIONS OF HORMONES

Hormones in general are the chemical regulators of enzyme systems. Many hormones have rather specific functions, such as the role of insulin in carbohydrate metabolism, or the action of secretin in stimulating the pancreas to secrete its thin alkaline fluid. The hormone of the thyroid gland accelerates oxidative enzyme systems. Many hormones play an important part in the regulation of growth and metabolism. The part played by the sex hormones in the development and maintenance of the reproductive system is well

The endocrine system

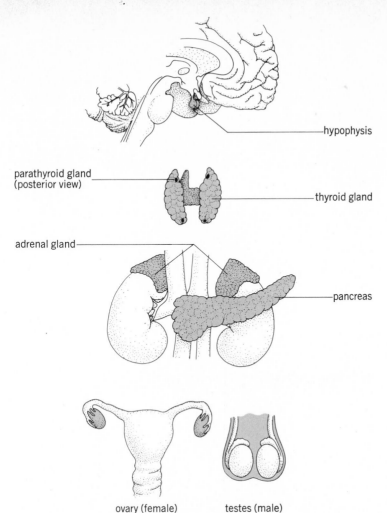

hypophysis

parathyroid gland
(posterior view)

thyroid gland

adrenal gland

pancreas

ovary (female)   testes (male)

FIGURE 20.1
Various glands performing endocrine
functions.

known. Some of the effects of hormones are less obvious, but they contribute
indirectly to personality, intellectual attainment, and physical ability.

Research workers have opened up many new fields of investigation.
There has been a search for hormone receptors and receptor sites on target
organs—for example, receptor sites on various tissues capable of binding
insulin. Releasing factors for major hormones have been extensively studied
and identified. In the forefront of investigation is the role of cyclic AMP
as an intracellular mediator for many hormones. Adenosine 3,5-cyclic monophosphate (cyclic AMP) is now well recognized as the intracellular "second
messenger" that mediates the hormonal effects within the cell; in other
words, these effects are not brought about directly by the original hormone.
Apparently first affecting the membrane of the cell and probably combining
with a receptor, it stimulates changes within the cells of the target organ
which ordinarily would be associated with the action of that particular
hormone. As we have noted before, the enzyme adenyl cyclase converts ATP
to cyclic AMP.

A second cylic nucleotide, guanosine 3′5′-monophosphate (cyclic

Functions of hormones

517

TABLE 20.1
Hormones of the hypophysis

Hormone	Target organ	Functional effects
Adenohypophysis		
Growth hormone (GH) (somatotropic)	Various tissues, skeletal, muscle	Growth of bones and other tissues
Thyrotropic hormone (TSH)	Thyroid gland	Stimulates the secretion of thyroxine
Adrenocorticotropic hormone (ACTH)	Adrenal cortex	Secretion of glucocorticoids
Gonadotropic hormones:		
In male:		
Follicle stimulating hormone (FSH) plus luteinizing hormone (LH)	Seminiferous tubules	Production of sperm
Luteinizing hormone	Interstitial cells of testes	Secretion of testosterone
In female:		
Follicle-stimulating hormone (FSH)	Follicles of ovaries	Maturation of ovarian follicles
Luteinizing hormone	Interstitial cells of ovaries	Stimulates secretion of estrogen and progesterone Formation of corpus luteum
Lactogenic hormone (prolactin)	Mammary glands	Production of milk in activated glands
Intermediate part (pars intermedia)		
Melanocyte-stimulating hormone (MSH)	Melanophores	Darkening of skin by pigmentation
Neurohypophysis		
Oxytocin	Uterus	Stimulates contraction
Vasopressin: antidiuretic hormone (ADH)	Kidney	Retention and reabsorption of water

GMP) may prove to be another intracellular regulatory agent, possibly antagonistic to cyclic AMP in bidirectionally controlled systems such as muscular contraction and relaxation or involved in a response to the action of certain hormones. The enzyme guanylate cyclase catalyzes the formation of cyclic GMP.

ENDOCRINE BALANCE Hormones are thought to accelerate or depress the activity of certain enzyme systems but not to initiate any action that is strictly their own. Probably no hormone or endocrine gland within the body is able to carry on its functions adequately without being affected by the secretions

TABLE 20.2
Survey of principal hormones

Gland	Hormone	Functional effects
Hypophysis (see Table 20.1)		
Thyroid	Thyroxine, T_4, and triiodothyronine, T_3	Accelerates metabolism; promotes normal growth
	Thyrocalcitonin	Regulates serum-calcium level
Parathyroid	Parathormone	Raises blood-calcium levels by stimulating calcium re-absorption from bone
Adrenal glands (see Table 20.3)		
Pancreas	Insulin	Lowers blood-sugar level; part of the sugar is stored as glycogen in liver and muscles
	Glucagon	Causes a rise in blood sugar by accelerating the conversion of liver glycogen to glucose
Kidneys	Erythropoietin	Stimulates the production of erythroblasts
Gonads Female: ovaries	Estrogen	Development of female reproductive system and secondary sexual characteristics
	Progesterone	Produced by the corpus luteum, it is concerned with the maintenance of the uterine lining, preparing it for implantation. Also development and maintenance of the placenta; development of mammary alveoli
Male: testes	Testosterone	Development of male reproductive system and secondary sexual characteristics
Hormones of the digestive tract (see Table 17.3)		

of other glands. There is a delicate balance between the endocrine glands that, in a normal individual, provides for normal growth and metabolism. Conversely, if one of the glands is thrown out of balance by disease, the other glands may be more or less affected.

While some glands are directly interdependent, others appear to be only remotely so. It would be helpful if all the glands could be discussed at once in order to stress these relationships, but for the sake of clarity it is better to consider them one at a time. Among the glands to be discussed are the thyroid gland, the parathyroid glands, the adrenal glands, and the pituitary gland or hypophysis. The thyroid and parathyroid glands develop in the neck; the adrenal glands are located on top of the kidneys; and the hypophysis is located below the forebrain (see Tables 20.1 and 20.2).

The hypophysis, or pituitary gland, has been called the *master gland* of all the endocrines because, through its tropic hormones, it exerts a regulatory effect over the activity of other endocrine glands. Actually there is an interdependence or endocrine balance between all the glands of internal

HYPOPHYSIS

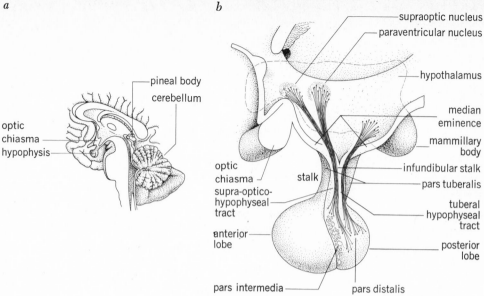

a

b

supraoptic nucleus

paraventricular nucleus

hypothalamus

pineal body

cerebellum

median eminence

mammillary body

optic chiasma

hypophysis

optic chiasma

stalk

supra-optico-hypophyseal tract

anterior lobe

infundibular stalk

pars tuberalis

tuberal hypophyseal tract

posterior lobe

pars intermedia

pars distalis

secretion. The use of the word *pituitary* is somewhat unfortunate, since it originally referred to a mucous secretion or phlegm. *Hypophysis* refers to the location and development of the gland under the brain. Both terms are in common use, but hypophysis is favored in medical and technical research.

The hypophysis is a gland about 1 centimeter in diameter along its anterior-posterior axis. It lies well protected in a depression of the sphenoid bone called the *sella turcica*. The most prominent parts of the gland are the anterior lobe, the posterior lobe, and the stalk, or infundibulum. There is also an intermediate lobe, which varies greatly in different species. The anterior and posterior lobes have different embryonic origins. The *anterior lobe (pars distalis)* arises as a diverticulum of Rathke's pouch, which in the embryo is a backward and upward extension of the oral epithelium. The ventral portion of the pouch thickens and later gives rise to glandular cells, which can be differentiated by appropriate stains. The anterior lobe formed in this way is also called the *adenohypophysis*, meaning glandular portion of the hypophysis (Figure 20.2).

There is a close association between the regulatory functions of the endocrine system and those of the central nervous system. This relationship is most clearly developed in the close association between the hypothalamus and the hypophysis. Small blood vessels in the lower portion of the hypothalamus join and extend down the stalk of the hypophysis (infundibulum), forming the hypothalamic-hypophyseal portal blood vessels supplying the adenohypophysis. The hypothalamus is the source of hypophyseal hormone—*releasing factors*, which are considered to be *neurohormones* rather than neurohumors. These neurohormones are released into the bloodstream of the hypothalamic-hypophyseal portal vessels, whereas neurohumors are released at nerve endings, where they activate a succeeding nerve-cell body. There is an individual releasing factor for each hormone, with possibly one or two exceptions. So, for example, we may speak of growth hormone–releasing factor or thyrotropin-releasing factor, and so on. The hypothalamus also produces release-inhibiting hormones. Two that are known are prolactin release–inhibiting hormone and melanocyte-stimulating hormone (MSH)–release inhibitor.

ANTERIOR LOBE, OR ADENOHYPOPHYSIS The epithelial tissue of the anterior lobe of the hypophysis contains at least three kinds of glandular cells. One type, the *chromophobe cells*, consists of small cells with nongranular cytoplasm and stains very poorly. There are two types of larger cells that take stains readily; they are called *chromophil cells*. Chromophil cells that show an affinity for acid stains are referred to as *acidophils*; those that stain with basic dyes are *basophils*. Investigators who have attempted to identify one type of cell as the source of one specific hormone have reached no definite conclusions. Chromophobe cells appear to be immature and may be simply precursors of chromophil cells.

The anterior lobe produces six hormones that are well verified (see Table 20.1). These are as follows:

A growth hormone
Two gonadotropic hormones: FSH and LH
An adrenocorticotropic hormone
A thyrotropic hormone
A lactogenic hormone

tropic hormone

A tropic hormone is one that influences another gland or organ. The part affected is often referred to as the *target gland* or *organ*. These hormones are all proteins and are inactivated by digestion if taken by mouth. They must therefore be carried by the blood to be effective in glandular therapy or in experimental research.

Removal of the entire gland (hypophysectomy) in a young mammal results in a series of degenerative changes but not necessarily in death. Many of the early experiments in extirpation resulted in extensive damage to the brain and made the effects more difficult to interpret. Most of the effects noted following total hypophysectomy can be obtained by removal of the anterior lobe alone.

Upon removal of the anterior pituitary in a young mammal the animal fails to grow and develop into a normal adult. The gonads do not develop; the animal remains sexually immature. Metabolism is depressed, resulting in increased accumulation of fat. Abnormalities occur in the thyroid gland and the adrenal cortex. Injection of anterior pituitary extract tends to promote growth and development of the gonads in such animals.

When normal young animals are treated with anterior pituitary extract, there is a tendency toward gigantism, overgrowth of skull bones, large feet, and early sexual maturity. The thyroid gland and adrenal cortex enlarge, with evidence of hypersecretion. The external genitalia may become larger than normal.

GROWTH HORMONE The process of growth is dependent upon a number of factors that determine the course of metabolism. Genetic and nutritional factors exercise an important influence on growth, while hormonal factors can regulate or modify. It is obvious that the growth factor in anterior pituitary secretion is not solely responsible for growth.

Basically, growth is indicated when there is an accumulation of protein in the tissues of the body. When the body fails to grow or loses weight, an increase in the excretion of nitrogen is noted. Growth hormone injected into animals retards the excretion of nitrogen and therefore favors the

Hypophysis

521

retention of protein for growth. Hypophysectomized animals excrete nitrogen in excessive amounts.

It has been known for some time that if the anterior lobe of the hypophysis and the entire pancreas are removed at the same time from an experimental animal, the animal does not develop diabetes mellitus. It was formerly thought that there must be a diabetogenic hormone, that is, one capable of producing diabetes, secreted by the anterior hypophysis as a distinct hormone. Experimentation has shown, however, that, if there is continued injection of growth hormone in large amounts into an animal, it will eventually develop permanent diabetes mellitus. Evidently it is the growth hormone that is diabetogenic under experimental conditions.

Skeletal growth is stimulated by daily injections of the growth hormone into young animals. Rats and dogs have been made to attain unusual proportions by this method. One of the abnormalities in human beings is concerned with an increase in the length of the bones of the arms and legs, resulting in heights of 7 or 8 feet. The term *gigantism* is used to indicate this condition, since the body proportions are different from those of normally tall persons. It is caused by a hypersecretion of the anterior lobe. Growth in height can occur only so long as the zones of growth in the long bones remain open. Another condition resulting from hypersecretion in the adult is called *acromegaly*. Overgrowth of bone is especially noticeable in the enlargement of the brow ridges, the lower jaw, and the hands and feet. The nose and lips thicken, the face loses its intelligent expression, and there is mental regression. Tumors of the anterior pituitary lobe are usually found to be responsible for hypersecretion in these cases (Figure 20.3).

Hyposecretion of the anterior lobe in infants and children produces pituitary dwarfism. Pituitary dwarfs, unlike cretins, have good features, they are normally intelligent, and their body proportions are good. They apparently do not obtain enough growth hormone to permit normal growth. Since the gonadotropic hormones are also lacking, they remain sexually infantile (Figure 20.4).

GONADOTROPIC HORMONES   We have noted that removal of the anterior lobe of the pituitary in a young animal results in failure of the gonads to develop; the animal remains sexually infantile. When hypophysectomy is performed on an adult animal, the gonads regress or atrophy. Introduction of anterior pituitary extracts produces marked enlargement of the ovaries of both normal and hypophysectomized female rats and other experimental animals. By careful experimental procedures investigators have been able to differentiate two gonadotropic hormones of the anterior pituitary. One is the *follicle-stimulating hormone (FSH)*, which in the female stimulates development of the follicle cells surrounding each maturing egg in the ovary (Figure 20.5). In the male it stimulates development of the seminiferous tubules in the testes, which produce the male sex cells, or sperm. The second hormone is the *luteinizing hormone (LH)*, which stimulates the development of corpora lutea in the ovary. Actually the combined action of both hormones is essential to the development of the ovum and to ovulation. LH in the male stimulates the interstitial cells of the testes to secrete testosterone.

The secretion of the follicle-stimulating hormone and the luteinizing hormone is regulated by a releasing hormone produced by the hypothalamus. The releasing hormone which activates both FSH and LH is a small polypeptide. Research investigation indicates that the hypothalamus depresses the secretion of gonadotropic hormones until puberty.

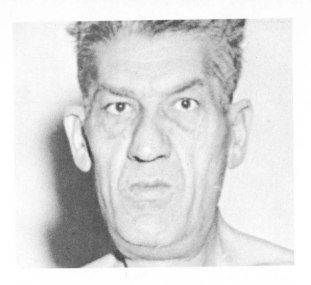

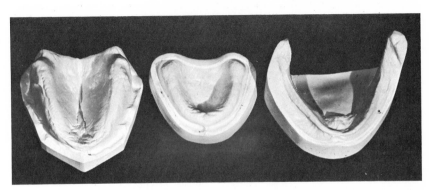

FIGURE 20.3
Acromegaly. (Above) Patient exhibiting coarse facial features and enlargement of jaws. (Below) Enlargement of the jaws in acromegaly, compared with normal plaster-cast model in center. (*E. Cheraskin and L. Langley, "Dynamics of Oral Diagnosis," The Year Book Medical Publishers, Inc., Chicago, 1956.*)

The corpora lutea are the yellow bodies of the ovary. After the ovum ruptures through the surface of the ovary (ovulation), the follicular cells, which formerly surrounded and nourished the ovum, change in appearance and function. They form a yellow glandular body called the *corpus luteum*, which then begins to function as an endocrine gland in its own right (see section on hormones of the gonads, further on in this chapter; refer also to Chapter 21).

The luteinizing hormone is thought to stimulate the development of the ovarian follicle before ovulation as well as to stimulate ovulation and to provide for the development of the corpus luteum. It is concerned also with the secretion of estrogen and with the secretion of progesterone by the corpus luteum.

In the male, LH is more appropriately called the *interstitial cell–stimulating hormone (ICSH)* because it stimulates the development of the small interstitial cells of the testes. These cells are located in groups surrounded by the sperm-producing cells (seminiferous tubule cells). The interstitial cells secrete the male sex hormone. The disparity in the names LH and ICSH arose when it was thought that they were two separate hormones; now it is known that they are identical hormones in either sex. The biological activity of extracts containing ICSH is usually tested on young or hypo-

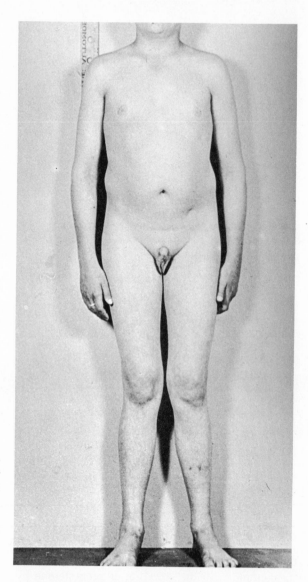

physectomized male rats, in which it causes enlargement of the seminal vesicles and the prostate gland.

ADRENOCORTICOTROPIC HORMONE Removal of the hypophysis in experimental animals is followed by a great reduction in the size and activity of the adrenal cortex. Enough activity remains, however, to maintain life. The adrenal medulla undergoes no change in size, and apparently its function is unaffected in hypophysectomized animals. When the adrenocorticotropic hormone is introduced into normal animals, it induces hypertrophy of the cortex and stimulates hyperfunction.

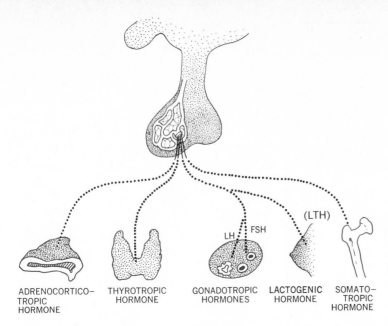

ADRENOCORTICO-
TROPIC
HORMONE

THYROTROPIC
HORMONE

GONADOTROPIC
HORMONES

LH FSH

LACTOGENIC
HORMONE

(LTH)

SOMATO-
TROPIC
HORMONE

FIGURE 20.5
Diagram illustrating some of the
regulatory effects of the anterior
lobe hormones.

There is a high concentration of ascorbic acid and cholesterol in the adrenal cortex. A single injection of adrenocorticotropin into experimental animals causes a rapid depletion of ascorbic acid concentration in the adrenal cortex. The cholesterol concentration is reduced more slowly as cholesterol is converted into adrenal glucocorticoids. The blood-serum cholesterol is also reduced by the administration of ACTH over a period of several days. Cholesterol is a source in the biosynthesis of adrenal cortical hormones as well as the sex hormones.

Melanocytes are stimulated by ACTH, producing a darkening of the skin. Activation of the enzyme adenyl cyclase produces more cyclic AMP from ATP (Table 20.1).

THYROTROPIC HORMONE (TSH)   After hypophysectomy the thyroid gland becomes reduced in size, and its secretion is greatly diminished. The adenohypophysis produces a thyroid-stimulating hormone thyrotropin or (TSH), which aids in maintaining the normal functioning of the thyroid gland. It can be demonstrated that TSH increases the blood circulation through the thyroid gland and that this, in turn, increases the production of the hormone *thyroxine*. There is also a lipolytic effect, that is, the breakdown of fats to fatty acids.

The hypothalamus gives rise to a thyrotropin-releasing hormone (TRH) which stimulates the adenohypophysis to release TSH.

The exophthalmos occurring in some cases of hyperthyroidism has usually been associated with the release of TSH, but apparently other factors are involved, possibly an unidentified hormone.

PROLACTIN   The development of the mammary glands (breast glands of the female) is stimulated by ovarian hormones, but their secretion, once they are developed, is controlled by a hormone secreted by the adenohypophysis and called *prolactin*. It is also a luteotropic hormone in rodents (mouse and rat) and is so named because it exerts a luteotropic influence on the corpus luteum of the ovary. Stimulation of the corpus luteum causes it to secrete the hormone progesterone. The luteotropic effect does not occur in the human being.

Hypophysectomy arrests the formation of milk if performed during a lactation period. A prolactin release–inhibiting neurohormone (PIH) is formed in the hypothalamus which inhibits the release of prolactin from the adenohypophysis. It would appear that the release–inhibiting hormone is in control except at the time of lactation, at which time suckling may initiate the release of prolactin.

MELANOCYTE-STIMULATING HORMONE (MSH) Melanocytes are pigment cells, most evident in lower vertebrates that are able to darken their skin by expansion of these cells. In man, the pigment melanin not only is found in static melanocytes but is distributed freely in the outer layer of the skin. MSH is found in the adenohypophysis and in the pars intermedia of those vertebrates having an intermediate lobe, where it is often called *intermedin*. The pars intermedia in man is poorly developed. The function of MSH in man is obscure. It is generally considered to have some effect on melanin production and darkening the skin. MSH release–inhibiting neurohormone (MIH) is apparently produced in the hypothalamus.

The posterior lobe, or *neurohypophysis*, arises as a downgrowth of the diencephalon part of the brain from the region that will later become the floor of the third ventricle. The stalk by which it is attached is the *infundibulum*. The base of the infundibulum is located behind the optic chiasma. The *neurohypophysis (pars nervosa)* has intimate connections with the hypothalamus through the infundibulum (Figure 20.2). The cells that compose the neurohypophysis resemble neuroglial cells amidst numerous nerve fibers; the cells are called *pituicytes*. There seems to be no doubt that the secretions of the posterior lobe arise in the hypothalamus. Droplets of secretion pass down nonmyelinated nerve fibers and are stored in the neurohypophysis.

NEUROHYPOPHYSEAL HORMONES The two principal hormones found in the posterior lobe are *oxytocin* and *vasopressin*.

Oxytocin Oxytocin induces marked contractions in the uterus, especially in the pregnant uterus. Its function in the male, if any, is not known. Oxytocin also causes the release of milk in the mammary glands of the nursing mother. The sucking stimulus initiates nerve impulses which are directed to the hypothalamus. These afferent impulses to the hypothalamus cause the neurohypophysis to release oxytocin into the bloodstream. In the mammary gland, numerous myoepithelial cells surrounding alveoli are stimulated to contract, ejecting milk into the ducts. The infant is then able to remove milk from the lactating gland by sucking. The synthesis and secretion of milk are governed by various hormones, among them, prolactin. Oxytocin effects the ejection of milk after it has been formed by glandular cells of the mammary glands.

Vasopressin The early experiments with vasopressin were performed on dogs, and the principal effect noted was a rise in blood pressure; hence the name vasopressin. However, this hormone has little effect on general blood pressure in human beings, although it may have some regional effects, such as contraction of the smooth muscle in the arterioles of the vasa recta surrounding the kidney tubules. There is also some evidence of a mild pressor effect following severe hemorrhage.

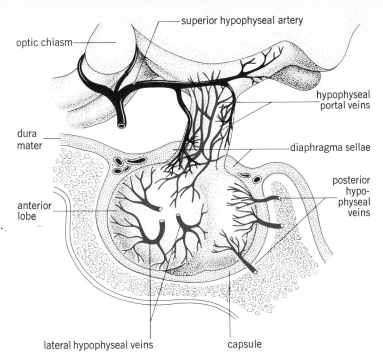

optic chiasm

superior hypophyseal artery

hypophyseal portal veins

dura mater

diaphragma sellae

posterior hypo-physeal veins

anterior lobe

lateral hypophyseal veins

capsule

FIGURE 20.6
Lateral view of the hypophysis illustrating the hypophyseal portal system.

The most evident function of the hormone relates to depressing the flow of urine from the kidneys, and in this respect it is known as the antidiuretic hormone (ADH). Decrease in the flow of urine is accomplished, apparently, by increasing the permeability of membranes in the distal convoluted tubules and collecting ducts so that more water is removed from the filtrate. The movement of water is passive, as we have seen. The action of ADH enables the kidney to reabsorb and conserve water.

When the posterior lobe of an animal is removed without injury to the hypothalamus, a condition develops in which the kidneys excrete great amounts of water (polyuria). The disease called *diabetes insipidus* is a similar condition, in which an unusual amount of urine of low specific gravity is excreted. The urine does not contain sugar as it does in diabetes mellitus, and the amount of urine excreted is much greater. Diabetes insipidus is considered to be a manifestation of posterior-lobe deficiency and can be alleviated by continued administration of antidiuretic hormone (ADH). Injury to the hypothalamus or to the nerve pathway in the infundibulum also causes polyuria to develop. A functional nerve pathway between the hypothalamus and the neurohypophysis seems to be essential to the production of ADH.

In summary, most recent evidence indicates that the neurosecretion of the posterior lobe arises in certain nuclei of the hypothalamus and that there are two hormones: vasopressin, or ADH, and oxytocin. The most important function of vasopressin is its antidiuretic action, causing increased reabsorption of water from kidney tubules. Oxytocin stimulates some smooth-muscle tissue, especially the muscle of the pregnant uterus. It also stimulates myoepithelial cells in the mammary glands that bring about a "let down" of milk in lactating glands.

Hypophysis

527

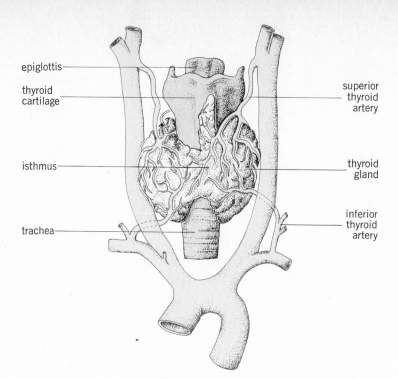

FIGURE 20.7
The thyroid gland, anterior view.

epiglottis

thyroid
cartilage

isthmus

trachea

superior
thyroid
artery

thyroid
gland

inferior
thyroid
artery

THYROID GLAND

The greater part of the thyroid gland arises from the floor of the pharynx. Its connection with the pharynx is soon lost, and the gland develops into a bilobed structure with a connecting portion anteriorly across the trachea just below the larynx (Figure 20.7). The gland grows during childhood and reaches its normal adult size at puberty. The thyroid is the largest of the glands that are entirely endocrine in function. It is larger in the female than in the male, but its size is affected by a number of factors. It enlarges if not enough iodine is available. There is a greater incidence of abnormal conditions of the thyroid in the female than in the male.

Histological preparations of the tissue of the gland show it to be composed of spherical sacs called *vesicles*, or *follicles*. The walls of the vesicles are composed of a single layer of cuboidal epithelial cells. Within the vesicle is a viscid colloidal fluid, which contains iodine. The colloid serves to store the thyroid hormone (Figure 20.8a).

The thyroid gland receives an abundant blood supply from the thyroid arteries, which branch from the external carotid and subclavian arteries. The arteries anastomose freely, and much of the blood volume appears to pass more or less directly into the thyroid veins. Each vesicle is surrounded by a dense capillary network.

THYROID HORMONES The major thyroid hormone is *thyroxine*, or tetraiodothyronine (T_4), but there is also *triiodothyronine* (T_3), which is physiologically very active but is present in the blood in only very small amounts. A third hormone of quite different chemical nature is *calcitonin*, which will be discussed later. It is evident that iodine forms a part of the thyroxine chemical compound. There are several steps in the synthesis of these two thyroid

The endocrine system

528

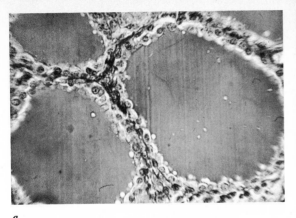

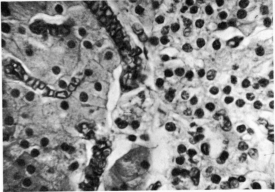

a   *b*

hormones. In the first place, ionic iodide is converted to elemental iodine. This is an oxidative reaction catalyzed by a tissue peroxidase. Elemental iodine is then combined with the amino acid tyrosine by enzymatic action to form monoiodotyrosine. But some tyrosine molecules combine with two atoms of iodine and form diiodotyrosine. A coupling enzyme aids in combining the mono and diiodo forms into the hormone triiodothyronine. Two molecules of diiodotyrosine form tetraiodothyronine, which is thyroxine.

Large vesicles of the thyroid gland contain the glycoprotein *thyroglobulin*, which is secreted by cells of the vesicles. The thyroid hormones are stored in the colloid of the vesicles, chemically attached to thyroglobulin. Before the hormones can enter the bloodstream they must be released from this large globulin, since it does not enter the blood.

FUNCTIONS OF THYROXINE   The most outstanding function of the thyroid hormone is its ability to accelerate metabolism. The effect appears to be at the cellular level and is probably due to the effect of the hormone in oxidative enzyme systems. The rate of energy exchange is affected; it is accelerated in hyperthyroidism and reduced in hypothyroidism. Growth is also affected, for hypothyroidism in the young results in retarded growth. Metabolism in patients with low basal rates can be greatly accelerated by the administration of thyroid substance.

One should keep in mind that normally the functions of the thyroid are merely regulatory. As we have seen, the adenohypophysis releases thyrotropin, which is the chief regulator of thyroid function. An increase in the metabolic rate inhibits the secretion of thyrotropin in a feedback mechanism which includes the hypothalamus. Much can be learned concerning the functions of the thyroid, however, by considering some of the abnormal conditions.

GOITER   Many types of enlargement of the thyroid gland may be called *goiter*. Simple uncomplicated goiter usually means an enlargement of the gland due to lack of iodine. It does not necessarily result in any endocrine disturbance, such as a change in the rate of metabolism, although a low rate of hormone production is indicated. The incidence of goiter is much higher in mountainous or desert regions, where there is a lack of iodine in the drinking water. Fresh-water areas remote from the sea are also apt to be deficient. Deficiency

Thyroid gland

529

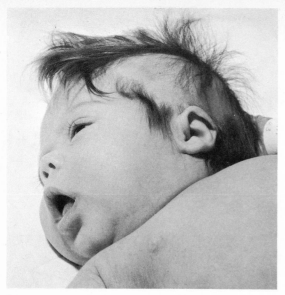 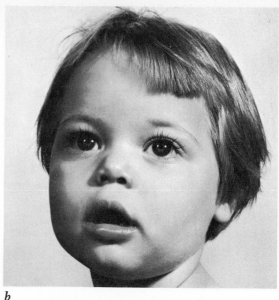

a   b

FIGURE 20.9

a A ten-week-old baby showing typical features of severe hypothyroidism. Note especially the fragile hair (which is falling out) and the thickened facial features. *b* The same child after 16 months of treatment. (*Courtesy of Dr. Milton S. Grossman.*)

in iodine is being overcome by the more general use of iodized table salt. Goiter occurs more frequently in women than in men. Adolescents and pregnant women should be especially careful to obtain an adequate supply of iodine in their food and drinking water. However, the incidence of goiter has become relatively low now that its cause is understood.

OTHER ABNORMAL CONDITIONS   An abnormal condition of the thyroid gland is responsible for three well-known endocrine states, namely, cretinism, myxedema, and exophthalmic goiter. The first two conditions are examples of hypothyroidism. The prefix *hypo-* indicates a deficiency of glandular secretion; the prefix *hyper-* refers to an oversecretion. Exophthalmic goiter is an example of hyperthyroidism.

Cretinism   Hypothyroidism manifests itself in many ways. Probably it is more accurate to limit the term *cretin* to those extreme cases resulting from failure of the thyroid to develop in intrauterine life. The brain of the fetus is retarded in its development, and the infant shows early signs of hypothyroidism. It fails to grow properly, the tongue thickens and often protrudes, the bridge of the nose fails to develop (saddle nose), and the child is mentally retarded. The treatment of this particular kind of cretinism may be unsatisfactory, largely because the developing brain of the fetus has been seriously injured. Administration of thyroid substance usually results in good physical growth if treatment is begun early in infancy, but mental development may be retarded. There are, however, many cases of hypothyroidism in infants and children that will yield to treatment, and the individuals will be greatly improved. Thyroid substance is one of the few hormones that can be taken by way of mouth; the earlier the treatment is begun, the better are the chances of normal development (Figure 20.9).

Myxedema   In some forms of hypothyroidism the skin becomes thick and puffy, resembling an edema; therefore the term *myxedema* was suggested as de-

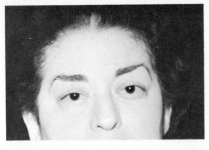

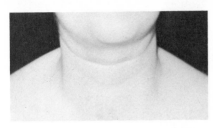

a   b

FIGURE 20.10

Response of a patient with myxedema to thyroid therapy: *a* Note puffy areas around eyes and thick upper lids, including heaviness of facial features. *b* Improvement in facial appearance after treatment. (*Courtesy of Robert Gittler, M.D.*)

scriptive of this condition. It is not a true edema or swelling, although there is an accumulation of a semifluid albuminous substance in the skin. Some types of hypothyroidism in children may be described as juvenile myxedema. Not all forms of adult hypothyroidism are myxedematous, especially in the beginning. A low basal metabolic rate is the most constant finding. The patient feels cold, because a low rate of metabolism results in decreased heat production. Physical changes and mental retardation develop slowly. The skin thickens and becomes dry, and the face loses its normal intelligent expression. If the individual is an adult, growth cannot be retarded; but otherwise myxedema resembles cretinism (Figure 20.10*a*).

The response to treatment with thyroxine is excellent and often dramatic. Within a few days the patient may appear alert, with improved speech and a higher basal metabolic rate. The facial expression slowly improves as the individual returns to normal mental and physical health (Figure 20.10*b*).

Hyperthyroidism The term *toxic goiter* includes various manifestations, but it always means a gland that is overactive in producing the hormone in excessive amounts. Unlike simple goiter, the gland may be only slightly enlarged, if enlarged at all. One form of toxic goiter is caused by an adenomatous growth (or tumor). The causal factors in this case would appear to be within the gland itself. In the case of exophthalmic goiter, or Graves' disease, the causal factors probably are not located within the gland. The gland is overactive because it is stimulated from some other source. Experimental evidence points to some unknown factor, probably associated with thyrotropic hormone.

Exophthalmic goiter is characterized by moderate enlargement of the gland, protrusion of the eyeballs, and a high basal metabolic rate. Ocular signs are not always present, but when the eyes are affected, the protruding eyeballs show too much of the white and the patient has a tense, frightened appearance (Figure 20.11). The patient is apt to be nervous and to have

Thyroid gland

531

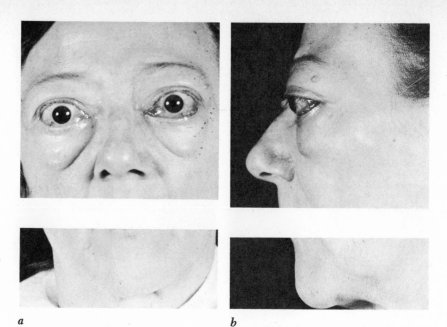

a   *b*

a high pulse rate. Since the basal metabolism is greatly increased, the hyperthyroid patient may eat a great deal and still lose weight. He will produce considerable body heat from heightened metabolic activity, feel warm and perspire readily. Muscular weakness may result in tremor, especially evident in the fingers.

Various methods of treatment attempt to suppress the activity of the gland. Drugs are used to reduce the activity of the thyroid, but unfortunately most of these drugs produce toxic side effects. Partial removal of the gland by surgery (subtotal thyroidectomy) after treatment with iodine and other drugs is considered to be the best treatment in most cases. Radioactive iodine can be administered as another form of treatment. It is rapidly taken up by the thyroid gland and emits rays that destroy portions of thyroid tissue, thus reducing the activity of the gland.

CALCITONIN One of the more recently discovered hormones is the polypeptide calcitonin, produced by interstitial or parafollicular "C" cells of the thyroid gland. The interstitial cells have an interesting biological history. They arise from the epithelial tissue of embryonic branchial pouches which develop into the postbranchial or ultimobranchial bodies of lower vertebrates (Figures 22.32 and 22.33). In the human thyroid as well as in that of most other mammals, cells from the embryonic ultimobranchial bodies become incorporated in the thyroid gland as interstitial cells. The hormone produced by these cells is not a part of the colloidal material within the vesicles. Calcitonin has also been found in the parathyroid and thymus glands, but the primary source is believed to be in the thyroid.

Calcitonin functions in regulating hypercalcemia. In cooperation with parathyroid hormone, which is hypocalcemic in function, these two hormones appear to be the chief regulators of the calcium level of the blood plasma. A negative feedback mechanism maintains a calcium homeostasis at a precise level. Skeletal structures provide a great reserve source of calcium.

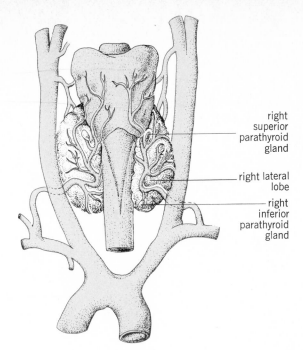

right
superior
parathyroid
gland

right lateral
lobe

right
inferior
parathyroid
gland

FIGURE 20.12
Thyroid and parathyroid glands,
posterior view.

PARATHYROID GLANDS

Partially embedded on the posterior surface of the thyroid gland are four
small yellowish or brown bodies known as the *parathyroid glands* (Figure
20.12). They are closely related to the thyroid in point of origin, for they
arise in the embryo from the entoderm of the third and fourth pharyngeal
pouches (Figures 22.32 and 22.33). The cells are in a compact mass and do
not closely resemble thyroid tissue (Figure 20.8*b*). Neither are they closely
related from a functional standpoint. The parathyroid glands are the smallest
of the compact endocrine glands, being only about the size of a cherry seed.
The hormone produced by the glands is called *parathormone*. It is primarily
concerned with the metabolism of calcium and phosphorus.

A most striking effect follows the removal of the parathyroid glands
from experimental animals, especially those of the order *Carnivora*. Muscular
tremors occur after a few days, progressing toward tetany and death. There
is a sharp drop in the calcium level of the blood, as well as a rise in the
phosphorus content. The tetany can be alleviated by the injection of an
extract of the parathyroid glands containing the hormone, by administration
of calcium, or by using certain drugs that enable the body to maintain the
calcium level of the blood.

PARATHYROID HORMONE   An extract of parathyroid glands has been prepared,
and the hormone has been isolated in pure form. It is a polypeptide protein
substance and plays a part in regulating the calcium level of the blood. It
cannot be given by mouth, since it would be subject to breakdown by
digestive proteases.

The parathyroid glands are an exception to the general rule in that
they are not controlled by any known secretion of the anterior pituitary

Parathyroid glands

533

gland. The concentration of calcium ions in the blood acts as a control over the secretion of the parathyroid glands. When blood calcium becomes low, the parathyroids are stimulated and the secretion of parathormone is increased. The blood-calcium level may be increased in several ways: (1) by increased absorption from the intestine, (2) by increased absorption through kidney tubules, and (3) by increased release of calcium stored in the bones. More phosphate is released as bone structure is broken down by osteoclast cells to obtain calcium, but at the same time the kidneys are stimulated to increase the excretion of phosphate so that a balance is maintained.

Almost all (about 99 percent) calcium stored in the body is in the skeleton, where it is found in the form of phosphates and carbonates. The calcium level of the blood serum is only about 10 milligrams per 100 milliliters. Calcium is present in protoplasm and in the extracellular fluid in minute amounts. It is absorbed from foods through the intestinal wall and eliminated chiefly by way of the intestinal tract, although a small amount appears in the urine.

Phosphorus, largely in the form of phosphates, is widely distributed throughout the body, but about 80 percent is found in the skeleton. Phosphates occur in various combinations in the blood, protoplasm, and extracellular fluid. They form a part of adenosine triphosphate, phosphocreatine, phospholipids, phosphoproteins, and many other organic compounds. Phosphates are obtained from foods and are eliminated largely by way of the kidneys.

HYPOPARATHYROIDISM Parathyroid insufficiency in man is not often of the acute type. As we have noted, complete extirpation of the glands in animals usually results in tetany and death. In people there is seldom a complete destruction of the glands, and tetany is more commonly of a chronic or latent type. Hyperexcitability of the peripheral neuromuscular system is a common feature of tetany. The muscles of the face contract when the facial nerve is stimulated by tapping and give the face a tense, sad, or crying expression. Pressure on the arm can cause the fingers to fold together in a characteristic fashion. The serum-calcium level is low, and the serum-phosphate concentration is high. The parathyroid hormone, though it will cause a rise in the serum-calcium level and a reduction in serum phosphate through urinary excretion, has not proved satisfactory for continuous treatment. Massive doses of calciferol (vitamin D_2) increase the calcium absorption from the intestine and, in so doing, help the body to maintain the serum-calcium level. Another sterol produced as an irradiation product of ergosterol is dihydrotachysterol (A.T.10). It resembles the action of parathyroid extract in the reduction of serum phosphate through increased urinary excretion. It also causes a moderate rise in serum-calcium level, resulting from increased absorption from the intestine. These effects vary with the amount of secretion available from the patient's parathyroid glands, the medication having a tendency to decrease the activity of the gland.

HYPERPARATHYROIDISM A greater than normal secretion of the parathyroid glands is commonly associated with a tumor or growth on one or more of the glands. Hypersecretion results in a loss of tone in the muscles and a disturbance in calcium and phosphate metabolism. The serum-calcium level is high and is partially maintained by the withdrawal of calcium from the skeleton. The skeleton may become greatly weakened through loss of calcium, and spontaneous fractures may occur. The bones often become greatly deformed, containing cysts and fibrous tissue. The excess calcium can be deposited in tissues other than bone, especially as stones in the kidneys. The serum-phosphate level is high also, and the excretion of both calcium and phosphate by the kidneys is increased. The surgical removal of excess parathyroid tissue appears to be the best treatment for hyperparathyroidism. Ordinarily the parathyroid glands are stable in the performance of their duties; fortunately hypoparathyroid and hyperparathyroid conditions as primary disturbances are rare (Table 20.2).

ADRENAL GLANDS

The adrenal gland is actually two glands in one; it has an outer portion, or cortex, and an inner portion, or medulla (Figure 20.13). The two parts are distinct in origin and function. The adrenal glands of sharks and similar fish are of interest, for they have a separate internal gland between the kidneys, which corresponds to the cortex, and a double row of bodies close to the chains of sympathetic ganglia, which are homologous to the tissue of the medulla. The cortical portion in the human embryo arises from mesoderm of the same general region that gives rise to the gonads. The medulla, on the other hand, is formed by the infiltration of neural-crest ectodermal cells, which have a common origin with cells of the chain of sympathetic ganglia (Figure 22.22). These strands of cells, which invade the mass of cortical tissue, stain dark brown with chromic acid and are called *chromaffin cells*. Eventually they form the entire middle, or medullary, portion of the gland. Preganglionic cholinergic fibers supply the medullary tissue by way of the splanchnic nerves. There are no true postganglionic fibers; the medullary cells probably act in this capacity.

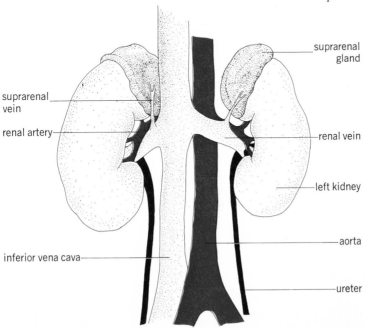

FIGURE 20.13
Adrenal glands (suprarenal glands) *in situ.*

suprarenal gland

suprarenal vein

renal artery

renal vein

left kidney

aorta

inferior vena cava

ureter

Adrenal glands

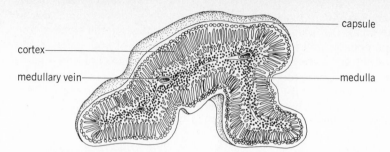

FIGURE 20.14

Human adrenal gland in schematic section to show cortex and medulla.

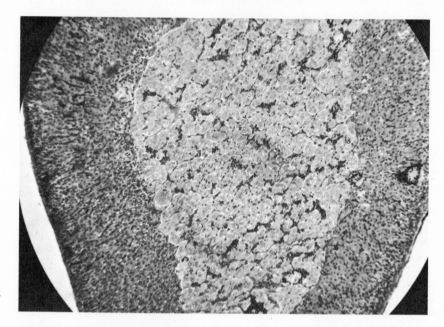

FIGURE 20.15

Microphotographic section through mouse adrenal gland to show differentiation between cortex and medulla. (*Courtesy of Edgar P. Jayne.*)

In man the glands lie on top of the kidneys and may be called *suprarenal glands*, but in most mammals they are separate from the kidneys, and therefore *adrenal* is the better term (Figure 20.14). In a section of a fresh gland, the cortex is yellow, while the medulla ia a red-brown color. The gland is enclosed by a capsule of fibrous connective tissue.

Microscopic examination reveals that the cortex is poorly differentiated into three zones, while the medulla consists chiefly of chromaffin cells arranged in irregular masses (Figure 20.15). The medulla receives a most abundant blood supply—a significant factor in view of its function. The cells of the cortex secrete hormones that are distinct from the hormones secreted by the medullary tissue. Since the hormone of the medulla is somewhat better known, we shall consider it first.

EPINEPHRINE and NOREPINEPHRINE Two closely related hormones are produced by the adrenal medulla. They are *epinephrine*, also known as adrenaline, and *norepinephrine* or noradrenaline. Both hormones are catecholamines derived from their precursor, dopamine. The two hormones have similar effects but not identical action. Epinephrine is produced much more abundantly than its counterpart. Release of the hormones is secured directly by stimulation of the splanchnic nerves, not by hormone control. Norepinephrine, it may be recalled, is the principal neurotransmitter in the sympathetic nervous system. Epinephrine has been shown to increase the level of cyclic AMP in many tissues, thereby adding to its effectiveness.

The action of epinephrine is almost identical with effects produced by stimulating the sympathetic nervous system. The blood pressure rises, largely because the smooth muscles in arterioles are stimulated to constrict; peristaltic movements of the intestine are inhibited; the pupil dilates; and the bronchial muscle relaxes. The similarity of action between the hormone and the sympathetic system is not so surprising when one recalls that the modified ganglion cells of the adrenal medulla migrate from the mass of tissue that forms the sympathetic ganglia. The action of epinephrine, therefore, is said to be *sympathomimetic;* that is, it mimics the action of the sympathetic system.

The theory that epinephrine comes to the aid of an animal in an emergency was proposed by Cannon and his associates. A considerable body of data was compiled in support of this theory. Though the evidence is largely indirect, it seems to indicate that many of the effects produced by an emergency situation are beneficial. Still, it is doubtful that epinephrine is essential for survival in an emergency. The theory, therefore, does not necessarily explain the function of the adrenal medulla. There is still some question that the adrenal glands play a primary role in enabling an animal to resist shock and stress, although this concept is seemingly well established.

Specific effects   When epinephrine is injected into the veins, there is a sudden rise in blood pressure, which is of short duration. The effect on blood pressure in man is much more striking than the effect produced in laboratory animals. Though many factors contribute to a rise in blood pressure, the most evident factor here is the constriction of peripheral arterioles. The great splanchnic capillary bed is also affected by vasoconstriction.

Not all parts of the circulatory system are affected in the same way. The coronary arterial system for example, is not constricted. The concentration of the solution injected and whether it is injected into a vein or into the abdominal cavity of the laboratory animal may be expected to vary the effect. In the normal animal, epinephrine increases the blood supply to the lungs and to the skeletal muscles.

Norepinephrine is primarily a peripheral vasoconstrictor, having little effect on heart rate but being capable of causing a sharp rise in blood pressure. Natural epinephrine as secreted by the adrenal medulla contains norepinephrine in small amounts. Cannon and his associates stated that when sympathetic postganglionic fibers are stimulated, a substance resembling epinephrine, but not identical with it, is liberated at the neuromuscular junction. They called the substance *sympathin.* Sympathin is now known as norepinephrine. This is the adrenergic substance released at postganglionic sympathetic nerve endings.

Use in surgery   Epinephrine and norepinephrine have numerous uses as vasoconstrictors in surgery and dentistry. One or the other may be included in a local anesthetic to keep the anesthetic from being rapidly carried away by the blood. Since it causes local vasoconstriction, the anesthetic is localized. Epinephrine reduces the bleeding in surgery; the larger vessels are tied off and the smaller ones are swabbed with the hormone to induce vasoconstriction. It is used also as a heart stimulant and to secure a temporary rise in blood pressure.

Adrenal glands

Use in allergies Epinephrine dilates the bronchioles and so finds a use in the treatment of bronchial asthma. It is also effective in other types of allergy such as hives, presumably because it regulates the permeability of capillaries and the retention of salts and water in the tissues.

Effect on respiration and metabolism The injection of epinephrine or increased secretion of the hormone increases oxygen consumption and the respiration rate, since it increases the basal metabolic rate. The blood sugar level is raised through the reduction of stored liver glycogen to glucose. Muscle glycogen is also utilized, but it appears in the blood as lactic acid rather than glucose. The lactic acid is carried to the liver and resynthesized to glycogen. It has been suggested that during starvation, when liver stores of glycogen have been depleted, the administration of epinephrine can increase liver glycogen by this method. Cigarette smoking is said to cause an increased secretion of epinephrine.

ABNORMAL FUNCTION The adrenal medulla is remarkably stable in its function, and abnormalities are rare. We have noted that insufficient secretion of epinephrine is not a serious inconvenience. Occasionally a tumor of the adrenal medulla produces an excess secretion. The commonest effect is a sudden, paroxysmal rise in the blood pressure to readings above 200 mm Hg, the blood pressure returning to fairly normal readings between attacks. Persistent hypertension is not characteristic, although it has been reported associated with tumors of the adrenal medulla. Surgical removal of the tumor usually results in a return to normal blood-pressure levels.

THE ADRENAL CORTEX The functions of the adrenal cortex are more complex than those of the medullary portion since so many organs and tissues are affected in various ways. The adrenal cortices are essential to life; the effects caused by their removal are very severe and invariably fatal unless essential hormones are replaced.

Perhaps the best way to investigate the function of the adrenal cortex is to consider the physiological changes that occur when the adrenal glands are removed from an experimental animal. The dog or cat, after such an operation, appears normal for several days; but later there is a loss of appetite, and signs of fatigue appear. Digestive disturbances follow, and the animal shows evidence of muscular weakness. Finally the blood pressure and the body temperature begin to fall, the animal exhibits extreme prostration, and death occurs with symptoms of shock. Physiological sequences are the drop in sodium- and chloride-ion concentrations in the blood and increased excretion of these ions by the kidneys. Animals from which the adrenal glands have been removed can be kept alive for some time by giving them increased amounts of table salt and replacement therapy, but this procedure does not completely correct other injurious effects of adrenal insufficiency.

Adrenocortical insufficiency renders both animals and man much more susceptible to the stress of disturbing conditions that may result in shock. It also renders the organism more susceptible to infections and to the onset of fatigue. Potassium ions in the blood are increased, while the excretion of water by the kidneys is decreased. Under these conditions the fluid content of the tissue cells increases as blood volume decreases. The blood pressure falls as the blood volume decreases, so that the kidneys are unable to function properly with a reduced blood supply; eventually a state of shock is produced.

In the fasting adrenalectomized animal the blood-sugar level becomes low, and the liver is depleted of its store of glycogen. The amount of nitrogen excreted is reduced also. If adrenocortical steroids are introduced in fairly large dosages, the blood-sugar level can be restored, with an increase in liver glycogen, and the excretion of nitrogen is increased. It has been suggested that the introduction of the hormone permits the fasting animal to derive carbohydrate from protein.

Nature of adrenocortical hormones The hormones secreted by the cortices of the adrenal glands may be divided into three groups: mineralocorticoids, glucocorticoids, and the sex hormones, androgens and estrogens.

Mineralocorticoids Aldosterone is the principal mineralocorticoid. Its functions evolve around the regulation of electrolytes, especially the reabsorption of sodium by the renal tubules and also the increased absorption of sodium by the intestinal epithelium. At the same time there is an increase in potassium loss from the kidneys. Water loss from the kidneys is decreased, causing a consequent increase in the volume of the blood plasma. The mineralocorticoids are of importance also in regulating the water balance in the tissues.

Glucocorticoids The more important are *cortisol* (*hydrocortisone*), *corticosterone*, and *cortisone*. Cortisol is by far the most active of the three hormones. As much as 95 percent of glucocorticoid effectiveness has been attributed to it. Cortisone is the least effective. Many functions have been assigned to these steroids. Among those that are well established by research investigation are (1) the glucocorticoids promote glycogen synthesis in the liver, which is in part related to an increase in gluconeogenesis, the production of glucose from fat and protein sources rather than from carbohydrates; (2) there is often a redistribution of fat deposits; (3) the hormones produce an anti-inflammatory response and therefore a tendency toward healing; and (4) they produce an increased resistance to injury and to physical "stress."

Inflammation, injury, and stress involve the hypothalamus and a hormone from the adenohypophysis as well as the adrenal cortical hormones. The hormone from the anterior pituitary gland is called the adrenocorticotropic hormone, or ACTH. The pain from an injury or the response to stress probably initiates a feedback mechanism to the hypothalamus. The hypothalamus, then, produces a releasing factor (adrenocortico-releasing-factor) which is conducted to the anterior lobe of the pituitary. The anterior pituitary then releases its adrenocorticotropic hormone (ACTH), which, in turn, stimulates the adrenal cortex to secrete its hormones.

The *sex hormones* of the adrenal cortex are normally produced only in trace amounts. It is only under very abnormal conditions that there is any evidence of their activity. Visible effects, if any, appear to be produced mainly by androgens and usually in females and young boys.

Addison's disease A well-known form of adrenocortical insufficiency is called *Addison's disease*. An English physician by the name of Addison, in 1855, described a condition associated with the destruction of the cortex of the adrenal glands, which has since come to bear his name. The condition is characterized by extreme lassitude with muscular weakness, weak heart action with lowered blood pressure, and digestive disturbances accompanied by loss of weight. Especially characteristic is a peculiar increase in the pigmentation of the skin and mucous membranes of the mouth (Figure 20.16). This increase is due to an increase in melanin, the normal pigment of the

Adrenal glands

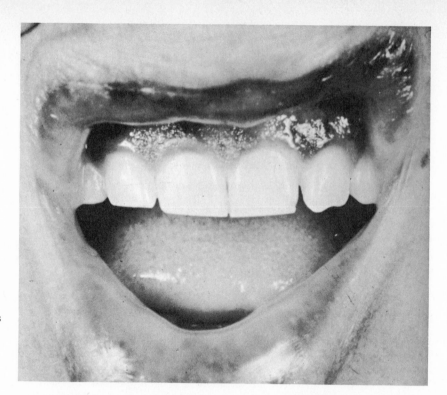

skin. The pigmentation may be evenly distributed, or it may be in blotches. It is greatest in those areas that normally contain pigment, such as the area around the nipples.

Deterioration of the adrenal cortex is commonly found in connection with Addison's disease. Often the cortex is destroyed by tuberculous lesions. The treatment of Addison's disease is not perfect; but with salt therapy and the use of cortical steroids, a great deal of progress has been made. Only a few years ago this condition was regarded as an often fatal disease. The outlook now is greatly improved.

Hyperactivity of the adrenal cortex One occasionally hears of children who have attained considerable sexual development by the time they are two or three years old. Such cases are usually associated with abnormal activity of the adrenal cortex. Often there is a tumor or growth causing enlargement. Young girls before puberty develop rapidly, with some indication of masculinization. Pubic and axillary hair may develop at an early age, and the distribution of body hair is diffuse or more like the masculine type. There may be some growth of hair on the face, especially on the chin. The clitoris usually becomes greatly enlarged, but the ovaries and uterus remain small and there is no menstruation.

Young boys appear much older than their actual age. They grow rapidly during the early years and show considerable muscular development. Pubic hair develops, even though the child may be only three years old. The penis becomes greatly enlarged, but the testes remain small and apparently do not produce mature sperm.

Hyperactivity of the adrenal cortex after puberty in the female is

TABLE 20.3

Hormones of the adrenal glands

General classification	Hormone	Functional effects
Cortical hormones		
Mineralocorticoids	Aldosterone	Increases sodium reabsorption by kidney tubules, increases potassium in urine, increases water retention
Glucocorticoids	Cortisol (hydrocortisone) Corticosterone Cortisone	Stimulates gluconeogenesis, increases protein breakdown, increases blood glucose and amino acids in blood plasma, anti-inflammatory response
Adrenal sex hormones	Androgens	Trend toward masculinity
	Estrogens	Trend toward feminine characteristics
Medullary hormones	Epinephrine	Increases blood supply to lungs and muscles, increases cardiac output, raises blood-sugar level
	Norepinephrine	Causes peripheral vasoconstriction; an adrenergic neurotransmitter

characterized by virilism or masculinization. There is often a great development of body hair with masculine distribution. Facial hair develops strongly, especially on the chin. The bearded lady in the circus side show may be suffering from abnormal growth of the adrenal cortex. The body build tends toward the masculine type, and the voice is deeper than normal for the female. The adrenogenital syndrome in the adult male is not well known.

Another type of hyperfunction of the adrenal cortices produces the condition known as *Cushing's disease*. It is characterized by redistribution of fat, often with great obesity, muscular weakness, skeletal weakness, and high blood pressure (Table 20.3).

HORMONES OF THE GONADS

The gonads are the reproductive organs that give rise to the sex cells. The ovaries of the female produce ova, or eggs; the testes of the male produce sperm, or spermatozoa. The gonads also function as endocrine glands, producing the sex hormones. These hormones exercise control over the development and function of the reproductive organs, including the development of the secondary sexual characteristics.

FEMALE SEX HORMONES Hormones secreted by various parts of the female reproductive system are as follows:

1 A group of estrogenic substances is secreted by the ovary. They are commonly called *estrogens*, or *female sex hormones*. One of them, *estradiol*, may be the true female sex hormone. It is the most potent of several naturally occurring estrogenic substances. The estrogenic hormones are so named because they have the ability to produce the estrus (or mating state) in female animals.

Hormones of the gonads

2 The corpus luteum of the ovary (Figure 22.11) produces a progestational hormone. The active principle is called *progesterone*. *Gestation* means pregnancy, and the progestational hormone is concerned with changes in the lining of the uterus that favor the implantation of the developing embryo and therefore pregnancy. The corpus luteum was described in the section on gonadotropic hormones and is considered further in Chapter 21.

The placenta also secretes estrogens and progesterone. In addition, it produces a gonadotropic hormone called *chorionic gonadotropin*. The chorion is the fetal membrane that forms the embryonic portion of the placenta. The placenta, in turn, is the nourishing organ that provides for the exchange of food and oxygen between the blood of the mother and the blood of the embryo. It is also the organ that provides for the elimination of carbon dioxide and excretory products formed by the embryo.

Functions of estrogenic substances The female sex hormones are responsible for the changes that occur in the female at puberty. These changes include the development of the internal and external genitalia and the appearance of the secondary sexual characteristics. The development of the breasts, pubic and axillary hair, and the feminine body contour with broadening of the pelvis are secondary sexual characteristics. There is also a change of voice in the maturing female. The change is not as striking as in the male, but most women lose the high thin voice of little girls and assume a lower voice of better quality. Menstruation—the flow of blood from the degenerating lining of the uterus—begins with the attainment of puberty.

Estrogen is found in high concentration in the follicular cells surrounding the developing ova. Estrogenic substances are not produced exclusively by the ovary; they can be recovered from the urine of both male and female. The urine of pregnant women contains more estrogen than the urine of nonpregnant females. It probably originates in the placenta during pregnancy. The urine of the stallion is a good source of estrogen, and estrogenic substances have also been isolated from plant sources.

Hormone role in menstrual cycle and pregnancy The human menstrual cycle is essentially a lunar-month, or 28-day, cycle in which about 4 days are concerned with menstruation. There follows a period in which the lining of the uterus is regenerated and the ovarian follicle matures. This is sometimes called the *recovery*, or *preovulatory*, *period*. It lasts until around the middle of the cycle. Ovulation, or the rupture of the egg from the ovary, usually occurs at approximately the thirteenth to fifteenth day of the 28-day cycle. The cells of the ruptured follicle change into the yellowish cells of the corpus luteum (or yellow body) of the ovary. The lining of the uterus is soft and thick—ready for implantation, if the ovum is fertilized. If the ovum is fertilized and implanted, the corpus luteum remains active during the early weeks of pregnancy before it begins to degenerate. If the ovum is not fertilized, changes occur in the lining of the uterus that lead to menstruation. The latter part of the period between ovulation and menstruation is called the *premenstrual period*. The corpus luteum undergoes degeneration during the latter part of the premenstrual period, and menstruation follows at the end of the cycle (Figure 20.17).

During the preovulatory period of the menstrual cycle (described in Chapter 21), the estrogen level of the blood rises; this is correlated with the growth and maturation of the ovarian follicle. The lining of the uterus also develops rapidly during this period. Anterior-pituitary hormones FSH and LH stimulate the development of the follicles.

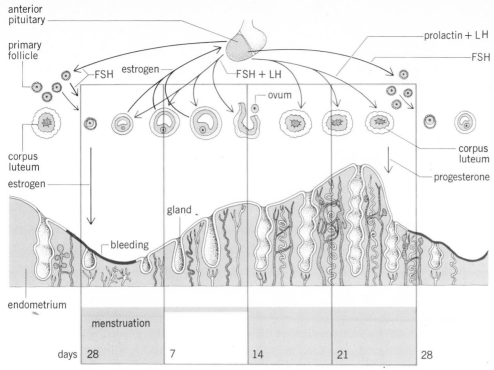

anterior pituitary
primary follicle
prolactin + LH
FSH
FSH
estrogen
FSH + LH
ovum
corpus luteum
progesterone
corpus luteum
estrogen
gland
bleeding
endometrium
menstruation
days | 28 | 7 | 14 | 21 | 28

FIGURE 20.17
Role of hormones in the menstrual cycle.

Progesterone is the hormone secreted by the corpus luteum. It seems to be dependent on estrogen for building up the lining of the uterus after menstruation but then is able to stimulate the maintenance of the glandular lining, once it is developed. Experimentally progesterone seems to play no part in the development of the secondary sexual characteristics. It does stimulate the glandular tissue of the breasts after the tubules have been developed under the influence of estrogen and pituitary hormones.

Progesterone plays an important part in pregnancy, for it is concerned with the maintenance of the secretory lining of the uterus after implantation. As we have seen, during the early part of pregnancy the corpus luteum remains functional and continues to secrete its hormone. Ovulation is suppressed, and menstruation does not occur, since the lining of the uterus is maintained in a state of proliferation and at a high degree of activity; hence the use of progesterone in birth control pills. Progesterone also appears to be essential for the development of the maternal portion of the placenta.

Progesterone, in the process of metabolism, is reduced to an inactive substance called *pregnandiol* and excreted in the urine. It can be converted readily to progesterone in the laboratory. It is of interest that the bull and the human male also excrete pregnandiol or a closely related substance. In this case it is considered to be secreted by the testis or the adrenal cortex.

A hormone called *relaxin* is secreted by the ovaries, especially during pregnancy and by the placenta. The corpus luteum of pregnancy probably contributes the greatest amount. The effects of relaxin vary in different species. It relaxes the symphysis pubis in some animals but not in others. In the female it softens and relaxes the cervix of the uterus before childbirth.

Hormones of the gonads

543

Experimentally, tissues must be exposed to estrogen before relaxin can be effective.

PLACENTAL HORMONES The placenta is primarily a structure that functions as an organ providing for an exchange of food materials and oxygen between the blood of the mother and the blood of the fetus. The placenta also functions as a temporary endocrine gland, producing estrogenic substances, progesterone, and chorionic gonadotropin. The secretion of estrogens by the placenta causes a very considerable rise in the estrogen level. The production of estrogens during pregnancy is much greater than during the normal monthly cycle. The high estrogen level inhibits the production of FSH; consequently, follicle production in the ovary is suppressed during pregnancy.

Progesterone also is secreted in much greater amounts by the placenta. It produces changes in the lining of the uterus favorable to the developing fetus. Estrogens and progesterone aid in developing the breasts for lactation.

Chorionic gonadotropin prevents the decline of the corpus luteum by stimulating it to enlarge and become more active. It also stimulates the corpus luteum to produce estrogens and progesterone in large quantities. During the third or fourth month of pregnancy the secretion of chorionic gonadotropin is suppressed, and the corpus luteum begins to deteriorate. The placenta is now secreting estrogen and progesterone in adequate amounts, so that the degeneration of the corpus luteum can proceed without any loss of hormones necessary to promote the development of the embryo.

Chorionic gonadotropin is obtained from pregnancy urine (P.U.) and from placentas. Unlike pituitary gonadotropins, human chorionic gonadotropin does not stimulate the growth of ovarian follicles. When injected into immature mice, rats, or isolated rabbits, it causes marked changes in the ovary, where it is predominantly luteinizing. Mature follicles may erupt, hemorrhage spots may occur in unruptured follicles, and atretic corpora lutea may be formed from the unruptured follicles. Changes in the ovary of immature or nonpregnant laboratory animals after injection with pregnancy urine are the basis for various tests for pregnancy. These tests on laboratory animals give accurate indications even when there is a pregnancy of only 2 weeks and before there are any outward signs.

More rapid pregnancy tests are also performed with certain species of African and South American toads or native male frogs. Tests of this nature take little time (about 2 hours), and each animal can be used again. Female toads extrude their eggs if there is a positive test for chorionic gonadotropin; with male toads or frogs a positive test is determined upon finding seminal discharge in their urine.

Human chorionic gonadotropin is used clinically to stimulate the descent of the testes, if the testes have failed to descend into the scrotum. The interstitial cells of the testes also are stimulated.

MALE SEX HORMONES *Androgenic substances* Several substances (steroids) that have masculinizing effects can be obtained from urine. These substances are called *androgens*. At first it was thought that these substances represented the male sex hormone, but they are now considered to be breakdown products of the hormone that retain some androgenic activity. A substance called *testosterone* has been isolated from the testes. It possesses great androgenic activity and is believed to be the male sex hormone.

Testosterone is produced by the interstitial cells of the testes. The cells

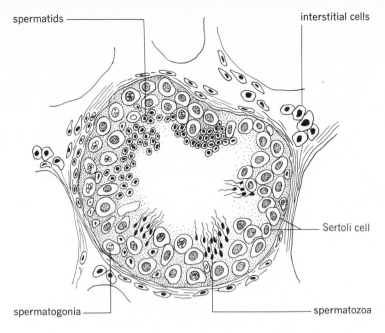

spermatids — interstitial cells

Sertoli cell

spermatogonia — spermatozoa

FIGURE 20.18
Cross section through a seminiferous (convoluted) tubule of the
testis. Note the location of the interstitial cells, which produce the
hormone testosterone.

that secrete the hormone are located in the tissue between the sperm-producing follicles. The interstitial cells are stimulated by ICSH of the pituitary. ICSH appears to be more effective when FSH is present also (Figure 20.18).

The male sex hormone is responsible for the development of the secondary sexual characteristics of the male, such as the beard, the pubic and axillary hair, the male body contour, and the deep voice. The male sex hormone influences the development of the primary sex organs and internal genitalia such as the prostate gland and the seminal vesicles. Gonadotropic hormones from the anterior pituitary are essential also.

Much of our knowledge of the action of the male sex hormone has been derived from the castration of animals. It has been known for centuries that castration of meat animals makes them more docile and easier to fatten. The meat is of better quality, and the muscles are not so tough. Castration of a young cockerel causes the bird to grow larger than a normal male because the closure of the zones of growth in the long bones is delayed. The castrate (or capon) then becomes a large and desirable meat animal. The loss of the sex hormone is especially evident in the appearance of the head. The large comb and wattles of the rooster fail to grow; instead the head tends to resemble that of the female, with a low pale comb.

Early castration, before puberty, in man prevents the expression of the secondary sexual characteristics. The voice does not change, facial hair growth is scanty, and the skin remains pale and soft. Eunuchs (or castrates) commonly grow tall, with long arms and legs, narrow shoulders, and a wide pelvis. Late castrates may not show much change in outward appearance. There is usually no change in the voice. Internally the seminal vesicles and prostate gland gradually regress.

OTHER SOURCES OF ANDROGENS AND ESTROGENS The adrenal cortex, as we have seen, is capable of producing both androgens and estrogens. We have noted

Hormones of the gonads

that tumors of the cortex often produce masculinizing effects. The ovaries also produce androgens, and the testes, estrogens. Androgens and estrogens can be recovered from the urine of either sex, but estrogens predominate in the female and androgens are more abundant in the male.

ADOLESCENT ACNE The eruptions of the skin, so common after puberty, may be primarily the result of androgenic activity. During adolescence the sebaceous glands of the akin become more active, producing an oily skin. These glands often become inflamed and eruptive. Castrates are not subject to acne unless they are treated with androgens.

Undoubtedly many factors influence the eruptions of the skin. Diet, the state of health, and the condition of the skin all have some bearing on the severity of the condition. Maintaining extreme cleanliness of the skin is recommended. In severe cases, a physician should be consulted. Medical treatment may include skin medication and vitamin or hormone treatments.

REJUVENATION It is almost an axiom that the introduction of a glandular extract will fail to stimulate the gland that normally is responsible for the secretion, because of a negative feedback mechanism. The introduced hormone very often has a depressing rather than a stimulating effect on the gland involved. It is unlikely that injections of androgens, or testicular grafts or implantations, will produce anything more than a transient beneficial effect on aging men. The power of suggestion is great, and claims of improved muscular strength and sexual vigor after treatment appear to be largely psychological rather than physical.

THYMUS In the upper part of the thorax above the heart is a bilobed structure called the *thymus* (Figure 20.19). It is pink and consists of lymphoid tissue. The thymus of meat animals is the throat sweetbread of the market. Relatively large in young children, it reaches its greatest size at puberty and then

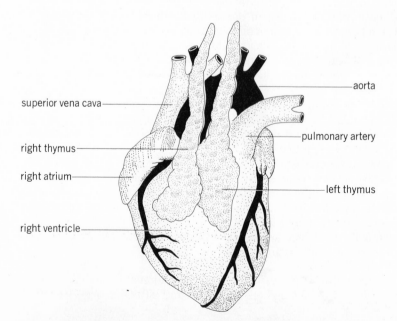

FIGURE 20.19
The thymus gland in a young adult.

superior vena cava

aorta

right thymus

pulmonary artery

right atrium

left thymus

right ventricle

The endocrine system

undergoes regression, the lymphoid tissue being largely replaced by adipose tissue. Numerous attempts to discover a secretion of the thymus have been unsuccessful. The administration of the thymus extracts has given inconclusive results. Like all lymphoid tissues, such as tonsils and adenoids, it is known to produce lymphocytes; perhaps it should be classified with such tissues rather than with the endocrine glands.

There are indications, however, that the thymus produces a factor that enables the lymph nodes and spleen to develop lymphocytes capable of acting as immunologically competent cells. The thymus plays a major role in establishing an immunological response.

It is also considered probable that the thymus produces a hormone which acts upon lymphocytes that are not producing antibodies and converts them into cells which have the ability to produce antibodies. The lymphocytes would most likely be converted into plasma cells capable of reacting to an introduced antigen. The hormone responsible, if present, has not been identified.

THE PINEAL GLAND

The pineal body (Figure 20.2) was originally a photoreceptor, a median eye in prehistoric reptiles. Often regarded as a vestigial structure, it now appears to be a valid endocrine structure, producing the hormone *melatonin*. The clearest picture of melatonin's function is found in those animals that possess active melanophores. Whereas the melanocyte-stimulating hormone (MSH) causes the skin of these animals to darken, melatonin gathers the melanophores together and lightens the color of the skin. In these lower vertebrate animals the pineal body responds to light waves received at the retina. Nerve impulses probably are carried to the gland by sympathetic nerves to cause the release of the hormone. The pineal function in man, however, remains largely obscure.

PROSTAGLANDINS

Prostaglandins are hormone-like substances found in various tissues of the body and having a wide spectrum of physiological functions. They are largely regulatory, affecting the action of smooth muscle, blood pressure, glandular secretion, and the female reproductive system, especially contraction of the uterus. Probably they are associated with cyclic AMP in conducting regulatory activities.

Hormones of the kidney and hormone-like substances have been previously discussed. We shall mention only the renin-angiotensin system, concerned with hypertension of renal origin; erythropoietin and its effect on red cell production; and neurohumors produced at nerve endings, in synapses, and in motor end plates.

Insulin and glucagon also have been considered in regard to hormones of the pancreas.

HORMONES ASSOCIATED WITH THE DIGESTIVE TRACT

Numerous hormones are associated with various digestive processes. The functions of the following hormones were considered in Chapter 17: enterogastrone, gastrin, secretin, pancreozymin, cholecystokinin, and enterocrinin (see Table 17.3).

SUMMARY   Endocrine glands are glands of internal secretion. They are ductless glands, and their secretions are called hormones.

The thyroid gland secretes the hormone thyroxine. Iodine is required in its formation. Simple goiter is an enlargement of the gland due to lack of iodine. Cretinism and myxedema are examples of hypothyroidism. Cretins fail to grow properly; they are also mentally retarded. Myxedema is adult hypothyroidism. Hyperthyroidism results from an overactive thyroid gland. The thyrotropic hormone of the hypophysis probably causes the gland to become overactive. One type of hyperthyroidism, called exophthalmic goiter, is characterized by a high metabolic rate and protrusion of the eyeballs.

The parathyroid glands consist of four small yellowish bodies embedded in the posterior side of the thyroid gland. These glands secrete parathormone, a hormone concerned with regulating calcium metabolism. Complete extirpation, especially in carnivorous animals, results in tetany and death. In man there is often a tensing of the muscles of the face, and the fingers fold together in a characteristic manner. Hyperparathyroidism is characterized by withdrawal of calcium from the skeleton. The serum calcium level is high. Spontaneous fractures may occur, and bones may become greatly deformed.

The adrenal glands are located on top of the kidneys and are composed of a medullary portion and a cortex. The medullary portion secretes epinephrine and norepinephrine. These hormones acting together exert a pressor effect on the heart, blood pressure, and smooth muscle. Norepinephrine is essentially a peripheral vasoconstrictor.

The adrenal cortex secretes a large number of chemical substances. The more important active corticosteroids are as follows: glucocorticoids, cortisone, and cortisol and the mineralocorticoid aldosterone. Glucocorticoids are concerned especially with carbohydrate metabolism; mineralocorticoids affect salt concentration and water balance.

Addison's disease is the result of adrenocortical insufficiency.

The adrenal cortex also secretes sex hormones. Hyperactivity of the adrenal cortex often involves premature sexual development in children.

The hypophysis is composed of two lobes, a posterior lobe and an anterior lobe. The posterior lobe produces two hormones, vasopressin and oxytocin. Vasopressin is the antidiuretic hormone (ADH). Oxytocin stimulates especially the smooth muscle of the pregnant uterus. The anterior lobe of the hypophysis produces six hormones: a growth hormone, two gonadotropic hormones, an adrenocorticotropic hormone, a thyrotropic hormone, and a lactogenic hormone. The gonadotropic hormones are a follicle-stimulating hormone (FSH) and a luteinizing hormone (LH). ACTH stimulates the adrenal cortex; the thyrotropic hormone regulates the secretion of the thyroid gland.

Hyperfunction of the anterior lobe can produce a condition of gigantism in a growing individual. In an adult the condition is called acromegaly. Hypofunction may result in pituitary dwarfism if it affects children.

The gonads produce sex hormones. Female sex hormones are estrogens and progesterone. The placenta also secretes these hormones and, in addition, chorionic gonadotropin. Estrogen stimulates the development of the secondary sexual characteristics of the female. Progesterone is secreted by the corpus luteum. Progesterone maintains the lining of the uterus until after ovulation during the regular monthly cycle. During pregnancy it maintains the lining of the uterus after implantation. Ovulation and menstruation are suppressed during pregnancy.

Chorionic gonadotropin prevents the decline of the corpus luteum during the first 3 or 4 months of pregnancy. Chorionic gonadotropin excreted in the urine of the pregnant female forms the basis for various pregnancy tests.

The male sex hormone is testosterone. It is secreted by the interstitial cells of the testes. Other androgens may be produced by the adrenal cortex. The adrenal cortex may also be a source of estrogens.

1 Name and locate the endocrine glands.
2 Prepare a set of experiments to prove that a certain gland produces a specific hormone.
3 What is meant by endocrine balance?
4 Why is it that the incidence of goiter is higher in some geographical regions than in others?
5 Discuss various conditions that can arise from abnormalities of the thyroid gland.
6 Explain the action of the parathyroid gland in the regulation of calcium and phosphorus metabolism.
7 Differentiate between the structure and function of the adrenal medulla and the adrenal cortex.
8 Discuss the practical uses of adrenal hormones.
9 Explain the function of the hypophysis as the master gland of all the endocrines.
10 Distinguish between the origin and function of the two lobes of the hypophysis.
11 List the hormones of the hypophysis and explain their functions.
12 Describe the changes that occur as a result of castration. Which hormones are concerned?
13 Discuss the function of the female sex hormones in relation to the menstrual cycle.
14 Would it be possible to devise a test that would indicate a pregnancy of only 2 weeks? How could such a test be performed?
15 Is it true that androgenic substances can be recovered from the urine of either sex? Where might these substances arise in the female?

SUGGESTED READING

ALIAPOULIOS, MENELAOS A., PAUL GOLDHABER, and PAUL MUNSON: Thyrocalcitonin Inhibition of Bone Resorption Induced by Parathyroid Hormone in Tissue Culture, Science, 151:330-331 (1966).
AXELROD, J.: Comparative Biochemistry of the Pineal Gland, Am. Zool., 10:259-267 (1970).
COPP, D. H.: Endocrine Regulation of Calcium Metabolism, Ann. Rev. Physiol., 32:61-86 (1970).
FRIEDEN, E., and H. LIPNER: "Biochemical Endocrinology of the Vertebrates, "Prentice-Hall, Inc., Englewood Cliffs, N.J., 1971.
GILLIE, R. B.: Endemic Goiter, Sci. Am., 224:92-101 (1971).
GREEN, R.: "Human Hormones," McGraw-Hill Book Company, New York, 1970.
GUILLEMIN, R. and R. BURGUS: The Hormones of the Hypothalamus, Sci. Am., 227:24-33 (1973).
HOLLANDER, C. S., et al.: Thyrotropin-releasing Hormone: Evidence for Thyroid Response to Intravenous Injection in Man, Science, 175:209-210 (1972).
MITNICK, M., and S. REICHLIN: Thyrotropin-releasing Hormone: Biosynthesis by Rat Hypothalamic Fragments in Vitro, Science, 172:1241-1243 (1971).
MULROW, P.: The Adrenal Cortex, Ann. Rev. Physiol., 34:409-424 (1972).
NETTER, F.: CIBA Collection of Medical Illustrations, vol. 4, "Endocrine System and Metabolic Diseases, "CIBA Corp., Newark, N.J., 1965.
ORCI, L., K. H. GABBAY, and W. J. MALAISSE: Pancreatic Beta-cell Web: Its Possible Role in Insulin Secretion, Science, 175:1128-1130 (1972).
PASTAN, IRA: Cyclic AMP, Sci. Am., 227:97-105 (1972).
PIKE, J. E.: Prostaglandins, Sci. Am., 225:84-92 (1971).
SCHALLY, A. V., et al.: Gonad-releasing Hormone: One Polypeptide Regulates Secretion of Luteinizing and Follicle-stimulating Hormones, Science, 173:1036-1038 (1971).
YATES, F. E., S. M. RUSSELL, and J. W. MARAN: Brain-adenohypophysial Communication in Mammals, Ann. Rev. Physiol., 33:393-444 (1971).

Suggested reading

unit nine

reproduction

chapter 21
the reproductive system

The reproductive system is unique among the organ systems; the organs of this system vary greatly between the sexes. Differentiation of the external genitalia is interesting from the standpoint of homologous structures. Male and female children do not differ remarkably in body form until they reach the age of puberty. At this time, under the influence of hormones, striking changes occur in several systems. The voice gradually changes in the male to a deeper masculine tone; the beard becomes a little stronger; pubic, axillary, and body hair develop; and the boy gradually assumes the characteristics of the adult male. The body form of the adult male develops increased musculature, with broader shoulders and narrow hips. The female at puberty develops a feminine contour due largely to deposition of subepidermal fat, and the mammary glands become larger. The internal and external genitalia approach maturity, and the gonads begin to produce mature sex cells (Figure 21.1).

FEMALE REPRODUCTIVE SYSTEM

The internal reproductive organs of the female are the ovaries, fallopian tubes (or oviducts), the uterus, and the vagina. The ova arise and develop in the ovaries; when they are mature they rupture from the surface of the ovary and pass down the fallopian tubes to the uterus. If the ovum is fertilized during its passage down the oviduct, the developing blastocyst becomes implanted in the lining of the uterus. If it remains unfertilized, it soon breaks

blastocyst

FIGURE 21.1
The male and female figures, in children and after puberty. Note that male and female children before puberty do not differ markedly in body form.

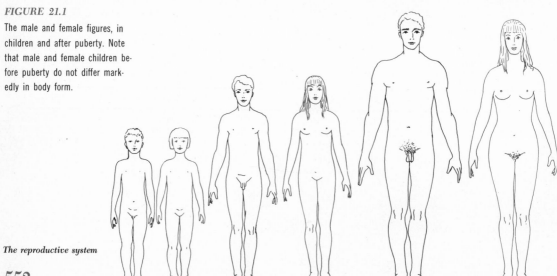

The reproductive system

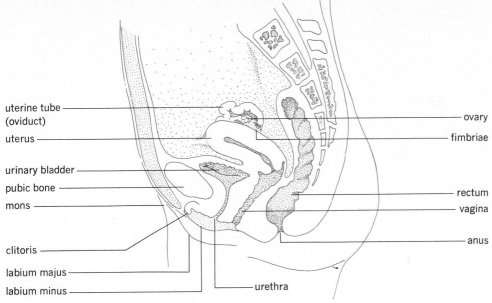

uterine tube (oviduct)

uterus

urinary bladder

pubic bone

mons

clitoris

labium majus

labium minus

urethra

ovary

fimbriae

rectum

vagina

anus

FIGURE 21.2

The female reproductive organs in sagittal section.

down and becomes lost in mucous secretions. The uterus leads into the vagina, which is a narrow passageway opening to the exterior (Figures 21.2 and 21.3).

OVARIES   The paired ovaries lie on either side of the uterus and below the fallopian tubes. They are oblong bodies 2.5 to 4 centimeters in length and about 1.5 to 2 centimeters in their anterior-posterior measurement. Their thickness is a few millimeters less than their depth. They lie posterior to a supporting fold of the peritoneum called the *broad ligament* and are attached to the uterus by ovarian ligaments.

The internal structure of the ovary consists of a connective-tissue framework, which supports the developing germ cells, muscle cells, blood

FIGURE 21.3

Female pelvic diaphragm, superior view.

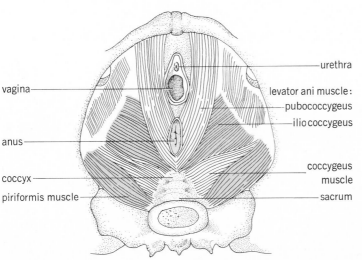

vagina

anus

coccyx

piriformis muscle

urethra

levator ani muscle:

pubococcygeus

iliococcygeus

coccygeus muscle

sacrum

Female reproductive system

vessels, and nerves. The cortex of germinal epithelium contains numerous germ cells and follicles in various stages of development (Figure 22.11). The ovary is covered with a delicate membrane of columnar epithelium.

The ova, as we have seen, develop within the ovarian follicle. The various stages of oogenesis are passed there, and the developing ovum in one of the more mature follicles is in reality a primary oocyte. Follicles develop under the influence of the follicle-stimulating hormone (FSH) and the luteinizing hormone (LH) originating in the pituitary gland. From puberty to the menopause, mature follicles approach the surface of the ovary and rupture mature ova through the surface at fairly regular monthly intervals in the process known as *ovulation*. It is assumed that the ovaries alternate in producing mature ova, but little is known about the regularity of the process. It is certain that occasionally more than one mature ovum is produced at the time of ovulation, as in the case of fraternal twins; but these ova could have been produced by one ovary. Of the thousands of potential ova found in the ovaries before puberty, only a few ever reach maturity. The mature follicle is 10 to 12 millimeters in diameter and bulges from the surface of the ovary. Ovulation occurs about the middle of the 28-day menstrual cycle, but the follicle cells persist, undergoing a transformation into the corpus luteum (Figure 20.17).

Corpus luteum The follicular cells, after ovulation, enlarge and increase in numbers so that the number of cell layers increases. The cavity of the old follicle becomes filled with blood, but the blood is gradually resorbed as new cell layers fill in the cavity. Connective tissue and blood vessels grow in from a connective-tissue layer surrounding the old follicle.

A yellowish thick-walled body called the *corpus luteum* replaces the old follicle (Figure 21.4). The cell cytoplasm contains a lipoidal substance known as *lutein*. It gives the cell mass a slightly yellowish color, especially after the corpus luteum is fully formed. In the period between ovulation and menstruation the corpus luteum secretes the hormones progesterone and estrogen, which apparently exert a sustaining influence on the lining of the uterus. If the ovum is not fertilized, the corpus luteum begins to degenerate toward the end of the menstrual cycle and menstruation follows. If the ovum is fertilized, the corpus luteum of pregnancy reaches the height of its development about the third month, after which it begins to degenerate. Probably the production of gonadotropic hormones following implantation of the blastocyst is responsible for maintenance of the corpus luteum. After the placenta is formed, it is capable of producing progesterone and estrogens; this may relieve the corpus luteum of its function. The corpus luteum degenerates slowly and is still present in the ovary at the time of childbirth.

Whether degeneration of the corpus luteum takes place in a monthly cycle or following a pregnancy, the cellular substance is replaced by fibrous connective tissue. The location of the old corpus luteum is marked by an area of white scar tissue in the ovary and is called the *corpus albicans* (Figure 22.11).

FALLOPIAN TUBES The tubes that conduct the ova from the ovaries to the uterus are usually called *oviducts* in animals; in man they are more commonly referred to as *fallopian tubes*, or *uterine tubes*. They lie in a horizontal plane above the ovaries (Figure 21.2). The distal ends near the ovaries flare out in a funnel-like fashion. The funnels bear fringed processes (*fimbriae*), which aid in guiding the ovum into the tube. The tube is not passive at the time

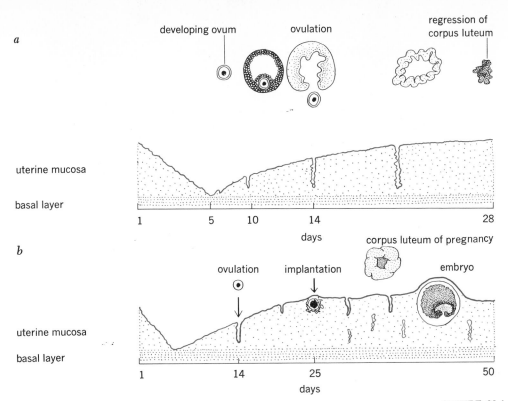

a

developing ovum   ovulation   regression of
                                        corpus luteum

uterine mucosa

basal layer

1   5   10   14   28
                        days

b

                                        corpus luteum of pregnancy

            ovulation   implantation   embryo

uterine mucosa

basal layer

1   14   25   50
                        days

FIGURE 21.4

The sequence of events during an
ordinary menstrual cycle and during pregnancy: *a* ordinary menstrual cycle, showing the development of uterine mucosa and its
deterioration at menstruation; *b*
development of the uterine mucosa
in the case of fertilization, implantation, and development of the
embryo.

of ovulation. The fimbriae are erectile, and at ovulation the fimbriae and
funnel move closer to the ovary to receive the ovum. A ciliated epithelium
lines the tube, and the beating of the cilia moves the ovum along toward
the uterus. Smooth muscles in the wall of the tube also aid in propelling
the ovum.

UTERUS   The single uterus is a thick-walled organ located in the upper part
of the pelvic region. Its function is to receive the blastocyst and to provide
protection and nourishment to the developing embryo and fetus after implantation. It is a small organ during childhood, but after puberty it is usually
about 3 inches long, nearly 2 inches wide, and about 1 inch thick. It is
somewhat larger after the first pregnancy. The cavity of the nonpregnant
uterus is always small (Figure 21.2).

The uterus is capable of great enlargement during pregnancy, extending
high into the abdominal cavity. An increase in the number of muscle fibers
and the lengthening of fibers permit the uterus to expand. The soft mucosal
lining is called the *endometrium.*

The position of the uterus varies, but it is usually tipped forward over
the urinary bladder (Figure 21.2). It is supported by the broad ligament and
the round ligament. The lower part of the uterus is more cylindrical in shape
and is called the *cervix.* Its external orifice opens into the vagina (Figures
21.5 and 21.6).

VAGINA   A canal leading from the vestibule of the external genitalia to the
cervix of the uterus is called the *vagina.* It is a muscular canal lined with

Female reproductive system

mucous membrane and capable of considerable distention. It lies almost at a right angle to the plane of the uterus, extending inward about 7 to 9 centimeters (Figure 21.2). The anterior wall is shorter than the posterior wall, which extends behind the cervix. The projection of the cervix into the vagina creates a pocket or fissure called the *fornix*, the posterior fornix being the deeper recess.

The vagina receives the penis of the male during sexual intercourse; a seminal emission releases sperm near the external orifice of the uterus. At childbirth the vagina becomes greatly distended to form the birth canal from the cervix to the exterior.

The external orifice of the virginal vagina is partially occluded by a fold of membrane known as the *hymen*. The hymen varies considerably in its shape and degree of extensibility. It may be distended or torn slightly at the first sexual intercourse, but it is not regarded as a very reliable sign of virginity.

EXTERNAL GENITALIA The female external genitalia consist of the *labia majora*, the *labia minora*, the *clitoris*, and the *vestibule* (Figure 21.5). The labia majora are two outer fleshy folds covered with pubic hair. They are continuous with the *mons pubis* above. The mons is an eminence of fat over the symphysis pubis and is covered with pubic hair. The labia minora are two membranous folds underneath and medial to the labia majora. They are red or pink in color and devoid of fat and do not bear pubic hair. At their upper extremity the labia minora extend around the clitoris, a structure homologous to the penis of the male. The clitoris contains erectile tissue, blood vessels, and nerves. It is a small structure 2 to 2.5 centimeters long but is largely embedded in tissue. Only the tip or glans portion protrudes, and it is ordinarily covered with membranes. The glans contains sensory receptors and represents an erogenous zone. The vestibule is the space bounded by the

FIGURE 21.5

Female external genitalia, anterior view.

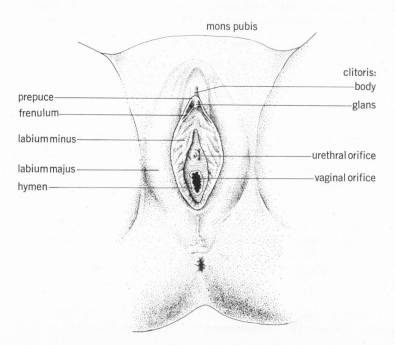

mons pubis

clitoris:
body
glans

prepuce
frenulum

labium minus

labium majus
hymen

urethral orifice
vaginal orifice

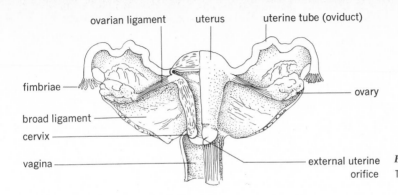

ovarian ligament   uterus   uterine tube (oviduct)

fimbriae —

broad ligament —

cervix —

vagina —

ovary

external uterine orifice

FIGURE 21.6
The uterus and appendages.

labia minora and the clitoris. It is evolved from the urogenital sinus of the embryo and contains the external orifices of the urethra and the vagina. The major vestibular glands, or Bartholin's glands, also empty into the vestibule. Their secretion is believed to function as a lubricant (Figure 21.6).

MAMMARY GLANDS   The mammary glands are modified skin glands that develop from two rows of differentiated ectodermal epithelium in the embryo. The so-called "milk line" extends from the axillae to the inguinal region in the embryo; in some animals a row of mammary glands develops from each milk line. The human breast glands develop in the pectoral region at the level of the fourth and fifth ribs. They are present in both sexes but develop under the influence of female sex hormones.

The female breasts remain relatively undeveloped until puberty, when the accumulation of fat adds materially to their size. The glandular portion does not mature and become secretory until the termination of pregnancy. Lactation is stimulated and maintained under the influence of an anterior-pituitary hormone, prolactin. The secretion of the mammary glands is at first a thin yellowish substance called *colostrum*. It contains nutrient materials, but the composition is different from milk. Within a few days milk is secreted, and lactation may continue for several months. Human milk is presumably best adapted for feeding infants. Though cow's milk is frequently substituted, it should be realized that the composition of cow's milk is somewhat different from that of human milk.

The breast, or mamma, is covered with thin, soft skin. At the apex is a nipple, which contains 15 to 20 depressions representing the individual openings of ducts. Surrounding the nipple is a pigmented circular area called the *areola*. The pigmentation varies with the complexion of the individual and deepens during pregnancy (Figure 21.7).

The glandular portion of the breast is composed of 15 to 20 lobes, each with an individual lactiferous duct opening through the nipple. Each lobe is subdivided into lobules with their ducts emptying into the larger lactiferous duct. The lobules, in turn, are compound glands composed of small glandular sacs (or alveoli). The glands regress during the later years of life, the alveoli being largely resorbed. The tubular structure remains, supported by connective and adipose tissue (Figure 21.7b).

Female reproductive system

PHYSIOLOGY   The female reproductive functions are greatly influenced by hormones, as discussed in Chapter 20. Also Chapter 22 considers the origin

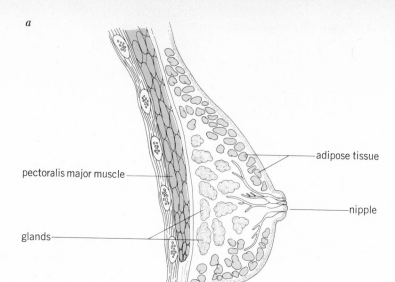

a

pectoralis major muscle

glands

adipose tissue

nipple

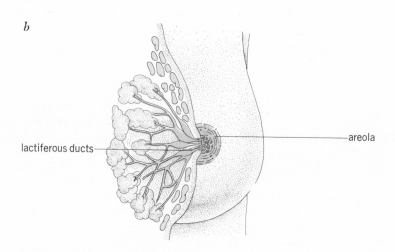

b

lactiferous ducts

areola

FIGURE 21.7
Mammary gland: *a* sagittal section; *b* partial anterior section.

and development of the germ cells and the reproductive organs. The ordinary menstrual cycle is as follows:

1st–4th day	Menstruation
5th–12th day	Preovulatory period
	Development of ovarian follicle and growth of endometrium
	Rise of estrogen level
13th–15th day	Ovulation
15th–20th day	Migration and breakdown of unfertilized ovum
	Development of corpus luteum
	High progesterone level
21st–28th day	Premenstrual period
	Regression of corpus luteum
	Progesterone level falls
	Deterioration of endometrium

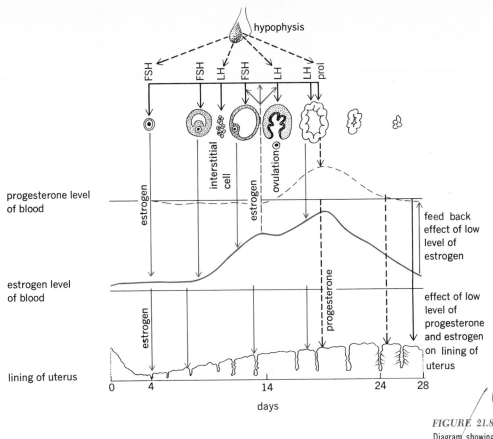

progesterone level
of blood

estrogen level
of blood

lining of uterus

days

feed back
effect of low
level of
estrogen

effect of low
level of
progesterone
and estrogen
on lining of
uterus

graafian follicle

FIGURE 21.8
Diagram showing influence of hormones on female reproductive cycle. A rise in blood-estrogen level
tends to inhibit production of FSH
by the anterior lobe of the hypophysis. The rise in blood-
estrogen level also stimulates secretion of LH and prolactin
(PROL). At the end of the monthly
cycle the low estrogen level stimulates renewed secretion of FSH by
a feedback mechanism. The low
level of estrogen and progesterone
is associated with deterioration of
the lining of the uterus, and menstruation follows.

If menstruation represents the beginning of the monthly cycle, then during the first to the fifth day the final deterioration of the soft mucous lining of the uterus and its removal in the menstrual flow of blood take place. During the preovulatory period the junctional layer of the mucosal lining gradually thickens and presents a soft, highly vascular bed for the implantation of the ovum, should it be fertilized. In the ovary a graafian follicle is growing, and the ovum is reaching maturity (Figure 21.4). The production of estrogen is high at this time. It should be stressed that although ovulation ordinarily occurs about the middle of the menstrual cycle, it cannot be said to occur precisely on the fourteenth day or necessarily within the 13- to 15-day period. A statistical average seems to indicate that ovulation most frequently occurs during this period. Ordinarily there are no outward signs of ovulation, and it is difficult to determine just when it does occur. There is, however, a slight rise in rectal temperature of about $\frac{1}{2}°$F following a slight drop below normal.

The period following ovulation is a secretory phase in which the endometrium continues to grow and is in a state of heightened activity. The coiled arteries are distended and the mucus-secreting glands enlarged. The corpus luteum is active, and blood progesterone and estrogen are at a high level (Figure 21.8). Toward the end of the period, if the ovum is not fertilized, a regression in the endometrium takes place. The circulation in the capillaries is greatly diminished during the latter part of the premenstrual period. Leukocytes begin to migrate into the area, and a breaking up of the super-

Female reproductive system

559

ficial tissue takes place; bleeding occurs from degenerating capillaries, and the menstrual period follows.

The functions of the ovary are directly influenced by gonadotropic hormones of the pituitary gland. FSH influences the development of ovarian follicles, and LH aids in stimulating ovulation and is concerned with the development of the corpus luteum. It appears that several hormones take part in the luteotropic process, rather than LH alone. The hormones thought to be involved, including LH, are prolactin, estrogens, FSH, and their releasing hormones. These hormones aid in maintaining the corpus luteum and enable it to secrete progesterone.

The action of female sex hormones on their target organs is generally considered to include the chemical binding of estrogen to receptor proteins. Whereas cyclic AMP serves as an intermediate or "second messenger" for many hormones, it has not been demonstrated that it serves to mediate either estrogens or androgens.

The fertilization of the ovum, the early cleavage stages, and the implantation of the blastocyst are discussed in the following chapter. If fertilization occurs and is followed by implantation, the corpus luteum grows larger and persists as the corpus luteum of pregnancy. Ovulation does not ordinarily occur during pregnancy, and menstruation is also suppressed. The blood estrogen level is high, suppressing the FSH of the pituitary. Progesterone is secreted first by the corpus luteum and later by the placenta. Progesterone stimulates the growth of the endometrium, producing a favorable environment for implantation and for the maintenance of pregnancy.

The discomfort that some women experience at the onset of the menstrual period is believed to be caused by hormonal disbalance. The corpus luteum regresses and stops secreting progesterone at this time. The blood level of progesterone is therefore low. The estrogen blood level is also low, since graafian follicles are stimulated to develop actively after menstruation. Ovarian hormone production is high during the premenstrual period, and women commonly find this to be a period of maximum efficiency.

ORAL CONTRACEPTION Oral contraceptives today are commonly referred to as "the pill." In the original experimentation the natural hormones progestin and estrogen were used but were not satisfactory, causing undesirable side effects. The pill now includes synthetic estrogen and progestin components and, in the lower doses used, produces comparatively few side effects in most women. These components inhibit FSH production, which normally causes ovulation to occur at monthly intervals. There is some danger of blood clotting, such as the condition known as thrombophlebitis associated with inflammation of the veins, usually leg veins, but the pill is an effective oral contraceptive.

The female reproductive cycle comes to an end when menstruation gradually ceases, during the *menopause*. This condition is reached ordinarily at an age of approximately forty-seven years. Hormone imbalance may lead to emotional disturbance for a time until an adjustment is accomplished.

MALE REPRODUCTIVE SYSTEM

TESTES Spermatogenesis takes place in the testes. The testes descend from an abdominal position before birth and come to lie in a sac called the *scrotum* (Figure 21.9). The scrotum is composed of two compartments; a median seam

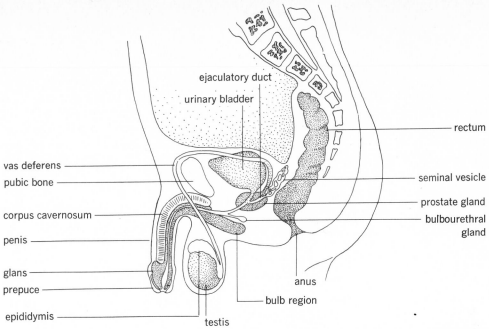

ejaculatory duct

urinary bladder

vas deferens

pubic bone

corpus cavernosum

penis

glans

prepuce

epididymis

testis

bulb region

anus

rectum

seminal vesicle

prostate gland

bulbourethral
gland

FIGURE 21.9
Male reproductive system in longitudinal section.

(or raphe) indicates where the embryonic scrotal folds have grown together. The skin is wrinkled into transverse ridges by the contraction of dermal and subcutaneous muscle fibers, and it bears a sparse coat of hair. Occasionally the testes fail to descend into the scrotum, a condition known as *cryptorchism*. Undescended testes are almost invariably sterile, although they produce the male sex hormone. Spermatozoa are very sensitive to heat, and it is generally believed that the temperature within the body cavity is unfavorable for spermatogenesis. When the body is subjected to a cold environment, the testes are drawn up close to the body; in a warm environment, muscular relaxation permits them to lie deep in the scrotum away from the body.

The two testes are essentially ovoid in shape, being a little flattened laterally. They are about 4 centimeters in the longer axis and 3 centimeters in width. The internal structure consists of compartments or lobules filled with seminiferous (convoluted) tubules. The tubules are lined with germinal epithelium, and it is there that spermatogenesis takes place (Figure 22.1). The germinal epithelium also contains *Sertoli cells*, which are modified columnar cells with large oval nuclei and prominent nucleoli. They are thought to secrete a nourishing fluid for spermatids which attach themselves to these cells (Figure 22.1).

Interstitial cells (*Leydig cells*) lie between the seminiferous tubules and produce the androgen testosterone. The adenohypophysis produces the gonadotropic hormone ICSH, which stimulates the interstitial cells to secrete testosterone (Figure 20.18).

The male sex hormone is responsible for the development of the secondary sexual characteristics of the male. These include the beard, the deep voice, and body hair patterns. The pubic-hair pattern of the male commonly resembles a triangle with the apex extending upward toward the umbilicus, whereas in the female pubic hair extends upward only to a horizontal line along the superior border of the mons veneris.

Male reproductive system

561

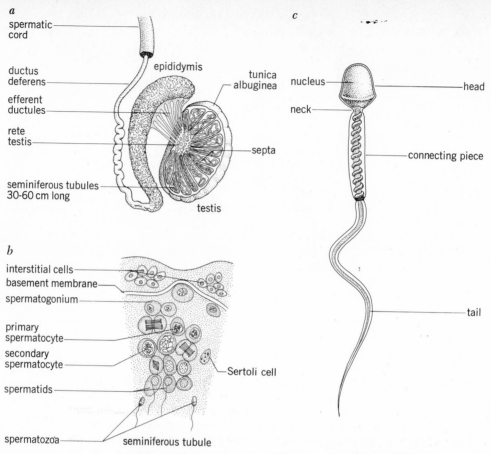

a

spermatic cord

ductus deferens

efferent ductules

rete testis

seminiferous tubules 30-60 cm long

epididymis

tunica albuginea

septa

testis

c

nucleus

neck

head

connecting piece

tail

b

interstitial cells

basement membrane

spermatogonium

primary spermatocyte

secondary spermatocyte

spermatids

spermatozoa

Sertoli cell

seminiferous tubule

FIGURE 21.10

Testes and spermatozoa: *a* section of the testis, lateral view; *b* section of seminiferous tubule illustrating spermatogenesis; *c* enlargement of spermatozoon showing structure.

The length of time required to produce mature spermatozoa from spermatogonia has been studied in various ways. Spermatogenesis in man occurs in four cycles and has been estimated to require 64 to 74 days.

EPIDIDYMIS The sperm at this stage are not motile but are propelled up through the convoluted seminiferous tubules into a network of fine tubules (the rete testis) and on into the efferent ducts of the epididymis. The epididymis is a body containing a tightly convoluted tubule and is located along the posterior surface of the testis (Figure 21.10). The tube is only about 0.4 millimeter in diameter, but it is 18 to 20 feet in length.

VAS DEFERENS The duct of the epididymis is continuous with a larger duct, the ductus deferens or vas deferens, which leads the sperm away from the testis. The vas deferens extends upward from the testis through the spermatic cord; it passes through the inguinal canal, over the pubic arch, and posteriorly over the urinary bladder to terminate in the ejaculatory duct. The roundabout position of the vas deferens is explained by the fact that the testes have changed their location. When the testes descend through the inguinal canal into the scrotum, they carry with them the vas deferens, blood vessels, lymph vessels, and nerves. From the abdominal inguinal ring to the testis these structures form the spermatic cord. The enclosing fasciae contain muscle fibers of the cremaster muscles, which aid in drawing the testes close to the body.

The vas deferens joins the duct from the seminal vesicle and enters the tissue of the prostate gland as the ejaculatory duct. The right and left ejaculatory ducts open into the urethra within the prostate gland. They are much smaller ducts than the vas deferens and only 2 centimeters long.

SEMINAL VESICLES   The seminal vesicles are lobulated sacs located at the posterior surface of the bladder. They secrete a fluid that forms a part of the semen. The fluid passes down a small duct and enters the ejaculatory duct. It is thought to contribute to the viability of the spermatozoa.

PROSTATE GLAND   A muscular and glandular organ, the prostate gland, is located below the bladder and anterior to the rectum. The base of the urethra passes through it. The prostatic secretion is alkaline, somewhat milky, and contributes to the odor of semen. The gland measures about 4 centimeters in its transverse or horizontal plane and about 3 centimeters in its vertical plane. The base of the urethra runs almost vertically through the anterior portion of the gland when the body is in a standing position. The lobules of the gland discharge their secretion through 20 to 30 small ducts, which open by minute pores into the urethra (Figure 21.9).

The prostate gland is somewhat unfortunately located, since it surrounds the base of the urethra. It tends to enlarge in older men and often constricts the urethra and makes it difficult to empty the bladder.

BULBOURETHRAL (COWPER'S) GLANDS   These are two small, yellow glands about the size of peas located in the bulb region at the base of the penis and emptying into the urethra from below. The secretion is a clear, mucoid fluid discharged during sexual stimulation. Much of the secretion precedes seminal emission, and it has been suggested that its function is to lubricate the urethra and glans penis as well as to neutralize the uric acid in the urethra before the spermatozoa pass through. A small amount of the secretion is contributed to the seminal fluid (Figure 21.9).

PENIS   The penis is the copulatory organ of the male. It is attached to the pubic arch and covered with skin that is continuous with the integument covering the scrotum. The body of the penis is composed of three longitudinal columns of *erectile tissue.* Two of these bodies are in the dorsolateral part of the penis and are called the *corpora cavernosa;* the third is midventral and contains the urethra (the *corpus spongiosum urethrae*). Erectile tissue is composed of blood spaces, which ordinarily are not distended with blood, the penis then being soft and flaccid. Sexual excitement causes blood to pour into these spaces faster than it is drained away by the veins. As a result the walls of the tissue become distended with blood, and the penis becomes hard and erect. It is in this condition that it is inserted into the vagina in the act of sexual intercourse. After sexual excitement has passed, blood is drained out of the erectile tissue and the penis becomes soft again. Erectile tissue is also present in the clitoris, as we have noted.

The corpus cavernosum urethrae is reflected back over the end of the penis like a cap and contains the vertical slit, which is the external orifice of the urethra. The smooth tip of the penis is the glans portion and is covered by loose skin called the *foreskin,* or *prepuce.* Sometimes the foreskin covers the glans too tightly or becomes adherent. *Circumcision* is an operation to remove the foreskin. The operation may be complete or partial. The area posterior to the glans contains modified sebaceous glands that secrete a soft, whitish substance, which soon deteriorates. Circumcision exposes the surface of the glans and makes it easier to cleanse the area where the secretion (*smegma*) has collected.

SEMINAL EMISSION   The amount of fluid in a single seminal emission averages 3 milliliters. The average number of spermatozoa is around 120 million per milliliter. Some 300 million spermatozoa in a seminal discharge is a remarkably large number, especially since only one is permitted to enter the ovum at fertilization. When the number of spermatozoa per milliliter drops below 60 million, the fertility of the individual is considered to be far below average.

Seminal emission is accomplished by stimulating the vas deferens, seminal vesicles, and prostate gland to pour their accumulated contents into the base of the urethra by way of the ejaculatory ducts. Skeletal muscles of the bulb region then contract, and with the relaxation of the urethral sphincter, the semen is ejected. The ejection is intermittent, coming in several waves. The ejection of semen is a reflex act involving both the parasympathetic and sympathetic divisions of the autonomic nervous system, but the basic muscular reactions are largely parasympathetic. A variety of sensations resulting from the ejection of semen constitutes the orgasm in the male. It is followed by relaxation of the reproductive organs and general lassitude. Orgasm in the female is essentially the same reflex phenomenon following sexual stimulation to a high degree, but there is no similar ejection of fluid.

The spermatozoa become highly motile upon ejection. If they are ejected in the vagina of the female, they move up through the uterus and the fallopian tubes at a surprising rate of speed. Recent estimates indicate that the sperm reach the upper part of the uterine tubes within 30 minutes after being deposited near the cervix of the uterus. The sperm may maintain their motility in various parts of the female genital tract for at least 50 hours. The fact that sperm are still motile, however, does not mean necessarily that they would be able to effect fertilization. The fertilizable life of the ovum has been estimated at 6 to 12 hours. It may be less than 6 hours. It would seem that the period of high fertility in the female is very closely linked to the time of ovulation.

Sexual intercourse is not necessary to maintain the reproductive organs in a good state of health; neither is masturbation. Healthy young men may experience occasional nocturnal emissions, commonly called *wet dreams*, which serve to eliminate excess accumulations of semen.

STERILIZATION

The effects of castration were discussed in Chapter 20, but sterilization by other means is often confused with castration. Sterilization in the male may be caused by accidental injury to the conducting ducts or by sectioning and tying off the vas deferens in the scrotum (vasectomy). In the female sectioning and tying off the uterine tubes near the uterus requires an abdominal operation. It is important to understand that an operation of this sort designed to produce sterility does not affect the gonads. They will continue to produce germ cells and sex hormones. Since the hormones are absorbed by the blood, there is no change in the availability of these hormones and therefore no change in sexual characteristics that they control. Sectioning of the conduct-

The reproductive system

ing tubes merely prevents the germ cells from becoming available for fertilization. Some of the venereal diseases can also cause sterility through injury to the conducting tubes or the gonads. Suppression of germ cells by chemical means probably will supersede surgical methods of sterilization in the future.

VENEREAL DISEASES

GONORRHEA   Venereal diseases are almost always acquired by sexual contact. There are several well-known venereal diseases; gonorrhea and syphilis are the most important. Gonorrhea is caused by pus-forming, round, diplococcal forms of bacteria. The organism, *Neisseria gonorrhoeae* (Figure 21.11), invades the urogenital tract, causing inflammation of membranes lining the passageways. The inflammation often causes closure of small ducts and results in sterility. If untreated, the disease may spread through blood or lymph to various parts of the body. It often localizes in the joints, causing an arthritis of gonorrheal origin.

Gonorrheal ophthalmia is an infection of the conjunctiva and can be acquired from recently contaminated towels or from the hands. It is a serious disease and can result in blindness unless promptly treated. Infants can acquire a similar infection of the eyes (*opthalmia neonatorum*) as the head of the newborn infant passes through an infected birth canal. Many states require that the eyes of newborn infants be treated for this disease whether or not there is any indication of gonorrhea in the mother. Such treatments have helped to reduce blindness in newborn infants.

SYPHILIS   Another serious venereal disease responsible for a great deal of human misery is syphilis. It causes many persons to become invalids for life and is responsible for many deaths. The causative organism, method of attack, and course of the disease are quite different from those of gonorrhea. The organism, *Treponema pallidum*, is a tightly coiled spirochete 6 to 14 microns

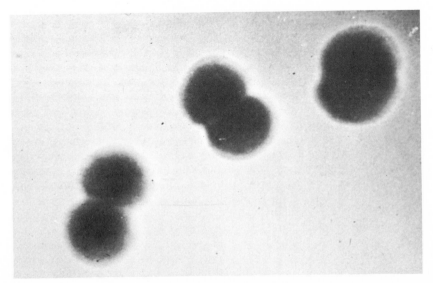

FIGURE 21.11
The causative organism of gonor-rhea, *Neisseria gonorrhoeae*. (*Courtesy of U. S. Department of Health, Education, and Welfare, Public Health Service.*)

Venereal diseases

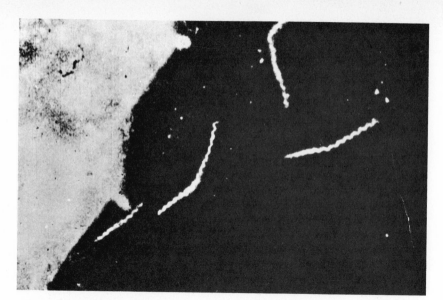

in length and only 0.3 micron in diameter (Figure 21.12). It is capable of entering the body through a soft, moist surface; the usual points of entry are on the surface of the penis in the male or on the external genitalia and in the vagina of the female. A lesion, called a *chancre*, develops at the point of entry after an incubation period of 2 to 4 weeks. If the lesion occurs within the vagina or on the cervix of the uterus, it may not be noticed at this stage. This is the primary stage of the disease. The ulcers are swarming with syphilitic spirochetes and are highly infectious. The ulcer disappears after about 6 weeks, and an uninformed person may think the disease has been cured or at least arrested.

The spirochetes then enter the bloodstream and spread all through the body in the second stage of the disease. During the next 6 to 12 weeks various symptoms of disease manifest themselves. Often a syphilitic rash appears on the skin, and lesions may appear in various parts of the body. These symptoms also subside after a time, and the disease may become latent for several months or several years.

The organism becomes localized during the third stage of the disease. The localization can affect the heart, aorta, brain, spinal cord, skeletal system, or other organs. Degenerative lesions of vital organs often cause death. Syphilitic lesions in the brain can cause insanity or paresis; in the spinal cord they can cause a kind of paralysis called *locomotor ataxia*. The heart and the arch of the aorta are commonly attacked.

There are several good diagnostic tests that determine whether or not a person has acquired the disease. Many states require a blood test before marriage, for if one of the partners in marriage has syphilis, the other partner will very likely become infected. An infected pregnant mother can transmit the disease to the fetus. Such infection is called *congenital syphilis* and is responsible for many stillbirths. If the effects of the disease are not evident in the newborn infant, symptoms may appear later.

Venereal disease must be treated promptly by a physician. Gonorrhea and syphilis can now be brought under control, especially if treatment is begun during the early stages. Penicillin has proved to be of great value in the treatment of these diseases, but no one should consider trying to treat himself. Prophylaxis is largely limited to the male but is apt to be ineffective. At any rate, one should keep in mind that gonorrhea in the male invades by way of the urethra, whereas syphilis is more likely to invade through the surface of the penis in its first stage; prophylaxis, to be effective, must include both these areas.

The internal organs of the female reproductive system are the following: ovaries, oviducts, uterus, and vagina. The ovaries lie on either side of the uterus and below the oviducts. Ova develop within follicles. As the follicles and ova mature, the larger follicles appear at the surface of the ovary. The ova rupture through the surface about the middle of the monthly cycle in the process known as ovulation.

After ovulation, the follicle is transformed into a yellowish body called the corpus luteum. It secretes progesterone and estrogens.

When the ovum is ruptured from the ovary, it passes down the oviduct. If fertilized, it passes through its early cleavage stages and becomes implanted in the soft lining of the uterus.

The uterus is a thick-walled, muscular organ, somewhat pear-shaped, with the neck or cervix projecting into the vagina. The vagina is the canal leading inward from the vestibule of the external genitalia. The external orifice of the virginal vagina is partially occluded by a membrane called the hymen. The labia majora, the labia minora, the clitoris, and the vestibule constitute the female external genitalia. The mammary glands are the breast glands.

If the ovum is fertilized and implanted, the corpus luteum persists and is very active through the first 3 or 4 months of pregnancy. The progesterone level is high, producing a favorable environment for pregnancy. After the first few months, the placenta produces enough estrogen and progesterone to supply the needs of the body; the corpus luteum then begins to regress. The placenta also secretes chorionic gonadotropin.

The spermatozoa of the male arise in the testes. The testes are ovoid organs, which lie in a sac called the scrotum. The copulatory organ of the male is the penis.

A greatly coiled tubule, the epididymis, leads sperm away from the testis into the vas deferens. The vas deferens passes through the inguinal canal and over the urinary bladder. Posterior to the urinary bladder, the vas deferens receives a duct from the seminal vesicle. It then enters the tissue of the prostate gland as the ejaculatory duct. The short right and left ejaculatory ducts open into the urethra. The urethra traverses the penis to the external orifice.

The seminal vesicles and the prostate gland secrete a portion of the seminal fluid. The prostate gland surrounds the base of the urethra. If it enlarges, it may constrict the urethra.

The two most important venereal diseases are gonorrhea and syphilis. The gonorrheal organism is a diplococcal form of bacteria. It is pus-forming and invades by way of the mucous membranes of the urogenital tract.

The causative organism of syphilis is a coiled spirochete, *Treponema pallidum*. There are three stages of the disease. In the primary stage a chancre develops at the point of entry after an incubation period of 2 to 4 weeks. In the second stage, the spirochetes enter the bloodstream and spread over the body; a syphilitic skin rash or mouth sores may appear during this stage.

The organisms become localized in various organs of the body during the third stage. The brain, spinal cord, heart, and arch of the aorta are commonly attacked.

QUESTIONS

1 Outline the menstrual cycle.
2 Discuss the changes that occur in the ovary and in the lining of the uterus following ovulation.
3 Trace the pathway of the sperm from their origin in the testes to the place where the ovum is fertilized.
4 Explain the mechanism that enables erectile tissue to function.
5 Discuss the changes that occur at puberty.
6 Approximately how many sperm are produced in a single seminal discharge? How many ova are produced during the monthly cycle?
7 How long do spermatozoa retain their motility? What is the estimated fertilization life of the ovum? Estimate the time of highest fertility in the female.
8 By what means does gonorrhea invade the body? How does syphilis enter? Describe the causative organism in each case.
9 How are newborn infants infected in the case of gonorrhea; in the case of syphilis?
10 Discuss the three stages in the development of syphilis.

SUGGESTED READING

ARMSTRONG, D. T.: Reproduction, *Ann. Rev. Physiol.,* 32:439–470 (1970).

CLARK, J. H., J. ANDERSON, and E. J. PECK, JR.: Receptor-Estrogen Complex in the Nuclear Fraction of Rat Uterine Cells during the Estrous Cycle, *Science,* 176:528–530 (1972).

HARRISON, R. J., and WM. MONTAGNA: "Man," Appleton Century Crofts, New York, 1969.

JAFFE, F. S.: Toward the Reduction of Unwanted Pregnancy, *Science,* 174:119–127 (1971).

JOHNSON, C. E.: "Human Biology: Contemporary Readings," Van Nostrand, Reinhold Co., New York, 1970.

LANGER, W. L.: Checks on Population Growth: 1750–1850, *Sci. Am.* 226:93–99 (1972).

LISK, R. D.: The Physiology of Hormone Receptors, *Am. Zoologist,* 11:755–767 (1971).

SCHWARTZ, N. B., and C. E. McCORMACK: Reproduction: Gonadal Function and its Regulation, *Ann Rev. Physiol.,* 34:425–472 (1972).

WILLIAMS-ASHMAN, H. G., and A. H. REDDI: Actions of Vertebrate Hormones, *Ann. Rev. Physiol.,* 33:31–82 (1971).

unit ten

developmental anatomy and physiology

chapter 22
human development

The gonads are the male and female generative organs. They are the primary sex organs, which produce the germ cells, or gametes. The male gonads, or testes, produce spermatozoa, and the female gonads, or ovaries, develop ova. The gametes are derived from undifferentiated germ cells and develop by a process called *gametogenesis*. The development of male gametes, or spermatozoa, is called *spermatogenesis;* in the female the development of ova is referred to as *oogenesis*.

GAMETOGENESIS

The gonads form in the early embryo from two parallel ridges of peritoneal epithelium called the *genital ridges*. Primordial germ cells become differentiated and appear in the area where the reproductive organs form. The exact origin of the primordial germ cells remains a biological problem. There are also various interpretations regarding the further development or fate of the primordial germ cells. Germinal epithelium containing primordial germ cells can be demonstrated in 10-millimeter human embryos, but it is questionable whether these cells give rise to mature sperm or ova. The problem is of especial interest in the female, where the ovary contains thousands of potential ova at birth. Are these the cells that form mature ova many years later, or are new germ cells formed constantly from the germinal epithelium? Many primordial germ cells degenerate, while others divide and become closely associated with the germinal epithelium of the gonads. At present there is no definite answer to the question.

The development of mature spermatozoa and ova is a complicated process involving nuclear and cytoplasmic changes of great significance. There are marked differences between the development of spermatozoa and ova, but here we shall emphasize the similarities in the behavior of the chromosomes. The process begins with the multiplication of cells derived from undifferentiated germ cells. The cells increase in number by mitotic division. These cells contain the complete, or diploid, number of chromosomes, which in a human being is 46. Traditionally, the number of chromosomes in man was thought to be 48, but more recent studies indicate that 46 is the correct number.

The developing germ cells pass through a period of growth and multiplication, after which certain cells in the male may be designated as *primary spermatocytes* and in the female, *primary oocytes*.

MEIOSIS

We have considered the behavior of the chromosomes during the process of mitosis in Chapter 3. Mitotic division provides for an equal number of chromosomes in each of the two resulting cells, whereas the primary objective of meiotic division is the reduction of the chromosome number to half

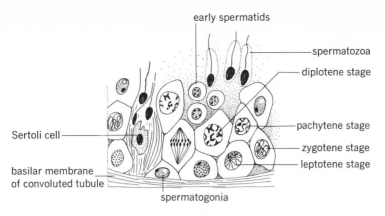

early spermatids

spermatozoa

diplotene stage

pachytene stage

zygotene stage

leptotene stage

Sertoli cell

basilar membrane
of convoluted tubule

spermatogonia

FIGURE *22.1*
Some of the stages of spermatogenesis (semidiagrammatic). (*After
various sources.*)

that existing in the original germ cells. The complete number of chromosomes is said to be the *diploid* condition, whereas the reduced number is *haploid*. During meiosis the chromosome number is reduced from the diploid condition in the undifferentiated germ cells to the haploid condition in the gametes.

PROPHASE   The chromosomal behavior in the long and varied prophase I of meiosis can be followed step by step, and the different stages are named. The first stage is the *leptotene stage*, in which the chromosomes change from the resting-cell condition and become evident as long, slender threads. The behavior of the chromosomes is best observed in the male; in a prepared section of the testis various stages of mitotic and meiotic divisions can be observed (Figure 22.1).

The *second*, or *zygotene*, *stage* is characterized by the pairing of homologous chromosome threads. The slender threads lie side by side in a process called *conjugation*, or *synapsis*. The paired chromosomes now appear as single "chromosomes"; but since they are really two, they are referred to as *bivalent*. The chromosomes in the diploid state consist of 23 pairs of homologous, or like, chromosomes. Of any one pair, one chromosome represents the paternal inheritance and the other, the maternal. Twenty-three individual chromosomes are contributed in the sex cells of each parent. When like chromosomes are associated in pairs, one is said to be the *homolog* of the other. The chromosomes ordinarily appear to be scattered throughout the nucleus, but in the zygotene stage like pairs lie together. Meiosis is a gradual process; as in mitosis, one stage gradually merges into another (Figure 22.2).

The *pachytene stage* is the third stage of the meiotic prophase, in which the chromosome threads shorten and thicken somewhat. Each chromosome thread is then double except at the *centromere* (kinetochore), and the pair is known as a *dyad*. Since the chromosome threads are side by side in conjugation, the two dyads divide longitudinally to form *tetrads*. Each chromosome thread at this stage is called a *chromatid*. A tetrad therefore may be considered to consist of a bivalent chromosome of four chromatids.

In the *diplotene*, or *fourth*, *stage*, the paired chromosomes tend to move away from each other in certain regions; but at some points they stay together and cross over each other, forming *chiasmata* (chiasma, singular, a crossing). The chiasmata are of great significance genetically, since it is in this region that the four chromatids of a tetrad often appear to exchange

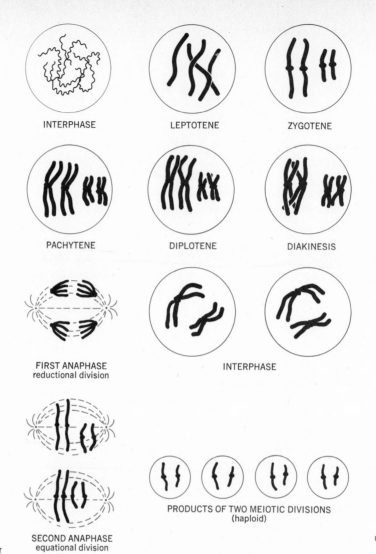

INTERPHASE LEPTOTENE ZYGOTENE

PACHYTENE DIPLOTENE DIAKINESIS

FIRST ANAPHASE
reductional division

INTERPHASE

SECOND ANAPHASE
equational division

MEIOSIS

PRODUCTS OF TWO MEIOTIC DIVISIONS
(haploid)

FIGURE 22.2
Various stages of meiosis in the development of spermatocytes or oocytes.

portions. This process is known genetically as *crossing over* and may bring about a new arrangement of groups of genes on the chromatids involved.

DIAKINESIS The chromosomes at the diplotene stage begin to shorten and become thicker. There may be rotation and twisting. The diplotene stage in oogenesis is an exception in that the chromatids elongate and become somewhat diffuse, but later they shorten and thicken as the chromatids enter a fifth stage, *diakinesis*, preparatory to the first meiotic division. The chiasmata tend to move toward the ends of the chromatids at diakinesis. Since the chromosomes stain deeply at this time, it is an important stage in studying and counting the chromosomes of various species. The nucleolus disappears during this stage, and the disappearance of the nuclear membrane marks the end of diakinesis and the beginning of the first metaphase.

Human development

572

METAPHASE I The bivalent chromosomes (tetrads) are distributed at the equatorial plate of the spindle.

ANAPHASE I AND THE FIRST MEIOTIC DIVISION The paired chromosomes move away from their homologs, often presenting a V-shaped appearance as they move toward opposite poles. The separation of the chromosome pairs reduces the number of the chromosomes to half in each resulting cell. The first meiotic division is a reductional division. The secondary spermatocyte or oocyte produced contains the haploid number of double-stranded chromosomes. The centromeres of the individual chromosomes have not divided.

TELOPHASE I The telophase is usually a short period preceding the initiation of the second meiotic division.

THE SECOND MEIOTIC DIVISION The secondary spermatocyte or oocyte passes through another prophase, metaphase, and anaphase. A spindle forms in this second metaphase, and the chromosomes become arranged near the equator. The second anaphase is characterized by the division of the centromeres and the separation of the chromatids. This is commonly called an *equational division*, since there is no further reduction of chromosome number. Each chromatid in the second meiotic division becomes a separate chromosome. The spermatid or ovum contains the haploid number of individual chromosomes.

ESSENTIAL DIFFERENCES BETWEEN MITOSIS AND MEIOSIS During mitosis there is no pairing of homologous chromosomes; the individual chromosomes replicate themselves, and the duplicates then separate during cell division. Each cell resulting from a mitotic division receives the same number of chromosomes, which have the same genetic composition as the parent cell (Figure 22.3).

Meiotic division is characterized by the pairing of homologous chromosomes during the first prophase and the subsequent separation of chromatids at anaphase II. This anaphase separation is the reductional division, which results in the production of secondary spermatocytes or oocytes containing the reduced, or haploid, chromosome number. The gametes, therefore, are haploid, and the diploid condition is restored at the time of fertilization of the ovum when the male and female pronuclei unite. As a result of crossing over during meiosis, the genetic constitution of the chromosomes may be changed, and the union of the sex cells at fertilization will bring about new combinations of characters.

THE BEHAVIOR OF THE SEX CHROMOSOMES IN GAMETOGENESIS

Although the sex chromosomes are of special interest, it should be remembered that all chromosomes contain hereditary characters called *genes*. The chromosomes exclusive of the sex chromosomes are called *autosomes*. In man there are 44 autosomes plus two sex chromosomes. The sex chromosomes of the male are designated XY; those of the female are XX. The total number of chromosomes may be indicated as 44 + X + Y or 44 + X + X (Figures 22.4 and 22.5).

The two X chromosomes of the female follow exactly the same pattern in meiosis as the autosomes. The net result provides the developing ovum

MITOSIS

PROPHASE
(chromosomes
become visible)

METAPHASE
(no pairing of
chromosomes)

ANAPHASE
(daughter
chromosomes
move toward
opposite poles)

LATE ANAPHASE
(complete number
of chromosomes
go to each
resulting cell)

MEIOSIS: first meiotic division

PROPHASE
(chromosome pairs
appear together)

METAPHASE
(bivalent chromosomes
on equatorial plate as
tetrads)

ANAPHASE
(paired chromosomes
move away from
their homologs)

LATE ANAPHASE
(the secondary
spermatocyte
or oocyte contains the
haploid number
of chromosomes)

MEIOSIS: second meiotic division

PROPHASE

METAPHASE

ANAPHASE
(separation of
chromatids)

LATE ANAPHASE
(each spermatid
or ovum receives
the haploid
number of individual
chromosomes)

FIGURE 22.3

A series of diagrams illustrating
the differences between mitosis
and the first and second meiotic
divisions.

with one X chromosome. The XY chromosomes of the male follow essentially
the same meiotic procedure, but pairing is limited to homologous parts of
the two chromosomes. Reduction division (when the chromosome pairs
separate) produces secondary spermatocytes that contain either an X or a
Y chromosome. The spermatids, and ultimately the spermatozoa, therefore
carry either X or Y chromosomes. At fertilization, when male and female
gametes unite their pronuclei to form the fertilized egg, or zygote, the
chromosome pairs will be restored. If the ovum is fertilized by a Y sperm,
the chromosome pair is represented as XY and the offspring will be male;
fertilization of the ovum by an X sperm restores the XX condition, and the
offspring will be female. Many animals, such as reptiles, birds, moths, and
butterflies, have a different pattern of sex determination. In these animals

Human development

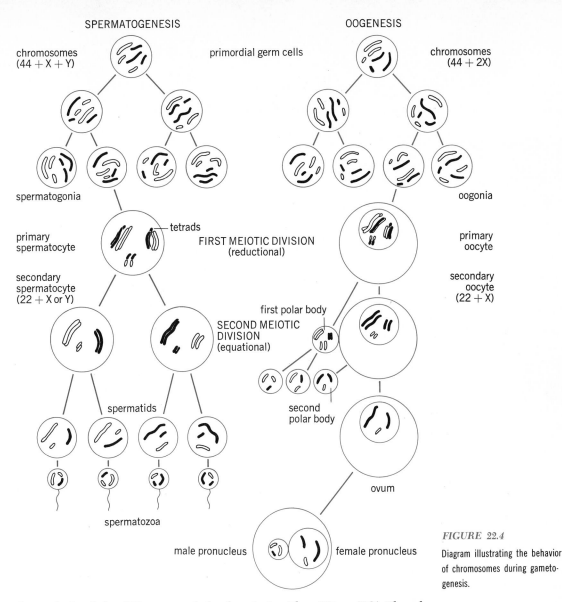

SPERMATOGENESIS

chromosomes
(44 + X + Y)

primordial germ cells

OOGENESIS

chromosomes
(44 + 2X)

spermatogonia

oogonia

primary
spermatocyte

tetrads

FIRST MEIOTIC DIVISION
(reductional)

primary
oocyte

secondary
spermatocyte
(22 + X or Y)

first polar body

SECOND MEIOTIC
DIVISION
(equational)

secondary
oocyte
(22 + X)

second
polar body

spermatids

ovum

spermatozoa

male pronucleus

female pronucleus

FIGURE 22.4

Diagram illustrating the behavior
of chromosomes during gametogenesis.

the male is of the XX type, and the female is either XY or XO. Though
the sex pattern is established by certain combinations of sex chromosomes,
sexual differentiation is influenced by hormones and possibly other factors
in the internal environment.

The behavior of the sex chromosomes is of special interest in regard
to understanding sex-linked characteristics. Two such common characteristics
are red-green color blindness and hemophilia. The genes for these conditions
are located on an X chromosome. Sons do not inherit the condition from
their father, since they do not receive an X chromosome from the male
parent. Daughters can receive the X chromosome (which carries the gene
responsible for the defect) from either parent. Males will be color blind or
have hemophilia if they receive the defective gene from their mother (see
Chapters 12 and 13). The method of inheritance of sex-linked characters

*The behavior of the sex
chromosomes in gametogenesis*

575

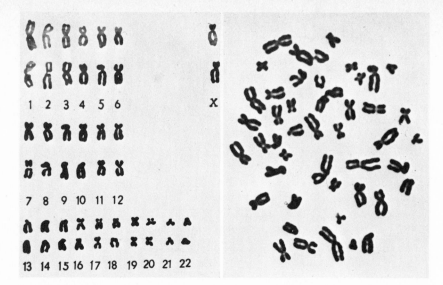

FIGURE 22.5

Chromosomes of a normal female. (*Courtesy of Earl H. Newcomer.*)

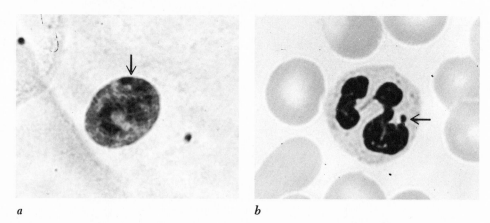

a   *b*

FIGURE 22.6

a Nucleus of normal female squamous epithelium showing Barr body (see arrow); *b* white blood cell of normal human female showing drumstick on nucleus (see arrow). ×1,400. (*Courtesy of Carolina Biological Supply Company.*)

affords additional proof that genes are located on chromosomes. The Y chromosome is thought to contain only a few genes, and in some animals the Y chromosome is described as "empty" or containing no genes (Figure 12.18).

Sex-chromatin differences in cells Certain cells of the female show a dark-staining body near the edge of the interphase nucleus. This sex chromatin, or *Barr body*, is related to the female X chromosome and is not found in male cells. Though present in various tissues, it is commonly demonstrated in cells scraped from the mucosal membrane lining the cheek. A similar structure is found in polymorphonuclear leukocytes of the female. This is a stalked body resembling a drumstick attached to one of the lobes of the nucleus. It is found in only a small number of these leukocytes and only in females. These examples of sex chromatin can be demonstrated in cells that are not undergoing mitosis (Figure 22.6*a* and *b*).

Abnormalities in chromosome number Occasionally the chromosome number varies from the normal 46. Chromosomes, of course, have to be observed when

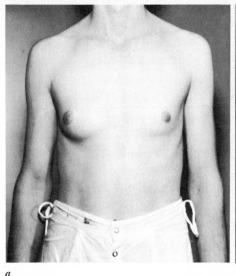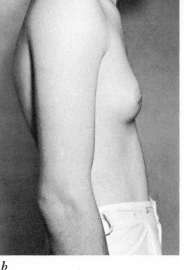

FIGURE 22.7
Klinefelter's syndrome. (*Courtesy of Armed Forces Institute, Washington, D. C.*)

a   *b*

the cell is undergoing division. The cells are usually prepared, stained, and mounted on a microscope slide. One chromosome abnormality is known as *Klinefelter's syndrome* (Figure 22.7). The individual is male, but his sex chromosomes are XXY, $2n = 47$. The testes are undeveloped; there is some enlargement of the breasts, as well as often some mental retardation.

In cases of *Down's syndrome (mongolism)* there are 47 chromosomes but the extra one is an autosome, not a sex chromosome. These individuals are of short stature, and their fingers and toes are short and thick. The face is rounded, and the degree of intelligence is very low (mongolian idiocy) (Figure 22.8).

Another abnormality is called *Turner's syndrome.* These individuals are female but have only one X chromosome. They are sexually infantile, with immature ovaries. They are usually quite short and have peculiar "webbed" necks. They are XO individuals, and their $2n$ chromosome number is 45.

Nondisjunction in the paired X chromosomes of the mother could readily give rise to Klinefelter's syndrome provided the ovum is fertilized by a Y sperm. Turner's syndrome probably arises when the female ovum does not contribute an X chromosome. If there is only a Y chromosome the ovum fails to develop.

SPERMATOGENESIS

The spermatozoa develop in the seminiferous tubules of the testes. There is a period of growth and multiplication of the diploid spermatogonia, and eventually some cells with enlarged nuclei can be recognized as primary spermatocytes (Figure 22.1). The primary spermatocyte undergoes a meiotic division, as previously described, producing two haploid secondary spermatocytes (Figure 22.4). A further equational division results in the production of four spermatids. Without further division the spermatids develop into mature spermatozoa. Mature spermatozoa are not produced until puberty, but from puberty through the long period of reproductivity in the male, they are produced in remarkably large numbers. It has been estimated that

Spermatogenesis

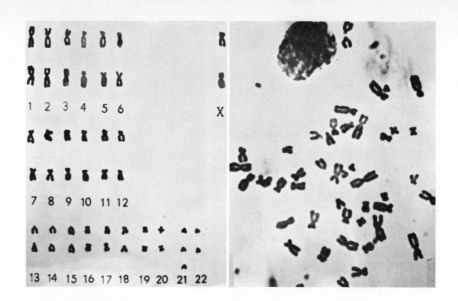

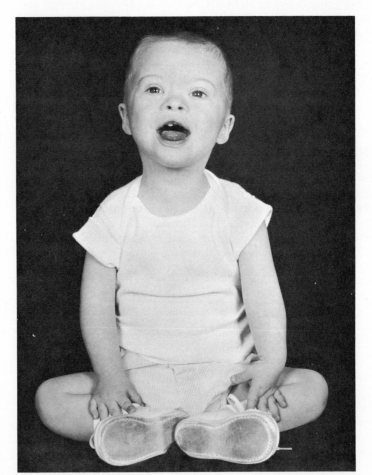

FIGURE 22.8

Chromosomes of a female with Down's syndrome (top). (*Courtesy of Earl H. Newcomer.*) Boy with Down's syndrome (bottom). (*Courtesy of Maurice Whittinghill, "Human Genetics and Its Foundations," Reinhold Publishing Corporation, New York, 1965.*)

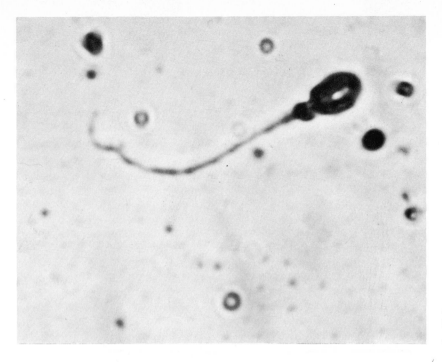

FIGURE 22.9

Human spermatozoon. (*Courtesy of Dr. L. B. Shettles.*)

there are 200 million to 500 million spermatozoa in a single seminal discharge. The spermatozoa are not motile in the testes and are found clustered around tall Sertoli cells (or sustentacular cells), which support them and perhaps nourish them.

SPERMATOZOA Human spermatozoa are microscopic in size, with a length of only 0.06 millimeter. The male gamete is not a typical cell; it is a motile specialized germ cell carrying the male chromatin material in a condensed condition. The spermatozoon consists of a head, neck, body, and tail. The head is oval in its widest aspect, but narrow and oblong in side view. The neck is very short, and the body is often called a connecting piece between the neck and the tail. The long tail is capable of violent swimming motions, which propel the sperm forward. The chief portion of the tail is enclosed in the same sheath that covers the head, neck, and body. The short end piece is the termination of the axial filament that extends through the body and tail (Figure 22.9).

Electron micrographs reveal much more of the minute structure of spermatozoa (Figure 22.10). The axial filament in cross section is shown to be composed of nine pairs of fibers arranged in a circumferential ring, with two fibers in the center. This is a characteristic structure of cilia and flagella. The mitochondrial sheath in the neck region contains rows of closely packed mitochondria. Proximal and distal centrioles are also present in the neck region.

The development of ova takes place in the ovaries; the process is called **OOGENESIS** *oogenesis.* The germ cells are derived from germinal epithelium in the fetus. At birth thousands of oogonia are located in the cortex of the ovary. Many

579

FIGURE 22.10

Electron micrographs of spermatozoa: longitudinal section of tail
portion in rat sperm, showing filaments and mitochondria; cross
section of tail portion, illustrating
the typical 9 + 2 arrangement of
filaments (inset). (*Courtesy of
James A. Freeman, "Cellular
Fine Structure," McGraw-Hill
Book Company, New York,
1965.*)

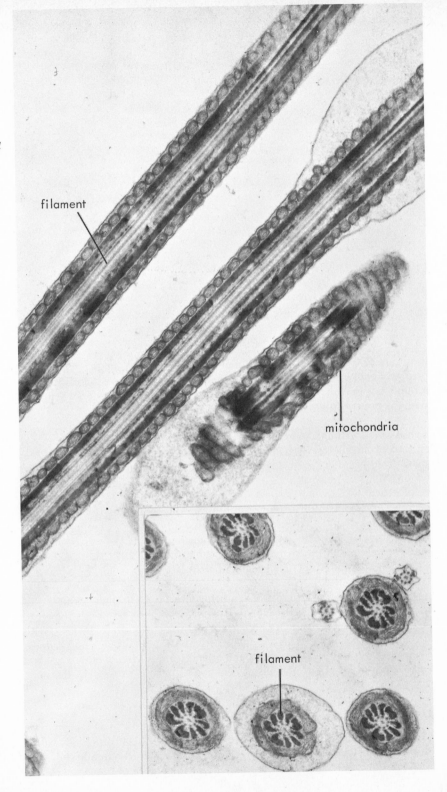

Human development

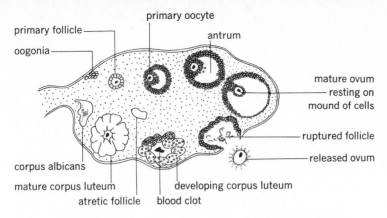

primary oocyte

primary follicle

oogonia

antrum

mature ovum resting on mound of cells

ruptured follicle

released ovum

corpus albicans

mature corpus luteum

developing corpus luteum

atretic follicle   blood clot

FIGURE 22.11

Schematic diagram of the ovary, illustrating the development of the ovum and its release from the follicle. (*Redrawn from Bradley M. Patten, "Human Embryology," McGraw-Hill Book Company, New York, 1968.*)

oogonia fail to develop, and it is generally accepted that oogenesis does not progress beyond multiplication of the oogonia until puberty. The reproductive period of women is approximately from fourteen to forty-seven years of age, and during this time the ovaries are actively producing mature ova. There is a tendency to reduce the number of mature ova and to conserve the amount of yolk. Some of the oogonia enlarge, and surrounding oogonia become modified into a protective sac of cells called a *primary follicle* (Figure 22.11). The enlarged oogonium surrounded by the cells of the primary follicle is then called a *primary oocyte*. It is diploid, and its chromosome number may be designated as 44 + X + X (Figure 22.4). Mammalian ova do not contain a great amount of yolk, and it is distributed throughout the cytoplasm. The nucleus is centrally located. The primary oocyte is conspicuously larger than the individual follicle cells that surround it.

The ovarian (graafian) follicle grows rapidly as the oogonium develops. It changes from a single layer of cells to a double layer and finally to a profusion of cells around the follicle and the oogonium. The oogonium is at first centrally located; with the rapid growth of the follicle, a fluid-filled cavity develops, and the oogonium lies in a mound of cells along the side.

MATURATION OF THE OVUM   The behavior of the chromosomes in meiosis has already been described. We observed that during spermatogenesis reduction division resulted in two secondary spermatocytes of equal size. Reduction division of the primary oocyte produces two secondary oocytes of unequal size. The secondary oocyte that is destined to become a mature ovum retains practically all the yolk but is able to discard half the bivalent chromosomes into a small degenerate cell commonly called a *polar body*. The polar body is actually a small oocyte, but since it is often found attached to the egg near the nucleus, it was called a polar body before its real identity was determined. Occasionally the polar body undergoes a second division, but usually it degenerates before such a division occurs.

The secondary oocyte and the follicle then develop rapidly and approach the surface of the ovary. The mature follicle is approximately 1 centimeter in diameter, but the developing ovum is only a little over 0.1 millimeter in diameter. The second maturation division in mammalian ova does not occur ordinarily until after the sperm enters the ovum. The mature follicle ruptures, and the developing ovum is washed out with the follicular fluid in the process called *ovulation*. After the egg is free from the ovary

Oogenesis

581

and usually after the entrance of the spermatozoon, the second maturation division occurs and the second polar body is formed. The second polar body and the mature ovum are haploid after an equational division of the chromosomes in which there is no further reduction but simply a separation of chromatids.

The nucleus of the ovum then becomes the female pronucleus and unites with the male pronucleus as fertilization is accomplished. Oogenesis resembles spermatogenesis in its general plan, but the final result is one mature ovum instead of four spermatids. The first polar body corresponds to a small secondary oocyte. If it were to divide, there would be three polar bodies and one mature ovum, comparable in number to the four spermatids formed in spermatogenesis. The meiotic behavior of the chromosomes is essentially the same in the ovum and spermatids, but the chromosome content may differ, in that the mature ovum always contains an X chromosome, whereas the spermatid may contain either an X or a Y sex chromosome.

HUMAN OVUM The ova of placental mammals, unlike the eggs of reptiles and birds, are minute and contain little yolk. Much of our knowledge concerning the development of the human ovum and its early embryology has been gained from the study of these processes in other mammals. The human ovum is so small that it is barely visible to the unaided eye. It is estimated to weigh about one-millionth of a gram. When the oocyte ruptures from the ovarian follicle, it is surrounded by numerous cells constituting the *corona radiata* (Figure 22.12). The limiting noncellular layer around the ovum is the transparent *zona pellucida*. Stained specimens may show early mitotic divisions of the chromosomes preparatory to budding off the first or second polar bodies.

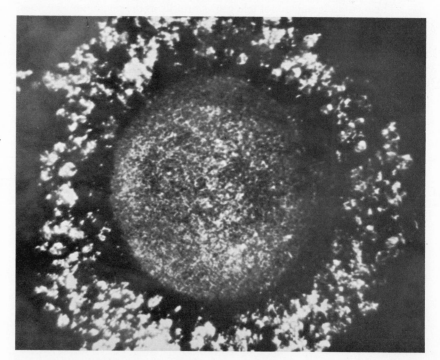

FIGURE 22.12
A living human oocyte surrounded by from 3,000 to 4,000 corona radiata cells, photographed in phase contrast. The oocyte was removed from an ovarian follicle by aspiration with a syringe and needle as it neared ovulation. Original diameter of oocyte proper, 100 microns. (*Courtesy of Dr. L. B. Shettles.*)

Human development

582

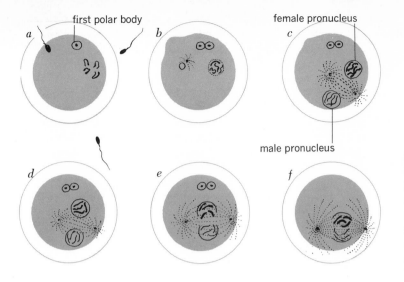

first polar body

female pronucleus

male pronucleus

FIGURE 22.13
Fertilization of the ovum and the formation of the first cleavage spindle, shown diagrammatically. (*Redrawn from William Patten in Bradley M. Patten, "Human Embryology," McGraw-Hill Book Company, New York, 1953.*)

FERTILIZATION

The intricate stages of gametogenesis reach a logical culmination at the union of male and female gametes in the process known as *fertilization*. The first meiotic division, with the extrusion of the first polar body, has been accomplished; ovulation has occurred, and the ovum starts on its journey down the oviduct. Fertilization takes place high in the oviduct, the spermatozoa traveling rapidly through the neck and body of the uterus and on into the oviduct (or uterine tube) to meet the descending ovum. The ovum is waved down the oviduct by the beating of cilia, which line the tube, and by the muscular contractions of the walls.

Only one sperm enters the egg, although there is evidence that the presence of numerous spermatozoa is necessary to ensure fertilization. The cells of the corona radiata are held together by an organic acid called *hyaluronic acid.* The individual sperm contains an enzyme called *hyaluronidase,* which is capable of breaking down hyaluronic acid, but the amount of enzyme contained in a single sperm is so minute that it is ineffective. If several thousand sperm surround the ovum, however, the cells of the corona can be loosened so that one sperm can successfully enter the egg and accomplish fertilization.

When the sperm enters the egg, the tail of the sperm is absorbed and the nucleus begins to move through the cytoplasm of the egg as the male pronucleus. The entrance of one sperm into the egg causes a change in the surface membrane, and no additional sperm can gain entrance. The entrance of the sperm provides the stimulus for the second maturation division. The second polar spindle is by then already formed in mammalian ova, and the second polar body is promptly extruded (Figure 22.13).

The male and female pronuclei then approach each other, a mitotic spindle forming between them. The nuclear membrane disappears, and the chromosomes are observed on the spindle where the chromosome pairs are restored. The process of fertilization is thus completed, and the fertilized ovum or zygote is restored to the diploid condition. It may be well to emphasize that, in restoring the chromosome pairs, the male parent has

Fertilization

583

contributed 23 chromosomes representing his own characteristics, and the female parent has been responsible for 23 chromosomes representing her line of descent. The zygote therefore contains a new arrangement of genes on the chromosomes never before duplicated in any other individual. The offspring destined to develop from the fertilized ovum will have a genetic constitution different from that of either parent and from anyone else in the world.

The union of the paired chromosomes also brings together the sex chromosome pairs. As we have seen, if the ovum is fertilized by a sperm bearing a Y chromosome, the paired chromosomes will be XY and the offspring will be male; if the sex chromosome pairs are reconstituted as XX, the offspring will be female. Thus sex determination is effected at fertilization. The union of the male and female pronuclei also provides the stimulus for further mitotic divisions of the zygote as it passes through a series of divisions known as *cleavage stages*.

CLEAVAGE The zygote enters a period of rapid mitotic divisions in which it passes through stages described as two-cell, four-cell, eight-cell stages, etc. The

FIGURE 22.14

a Living human ovum, minus corona radiata cells, surrounded by spermatozoa. The outer noncellular layer which acts as a limiting membrane is the zona pellucida. Here, under phase-contrast microscopy, it appears as the outer dark zone. Original magnification, ×400. *b* Living human ovum with two polar bodies. The corona radiata is absent. *c* Living two-cell stage, surrounded by the zona pellucida. One polar body can be seen. The difference in the size of the two cells appears exaggerated because they are at different focal levels. Phase contrast; original magnification, ×100. *d* Dividing human ovum. Phase-contrast microscopy; original magnification, ×200. (*Courtesy of Dr. L. B. Shettles.*)

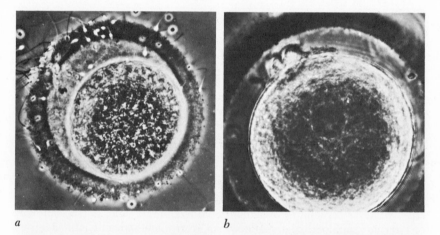

a   b

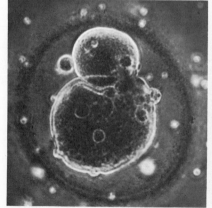
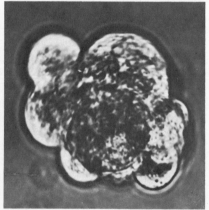

c   d

Human development

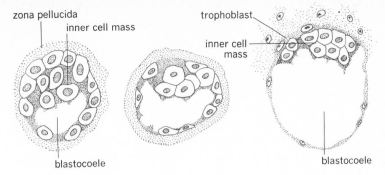

FIGURE 22.15
Early development of mammalian blastocyst (blastodermic vesicle). (*Redrawn from Bradley M. Patten, "Human Embryology," McGraw-Hill Book Company, New York, 1968.*)

mammalian ovum has been described as a very small germinal cell, but it is large compared with the average size of somatic cells. Successive cleavages produce cells of smaller size so that the developing mass of cells is but a little larger than the original zygote. The pattern of cleavage varies with ova of various kinds, but cleavage planes tend to be more complete in ova containing a small amount of yolk.

Most of our knowledge concerning cleavage in the human zygote has been derived from observing cleavage in other mammals. Detailed study of cleavage in the monkey ovum has presented an excellent picture of the probable course of events in man. The cells derived by cleavage are called *blastomeres* (Figure 22.14).

The cells continue to divide within the zona pellucida, which acts as a limiting membrane, and finally there is formed a solid ball of cells called *morula* because of a fancied resemblance to a mulberry. The cells of the morula continue to multiply and form a hollow ball of cells called a *blastula*. The mammalian blastula is a specialized one and is more properly called a *blastodermic vesicle*, or *blastocyst*. It consists of a thin layer of cells, the *trophoblast*, forming a cavity filled with fluid and containing an inner cell mass (Figure 22.15).

THE BLASTODERMIC VESICLE   The cells of the trophoblast surround an *inner cell mass*, which is destined to form the embryo itself. The inner cell mass forms at the *animal pole* of the blastodermic vesicle, where it soon flattens and becomes the blastoderm from which the germ layers of the embryo are derived. The trophoblast plays no direct part in the formation of the embryo; it is concerned with the development of fetal membranes in the fetal portion of the placenta. The ovum passes through its cleavage stages as it moves down the oviduct, but the blastodermic-vesicle stage is reached in the uterus just before implantation (Figure 22.16). The zona pellucida disappears at the time of formation of the blastodermic vesicle, and the trophoblastic cells form the attachment with the uterine epithelium.

IMPLANTATION

Little is known about the implantation of the blastocyst in man, but there is every reason to believe that it follows the same general plan observed in other mammals. During implantation the blastodermic vesicle sinks into the soft lining of the uterus and later becomes firmly attached (Figure 22.17). The minute blastodermic vesicle of the monkey, only 0.5 millimeter in

FIGURE 22.16

Ovulation and the passage of the fertilized ovum through the oviduct. Various cleavage stages are shown diagrammatically. Finally the blastocyst is shown at the beginning of implantation. (*Redrawn from W. J. Hamilton, J. D. Boyd, and H. W. Mossman, "Human Embryology," 2d ed., W. Heffer and Sons, Ltd., Cambridge, England, 1952.*)

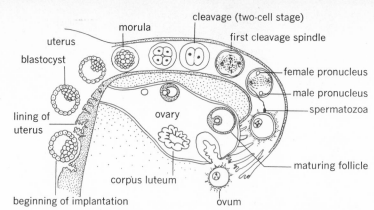

FIGURE 22.17

a Surface view of the lining of the uterus, showing the implantation site of a normal 7-day ovum (Carnegie No. 8225). The implantation site is pale gray and is a little less than 1 millimeter in diameter. Observe the prominent openings of the glands in the uterine lining and the wrinkling of the surface which is consistent and typical of the aging lining at the twenty-second day of the menstrual cycle. *b* A profile view of a 12-day ovum (Carnegie No. 8330), showing the elevation of the blastocyst from the surrounding endometrium. [*From Hertig, Rock, and Adams, Am. J. Anat.,* **98:**435–494 (1956); *courtesy of Carnegie Institution of Washington.*]

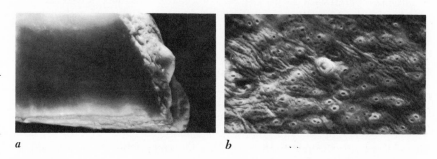

a   *b*

diameter, is known to reach its implantation stage on the ninth day. It usually becomes implanted in the upper part of the uterus on either the dorsal or ventral surface, as does the human blastocyst. It is known that the human blastocyst sinks much more deeply into the lining of the uterus than does that of the monkey. The human blastocyst becomes completely covered by the uterine lining, but the monkey blastocyst remains exposed. The lining of the human uterus thickens below the implanted blastocyst, and cells of the trophoblast grow down into it. Later these processes become long and slender and are called *villi*. They function in the exchange of products of metabolism between the blood of the mother and that of the embryo (Figure 22.17).

GERM LAYERS AND THEIR DERIVATIVES

The differentiation of the three germ layers and their subsequent location in typical positions characterize *gastrulation* in mammals. Gastrulation in the human embryo is specialized, but the essential features can be interpreted from studies of the gastrula stages of other mammals and birds.

ENTODERM   In mammals the lower part of the inner cell mass produces a layer of cells called the *entoderm*. The entoderm, at first a thin layer of cells, extends rapidly to line the primitive gut, or the primitive cavity known as the *archenteron*. The entoderm eventually forms the epithelial lining of the digestive tract and the glands associated with it, the respiratory tract, and part of the urogenital system.

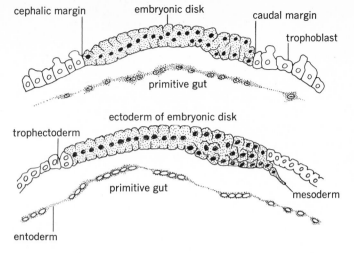

FIGURE 22.18
Longitudinal section of a mammalian embryonic disk. The three germ layers are shown also. (Redrawn from Bradley M. Patten, "Human Embryology," McGraw-Hill Book Company, New York, 1953.)

ECTODERM The remaining cells of the inner cell mass form a layer of cells called the *embryonic disk*. The dorsal portion forms the *ectoderm* of the embryonic disk (Figure 22.18). The ectoderm eventually gives rise to the outer covering of the body, the epidermis, with its derivatives such as hair and fingernails. It also gives rise to the epithelial lining of the mouth, nasal cavity, and anal canal. An infolding of the dorsal ectoderm forms the dorsal nerve cord and its derivatives. The lens of the eye and the enamel of the teeth are also of ectodermal origin.

MESODERM The intermediate layer between the entoderm and the ectoderm is the *mesoderm*. Its origin varies considerably in different animals. In the mammal a thickening at the caudal margin of the embryonic disk indicates the origin of this germ layer. It spreads rapidly and finally divides into two layers: a dorsal layer of somatic mesoderm and a ventral layer of splanchnic mesoderm. The space between the two layers of mesoderm is the *coelom*, or body cavity. The somatic mesoderm becomes closely associated with the ectoderm to form the musculature and connective tissue of the body wall covered by the epidermis (ectodermal), whereas the splanchnic mesoderm combines with the entoderm to form the wall of the digestive tract lined with an entodermal epithelium (see Figure 22.23).

Mesodermal derivatives include the musculature, skeletal, visceral, and cardiac; the structural elements, such as bone and the connective tissues; the blood, bone marrow, and lymphoid tissues; the epithelial lining of body cavities, pleural, cardiac, and peritoneal; as well as the lining of the kidneys, ureters, ducts of the reproductive system, and gonads.

EXTRAEMBRYONIC, OR FETAL, MEMBRANES

The early cellular layers develop far beyond the region of the embryonic disk and are known as *extraembryonic layers*. They do not contribute directly to the formation of the embryo itself but are concerned with the formation of extraembryonic or fetal membranes. These membranes are the *amnion*, *yolk sac*, *allantois*, *chorion*, and *placenta*.

587

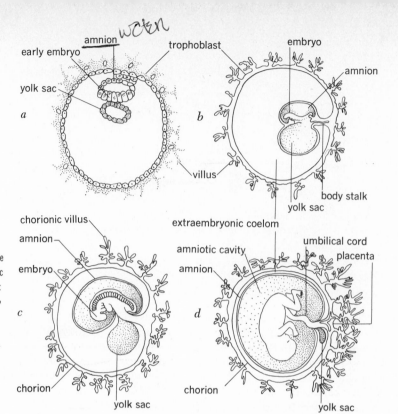

early embryo — amnion wcen — trophoblast — embryo

yolk sac

amnion

a

b

villus

body stalk

yolk sac

extraembryonic coelom

chorionic villus

amnion

amniotic cavity

umbilical cord

embryo

amnion

placenta

c

d

chorion

yolk sac

chorion

yolk sac

FIGURE 22.19

Diagrammatic representation of the development of the extraembryonic membranes in the human embryo: *a* an embryo of about 12 days; *b* embryo of about 16 days; *c* embryo of about 28 days; *d* embryo of about 12 weeks. (*After Bradley M. Patten, "Human Embryology," McGraw-Hill Book Company, New York, 1968.*)

AMNION   The amnion is a thin protective membrane containing amniotic fluid. The embryo develops in this clear, watery fluid, which affords protection against mechanical injury. The amnion expands as the fetus grows until it fills the extraembryonic cavity. It commonly ruptures just before childbirth, releasing the fluid. Individuals said to be "born with a veil" are those who are born with the amnion or a portion of it covering the face.

YOLK SAC AND ALLANTOIS   The ova of many animals contain large amounts of yolk, and the yolk sac is therefore large and of great importance. The human ovum contains very little yolk, and the yolk sac is evident only at an early embryonic stage. As the fetus develops, the yolk sac becomes incorporated in the umbilical cord, where it persists as a degenerate structure.

The allantois, like the yolk sac, is an important membrane in the embryonic development of many animals, especially birds and reptiles. In these animals it has an excretory function, serving as a temporary storage place for urine. Combining with the chorion, it serves as a respiratory organ within the egg. The placenta takes over the excretory and respiratory functions of the human embryo, and the allantois remains minute and rudimentary. The base of the allantois, however, gives rise to the urinary bladder.

CHORION   The chorion is a much thicker membrane than the amnion. It limits the extraembryonic cavity and forms the embryonic portion of the placenta. At an early stage in the human embryo a layer of mesodermal cells is added to the ectodermal trophoblast (*trophectoderm*), and this outer layer of the

Human development

588

blastocyst is called the *trophoderm*. The trophoderm combines with the allantois to form the chorion. The chorion as an outer membrane develops into a highly vascular membrane, capable of functioning in the interchange of the products of metabolism (Figure 22.19).

PLACENTA Small branching processes develop over the outer surface of the chorion. These processes are vascular and are called *chorionic villi*. They cover the chorion at first, but later they are best developed at the place where the chorion makes contact with the lining of the uterus. The mucosal lining of the uterus (endometrium) becomes greatly altered during pregnancy; it is eventually cast off after the child is born. It is therefore called the *decidua*, as trees that shed their leaves in the fall are called *deciduous* trees. The portion of the decidua where the chorion is attached is called the *decidua basalis*. It is there that an interchange of oxygen and food from the mother and excretory products from the fetus takes place.

The placenta is composed of two portions, even though the two parts seem to be closely associated. The chorionic villi become elaborately branched (chorion frondosum) in the area opposite the decidua basalis, and this portion of the chorion is the fetal placenta. The decidual basalis becomes filled with blood sinuses into which the chorionic villi of that region project. The decidua basalis is therefore the maternal portion of the placenta. There is no exchange of blood between the two portions. The fetus soon develops its own blood and circulatory system, and the fetal heart pumps blood out through the villi of the fetal placenta. The mother's blood fills the blood spaces and bathes the chorionic villi with blood. The maternal placenta derives its blood supply from the uterine arteries, and the blood is returned by way of uterine veins (Figure 22.20).

The mature human placenta is a circular disk about 8 inches in diameter

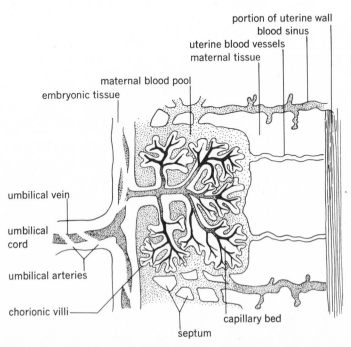

portion of uterine wall
blood sinus
uterine blood vessels
maternal tissue
maternal blood pool
embryonic tissue
umbilical vein
umbilical cord
umbilical arteries
chorionic villi
septum
capillary bed

FIGURE 22.20
Diagrammatic representation of a portion of the placenta, illustrating the relationship between fetal and maternal tissues.

Extraembryonic, or fetal, membranes

589

and nearly an inch thick. It becomes detached from the uterus at parturition and accounts for a large part of the afterbirth.

The mother, from her food, provides nourishment for the developing fetus. The food—digested, and absorbed by the blood—diffuses through the membrane of the chorionic villi. Once absorbed into the blood of the fetus, simple food molecules are rebuilt into complex molecules. Oxygen from the mother's blood diffuses into the blood of the embryo, while carbon dioxide is exchanged. The waste products of metabolism, such as urea, diffuse out of the embryonic bloodstream and are disposed of by the excretory organs of the mother.

The placental membranes present an effective barrier against bacteria, and the fetus is usually well protected against diseases of bacterial origin. The viruses and some active blood-borne diseases, such as syphilis, can affect the fetus.

In review, it should be mentioned that the placenta produces estrogenic hormones, progesterone, and chorionic gonadotropin.

UMBILICAL CORD The connection between the fetus and the placenta is established by way of the umbilical cord. Its fetal attachment is at the umbilicus (or navel). The cord is not an outgrowth of the body wall of the fetus; it is formed by the amnion. The first attachment of the embryo, the body stalk, and the yolk sac are pulled together and become incorporated within the cord. The umbilical blood vessels consist of two arteries and a single large vein. A mucous tissue that is jellylike in consistency (Wharton's jelly) is peculiar to the umbilical cord. The cord at the full term of pregnancy is coiled and twisted; it measures about $\frac{1}{2}$ inch in diameter and is nearly 2 feet long. It attaches near the center of the placenta, and after the child is born, it is shed with the placenta as part of the afterbirth.

TWINNING Twins can be produced in various lower animals as early as the two-cell stage of the developing embryo. If two organizing centers are established, two embryos can develop separately and produce identical twins. In the higher mammals, including man, early separation of the blastomeres is unlikely because cleavage takes place within the zona pellucida. Twinning therefore is more likely to occur at the blastocyst stage as a result of divisions occurring in the inner cell mass. Separate organization centers can develop to provide for the development of two embryos from a single ovum. Human identical twins are single-ovum twins, are always of the same sex, and look very much alike. They develop with a single chorion and placenta (Figure 22.21). They are of scientific interest, since they represent two individuals with apparently identical chromosomes. This poses a question for scientists: How much can two individuals vary who have the same chromosomal constitution? It is well known that identical twins usually resemble each other closely in appearance but can develop different capabilities. Occasionally identical twins are not completely separated. The two individuals may be joined at various places and may have organs in common. Such twins do not ordinarily live long, but some, such as those joined at the base of the spine, have lived to old age.

Fraternal twins occur when two ova are fertilized and both develop. They develop in individual chorions, and each has its own placenta. Such

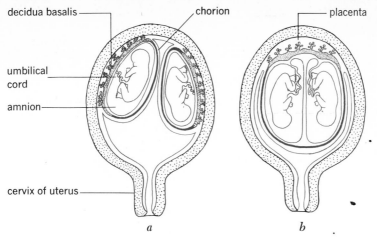

decidua basalis

chorion

placenta

umbilical cord

amnion

cervix of uterus

a

b

FIGURE 22.21
Diagram illustrating the differences
in the fetal membranes between
fraternal and identical twins: *a*
fraternal twins with entirely separate membranes; *b* identical twins
with common chorion and placenta.
(*Redrawn from J. Kollmann,
"Handatlas der Entwicklungs-
geschichte des Menschen,"
Fischer, Jena, 1907.*)

twins may be of opposite sex and have only a family resemblance to each other.

Triplets, quadruplets, and quintuplets may be identical or combinations of identical and fraternal twins with single-ovum individuals.

Development of cylindrical form We have described the embryo in the spherical form of the blastocyst and then as a flat embryonic disk; now it becomes necessary to visualize a transformation into a cylindrical shape. The dorsal nerve cord and the primitive digestive tract are essentially tubes within the cylindrical form. The nerve cord begins to appear as a longitudinal groove in the middorsal ectoderm (Figure 22.22). The groove deepens and finally becomes pinched off as a tubular nerve cord. Directly below it is a rodlike

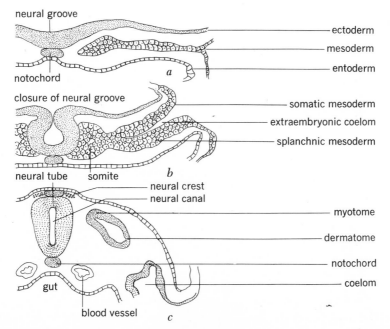

neural groove

ectoderm

mesoderm

entoderm

notochord

a

closure of neural groove

somatic mesoderm

extraembryonic coelom

splanchnic mesoderm

neural tube   somite

b

neural crest
neural canal

myotome

dermatome

notochord

coelom

gut

blood vessel

c

FIGURE 22.22
Diagrammatic cross section showing the invagination of the neural
groove and the position of the
three germ layers in the mammalian embryo. Note the two layers of
the mesoderm with the coelom, or
body cavity, forming between
them.

Twinning

591

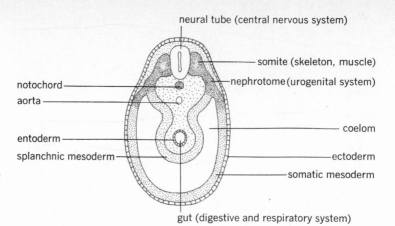

neural tube (central nervous system)

somite (skeleton, muscle)

nephrotome (urogenital system)

notochord

aorta

entoderm

splanchnic mesoderm

coelom

ectoderm

somatic mesoderm

gut (digestive and respiratory system)

FIGURE 22.23

Transverse section of a vertebrate embryo, showing the development of somite, nephrotome, coelom, and gut. (*After Leslie B. Arey, "Developmental Anatomy," W. B. Saunders Company, Philadelphia, 1965.*)

structure derived from mesoderm. It is the *notochord*, a supporting structure common to all vertebrate embryos. The digestive tube forms with an epithelial lining derived from entoderm (endoderm) and a muscular wall derived from the splanchnic layer of mesoderm. The oral and anal cavities invaginate, pulling in surface epithelium, with the result that these cavities are lined with an epithelium of ectodermal origin.

The spherical stage of the embryo through the blastocyst is passed in the first 2 weeks; the flat embryonic disk is developed early in the third week; and at about $3\frac{1}{2}$ weeks the embryo is assuming a cylindrical form. The lateral and ventral parts of the body wall are completed by a downgrowth of somatic mesoderm (somatopleure) on either side. The two downgrowths meet and fuse along a midventral line to form the coelom, or body cavity (Figure 22.23).

DEVELOPMENT OF THE HUMAN EMBRYO

The human embryo at the end of 1 month's growth has attained a length slightly less than 4 millimeters. Probably the most characteristic formation at this stage is the development of muscle segments, called *somites*, on either side of the neural tube. There are 30 pairs of somites at this stage, representing the beginning of the skeletal musculature (Figure 22.24). The neural tube has enlarged and has undergone flexure (that is, bending) in the early formation of the brain. A definite tail is present and is characteristic of human embryos. The heart and liver are developing and can be observed externally as bulges or prominences along the ventral side of the trunk.

The neck region contains a number of pharyngeal arches with clefts between them. The clefts are gill clefts or gill slits in fish, but, of course, no gills form in the human embryo. The arches form structural elements of the face and pharynx, and the pouches that develop in the clefts give rise to epithelial linings and glandular structures associated with the pharynx. The first arch, from which the lower jaw is derived, is called the *mandibular arch*. The second arch contributes to the construction of the hyoid bone and is called the *hyoid arch*. Arches 3, 4, and 5 are rudimentary but take part in the formation of some of the cartilages of the larynx (Figure 22.32). The pouch that forms in the first, or hyomandibular, cleft gives rise

Human development

592

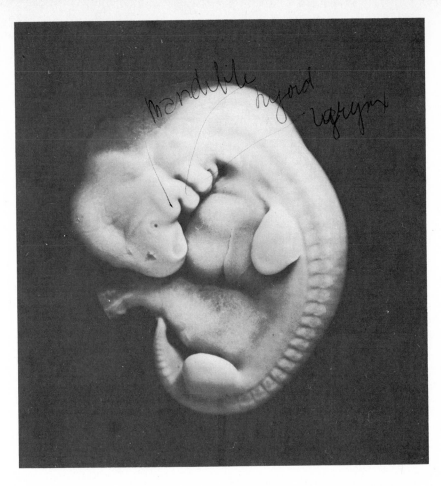

handwritten annotations on image: mandible, hyoid, larynx

FIGURE 22.24

Human embryo (Carnegie No. 6502) about 5 weeks after fertilization. Greatest length, 6.7 millimeters. [*From G. L. Streeter, Carnegie Inst. Wash. Contrib. Embryol.,* **31**:199 (1945); *courtesy of Carnegie Institution of Washington.*]

to the eustachian tube and the cavity of the middle ear. The second branchial pouch eventually takes part in the formation of the tonsils; the third and fourth pouches contribute to the formation of the parathyroid and thymus glands. The fifth pouch ordinarily does not persist in the human embryo (Figure 22.33).

LIMB BUDS The arm and leg buds make their appearance in the fifth week of the embryo. The arm buds are slightly more advanced in their development. Both arm and leg buds elongate rapidly and during the sixth week terminate in little paddlelike expansions, which will form the hands and feet (Figure 22.25). Fingers and toes develop during the seventh and eighth weeks. The nails develop slowly, but as early as the third month there is a thickening of the epithelium indicating where the nail will develop.

FORMATION OF THE FACE The head of the embryo appears very large when compared with the size of the trunk region. In the embryo of 5 weeks the head seems very elongate, in part because of considerable space devoted to the formation of the pharyngeal clefts. The face is in the very early stage of formation and does not have the appearance of a face. The rapid development of forebrain and midbrain is indicated by bulges in the cephalic region.

Development of the human embryo

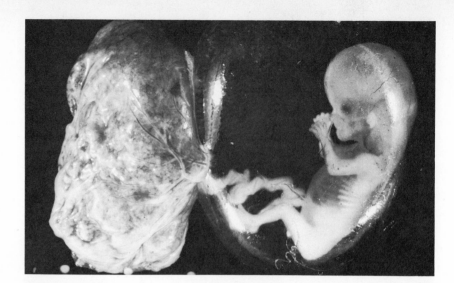

FIGURE 22.25
Human fetus of 8½ weeks within the amnion. The umbilical cord leads outward from the fetus to the placenta. (*Courtesy of David Dreizen, M.D.*)

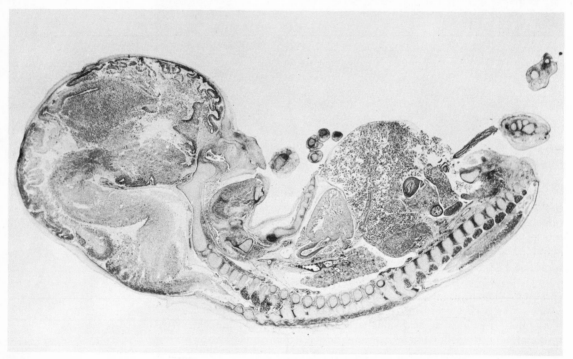

FIGURE 22.26
Normal human fetus of 10 weeks, longitudinal section. Note flexure at base of brain; the lung, heart, and typical large fetal liver below. (*Courtesy of Percy Brooks.*)

Flexures in the body and neck regions cause the face to lie just above a thoracic prominence that indicates the developing heart (Figure 22.26).

It is difficult to study the face in this position, but if the embryo is straightened out, the face appears as a series of folds centered about the early formation of the mouth and nose (Figure 22.27). The mouth is formed in the cleft between the maxillary and mandibular processes. The cleft is very extensive in the embryo of 6 weeks but becomes greatly reduced with

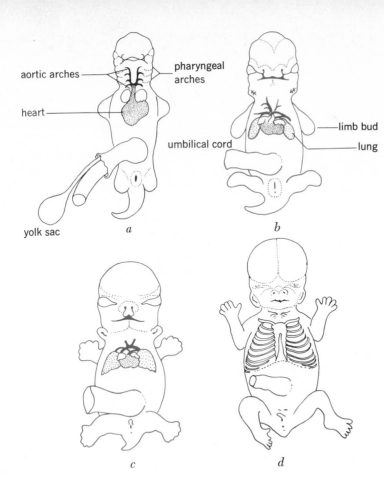

aortic arches

pharyngeal arches

heart

limb bud

lung

umbilical cord

yolk sac

a

b

c

d

FIGURE 22.27

Frontal views of four young human embryos, drawn without regard to body curvature: *a* embryo of 5 weeks; *b* embryo of 6 weeks; *c* embryo of 7 weeks; *d* embryo of 8 weeks. Note development of face and appendages in the series of embryos. (*Redrawn from William Patten in Bradley M. Patten," Human Embryology," McGraw-Hill Book Company, New York, 1953.*)

the formation of lips and cheeks. The nasal pits are at first far apart, separated by a broad median nasal process. As the nose becomes elevated, the nasal pits move closer together. The maxillary processes grow forward from either side until they meet and fuse with the median nasal process in the formation of the upper lip. Malformations such as harelip occur when these processes fail to unite properly.

EYES An evagination of the lateral wall of the forebrain is called the *optic vesicle*. It forms very early in the human embryo, as early as $3\frac{1}{2}$ to 4 weeks. The optic vesicle forms a two-layered cup, the optic cup, as it grows out to meet the surface ectoderm. Just above the optic cup, the surface ectoderm thickens and invaginates into the cup as a hollow lens vesicle. Later it becomes pinched off from the surface ectoderm to form the lens of the eye (Figure 22.28). The lens is nearly spherical when it is first formed, but it gradually flattens and becomes biconvex. The cells multiply until the cavity of the lens is completely filled. The growing lens is supplied with blood from the hyaloid artery, which normally disappears before birth.

The layer of the optic cup closest to the lens becomes the neural retina, and the deeper layer forms the pigment back of it. The optic-nerve fibers form from ganglion cells in the retina and grow back through the optic stalk as the optic nerve.

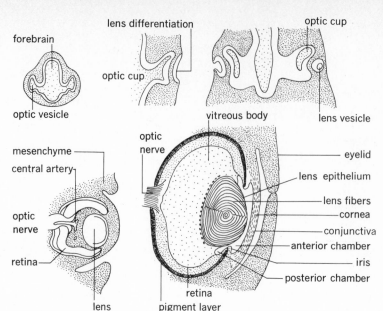

FIGURE 22.28

The development of the eye in the mammalian embryo. The lens is enclosed in an external noncellular capsule (not shown). The lens epithelial cells are located beneath the capsule. Epithelial cells elongate and form lens-fiber cells. [*After Lewis in Barry J. Anson (ed.), "Morris' Human Anatomy," 12th ed., McGraw-Hill Book Company, New York, 1966.*]

Labels in figure: forebrain, lens differentiation, optic cup, optic cup, optic vesicle, vitreous body, lens vesicle, mesenchyme, optic nerve, central artery, eyelid, lens epithelium, lens fibers, cornea, optic nerve, conjunctiva, anterior chamber, retina, iris, posterior chamber, lens, retina, pigment layer

In the field of experimental embryology, Coulombre and others have performed some very interesting experiments reversing the lens in experimental animals so that the lens epithelium, instead of facing the cornea as it normally does, is turned to face the neural retina. The epithelial cells differentiate into a new set of lens fibers on the posterior side of the lens, and a new epithelium is formed anteriorly. The most acceptable interpretation of these results would indicate that the neural retina produces a factor that induces the differentiation of the lens. Ribonucleoprotein particles found in the cells of the optic vesicle may eventually prove to be the inducer substance, although as yet this factor remains unknown.

The eyes of the embryo of 6 weeks are indicated by the optic vesicle outlined under a thin covering of ectoderm. They are located far around on the side of the head. With the development of the face at 7 weeks the eyes are in a forward position though still somewhat far apart. The formation of eyelids at this time emphasizes the outlines of the eyes. The eyelids meet and fuse during the ninth week and remain closed until the seventh month of intrauterine life. The human infant is not born with the eyes closed as many mammals are.

DEVELOPMENT OF THE EAR The early formation of the ear is indicated by a thickening of the ectoderm on either side of the neural groove, which deepens to form the auditory pit. The auditory pit is present in the embryo of 4 weeks. It deepens and becomes a closed sac—the auditory, or otic, vesicle. Later, outgrowths from the vesicle give rise to the membranous labyrinths of the cochlea, the vestibule, and the semicircular canals of the inner ear.

The first pharyngeal pouch forms the tympanic cavity of the middle ear and the eustachian tube, whereas the first branchial groove forms the external auditory canal. The partition separating the two acquires an inner layer of mesoderm and becomes the tympanic membrane. It is lined with

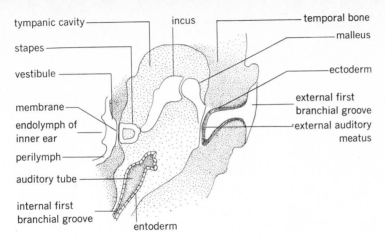

FIGURE 22.29
The development of the ear in the embryo. (*After G. S. Dodds, "The Essentials of Human Embryology," John Wiley & Sons, Inc., New York, 1946.*)

entoderm from the pharyngeal pouch and covered with ectoderm from the external depression of the first branchial groove (Figure 22.29).

The external ear arises from several growth centers located around the mandibular and hyoid gill clefts. In the embryo of 6 weeks little more than the growth centers are visible, but during the seventh and eighth weeks the external ear begins to take shape (Figure 22.27). The ears form very low on either side of the head. Their position may not seem so strange when one considers that they are formed in connection with pharyngeal grooves and pouches. Even in the newborn infant the tympanic membrane lies close to the surface at the distal end of a very shallow external auditory canal.

SKIN AND ITS DERIVATIVES   The skin and its derivatives provide a protective covering for the body. As we have seen in earlier chapters, the skin plays an important part in the regulation of body temperature. The receptors for the sense of touch, pressure, pain, and temperature are largely located in the skin. Adipose tissue, as a subepidermal layer, provides protection, insulation, and a reserve store of energy materials.

The integument of the embryo is, at first, a single layer of cuboidal epithelium, derived from the ectoderm, over a layer of mesenchyme of mesodermal origin. The epithelial layer soon becomes stratified and develops into several specialized layers common to the epidermis (Figure 22.30). The deeper layer of mesenchyme gives rise to a layer of fibroelastic connective tissue called the *dermis* or *corium*. Connective-tissue fibers attach the corium to the superficial and deep fascia covering skeletal muscles, tendons, and bones.

The ridges and whorls on the finger tips and on the soles of the feet present a characteristic pattern for a given individual that will remain constant throughout life and can be used for identification.

NAILS AND HAIR   The nails and hair are derivatives of the skin. The nails arise from thickenings of the epidermis at the ends of the fingers and toes. The nails develop during the third month and grow slowly, reaching the finger tips during the ninth month. The toenails are a little later in attaining completion.

The hair follicle originates as a downgrowth of epidermal cells into

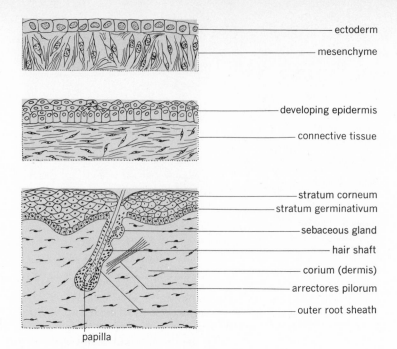

ectoderm
mesenchyme

developing epidermis
connective tissue

stratum corneum
stratum germinativum
sebaceous gland
hair shaft
corium (dermis)
arrectores pilorum
outer root sheath

papilla

FIGURE 22.30

Stages in the development of the skin in the embryo.

the dermis below. The invading cells become club-shaped and invaginated at the lower end, where mesodermal tissue pushes in to form the papilla. The hair shaft begins to form at the base of the downgrowth and grows outward to penetrate the skin surface. The hair follicle develops at an angle to the surface of the skin, and along its lower side are two outgrowths of cells. The upper one is the primordium of a sebaceous gland, which produces an oily secretion at the base of the hair. The lower outgrowth is called the *epithelial bed* and is associated with the rapidly growing hair root (Figure 22.30). Also attached on the lower side of the hair follicles are the arrectores pilorum muscles, which pull on the bases of the hairs and cause small elevations in the skin (gooseflesh) as a reaction to cold or fear.

The first body-hair covering of the fetus is a soft, downy coat called the *lanugo*. It appears about the seventh month and is shed before birth. The body hair that replaces it is somewhat coarser but remains delicate and inconspicuous in children and in the mature female to a certain extent. The changes associated with puberty produce coarser body hair in both sexes, including pubic and axillary hair and also the beard in the male.

SWEAT GLANDS   The sweat glands arise from an invagination of epidermal cells into the dermis. The cylindrical downgrowth is solid at first, but later a cavity develops, which then forms a long hollow tube. The base of this tube coils extensively in the mature gland. Glands of the axilla, eyelids, and external auditory canals are regarded as specialized sweat glands. The sweat glands begin to form in the fourth month and are well established in the seventh month.

SKIN COLOR AS AN EXAMPLE OF MULTIPLE-FACTOR INHERITANCE   We have observed that individuals of the so-called "white" race vary in skin color, as between

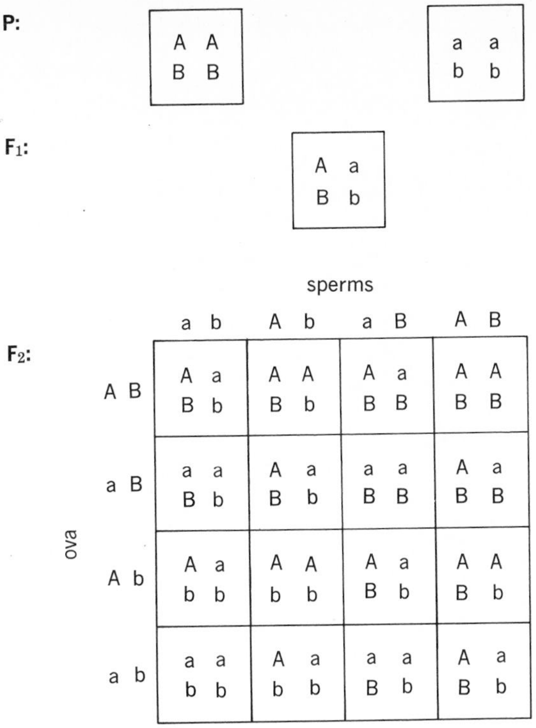

P:

| A A B B | | a a b b |

F₁:

| A a B b |

sperms

	a b	A b	a B	A B
A B	A a B b	A A B b	A a B B	A A B B
a B	a a B b	A a B b	a a B B	A a B B
A b	A a b b	A A b b	A a B b	A A B b
a b	a a b b	A a b b	a a B b	A a B b

ova

F₂:

FIGURE 22.31

Multiple factors affecting skin color. The parent designated *AABB* has the greatest amount of pigmentation; the parent designated *aabb* has the least amount. The F₂ shows the 16 possible combinations.

blonds and brunettes. Negroid peoples also vary in skin color from very light to very dark. The phenotypic appearance is further complicated by the effect of sun and wind. Melanocytes secrete the pigment melanin which gives color to the skin and hair.

C. B. Davenport[1] in 1913 made extensive studies of skin color inheritance among Negroes in Jamaica. His dihybrid hypothesis explained his findings rather well in this limited geographical area, but more extensive studies favor a more complex, multiple-factor interpretation. R. Ruggles Gates[2] in 1949 suggested that a three-factor inheritance for skin color provided a more adequate explanation than a two-factor hypothesis.

Even though skin color is determined by multiple factors, the simple dihybrid hypothesis presents the facts fairly well and can be used as an example. We must consider that the two-factor pairs operate without dominance and that their color effects are cumulative.

If we assume that the "full-blooded" Negro-type individual can be designated by the symbols *AABB*, then the fair-skinned white would be indicated as *aabb*, the latter symbol meaning very little pigmentation. The F₁ or first filial generation of a cross between these two genetic individuals would produce mulatto children of the genotype *AaBb*.

Now from a genetic standpoint, an F₂ mating of two individuals of this *AaBb* genotype would produce only 1 out of 16 of the genotype aabb for white skin and similarly only 1 in 16 would be likely to be of the genotype *AABB*, or black-skinned (see Figure 22.31). The remaining zygotes of the

[1] C. B. Davenport: Heredity of Skin Color in Negro-White Crosses, *Carnegie Inst. Wash., Publ.* 188, 1913.
[2] R. Ruggles Gates: "Pedigrees of Negro Families," McGraw-Hill Book Company, New York, 1949.

*Development
of the human embryo*

16 possible combinations would produce individuals of various degrees of pigmentation. Six of the 16 would have two genes for pigmentation *AAbb*, *aaBB*, or *AaBb* and would have essentially the same degree of pigmentation as their parents. Notice, however, that 4 of the 16 would have three genes for pigmentation, *aABB* or *AABb*, and would be darker than their parents, and 4 out of 16 would have only one gene for pigmentation, *aaBb* and *Aabb*, and would be of lighter complexion than the parents.

Davenport's hypothesis probably is oversimplified, and it is now considered that more than two pairs of genes are involved in producing pigmentation. In a marriage between Negro and white, if the white individual carries no genes for pigmentation (genotype *aabb*), then no child in this family can be darker than the parent showing the greatest amount of pigmentation.

GROWTH   The first 2 months of embryonic life provide the general external features and body form of the mature infant. The basic plan of the organ systems is worked out, and at about the third month the term *fetus* rather than *embryo* is used to describe the developing form. The period of the fetus from the third through the ninth month is characterized by a great increase in size and by the maturing of the organ systems. The head size remains large in proportion to the rest of the body, and the neck is very short. The pelvis and the legs remain small compared with the development of the upper part of the body. During the ninth month the fetus attains a crown-to-heel length of approximately 18 inches and should weigh about 7 pounds.

DEVELOPMENT OF BLOOD   We have stated that the embryo develops its own blood and is not dependent on the mother for the formation of blood cells. The first indication of an embryonic blood supply is the formation of clusters of blood cells in mesodermal blood islands of the yolk sac. Centers for blood formation also give rise to endothelial tubes similar to those that line blood vessels. There are various sites of blood formation; the early yolk-sac center soon is replaced by centers in the body mesenchyme and later by those in the liver and spleen. When bone is formed, the red marrow of bone is the principal blood-forming center.

The mesenchyme is a tissue derived from mesoderm and possessing great potential capabilities for giving rise to other tissues, such as connective tissues, bone, muscle, reticuloendothelial tissue, and blood. Wandering cells of mesenchyme form the blood islands and give rise to various kinds of blood cells. The first recognizable blood cell from which other types of blood cells are thought to originate is called a *hemocytoblast*. The first corpuscles to appear in the embryonic blood are not identical with mature erythrocytes. They retain their nucleus and more closely resemble the immature red cell, or *erythroblast*. The hemocytoblast is thought to be the parent cell from which the red cells and various kinds of white cells arise. Very large cells called *megakaryocytes* may arise directly from undifferentiated mesenchyme cells. They are of special interest, since it is thought that they may give rise to the blood platelets through fragmentation of their cytoplasm. It is the hormone erythropoietin that stimulates red cell formation. The hormone now has been isolated in pure form.

DEVELOPMENT OF BLOOD VESSELS AND THE HEART Blood vessels arise as simple endothelial tubes forming a capillary network. Some of these tubes enlarge to form the definitive pathways of the larger arteries and veins. At first it is not possible to distinguish structurally between arteries and veins, but later, when connective and muscular tissue form the walls around the endothelial tube, they can be identified.

The mammalian heart forms from paired areas of splanchnic mesoderm, which lengthen into two strands of cells. The strands become hollow tubes, and as they grow caudad, they unite to form a single endothelial tube. The endothelial tube is essentially the developing endocardial lining of the heart. A layer of mesoderm external to the endothelial tube gives rise to the muscular wall of the heart, the myocardium, and to its covering membrane, the epicardium.

The single tubular heart turns on itself and eventually forms two atria and two ventricles. As we have seen in Chapter 14, the wall separating the two atria is not complete. An opening, the foramen ovale, permits blood to pass directly from the right atrium to the left during the period of fetal circulation. It is also possible for the blood to pass from the left pulmonary artery into the aorta by way of the ductus arteriosus and thus bypass the lung (see Chapter 14).

The heart begins to develop from paired primordia of splanchnic mesoderm about the third week. Human embryos as early as the fourth week show an external cardiac prominence above the developing heart. The end of the fourth week finds the four chambers of the heart developing, although the heart is still functioning as a single contractile tube. The completion of the chambers of the heart and the development of the valves must be accomplished while the heart is at work maintaining the fetal circulation. Furthermore, the fetal heart must be able to adapt itself quickly at birth to support the added burden of pulmonary circulation. This change is accomplished by the gradual closing of the foramen ovale and the constriction of the ductus arteriosus.

RETICULOENDOTHELIAL TISSUE Reticular tissue is a loose fibrous tissue found in lymphoid tissues, liver, spleen, and bone marrow. Arising from mesenchyme, it is closely associated with the endothelial lining of blood vessels and with blood-forming cells in the embryo. Reticuloendothelial tissue gives rise to wandering phagocytic cells such as macrophages and to fixed phagocytic cells lining the passageways of lymph nodes, spleen, and sinusoids of the liver. The phagocytic cells of these organs are capable of removing foreign matter from the blood; the lymph nodes, spleen, and liver, therefore, have been called the *primary filters* of the body.

EMBRYONIC DIGESTIVE SYSTEM The embryonic digestive tube is essentially a blind tube of entoderm as it is first formed. The muscular layer that surrounds the tube is formed secondarily from splanchnic mesoderm. An oral cavity (or stomodeum), which will later become the mouth, invaginates from the anterior end of the embryo to meet the anterior end of the entodermal tube. The surface ectoderm is pulled in with this invagination and lines the anterior part of the oral cavity. In the same way a posterior ectodermal invagination pushes in to meet the entodermal tube near its caudal end. The caudal end of the entodermal tube forms a common posterior opening called the cloaca. In mature animals that retain the cloaca, such as birds and reptiles, the

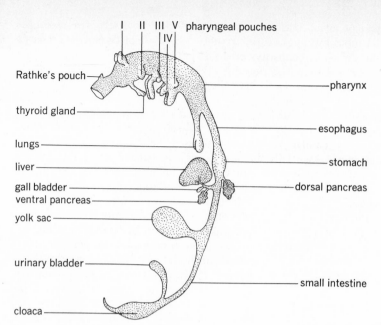

FIGURE 22.32
Derivatives of the embryonic digestive tract, shown diagrammatically.
(After C. D. Turner, "General Endocrinology," W. B.
Saunders Company, Philadelphia, 1971.)

urinary and reproductive ducts empty into this common opening. The human embryonic cloaca soon divides to form the rectum dorsally and the urinary bladder and urogenital sinus ventrally. The urethral and anal canals are soon established, and the cloaca disappears. The anal canal by virtue of its origin is lined with an epithelium of ectodermal origin.

The pharynx region has been discussed from the standpoint of its derivation. Certainly the structures derived from the pharyngeal pouches and gill clefts constitute a most interesting part of the history of embryonic development (Figures 22.32 and 22.33).

The pharynx itself is a soft tube in back of the mouth leading to the esophagus. The tonsils lie embedded in its lateral walls. The esophagus is at first a short tube leading to a slight enlargement in the digestive tube, the stomach (Figure 22.32). During the fifth week the stomach enlarges and begins to assume the shape of the mature organ. It is originally located rather high in the body cavity, but during the sixth and seventh weeks the body elongates and the stomach appears closer to its permanent location. The esophagus elongates at this time also.

Liver The liver arises as a diverticulum below the stomach from the region of the intestine destined to become the duodenum. The entodermal diverticulum grows into a thick-walled vesicle from which the liver tubules and hepatic ducts arise. The entodermal ducts grow into splanchnic mesoderm, which provides the connective tissue of the liver and its capsule. The posterior portion of the diverticulum gives rise to the gallbladder and the cystic duct. The fetal circulatory plan provides that food-laden blood from the placenta shall pass through the liver. The liver becomes proportionately very large in the fetus, accounting for around 10 percent of the body weight in the fetus of 9 weeks. The liver is still large at birth but represents approximately only 5 percent of the body weight, while in the adult the proportional weight shrinks to 2 to 3 percent. The liver eventually becomes located below

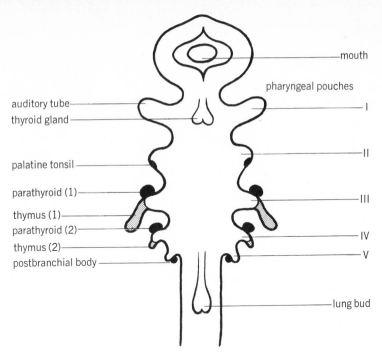

auditory tube
thyroid gland

palatine tonsil

parathyroid (1)

thymus (1)
parathyroid (2)
thymus (2)
postbranchial body

mouth

pharyngeal pouches

I

II

III

IV

V

lung bud

FIGURE 22.33

Derivatives of the embryonic pharynx, shown diagrammatically.

the diaphragm largely on the right side. The stomach lies to the left and is partially covered by the liver.

Pancreas Two diverticula from the primitive entodermal tube give rise to the pancreas. One evagination from the dorsal wall becomes the dorsal pancreas, and the other evagination on the opposite side becomes the ventral pancreas. The duct of the ventral pancreas is associated with the common bile duct. The lengthening of the bile duct and the growth and flexure of the duodenum bring the ventral pancreas to a position directly below the base of the dorsal pancreas, and the two parts fuse in the embryo of 7 weeks. The greater part of the mature gland is derived from the dorsal pancreas; the ventral pancreas forms only the lower basal portion. The dorsal pancreatic duct joints the ventral duct in a manner to retain the base of the ventral duct and its primitive connection with the bile duct. The single duct of the mature pancreas joins the common bile duct in the *ampulla of Vater* and empties the pancreatic fluid into the duodenum along with bile from the liver.

Intestine The intestine is at first a straight tube extending from the stomach to the cloaca. It is held in place by a connective-tissue sheath called the *mesentery*. Anteriorly in the duodenal region there are both a dorsal and a ventral mesentery, dividing the body cavity, or coelom, into right and left portions (Figure 22.34). Posteriorly the right and left coelom forms a common cavity. The coelom itself arises as a narrow cavity between layers of somatic and splanchnic mesoderm (Figure 22.22). The transverse septum, which later becomes a part of the diaphragm, separates the pleural and cardiac cavities from the abdominal cavity. The mesentery is continuous with the peritoneum lining the abdominal cavity. The lungs and heart occupy the thoracic cavity, and the intestine comes to lie in the abdominal cavity.

Growth

603

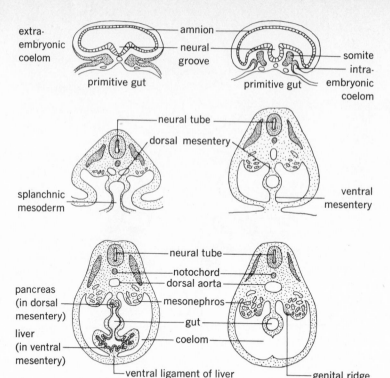

FIGURE 22.34

Transverse sections of embryos, illustrating early stages in the development of the coelom and mesenteries, shown diagrammatically. (*Redrawn from Bradley M. Patten, "Human Embryology," McGraw-Hill Book Company, New York, 1953.*)

The intestine lengthens, forms a loop, and then grows rapidly to form so many coils that the small abdominal cavity apparently cannot contain them. A loop of the intestine pushes out into the umbilical cord at about 5 weeks. The abdominal cavity continues to enlarge, and in the embryo of 10 weeks the extruded portion is pulled back through the umbilical ring. The caudal part of the intestine, which later will form the large intestine, is actually of smaller diameter than the small intestine in the early embryo. At a later stage in its development, as in the fetus of 5 months, it begins to resemble the large intestine of the newborn infant.

After the fourth month the intestine begins to fill with a greenish mixture called *meconium*. It is composed of sloughed-off epithelial cells, mucus, and bile to which are added sebaceous secretions and lanugo hairs swallowed with the amniotic fluid. The contents of the intestine are sterile before birth but acquire a bacterial flora after feeding begins. The greenish meconium is voided from the intestine within 3 or 4 days after birth.

RESPIRATORY SYSTEM   The area immediately posterior to the pharyngeal pouches gives rise to a laryngotracheal ridge on its ventral surface in very young embryos. The primordial outgrowth of the larynx and trachea then arises as a bud from the primitive gut in this region. The tracheal bud elongates and branches, forming bronchial buds. The whole structure is bilobed and is commonly called the *lung bud* (Figures 22.32 and 22.33). The bronchial buds, by continued branching, form the entire respiratory tree, the bronchial tubes, the bronchioles, and the alveoli. The bronchial buds are entodermal, but they grow into a mass of mesenchyme from which will arise the supporting tissues of the lungs and bronchial tree; only the epithelial lining of the passageways remains entodermal in origin. The lung buds grow out dorsally and on either side of the heart into the portion of the body cavity that will later become the pleural cavities.

The lungs are small at birth, since they are never fully expanded with air. Though respiratory movements can take place before birth, the lungs are not functional until the newborn infant takes its first gasp of air. The lung tissue may not become completely inflated until several days after birth.

SKELETAL AND MUSCULAR SYSTEMS   The development of connective tissue and bone has been discussed in previous chapters. The supporting tissues arise from mesenchyme, and bone begins to form as early as the sixth week. The embryo of 9 weeks begins to develop the membrane bones of the skull, while cartilage is replaced by bone in the long bones, such as the ribs and bones of the appendages. Even at birth the bones of the skull are incomplete, and the long bones are largely cartilage at the joints.

Muscle   Visceral muscle develops from the mesenchyme that envelops primitive epithelial tubes, such as the primitive digestive tube, and its evaginations. The musculature of the digestive tube arises from the splanchnic layer of mesoderm (Figure 22.23). Cardiac muscle also arises from splanchnic mesoderm in the region where the primitive heart tubes develop. It forms the muscular wall of the heart, the myocardium. Skeletal muscles of the trunk are derived from the myotomes of segmentally arranged somites so characteristic of early embryos (Figure 22.24). Muscles of the head and face and some of the muscles of the neck develop from the mesenchyme of the branchial arches. The origin of the musculature of the appendages from somites near the limb-bud region has been questioned, but muscle tissue appears to grow into the limb buds from the ends of myotomes at limb-bud levels. The musculature soon loses its earlier segmental arrangement, and in the embryo of 8 weeks individual muscles are already discernible.

UROGENITAL SYSTEM   In Chapter 19, we discussed the transient nature of the pronephric and mesonephric kidneys in the human embryo. We traced the development of the metanephric kidney and indicated that in the male the old mesonephric duct is converted into a sperm duct. There is such a close relationship between the ducts of the two systems that they are commonly considered as the urogenital system.

The urogenital system arises from a ridge of tissue lateral to the somites, called the *urogenital ridge* (Figure 22.23). Nephrogenic tissue from this region gives rise to the kidneys and their ducts. The genital ridge (gonadal ridge) forms along the midventral surface of the mesonephros and gives rise to the gonads (Figure 22.34). The testes and the ovaries are formed high in the abdominal cavity and later assume a lower position. The testes descend into the scrotum usually before birth, and the ovaries come to lie on either side of the uterus and under the oviduct.

The metanephric kidney develops with a new duct, the ureter. The old mesonephric duct in the male is then converted into the epididymis and the vas deferens—ducts that lead the sperm away from the testes. The mesonephric ducts do not persist in the female except as vestigial structures. Newly formed Müllerian ducts develop into the oviducts of the female adult, which conduct the ova from the ovary. They are never a part of the urinary system but are purely reproductive in function (Figure 19.3).

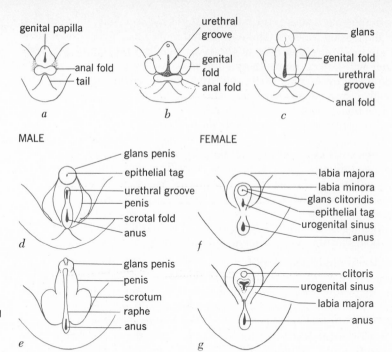

genital papilla

urethral groove

anal fold
tail

genital fold
anal fold

glans

genital fold
urethral groove

anal fold

a

b

c

MALE

FEMALE

glans penis
epithelial tag
urethral groove
penis
scrotal fold
anus

d

labia majora
labia minora
glans clitoridis
epithelial tag
urogenital sinus
anus

f

glans penis
penis
scrotum
raphe
anus

e

clitoris
urogenital sinus
labia majora
anus

g

FIGURE 22.35
Development of the external genitalia in the human embryo: *a, b, c* undifferentiated stages; *d, e* differentiation of the male external genitalia; *f, g* differentiation of the female external genitalia.

EARLY DEVELOPMENT OF EXTERNAL GENITALIA The external genitalia begin to develop during the sixth and seventh weeks, but it is not possible to tell the sex of the embryo at this time because the external genitalia are in an undifferentiated stage (Figure 22.35). As early as the fifth week, before the cloacal opening divides into a urogenital sinus and rectum, a slight elevation called the *urogenital tubercle* (papilla) forms above the cloacal opening. The tubercle develops a distal glans portion and will become the penis of a male embryo or the clitoris of a female. A median urethral groove below the glans is bordered by two ridges called the *genital folds*. Lateral to the genital folds, as development of the genitalia progresses, appear two additional elevations known as *genital swellings*. They form the scrotal fold of the male or the labia majora of the female (Figure 22.35).

Male genitalia In the male embryo the penis lengthens, and by an infolding of the urethral (genital) folds, the urethra becomes incorporated in the penis. The closing of the urethral groove leaves a seam (or raphe) along the ventral side of the penis. The genital swellings enlarge and fuse to form the scrotum, the median scrotal raphe being continuous with that of the penis. These changes occur in the embryo around the tenth to the twelfth week, but the testes do not descend into the more mature scrotum until the eighth month.

Female genitalia It is evident that the clitoris of the female is the homolog of the penis of the male, since both are derived from the genital tubercle. The urethral groove does not extend into the clitoris in the embryo of 12 weeks; it opens into the urogenital sinus (vestibule). The urethra of the female is shorter than that of the male, and in the mature reproductive system it opens directly to the exterior through the urethral orifice.

The female external genitalia do not differentiate as greatly from the embryonic condition as do the male genitalia. The clitoris develops slowly and does not become large. The urethral folds become the inner membranous folds, or *labia minora*. The genital folds develop into the outer folds, or *labia majora* (Figure 22.35). The vagina is at first a slender tube opening into the urogenital sinus or vestibule, but toward the end of the fetal period it enlarges and opens separately to the exterior. The entrance to the vagina is partly occluded by a thin membrane called the *hymen*.

The full-term fetus grows rapidly in the last 3 months, and the uterus expands high into the abdominal cavity. The fetus changes its position many times, but just before birth it commonly assumes a position with the head down at the opening into the cervix of the uterus. The date of childbirth, or *parturition*, is commonly calculated as 280 days from the beginning of the last menstrual period rather than approximately 9 months after conception. The date of the last menstrual period usually can be determined more accurately, but by any method of calculation the baby may be born a little earlier or a little later than the estimated date.

At childbirth the cervix gradually dilates, and the fetal membranes rupture. Rhythmic contractions of uterine musculature propel the fetus through the dilated cervix (Figures 22.36 to 22.39). The uterine contractions are involuntary but are supplemented by voluntary contractions of the abdominal muscles of the mother. The dilation of the cervix is usually a slow process; when this is accomplished, the obstetrician considers that the first stage of labor is completed. During the second stage the fetus starts to move through the birth canal; if the head is to appear first, the crown of the head is "presented" at the bulging perineum. The emergence of the infant into a new world brings about rapid changes in its physiological activities. The newborn child is now freed from its dependence on the placenta and must begin to breathe for itself. Pulsations cease in the umbilical cord, and it is cut and tied off. The heart adjusts itself to assume the added work of active pulmonary circulation. Later, food must be taken by way of the mouth for the first time. The new individual is well adapted to survive, but many growth processes are still incomplete at birth.

The uterus, by further contractions, expels the loosened placenta and

CHILDBIRTH

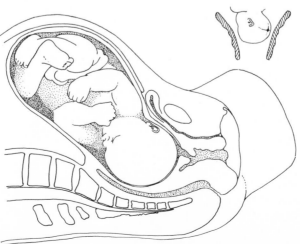

FIGURE 22.36
The beginning of childbirth. The cervical canal still is not dilated. The inset shows the position of the head in relation to the pelvis, frontal view. (*After Bryant and Overland, "Woodward and Gardner's Obstetric Management and Nursing," F. A. Davis Company, Philadelphia, 1964.*)

Childbirth

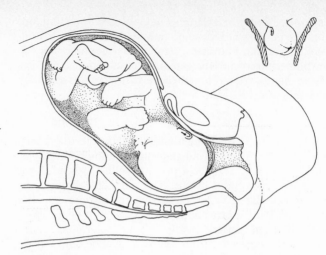

FIGURE 22.37
Early stages of childbirth. The cervix has dilated, and the membranes have broken. Note flexion and descent of head in inset. (*After Bryant and Overland, "Woodward and Gardner's Obstetric Management and Nursing," F. A. Davis Company, Philadelphia, 1964.*)

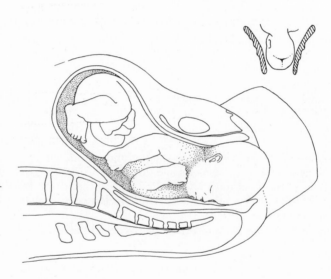

FIGURE 22.38
The progress of childbirth. The head has reached the pelvic floor, and the occiput has rotated so that it faces the pubic symphysis. Note the position of the head in inset. (*After Bryant and Overland, "Woodward and Gardner's Obstetric Management and Nursing," F. A. Davis Company, Philadelphia, 1964.*)

FIGURE 22.39
The progress of childbirth. The head is presented: first the occiput, then the brow, and finally the face. (*After Bryant and Overland, "Woodward and Gardner's Obstetric Management and Nursing," F. A. Davis Company, Philadelphia, 1964.*)

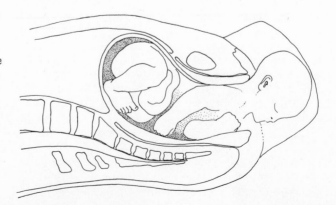

decidual membranes as the *afterbirth*. The uterine mucosa then undergoes a period of repair and regeneration similar to that following menstruation. The period of repair varies, but commonly menstruation is not resumed until about 3 months after parturition. Ovulation can occur within this period, however, and a new pregnancy can be started.

Gametogenesis means the development of the gametes, or germ cells. Undifferentiated germ cells multiply by mitotic division. After a period of growth and multiplication, certain male germ cells can be identified as primary spermatocytes and similar cells in the female as primary oocytes.

During meiosis, unlike in mitosis, the chromosome number is reduced from the diploid condition in undifferentiated germ cells to the haploid condition of the gametes. The prophase consists of several stages, as follows: leptotene, zygotene, pachytene, and diplotene stages and diakinesis. The chromosomes then enter metaphase I and anaphase I of the first meiotic division. The first meiotic division is a reduction division; the secondary spermatocytes or oocytes contain the haploid number of chromosomes.

Telophase I is a short period, and the spermatocytes or oocytes start the second meiotic division. The prophase and metaphase are typical of a mitotic division. Anaphase II is characterized by the division of the centromeres and the separation of the chromatids. This is the equational division; there is no further reduction in number, and each chromatid becomes a separate chromosome.

There are 46 chromosomes in man: 44 plus X and Y sex chromosomes in the male, 44 plus two X chromosomes in the female. The development of sperm cells is called spermatogenesis; the development of ova is called oogenesis. Fertilization of the ovum occurs upon the union of male and female pronuclei. The fertilized ovum is a zygote.

The zygote passes through cleavage stages, first forming a solid ball of cells, which later becomes a hollow ball, or blastula. The mammalian blastula is specialized and is called a blastodermic vesicle, or blastocyst.

The zygote passes through its cleavage stages as it moves down the oviduct. By the time it reaches the uterus, it is a blastodermic vesicle; it becomes implanted at this stage.

Three germ layers are differentiated; they are the ectoderm, mesoderm, and entoderm. The ectoderm gives rise to the skin and its derivatives, the dorsal nerve cord and its derivatives, and the epithelial lining of the mouth and anal canal. Mesodermal derivatives include muscle tissue, bone, connective tissues, blood, bone marrow, and lymphoid tissues. The entoderm forms the delicate epithelial linings of the digestive tract and the glands associated with it, as well as the lining of the respiratory tract and parts of the urogenital system.

The principal fetal membranes are the amnion, chorion, and placenta. The amnion is a thin, protective membrane, which immediately surrounds the embryo. It contains the amniotic fluid. The chorion limits the extraembryonic cavity and forms the embryonic portion of the placenta. The placenta is composed of a maternal portion and an embryonic portion. In the placenta the exchange of oxygen and carbon dioxide occurs. Nutritive materials are transferred across lymph spaces from the mother's blood to the fetal circulation. The placenta also functions as an excretory organ for the fetus, removing waste products from the fetal circulatory system.

The human embryo of about 4 weeks is about 4 millimeters long and exhibits gill slits; muscle segments, or somites; limb buds; and a tail. The face begins to form at 5 weeks by an infolding from either side. The eyes are indicated at 6 weeks by the formation of the optic vesicles. The auditory pit is present at 4 weeks. The first branchial groove forms the external auditory canal; the first pharyngeal pouch forms the tympanic cavity of the middle ear and the auditory tube.

The first indication of an embryonic blood supply comes from the formation of blood islands in the yolk sac. The heart and circulatory system develop quickly to supply the needs of the embryo. The heart, until about the fifth week, functions as a single pulsating tube.

Organs and glands of the digestive system arise as diverticuli or enlargements of the digestive tube. The larynx, trachea, and lungs arise as outgrowths from the primitive gut just posterior to the pharyngeal pouches. The whole bilobed structure is commonly called the lung bud.

The urogenital system arises from the urogenital ridge. The pronephric and mesonephric kidneys are transient and give way to the development of the metanephric or permanent type of kidney. The metanephric kidney develops a new duct (the ureter).

The external genitalia develop from an undifferentiated stage consisting of a genital tubercle and two ridges called genital folds. The tubercle forms the penis of the male or the clitoris of the female. Genital swellings of the original genital folds become the scrotum of the male or the labia majora of the female.

The fetus grows rapidly during the last 3 months and becomes heavier. The date of childbirth is calculated as 280 days from the beginning of the last menstrual period.

At the beginning of parturition, the cervix gradually dilates, and the amnion commonly ruptures. Rhythmic uterine contractions propel the fetus through the dilated cervix and into the birth canal. After childbirth, further uterine contractions expel the placenta and decidual membranes as the afterbirth.

QUESTIONS

1 Trace the development of spermatozoa or ova through various stages of gametogenesis.
2 Discuss the behavior of the sex chromosomes in gametogenesis.
3 Just when is fertilization of the ovum completed?
4 Where does fertilization occur?
5 Outline the various stages of cleavage.
6 Describe the blastodermic vesicle, or blastocyst.
7 What is meant by implantation?
8 Discuss the origin of the germ layers and their derivatives.
9 Explain the function of each of the fetal membranes.
10 Describe the structure, function, and content of the umbilical cord.
11 Distinguish between identical and fraternal twins.
12 Describe the formation of the face.
13 Discuss the structure and function of the skin and its derivatives.
14 How does the heart arise?
15 List the derivatives of the primitive digestive tube.
16 Explain the changes that occur in the development of the urogenital system.
17 Describe the early differentiation of the external genitalia.
18 What changes in the physiology of the newborn infant must take place at birth?

ADAMS, FORREST H.: Fetal and Neonatal Cardiovascular and Pulmonary Function, *Ann. Rev. Physiol.*, 27:257-284 (1965).

ALSTON, R. E.: "Cellular Continuity and Development," Scott, Foresman and Company, Glenview, Ill. 1967.

COULOMBRE, ALFRED J., and JANE L. COULOMBRE: Lens Development: Fiber Elongation and Lens Orientation, *Science*, 142:1489-1490 (1963).

HELLER, C. G., and YVES CLERMONT: Spermatogenesis in Man: An Estimate of Its Duration, *Science*, 140:184-185 (1963).

MONTAGNA, WM.: The Skin, *Sci. Am.*, 212:56-66 (1965).

MOORE, J. A.: "Heredity and Development," Oxford University Press, New York, 1963.

PAPACONSTANTINOU, JOHN: Molecular Aspects of Lens Cell Differentiation, *Science*, 156:338-346 (1967).

SPEMANN, H.: "Embryonic Development and Induction," Yale University Press, New Haven, Conn., 1938.

STREETER, G. L.: Developmental Horizons in Human Embryos, *Contrib. Embryol.*, 32:133-203; *Carnegie Inst. Wash. Publ.* 575, 1948.

WADDINGTON, C. H.: "Principles of Development and Differentiation," The Macmillan Company, New York, 1966.

reference books

GROSS ANATOMY TEXTBOOKS AND ATLASES

ANSON, B. J.: "Atlas of Human Anatomy," 3d ed., W. B. Saunders Company, Philadelphia, 1963.

———— (ed.): "Morris' Human Anatomy," 12th ed., McGraw-Hill Book Company, New York, 1966.

CROUCH, JAMES E.: "Functional Human Anatomy," 2d ed., Lea & Febiger, Philadelphia, 1972.

CUNNINGHAM, D. J.: in G. J. Romanes (ed.), "Textbook of Anatomy," 10th ed., Oxford University Press, Fair Lawn, N.J., 1964.

EDWARDS, L. F., and G. R. L. GAUGHRAN: "Concise Anatomy," 3d ed., McGraw-Hill Book Company, New York, 1971.

GRAY, H.: in C. M. Goss (ed.), "Anatomy of the Human Body," 29th ed., Lea & Febiger, Philadelphia, 1973.

SOBOTTA, J., and F. H. J. FIGGE: "Atlas of Descriptive Human Anatomy," 8th ed., Stechert-Hafner, Inc., New York, 1963.

WISCHNITZER, SAUL: "Outline of Human Anatomy," McGraw-Hill Book Company, New York, 1963.

CELLULAR BIOLOGY

ALLEN, JOHN M. (ed.): "The Molecular Control of Cellular Activity," McGraw-Hill Book Company, New York, 1962.

BAKER, J. J. W., and G. E. ALLEN: "Matter, Energy and Life," 2d ed., Addison-Wesley Publishing Company, Inc., Reading, Mass., 1970.

BOURNE, G. H.: "Division of Labor in Cells," 2d ed., Academic Press, Inc., New York, 1970.

De ROBERTIS, E. D. P., W. W. NOWINSKI, and F. A. SAEZ: "Cell Biology," 5th ed., W. B. Saunders Company, Philadelphia, 1970.

FAWCETT, DON W.: "The Cell; An Atlas of Fine Structure," W. B. Saunders Company, Philadelphia, 1966.

FREEMAN, JAMES A.: "Cellular Fine Structure," McGraw-Hill Book Company, New York, 1964.

GIESE, A. C.: "Cell Physiology," 3d ed., W. B. Saunders Company, Philadelphia, 1968.

LOEWY, A. G., and PHILIP SIEKEVITZ: "Cell Structure and Function," 2d ed., Holt, Rinehart and Winston, Inc., New York, 1969.

McELROY, WILLIAM D.: "Cell Physiology and Biochemistry," 3d ed., Prentice-Hall, Inc., Englewood Cliffs, N.J., 1971.

NASS, GISELA: "The Molecules of Life," McGraw-Hill Book Company, New York, 1970.

PORTER, KEITH R., and MARY A. BONNEVILLE: "An Introduction to the Fine Structure of Cells and Tissues," 4th ed., Lea & Febiger, Philadelphia, 1973.

SMITH, C. U. M.: "The Architecture of the Body," Faber & Faber, Ltd., London, 1964.

SWANSON, CARL P.: "The Cell," 3d ed., Prentice-Hall, Inc., Englewood Cliffs, N.J., 1969.

HISTOLOGY

AREY, L. B.: "Human Histology," 3d ed., W. B. Saunders Company, Philadelphia, 1968.

BLOOM, W., and D. W. FAWCETT: "A Textbook of Histology," 9th ed., W. B. Saunders Company, Philadelphia, 1968.

DiFIORE, M. S. H.: "An Atlas of Human Histology," 3d ed., Lea & Febiger, Philadelphia, 1967.

FINERTY, JOHN C., and E. V. COWDRY: "A Textbook of Histology," 5th ed., Lea & Febiger, Philadelphia, 1960.

GREEP, R. O. (ed.): "Histology," 3d ed., McGraw-Hill Book Company, New York, 1973.

HAM, A. W., and T. S. LEESON: "Histology," 5th ed., J. B. Lippincott Company, Philadelphia, 1965.

WINDLE, W. F.: "Textbook of Histology," 4th ed., McGraw-Hill Book Company, New York, 1969.

PHYSIOLOGY

BECK, W. S.: "Human Design," Harcourt, Brace, Jovanovich, Inc., New York, 1971.

BEST, C. H., and N. B. TAYLOR: "The Physiological Basis of Medical Practice," 8th ed., The Williams & Wilkins Company, Baltimore, 1966.

CROUCH, J. E., and J. R. McCLINTIC: "Human Anatomy and Physiology," John Wiley & Sons, Inc., New York, 1971.

DAVSON, HUGH, and M. GRACE EGGLETON (eds.): "Starling and Evan's Principles of Human Physiology," 13th ed., Lea & Febiger, Philadelphia, 1962.

GROLLMAN, SIGMUND: "The Human Body," 2d ed., The Macmillan Company, New York, 1969.

GUYTON, A. C.: "Function of the Human Body," 3d ed., W. B. Saunders Company, Philadelphia, 1969.

————: "Textbook of Medical Physiology," 4th ed., W. B. Saunders Company, Philadelphia, 1971.

————: "Basic Human Physiology," W. B. Saunders Company, Philadelphia, 1971.

JACOB, S. W., and C. A. FRANCONE: "Structure and Function in Man," 2d ed., W. B. Saunders Company, Philadelphia, 1970.

LANGLEY, L. L.: "Physiology of Man," 4th ed., Van Nostrand Reinhold Company, New York, 1971.

————: "Review of Physiology," 3d ed., McGraw-Hill Book Company, New York, 1971.

————, I. R. TELFORD, and J. B. CHRISTENSEN: "Dynamic Anatomy and Physiology," 3d ed., McGraw-Hill Book Company, New York, 1969.

RUCH, T. C., and HARRY D. PATTON (eds.): "Physiology and Biophysics," 19th ed., (Howell-Fulton), W. B. Saunders Company, Philadelphia, 1965.

TUTTLE, W. W., and BYRON A. SCHOTELIUS: "Textbook of Physiology," 16th ed., The C. V. Mosby Company, St. Louis, 1969.

VANDER, A. J., J. H. SHERMAN, and D. S. LUCIANO: "Human Physiology," McGraw-Hill Book Company, New York, 1970.

WOOLDRIDGE, DEAN E.: "The Machinery of Life," McGraw-Hill Book Company, New York, 1966.

EMBRYOLOGY

AREY, L. B.: "Developmental Anatomy," 7th ed., W. B. Saunders Company, Philadelphia, 1965.

BALINSKY, B. I.: "An Introduction to Embryology," 3d ed., W. B. Saunders Company, Philadelphia, 1970.

BERRILL, N. J.: "Developmental Biology," McGraw-Hill Book Company, New York, 1971.

HAINES, R. W., and A. MOHIUDDIN: "Handbook of Human Embryology," The Williams & Wilkins Company, Baltimore, 1965.

PATTEN, BRADLEY M.: "Foundations of Embryology," 3d ed., McGraw-Hill Book Company, New York, 1968.

SPEMANN, HANS: "Embryonic Development and Induction," Stechert-Hafner, Inc., New York, 1962.

WILLIAMS, P. L., C. P. WENDELL-SMITH, and SYLVIA TREADGOLD: "Basic Human Embryology," J. B. Lippincott Company, Philadelphia, 1966.

GENETICS

BONNER, DAVID M., and STANLEY E. MILLS: "Heredity," 2d ed., Prentice-Hall, Inc., Englewood Cliffs, N.J., 1964.

DeBUSK, A. GIB: "Molecular Genetics," The Macmillan Company, New York, 1968.

GARDNER, E. J.: "Principles of Genetics," 3d ed., John Wiley & Sons, Inc., New York, 1968.

KING, R.: "Genetics," 2d ed., Oxford University Press, Fair Lawn, N.J., 1965.

LEVINE, R. P.: "Genetics," 2d ed., Holt, Rinehart and Winston, Inc., New York, 1968.

McKUSICK, V. A.: "Human Genetics," Prentice-Hall, Inc., Englewood Cliffs, N.J., 1964.

MOODY, P. A.: "Genetics of Man," W. W. Norton & Company, Inc., New York, 1967.

MOORE, JOHN A.: "Heredity and Development," 2d ed., Oxford University Press, Fair Lawn, N.J., 1972.

SINNOTT, E. W., L. C. DUNN, and T. DOBZHANSKY: "Principles of Genetics," 5th ed., McGraw-Hill Book Company, New York, 1958.

SRB, ADRIAN M., RAY D. OWEN, and ROBERT S. EDGAR: "General Genetics," 2d ed., W. H. Freeman and Company, San Francisco, 1965.

STERN, CURT: "Principles of Human Genetics," 2d ed., W. H. Freeman and Company, San Francisco, 1960.

SUTTON, H. ELDON: "Genes, Enzymes and Inherited Diseases," Holt, Rinehart and Winston, Inc., New York, 1961.

WALLACE, BRUCE: "Chromosomes, Giant Molecules and Evolution," W. W. Norton & Company, Inc., New York, 1966.

WHITTINGHILL, MAURICE: "Human Genetics and Its Foundations," Reinhold Publishing Corporation, New York, 1965.

WINCHESTER, A. M.: "Genetics," 3d ed., Houghton Mifflin Company, Boston, 1966.

Reference books

ENDOCRINOLOGY   FRIEDEN, E., and H. LIPNER: "Biochemical Endocrinology of the Vertebrates," Prentice-Hall, Inc., Englewood Cliffs, N.J., 1971.

GREENE, R.: "Human Hormones," McGraw-Hill Book Company, New York, 1970.

TURNER, C. D., and J. T. BAGNARA: "General Endocrinology," 5th ed., W. B. Saunders Company, Philadelphia, 1971.

WILLIAMS, ROBERT H.: "Textbook of Endocrinology," 4th ed., W. B. Saunders Company, Philadelphia, 1968.

ZARROW, M. X., J. M. YOCHIM, and J. L. McCARTHY: "Experimental Endocrinology," Academic Press, Inc., New York, 1964.

BIOCHEMISTRY   BAKER, J. J. W., and GARLAND E. ALLEN: "Matter, Energy and Life," 2d ed., Addison-Wesley Publishing Company, Inc., Reading, Mass., 1970.

JELLINCK, P. H.: "Biochemistry," Holt, Rinehart and Winston, Inc., New York, 1963.

————: "The Cellular Role of Macromolecules," Scott, Foresman and Company, Glenview, Ill., 1967.

KARLSON, P.: "Introduction to Modern Biochemistry," 2d ed., Academic Press, Inc., New York, 1965.

KAY, ERNEST R. M.: "Biochemistry: An Introduction to Dynamic Biology," The Macmillan Company, New York, 1966.

KLEINER, ISRAEL S., and JAMES M. ORTEN: "Biochemistry," 7th ed., The C. V. Mosby Company, St. Louis, 1966.

MAZUR, A., and B. HARROW: "Textbook of Biochemistry," 10th ed., W. B. Saunders Company, Philadelphia, 1971.

BIOLOGY AND ZOOLOGY

COCKRUM, E. LENDELL, WILLIAM J. McCAULEY, and NEWELL A. YOUNGGREN: "Biology," W. B. Saunders Company, Philadelphia, 1966.

GOIN, C. J., and O. B. GOIN: "Man and the Natural World," The Macmillan Company, New York, 1970.

HARDIN, G.: "Biology: Its Principles and Implications," 2d ed., W. H. Freeman and Company, San Francisco 1966.

JOHNSON, C. E.: "Human Biology: Contemporary Readings," Van Nostrand Reinhold Company, New York, 1970.

JOHNSON, W. H., LOUIS E. DELANNEY, E. C. WILLIAMS, and T. A. COLE: "Principles of Zoology," Holt, Rinehart and Winston, Inc., New York, 1969.

PHILLIPS, E. A.: "Basic Ideas in Biology," The Macmillan Company, New York, 1971.

SIMPSON, G. G., and WM. S. BECK: "Life, An Introduction to Biology," 2d ed., Harcourt, Brace & World, Inc., New York, 1965.

STORER, T., ROBERT USINGER, R. C. STEBBINS, and J. W. NYBAKKEN: "Elements of Zoology," 5th ed., McGraw-Hill Book Company, New York, 1972.

VILLEE, C. A.: "Biology," 6th ed., W. B. Saunders Company, Philadelphia, 1972.

————, WARREN F. WALKER, Jr., and F. E. SMITH: "General Zoology," 3d ed., W. B. Saunders Company, Philadelphia, 1968.

WEISZ, P. B.: "The Science of Biology," 4th ed., McGraw-Hill Book Company, New York, 1971.

————: "The Science of Zoology," McGraw-Hill Book Company, New York, 1966.

———— (ed.): "The Contemporary Scene: Readings," McGraw-Hill Book Company, New York, 1970.

WINCHESTER, A. M.: "Biology and Its Relation to Mankind," 4th ed., D. Van Nostrand Company, Inc., Princeton, N.J., 1969.

glossary

abduction. The movement of a part away from the midline or axis of the body.

accommodation. The adjustment or focusing of the eye for vision at different distances.

acetabulum. The round depression, or cavity, of the hip bone (os coxae) which receives the head of the femur.

acetylcholine. The acetyl ester of choline. A substance considered to be secreted at cholinergic nerve endings. Probably the transmitter substance at the synapse of cholinergic nerve fibers.

acid. A compound capable of dissociating in aqueous solution to form hydrogen ions.

acromegaly. An endocrine condition characterized by overgrowth of bones of the face and extremities.

acromion. A process of the scapula.

ACTH. Adrenocorticotropic hormone from the anterior hypophysis.

actin. A muscle protein which takes part in contraction.

active transport. The movement of materials through a membrane against a gradient and requiring expenditure of energy.

actomyosin. A muscle protein complex composed of actin and myosin.

adduction. The movement of a part toward another part or toward the midline of the body.

adenine. A purine base.

adenohypophysis. The glandular anterior lobe of the pituitary gland.

adenoid. An enlarged lymphoid growth in the nasopharynx.

adipose. Fatty tissue; fat.

adrenal (suprarenal) gland. An endocrine gland located on the superior border of the kidney in man.

adrenaline. See epinephrine.

adrenergic. Certain nerve fibers releasing an adrenaline-like, i.e., epinephrine-like, substance as a transmitter. Especially the terminal filaments of most sympathetic postganglionic neurons.

afferent. Leading toward, as sensory neurons. Vessels that progress toward or enter an organ.

albinism. A congenital condition characterized by lack of pigment in skin, hair, and iris.

aldosterone. A mineralocorticoid of the adrenal cortex.

alimentary. Pertaining to food or aliment, as the alimentary tract of the digestive system.

alkaline. A substance or solution containing more hydroxyl than hydrogen ions.

allantois. An extraembryonic membrane arising as an outgrowth of the embryonic hindgut.

allele. One of a pair of genes. Each has the same locus on homologous chromosomes.

alveolus. The bony socket of a tooth or the alveolar border of the jaw bone. A pouch in the air sac of the lung.

amacrine cells. Retinal neurons with long lateral processes.

ameba. A colorless, unicellular, protozoan organism, which constantly changes its form, progressing by means of pseudopodia. Ameboid, resembling an ameba.

amino acid. An organic compound with the basic formula NH_2—R—COOH.

amnion. An embryonic membrane filled with amniotic fluid, which immediately encloses the embryo.

amphoteric. Applied to fluids that possess qualities of both alkalies and acids.

ampulla. A dilatation of a canal or duct.

amylase. (Greek, *amyl*, starch.) Any starch-digesting enzyme that hydrolyzes starch to sugar.

amyloid. Starchy or starchlike.

amylopsin. The starch-digesting ferment of pancreatic fluid.

anastomosis. The opening of one vessel into another. Literally, to bring to a mouth.

ancon. The elbow.

anconeus. A muscle of the elbow joint.

androgen. A male sex hormone that influences the development of secondary sex characteristics in the male.

angiotensin (angiotonin). A polypeptide formed by the action of the proteinase renin on renin substrate. It causes a rise in blood pressure.

angstrom. A unit of length; angstrom = 0.0001 micron = 1×10^{-7} millimeter.

anlage. The embryonic primordium from which a body part or an organ develops. A blastema.

annulus. A ring-shaped opening. Annulus ovalis, the margin of the foramen ovalis of the fetal heart.

anode. The positive pole of a battery.

antebrachium. The forearm.

antecubital. Applied to the space in front of the elbow.

antibody. A protective substance formed to react with foreign substances or antigens that may be introduced into the body.

antigen. A substance that stimulates the production of antibodies or reacts with them.

antitoxin. A substance in the blood capable of neutralizing a specific toxin.

Glossary

615

antrum. A cave. The cavity or sinus in the maxilla is called the antrum of Highmore.

aorta. The largest of all the arteries in the body.

apnea. Suspension of breathing.

apoenzyme. The protein part of an enzyme.

aponeurosis. A layer of strong white fibrous tissue.

aqueous. Watery.

arachnoid. Like a spider's web, for fineness. One of the membranes of the brain and spinal cord.

archenteron. The primitive or embryonic digestive cavity.

areolar. Having little spaces.

arrectores pilorum. Cutaneous muscles attached to the bases of hairs. Arrectores pilorum cause the hairs to "stand on end."

arteriosclerosis. Hardening of the arteries.

artery. A vessel carrying blood away from the heart.

arthrosis. A joint or articulation.

asphyxia. A condition in which the blood is deprived of oxygen.

assimilation. The taking up of nutriment by the body tissues, in such a manner that it becomes a part of them.

atlas. The first cervical vertebra, upon which the skull rests, from the fabled giant who bore the globe upon his shoulders.

ATP. Adenosine triphosphate.

ATPase. Adenosine triphosphatase.

atrium. A hall; a chamber of the heart where blood enters.

atrophy. Wasting. Reduction in size, often with degeneration.

Auerbach's plexus. A sympathetic nerve network in the wall of the intestine.

auricular. Shaped like, or belonging to, an ear.

axilla. The armpit.

axis. The second cervical vertebra. Named because of the pivot around which the atlas revolves (like a wheel around an axis).

axon. The efferent fiber of a nerve cell.

azygos. Without a yoke. The name of certain vessels or nerves that are not in pairs.

Bartholin, glands of. Glands located on either side of the vaginal orifice. Vestibular glands.

basophil. A granular leukocyte. The granules and nucleus stain blue with basic or alkaline dyes.

biceps. Having two heads, as the biceps femoris, biceps brachii.

bicuspid. Having two points or cusps. A bicuspid tooth.

blastocoele. The cavity within the blastula.

blastocyst. A type of blastula characteristic of Mammalia.

blastoderm. The primitive germ layer, which gives rise to the primary germ layers.

blastodermic vesicle. The blastocyst.

bolus. A rounded mass, as a bolus of food in the intestine.

Bowman's capsule. The invaginated distal portion of a uriniferous tubule, which contains the glomerulus.

brachialis. Belonging to the arm, or brachium.

Broca's area. The motor speech center in the brain.

bronchus. (Plural, bronchi.) An air tube. The smallest air tubes are called bronchioles.

buccinator. From a word meaning trumpet. The blowing or trumpeting muscle.

bulbourethral glands. Cowper's glands, the ducts of which open into the urethra.

bundle of His. The atrioventricular, neuromuscular bundle of the heart.

bursa. Literally, a purse. The bursae are small sacs containing fluid and are found in the fascia under skin, or muscles, or tendons.

calcaneus. The heel bone. The tendo calcaneus, or tendo Achillis, is attached to the calcaneus.

calcitonin. A thyroid hormone.

calculus. A stonelike body formed in some fluid of the body. Renal calculus, in the kidney; biliary c., in the gallbladder, etc.

callus. A thickened portion of the skin. The material thrown out (provisional callus) for the repair of fractured bone, to become the permanent callus when the bone is completely ossified.

calorie, large. A term referring to the amount of heat required to raise 1 kilogram of water from 15 to 16°C. A kilogram calorie.

cancellous. Resembling lattice work. A cancellous or spongy bone.

canine. Resembling a dog. Canine teeth, like a dog's long, pointed teeth.

canthus. (Plural, canthi.) The angle at the meeting of upper and lower eyelid.

capillary. Resembling a hair in size.

capitellum or capitulum. A little head, an eminence on the lower extremity of the humerus.

capsule. A structure that encloses an organ or part. (The capsule of a joint.)

carbohydrate. An organic substance composed of carbon, hydrogen, and oxygen, as in sugars and starches.

carboxyl. The COOH group characteristic of organic acids.

cardiac. Referring to the heart or cardia.

caries. Decay of bone or teeth. Dental caries.

carotene. A yellowish carotenoid, precursor of vitamin A.

carotid. The name of the large arteries of the neck, once thought to cause sleep.

casein. A milk protein, precipitated by the action of rennin or acids.

cast. An albuminous structure molded in tubular form. Renal casts.

castration. Removal of the gonads.

cataract. A condition in which the lens of the eye becomes clouded or opaque.

cathode. The negative pole of a battery.

cauda equina. A horsetail. The name given to the bundle of spinal nerves in the lower portion of the spinal canal.

cecum. Blind. The blind pouch at the beginning of the large intestine.

celiac (coeliac). Pertaining to the celia or belly.

center. In the nerve system, a center is a collection of gray cells. The central nervous system comprises the brain and spinal cord, which contain the large nerve centers, e.g., respiratory center.

centrifugal. Referring to a force that is exerted from the center outward.

centriole. A minute body in the centrosome. During mitotic cell division, a centriole is found at either end of the spindle.

centripetal. Applied to a force that seeks a center.

cephalic. (Greek, *kephalē*, head.) Toward the head. Referring to the head.

cerebellum. Little brain. Lower, posterior, coordinating portion of brain.

cerumen. The wax of the ear.

cervix. Neck. Cervical, belonging to or resembling a neck. The cervix of the uterus.

chemoreceptor. A receptor reacting to chemical stimuli.

chiasma. A crossing or decussation of nerve fibers within the central nervous system. The optic chiasma.

chief cells. Cells of the gastric mucosa that secrete pepsinogen.

chlorophyll. The green coloring matter in plants that is concerned with photosynthesis.

choana. A funnel. The choanae are the posterior openings from the nose into the pharynx.

cholesterol. A sterol commonly present in animal fats but also in the protoplasm of cells. An important constituent of nervous tissue, blood, and bile.

choline. A base, commonly a constituent of phospholipids. Considered to be a vitamin of the B complex. See also acetylcholine.

cholinergic. Referring to nerve fibers that, when stimulated, release acetylcholine.

cholinesterase. An enzyme that catalyzes the hydrolysis of acetylcholine.

chorda tympani. A branch of the facial (VIIth cranial) nerve that passes through the tympanic cavity to unite with the lingual branch of the trigeminal (Vth cranial) nerve.

chordae tendineae. Tendinous cords attaching the heart valves to the papillary muscles.

choroid. The intermediate, vascular coat of the eyeball.

chromaffin. Cells having an affinity for, and staining deeply with, chromion salts. Usually refers to cells of the adrenal medulla or other cells of sympathetic origin.

chromatin. Material within the nucleus that stains deeply with basic dyes.

chyle. Lymph containing absorbed fat.

chyme. Acid, partially digested food material as it leaves the stomach.

cicatrix. A scar. It is formed of fibrous connective tissue.

cilia. Eyelashes. Ciliated, having tiny hairlike projections, as ciliated epithelium.

ciliary. The ciliary region of the eye presents radiating lines, caused by folds of the tissues composing it.

circumduction. Leading around. This is the motion made when a part is moved around in a circle, one end being stationary. The extremities, the digits, and the head can be circumducted.

circumflex. To bend around. Circumflex arteries wind around the arm or thigh.

circumvallate. Walled around. The circumvallate papillae at the base of the tongue are encircled by a ridge.

cisterna chyli. Chyle cistern. An enlargement at the base of the thoracic duct.

clavicle. The clavicula, which resembles a Roman key. Bone of the shoulder girdle.

climacteric. A time of life when the system is believed to undergo marked and permanent changes; usually applied to the time of the cessation of menstruation, the menopause.

clitoris. A small erectile organ in the upper part of the vulva, the homolog of the male penis.

coagulation. The clotting of blood.

coccyx. A cuckoo's beak. The bone at the end of the spinal column, named from its shape.

cochlea. A conch shell. A cavity of the internal ear resembling a snail shell in form.

coelom. The embryonic body cavity formed between somatic and splanchnic mesoderm.

coenzyme. The nonprotein or prosthetic group of an enzyme.

collagen. A protein found especially in white fibrous tissues, cartilage, and bone. On being boiled, it produces gelatin.

collateral. A side branch of an axon. Collateral circulation is secured by the union of branches of two vessels, whereby the main current of fluid may be carried by this side route if necessary.

commissure. A joining. A commissure connects right and left parts, as the commissures of the brain and spinal cord.

communis. Common.

concha. A shell.

condyle. A rounded eminence of bone.

congenital. Existing at or before birth.

conjunctiva. Connecting. The mucous membrane that lines the undersurfaces of the eyelids and covers the anterior surface of the eyeball.

convoluted. Twisted.

coracoid. Like a crow's beak. The coracoid process of the scapula.

corium. Leather. The deep portion of the skin from which leather is made.

cornea. The transparent anterior portion of the scleroid coat of the eye.

cornua. Plural of cornu, a horn.

coronal, coronoid. Pertaining to, or resembling, a crown.

coronary. The coronary arteries encircle the crown of the heart.

corpus callosum. The transverse commissure of the cerebral hemispheres.

corpus luteum. Yellow body. The tissue formed in a ruptured graafian follicle of the ovary.

cortex. Bark. The superficial layer, as the cortex of the brain.

cortisone. An adrenal cortical extract, 17-hydroxy-11-dehydro-corticosterone.

costal. Relating to a rib or costa. Costal cartilage.

coxae. Plural of coxa, the hip; also the genitive form, as os coxae, the bone of the hip.

cranium. The part of the skull that contains the brain.

cretinism. The condition of a cretin or undeveloped person, both mentally and physically, resulting from deficient activity of the thyroid gland.

cribriform. Resembling a sieve. Cribriform plate of the ethmoid bone.

cricoid. Like a ring. The cricoid cartilage of the larynx is shaped like a seal ring.

crista. A crest. The crista galli of the ethmoid bone. The crista acustica, a ridge of sensory hair cells in the ampulla of a semicircular canal. The cristae of mitochondria.

crucial. Like a cross. The crucial ligaments cross each other.

crural. Belonging to or like the lower extremity, from crus, a leg; the crura (or legs) of the diaphragm. The crura cerebri, or cerebral peduncles, descending nerve tracts from the cerebral hemispheres.

cryptorchism. A condition referring to the failure of the testes to descend.

crypts of Lieberkühn. Tubular glands at the bases of villi in the intestine.

cumulus oophorus. A mound of cells supporting the maturing ovum within the follicle.

cyclic AMP. Adenosine 3'5'-monophosphate.

cystic. Relating to a cyst, or a sac containing fluid (cystic duct). A cystic ovary has cysts developed from its substance.

cytochromes. Heme proteins involved in cellular respiration.

cytoplasm. The protoplasm of the cell exclusive of the nucleoplasm.

Glossary

618

cytosine. A pyrimidine base present in nucleic acid.

deamination. The removal of an amino, NH_2, group from an amino acid or from another organic compound.

decidua. (Latin, *deciduus*, that which falls off.) The mucous membrane lining the uterus, especially the part that is shed at menstruation or following parturition as a part of the afterbirth.

decussation. A crossing, as of nerve tracts. Decussation of the pyramidal tracts.

deglutition. The act of swallowing.

dehydrogenase. An enzyme that catalyzes the oxidation of a substrate by the removal of hydrogen.

deltoid. Shaped like the Greek letter delta, Δ.

dendrite. A process of a neuron carrying impulses toward the cell body.

dental. From dens, a tooth, belonging to a tooth. Dentated. Toothed.

dentine. The sensitive substance of the tooth between the enamel and the pulp.

dentition. The eruption or cutting of the teeth. The kind, number, or arrangement of the teeth.

desmosome. An electron-dense zone of attachment between apposing plasma membranes.

diabetes insipidus. A condition due to malfunction of the hypothalamus and characterized by excessive thirst and excessive secretion of urine of low specific gravity.

diabetes mellitus. A condition in which carbohydrates are not oxidized properly by the tissues. Usually insulin is lacking. There is an excess of sugar in the blood, and sugar is present in the urine.

diapedesis. The passing of blood cells through the walls of capillaries.

diaphoretic. An agent that increases the amount of perspiration.

diaphragm. A wall across a space. The muscle that separates the cavity of the thorax from that of the abdomen.

diaphysis. The greater part of the shaft of a bone.

diarthrosis. A movable joint.

diastole. A Greek word meaning a drawing apart. The dilatation of the chambers of the heart.

digastric. Double-bellied, as the digastric muscle.

digit. A finger or toe.

distal. Farthest from the head or trunk.

diuretic. An agent that increases the quantity of urine.

diverticulum. An out-pocketing or sac; the cecum.

DNA. Deoxyribonucleic acid.

dopa. Considered to be a possible neurotransmitter.

dopamine. 3,4-Dihydroxyphenylalanine.

dorsal. Belonging to the dorsum, or back.

ductus arteriosus. A blood vessel of the fetal circulatory system that diverts blood from the left pulmonary artery to the aorta.

duodenum. Meaning twelve. The duodenum is 12 fingerwidths long.

dura mater. The fibrous outer membrane of the brain and spinal cord.

dyspnea. Difficult breathing.

ECG. A record of heart action made by an electrocardiograph.

ectoderm. The outermost germ layer of the embryo.

edema. Swelling caused by effusion of serous fluid into areolar tissues.

EEG. Electroencephalogram. A record of rhythmic brain waves, largely from the cerebral cortex, indicating. the electrical activity of the brain.

efferent. Meaning away from. Efferent vessels leave organs. Efferent or motor neurons.

electrolyte. A substance that ionizes in solution and is capable of conducting an electric current.

electron. The smallest unit of negative electricity.

eliminate. To excrete substances that are useless.

embolism. The blocking of a blood vessel by the formation of a clot.

embryo. An organism in the early stages of development, especially before the third month in human beings.

emesis. Vomiting.

emetic. An agent that induces vomiting.

emission. A seminal discharge.

endo-. Within. Endocardium, within the heart. Endothelium, the epithelial lining of circulatory organs.

endocrine gland. A gland of internal secretion. A ductless gland.

endoderm. The innermost germ layer of the embryo (entoderm).

endolymph. The fluid contained in the membranous labyrinth of the inner ear.

endometrium. The mucous membrane lining the uterus.

endomysium. The sheath of a muscle fiber.

endoplasmic reticulum. A series of tubules and vesicles forming a network, granular or agranular, throughout the cytoplasm of the cell.

endosteum. The lining of medullary canals in long bones.

enema. A fluid introduced into the rectum.

ensiform. Sword-shaped. The process of the sternum.

enteric. Pertaining to the enteron or intestine.

enterokinase. An enzyme of the small intestine that changes trypsinogen to active trypsin.

enzyme. An organic compound produced by living cells, which acts as a catalyst in chemical reactions. Digestive enzymes.

eosinophil. A granular leukocyte whose granules stain eosin red with acid dyes.

epi-. Upon, as epicondyle, epidermis, epiglottis.

epidermis. The outer or ectodermal layer of the skin.

epididymis. A group of coiled tubules on the testis, continuous with the tubules of the testis and with the vas deferens.

epimysium. The connective-tissue muscle sheath.

epinephrine. The hormone of the adrenal medulla.

epiphysis. A part of a bone that is formed independently and joined later to complete the whole bone.

epithelial. Pertaining to epithelium.

epithelium. A tissue that forms the epidermis and the lining of ducts and hollow organs.

EPSP. Excitatory postsynaptic potential.

erythroblast. A nucleated red bone-marrow cell, a precursor of the mature erythrocyte.

erythrocyte. A red blood cell.

erythrocytopoiesis. The production and development of erythrocytes. Synonym: erythropoiesis.

erythropoietin. A hormone that stimulates the formation and development of erythrocytes.

esophagus. From a Greek word meaning to carry food. The esophagus transmits food from pharynx to stomach.

ester. A compound formed from an alcohol and an acid by the elimination of water.

esterase. An enzyme that catalyzes the hydrolysis of an ester into an alcohol and an acid.

estrogen. An estrus-producing hormone. A female sex hormone.

ethmoid. Sievelike. The ethmoid bone has many openings on its surface.

Eucaryota. Organisms with distinct nuclei. Protista.

eunuch. A male castrate.

eversion. Turning outward. To evert an eyelid is to fold it back so as to expose the interior surface.

excretion. A waste substance to be removed from the body. The process of removing waste products from the tissues.

extension. Stretching out or extending. (Bending backward is overextension.)

extirpation. The complete removal of a part from the body.

extrinsic. Originating outside a structure. Extrinsic muscles of the eyeball.

exudate. A collection of material that has filtered through the walls of vessels into surrounding tissues.

FAD. Flavin adenine dinucleotide.

falciform. Sickle-shaped.

fallopian tube. Uterine tube or oviduct.

falx. A sickle.

fascia. (Plural, fasciae.) A band. The tissue that binds organs or parts of organs together.

fasciculus. In the nervous system, a nerve tract.

fasciculus cuneatus. A laterodorsal afferent (ascending) tract of the spinal cord.

fasciculus gracilis. A mediodorsal afferent tract of the spinal cord. Proprioceptive from the lower part of the trunk and from the legs.

fauces. (Latin, *faux*, throat.) Isthmus of, the space bounded by the soft palate, tonsils, and tongue. Pillars of, the folds connecting the soft palate with the tongue and pharynx. (The tonsil is between the pillars of either side.)

feedback mechanism. A regulatory arrangement whereby a portion of the output of a system is recycled back into the system to control any further output. Negative feedback.

femoral. Belonging to the femur or thighbone.

fetus. The mammalian organism in the later stages of development, especially after the beginning of the third month.

fibrin. Protein threads that form the framework of a blood clot.

filiform. Threadlike in shape, slender; as filiform papillae of the tongue.

fimbria. A fringe; fimbriated, having a fringelike appearance.

fissure. A cleft or groove, as a fissure of the brain surface.

fistula. A tubelike passage formed by incomplete closure.

flavus. Yellow.

flexion. Bending. Flexure, a bend.

follicle. A very small sac (or bag) containing a secretion.

fontanel. A little spring. A membranous spot in the infant's skull; the name suggested by the rising and falling caused by the child's respirations.

foramen. (Plural, foramina.) An opening.

foreskin. The prepuce.

fornix. A vaultlike space. An arch of nerve fibers below the corpus callosum.

fossa. A depression or concavity.

fourchette. A little fork. The fold of mucous membrane at the posterior ends of the labia majora.

fovea. A small pit. The fovea centralis is a tiny depression in the macula lutea of the retina.

frenum. A curb or bridle. The frenum linguae is the fold of mucous membrane attaching the tongue to the floor of the mouth.

FSH. The follicle-stimulating hormone secreted by the anterior lobe of the hypophysis.

fundus. The base.

fungiform. Shaped like a fungus or mushroom.

fusiform. Spindle-shaped.

GABA. Gamma-aminobutyric acid. Considered to be a possible synaptic transmitter.

gamete. The mature ovum or spermatozoon.

gametogenesis. The origin and development of the gametes.

ganglion. (Plural, ganglia.) A group of nerve-cell bodies usually outside the brain and spinal cord.

gaster. The stomach. Gastric, belonging to the stomach, or gaster.

gastrocnemius. The belly of the leg. The prominent muscle of the calf of the leg.

gene. A hereditary unit having a definite location on a chromosome.

genotype. Having the same genetic constitution.

genu. A knee.

gigantism. A condition of abnormal growth involving excessive secretion of the anterior lobe of the hypophysis.

gingiva. The gum. The tissue that encloses the neck of the tooth and covers the jaw bone.

glabella. A little smooth space. The smooth space between the eyebrows.

gladiolus. A little sword. The body of the sternum.

gland. A collection of cells that can form a secretion or an excretion.

glans. The head of the clitoris or penis.

glaucoma. A condition of the eye characterized by a marked increase in intraocular pressure.

glenoid. Having the form of a shallow cavity. Belonging to a cavity.

globulin. A group of animal and plant proteins. Alpha, beta, and gamma globulins are fractions of blood-serum globulin. Gamma globulin contains antibodies.

glomerulus. The minute coiled mass of capillaries within a Bowman's capsule of the kidney.

glossopharyngeal. Belonging to the tongue and pharynx.

glottis. The upper opening of the larynx. Epiglottis, the leaf-shaped cartilage upon the upper border of the larynx.

glucagon. A hormone produced by alpha cells of the pancreatic islet group. The hormone causes a rise in blood-glucose levels.

glucokinase. An enzyme of the liver that catalyzes the phosphorylation of glucose.

gluconeogenesis. The formation of glucose from noncarbohydrate materials such as protein.

glucose. A monosaccharide sugar, $C_6H_{12}O_6$.

gluteal. Referring to the gluteus, or buttock.

glycogen. A polysaccharide, $(C_6H_{10}O_5)_n$, formed from glucose and found in various tissues but stored especially in the liver and muscles.

glycogenesis. The formation of glycogen from glucose in the liver.

glycogenolysis. The breakdown of glycogen to glucose.

glycolysis. The breaking down of glucose in the tissues, to pyruvic or lactic acid, by enzymatic action.

glyconeogenesis. The formation of glycogen from noncarbohydrate sources such as from amino acids.

glycosuria. The presence of an abnormally high proportion of sugar in the urine.

goblet cells. Mucus-secreting cells in the lining of the intestine.

goiter. The enlargement of the thyroid gland.

golgi apparatus. A smooth membranous structure in the cytoplasm of animal cells, usually located near the nucleus.

gonad. A reproductive organ. The testis or the ovary.

gonadotropic. A hormone secreted by the anterior hypophysis that influences the development and maintenance of the gonads.

graafian follicle. An ovarian follicle.

gram molecular weight. The weight of a substance in grams equivalent to its molecular weight.

groin. The depressed area of the abdomen adjacent to the thigh.

guanine. A nitrogenous base, a purine.

gustatory. Associated with the sense of taste.

gyrus. (Plural, gyri.) A circle. Convolutions of the brain cortex.

haploid. The reduced number of chromosomes after meiosis in the maturing germ cells.

haversian. Name applied to the central canals in bone tissue, from the English anatomist Havers.

helicotrema. The confluence at the apex of the cochlea between the scala tympani and the scala vestibuli.

helix. A spiral form.

hematocrit. The percentage by volume of blood cells in a unit volume of blood.

hemoglobin. The oxygen-carrying substance of red blood cells, to which their color is due.

hemolysis. Destruction of red blood dells.

hemophilia. A sex-linked hereditary condition characterized by reduced clotting ability of the blood, resulting in prolonged bleeding.

hemopoiesis. The formation of blood.

hemorrhoidal. From a word meaning flowing with blood. Pertaining to a hemorrhoid, or pile.

Henle's loop. A specialized turn in the uriniferous tubule.

heparin. A substance found in the liver that inhibits the coagulation of blood.

hepatic. Referring to the liver, or hepar.

hernia. The protrusion of a part of an organ through an opening.

heterozygous. Having different alleles representing a given character.

hilum. Literally, a little thing. Applied to the depression where vessels enter and leave an organ.

homeostasis. The maintenance of a steady state in the internal environment.

homologous. Having the same origin, development, and structure.

homozygous. Having common alleles representing a given character or referring to the whole individual.

hormones. Chemical substances formed in endocrine glands and conveyed by the blood to other organs, to influence their activity.

hyaline. Resembling glass, clear. Hyaloid has a similar meaning.

hydration. Saturating with water.

hydrocephalus. A collection of fluid either within the ventricles or outside the brain.

hymen. A membrane partially covering the orifice of the vagina.

hyoid. U-shaped, as the hyoid bone.

hyperglycemia. An excess of sugar in the blood.

hypermetropia. Farsightedness.

hyperplasia. Excessive formation of tissue.

hypertrophy. Overgrowth. Derived from two Greek words meaning too much nourishment.

hypochondrium. Under the cartilage. The hypochondriac region is under the cartilages of ribs.

hypogastric. Under the stomach.

hypoglossal. Under the tongue.

hypophysis. (Greek, growing under.) An endocrine gland located under the hypothalamus. The pituitary gland.

hypothalamus. An area that is located in the floor of the diencephalon.

ileum. A roll or twist; the portion of small intestine that appears rolled or convoluted.

ilium. The upper portion of the hipbone, or os coxae.

incisor. A cutting instrument. The front teeth are incisors.

index. Indicator. The first finger, named from its common use.

infra-. Beneath.

infundibulum. A funnel-shaped space or part. The stalk of the pituitary gland.

inguinal. Belonging to or near the thigh, or inguen. Inguinal canal.

inhibition. The restraining or stopping of normal action.

insertion. The attachment of a muscle to a bone at the more freely movable end.

instep. The arch of the foot, dorsal aspect.

insulin. The hormone of the pancreas secreted by the islet cells (of Langerhans).

inter-. Between, as intercostal, between the ribs; intercellular, between the cells, etc.

intercalated. Placed between, as the electron-dense areas at apposing plasma membranes in cardiac tissue.

intermediary metabolism. The metabolism of cells or body fluids after absorption and before excretion.

interstitial tissue. Tissue located in the interspaces of a structure. The endocrine tissue of the testis.

intima. The lining of blood vessels.

intrinsic. Located within a part or organ.

inversion. A turning in, as inversion of the eyelashes; inversion of the foot.

in vitro. In glass; referring to a process or reaction carried on in a test tube, that is, outside the body.

in vivo. Referring to a process or reaction carried on within a living organism.

involution. The changing back to a former condition, of an organ that has fulfilled a function, as the involution of the uterus after parturition.

IPSP. Inhibitory postsynaptic potential.

iris. A circle or halo of colors. The colored circle behind the cornea of the eye.

ischium. The lowest part of the hipbone, or os coxae.

islands of Langerhans. Small islet cells of the pancreas, which secrete the hormone insulin.

isometric. The same measure. A type of contraction wherein the tension increases but the length of the muscle remains essentially the same.

isotonic. Having the same osmotic pressure as another fluid taken as a standard. Contraction of muscle that permits the muscle to shorten.

isotopes. Elements that have the same number of atoms but different atomic masses.

jaundice. A condition characterized by yellowing of tissues caused by the absorption of bile pigments into the blood.

jejunum. Empty. The second portion of the small intestine, usually found empty.

jugular. Belonging to the neck, or jugulum.

karyokinesis. Mitosis.

keratitis. Inflammation of the cornea.

kidney. An important organ of elimination or excretion, in which the urine is formed.

kinase. An activator of a zymogen to form an enzyme.

kinesiology. The science of muscular movement.

kinetic. Referring to motion. Kinetic energy.

kinetochore. The constriction in the chromosome indicating the attachment of the spindle fiber.

Kupffer cells. Fixed macrophages lining sinusoids of the liver.

kymograph. An instrument consisting of a revolving drum covered with paper on which a record of some physiological activity can be traced.

kyphosis. Hunchback. An exaggerated dorsal spinal curvature in the thoracic region.

labia majora. The large outer folds of the female external genitalia.

labia minora. The small inner membranous folds of the female external genitalia.

labium. (Plural, labia.) A lip.

lacrimal. Having to do with tears, or lacrymae, as the lacrimal gland.

lactase. An enzyme that acts upon lactose.

lactation. The secretion of milk by the mammary glands.

lacteal. Like milk (from *lac*, milk). The lacteals are lymph vessels that carry the milky-looking chyle.

Glossary

622

lactic acid, $C_3H_6O_3$. An acid formed from carbohydrates, as in sour milk.

lactose. Milk sugar.

lacuna. A minute cavity, as in cartilage and bony tissue.

lambdoid. Resembling the Greek letter lambda, λ.

lamella. A little plate or thin layer.

lamina. A plate or layer.

lanugo. The fine downy hair that covers the fetus at about 5 months.

larynx. The part of the air passage extending from the base of the tongue to the trachea. The cartilaginous organ of the voice.

latissimus. Broadest. Latissimus dorsi, broadest muscle of the back.

lens. A glass or crystal, curved and shaped to change the direction of (or refract) rays of light.

lentiform. Shaped like a lens.

lesion. An injury to a tissue that changes its structure or function.

leukemia. A pathological condition of the blood-forming organs characterized by an uncontrolled production of leukocytes.

leukocyte. A white cell of the blood or lymph. Leukocytosis, an increase in the number of leukocytes.

leukopenia. A decrease in the number of leukocytes.

levator. A lifter. Levator palpebrae, lifter of the eyelid.

linea. A line.

linea alba. A median ventral line on the abdomen indicating the line of junction of the tendons of the external oblique muscles.

linea aspera. A rough line along the posterior surface of the femur.

lingual. Belonging to the tongue.

lipase. A digestive enzyme that acts upon fats.

lobule. A little lobe.

lumbar. Belonging to the loin. The lower part of the back.

lumen. The cavity of a hollow tube or organ.

luteal hormone. The hormone secreted by the corpus luteum of the ovary.

lymph. The clear fluid of the lymphatic system and tissue spaces.

lymph node. A nodule of lymphoid tissue occurring along lymphatic vessels.

lymphocyte. A nongranular leukocyte arising in lymphoid tissues.

lysosomes. Minute cell particles containing hydrolytic enzymes.

lysozyme. An enzyme found especially in tears. It is mildly antiseptic.

macrophage. A large phagocytic cell of the reticuloendothelial system. Not a leukocyte.

macula. A spot. Macula lutea, yellow spot.

major. Greater or larger.

malar. Belonging to the cheek.

malleolus. A little hammer. The two malleoli are processes located at the lower extremities of the tibia and fibula.

malleus. Mallet-shaped. The ear bone that is attached to the tympanic membrane.

maltase. A digestive enzyme acting upon maltose.

mammary. Pertaining to the breast.

mandible. (Latin, *mandere*, to chew.) The lower jaw bone.

manubrium. A handle. The first part of the sternum.

masseter. A chewer. One of the muscles of mastication or chewing.

mast cell. A large granular cell found in connective tissues.

mastitis. Inflammation of the breast, or mammary glands.

mastoid. Shaped like a breast.

maxilla. The jawbone. Applied to the upper jawbone.

meatus. A passageway.

medial. Toward the middle line.

median. Middle, as the median line of the body.

mediastinum. The space in the middle of the thorax.

medulla. Marrow. The central part of a gland or organ. Medulla oblongata. The posterior part of the brain.

medullary. Pertaining to, or like, marrow. The medullary canals contain marrow.

megakaryocytes. Giant cells found in bone marrow. They are thought to produce blood platelets by fragmentation.

melanin. A dark pigment.

melanocyte. A pigment cell containing melanin.

meninges. Membranes. Membranes of the brain and spinal cord.

meningitis. Inflammation of the meninges.

menopause. The cessation of menstruation at the close of the reproductive period.

menstruation. A periodic discharge of blood from the genital canal of a woman, associated with changes in the lining of the uterus.

mesenchyme. Embryonic connective tissue.

mesentery. From two Greek words, meaning middle and bowel. (The mesentery supports the intestine from the posterior abdominal wall.)

mesoderm. The middle germ layer of the embryo.

metabolism. The chemical changes associated with the assimilation of energy materials into cell protoplasm and the elimination of the waste products of cellular activity. Life processes.

metastasis. (Greek, to transpose). The translocation of a disease from its primary focus to some other part of the body through the medium of blood or lymph.

micron. 1 μ. It is 0.001 millimeter.

microvilli. Minute fingerlike projections on the luminal surface of certain epithelial cells, formed as modifications of the plasma membrane.

minimus. Least or smallest.

minor. Lesser.

mitochondria. Minute bodies in the cytoplasm of the cell.

mitral. Resembling a miter in outline. Mitral valve. The left atrioventricular valve of the heart.

molar. Like a millstone, or mola. The molar teeth grind the food.

molar solution. A solution containing in one liter as many grams of the substance (solute) as the molecular weight of the solute.

mucous. Containing or resembling mucus. Mucosa, a mucous membrane.

mucus. A thick clear fluid secreted by the cells of mucous membranes.

myelin sheath. The white, fatlike inner covering of a myelinated nerve fiber.

myenteric plexus. A network of sympathetic fibers in the wall of the intestine.

myocardium. The muscular wall of the heart.

myoglobin. A muscle protein capable of short-term storage of oxygen.

myopia. Nearsightedness.

myosin. One of the principal proteins of muscle.

myotatic reflex. A stretch reflex of muscle.

myotic. A drug causing constriction of the pupil.

myxedema. A condition resulting from hypothyroidism.

naris. The nostril.

navicular. Boat-shaped, as the navicular, or scaphoid, bone.

necrosis. The death of a portion of tissue, while still surrounded by living structures.

neural. Pertaining to nerves. The neural axis is the spinal cord. The neural canal is the spinal canal. The neural cavity contains the brain and spinal cord.

neurilemma. The outermost membrane covering a nerve fiber.

neuroglia. The ectodermal connective tissue of the brain and spinal cord.

neuron. A unit of the nerve tissues. It consists of a cell body, dendrites, axon, and terminal filaments.

neutrophil. A granular leukocyte staining with neutral dyes.

Nissl substance. Dark-staining material in the cytoplasm of nerve cells.

nodes of Ranvier. Constrictions in the myelin sheath of nerve fibers.

nucha. The nape of the neck.

nucleolus. A smaller nucleus within the nucleus of a cell.

nucleus. A small round body near the center of a cell enclosed in the nuclear membrane. The vital part of a nucleated cell. A group of cell bodies within the central nervous system.

nutrient. Nourishing.

nutrition. The process of nourishing the cells of living tissues.

nystagmus. Abnormal, involuntary, rhythmic oscillation of the eyeballs.

occipital. Belonging to the back of the head, or the occiput.

odontoid. Resembling a tooth in shape.

olecranon. The large process at the upper end of the ulna. The head of the elbow.

oligodendrocytes. A type of neuroglial cell found along neurons and capillaries of the central nervous system. Synonym: oligodendroglia.

omentum. A fold of peritoneum connected with the stomach.

omos. The shoulder. Omohyoid, belonging to shoulder and hyoid bone, as the omohyoid muscle.

oocyte. An immature ovum.

ophthalmic. Belonging to the eye or ophthalmos.

orbicular. Ring-shaped. A ligament or muscle that resembles a little circle.

organ. A structure designated for a particular function or use.

organ of Corti. An organ of hearing located in the cochlear duct of the cochlea. The spiral organ.

orifice. An aperture or opening.

os. A bone. Ossicle, a little bone.

os. A mouth.

osmosis. The movement of substances of different concentrations in solution through a semipermeable membrane.

osseous. Bony.

ossification. The formation of bone.

osteoblasts. Cells concerned with bone formation. The precursors of osteocytes.

osteoclasts. Large multinucleate cells thought to be involved in the breakdown of bony tissue.

osteogenic. Referring to the development of bony tissue.

osteology. The science that treats of bones.

outlet. The inferior opening, or strait, of the pelvis.

oviduct. The fallopian or uterine tube.

ovulation. The rupture of the ovum from the ovary.

ovum. (Plural, ova.) Female germ cell.

oxytocin. A hormone of the pituitary gland stimulating the contraction of uterine muscle.

pachytene. The stage in meiosis that follows synapsis. The homologous chromosome threads appear shorter and thicker.

pacinian corpuscles. Tactile or pressure receptors in the skin, mesentery, and other tissues.

palpebra. An eyelid. Palpebral fissure, the fissure between the eyelids.

pancreas. A digestive gland located below the stomach in the loop of the duodenum.

papilla. (Latin, nipple). A soft conic eminence.

papilla of Vater. An enlargement at the common opening of the pancreatic and bile ducts into the duodenum.

parenteral. Other than by way of the intestine or alimentary tract.

parietal. Referring to the wall of a body cavity or an outer membrane lining the wall, as opposed to an inner or visceral membrane. The parietal bones of the skull.

parietal cells. Found along the border of fundic glands of the stomach. They produce a substance that is later converted to hydrochloric acid.

parotid. Near the ear. The parotid salivary gland is below the external ear.

parturition. The act of bringing forth, or giving birth to, young.

patella. A little pan. The sesamoid bone in front of the knee joint; the kneecap.

pectoral. Connected with the breast, as pectoral muscles.

pedicle. A little foot. Peduncle has a similar meaning.

pelvis. A basin. The cavity in the lowest part of the trunk.

penis. External genital organ of the male.

pepsin. The protease of the stomach.

peptic cells. Also called chief cells of the stomach. They secrete pepsinogen.

pericardium. The membrane that encloses the heart.

perichondrium. The nourishing membrane that covers cartilage.

perilymph. The lymph of the perilymphatic spaces between the membranous and bony labyrinths of the inner ear.

perimysium. A connective tissue around small bundles of muscle fibers.

perineal. Pertaining to the perineum, that region of the body in front of the anus.

periosteum. The nourishing membrane around bone.

peristalsis. (From two Greek words, meaning around and constriction.) The intestinal movements that propel the food.

peritoneum. (From two Greek words, meaning around and to stretch.) The serous membrane around abdominal organs. The lining of the abdominal cavity.

peroneal. Relating to the fibula. Peroneal nerves supply muscles on the fibula.

petrous. Hard, like a rock.

Peyer's patches. Aggregations of lymph nodules in the mucous lining of the ileum.

phagocyte. White blood cells having the power to take microorganisms into their substance and to digest them.

phalanges. Plural of phalanx, a body of troops drawn up closely together. The fingers and toes.

phallus. The penis.

pharynx. That part of the food passage that connects the mouth and esophagus. The upper part is the nasopharynx, an air passage.

phenotype. A group of individual organisms that look alike but differ in their genetic constitution.

phenylketonuria. A condition resulting from inability of an individual to oxidize phenylalanine to tyrosine.

phlebitis. Inflammation of a vein.

phrenic. Pertaining to the phren, or diaphragm, as the phrenic nerves.

pia mater. Tender mother. The delicate membrane that bears the blood vessels of brain and spinal cord. The innermost membrane of the meninges.

pigment. Coloring matter.

pineal body. A cone-shaped body arising from the roof of the diencephalon.

pinocytosis. The process by which extracellular fluid and certain substances such as protein molecules are taken into the cell by invagination of the plasma membrane.

pituitary gland. An endocrine gland that lies beneath the brain in the sella turcica of the sphenoid bone. The hypophysis.

placenta. A membranous structure that provides the exchange of food materials and oxygen between the blood of the mother and that of the embryo or fetus. Waste products of fetal metabolism are removed through the placenta.

plantar. Belonging to the sole of the foot.

plasma. The name given to the fluid portion of circulating blood.

platysma. Broad. Platysma muscle.

pleura. A side. The name of the serous membrane that lines the thorax and covers the lungs.

plexus. A network. An arrangement in which vessels or nerves appear to be woven together.

pneumogastric. Referring to the lungs and stomach.

polycythemia. A condition in which there is a marked increase in the number of erythrocytes.

polymorphonuclear. Having nuclei of various shapes, such as granular leukocytes.

poples. A space behind the knee (popliteal space).

popliteal. Belonging to the back of the knee.

porta. A gate. The portal vein enters the porta or gate of the liver.

prehension. Taking hold of.

premolar. Applied to the teeth that stand immediately in front of the molars.

prepuce. The foreskin of the penis or clitoris.

presbyopia. The farsightedness of aging, due to diminished elasticity of the lens.

pressor. Producing a rise in blood pressure.

process. In anatomy, a projection, a prominence, or an outgrowth.

proctodeum. The invaginated portion of the hindgut, which is lined with ectoderm.

progesterone. The hormone of the corpus luteum.

prolactin. A lactogenic hormone of the anterior pituitary gland that stimulates the production of milk in nursing mothers.

pronation. Literally, bending forward. The position of the hand when the thumb is toward the body. The act of turning the hand palm downward, or in the prone position.

prostaglandins. Hormone-like substances present in various tissues and having a wide range of physiological functions.

prostate. From Greek words meaning to stand before. The prostate gland is in front of the neck of the bladder and surrounds the base of the urethra in the male.

prothrombin. The inactive form of thrombin.

protoplasm. The essential living matter of all cells.

protuberance. A knoblike projection.

proximal. Nearest to the trunk, median line, or center.

psychic. Pertaining to the mind.

pterygoid. Wing-shaped.

ptosis. (Greek, a falling.) Prolapse or lowering of an organ or part.

ptyalin. A salivary amylase.

puberty. The age at which the reproductive organs become functional.

pubes. The anterior portion of the os coxae, the pubic bones. The hairy region above the pubic bones.

pudendum. The external genitalia, especially of the female.

pulmonary. Pertaining to the lung, or pulmo.

Purkinje cells. Neurons of the cortex of the cerebellum with great arborization of the dendrites.

pyloris. The lower opening of the stomach into the duodenum.

pyramidal. Shaped like a pyramid. Pyramidal cells of the motor area of the cerebral cortex.

pyramidal tracts. Corticospinal nerve fibers from the cerebral motor area and extending downward into the spinal cord.

pyridoxine. Vitamin B_6.

quadrate. Four-sided; square or rectangular.

quadri-. A combining form meaning four.

quadriceps. Four-headed.

quadrigeminal. Consisting of four parts.

quinti-. Referring to the fifth.

racemose. Resembling a cluster of grapes.

rachitis. Rachis, the vertebral column. A vitamin D deficiency, rickets.

radius. A rod or spoke. The lateral bone of the forearm.

ramus. A branch, as the ramus of the mandible.

raphe. A seam. The union of two parts in a line, like a seam.

reaction. Response to a stimulus or test. The iris reacts to the stimulus of light. Urine reacts to the litmus test.

receptor. The specialized end organ of an afferent neuron.

recession. Withdrawal, as the margin of the gums from the teeth.

rectus. Straight, as rectus muscles.

refractory period. The period of reduced irritability, as in muscle and nerve.

renal. Pertaining to the kidney.

renin. A protease from the kidney that releases angiotensin from angiotensinogen.

rennin. A milk-coagulating enzyme of the gastric juice.

rete. A net.

reticular. Resembling a network.

retina. A net. The complicated inner nerve coat of the eye.

retinal. A rod cell pigment. (Retinene.)

retro-. Prefix meaning behind. Retroperitoneal, behind the peritoneum.

Rh factor (rhesus factor). An antigen known to occur on the red cells of about 85 percent of the white population of the United States and Great Britain.

rhinencephalon. The olfactory part of the brain.

rhino-. A combining form referring to the nose.

riboflavin. One of the vitamins of the vitamin B complex.

ribose. A five-carbon sugar, a pentose.

ribosomes. Particles consisting of ribonucleic acid, commonly found on the endoplasmic reticulum.

rickets. A vitamin D deficiency causing malformation of bone and cartilage.

rigor mortis. Rigidity of death. The muscular stiffness that occurs after death.

RNA. Ribonucleic acid.

rod cells. Visual cells of the retina adapted for night vision.

roentgenogram. An x-ray photograph.

Rouget cells. Supporting cells on capillary walls.

rugae. (Plural of ruga.) Folds. Wrinkles.

saccade. Rapid eye movements.

saccule. A little sac.

sacral. Relating to the sacrum, or the bone that protects the pelvic organs, which were held sacred by the ancients.

sagittal. Like an arrow, straight. The straight suture of the skull. A plane that divides the body into right and left portions. Midsagittal, a median longitudinal plane.

saline. Salty.

saliva. The mixed secretions of glands of the mouth and salivary glands.

saphenous. Manifest or plainly seen. The large superficial vein on the medial side of the lower extremity and the longest vein in the body.

sarcolemma. The delicate sheath around a muscle fiber.

sarcomere. In muscle, a segment limited by a Z membrane at either end.

sarcoplasmic reticulum. A network of minute channels within the striated muscle fiber.

sartorius. (Latin, *sartor*, a tailor.) An anterior muscle of the thigh.

Schwann cell. A type of cell that covers myelinated and unmyelinated fibers of vertebrate peripheral nerves.

sciatic. Pertaining to the ischium. The sciatic nerve.

sclerotic. Hard. The sclerotic is the tough fibrous coat of the eye; the sclera.

scoliosis. Lateral curvature of the spine.

scrotum. The pouch containing the testes.

scurvy. A vitamin-deficiency condition due to lack of vitamin C.

sebaceous. Applied to the glands that produce the oil, or sebum, of the skin.

secretin. An intestinal hormone that stimulates the pancreas.

secretion. A substance formed by glandular cells.

sella turcica. Turk's saddle. A saddle-shaped depression in the sphenoid bone.

semen. The secretion of male reproductive glands containing spermatozoa.

semilunar. Shaped like a half-moon.

seminal vesicle. A secretory gland of the male reproductive system.

septum. A partition.

serous. Of the nature of serum, a thin watery fluid derived from the blood.

serrated. Having teeth like a saw.

Sertoli cells. Modified cells present in male germinal epithelium.

serum. The fluid portion of the blood after clotting has taken place.

sesamoid. Resembling a grain in form. Applied to small nodules of bone sometimes found in tendons.

shaft. The main portion of a long bone.

sigmoid. Curved like the letter S. As the sigmoid (or transverse) sinus; the sigmoid colon.

sinoatrial node. A group of nerve cells and fibers in the wall of the right atrium near the opening of the superior vena cava, which functions as a pacemaker for the heartbeat.

sinus. A hollow space, or cavity. A cavity within a bone. A cavity containing blood, as a venous sinus.

soluble. That which can be dissolved or made into a solution.

somatic. Pertaining to the body. Somatic cells, body cells exclusive of germ cells.

specific gravity. The weight of a substance, judged in comparison with an accepted standard. In the case of urine, the standard is an equal volume of distilled water—at greatest density.

spermatozoa. Male germ cells.

sphenoid. Wedge-shaped. The sphenoid bone.

sphincter. A muscle that closes an orifice.

sphygmomanometer. An instrument for measuring blood pressure.

spirometer. An instrument for determining the amount of air respired.

splanchnic. Pertaining to the viscera or internal organs.

squamous. Shaped like a scale.

stenosis. The narrowing or contraction of a passageway.

stereognosis. The faculty of recognition of objects by the sense of touch.

sternum. Breastbone.

stimulus. That which excites activity or function.

stomodeum. The embryonic ectodermal invagination that leads to the formation of the mouth.

stratum. A layer.

striated. Striped. The alternate light and dark bands of striated muscle.

sty. An inflammation of a sebaceous gland of the eyelid.

styloid. Pointed, like the stylus that was used for writing in ancient times.

sub-. Prefix meaning below or under.

subarachnoid space. The space beneath the arachnoid membrane filled with cerebrospinal fluid.

subcutaneous. Under the skin.

submucosa. A layer of fibrous connective tissue beneath or adjacent to a mucous membrane.

subserous. Under a serous membrane.

substrate. The substance acted upon by an enzyme.

sucrase. An enzyme that acts upon sucrose.

sudoriferous. Bearing sweat, as sudoriferous glands. (Sudoriparous has the same meaning.)

sulcus. (Plural, sulci.) A furrow or groove. The depressions between the convolutions of the brain.

super-. Prefix meaning above.

supercilium. The eyebrow, or prominence above the eyelashes. (Adj., superciliary.)

supination. The attitude of one lying on the back. The position of the hand when the little finger is next to the body, or turning the palm upward.

suppuration. The formation of pus.

supra-. A prefix meaning above.

suprarenal. Above the kidney. Adrenal. An endocrine gland located on the superior surface of the kidney.

surfactant. A lipoprotein which lowers the surface tension in lung alveoli.

sustentaculum tali. A process of the calcaneus (heelbone) that supports the talus.

suture. (Latin, *sutura*.) A seam. The joints of the cranium are sutures.

sympathomimetic. To mimic the action of the sympathetic nervous system. To cause physiologic actions similar to those produced by the sympathetic nervous system.

symphysis. A joining of two bones especially in a sagittal plane, as the symphysis of the mandible and the pubic symphysis.

synapse. The region where the end knobs of the terminal filaments of one neuron come in close physiologic relationship with the dendrites and cell body of a succeeding neuron.

synapsis. The conjugation of homologous chromosomes during meiosis.

synaptic cleft. A space of about 100 to 200 angstroms between the presynaptic and postsynaptic membranes.

synarthrosis. An immovable joint.

syncytium. A multinucleate mass of protoplasm.

syndrome. A typical set of conditions that characterize a deficiency or disease.

synergic, synergetic. Two or more agents acting as one. The cooperative action of certain muscles.

synovia. A fluid resembling the raw white of an egg, found in joint cavities, bursae, and tendon sheaths.

synovial membrane. A membrane lining a joint cavity, a bursa, or tendon sheath and concerned with the secretion of synovial fluid.

systole. (Greek, contraction.) The contraction of the chambers of the heart.

tactile. Referring to the sense of touch.

taeniae coli. Three tapelike longitudinal muscle bands of the colon.

talus. The ankle bone upon which the tibia rests, also called astragalus.

tendo Achillis. The tendon of Achilles, or calcaneus tendon. The tendon of leg muscles attached to the calcaneus or heelbone by which Achilles was held when his mother submerged him in the river Styx to render him invulnerable. Only the heel remained vulnerable.

tensor fasciae latae. A muscle that tenses the fascia of the thigh.

tentorium. A tent. The tentorium cerebelli (of the cerebellum) covers the cerebellum.

teres. Round. (*Ligamentum teres*, round ligament.)

testes. The male reproductive glands that produce spermatozoa.

testosterone. The male sex hormone secreted by the interstitial cells of the testis.

tetanus. (Physiol.) A sustained contraction of muscle especially when produced experimentally.

thalamus. (Greek, a bed.) The optic thalamus is in the base or bed of the brain.

thenar. Relating to the palm or sole.

thiamine hydrochloride. An essential vitamin of the vitamin B complex.

threshold stimulus. The least strength of stimulus that will cause a reaction. A minimal stimulus.

thorax. The chest. The portion of the trunk that contains the heart and lungs.

thrombin. An enzyme acting upon fibrinogen to produce fibrin.

thrombocyte. A blood platelet.

thromboplastin. A thromboplastic substance which, along with calcium ions and other factors, converts prothrombin to active thrombin.

thrombus. A blood clot formed within the heart or blood vessels, as in coronary thrombosis.

thymus. A lymphoid structure located beneath the sternum in the mediastinum.

thyrocalcitonin. A hypocalcemic factor; a regulator of calcium in blood plasma.

thyroid. Shield-shaped. The thyroid gland, an endocrine gland.

thyrotropin. An anterior-pituitary hormone that regulates the secretion of the thyroid gland.

thyroxine. A hormone of the thyroid gland.

tidal air. The amount of air inspired or expired during quiet breathing.

tissue. A group of cells of similar origin, structure, and function.

tonus. A state of mild contraction exhibited by muscle tissue.

torticollis. Twisted neck, wryneck.

trabeculae. Little beams. The cross bands of connective tissue that support soft structures, as in the spleen.

trapezius. A muscle of the back.

trauma. A wound or injury.

triceps. Three-headed.

trigone. A space or surface having three angles or corners.

trochanter. From a word signifying a wheel. (The muscles that are attached to the trochanters roll the femurs.) A bony process of the femur.

trochlea. A pulley. A trochlear surface is a grooved convexity, as the trochlea of the humerus. A ring of connective tissue in the upper margin of the orbit.

trochlear. Pertaining to a pulley. The trochlear cranial nerve innervates the superior oblique (pulley) muscle of the eye.

tropomyosin. A muscle protein.

trypsin. The enzyme of the pancreas that digests proteins. A protease.

tuber. A swelling or bump.

tubercle. A small projection like a swelling.

tuberosity. A large projection on a bone.

turbinated. Rolled, like a scroll. Turbinate bones, the nasal conchae.

tympanum. Pertaining to the middle ear or to the eardrum. Tympanic cavity. Tympanic membrane.

tyrosine. An amino acid.

ubiquinone. Coenzyme Q, a factor in electron transport.

ulna. A cubit; the elbow. The longer bone in the medial side of the forearm.

umbilicus. (Latin, *umbo*, the elevated or depressed point in the middle of an oval shield.) The navel.

uncinate. Hooked. A process shaped like a hook.

ungual. Belonging to the nail.

uracil. A constituent of nucleic acids; a pyrimidine base.

urea. $CO(NH_2)_2$. A substance representing the chief nitrogenous product of tissue waste.

ureter. The duct of the kidney, which conveys urine to the bladder.

urethra. The passage through which urine is expelled from the bladder.

uterus. A pear-shaped, muscular organ of the female reproductive system, in which the fetus develops.

utriculus. A membranous sac in the vestibule of the inner ear, connected with the semicircular canals.

uvula. The median, posterior tip of the soft palate.

vagina. A sheath. The passageway from the uterus to the external orifice.

vagus. The Xth cranial nerve.

vallate. Situated in a cavity surrounded by a ridge. The vallate papillae of the tongue.

valvulae conniventes. Little valvelike folds. Seen on the mucous coat of the small intestine.

varicose veins. Abnormally swollen and tortuous veins.

vasa efferentia. Tubules that lead from the testis into the vas deferens.

vasa recta. The capillary network around kidney tubules.

vascular. Having many blood vessels.

vas deferens. The efferent duct of the testis.

vasomotor. Literally, vessel mover. Applied to the nerves that dilate blood vessels or contract them, or vasodilators and vasoconstrictors.

vasopressin. A posterior-pituitary antidiuretic hormone.

velum. The veil, or soft hanging portion of the palate or roof of the mouth.

vena cava. A large, hollow vein.

venesection. Cutting a vein.

ventral. Toward the front of the body, as the ventral cavity.

ventricle. A cavity in the brain or in the heart.

venule. A very small vein.

vermiform. Worm-shaped. The vermiform appendix.

vertebrae. From a Latin word meaning to turn. Certain movements of the vertebrae turn the body from side to side. The bones of the spinal column.

vertex. The crown of the head.

vesicle. A liquid-filled sac or cavity.

vestibule. A cavity of the internal ear through which impulses are transmitted to auditory and vestibular nerves.

villus. The villi of the intestine are hairlike in shape and belong to the mucous coat. Vascular, fingerlike processes of the chorion.

viosterol. Activated ergosterol. Vitamin D_2. An antirachitic compound.

viscus. (Plural, viscera.) An internal organ of the head or trunk.

vitamin. An organic compound usually present in minute amounts in foods and essential for growth and nutrition.

vitreous. Glassy. The vitreous humor resembles glass in appearance. The vitreous layers of the skull are brittle like glass.

volar. Belonging to the palm.

vulva. The external genitalia of the female.

Wernicke's area. An area located in the temporal lobe of the brain close to the auditory area. Injuries to this area result in failure to understand the meaning of spoken words.

Wharton's jelly. Mucoid connective tissue of the umbilical cord.

white matter. Nerve tissue composed chiefly of nerve fibers.

Wormian bones. Small supernumerary bones in the sutures of the skull.

xantho-. A combining form meaning yellow.

xanthophyll. A yellow pigment found in plants.

X chromosome. One of the sex chromosomes.

xero-. A combining form meaning dry.

xerophthalmia. A vitamin A deficiency condition characterized by a dry and thickened conjunctiva.

xiphoid. Sword-shaped. The third piece of the sternum is the xiphoid, or ensiform, process.

Y chromosome. A male sex chromosome in man.

yellow spot. The macula lutea of the retina.

yolk. Nutritive material of the ovum.

yolk sac. An extraembryonic membrane containing the yolk if yolk is present.

Z line. A narrow zone of dense material present in skeletal-muscle tissue.

zein. A prolamine of maize.

zona pellucida. A transparent, noncellular, secreted layer surrounding the ovum.

zoology. The science of animal life.

zygapophysis. An articular process of a vertebra.

zygoma. A yoke. The arch of bone at the side of the face formed between zygomatic and temporal bones.

zygote. The fertilized ovum before cleavage.

zymase. An enzyme present in yeast concerned with alcoholic fermentation.

zymogen. The inactive precursor of an enzyme.

index

Page numbers in italic indicate figures.

A band of muscles, 123, 127
Abdomen, cavity, 425–426, 604
Abdominal aorta, 358, 361, 365, 369
    principal branches, 360
Abducens nerve, 229, 231, 285, 295
Abductor muscles, 181
Absorption of food, 59, 445–448
Accelerator globulin (factor V), 334
Accessory nerve, 231, 233, 248, 249,
    251
Accommodation reflex, 303
Acetabulum, 110
Acetic acid, 11, 407, 450, 458, 476
Acetoacetic acid, 434, 444
Acetone, 444
Acetylcholine, 128, 138, 194
    breakdown, 128
    function as a choline, 512
    and para sympathetic fibers, 258,
        353
    specific effects, 138
    as transmitter, 194, 457
Acetylcholinesterase, 128
Acetylcoenzyme A, 407, 470, 476
Achilles tendon (calcaneus tendon),
    176, 178, 179
Acid solutions, 10
Acidophils, 521
Acids, 10–12
Acne, 546
Acoustic (vestibulocochlear) nerve,
    232, 233, 272
Acromegaly, 522, 523
Acromion process, 103
ACTH, 524, 525, 539
Actin, 125
Action potential, 196, 197
Actomyosin, 124
Adam's apple, 97
Adaptation: to light, 298–299
    olfactory, 266–267
Addison's disease, 539–540
Adductor longus muscle, 172
Adductor magnus muscle, 170
Adductor muscles, 172
Adenine, 21, 30, 36, 37, 38
Adenoids, 416
Adenosine, 22, 23, 36, 138
Adenosine diphosphate, 21
    (See also ADP)
Adenosine monophosphate, 21, 22,
    36
Adenosine triphosphate, 26
    (See also ATP)
ADH, 527
Adipose tissue, 67–70
    cells and fibers, 67
ADP, 22–23, 25, 127, 129–131, 442
    and ATP, 22–23, 25, 127,
        129–131, 477
    and ATP phase, 129–131
    in glycolysis, 407, 476, 477

ADP-ATP system, 127, 129, 436–440,
    516–517
Adrenal cortex, 470, 535, 536,
    538–541, 545–546
    hormones, 524–525, 539–541
    hyperactivity, 540–541
    and medulla, 535–536
    pigment, relation to, 470
Adrenal glands, 516, 535–536,
    537–538
    and adrenergic fibers, 255
    and carbohydrate metabolism, 437
    effect on heart rate, 369
    and endocrine balance, 519
    hormones, 337–541
    and internal secretion, 516
    medulla, 535–538
Adrenalin, 536
Adrenergic fibers, 255
Adrenergic substance, 353
Adrenocortical hormones, 539–541
Adrenocorticotropic hormone
    (ACTH), 521, 524, 525, 539
Aerobic respiration, 406–407
Aeroembolism, 410
Afterbirth, 589–590, 607
Afterimage: negative, 308–309
    positive, 308–309
Agglutination, 337, 351
Agglutinins, 319, 332, 337–351
AHF, 334–336
Air: composition, 403–404
    residual, 404
    tidal, 403
Albinism, 71, 289–290
Aldosterone, 508, 539
Alimentary canal, 416
Alkaline reserve, 319, 444
Alkaline solution, 10
Alkaline tide, 506
All-or-nothing principle, 197–198
Allantois, 496, 587–588
Allergy, 320
Alveolar air, 405
Alveolar duct, 394
Alveolar glands, 56, 61
Alveolar process of maxilla, 394,
    395, 419
Alveoli, 394, 395, 419
Ametropia, 305
Amino acids, 407
    in blood plasma, 322
    and DNA, 42–43
    essential, 459–460, 485
    of food proteins, 458
    glycogenic, 476–477
    ketogenic, 476–477
    and RNA, 41–42
    transport through cell membranes,
        26, 41
Aminoacetic acid, 458
Ammonia, 450

Ammonium salts, 323
Amnion, 588
    and fetus, 591, 594, 604
    and umbilical cord, 590
Amniotic cavity, 588
Amniotic fluid, 497, 588, 604
AMP, 21, 22, 36
Amphetamines, 226, 536
Ampulla: of semicircular canals,
    270, 276
    of Vater, 441, 603
Amylase: pancreatic, 432, 433
    salivary, 422, 423
Anabolism, 29, 474
Anal canal, 449, 451, 602
Anal fold, 606
Anaphase, 43, 44, 45, 46, 573
Androgenic substances, 544
Androgens, 539, 544–545
Anemia, 324–325
    pernicious, 325, 471
    and radiation, 342
    treatment of, 444, 461, 471
Angiotensin, 508
Angiotensinase, 508
Angiotensinogen, 508
Animal pole, 585
Anion, action in electric field, 9
Anisotropic (A) bands, 123, 125,
    127, 229
Annulus of Zinn (Zinn's ring), 284
Anode, 9
Anterior chamber of eye, 287
Anterior commissure of brain, 229
Antibodies, 319
    and agglutination, 337–341
    relation to serum globulin, 323
Anticoagulant of blood, 330, 336
Antidiuretic hormone (ADH), 527
Antigens, 319, 331, 337–341
Antihemophilic factor (AHF), 334,
    348
Antithromboplastin, 336
Antitoxins, 310
Anus, 449, 606
Aorta, 374, 352, 354–355, 356–361,
    368–369
    abdominal, 357, 360, 361, 369
    arch, 347, 357–358, 359, 360,
        368, 369
    ascending, 357
    descending, 357
    principal branches, 357
    thoracic, 357, 359
Aortic arches of embryo, 595
Aortic bodies, 360, 377–378
Aortic semilunar valves, 349
Aortic sinuses, 357
Aphasia, 220
Apocrine glands, 61–62
Aponenzyme, 23
Aponeurosis, 69, 142, 148

Apoplexy, 218
Appendicitis, 450
Appendicular skeleton, 103, *104*, *105*
Appendix, vermiform, 417, 449
Aqueduct of Sylvius (cerebral
    aqueduct), *216*, 225, 229
Aqueous humor, 210, 286, *287*, 298,
    301
Arachnoid membrane, 208, 211
Arachnoid villi, 211
Arbor vitae, 229
Archenteron, 586
Arches of foot, 115, 116
Area centralis, 299
Areola, 577
Areolae, 65
Areolar tiissue, 64, 65, 67
Arm: bones, 103, *104*–108
    muscles, 154, *155*–*156*, 157
Arrectoris pilorum muscles, 598
Arterial blood, 403
    (*See also* Blood)
Arterial circle of Willis, 359
Arteries, 355–357, *358*, 373, *374*
    abdomen, *360*–361
    arm, 357, *358*, 360
    coronary, 346–347, 353
    head and brain, 358–359
    lower extremities, 362
    pelvis, 367
    principal trunk, 357
    pulmonary, 347, 348, *353*, 357
    structure, 355–356
    systemic, 357–*358*, 359
    thorax, 359
Arterioles, 346–347, *373*, 378
    kidney, 499
Arteriosclerosis, 382
Arthritis, 118
Articulations, 118, 119
Artificial respiration, 410, *411*, 413
    American Red Cross, 412
    Holger-Nielsen method, 411
    mouth-to-mouth technique,
        411–412
    Schafer method, 410
Ascorbic acid (vitamin C), 462, 472
Aster, 45
Astigmatism, 287–288
Astragalus, 112
Astrocytes, 213–215
Atherosclerosis, 382
Atlas, 98–*99*
Atomic mass, 7–8
Atomic numbers, 3–*4*
Atomic structure, *4*–5
Atoms, 3–5
ATP, 18–26, 28
    and actomyosin, 124
    adenosine formation, 138
    and carbohydrate breakdown, 24
    in cardiac tissue, 139
    dephosphorylation, 126
    energy release, 127–130
    formation of, 29, 46
    and membrane transport, 18, 130,
        197, *407*–410
    myosin, 129–131
    production, 122
    relation to RNA, 41
    and vitamin A, 476, 501

ATPase, 19, 126–127, 129
Atria of heart, 346–350, 352–*371*
    contraction, 347–348, 350–352
    musculature, 349–352
Atrioventricular bundle, 350–*351*,
    352
Atrioventricular (AV) node, 350–*351*,
    353
Atrioventricular (AV) valves, 346,
    347, *349*
Auditory meatus, *91*, 268
Auditory nerve, 267, 272
Auditory pit, 596–597
Auditory tube, *268*, 269–270, *597*,
    602
Auditory vesicle, 596–597
AUG, 43
Autonomic circulatory effects, 247,
    252
Autonomic nervous effects, summary,
    252–254
Autonomic nervous system, 241–260,
    349, 422, 564
    parasympathetic, 243, 248–252,
        353–422
    sympathetic, 241–243, *244*–247,
        303–304
Autosomes, 573–577
Avidin, 470
Axial skeleton, *89*–93, *94*–95, *104*–*105*
Axillary artery, *358*, 360
Axillary vein, 364, *366*
Axis, *99*
Axolemma, 191, 192
Axon, 188–189, *190*–*191*
    structure of process, 29, *30*
Axon hillock, 189, 191
Azygos vein, 364, 387, 395

Backbone, 98–*100*, 101
Bacteriolysins, 319
Bainbridge reflex, 377
Barr body, 576–577
Bartholin's glands, 577
Basal ganglia, 223
Basal metabolic rate (BMR), 480,
    484, 531–532
Basal metabolism, 476–477,
    480–485
    apparatus, 480–*481*
Bases, 10–12
    nitrogenous, 36–38
Basilar artery, 323
Basilar membrane, *270*, 271–275,
    364–365, *571*
Basilic vein, 365
Basophils, 330, 521
Beat of heart, 346–352
Bell-Magendie law, 199
Beriberi, 465–466
Beta-hydroxybutyric acid, 444
Bicarbonate buffer system, 12
Bicarbonate-ion concentration, 406
Biceps brachii muscle, 108, *122*,
    143, 154, *156*
Biceps femoris muscle, *170*, 172, *176*
Biceps muscle, 143, 417–419
Bicuspid teeth, 418, 419
Bicuspid valves, 347
Bile, 434, 438, 441, 457, 474,
    603, 604

Bile duct, 417, 432, 603
    common, *416*, 432, *441*, 442, 603
    development, 603
    interlobular, 441
Bile pigments, 441
Bile salts, 433, 440, 441, 448
Bilirubin, 441
Biliverdin, 441
Binocular vision, 305–306
Bios factors, 470
Biotin, 470
Bipolar cells of retina, *291*
Bipolar nerve cells, 188–*189*
Blacktongue, 468
Blastocoele, *585*
Blastocyst, 552, 554, 585–*586*
Blastoderm, *585*–*586*
Blastodermic vesicle, 585
Blastomeres, *584*–585
Blastula, 585
Blind spot, *287*, *293*, 294
Blood, 318–358
    agglutination, 337, *339*–341
    cardiac cycle, 351–355
    cardiac output, 375–376
    cell count, 323–327, *332*–*333*
    cells, 322–325, *326*–*332*, 333, 576
    circulation (*see* Circulation)
    clotting mechanism, 334–335,
        336, 444, 457, 474
    defibrination, 335
    development, 600–601
    effect of calcium ions, 334
    functions, 318–320
    grouping, 337, *338*–*339*, 341–342
    platelets, *323*, 330, 333, 557
    pressure, 375–377, *378*–*379*,
        380–382, *383*–*384*
        and absorption pressure, 378
        in arteries and blood vessels,
            372–373, 374–377,
            378–379, 380–382
        diastolic phase, 379
        and effective hydrostatic
            pressure, 378
        and effective osmotic
            pressure, 378
        and filtration pressure, 378
        measurement, 379–389
        skeletal muscles, 142
        systolic phase, 379–380
        values, 380–382
    proteins, 444
    serum, 336, 337
    sugar, 431, 434, 442
        level, 539
    transfusions, 336–*337*, 338, 340
    typing, 338–339
    velocity: in arteries, *372*
        in capillaries, 372–*373*
        in veins, 372–373
    vessels, 142, 346–353, *354*–*355*,
        356–*358*
        development, 600–601
    volume, 320
Blue babies, 369
Blue line, 423
BMR, 480, 482, 531, 532
Body temperature control, 241,
    372–374, 376
Bone marrow, 75–77

Bones, 75–82, 88–120
  arm, *103–104, 107–108*
  composition, 75
  cranial, *89–94*
  facial, 92–97
  foot, 113, 115–116
  growth, 76–*82*, 89, 605
  hand, 106, 109, 115
  leg, 111, *112, 113*, 115
  neck, 97, *98*
  pelvic girdle, 109, *110, 111*
  shoulder girdle, 103–*104*, 105
  skeleton, *104, 105*
  skull, 89–90, 91–*92*, *93–94*
  structure, *75–77*, *242*
  thorax, 100–101
  trunk, 98–*99*, *100*, 101, *104–105*
  wrist, 108–*109*
Bony labyrinth, 270
Bowman's capsule, 497–*498*
Brachial artery, *358*
Brachial plexus, 238
Brachial vein, *366*
Brachialis muscle, 143, 154–*155*, *156*
Brachiocephalic artery, 357, *358*
Brachiocephalic vein, 363, *366*
Brachioradialis, 155
Brain, 208, *210, 211*–239
  action of chemical compounds,
    224, 225
  development, 208–212
  divisions, primary and secondary,
    210
  structures, 210, *222, 225, 231, 242*
  ventricles, 210
Branchial groove, *597*
Breathing, *398*–399
Bronchi, *394*–395
Bronchial arteries, 359, 395
Bronchial buds, 604
Bronchial tubes, *394*–395
Bronchial veins, 395
Bronchiole, 394
Bronchitis, 394
Buccal glands, 421
Buccinator muscle, *144*, 145
Buffers, 12–14, 318
  bicarbonate, 10, 12
  phosphate, 10–13
  protein, 10–14
Bulbourethral glands (Cowper's),
    516, 563
Bundle of His (neuromuscular
    bundle), 349
Bunion, 116
Bursae, 75, 112, 117, 183
Bursitis, 112, 117
Butyric acid, 450

Caffein in relation to nerve impulses,
    506
Caisson disease, 410
Calcaneus, *115*
Calcaneus tendon, *178*
Calciferol, 465, 534
Calcitonin, 528, 532
Calcium, 416, 464–465
  level in blood, 529, 533–535
  metabolism, 533–535
  as relaxing factor, 130
Calcium ions, 334

Caloric valve of foods, 478–479
Calories, 131, 478–488
  daily total, 482–486
  gram, 477
Calorimetry, 477–479
Calyx of kidney, 497–*498*
Canal of Schlemm, *287–289*
Canaliculi, 75–76, *77*
Cancellous bone, 75
Cancer, 52, 342
Canine teeth, *418*
Capillaries, 346–347, *355*–357,
    365, 372
  a-v, 372, *373*
  movement of fluids through
    membrane, *378*
  wall, 356
Capitulum of radius, 108
Capon, 545
Carbaminohemoglobin, 405
Carbohydrates, 434–437, 455–456
  metabolism, 434–437
  oxidation, 477
Carbon compounds, 7
Carbon dioxide: in cellular
    metabolism, 22–23
  in digestion, 443
  in photosynthesis, 454
  in respiration, 400–401, 403–404,
    405–407, 408, 479
Carbonates, 534
Carbonic acid, 13, 405
Carbonic anhydrase, 405
Carboxylase, 466
Cardiac cycle, 351–*352*
Cardiac impression, *396*
Cardiac inhibitory center, 230
Cardiac output, 352
Cardiac plexus, 233, 245, *247*
Cardiac tissue, 138–*139*, 350
Cardiac veins, 346–*349*, *353*, 363
Carotid arteries, *358*, 359, *360*
Carotid body, *358*, *400*
Carotid sinus, 359
Carotine, 71
Carpal bones, 108, *109*
Cartilage, 72, *73*, *74*, 605
  elastic, 72, *73*
  hyaline, 72, *73*
  matrix, 72, *73*
  white fibrocartilage, 72, *73*
Cartilage bone, 79, *80*
  development, 79, *80*
  endochondral, 79
  perichondral, 79
Caruncle, 283
Casein, 430
Castration, 545
Catabolism, 29, 474
Catalyst, 23
Cataract, 290, *300*
Cathode, 9
Cation, 9, 10
Cauda equina, 233
Caval veins, 346, 377
Cecum, *417, 449*
Celiac artery, *358–359, 361*
Celiac plexus, 245, *247*
Cell, 2–3, 28–53
  horizontal, 291
  mitotic division, 43–*44*, 45–46

physiology, 28–29
  structure, 29
Cell membrane, 2, 29
  movement through, 15–22
Cellular coenzymes, 23–26
Cellular enzymes, 23–24
Cellular metabolism, 20–26, 28–29,
    401–*410*
Cellular respiration, 19–22, 406–409
Cellulose, 450, 454, 456, 477
Cementum, 419, 420
Central artery, *596*
Central canal, *202*, 230, 233, *235*
Central sulcus, *217–218*, 229
Central tendon, 150
Centriole, 30, 34–35
Centromere, *44–45*, 46, 571, 573
Centrosome, 34
Centrum, 98
Cephalic vein, 363, 364, *366*
Cephalin, 334, 457
Cerebellar peduncles, 229
Cerebellum, *209*, 228–230, *231*
  functions, 228–230
Cerebral aqueduct, *209, 216*, 227,
    229
Cerebral arteries, 218, 223–224,
    227, 359
Cerebral cortex, 205, 215, 216, 218
Cerebral hemorrhage, 218, 382
Cerebral peduncles, *219*, 227, 229
Cerebrospinal fluid, 202, 211–212
Cerebrospinal tract: lateral, *219*
  ventral, *219*
Cerebrum, *209*, 215–*216*, *217–219*,
    220–221, 223–224, 227–228
  lobes, *216–217*
  localization, *217–218*, *219*
  memory processes, 223–224,
    227–228
Ceruminous glands, 62
Cervical canal, *607*
Cervical ganglia, 422
Cervical nerves, 237–238
Cervical plexus, 238
Cervical veins, 364
Cervix, 555, 556, *557*, *591*, *607, 608*
Chambers of heart, 346–347, *349*
Chancre of syphilis, 566
Cheekbones, 95
Chemical elements, 3–5
Chemoreceptors, 359, 399–*400*,
    401–*402*
  mechanism, 377
Chiasmata, 571–572
Chief cells, *428*
Childbirth, 607, *608–609*
Chloride ions: in blood, 538
  movement through nerve cell
    membranes, 197
Chloride shift, 405, *406*
Chlorides, 450
Chlorophyll, 454
Cholecystokinin, 440, 547
Cholesterol, 322, 323, 382,
    439–*441*, 457, 525
Cholic acid, 441
Choline, 457, 469–472
Cholinergic fibers, 255
Cholinesterase, 194, 255, 261
Chorda tympani, 232, 269, 422

Chordae tendineae, 346, *347*
Chorion, 542, 587, *588–589*, 590, *591*
Chorion villi, *589–590*
Chorionic gonadotropin (APL), 542, 544, 590
Choroid, *287–288*, 289
Choroid plexus, 222, 228
Chromaffin cells, 535
Chromatids, 51, 571, 572, *574*, 582
Chromatin, 46
Chromophil cells: acidophils, 521
  basophils, 330, 521
Chromophobe cells, 521
Chromosomes, 35, 39–41, 570–584
  crossing over, *52*
  division, 43, *44–45*, 46
  homologous, 48, 49, *50*, *51*
Chyle cistern, 386, 387
Chylomicrons, 449
Chyme, 428
Chymotrypsin, 432
Cilia, 57, *60*
Ciliary body, 287, 288, *289*, 290
Ciliary ganglion, 290
Ciliary muscle, 288, *289*, 290
Ciliary nerve, 290
Ciliary process, 288
Ciliary ring, 288
Circle of Willis (arterial circle), *359*
Circulation: hepatic portal system, *365*
  physiology, 372–390
  pulmonary, 346–347
  systemic, 355–364
Circumcision, 563–564
Cisterna chyli, 386–*387*
Citrates, 33, 406–407, 470
Citric acid cycle, *24*, 33, 407, 477
Clavicle, 101, 103
Cleavage, 585, *586*, 590
Cleft palate, 94
Clitoris, *553*, 556, 557, *606*
Cloaca, 601–603
Clotting of blood, 334–336, 444, 457, 473, 474
Cobalt compounds, 471
Cocarboxylase, 24, 466
Coccygeal nerve, 237
Coccygeal vertebra, 98
Coccygeus muscle, 168–169
Coccyx, 98, *99*, 100
Cochlea, 232, *268*, *270–272*, 275, 596
Cochlear duct (scala media), *270–272*
Cochlear nerve, *272*
Codons, 43
Coelom, 494, 587, *591*, *592*
  development, 495, 592
Coenzymes, 23–26
  ATP, 18–19, 25
  coenzyme A, 25, 406–407, 444, 470
Colic valve, 445
Collagen, 64–*66*
Collagenous fibers (white), 64, *65*, *66*, 68, 69, *74*
Collarbone, 103
Colliculi, 604
Colon, *417*, 449–451
  development, 604

Color blindness, 310–*311*, 575
  inherited, 311, 575
Color vision, 307–309
Colostrum, 557
Columnar epithelium, *57*, 58, *59*, 446
Coma, 443
Communicating arteries, 323
Compounds, chemical, 13, 14
Concha, nasal, *90*, 92, 95–96
Condition of nerve impulse, 195–*197*, 198
Conditioned reflex, 205–206, 422, 430
Condyle, occipital, 88
Condyloid process, 92
Cone cells, *292*, 294–295, 298–*299*, 300
  spectral sensitivity, *308*, 309
Conjugation, 571
Conjunctiva, 283–284, *287*, 289, 565, *596*
Conjuctivitis, 284
Connective tissues, 64–82, 441, 597, 598
  adipose, 67–*68*
  aponeuroses, 69
  areolar, *65*, 67, 446
  bone, 75–82
  cartilage, 72–75
  elastic, 70
  fasciae, 69
  fibrous, 70
  integument, 70–72
  reticular, 69, 70
Contraction (muscle), 124–129, 137
Convergence of eyes, 199, 304–305
Convolutions of brain, 215
Convulsions, 437, 443
Coracobrachialis muscle, 154, *156*
Coracoid process, 103
Corium, 70–71, 596, 597, *598*
Cornea, 285–*287*, 289
  curvature, 287–288
  drainage, 301
  nerve supply to, 287
Corona radiata, *582*
Coronal suture, *89*, *93*
Coronary arteries, 347, *355*, 357, *358*
Coronary sinus, 355, 363
Coronoid process, 92, 106–107
Corpora quadrigemina, 288
Corpus albicans, 554, *581*
Corpus callosum, 216, *229*
Corpus cavernosum, *561*, 563
Corpus luteum, 523–524, 542–544, 554–559, 560, *581*
Corpus striatum, 220, 225
Corrugator muscle, 144
Cortex: cerebellum, 228–229
  cerebrum, 215–216, 220, 223–224, 227–228
  kidney, *497–498*
Corticospinal tracts, *235–236*
Corticosterone, 539
Cortisone, 539
Covalence, 5–6
Cowper's glands, 563
Cranial bones, *89–92*, 94
Cranial nerves, 225, 230–*232*, 233, 238, 249
Craniosacral system, 242, *245*

Creatine, 323, 506
Creatine kinase, 130
Creatine phosphate (CP), 130
Creatinine, 323, 501, 506, *507*
Creatinuria, 507
Cretinism, 530–*531*
Cribriform plate of ethmoid bone, 92
Cricoid cartilage, 79
Crista ampullaris, 276, *277*
Crista galli, *79*, *92*
Cristae, 32, *34*
Cross eyes, 306
Crossing over (chromosomes), 47–49, *52*, 571–572
Cruciate ligament, *175*
Cryptorchism, 561
Cuboid bone, 112–113
Cuboidal spithelium, *30*, *57*, 58
Cuneiform bones, 112–113
Cupula, 276–*277*, 279
Curds, 430
Cushing's disease, 541
Cuspid teeth, 418
Cutaneous cerebral sensory area, *217–220*
Cyanocobalamine, 471
Cyanosis, 326
Cysteine, 458, 461
Cystic duct, *441*
  development, 602
Cystic vein, 367
Cytochrome oxidase, 23, 408
Cytochrome series, 23, 408–409
Cytochromes, 23, 408
Cytoplasm, 32–34
Cytosine, *36*, 38

Dark adaptation, 298
Deamination, 444, 460
Decarboxylases, 23–24
Decarboxylation, 407
Decidua, 589
Decidua basilis, 589, *591*
Deciduous teeth, 417–419
Decompression, 410
Defibrination of blood, 335
Dehydrocholesterol, 465
Dehydrogenases, 21, 23, 24, 408
Dehydrogenation, 19, 21
Deltoid muscle, 154–155
Dendrites, 188–*189*, 190
Dental canaliculi, 420
Dentate nuclei, 229
Dentine, *420*
Deoxycorticosterone, 539
Deoxyribonucleic acid, 36
  (See also DNA)
Deoxyribose, 39
Depolarization, 128, 129, 199
Dermatome, *591*
Dermis, *62*, 67, 70–72, *597–598*
Desmosomes, 59, 61
Dextrins, 423, 433, 456
Diabetes insipidus, 505, 527
Diabetes mellitus, 434–435, 443, 444, 504, 505, 521–522, 527
Diakinesis, *572*
Dialysis, 17
Diapedesis, 329
Diaphragm, 101, 150–*151*, 392
  development, 602–603

Diaphysis, *80–82*
Diastole, 351
Diencephalon, 208–*209, 210,* 222
Diet: foodstuffs, 454–460
  inorganic requirements, 460–474
    chemical elements and salts, 460–461
    vitamins, 461–474
Dietary allowances, 484–487
Diffusion: through membranes, 15–16
  of nerve impulses, 200–201
Digestive system, 415–453, 601–604
  development, 209, 601–602
  enzymes and fluids, 416, 421–451
  regions, 416–452
Dihybrid cross, *47, 48*
Dihydrotachysterol (A.T.10), 534
Dilator muscle, 290
Diodrast, 508
Dipeptidase, 428
Diphosphopyridine nucleotide (DPN), 21
Diploid number of chromosomes, 570–571, 573–575, 581, 583–584
Diplotene stage of meiosis, *572*
Disaccharides, 455, 456
Dislocations, 116
Diuresis, 505, 506
DNA, 35–*37,* 39, 41–42, 45–46, 50–51, 53
  function, 38, 39, 41, 42, 45–46, 50–51, 53
  and genetic code, 36–37, *42*
  and polymerization of nucleotides, 53
  relationship to RNA, 38–40
  structure, *39*
DOPA, 225
Dopamine, 225
Dorsal aorta, *604*
Dorsal pancreas, *602,* 603
Dorsalis pedis artery, *358,* 362
Down's syndrome, 577–*578*
DPN, 21
Ductus arteriosus, *369,* 601
Ductus endolymphaticus, *270,* 276
Ductus venosus, *369*
Duodenum, *417, 426,* 431–433, 440–*441, 443*
  development, 603
Dura mater, *208,* 210–*211, 212, 270*
Dyads, *152, 571*
Dyspnea, 403

Ear, 72, 267–281
  development, *597*
  external, 72, 267, *268*
  internal, 268, 269, *271–272*
  middle, *268–270*
Eardrum, 267
Eccrine glands, 61, *62*
Ectoderm, *587, 591,* 596, *597, 598*
Ectoplasm, 32
Edema, 67, 319, 378–379, 388, 509
Ejaculatory duct, *516,* 562–563
Elastic cartilage, 72, *73, 74*
Elastic connective tissue, 70, 356–357
Elastic fibers, *65, 73*
Elbow joint, 107

Electrocardiogram, 353–*354*
Electrolyte, 9
Electron acceptors, 9, 408
Electron carriers, 408, 409
Electron donors, 9
Electron microscope, 2, 17, *29–31,* 35
Electron transport system, 18–19, 23, *25,* 406, 408, *409*
Electrons, 3–6, 8–10
  and photosynthesis, 20
  transfer, 9
Elements, chemical, 3–5
Embden-Meyernof pathway, 406, 476
Embolism, 335
Embolus, 336
Embryo, 555–557, 570–*588,* 589–591, *592–593,* 594–*596,* 597–607
Embryonic disk, *587,* 591
Emmetropia, 305
Emotional states, 241
Emphysema, 403
Emulsification of fats, 433, 441
Enamel, 419–*420*
End organs, *194*
  of Ruffini, *261–262*
End plate, *129*
End-plate potential, *129*
Endocarditis, 348–349
Endocardium, 348–349
Endochondral bone, 79
Endocrine glands, 516, *517–519*
Endocrine system, 516–548
Endolymph, 270, *597*
Endolymph space, *276*
Endolymphatic sac, *270*
Endometrium, 555, 559, 560, 589
Endoneurium, *195*
Endoplasm, 32
Endoplasmic reticulum, 18, 30, *31,* 32
Endosteum, *80*
Endothelial cells, 356
Endothelial tissue, 56
Endothelial tubes, 601
Endothelium, 63
Energy, 3–4, 407, 409, 444, 454–456, 474–477
  cellular, 24–26
  expenditures, 489
  requirements, 482–487, 489
  ultilization, 21–22
Ensiform process, 101
Enteric plexus, 251–252
Enteric system, 446
Enterocrinin, 447, 547
Enterogastrone, 431, 547
Enterokinase, 433
Entoderm, 586, *587,* 592
Enuresis, 510, 511
Enzymes, 454
  cellular, 23–24, 35
  digestive, 416, 421–430–452
  in lymph, 382
Eosinophilia, 332
Eosinophils, 70, 330, 332
Ependyma, 212
Epicranial aponeurosis, 145
Epicranial muscle, 144
Epidermis, *61,* 70–71, 597
Epididymis, *496, 516,* 562

Epiglottis, 72, *392, 393,* 424
Epinephrine, 138, 285, 536
  specific effects, 138, 537–538
Epiphysis, 80, *82*
Epiploic arteries, 360
Epithalamus, 222
Epithelial bed, 598
Epithelial membranes, 62–63, 85
Epithelium, types, *30,* 56, *57, 60*
EPSP, 198
Erector spinae muscle, 154
Ergastoplasm, 31–32
Ergosterol, 465
Erythroblastosis fetalis, 340–341
Erythroblasts, *30,* 322–*324,* 325–327, 337–344
  maturation factor, 325, 471
  number, 322, 323, 327
  plasma membrane, 29–30
  relation to hemoglobin, 323–327
Erythropenia, 333
Erythropoietin, 323, 600
Esophageal arteries, 323
Esophageal plexus, 233, 251
Esophagus, 416, *417,* 423, *425, 426*
  development, *602*
Estradiol, 542
Estrogens, 539, 541–546, 554, 558–560, 590
Ethmoid bone, *89, 92, 93*
Ethmoidal sinuses, 92
Eukaryote cells, 43
Eustachian tube, 72, 267, *268,* 269
Excitation (muscular), 124–129
Excitatory postsynaptic potential (EPSP), 198
Excretion of embryo, 589–590
Excretory system, 494–513
Exercise effects, 402–403
Expiration, 150, 397–398
Expiratory center, *397,* 398
Expiratory reserve, 403–404
Extensor digitorum brevis muscle, 181
Extensor digitorum longus muscle, *175–176*
Extensor hallucis longus muscle, *175–176*
Extensors, 143
  foot, 175–178
  hand and wrist, 156–157
External auditory canal, 596–*597*
External auditory meatus, 597
External ear, 596–597
External genitalia, *606,* 607
External oblique muscle, 164, *165*
External occipital protuberance, *105*
Exteroceptors, 259
Extraembryonic cavity, *588*
Extraembryonic coelom, *588,* 591
Extraembryonic layers, 587–589
Extrapyramidal system, 236
Extrinsic eye muscles, 285, 293
Extrinsic food factor, 471
Extrinsic muscles of tongue, 146
Eye, 283–313
  accessory structures, 283–284
  blindspot, 293
  development, 595–596
  extrinsic muscles, 284–285, *286*
  lacrimal structures, *284*
  vision, 301, *302,* 306, 307

Eye sockets, *90*
Eyeball, *285–286*, 287–*289*, 290
  fluids within, 300–301
  refracting media, 300
Eyebrows, 283
Eyelashes, 283
Eyelids, 144, 283, 284, 595–596

Face of embryo, 592–595
Facial bones, *92, 93–96, 97*
Facial nerve, 229, *231*, 232, 248,
  249, 263, 269, 421
Facial vein, 421
FAD, *407*, 408, 467
FADH$_2$, *407*, 408
Fallopian tube (uterine tube),
  552–555
Falx cerebri, 92
Far point, 304
Farsightedness (hyperopis), *305*
Fascia bulbi (Tenon's capsule), 285
Fascia lata, *170*
Fasciae, 69, *142*
Fasciculus cuneatus, *235*
Fasciculus gracilis, *235*
Fat cells, *67*, 75
Fatigue: muscle, 134, 136
  nerve, 200
Fats: carried in blood, 322–323
  digestion, 433, 440, 476
  metabolism, 408
  oxidation, 477–478
  storage in body, 456–457
Fatty acids, 322, 433, 440, 441, 457
Fauces, 423
Fecal matter, 450–451
Femoral artery, *358*, 362
Femoral hernia, *165*, 167
Femoral nerve, 238
Femur, 75, 111, *114, 119*
Fertilization, 552–553, 573, *583–584*
Fetal circulation, 340–341, 362,
  367–*369*, 370, 601–602
Fetal membranes, 585–587,
  588–*589*, 590–591
Fetal skeleton, 82
Fetus, 340, 341, 495, 497, 544,
  589, 590, 600–609
Fibers: muscle, 122–*123, 124–125,
  127, 128, 129,* 142
  nerve, 191–*193,* 194–*195*
  type of tissue, 64, *65,* 66–*69*
Fibrin, 335
Fibrinogen, 322, 335–338, 438, 444
Fibrinolysin, 336
Fibroblasts, 64, *65,* 67
Fibrocartilage, 72–73, *74*
Fibrocytis, 64, *65*
Fibrous connective tissue, 68, *69*
Fibrous tunic of eye, 287–288
Fibula, 82, 112, *113,* 114, *119,* 120
Filtrate, 26
Filtration, 26
Fimbriae, 554, *557*
Finger bones, 108, *109*
Fissures of brain, 215–216
Flatfoot, 115, *116*
Flavin adenine dinocleotide (FAD),
  407–408, 467
Flavins, 467
Flavoproteins, 408, 467

Flexor digitorum brevis muscle, 178,
  181
Flexor digitorum longus muscle, 177,
  179, 181
Flexor hallucis brevis muscle, 177,
  180, *181*
Flexor hallucis longus muscle, 177,
  180, 181
Flexors, 143
  foot, 180–181
  hand and wrist, 155–156
Folic acid, 471–472
Follicle stimulating hormone (FSH),
  522, 528, 542, 544, *559*
Follicles of thyroid, 528
Fontanels, 79, 89
Foods: caloric value, 479
  classification, 455
  specific dynamic action, 482–485
Foot: arches, *116*
  bones, *115*
  muscles, 178–180, *181, 182,* 183
Foot plates, *214*
Foramen, 89
  mental, *90,* 92, *93*
  ovale, 368, *369,* 601
Foramen magnum, 90, *95*
Forebrain, 208–*209*
Fornix, *229,* 556
Fossa, 89
Fourth ventricle, 216, 228–230
Fovea centralis, *287, 289, 293, 299*
Foveola, *299*
Fractures, type, 116, *117*
Frontal bone, *79, 89, 92, 93*
Frontal lobe of brain, *213, 216–218*
Frontal process of maxilla, 94
Frontal sinus, 89–90, *92*
Frontalis muscle, *145*
Fructose, 448, 455, 456, 507
FSH, 522, 528, 542, 544, 554, *559*
Fumaric acid, 407, 408
Fundus of stomach, *426*

Galactose, 448, 455
Gallbladder, *417, 438, 439–441*
  development, 602
Gallstones, 441
Gametes, 570–573, 578
Gametogenesis, *507,* 570, 573–*575,*
  576–577
Gamma globulin, 319, 320
Ganglia, 227, *292,* 306–307, 537
Gastric acidity, 431
Gastric artery, 360, *361*
Gastric digestion, 430–431
Gastric ulcers, 431
Gastric vein, 366, 547
Gastrin, 431
Gastrocneminus muscle, *175, 176,*
  177, 178
Gastroepiploic arteries, 360
Gastroepiploic vein, 366
Gastrulation, 586
Generator current, 261
Genes, 47–53, 289–290, 571–577,
  583–584
Genetic code, 36–43, 47, 53
Geniculate bodies, 272
Genital folds, *606,* 607
Genital ridge, 517, *604,* 605

Genitalia, 556, 606, 607
  development, *606,* 607
  female, *556*
Genohyoid muscle, *393*
Genotype, 49
Genu, *229*
Germ cells, 29
Germ layers, 586–587
Gestation, 542–545
Gigantism, 521–522
Gill clefts, 592, 597, 602
Gingiva (gum), 419, 420
Glands: alveolar, 61
  endocrine, 516–547
  exocrine, 516
Glans, *516,* 563–564, *606*
Glaucoma, 301
Glenoid fossa, 103, *106*
Glial cells, 215
Globin, 323
Globulin, 334, 336
Glomerular capsule, 449–*500*
Glomerulus, *495,* 498, 499, *500,* 501
Glossopharyngeal nerve, 232, 233,
  248, 249, *260,* 263
Glottis, 392, *393*
Glucagon, 434
Glucocorticoids, 539
Gluconeogenesis, 443, 444, 539
Glucose, 367, 423, 442, 448,
  455–456, 507
  in blood, 323, 434, 436, 507
  oxidation, 20–21, 131
  phosphorylation, 18, 23–24
  transport through cell membrane,
  437
Glutamic acid, 476
Gluteus maximus muscle, 167, 169,
  170, 171
Glycerol, 407, 433, 448, 456, 476
Glycine, 441, 458
Glycogen, 23, 24, 130, 131, 323,
  367, 406, 436, 442, 455, 474,
  476, 538, 539
Glycogenesis, 442, 443
Glycogenolysis, 443
Glycolysis, 23, 24, 131, 406, *407,*
  476
Glycosuria, 434, 436, 437, 504
Goblet cells, *57,* 58, *61*
Goiter: exophthalmic, 530, 531, *532*
  simple, 529
  toxic, 531
Golgi complex, 32
Gonadotropic hormones, 521, 522,
  525, 544–545, 554
Gonads, 522, 535, 541, 570
  early development, 605
Gonorrhea, *565,* 566
Gonorrheal ophtalmia, 565
Graafian follicle, 559, 581
Gracilis muscle, 172
Granulated lids, 284
Graves' disease, 531
Gravity effect, 381–382
Gray matter, 189
  brain, *211,* 215–216, *217,* 228–229
  dorsal horn, *202, 212*
  spinal cord, *201–202,* 212, 214
  ventral horn, 212
Gray ramus communicans, 244, *246*

Greater omentum, 426–427
Growth of cells, 29
Growth hormone, 521–522
Guanine, 36, *38*
Gum (gingiva), 419, 420
Gut, 591, 592, *604*

H band of muscle, 123, *125, 126,* 127
Hair cells, 272, 273, *277*
Hair development, 597–598
Hair follicle, *61,* 62, 597–598
Hair shaft, *598*
Hallucinations, 224
Hand: bones, 108, *109, 115*
 muscles, 156–157
Haploid number of chromosomes, 571–573
Harelip, 94
Haversian canals, 75–76
Haversian system, 75
Headache, 227
Hearing, 266–277
 physiology, 274–275
 place theory, 274
 traveling-wave theory, 274–275
Heart, 346–355
 blood vessels to and from, 346–371
 cavities, 348
 chambers, 346, *347, 348,* 349, *350, 351, 352, 354*
 circulation through, 346–352
 contraction, 347, 349–352
 cycle, 351–*352*
 development, 600–601
 musculature, 138–139, *348,* 349, *350,* 351, 352
 nerve supply to, 349
 sounds, 352
 valves, 346–*349, 352*
 (*See also under* Cardiac)
Heart murmur, 348–349
Heart rate, 353–354
Heat production, 131, 474, 476, 488
Helicotrema, 270, *272*
Hematin, 416
Hematocrit, 327
Heme, 323
Hemiazygos vein, 364, *387*
Hemocytoblast, 600
Hemocytometer, 327
Hemoglobin, 13, 322–327, 404–405, 441, 442, 457, 461, 471
Hemophilia, 336, 575
Hemoproteins, 408
Heparin, 70, 330, 335, 438, 444
Hepatic artery, 360, *361*
Hepatic duct, *440, 441*
Hepatic plexus, *251*
Hepatic portal system, 365–367
Hepatic veins, 365–*366, 367, 369*
Hering-Breuer vagal reflex, 399
Hernia, types, *165,* 167
Hexokinase, 23, 476
High blood pressure, 382–383
Hilus, 384, *385,* 497
Hindbrain, 208, *209,* 210
Hipbones, 109–111
Hippocampus, 110, 220, 266
Histamine, 330, 431

Histones, 35
Homeostasis, 254
Horizontal cells, 291
Hormones, 516–549
 and blood plasma, 322
 function, 433, 434, 440, 543
Humerus, 103, *104,* 107, 122
Hunchback, 100
Hyaline cartilage, 72–*73*
Hyaloid artery, 291, 595
Hyaloid canal, *287*
Hyaluronic acid, 583
Hyaluronidase, 583
Hydrochloric acid, 11, 12
Hydrochloric acid secretion, 429–431
Hydrogen-ion concentration, 10–12, 405
 buffer action, 12–13
Hydrolysis, 455–456
Hymen, 556, 607
Hyoid arch, 592
Hyoid bone, *79, 97, 393*
Hyperglycemia, 434, 437, 443
Hyperinsulinism, 436–437
Hyperopia, *305*
Hyperparathyroidism, 534–535
Hyperpnea, 400–401
Hypertension, 382–383
Hyperthyroidism, 437, 525, 530–532
Hypertonic solutions, 16, 450
Hypodermis, 70
Hypogastric artery, *361,* 362, 369
Hypogastric plexus, *247*
Hypogastric vein, 365
Hypoglossal nerve, 233, 424
Hypoglycemia, 436–437, 443
Hypoparathyroidism, 534
Hypophysectomy, 526
Hypophysis (pituitary body), 91, 223, 517, 519–*520,* 521–527
Hypothalamus, 215, 223, 228, *229,* 257, *520,* 522, 525, 527, 539
Hypothyroidism, *530*
Hypotonic solutions, 16

I band filaments, 123, 124, *125–127*
ICSH, 523, 545
Ileocecal valve, 431–432, 445, *449*
Ileum, 417, 432, 445–448, *449*
Iliac arteries, *358, 361,* 362, 369
Iliac spine, *110, 171*
Iliac veins, 356, *366, 367*
Iliacus muscle, 167–*168*
Iliofemoral ligament, *168*
Iliopsoas muscle, 167
Ilium, 168, 169
Implantation, *555, 586*
Incisive fossa, *95*
Incisive suture, *95*
Incisor teeth, *95,* 418
Incus, 91, *267, 268, 272,* 597
Indole, 450
Infectious mononucleosis, 333
Inferior colliculi, 272
Inferior mental nerve, 92
Inferior nasal conchae, *90,* 92, 95–96
Inferior oblique muscle, 285, *286*
Inferior rectus muscle, 285, *286*
Inferior vena cava (*see* Vena cava, inferior)
Infraorbital foramen, 90

Inguinal canal, 164, 167, 168
Inguinal hernia, *165,* 167
Inguinal ligament, 164, *165,* 171
Inguinal ring, 164, *165*
Inhibitions (physiological): post-synaptic (IPSP), 198
 presynaptic, 199
Initial heat, 131
Innominate (brachiocephalic) artery, 357, *358, 360*
Innominate veins, *363–364,* 366
Inorganic diet needs, 460–474
Inositol, 469, 470
Insensible perspiration, 478
Inspiration, 149, 397–398
Inspiratory capacity, 404
Inspiratory center, 398–399
Insula, 216–*217*
Insulin, 434–437
 protamine zinc form, 436
Insulin shock, 437, 443
Intercalated disk, 138–139
Intercostal arteries, 359–360
Intercostal muscles, 150
Interlobular arteries of liver, 438–439
Interlobular veins, 439
Intermediolateral column, 202
Internal environment, 321
Internal mammary veins, 364
Internal oblique muscle, 165
Internal pterygoid muscles, 146
Internal vesicle sphincter, 510
Interoceptors, 260
Interphase, 43, *44,* 45, 572
Interpreting center of special senses, 259–260
Interstitial-cell-stimulating hormone (ICSH), 523, 545
Interstitial cells, 545
Interventricular foramen, *216*
Intestine, 416, *417,* 432, *433–448, 449–485*
 development, *209,* 603–604
 function, 445–452
Intima, 356
Intrinsic factor, 471
Intrinsic muscles of tongue, 146
Inulin, 507–508
Inversion of image, *302*–303
Inverted retina, 291
Iodine, 528, 529
 use of radioactive, 532
Iodopsin, 299, 463
Iodopyracet, 508
Ionization, 8–10
Ions, 8–13
 movement through cell membrane, 18, 196, 197
IPSP, 198
Iris, 288, 289, *596*
Iron, 409, 457, 461
Irradiation, 465
Irritability, 28
Ischium, *110*
Island of Reil (insula), 216, *217*
Islet cells of Langerhans, *434*
Isoagglutinins, 337
Isocitric acid, 407
Isometric contraction, 135
Isotonic contraction, 135
Isotonic solutions, *16*

Isotopes, 7–8
Isotropic (1) band, 123, *125*, *126*

Jaundice, 441–442
Jejunum, *417*, 431–432, 445–482
Joints: ball-and-socket, 103, 119
    classification, 118–119
    hinge, 107
    immovable, 89, 118
    pivot, 108
Jugular veins, 363, *366*

Keratin, 71
Ketoglutaric acid, 407, 476
Ketones, 434, 444
Ketosis, 444
Kidney, 444, 494–513, 534
    artificial, 508–509
    blood supply, 497, *498*, 499–*500*,
        501–502
    development, 496
    function, 494, 500–502
    internal morphology, 494–*497*,
        498–*502*
Kinetochore, 571
Klinefelter's syndrome, 577
Knee joint, *113*, 130
Kneecap, 113, 130
Krebs cycle (*see* Tricarboxylic acid
    cycle)
Kupffer cells, 70, 439
Kymograph, 132
Kyphosis, 100

Labia majora, *553*, 556, 606, 607
Labia minora, *553*, 556, *666*
Labyrinth of ear, 269–270
Lacrimal bone, *79*, 80, *90*, *92*, *93*, 95
Lacrimal canals, 284
Lacrimal ducts, *284*
Lacrimal gland, 249, *284*
Lacrimal lake, 283
Lacrimal sac, 285
Lacrimal secretion, 284
Lactation, 557
Lacteal, 386, 446–448
Lactic acid, 130, 538
Lactiferous duct, 557
Lactoflavin, 467
Lactogenic hormone (LTH), 521, 522,
    524, 560
Lactose, 455
Lacunae, 72, 74, 75, 77
Lambdoid suture, *89*
Lamellae, 75, *77*
Lanugo, 598
Large intestine, 416, 417, 449–450
    development, 603–604
Laryngitis, 392
Laryngotracheal ridge, 604
Larynx, 392, *393*, *394*, 424
    development, 604
Latent period, 133
Lateral fissure of brain, *216*, *217*
Lateral sacs, 128, 129
Lateral ventricles of brain, 208, *209*,
    210, *216*
Latissimus dorsi muscle, *153*–154
Law of the heart, 377
Lecithin, 457, 472

Lens (eye), 286–287, 290–291
    accommodation, 302–304
    embryonic development, 290–291,
        595, *596*
    refractive index, 300
Lens epithelium, *596*
Lens fibers, 290, *596*
Lens vesicle, *596*
Leptotene stage of meiosis, 571–572
Lesser omentum, 426–427
Leukemia, 33
Leukocytes, 327–333, 382, 559–560,
    576
    function, 331–333
    radiation hazards, 342
Leukocytosis, 333
Leukopenia, 333
Leukotaxine, 332
Levator ani muscle, 168–169
Levator palpebrae superioris muscle,
    144
Levers of body, 143
LH, 521–523, 542, 554, *559*, 569
Ligaments, 69
Light adaptation, 298–299
Light reflex, 303–304
Limb buds, 593
Limbic system, 220, 223
Linea alba, 164–165
Lingual tonsils, 416
Linoleic acid, 457, 485
Linolenic acid, 457, 485
Lipase, 430, 432, 433
    pancreatic, 441, 446
Lipids, 18, 437, 444, 456–457
Lipoids, 457
Litmus reaction, 12
Liver, 417, 438–*439*, 440–*441*,
    442–444, 601
    development, *209*, 592, *602*–603
    functions, 440–479
    structures, 439–445
Liver sinusoids, 445
Locomotor ataxia, 237, 566
Loop of Henle, 498–500
Lordosis, *100*
LTH, 521, 522, 560
Lumbar arteries, 361
Lumbar puncture, 211
Lumbar vein, 364
Lumbar vertebrae, 98
Lumbosacral plexus, 238
Lung bud, *209*, 604
Lung reflex, 399
Lungs, 394–*396*, 511
    development, *602*, 604–605
    respiratory gases in, 404–407
    water loss, 511
Lutein, 554
Luteinizing hormone (LH), 521–523,
    542, 554, *559*, 560
Luteotropin (LTH), 521, 522, 524,
    525, *559*, 560
Lymph, 287, 378
    follicles, 388
    nodes, 331, 384, *385*–*386*, 387,
        601
    nodules, aggregated, 388
    sinuses, *385*
Lymphagogues, 387
Lymphatic capillaries, 382, 383, *384*

Lymphatic duct, 385–*387*
Lymphatic plexus, 382, 384
Lymphatic system, 322, 382–388,
    448
Lymphatic vessels, 142, 383–385
Lymphocytes, *30*, 69–70, 327,
    330–333, 342–344, 382, *385*,
    547
Lymphocytosis, 333
Lymphoid tissue, 388, 416
Lysergic acid diethylamide, 224–225
Lysosomes, 31, 32
Lysozymes, 284

M substance, 124
Macrophages, 70, 331, 439, 444, 601
Macula, 276, *277*–*279*
Macula lutea, *293*, 299
Magnesium, 460
Malar bones, 95
Malaria, 326
Malic acid, *407*
Malleolus: lateral, 112, *114*
    medial, 112, *114*
Malleus, 91, 267–*268*, 597
Maltase, 423
Maltose, 423, 455, 456
Mammary glands, 524, 557, *558*
Mammary veins, 364
Mammillary bodies, *231*
Mandible, *79*, *89*, *90*, *92*, *93*
Mandibular arch, 592
Mandibular fossa, 95
Mandibular nerve, 231–232
Mannitol, 507
Mannose, 448
Manubrium, 101, 103
Marey's law, 377
Marrow, 75–76, 342
    (*See also* Bones)
Marrow cavity, 79–80
Masseter muscle, 144–146
Mast (basophilic) cells, 70, 335
Master gland, 519
Mastoid fontanel, 89
    process, 90–91, 95
Mastoiditis, 91
Matrix, 64
    cartilage, 72, *73*
Matter, 3, 13
Maturation division, 583
Maxilla, *79*, *89*, *90*, *92*–94, 95
    processes, 94–95
Maxillary nerve, 231–232
Maxillary sinuses, 94
Meconium, 604
Mediastinal arteries, 359
Mediastinum, 346, *396*
Medulla, 218, *219*, *229*, 230–233,
    376
Medullary canal, 75
Megakarocytes, 333, 600
Meibomian glands, 283
Meiosis, 570–573
    first meiotic division, 46, 571, *572*,
        573, *574*, 575
    second meiotic division, 572, 573,
        574, *575*
Meissner's corpuscle, 194, 260, *261*
Melanin, 71, 289, 539

Membrane:
 basilar, 274
 cell, 2, 18, *29*, 128, 274
 plasma, 128
 of Reissner, 270
Membrane bone, 76–77
Membrane system, 32
Membranous canal, *276*
Membranous labyrinth, 596
Memory processes, 223–224
Mendel, Gregor, 47
Meninges, 208, *211–212*
Meningitis, 212
Menopause, 554, 560
Menstrual cycle, 542, 554, *555*, 558–559
Menstrual period, 607
Menstruation, 542–543, 554, 559–560, 609
Mental foramen, 90, 92, *93*
Mercury manometer, 379
 sphygmomanometer, *380*
Meromysins, 140
Mesaxon, 191
Mesencephalon, 208, *209, 210*
Mesenchyme, 64, *596, 597, 598, 600, 601, 605*
Mesenteric arteries, *358, 361, 369*
Mesenteric intestine, 432
Mesenteric veins, 366, *367*
Mesentery, 365, 426, 432, 450, *604*
Mesoderm, *587, 591, 592,* 596–597, *604*
Mesonephric duct, 495, *496,* 605
Mesonephros, 494–495, *496, 604,* 605
Mesothelium, 56, 63
Metabolism, 474–488
 cellular, 15–26, 28–29, 406–409
 and hormones, 529
Metacarpal bones, 108, 109
Metanephric kidney, *496,* 605
Metanephros, 494–496
Metaphase, 43, *44, 45,* 46, 573, *574*
Metastasis, 52
Metatarsal bones, *115*
Metatarsus, 113
Metencephalon, 208, *209, 210,* 228
Methane, 450
Methionine, 416
Microgliocytes, 214, *215*
Microsome faction, 34
Microvilli, 29, 59, *60*
Micturition, 510
Midbrain, 208–*209,* 210, 227
Milk line, 557
Mineral salts, 460–461
Mineralocorticoids, 539
Mitochondria, 18, *31–33,* 34, *127,* 406, 444, 578, *580*
 ATP production, 122
Mitosis, 43–46, 570–571, *574,* 576, 587
Mitral cells, 266
Mitral (bicuspid) valve, *349*
Modiolus, 270, 272
Molar teeth, *418,* 419
Molecules, *5–6*
Mongolism, *576, 577, 578*
Monocytes, 70, 300, 327, 331
Monohybrid cross, 47

Mononucleosis, 333
Monosaccharides, 448, 455
Mons pubis, 556
Morula, 585, *586*
Motility, 28–29
Motion sickness, 276
Motor area of cerebrum, *217–219, 229*
Motor end plates, *193, 202*
Motor unit, *132,* 140
Mouth, 416–417
Mucin, 284, 422, 423, 428, 440
Mucosa, 97
Mucous cells, *428*
Mucous membranes, 62–63
Müllerian ducts, 605
Müller's fibers, *292*
Multicellular glands, 50–63
Mumps, 421
Muscle physiology, 121–127, *129–137,* 139
Muscle tissue, 121–140
 cardiac, 138–*139*
 skeletal, 132–133, *134,* 136–137
 smooth, 136–137
 visceral, 136
Muscle zones and filaments, 123, *124–126, 132–136*
Muscles, 121–140, 142–186
 abdominal wall, 164, *165,* 166
 back, *151–153,* 154
 foot, 180, *181–182,* 183
 head, face, and neck, 138–*139, 145–146,* 147–148
 hip, *168, 170,* 171
 leg, *175, 176, 177–179*
 pelvis, *169*
 shoulder and arm, *155–156, 157–159, 160, 161, 162, 163,* 164, *177*
 shoulder girdle and thorax, 148, *149,* 150–*151*
 thigh, 171, 172, *173–174,* 175
Muscular contraction, 124–*125, 126, 127,* 128, *129,* 130
Mutations, 49–52
Myasthenia gravis, 194
Myelencephalon, 208, *209, 210*
Myelin, 192–193
Myelin sheath, 189, *190, 191, 192*
Myeloblast, 328
Myenteric plexus, 251, *252*
Myocardium, 349, 601, 605
Myofibrils, 122–123, 127–128, 130, 140
Myoglobin, *132,* 139
Myoneural junction, *129,* 140, 193, 194
Myopia, 305–306
Myosin, 124–128
Myotatic reflex, 204–205
Myotome, 591
Myxedema, 530–*531*

NAD, 20, 21, 23, 34, *407–408,* 467, 469, 477
NADP, 20, 21, 407, 408, 467
NADPH, 20
Nails, 597
Nasal bones, *79, 89, 90, 92, 93,* 95
Nasal cavity, 95, 392
Nasal concha, *90,* 92, 95–96

Nasolacrimal ducts, *284*
Nasopharynx, *393, 417,* 423
Navel, 367
Navicular bone, *115*
Negative feedback, 254
Neisseria gonorrheae, 565
Nephritis, 508–509
Nephron, 497, *500*
Nephrotomes, *495, 592*
Nerve cells, 188, *189–190,* 191, 214
Nerve endings, 260–261
Nerve fibers, 191, *193, 195*
Nerve impulses, 128, 129, 195–198
 conduction, 195–*197,* 198
 convergence, 200–*201*
 diffusion, 201
 divergence, *201*
 refractory period, 199–200
Nerve plexuses, 238, 245, 247, 251–*252*
Nerve tracts: brain, 216–218, *219*
 spinal cord, *219,* 233–235, 236–237
Nerves, 194–*195*
 autonomic (*see* Autonomic nervous system)
 cranial, 230–*231, 232–233,* 238
 spinal, 237–*238,* 239
Nervous system, 188–206, 208–238, 241–257
 (*See also* Autonomic nervous system; Parasympathetic nervous system)
Neural canal, *495, 591*
Neural crests, *209,* 495
Neural groove, *209, 591, 604*
Neural pathway, 199, 259, 262
Neural plate, *209*
Neural tube, *209,* 495, 519, *592, 604*
Neurilemma, 195
Neurofibrils, 189
Neuroglia, 213–*214,* 215
 astrocytes, *214*
Neuromuscular bundle, 349–351
Neuromuscular junction, 128
Neuronal circuits, 224
Neurons, *30,* 188–*189,* 190–191, *249*
 axons, 188–*191, 192*
 cell bodies, 188–189, 233, 235
 dendrites, *189–190,* 436
Neutral Protamine Hagedorn (NPH), 436
Neutrons, 3–*4,* 5
Neutrophils, 30, 327–330, 332, 356
NH_2 group in deamination, 443–444
Niacin (nicotinic acid), 462, 467–469
Nicotinamide, 469
Nicotinamide adenine dinucleotide, 20
 (*See also* NAD)
Nicotinamide adenine dinucleotide phosphate, 20
 (*See also* NADP)
Nicotine, 469
Nicotinic acid amide, 467–469
Nictitating membrane, 283
Night blindness, 297
Nissl substance, 188–189, *190,* 243
Nitrogen, 479, 485, 521, 539
Nodes of Ranvier, *190–191*
Norepinephrine, 38, 225, 255, 536–537

Nose, 263–265, 266, 267, 392
  development, 594–595
Notochord, 209, 495, 519, 592, 604
NPH, 420
Nuclear membrane, 31, 35
Nuclear pore, 31, 35
Nucleic acids, 35–36, 37–38
Nucleolus, 31, 35, 39, 572
Nucleoplasm, 28, 31, 35
Nucleoproteins, 506
Nucleotides, 35–38
Nucleus: atom, 3–4, 5
  cell, 31, 35
  nerve cell bodies, 235
Nucleus cuneatus, 235
Nucleus gracilis, 235
Nystagmus, 278–279

Obturator foramen, 110
Occipital bone, 79, 89, 90, 93, 95
Occipital condyles, 90, 95
Occipital fontanel, 89
Occipital lobe, 216–217
Occipitalis muscle, 145
Occipitofrontalis muscle, 144
Occipitoparietal suture, 93
Oculomotor nerve, 227, 231, 248,
  285, 286, 287
Odontoblasts, 420
Odontoid process of axis, 98
Olecranon fossa, 106
Olecranon process, 103, 106, 122
Oleic acid, 456
Olfactory area, 263–265
Olfactory bulb, 210, 231, 238, 265,
  266
Olfactory cells, 265, 266
Olfactory nerve, 230, 265
Olfactory receptors, 265
Olfactory tract, 231, 266–267
Oligocytes, 192
Oligodendrocytes, 214–215
Oligodendroglia, 192
Omentum, 63, 427
Oocyte, 554, 582
  primary, 570, 575, 581
  secondary, 573, 575, 581
Oogenesis, 554, 570, 572, 575,
  578–582
Oogonia, 575, 579, 581
Ophthalmic artery, 358
Ophthalmic nerve, 231
Opsonins, 319
Optic chiasma, 231, 238, 294, 303
Optic cup, 595, 596
Optic disk, 287, 293
Optic nerve, 230–231, 287, 289,
  293–294, 595, 596
Optic thalamus, 223
Optic tract, 230–231, 294
Optic vesicle, 595, 596
Orbicularis oculi muscle, 145, 283
Orbicularis oris muscle, 145
Orbitosphenoid bone, 79, 90
Organ of Corti, 232, 272
Organic compounds, 7
Orgasm, 564
Osmosis, 15–16
Osmotic pressure, 16, 323–333
Osseous labyrinth, 270
Osseous tissue, 75–81

Osteoblasts, 77
Osteoclasts, 80
Osteocytes, 75, 76
Osteogenic layer, 76
Osteon, 76
Otic ganglion, 248–249
Otic vesicle, 596
Otoliths, 276
Ova, 29, 30, 541–542, 552–554,
  555, 559, 564, 570
  chromosomes, 37, 38
  cleavage, 583, 584–585, 586
  fertilization, 583–584
  meiosis, 570–573, 574–575
  oogenesis, 578–581, 582
Oval window of ear, 268, 279
Ovarian arteries, 358, 361
Ovarian follicles, 523, 542, 544, 554,
  581–583
Ovarian veins, 365, 366
Ovaries, 516, 517, 523, 552–554,
  570, 577
  effect of sex hormones, 541–544
  oogenesis, 578–581
Oviducts, 554, 583, 586, 605
Ovulation, 523, 542, 543, 553–555
  558–560, 564, 581, 586, 609
Oxaloacetic acid, 407–408, 470, 477
Oxalosuccinic acid, 407
Oxidases, 23–24, 408
  cytochrome, 25
Oxidation, 20–23, 130–131,
  406–409
Oxidoreductase, 23
Oxygen, 403–404, 479–482
Oxygen debt, 131–132
Oxyhemoglobin, 13, 323, 404
Oxytocic effect, 526
Oxytocin, 518, 526, 527

PABA, 469
Pachytene stage: of meiosis, 571
  of spermatogenesis, 571
Pacinian corpuscle, 261
PAH, 508
Pain sensation, 255–256, 259–262
Palate, 94, 393, 416, 423
Palatine arches, 416
Palatine bones, 92, 94–95
Palatine process of maxilla, 94–95
Palatine tonsils, 416–417, 602
Paleocortex, 220
Palmar surface, 82
Palmitic acid, 456
Pancreas, 360, 416, 417, 432–435,
  441, 516, 517, 552
  development, 603, 604
Pancreatic duct, 432, 435, 441
Pancreatic veins, 365–366
Pancreozmin, 434, 547
Pantothenic acid, 25, 406, 469–470
Papillae, 57, 260, 264
Para-aminobenzoic acid (PABA), 469
Para-aminohippuric acid (PAH), 508
Paracasein, 531
Paranasal sinuses, 97, 98
Parasympathetic fibers, 232
Parasympathetic ganglia, 245
Parasympathetic nervous system,
  242, 245, 248–252, 253–254,
  422, 454

Parasympathetic plexuses, 252
Parathormone, 533
Parathyroid glands, 533, 535
  development, 593, 603
Parathyroid hormone, 533–534
Parathyroid tissue, 529
Parietal bone, 79, 89, 90, 93
Parietal cells, 428
Parietal lobe, 216, 217
Parietal pleura, 397
Parieto-occipital fissure, 216
Parotid duct, Stenson's, 421, 422
Parotid gland, 421
Pars intermedia, 202
Parturition, 590, 607–609
Past pointing, 279
Patella, 111
Patellar ligament, 171
Pectoralis major muscle, 148,
  149–150, 151
Pectoralis minor muscle, 149–150
Pellagra, 467–469
Pellagra-preventive factor (PP), 110,
  114, 467–469
Pelvic girdle, 109, 110–112
Pelvic plexus, 247
Pelvis, 110, 111
Penis, 376, 561, 563–564, 607
Pepsin, 428–429, 431, 433
Pepsinogen, 428–429
Peptidases, 432–433
Peptides, 432–433
Peptones, 428, 433, 448
Pericardial arteries, 359
Pericardial cavity, 347–348
Pericardial fluid, 348
Pericardium, 63, 347–348, 396
Perichondrium, 72, 73, 79
Perilymph, 270, 597
Perilymphatic space, 270, 276
Perineum, 169
Perineurium, 194, 195
Periodic table, 3–4, 5
Periodontal membrane, 420
Periosteal buds, 80
Periosteum, 76–77, 79, 116, 142,
  211, 419
Peripheral receptors, 194, 204
Peripheral system, 208
Peristalsis, 136, 424, 427, 446,
  509–510
Peristaltic waves, 446, 450–451
Peritoneum, 63, 426, 438, 450
Peritonitis, 426, 450
Perivitelline space, 584
Permeases, 19
Pernicious anemia, 368, 471
Peroneal artery, 358, 362
Peroneus brevis muscle, 176
Peroneus longus muscle, 175
Peroneus tertius muscle, 176
Perspiration, insensible, 485
Peyer's patches, 388
pH, 10, 433
  of blood, 13, 318–319
Phagocytes, 329, 332, 343, 439, 601
Phagocytosis, 34, 329
Phalanges: foot, 115
  hand, 108, 109, 115
Pharyngeal pouches, 596, 602, 604
Pharyngeal tonsils, 416

Pharynx, *393*, 416, 417, 424, 592, *602*
  laryngopharynx, *417*
  nasopharynx, *393*, *417*
  oropharynx, *417*
Phenotype, 49
Phenylalanine, 485–486
Phenylketonuria, 485–486
Phenylpyruvic acid, 486
Phosphate bonds, 21–22, 476
Phosphate buffers, 12–13
Phosphates, 437, 476, 534, 535
  ADP, ATP, 21–22
  DNA, RNA, 36, 37, 41
Phosphatides, 323
Phosphoglyceric acid, 407, 476
Phospholipids, 323, 334, 457, 471, 472
Phosphoric acid, 21, 22, *39*, 466
Phosphorus, 457, 460, 464, 465, 533, 534
Phosphorylases, 23, 24, 407, 476
Phosphorylation, 23, 34, 409
Photosynthesis, 20, 454, 479
Phrenic arteries, 361
Phrenic nerve, 238
Physiological saline solution, 15–16, 327
Phytic acid, 470
Pia mater, 208, *211*, *212*, 228
Pigmentation in disease, *540*
Pigments of skin and hair, 71, 470, 598–599
Pineal body, 222, *229*
Pinkeye, 284
Pinocytic vesicles, 30, *31*
Pinocytosis, 30, 34
Pituitary body, 91
Pituitary dwarfis, 522, *524*
Pituitary gland, *229*, 437, 519–525, 539
Placenta, *588*, *589*–590, *594*, 607
  fetal circulation, 362, 367–368, *369*
  secretion of sex hormones, 542–544, *555*
Plantar arteries, *358*, 362
Plantar surface, 181
Plasma, 320–323, 336–*339*, 404–405
  cofactor, 335
  composition, 320–323
  membrane, 128
Plasma-protein osmotic pressure, 378
Plasminogen, 336
Platelets of blood, *323*, 333, 335, 600
Platysma muscle, *145*, 146
Pleura, 63, *396*
Pleural cavity, 63, *396*
Pleurisy, 397
Plica semilunaris, 283
Pneumotaxic center, *398*, 399
Pneumothorax, 397
Polar bodies, 575, 578–582, *584*
  first, *575*, 582, *583*
  second, 575, 582
Poliomyelitis, 398
Polycythemia, 326–327
Polyneuritis, 466
Polyribosomes, 40

Polysaccharides, 454, 456
Polysomes, 40
Polyuria, 527
Pons, *209*, *210*, *219*, 228, *229*, 232
Popliteal artery, *358*, 362
Popliteal vein, 365
Pores, nuclear, *31*, 35
Portal system, 365–*367*, 441, 442
Portal vein, 366–*367*, *369*, 439
Posterior chamber of eye, *287*
Postganglionic neuron, 243, 244, *246*, *249*
  fibers, 243, *246*, 248, *249*, 255
Postsynaptic membrane, 224
Postsynaptic neuron, 199
Posture, 100
Potassium, 460, 461
Potassium ions: in blood, 538
  movement through cell membrane, 18, 197
Poupart's ligament, 164
PP factor, 468–469
Precapillary sphincter, 372
Preganglionic neuron, 243, *246*, 248
  fibers, 243, *246*, 248, *249*
Pregnancy, 340, 542–544, 554–557, 559–560
  tests, 544
  urine (P.U.), 544
Pregnandiol, 543
Premenstrual period, 542, 560
Premolar teeth, *418–419*
Preovulatory period, 542, 559
Prepuce, *561*, 563
Presbyopia, 304
Pressoreceptor mechanism, 376–377, 400–401
Pressoreceptors, 359, 376–377, 400–401
Presynaptic inhibition, 199
Primary filters of the body, 601
Primary follicles, 581
Primary motor area of brain, 217
Primitive gut, *587*, 604
Proctodeum, *209*
Profibrinolysin, 336
Progesterone, 539, 542–544, 554, 559–560, 590
Prolactin, 525, 557
Promine, 547
Pronephric duct, *495*
Pronephric tubules, *495*, *496*
Pronephros, 494
Pronuclei, 574, *575*, 582, *583*, *586*
Prophase, *43–44*, 45–46, 571–*572*, *574*
Proprioceptors, 259
Prorennin, 430
Prostate gland, *561*, 563–564
Proteases, 23
Protein: collagen, 64
  keratin, 71
Protein buffers, 12–13
Proteinase, 508
Proteins, 323, 457–460, 521
  classification, 457, 460
  metabolism, 444
  oxidation, 478, 479
  synthesis, 35, 36, 40–45, 53
Proteinuria, 504
Proteoses, 428

Prothrombin, 333, 334, 428
Protons, 3–*4*, 5
Protoplasm, 17, 28
Provitamin D substances, 464–465
Provitamins, 462
Pseudostratified epithelium, *57*, 58
Psoas major muscle, *167*–168
Psoas minor muscle, 167
Pteroylglutamic acid, 469–471
Pterygoideus medialis, 146
Ptyalin, 423
P.U., 544
Pubic arch, *110*–111
Pubic hair, 556, 564
Pubis, *110*–111
Pulmonary artery, 347, *352*, 357, *369*, 395
Pulmonary capillaries, *395*
Pulmonary circulation, 357
Pulmonary edema, 403
Pulmonary pleura, 404
Pulmonary plexus, 233, *251*
Pulmonary semilunar valves, 347, *349*, *352*
Pulmonary veins, 347, *349*, 352, 357, 363, 395
Pulp cavity, 420
Pulse, 397
Puncta lacrimalia, *284*
Pupil of eye, 288
  action, 303–304
  response to light, 303–304
Purines, 36–38
Purkinje cells, 188, *189*, 228–229
Purkinje effect, 298
Purkinje network of heart, 350
Pus, 332
Pyloric antrum, 426
Pyloric valve, 427
Pylorus, *426*, 431, 432
Pyorrhea, 421
Pyramidal cells, 189
Pyramidal tracts, crossing, *219*, 235–236
Pyramidalis muscle, 165
Pyridoxine (vitamin B_6), 469
Pyriform cortex, 220
Pyrimidines, 36–38
Pyrophosphate, 466
Pyruvic acid, 21, 23, 24, 130, 406, 407, 466, 476, 477

Quadratus femoris muscle, 111, 120, *171*
Quadratus lumborum muscle, 150, *167*, 168

Radial artery, *358*, 359
Radial notch, *103*, 108
Radiation hazards, 342
Radioactive isotopes, 8
Radius, *103*, 108, *109*
Rami, 92
Rathke's pouch, 520, *602*
Receptors, 204, 259–262
  auditory, *270*, 271–272
  eye, 283, 294–*295*, *301*
  naked nerve endings, 194
  olfactory, *265*
  pain, 259–260
  peripheral, 188, *194*, 204

sensory, 194, 204, 259–260, 359
  taste, 262–*264*
Recovery heat, 131
Recovery phase, 131
Rectum, *449*, 450, *553*, 602, 606
Rectus abdominus muscle, *165*–166
Rectus femoris muscle, *171*, 174
Rectus muscles of eye, 284–*286*
Red blood cells, 30, 322–*323*,
  324–327, 330, 336–344, 600
Red bone marrow, 75, 322–323, 333
Red nuclei, 227–228
Reduced hemoglobin, 13
Referred pain, 256
Reflex activity, 204–206
Reflex arc, 202, 204–205
Reflex center, 228
Refracting media of eye, 300
Refractory period: in muscle
    contraction, 134–135
  in nerve impulses, 199–200
Regulatory systems, 254
Rejuvenation, 546
Relaxation (muscular), 130–131
Relaxin, 543
REM, 306–307
Renal arteries, *358*, 361, *497*, *498*,
  500
Renal clearance, 506–508
Renal corpuscle, 499
Renal pelvis, 495, 497, *498*
Renal veins, 365, *366*, *497*, *498*
Renin, 508
Rennin, 429–430
Reproductive system, 552–568
  female, 552–*553*, 554–560
  male, 560–561, *562*–565
Respiration, 397–413
  aerobic, 406
  artificial, 410–*412*
  cellular, 19–22, 34, 406–409
Respiratory center, 230, *398*
Respiratory gases, 404–405
Respiratory movements, 398–400
Respiratory quotient, 479–480
Respiratory rate, 397, 400–*401*, *402*
Respiratory system, 392–413
  development, 604–605
Respiratory transport, 404–405
Reticular formation of nerve fibers,
  228
Reticular tissue, 69–70
Reticuloendothelial cells, 70
Reticuloendothelial tissue, 600–601
Retina, 286–*287*, *289*, 291–*292*, *293*,
  594, 595, *596*
Retinal artery, 293
Retinal image, 300
Retinal veins, 293
Retine, 547
Retinene, 563
Rh factor, 339–341
Rhinecephalon, 220
Rhodopsin, 245–297, 463
Riboflavin, 408, 467, *468*
Ribonucleic acid, 39–41
  (*See also* RNA)
Ribose, *39*
Ribosomes, *32*, 40–41
Ribs, *101*, 102, *105*

Rickets, 116–117, *464*–465
Ringer-Locke solution, 321–322
Ringer's solution, 321
RNA, 35–36, 38–*42*, 189, 224
  function, 36
  messenger, 40–42
  relationship to DNA, 38–40, 42
  synthesis, 35
  transfer, 41–42
Rocky Mountain spotted fever, 471
Rod and cone cells, 259, *292*, 294,
  295, 298, *299*
Rod cells, 291, *292*, 294–*295*, *296*,
  297, 298, 463
Rolandic fissure, *217*, 229
Round window of ear, 268, *270*, 272
Rubrospinal nerve tracts, *235*, 236
Rupture, 165, 167

Saccade, 306
Saccule, *270*, 276–277
  basal, 32
  distal, 32
Sacral arteries, 358, 361
Sacral autonomics, 251
Sacral canal, 98
Sacroiliac joint, 110, 111
Sacrospinalis muscle, 154
Sacrum, 98, *110*
Sagittal suture, 89, *213*
Saline solution, 16–17
Saliva, 417, 421–422
  composition, 422
  excretory function, 423
  role in digestion, 423
Salivary amylase, 423
Salivary glands, *417*, *421*–422, 423
  functions, 422
  innervation, *250*
Salts, 12, 503–505
Saphenous vein, 364
Sarcolemma, 122
Sarcomeres, 122–124, *125*
Sarcoplasm, 122
Sarcoplasmic reticulum, *127*, *128*,
  129, 140
Sarcosomes, 122, *127*
Sartorius muscle, *171*, 172
Satellite cells, 214
Scala media (chochlear duct),
  270–271
Scala tympani, *270*, 272
Scala vestibule, *270*, 272
Scapulae, 106
Schwann cells, *191*, *192*, 214
Sciatic nerve, 238
Sciatica, 238
Sclera, *287*, *289*
Scoliosis, 100
Scrotal fold, 606
Scrotum, 560–561, *606*
Scurvy, 472
Sebaceous glands, 61, *62*, 598
Sebum, 62
Secondary sex characters, 541–542,
  545
Secretin, 433–434, 547
Secretory cells, 32
Selective permeability, 15–19
Sella turcica, *79*, 91, *92*, *393*, 520
Semen, 563–565

Semicircular canals, 232, *268*, *270*,
  275, *276*–278
Semilunar valves: aortic, *349*, *352*
  pulmonary, 347, *349*, *352*
Semimembranosus muscle, *170*, 172
Seminal emission, 564
Seminal vesicle, *561*, 563
Seminiferous tubules, 522
Semitendinosus muscle, 170, 172
Sense organs (*see* Special senses)
Sensory receptors, 194, 204,
  259–260, 359
Sensory unit, 259
Septum pellucidum, 229
Serotonin, 224
Serous membranes, 62–63
Serratus anterior, 150, *165*
Sertoli cell, *571*, 578
Serum (*see* Blood)
Serum globulin, 323, 334, 438, 444
Sesamoid bones, 111, 181
Sex chromatin, 576
Sex chromosomes, 573–*575*,
  576–584
Sex hormones, 392, 441, 454, 455,
  539, 541–547
  chorionic gonadotropin, 542, 544,
    589
  estrogen, 539, 541–546, 557–*559*,
    560, 590
  progesterone, 539, 542–544, 554,
    558–560, 590
  testosterone, 544–545, 564
Sex linkage, 311, 325–326
Shin bone, 112
Sickle-cell anemia, 325–326
Sinoatrial (SA) node, 350–352
Sinus nerve, 359–*360*
Sinus venosus, 287, *289*
Sinuses, 89–97, *98*
Sinusitis, 97
Sinusoids, 367
Skatole, 450
Skeletal muscles, 122–*124*, 132,
  142–185
Skeletal system, early development,
  101, *104*, *105*)
  (*See also* Bones)
Skin, 62, 70–71, 597–*598*
Skin color, 598–*599*, 600
Skull, *79*, 89, *90*–*95*, 96, 97
Small intestine, 416, *417*, 431–432,
  445–448
  development, 603
  (*See also* Duodenum; Ileum;
    Jejunum)
Smegma, 602
Sodium chloride, 460–461
  ionization, 9, 18
Sodium glycocholate, 441
Sodium ions: in blood, 538
  movement through cell membrane,
    18, 196, 197
Sodium taurocholate, 441
Solar plexus, 245, *247*
Soleus muscle, *175*, 177
Solute, 15–16
Solutions, 15–16
Solvent, 15–16
Somatic mesoderm, *591*–592
Somites, 494, *495*, *591*–592

Spastic paralysis, 238
Special senses, 259–280
    equilibrium, 277–280
    hearing, 267–277
    skin sensations, 260–262
    smell, 263–265
    taste, 262–263
    vision, 283–313
Spectral sensitivity, 308–309
Sperm, 541, 556, 562
    (See also Spermatozoa)
Spermatic arteries, 358, 361
Spermatic veins, 366
Spermatids, 545, 571, 572, 573, 574, 575, 577, 582
Spermatocytes, 571, 572, 573, 574, 575, 577, 581
Spermatogenesis, 560–561, 570, 571, 575, 577, 578
Spermatogonia, 545, 561–562, 571, 575, 577
Spermatozoa, 28, 30, 541, 545, 561, 564, 578, 579, 580
    chromosomes, 37, 38, 573–575, 576–577, 578
    fertilization, 583–584, 585–586
Sphenoid bone, 79, 90, 91, 92, 93
Sphenoidal fontanel, 89
Sphenoidal sinuses, 91, 92
Sphincter muscles, 144–145, 440, 445, 446, 451
    cardiac, 424, 459
    pyloric, 426
Sphincter pupillae, 290
Sphygmomanometer, 379
Spinal artery, 359
Spinal canal, 202, 230, 233, 234
Spinal cord, 98, 201, 202, 203, 208, 212, 233–235, 236–239, 242, 243, 246
    conduction pathways, 234–235
    development, 208–209, 210
    injuries, 237
Spinal ganglia, 204, 212, 238
Spinal nerves, 203, 204, 237–238, 239
Spindle, mitotic, 583
Spinothalamic tracts, 235
Spiral lamina, 270
Spiral organ of Corti, 232, 271, 272
Spirochete of syphilis, 565–566
Splanchnic mesoderm, 587, 591, 592, 601, 604
Splanchnic nerves, 238, 376, 378, 432, 501, 535
Spleen, 330, 342–344, 427, 449
    and bile pigments, 441
    function, 342–344
    and lymphoid tissues, 388, 601
    and splenic vein, 366–367
Splenial muscles, 152, 343–344
Splenic artery, 360, 361
Splenic pulp, 343
Splenic sinuses, 343
Splenic vein, 363–366, 367
Splenius capitus, 145, 148
Sprain, 116
Squamous epithelium, 30, 56–57
Stapedius, 267
Stapes, 91, 268, 270, 272

Starch, 455–457
    digestion, 423–433
Starvation, 443, 444
Statoconia, 276
Steady state, 132
Stearic acid, 456
Stenosis, 349
Stereoscope, 306
Stereoscopic vision, 305–306
Sterilization, 564
Sternocleidomastoid muscle, 145, 146, 358
Sternum, 100–104
Steroids, 538–540
Sterols, 457
Stomach, 416, 417, 426–427, 428–431
    development, 209, 601–602
Stomodeum, 209
Strabismus, 306
Stratum corneum, 598
Stratum germinativum, 598·
Stretch reflex, 204–205
Stroke, 382
Strontium-90 half life, 342
Styloid process: of temporal bone, 79, 95, 97
    of ulna, 103, 106
Subarachnoid space, 208
Subclavian arteries, 357, 358, 359
Subclavian vein, 363–365, 366, 387
Subcutaneous inguinal ring, 164, 165
Sublingual gland, 249, 421
Submandibular ducts, 421
Submandibular gland, 249, 421
Subminimal stimuli, 134
Submucosal plexus, 251, 252
Substantia nigra, 225
Succinic acid, 407, 408, 477
Sucrase, 456
Sucrose, 455
Sulci of brain, 215
Sulfates, 450
Sulfur, 457, 461
Sulfuric acid, 11
Summation, 134, 198
    spacial, 198–199
    of subliminal stimuli, 134
    temporal, 199
Superior oblique muscle, 285, 286
Superior vena cava (see Vena cava, superior)
Supination, 154
Suppuration, 332
Supraglenoid tuberosity, 154
Supraorbital foramen, 90
Supraorbital ridges, 90
Suprarenal glands, 536
Suprarenal veins, 364
Sutures, 89
    coronal, 89, 90, 93
    incisive, 95
    intermaxillary, 95
    lamboidal, 89
    occipitoparietal, 93
    sagittal, 89
    transverse palatine, 95
Swallowing act, 417, 423–424
Sweat glands, 61–62, 254
Sylvian fissure, 216

Sympathetic ganglia, 238, 246, 247, 290, 422
Sympathetic nervous system, 241, 243–244, 245, 248, 303, 353, 376
Sympathetic plexus, 245–247
Sympathin, 537
Symphysis pubis, 110–111
Synapse, 198–200
    of amacrine cells, 291–292
    in vision, 306
Synapsis, 51, 224, 571
Synaptic cleft, 193
Synaptic knobs, 193, 198–199
Synovial bursae, 117, 142
Synovial cavity, 142
Synovial fluid, 119, 142
Synovial joints, 118
Synovial membranes, 75, 142
Synovial sheaths, 181
Syphilis, 565–567, 590
Systematic circulation, 357–371
Systole, 351–352

Tabes dorsalis, 237
Taenia coli, 449
Tail, 592
Talus, 115
Tapetum, 288
Target gland, 521
Tarsal glands, 283
Tarsus, 112–113
Tartar, 423
Tartrates, 450
Taste buds, 262–263
Taurine, 441
TCA cycle, 24
Tectum, 226
Teeth, 416, 417–418, 419–420
    classification, 418–420
    deciduous, 418–420
    structure, 419–420
Telencephalon, 208–210
Telodendria, 191
Telophase, 43–44, 45, 46, 573
Temperature regulation, 372–375, 376
Temporal bone, 79, 89, 90, 93, 597
Temporal lobe, 216, 217, 231, 238, 267, 273
    and word understanding, 272–273
Temporal muscle, 144, 146
Tendinous tissue, 69
Tendon sheath, 142
Tendons, 69, 88, 142
Tenon's capsule, 285
Tensor faciae latae, 172
Tensor tympani muscle, 268
Terminal filaments, 191, 193
Testes, 421, 516, 522–523, 544–545, 560–561, 563, 564, 578
    development, 605
Testosterone, 544, 564
Tetanus, 134–135
Tetrads, 49, 571, 574, 575
Thalamus, 219, 223, 229, 294
Thiamine, 24, 406–409, 461–462, 465–466
Third ventricle, 216, 299, 294
Thoracic aorta, 358, 359
Thoracic cage, 102

Thoracic cavity, *397*
Thoracic duct, *387–388*
Thoracic vertibrae, 98, *99*–101, *105*
Thoracolumbar nervous sysem, 241, 243–*244*, 245–249
Thorax, 100–102
Threshold stimuli, 197, 198
Thrombin, 334, 335
Thromboplastin, 334
Thrombosis, coronary, 335
Thrombus, 335, 349
Thymine, 36, 37
Thymus gland, 331, 338, 546–547
    development, 593
Thyroid arteries, 528
Thyroid cartilage, *79, 393, 394*
Thyroid glands, 437, 461, 482, *517*, 525, 528–532, 602
    development, *602, 603*
Thyroid tissue, *529*, 532
Thyroid veins, *366*, 528
Thyroidectomy, 532
Thyrotrepic hormones, 521, *525*
Thyrotropin, 529
Thyroxine, 353, 459, 461, 528, 531
Tibia, *82*, 112, *113–115*, 174, *178, 179*, 180
Tibial arteries, *358*, 362
Tibial veins, 365, *366*
Tibialis anterior muscle, *175, 176*, 178
Tibialis posterior muscle, 177, 180, *181*
Tissues: connective, 64–85
    epithelial, 56–63
    reticuloendothelial, 600, 601
Tocopherols (vitamin E), 473
Tongue, *264, 393*, 416, 417
Tonus: smooth muscle, 139
    visceral muscle, 137
TPN, 21
Trabeculae, *76*
Trace mineral, 461
Tracers, 461
Trachea, 392–393, *394*, 424
    development, 604
Tranquilizers, 225
Transfusion of blood, 336–340
Transport across membranes, 17–19, 196–197
    amino acids, 18
    glucose, 18, 437
Transversus abdominis muscle, 165, 166
Trapezius muscle, *126, 145*, 151–*152*, 153
Traveling-wave theory, 274
Treponema pallidum, 565–566
Treppe, 133
Tricarboxylic acid cycle, 24, 25, 33, 130, *407*, 408, 444, 470, 476
Triceps brachii muscle, *122*, 143, 155, 156
Trichinella, 330, 332
Trichinosis, 330
Tricuspid valve, 346, *349*
Trigeminal nerve, 229–231, 264–265, 424
Triglycerides, 449
Trigone, 510
Triiocothronine, 528

Triosephosphate, 407, 476
Triphosphopyridine nucleotide (TPN), 21
Trochlea, 285, *286*
Trochlear nerve, 231, 232, 285, *286*
Trochlear notch, 107
Trophectoderm, *587*
Trophoblast, *585, 587*, 588
Trophoderm, 588
Tropic hormones, 521
Tropomyosin, 125, 129, 140
Troponin, 125, 129, 140
Trypsin, 432–433
Trypsinogen, 432
Tubercle, 88
Tuberosities, 103, 104
Turner's syndrome, 577
Twinning, 590–*591*
Tympanic cavity, 596–*597*
Tympanic membrane, 267, *268*, 272, 596–597
Tympanic ring, *79, 89*
Typhus, 471
Tyrosine, 486, 488

Ubiquinone, 408, 473
Ulcers, gastric, 431
Ulna, 106, *107*
Ulnar artery, *358*, 359
Ultraviolet light, 465
Umbilical artery, 362, 367, 368, *369, 589*
Umbilical cord, 362, 367, 588, *589*, 590, *591*, 594
Umbilical hernia, 167
Umbilical ligament, 362
Umbilical ring, 604
Umbilical vein, 367, 368, *369, 589*, 590
Umbilicus, 367, 590
Unit membrane of cell, 18, *31*
Uracil, 39
Urea, 323, 422, 444, 460, 484, 506, *507*
Urease, 506
Ureter, 496, *497, 498*, 509
Urethra, 494, *496*, 510, 563, 602
Urethral groove, *602*, 606
Uric acid, 323, 506
Urinalysis, 511
Urinary system, *497*, 509–511, *553*
    development, 494–497, 588, *602*
Urine, 434, 444, 460, 479, 485, 494, 497, 501, *503*, 510, 588
Urobilin, 506
Urobilinogen, 440
Urochrome, 506
Urogenital ridge, 536
Urogenital sinus, 602
Urogenital system, 494–497
    early development, 494–496, 605
Urogenital papilla, *606*
Urogenital tubercle, 606
Uterine arteries, *589*
Uterine tubes, 554
Uterine veins, 365, 583, *589*
Uterus, 552–*553*, 554–560
    effect of sex hormones, 541–544
    in female reproductive cycle, 552–555, 556–*559*

in fertilization and pregnancy, 583, 585, *586*, 587–*588*, 589–590
Utricle, *270*, 276, *278*
Uvula, 94, *393*, 416, 423

Vagina, *553*, 555–*557*, 607
Vagus nerve, 230, 233, 248–250, *360*, 432, 440
Valence, 6, 9
Valves: of heart, 346–*349*, 351–352
    in veins, *356*, 363, 373–374
Varicose veins, 365
Vas deferens, *561*, 562–563, 605
Vasa recta, 500
Vascular system, 357–371
    at birth, 368–369
Vasoconstriction, 249, 372–378, 422, 537
Vasoconstrictor center, 230, 375–376
Vasoconstrictor nerves, 375
Vasodilation, 247–248, 372–*373*, 375–377, 422
Vasodilation center, 376
Vasomotor nerves, 375–376
Vasopressin, 527
Vastus intermedius muscle, 171
Vastus lateralis muscle, *171*
Vastus medialis muscle, *171*
Veins, 346–347, 355–357, 363–368, 373
    fetal circulation, 367–368, *369*
    head and neck, 363, *366*
    heart, 346–*349, 355*, 363
    hepatic portal system, 365–*367*, 369
    lower extremity, 364–*366*
    lungs, 363
    pelvis and abdomen, 364–366
    upper extremity, 363–365
Vena cava: inferior, 349, *355*, 365, *366*–368, *369*, 439
    superior, 349, *351, 355*, 364, *366*, 369
Venereal diseases, 565–567
Venous blood, 404
Venous pressure, 377
Venous sinus, 211
Venous system, 363–365, *366*, 367
Ventral mesentery, 603
Ventricles: brain, 208, *216*
    heart, 346–348, *349*, 350–352
Venules, 346, 347, *373*, 378
Vermiform appendix, *417*, 449
Vermis, 228
Vertebrae, 98–104
    cervical, 98–100
    coccygeal, 98
    lumbar, 98
    sacral, 98
    thoracic, 98, 99, 100–*104*, *105*
Vertebral artery, 98
Vertebral column, 98–*99*, 100, *234*, 235
    curvatures, *99*–101
Vertebral foramen, 98
Vertebral veins, 364
Vertebronchondral ribs, 101
Vesicles of thyroid, 528
Vestibular apparatus, 275–280
Vestibular membrane, *271*

Vestibular nerve, 232, 276
Vestibule, *268*, 269–270, *277*,
    556–557, 596–*597*
Vestibulocochlear (acoustic) nerve,
    232, 272, 276–288
Vestibulospinal tract, *235*, 236–237
Villi, 586–589
    chorionic, 588, 590
    intestinal, 386, *447*, 449
Viosterol, 465
Visceral afferent fibers, 255–256
Visceral afferent impulses, 256
Visceral muscle, 136
Vision, 301–313
    accommodation, 302–303
    color, 307–309
Visual area, 302–303
Visual interpreting area, *217*
Visual purple, 295, 463
Visual yellow, 463
Vital capacity, 404
Vitamin, 461–475
Vitamin A, 295–298, 462–465, 473
    deficiency, 463–465
Vitamin B, 25, 445–446, 451,
    465–467
Vitamin B$_1$, 466, 467
Vitamin B$_2$, 465, 467, *468*
Vitamin B$_6$, 469
Vitamin B$_{12}$, 325, 444–445
Vitamin C, 472

Vitamin D, 117, 441, 461, 463–465,
    475
Vitamin D$_2$, 464–465, 534
Vitamin D$_3$, 464–465
Vitamin E (torcopherols), 473
Vitamin G, 467
Vitamin H, 470
Vitamin K, 451, 467, 473, 474
Vitamin K$_1$, 473–474
Vitamin K$_2$, 473–474
Vitreous body, 286, 287, 291,
    300–301
Vocal folds, 392, 393
Voice, 392
Volkmann's canal, 77
Voluntary hyperventilation, 402
Vomer, *90, 92*, 95

Water: chemical make-up, 5–6
    in excretion, 504–505
    of oxidation, 485
    pH, 10
Water loss, 485
Wharton's duct, *421*
Wharton's jelly, 367, 590
Whey, 430
White blood cells, 327–333
    classification, 328
White cell count, 332–333
White matter: brain, 229
    spinal cord, 201–*202*, *212*

White ramus communicans, 243,
    246
Wisdom teeth, 418–419
Wormian bones, 88
Wrist bones, 108, *109*

X chromosome, 573–*575*, 576–577,
    582
Xerophthalmia, 463
Xiphoid process, 101, 102

Y chromosome, 573–*575*, 576–577,
    582, 584
Yellow marrow, 75
Yellow spot (macula lutea), 293, 299
Yolk, 588
Yolk sac, *588*, 590, *595*, 600

Z line of muscle, 124, *125, 126, 127*
Zinn's ring, 284
Zona pellucida, 582, *584*, 585, 590
Zones of muscles, *123, 125, 126, 127*
Zygomatic arch, *79*, 90, 95
Zygomatic bone, *90*, 95, 145
Zygomatic process: maxilla, 94, 95
    temporal bone, 90, *93*
Zygomaticus muscle, *144, 145*
Zygote, 574, 584, 585, 599–600
Zygotene stage of meiosis, *571, 572*
Zymogen granules, 428